We look forward to welcoming you
into our award-winning marinas

Join over 4000 happy berth holders across our
9 destinations including Marinas, Dry Stacks
and Boatyards.

1 Largs Yacht Haven
2 Troon Yacht Haven
3 Neyland Yacht Haven
4 Fambridge Yacht Haven
5 Lymington Yacht Haven
6 Haven Quay – Dry Stack, Lymington
7 Plymouth Yacht Haven
8 Yacht Haven Quay – Dry Stack, Plymouth
9 Jachthaven Biesbosch, The Netherlands

Call **01590 677071** or visit **yachthavens.com**

Yacht Havens

Poole...

The next page in your adventure

- The Jurassic Coast
- Brownsea Island
- Dining out
- Enterainment
- Fireworks
- **Plus much more!**

HOME OF THE
**POOLE HARBOUR
BOAT SHOW**

OUR FACILITIES:

PERMANENT BERTHS
It's in a private position that makes the most of the views and gorgeous sunsets, yet it's still close to Poole's historic quay, old town and vibrant shopping centre.

- 75 permanent berths
- Superyacht berths
- Floating docks for jet skis and RIBs up to 6.1m
- 24 hour security
- Deep water: 2.5 - 6m
- Water taxi service, parking

VISITOR MARINA
Use your boat as a holiday home; entertain family, friends, colleagues or customers onboard; sail the stunning Jurassic Coast.

Enjoy all the attractions of Poole, Bournemouth and beautiful Dorset. A warm welcome always awaits!

- 125 visitor berths all year for vessels up to 70m in length and up to 4.5m draft
- Swinging moorings

SWINGING MOORINGS
Relax with a glass of wine, on a sunny afternoon, on your own swinging mooring in Poole Harbour overlooking Brownsea Island. Away from the madding crowd, these offer you ultimate privacy, peace & tranquillity.

POOLE QUAY
Boat Haven
PORT OF POOLE
Marina

Poole Town Quay, Poole,
Dorset BH15 1HJ t: 01202 649488

poolequayboathaven.co.uk
VHF Channel 80 call sign "Poole Quay Boat Haven"

REEDS
CHANNEL
ALMANAC
2019

EDITORS **Perrin Towler and Mark Fishwick**

Free updates are available at www.reedsalmanacs.co.uk

IMPORTANT SAFETY NOTE AND LEGAL DISCLAIMER

This Almanac provides basic navigational data for planning and executing passages. The tidal prediction data has been reproduced by permission of national hydrographic offices. Chartlets illustrate items in the text, orientate the user and highlight key features; they should not be relied on for navigational purposes and must always be used in conjunction with a current, corrected navigational chart. Any waypoint or position listed in this Almanac must first be plotted on the appropriate chart to assess its accuracy, safety in the prevailing circumstances and relevance to the Skipper's intentions.

Navigational guidance or suggestions are based on the accumulated experience of editors, agents, harbour masters and users. They are generic and in compiling a passage plan or pilotage notebook all other available publications and information should be consulted. They take no account of the characteristics of individual vessels nor the actual or forecast meteorological conditions, sea or tidal state. These need to be checked with appropriate local authorities for the intended area of operation prior to departure.

While every care has been taken in compiling the Almanac, it is a human endeavour with many contributors. Despite rigorous checking there may be inadvertent errors or inaccuracies, and omissions resulting from the time of notification in relation to the publication date. To the extent that the editors or publishers become aware of these, corrections will be published on the website www.reedsalmanacs.co.uk (requires registration). Readers should therefore regularly check the website between January and June for any such corrections. Data in this Almanac is corrected up to Weekly Edition 25/2018 of Admiralty Notices to Mariners.

The publishers, editors and their agents accept no responsibility for any errors or omissions, or for any accident, loss or damage (including without limitation any indirect, consequential, special or exemplary damages) arising from the use or misuse of, or reliance upon, the information contained in this Almanac.

The use of any data in this Almanac is entirely at the discretion of the Skipper or other individual with responsibility for the command, conduct or navigation of the vessel in which it is relied upon.

ADVERTISEMENT SALES
Enquiries about advertising space should be addressed to:
adlardcoles@bloomsbury.com

Almanac manager Chris Stevens

Reeds Nautical Almanac
An imprint of Bloomsbury Publishing Plc
50 Bedford Square, London, WC1B 3DP
1385 Broadway, New York, NY 10018, USA

www.bloomsbury.com

Tel: +44 (0)207 631 5600
info@reedsalmanacs.co.uk
editor.britishisles@reedsalmanacs.co.uk
editor.continental@reedsalmanacs.co.uk
www.reedsalmanacs.co.uk

REEDS, ADLARD COLES NAUTICAL and the Buoy logo are trademarks of Bloomsbury Publishing Plc

First published 2018

© Nautical Data Ltd 1980-2003
© Adlard Coles Nautical 2004–2018
Cover photo © Getty

Reeds Channel Almanac 2019 – ISBN 978 1 4729 5755 9

British Library Cataloguing-in-Publication Data

A catalogue record for this book is available from the British Library.

Library of Congress Cataloguing-in-Publication data has been applied for.

Bloomsbury Publishing Plc makes every effort to ensure that the papers used in the manufacture of our books are natural, recyclable products made from wood grown in well-managed forests. Our manufacturing processes conform to the environmental regulations of the country of origin.

To find out more about our authors and books visit www.bloomsbury.com. Here you will find extracts, author interviews, details of forthcoming events and the option to sign up for our newsletters.

Typeset in Frutiger

Printed and bound in the UK by Bell & Bain Ltd

Because it's not always
plain sailing

Call us for a quotation
+44 (0) 1752 223656

The unique Pantaenius Yacht Scheme clauses, designed by yachtsmen for yachtsmen, offer some of the broadest cover available. Why not join the 100,000 boat owners worldwide who enjoy the peace of mind a Pantaenius policy provides.

PANTAENIUS
Sail & Motor Yacht Insurance

Plymouth · pantaenius.co.uk

Reference Contents

Navigational Contents

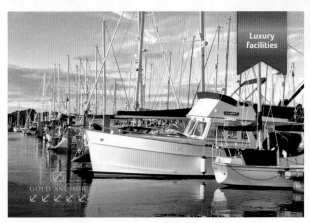

transeurope
MARINAS

VISITORS BERTHING 50% DISCOUNT

CRUISE AMONGST OVER 75 MEMBER MARINAS

1. Castlepark Marina
2. Greystones Harbour Marina
3. Malahide Marina
1. Bangor Marina
2. Rhu Marina
3. Troon Yacht Haven
4. Royal Quays Marina
5. Whitehaven Marina
6. Fleetwood Haven Marina
7. Liverpool Marina
8. Deganwy Marina
9. Conwy Quays Marina
10. Neyland Yacht Haven
11. Penarth Quays Marina
12. Upton Marina
13. Portishead Quays Marina
14. Mylor Yacht Harbour
15. Mayflower Marina
16. Dart Marina
17. Poole Quay Boathaven

18. Buckler's Hard Yacht Harbour
19. Town Quay Marina
20. Universal Marina
21. Cowes Yacht Haven
22. Royal Clarence Marina
23. Emsworth Yacht Harbour
24. Birdham Pool
25. Dover Marina
26. Gillingham Marina
27. Fambridge Yacht Haven
28. Tollesbury Marina
29. Fox's Marina
30. Brundall Bay Marina
31. Hull Marina
32. Beaucette Marina
1. Marina Den Oever
2. Jachthaven Waterland
3. Jachthaven Wetterwille
4. Marina Port Zélande
5. Jachthaven Biesbosch
6. Delta Marina

1. Puerto Deportivo Gijón
2. Marina Muros
3. Marina Combarro
4. Nauta Sanxenxo
5. Marina Davila Sport
6. Marina La Palma
7. Puerto Calero Marina
8. Marina Alcaidesa
9. Pobla Marina
10. Port Ginesta
1. Douro Marina
2. Marina Portimão
3. Quinta do Lorde Marina
1. Porto Romano
2. Venezia Certosa Marina
3. Marina del Cavallino
1. Marina Punat
1. Kos Marina
1. Marina de Saïdia
1. VVW Nieuwpoort

1. Dunkerque
2. Port Saint-Valéry
3. Fécamp
4. Le Havre Plaisance
5. Dives-Cabourg-Houlgate
6. Ouistreham/Caen
7. Granville
8. Saint Quay Port d'Armor
9. Perros-Guirec
10. Roscoff
11. Marinas de Brest
12. Douarnenez-Tréboul
13. Loctudy
14. Port La Forêt
15. Concarneau
16. Ports de Nantes
17. Île d'Yeu / Port Joinville
18. La Rochelle
19. Port Napoléon

Madeira

Canarias

Annual berth-holders based in TransEurope Marinas can benefit from a 50% visitor's berthing discount for up to five days per year in each member marina using their TransEurope membership card. For a current list of members and further information, please visit **www.transeuropemarinas.com**

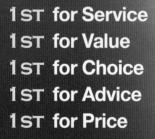

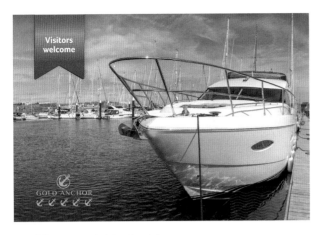

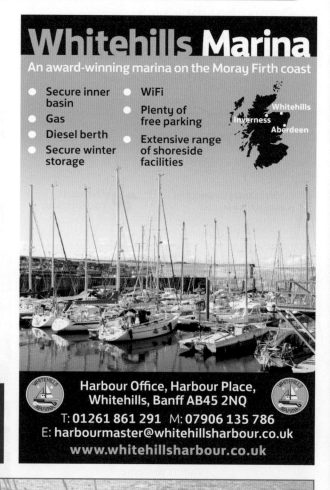

Reference data

0.1   THE ALMANAC

- **Acknowledgements**

The Editors thank the many official bodies and individuals for information and advice given in the compilation of this Almanac. These include: UKHO, Trinity House, HM Nautical Almanac Office, SHOM, HMSO, HM Revenue & Customs, Meteorological Office, BBC, IBA, MCA, RNLI, ABP, Harbour Masters and our many Harbour Agents.

- **Permissions**

Chartlets, tidal stream diagrams and curves are reproduced from Admiralty charts and publications (ALL, ATT and ALRS) by permission of the UKHO (Licence No GB DQ – 001 – Adlard Coles) and the Controller of HMSO.

UK and foreign tidal predictions are supplied by and with the permission of UKHO and by the French HO (SHOM) with permission to use the tidal predictions stated: Dunkerque, Dieppe, Le Havre, Cherbourg, St Malo, Brest; and Brest tidal coefficients (2018-131, 2018-133).

The tidal stream diagrams in 1.3 are printed by kind permission of the Royal Cruising Club.

Extracts from the *International Code of Signals, 1969,* and *Meteorological Office Weather Services for Shipping* are published by permission of the Controller of HMSO.

Ephemerides are derived from HM Nautical Almanac by permission of HM Nautical Almanac Office and the Council for the Central Laboratory of the Research Councils.

- **Disclaimer**

No National HO has verified the information in this product and none accepts liability for the accuracy of reproduction or any modifications made thereafter. No National HO warrants that this product satisfies national or international regulations regarding the use of the appropriate products for navigation.

Chartlets in this Almanac are not intended to be used for navigation. Always consult fully corrected official charts. See 0.31, Harbour Information, for more details.

- **Improvements**

Suggestions, however minor, for improving or correcting this Almanac are always welcome, especially if based on personal experience. All will be carefully considered. Please send your comments by email, if possible, direct to the relevant Editor (see below) or to info@reedsalmanacs.co.uk. Otherwise, a note to Adlard Coles Nautical (see page v) will be forwarded as necessary.

Perrin Towler (editor.britishisles@reedsalmanacs.co.uk) is responsible for Areas 1 to 3 (the south coast of the UK); Mark Fishwick (editor.continental@reedsalmanacs.co.uk) is responsible for Areas 4 to 7 (the north coast of France). We share the compilation of the Reference Data chapter.

- **Notifying errors**

Although every care has been taken in compiling this Almanac, errors may still occur. Please let us know if you spot any.

- **Harbour Agents**

Our Harbour Agents provide invaluable local information which may not appear in official sources. If this is of interest and you would like to earn a free copy of Reeds Nautical Almanac please apply to the relevant editor, giving brief details of your experience and the area(s) you would be able cover.

- **Sources of corrections**

This Almanac is corrected to Weekly edition No. 25/2018 of Admiralty Notices to Mariners.

Corrections to Admiralty charts and publications can be downloaded from the UKHO website or obtained from Admiralty Chart Agents (ACA) and certain Port Authorities.

- **Updates**

Free monthly updates, from January to June, can be downloaded at www.reedsalmanacs.co.uk. Please register online.

0.2   SYMBOLS AND ABBREVIATIONS

The following more common Symbols and Abbreviations feature in Reeds Almanacs and in some Admiralty charts and publications. Those symbols in the tinted box immediately below are frequently used within the text describing each port.

Symbol	Meaning
	Alongside berth
©	Automatic telling machine (ATM), cashpoint
®	Bank
	Bottled gas available, Calor Gas, Camping Gaz, or Kosangas
	Boatyard
	Boat hoist (+tons)
	Crane (+ tons)
	Chandlery
	Customs
	Diesel (supply by hose)
	Diesel (in cans)
	Electrical repairs
©	Electronic repairs
	Fresh water supply
	Food shop/supermarket
Ⓗ	Hospital
≠	In transit with, ie ldg marks/lts
@	Internet café/access
	Landing place/steps/ladder
	Laundry facilities available
LB,	Lifeboat, inshore lifeboat
	Light float, minor
	Lt float, major; Light vessel or Lanby
	Licensed bar, Public house, Inn
	National Coastwatch Institution (NCI)
	Marine engineering repairs
	Mooring buoy
	Petrol (supply by hose)
	Petrol (in cans)
⊠	Post Office
	Pump out facility
✕	Restaurant/cafe
	Rigger
	Sailmaker
	Shipwright (esp wooden hulls)
	Shower
	Shore power (electrical)
	Slipway
Ⓥ	Visitors' berths
	Visitors' mooring
	Wind turbine
	Weather forecast available
	Yacht Club/(YC), Sailing Club(SC)

abm	Abeam
ABP	Associated British Ports
ACA	Admiralty Chart, AC Agent
ACN	Adlard Coles Nautical (Publisher)
Aff Mar	Affaires Maritimes

AIS	Automatic Identification System
aka	Also known as
ALL	Admiralty List of Lights
ALRS	Admiralty List of Radio Signals
Al	Alternating light
ANWB	Association of road & waterway users (Dutch)
ATT	Admiralty Tide Tables
ATT	Atterisage (landfall/SWM) buoy
Auto	Météo Répondeur Automatique
B.	Bay, Black
BE	Belgian chart
Bk	Broken (nature of seabed)
Bkwtr	Breakwater
BMS	Bulletin Météorologique Spécial (Strong wind/Gale warning)
	Bn, bcn(s) Beacon, beacon(s)
BSH	German Hydrographic Office/chart(s)
BST	British Summer Time (= DST)
Bu	Blue
By(s)	Buoy, buoys
C.	Cape, Cabo, Cap
c	Coarse (sand; nature of seabed)
ca	Cable (approx 185m long)
Cas	Castle
CD	Chart datum (tidal)
CEVNI	Code Européen de Voies de la Navigation Intérieure (inland waterway signs etc)
cf	Compare, cross-refer to
CG	Coast Guard, HM Coastguard (in the UK)
CGOC	Coast Guard Operations Centre (UK)
chan	Channel (navigational)
Ch	Channel (VHF)
Ch, ⌖	Church
Chy	Chimney
Co	Coral (nature of seabed)
Col	Column, pillar, obelisk
CPA	Closest Point of Approach
CROSS	Centre Régional Opérationnel de Surveillance et Sauvetage (= MRCC)
CRS	Coast Radio Station
CRT	Canal and River Trust
C/S	COSPAS/SARSAT (satellite)
Cy	Clay (nature of seabed)
Dec	Declination (of the Sun) abbr: December
Defib	Automated External Defibrillator
dest	Destroyed
DF, D/F	Radio Direction Finding
DG	De-gaussing (range)
DGPS	Differential GPS
Dia	Diaphone (fog signal)
Dir Lt	Directional light
discont	Discontinued
DLR	Dockland Light Railway
Dn(s)	Dolphin(s)
DR	Dead Reckoning or Dries (secondary tides)
DSC	Digital Selective Calling
DST	Daylight Saving Time
DW	Deep Water (route)
DYC	Dutch Yacht Chart(s)
DZ	Danger Zone (buoy)
E	East
ECM	East cardinal mark (buoy/beacon)
ED	Existence doubtful. European Datum
EEA	European Economic Area
Elev	Elevation
Ent	Entrance, entry, enter
EP, △	Estimated position
ETA	Estimated Time of Arrival
ETD	Estimated Time of Departure
F	Fixed light
f	Fine (eg sand; nature of seabed)
F&A	Fore and aft (berth/mooring)
FFL	Fixed and Flashing light
Fl	Flashing light
FM	Frequency Modulation
Foc	Free of charge
Fog Det lt	Fog Detector light
Freq, Fx	Frequency
FS	Flagstaff, Flagpole
ft	Foot, feet
Ft,	Fort
FV	Fishing vessel
G	Gravel (nature of seabed), Green
GC	Great Circle
GDOP	Geometric Dilution of Precision (GPS)
GHA	Greenwich Hour Angle
GLA	General Lighthouse Authority
GMDSS	Global Maritime Distress & Safety System
grt	Gross Registered Tonnage
Gy	Grey
H, h, Hrs	Hour(s)
H−, H+	Minutes before, after the whole hour
H24	Continuous
Hd,hd	Headland, head (breakwater/mole)
HAT	Highest Astronomical Tide
HF	High Frequency
HFP	High Focal Plane (buoy)
HIE	Highlands & Islands Enterprise
HJ	Day service only, sunrise to sunset
HM	Harbour Master
HMRC	HM Revenue & Customs
HMSO	Her Majesty's Stationery Office
HN	Night service only, sunset to sunrise
HO	Office hours, Hydrographic Office
(hor)	Horizontally disposed (lights)
hPa	Hectopascal (= 1millibar)
HT	High Tension (overhead electricity line)
HW	High Water
HX	No fixed hours
IALA	International Association of Marine Aids to Navigation and Lighthouse Authorities
IDM	Isolated Danger Mark (buoy/beacon)
IHO	International Hydrographic Organisation
IMO	International Maritime Organisation
INMARSAT	International Maritime Satellite Organisation
intens	Intensified (light sector)
IPTS	International Port Traffic Signals
IQ	Interrupted quick flashing light
IRPCS	International Regulations for the Prevention of Collisions at Sea
Is, I	Island, Islet
ISAF	International Sailing Federation
Iso	Isophase light
ITU	International Telecommunications Union
ITZ	Inshore Traffic Zone (TSS)
IUQ	Interrupted ultra quick flashing light
IVQ	Interrupted very quick flashing light
kn	Knot(s)
kW	Kilowatts
L	Lake, Loch, Lough
Lat	Latitude
LAT	Lowest Astronomical Tide
Lanby, ⌑	Large automatic navigational buoy
Ldg	Leading (light)
LF	Low frequency
L Fl	Long flash

LH	Left hand
LNG	Liquefied Natural Gas
LNTM	Local Notice To Mariners
LOA	Length overall
Long, lng	Longitude
LPG	Liquefied Petroleum Gas
LT	Local time
Lt(s), ☆ ☆	Light(s)
M	Moorings. Nautical (sea) mile(s). Mud
m	Metre(s)
Mag	Magnetic. Magnitude (of star)
Mb, mb	Millibar (= 1 hectopascal, hPa)
MCA	Maritime and Coastguard Agency
Met/Météo	Meteorology/Météorologie (weather)
MHWN	Mean High Water Neaps
MHWS	Mean High Water Springs
MHz	Megahertz
ML	Mean Level (tidal)
MLWN	Mean Low Water Neaps
MLWS	Mean Low Water Springs
MMSI	Maritime Mobile Service Identity
Mo	Morse
Mon	Monument. Abbrev: Monday
MRCC	Maritime Rescue Co-ordination Centre
MRSC	Maritime Rescue Sub-Centre (not in the UK)
MSI	Maritime Safety Information
N	North
Navi	Navicarte (French charts)
NB	Nota Bene. Notice Board
NCM	North Cardinal Mark (buoy/beacon)
ND	No Data (secondary tides)
NGS	Naval Gunfire Support (buoy)
NM	Notice(s) to Mariners
NMOC	National Maritime Operations Centre (HMCG)
np	Neap tides
NP	Naval Publication (plus number)
NRT	Net registered tonnage
NT	National Trust (land/property)
Obscd	Obscured
Obstn	Obstruction
Oc	Occulting light
ODAS	Ocean Data Acquisition System (buoy)
Or	Orange (see also Y)
OT	Other times
P	Pebbles
(P)	Preliminary (NM)
PA	Position approximate
Pax	Passenger(s)
PC	Portuguese chart
PD	Position doubtful
PHM	Port-hand Mark (buoy/beacon)
PLA	Port of London Authority
Pos	Position
Prog	Prognosis (weather charts)
prom	Prominent
PSSA	Particularly Sensitive Sea Area
Pt(e).	Point(e)
Pta	Punta (point)
Q	Quick flashing light
QHM	Queen's Harbour Master
qv	Refer to (quod vide)
R	Red. River.
Racon	Radar transponder beacon
Ramark	Radar beacon
RCD	Recreational Craft Directive
RG	Emergency RDF station
RH	Right hand
Rk, Rky	Rock, Rocky (nature of seabed)
RNLI	Royal National Lifeboat Institution
ROI	Republic of Ireland
R/T	Radiotelephony
Ru	Ruins
RYA	Royal Yachting Association
S	South, Sand (nature of seabed)
S, St, Ste	Saint(s)
SAMU	Service d'Aide Médicale Urgente (ambulance)
SAR	Search and Rescue
SBM	Single buoy mooring
SC	Sailing Club. Spanish chart
SCM	South Cardinal Mark (buoy/beacon)
SD	Sailing Directions,Semi-diameter (of sun) Sounding of doubtful depth
Sh	Shells (nature of seabed). Shoal
SHM	Starboard-hand Mark (buoy/beacon); Simplified Harmonic Method (tides)
SHOM	Service Hydrographique et Océanographique de la Marine (FrenchHO/Chart)
Si	Silt (nature of seabed)
SIGNI	Signalisation de la Navigation Intérieure
SMS	Short Message Service (mobile texting)
so	Soft (eg mud; nature of seabed)
SOLAS	Safety of Life at Sea (IMO Convention)
Sp	Spire
sp	Spring tides
SPM	Special Mark (buoy/beacon)
SR	Sunrise
SRR	Search and Rescue Region
SS	Sunset. Signal Station
SSB	Single Sideband (radio)
St	Stones (nature of seabed)
Stbd	Starboard
subm	Submerged
SWM	Safe Water Mark (buoy/beacon)
sy	Sticky (eg mud; nature of seabed)
(T), (Temp)	Temporary
tbc	To be confirmed
tbn	To be notified
TD	Temporarily Discontinued (fog signal)
TE	Temporarily Extinguished (light)
tfn	Till further notice
Tr, twr	Tower
T/R	Traffic Report (route notification)
TSS	Traffic Separation Scheme
uncov	Uncovers
UQ	Ultra Quick flashing light
UT	Universal Time (= approx GMT)
Var	Variation (magnetic)
(vert)	Vertically disposed (lights)
Vi	Violet
vis	Visibility, visible
VLCC	Very large crude carrier (Oil tanker)
VNF	Voie Navigable de France (canals)
VQ	Very Quick flashing light
VTS	Vessel Traffic Service
W	West. White
WCM	West Cardinal Mark (buoy/beacon)
Wd	Weed (nature of seabed)
wef	With effect from
WGS	World Geodetic System (GPS datum)
wi-fi	Wireless Fidelity (internet access)
WIG	Wing in ground effect (craft)
WIP	Work in progress
Wk, ⌐ ⊕	Wreck (see also Fig 1(1))
WMO	World Meteorological Organisation
WPT, ⊕	Waypoint
WZ	Code for UK coastal navigation warning
Y	Yellow, Amber, Orange

0.3 PASSAGE PLANNING FORM

DATE:........................ FROM: .... TO: .... DIST: ....nm

ALTERNATIVE DESTINATION(S): ....

WEATHER FORECAST: ....

....

FORECASTS AVAILABLE DURING PASSAGE: ....

....

TIDES

DATE:.........................	DATE:.........................	DATE:.........................
PLACE:........................	PLACE:........................	PLACE:........................
HW	HW	HW ....
LW	LW	LW ....
HW	HW	HW ....
LW	LW	LW ....
COEFFICIENT:	....	....
HEIGHT OF TIDE AT:	....	....
.... hrs ....m	.... hrs ....m	......................hrs ....m

DEPTH CONSTRAINTS: ....

TIDAL STREAMS AT: ....

TURNS .... AT .... TOTAL SET (FM .... TO ....): ....° ....M

TURNS .... AT .... TOTAL SET (FM .... TO ....): ....° ....M

NET TIDAL STREAM FOR PASSAGE: ....° .... M

ESTIMATED TIME: ....hrs ETD: .... ETA: ....

SUN/MOON SUNRISE: .... SUNSET: ....

MOONRISE: .... MOONSET: .... PHASE: ....

WAYPOINTS

NO	NAME	TRACK/DISTANCE (TO NEXT WAYPOINT)
....	....	.... / ....
....	....	.... / ....
....	....	.... / ....
....	....	.... / ....
....	....	.... / ....

DANGERS   CLEARING BEARINGS/RANGES/DEPTHS

....

....

LIGHTS/MARKS EXPECTED ....

....

....

....

COMMUNICATIONS PORT/MARINA .... VHF .... ☎ ....

PORT/MARINA .... VHF .... ☎ ....

NOTES (CHARTS PREPARED & PAGE NUMBERS OF RELEVANT PILOTS / ALMANACS / ETC): ....

....

....

0.4 PASSAGE PLANNING

All passages by any vessel that goes to sea *must* be planned. This is defined as proceeding beyond sheltered waters. Full passage planning requirements may be found in Chapter V of the International Convention for Safety of Life at Sea (SOLAS), but more digestible guidance for small craft is in the MCA's Pleasure Craft Information Pack at: **https://www.gov.uk/government/uploads/system/uploads/attachment_data/file/282631/pleasure_craft_information_packdec07-2.pdf.**

Although the passage plan does not have to be recorded on paper, in the event of legal action a written plan is clear proof that the required planning has been completed. A suggested passage planning form is on the previous page. When completed this would constitute a reasonable passage plan. The blank form may be photocopied and/or modified.

Although spot checks on small craft are unlikely, the MCA could, following an accident or incident, take action under the Merchant Shipping Act if it could be proved that the skipper did not have a reasonable passage plan.

All passage plans should at least consider the following:

- **Weather.** Check the weather forecast and know how to get regular updates during the passage.
- **Tides.** Check tidal predictions and determine if there are any limiting depths at your port of departure, during the passage and at the port of arrival (and at alternative ports, if applicable). Tidal streams will almost certainly affect the plan.
- **Vessel.** Confirm she is suitable for the intended trip, is properly equipped, and has sufficient fuel, water and food on board.
- **Crew.** Take into account your crew's experience, expertise and stamina. Cold, tiredness and seasickness can be debilitating – and skippers are not immune.
- **Navigation.** Make sure you are aware of all navigational dangers by consulting up to date charts, pilot books and this Almanac. Never *rely* on GPS for fixing your position.
- **Contingency plan.** Consider bolt holes which can be entered *safely* in an emergency.
- **Information ashore.** Make sure someone ashore knows your plans, when they should become concerned and what action to take if necessary. Be sure to join the Coastguard Voluntary Identification Scheme.

0.5 POSITIONS FROM GPS

GPS uses the World Geodetic System 84 datum (WGS84). With the exception of much of the coast of Ireland and the west coast of Scotland, Admiralty charts of UK waters are now to the ETRS 89 datum. Harbour chartlets reflect this.

If the chart in use is not referenced to WGS84, positions read from the GPS receiver must be converted to the datum of the chart in use. This is printed on the chart and gives the Lat/Long corrections to be applied. They can be significant. There are two options:

- Set the receiver to WGS84. Before plotting positions, manually apply the corrections given on the chart. This option is advised by UKHO.
- Set the receiver to the datum of the chart in use; the datum corrections will be applied by the receiver's software. This method is not the most accurate due to the random nature of the differences.

0.6 VHF COMMUNICATIONS
Radio Telephony (R/T)

VHF radio (Marine band 156·00–174·00 MHz) is used by most vessels. Range is better than the line of sight between aerials, typically about 20M between yachts and up to 65M to a shore station; always invest in a good aerial, as high as possible.

VHF sets may be **Simplex**, ie transmit and receive on the same frequency, so only one person can talk at a time. **Semi-Duplex** (most modern sets), transmit and receive on different frequencies, or **full Duplex**, ie simultaneous Semi-Duplex, so conversation is normal, but two aerials are needed.

Marine VHF frequencies are known by their international channel number (Ch), as shown below.

Channels are grouped according to three main purposes, but some have more than one purpose.

Public correspondence: (via Coast radio stations)
Ch 26, 27, 25, 24, 23, 28, 04, 01, 03, 02, 07, 05, 84, 87, 86, 83, 85, 88, 61, 64, 65, 62, 66, 63, 60, 82, 78, 81.
All channels can be used for Duplex.

Inter-ship:
Ch 06*, 08*, 10, 13, 09, 72*, 73, 69, 77*, 15, 17.
These are all Simplex channels. * for use in UK.

Port Operations:
Simplex: Ch 12, 14, 11, 13, 09, 68, 71, 74, 69, 73, 17, 15.
Duplex: Ch 20, 22, 18, 19, 21, 05, 07, 02, 03, 01, 04, 78, 82, 79, 81, 80, 60, 63, 66, 62, 65, 64, 61, 84.

The following channels have one specific purpose:

Ch 0 (156·00 MHz): SAR ops, not available to yachts.

Ch 10 (156·50 MHz), **62** (160·725 MHz), **63** (160·775 MHz) and **64** (160·825 MHz): MSI broadcasts. The optimum channel number is stated on Ch 16 in the announcement prior to the broadcast itself.

Ch 13 (156·650 MHz): Inter-ship communications relating to safety of navigation; a possible channel for calling a merchant ship if no contact on Ch 16.

Ch 16 (156·80 MHz): Distress, Safety and calling. Ch 16, in parallel with DSC Ch 70, will be monitored by ships, CG rescue centres (and, in some areas, any remaining Coast Radio Stations) for Distress and Safety until further notice. Yachts should monitor Ch 16. After an initial call, stations concerned **must** switch to a working channel, except for Distress and Safety matters.

Ch 67 (156·375 MHz): Small craft safety channel used by all UK CG centres, accessed via Ch 16. Note: uniquely, also used in Wight/Solent area to make initial contact with NMOC, call sign **Solent Coastguard** to avoid congestion on Ch 16.

Ch 70 (156·525 MHz): Digital Selective Calling for Distress and Safety purposes under GMDSS.

Ch 80 (157·025 MHz): Primary working channel between yachts and UK marinas.

Ch M (157·85 MHz): Secondary marina working channel. For use in UK territorial waters only.

Ch M2 (161·425 MHz): for race control, with Ch M as stand-by. YCs often use Ch M2. For use in UK waters only.

Your position should be given as Lat/Long or the vessel's bearing and distance *from* a charted object, eg 'My position 225° Nab Tower 4M' means you are 4M SW of the Nab Tower (*not* 4M NE). Use the 360° True bearing notation and the 24-hour clock (0001–2359), specifying UT or LT.

0.7 DISTRESS CALLS

Distress signal - MAYDAY

Distress only applies to a situation where a *vessel or person is in grave and imminent danger and requires immediate assistance*. A MAYDAY call should usually be sent on VHF Ch 16 or MF 2182 kHz, but any frequency may be used if help would thus be obtained more quickly.

Distress, Urgency and Safety messages from vessels at sea are free of charge. A Distress call has priority over all other transmissions. If heard, cease all transmissions that may interfere with the Distress call or messages, and listen on the frequency concerned.

Brief your crew so they are all able to send a Distress message. The MAYDAY message format (below) should be displayed near the radio. Before making the call:

- Switch on radio (check main battery switch is ON)
- Select HIGH power (25 watts)
- Select VHF Ch 16 (or 2182 kHz for MF)
- Press and hold down the transmit button, and say slowly and distinctly:

- **MAYDAY MAYDAY MAYDAY**
- **THIS IS** ....
  (name of boat, spoken three times)
- **MAYDAY** ....
  (name of boat spoken once)
- **CALLSIGN / MMSI Number** ....
  (Following a DSC alert)
- **MY POSITION IS** ....
  (latitude and longitude, true bearing and distance *from* a known point, or general location)
- **Nature of distress** ....
  (sinking, on fire etc)
- **Help required** ....
  (immediate assistance)
- **Number of persons on board** ....
- **Any other important, helpful information** ....
  (you are taking to the liferaft; distress rockets are being fired etc)
- **OVER**

On completion of the Distress message, release the transmit button and listen. The boat's position is of vital importance and should be repeated if time allows. If an acknowledgement is not received, check the set and repeat the Distress call.

Vessels with GMDSS equipment should make a MAYDAY call on Ch 16, including their MMSI and callsign, *after* sending a DSC Distress alert on VHF Ch 70 or MF 2187·5 kHz.

0.7.1 MAYDAY acknowledgement

In coastal waters an immediate acknowledgement should be expected, as follows:

MAYDAY ....
(name of station sending the Distress message, spoken three times)

THIS IS ....
(name of station acknowledging, spoken three times)

RECEIVED MAYDAY

If you hear a Distress message, write down the details and, if you can help, acknowledge accordingly - but only after giving an opportunity for the nearest Coastguard station or some larger vessel to do so.

0.7.2 MAYDAY relay

If you hear a Distress message from a vessel, and it is not acknowledged, you should pass on the message as follows:

MAYDAY RELAY ....
(spoken three times)

THIS IS ....
(name of vessel re-transmitting the Distress message, spoken three times), followed by the intercepted message.

0.7.3 Control of MAYDAY traffic

A MAYDAY call imposes general radio silence until the vessel concerned or some other authority (eg the nearest Coastguard) cancels the Distress. If necessary the station controlling Distress traffic may impose radio silence as follows:

SEELONCE MAYDAY, followed by its name or other identification, on the Distress frequency.

If some other station nearby believes it necessary to do likewise, it may transmit:

SEELONCE DISTRESS, followed by its name or other identification.

0.7.4 Relaxing radio silence

When complete radio silence is no longer necessary, the controlling station may relax radio silence as follows, indicating that restricted working may be resumed:

MAYDAY

ALL STATIONS, ALL STATIONS, ALL STATIONS

THIS IS ....
(name or callsign)

The time ....

The name of the vessel in distress ....

PRUDONCE

Normal working on the Distress frequency may then be resumed, having listened carefully before transmitting. Subsequent calls from the casualty should be prefixed by the Urgency signal (0.8).

If Distress working continues on other frequencies these will be identified. For example, PRUDONCE on 2182 kHz, but SEELONCE on VHF Ch 16.

0.7.5 Cancelling radio silence

When the problem is resolved, the Distress call must be cancelled by the co-ordinating station using the prowords SEELONCE FEENEE as follows:

MAYDAY

ALL STATIONS, ALL STATIONS, ALL STATIONS

THIS ....(name or callsign)

The time ....

The name of the vessel in distress ....

SEELONCE FEENEE

0.8   URGENCY AND SAFETY CALLS

Urgency signal - PAN PAN

The radio Urgency prefix, consisting of the words PAN PAN spoken three times, indicates that a vessel, or station, has *a very urgent message concerning the safety of a ship or person*. It may be used when urgent medical advice is needed.

This is an example of an Urgency call:

> PAN PAN, PAN PAN, PAN PAN
>
> ALL STATIONS, ALL STATIONS, ALL STATIONS
>
> THIS IS YACHT SEABIRD, SEABIRD, SEABIRD
>
> Two nine zero degrees Needles lighthouse two miles
>
> Dismasted and propeller fouled
>
> Drifting east north east towards Shingles Bank
>
> Require urgent tow
>
> OVER

PAN PAN messages take priority over all traffic except Distress, and are sent on Ch 16 or 2182 kHz. They should be cancelled when the urgency is over.

If the message is long (eg a medical call) or communications traffic is heavy, it may be passed on a working frequency after an initial call on Ch 16 or 2182 kHz. At the end of the initial call you should indicate that you are switching to a working frequency.

If you hear an Urgency call react in the same way as for a Distress call.

0.8.1   Safety signal - SÉCURITÉ

The word SÉCURITÉ (pronounced SAY-CURE-E-TAY) spoken three times, indicates that the station is about to transmit an important navigational or meteorological warning. Such messages usually originate from a CG Centre or a Coast Radio Station, and are transmitted on a working channel after an announcement on the distress/calling channel (Ch 16 or 2182 kHz).

Safety messages are usually addressed to 'All stations', and are often transmitted at the end of the first available silence period. An example of a Sécurité message is:

> SÉCURITÉ, SÉCURITÉ, SÉCURITÉ ....
>
> THIS IS ....
> (CG Centre or Coast Radio Station callsign, spoken three times)
>
> ALL STATIONS ....
> (spoken three times) followed by instructions to change channel, then the message.

0.9   GMDSS

The Global Maritime Distress and Safety System (GMDSS) came into force in 1999. Most seagoing vessels over 300 tons are required by SOLAS to comply with GMDSS, but it is not compulsory for yachts. However, it is important that the principles of the system are understood, and you should at least consider fitting compliant equipment depending on your cruising area. Full details may be found in ALRS Vol 5 (NP 285).

0.9.1   Purpose

GMDSS enables a coordinated SAR operation to be mounted rapidly and reliably anywhere at sea. To this end, terrestrial and satellite communications and navigation equipment is used to alert SAR authorities ashore and ships in the vicinity to a Distress incident or Urgency situation. GMDSS also promulgates Maritime Safety Information (MSI).

0.9.2   Sea areas

The type of equipment carried by a vessel depends of her operating area. The four GMDSS Areas are:

A1	An area within R/T coverage of at least one VHF Coastguard or Coast radio station in which continuous VHF alerting is available via DSC. Range: 20–50M from the CG/CRS.
A2	An area, excluding sea area A1, within R/T coverage of at least one MF CG/CRS in which continuous DSC alerting is available. Range: approx 50–250M from the CG/CRS.
A3	An area between 76°N and 76°S, excluding sea areas A1 and A2, within coverage of HF or an Inmarsat satellite in which continuous alerting is available.
A4	An area outside sea areas A1, A2 and A3, ie the polar regions, within coverage of HF.

In each Area, in addition to a Navtex receiver, certain types of radio equipment must be carried by GMDSS vessels: In A1, VHF DSC; A2, VHF and MF DSC; A3, VHF, MF and HF or SatCom; A4, VHF, MF and HF.

0.9.3   Digital Selective Calling (DSC)

GMDSS comprises 'sub-systems' which are coordinated through Maritime Rescue Coordination Centres (MRCC) to ensure safety at sea. DSC is one of the 'sub-systems' of GMDSS. It uses terrestial communications for making initial contact and, in a distress situation, provides the vessel's identity, nature of distress and position (entered manually or automatically if linked with the GPS). In all DSC messages every vessel and relevant shore station has a 9-digit Maritime Mobile Service Identity (MMSI) which is in effect an automatic electronic callsign. Dedicated frequencies are: VHF Ch 70, MF 2187·5 kHz. A thorough working knowledge of the following procedure is needed.

A typical VHF/DSC Distress alert might be sent as follows:

- Briefly press the (red, guarded) Distress button. The set automatically switches to Ch 70 (DSC Distress channel). Press again for 5 seconds to transmit a basic Distress alert with position and time. The radio then reverts to Ch 16.

- If time permits, select the nature of the distress from the menu, eg Collision, then press the Distress button for 5 seconds to send a full Distress alert.

A CG/CRS should automatically send an acknowledgement on Ch 70 before replying on Ch 16. Ships in range should reply directly on Ch 16. When a DSC Distress acknowledgement has been received, or after about 15 seconds, the vessel in distress should transmit a MAYDAY message by voice on Ch 16, including its MMSI.

0.10 HM COASTGUARD

DOVER COASTGUARD
50°08'N 01°20'E. DSC MMSI 002320010
Langdon Battery, Swingate, Dover CT15 5NA.
☎ 01304 210008. 📠 01304 202137.
Area: Beachy Head to Reculver Towers.
Operates Channel Navigation Information Service (CNIS)

NATIONAL MARITIME OPERATIONS CENTRE (NMOC)
50°51'·50N 01°14'·90W, DSC MMSI 002320011
Kites Croft, Fareham, PO14 4LW
☎ 02392 552100 (Note: continues to accept calls on the local numbers of former CG stations at Portland (☎ 01305 760439) and Solent (☎ 02392 552100). 📠 02392 554131.
Make initial routine calls using call sign *Solent Coastguard* on Ch 67 (H24) to avoid congestion on Ch 16.
Area: Topsham (River Exe) to Beachy Head.

†*FALMOUTH COASTGUARD
50°09'N 05°03'W. MMSI 002320014
Pendennis Point, Castle Drive, Falmouth TR11 4WZ.
☎ 01326 317575. 📠 01326 318342.
Area: Marsland Mouth (near Bude) to Topsham (River Exe).
† Monitors DSC MF 2187·5 kHz. * Broadcasts Gunfacts/Subfacts.

0.11 CHANNEL ISLANDS COASTGUARD

Guernsey and Jersey Coastguard stations direct SAR operations in the North and South of the Channel Islands area respectively. Close liaison is maintained with adjacent French SAR authorities and a Distress situation may be controlled by the Channel Islands or France, whichever is more appropriate. For example, to avoid language problems, a British yacht in difficulty in French waters may be handled by St Peter Port or Jersey; and vice versa for a French yacht.

GUERNSEY COASTGUARD (CRS)
49°27'·0N 02°32'·0W   DSC MMSI 002320064
☎ 01481 720672   📠 01481 56432   guernseycoastguard@gov.gg   VHF Ch 16, **20**, 62, 67.
Area: The Channel Islands Northern area.

JERSEY COASTGUARD (CRS)
49°10'·9N 02°06'·8W   DSC MMSI 002320060
☎: 01534 447705   📠 01534 447799   jerseycoastguard@ports.je   VHF Ch 16, 25, 67, **82**.
Area: The Channel Islands Southern area.

0.12 NATIONAL COASTWATCH INSTITUTION

The National Coastwatch Institution (NCI) operates on dedicated **VHF Ch 65**. Call sign: name of selected station, (see full list at: www.nci.org.uk) plus abbreviation NCI, e.g: *Froward Point NCI*. Do **not** call on Ch 16 or use call sign *National Coastwatch* which can potentially be confused with Coastguard. There are 50 operational stations (2018) around the English and Welsh coast. Manned by volunteers keeping a visual watch during daylight hours throughout the year, they are able provide, on request, details of actual weather conditions and sea state at their individual coastal locations, radio checks and local knowledge.

0.13 FRANCE – CROSS/SNSM

CROSS (Centres Régionaux Opérationnels de Surveillance et de Sauvetage) provide an all-weather presence H24 along the French coast in liaison with foreign CGs. CROSS is an MRCC, and as well as fishery and anti-pollution work, coordinates SAR operations involving the French lifeboat service, SNSM (Société National de Sauvetage en Mer).

All centres monitor VHF Ch 16 and DSC Ch 70 and co-ordinate SAR on Ch 15, 67, 68, 73; DSC Ch 70. They can be contacted: by phone; see 0.12.1; via the National Gendarmerie, Affaires Maritimes or a Semaphore station. SNSM tel nos are given in Chapters 2 and 3 for most ports. A hefty charge may be levied if a lifeboat is called out to attend a vessel which turns out to be not in distress.

CROSS also monitors the TSS in the Dover Strait, off Casquets and off Ouessant using, for example, callsign *Corsen Traffic*.

For medical advice call the appropriate CROSS which will contact a doctor or Service d'Aide Médicale Urgente (SAMU). ⚠ Note: SAMU (Ambulance) will respond quicker than a doctor to a medical call-out in a hbr or marina. Just dial 15.

0.13.1 CROSS locations/contact details

Any CROSS can be telephoned simply by dialling **112**.

CROSS Gris-Nez 50°52'N 01°35'E
MMSI 002275100 ☎ 03·21·87·21·87; 📠 03·21·87·78·55 gris-nez@mrccfr.eu (Belgian border–Cap d'Antifer).

CROSS Jobourg 49°41'N 01°54'W
MMSI 002275200 ☎ 02·33·52·16·16; 📠 02·33·52·78·23 jobourg@mrccfr.eu (Cap d'Antifer–Mont St Michel).

CROSS Corsen 48°24'N 04°47'W
MMSI 002275300 ☎ 02·98·89·31·31; 📠 02·98·89·65·75 corsen@mrccfr.eu (Mont St Michel–Pointe de Penmarc'h).

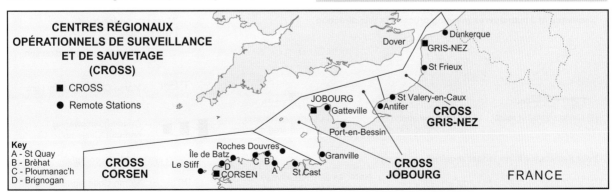

Fig 0(1)   FRANCE - CROSS areas

Fig 0(2)   Distress and life saving signals

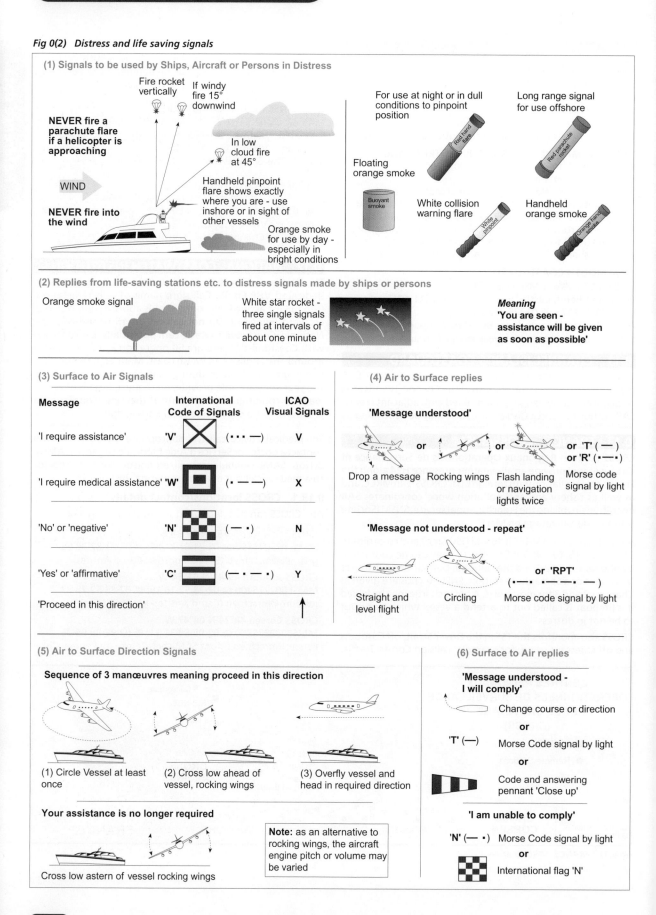

(1) Signals to be used by Ships, Aircraft or Persons in Distress

NEVER fire a parachute flare if a helicopter is approaching

Fire rocket vertically

If windy fire 15° downwind

In low cloud fire at 45°

WIND

NEVER fire into the wind

Handheld pinpoint flare shows exactly where you are - use inshore or in sight of other vessels

Orange smoke for use by day - especially in bright conditions

For use at night or in dull conditions to pinpoint position

Floating orange smoke

Buoyant smoke

White collision warning flare

Long range signal for use offshore

Red hand flare

Red parachute rocket

White pinpoint

Handheld orange smoke

Orange hand smoke

(2) Replies from life-saving stations etc. to distress signals made by ships or persons

Orange smoke signal

White star rocket - three single signals fired at intervals of about one minute

Meaning
'You are seen - assistance will be given as soon as possible'

(3) Surface to Air Signals

Message	International Code of Signals		ICAO Visual Signals
'I require assistance'	'V'	X (· · · —)	V
'I require medical assistance'	'W'	■ (· — —)	X
'No' or 'negative'	'N'	(— ·)	N
'Yes' or 'affirmative'	'C'	(— · — ·)	Y
'Proceed in this direction'		↑	

(4) Air to Surface replies

'Message understood'

Drop a message   or   Rocking wings   or   Flash landing or navigation lights twice

or 'T' (—)
or 'R' (· — ·)
Morse code signal by light

'Message not understood - repeat'

Straight and level flight   Circling

or 'RPT'
(· — · · — · —)
Morse code signal by light

(5) Air to Surface Direction Signals

Sequence of 3 manœuvres meaning proceed in this direction

(1) Circle Vessel at least once

(2) Cross low ahead of vessel, rocking wings

(3) Overfly vessel and head in required direction

Your assistance is no longer required

Cross low astern of vessel rocking wings

Note: as an alternative to rocking wings, the aircraft engine pitch or volume may be varied

(6) Surface to Air replies

'Message understood - I will comply'

Change course or direction

or

'T' (—)   Morse Code signal by light

or

Code and answering pennant 'Close up'

'I am unable to comply'

'N' (— ·)   Morse Code signal by light

or

International flag 'N'

0.14 NAVTEX

NAVTEX is the prime method of disseminating MSI to at least 200 miles offshore. A dedicated aerial and receiver with an LCD screen (or integrated printer) are required. The user selects which stations and message categories are recorded for automatic display or printing.

Two frequencies are used for broadcasts. On the international frequency, 518kHz, messages are always available in English with excellent coverage of Europe. Interference between stations is minimised by scheduling time slots and and by limiting transmission power. NAVTEX information applies only to the geographical area for which each station is responsible.

On the national frequency 490kHz (for clarity, shown in red throughout this chapter) the UK issues inshore waters forecasts and coastal station actuals. Elsewhere it is used mainly for transmissions in the national language. 490khz stations have different identification letters from 518kHz stations.

NAVTEX is convenient and particularly useful when preoccupied handling your vessel as you will not miss potentially important information.

0.14.1 Message numbering

Each message is prefixed by a four-character group:

The first character is the code letter of the transmitting station (eg **E** for Niton).

The 2nd character is the message category, see 0.13.2.

The third and fourth are message serial numbers, running from 01 to 99 and then re-starting at 01.

The serial number 00 denotes urgent messages which are always printed.

Messages which are corrupt or have already been printed are rejected. Weather messages are dated and timed. All Navtex messages end with NNNN.

0.14.2 Message categories

A*	Navigational warnings
B*	Meteorological warnings
C	Ice reports
D*	SAR info and Piracy attack warnings
E	Weather forecasts
F	Pilot service
G	AIS
H	LORAN
I	Spare
J	SATNAV
K	Other electronic Navaids
L	Navwarnings additional to **A**
M-U	Spare
V-Y	Special services – as allocated
Z	No messages on hand

* These categories cannot be rejected by the receiver.

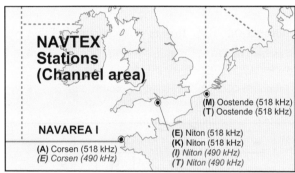

Fig 0(3)   NAVTEX stations – Navarea I

0.14.3 UK 518 kHz stations

The times (UT) of weather messages are in bold; the time of an extended outlook (a further 2 or 3 days beyond the shipping forecast period) is in italics.

E	–	**Niton**	*0040*	0440	**0840**	1240	1640	**2040**
		Thames clockwise to Fastnet, excluding Trafalgar.						

0.14.4 UK 490 kHz stations

These provide forecasts for UK inshore waters (to 12M offshore), a national 3 day outlook for inshore waters and, at times in bold, reports of actual weather at the places listed below. To receive these reports select message category 'V' (0.13.2) on your NAVTEX receiver. Actual Met data includes: Sea level pressure (mb), wind direction and speed (kn), weather, visibility (M), air and sea temperatures (°C), dewpoint temperature (°C) and mean wave height (m).

I	–	**Niton** The Wash to St David's Head	**0120**	0520	**0920**	**1320**	1720	**2120**

Sandettie Lt V, Greenwich Lt V, Solent, Hurn airport, Guernsey airport, Jersey airport, Portland, Channel Lt V, Plymouth, Culdrose, Seven Stones Lt V, St Mawgan and Roches Point (Cork).

0.14.5 NAVTEX coverage abroad

Selected NAVTEX stations in Metareas I and II, with their identity codes and transmission times are listed below. Times of weather messages are shown in **bold**. Gale warnings are usually transmitted 4 hourly.

METAREA I (Co-ordinator – UK)			Transmission times (UT)					
K	–	**Niton**	0140	0540	0940	1340	1740	2140
V	–	**Oostende**, Belgium	0330	0730	1130	1530	1930	2330
T	–	**Oostende**, Belgium (Note)	0310	**0710**	1110	1510	**1910**	2310
B	–	**Oostende**, Belgium (Mostly Dutch)	0010	0410	0810	1210	1610	2010

Note   Forecasts and strong wind warnings for Thames and Dover, plus NavWarnings for the Belgian coast.

METAREA II (Co-ordinator – France)								
A	–	**Corsen**, Le Stiff, France	**0000**	0400	0800	**1200**	1600	2000
E	–	**Corsen**, Le Stiff, France (In French)	0040	0440	0840	1240	1640	2040

0.15 MSI BROADCASTS BY HM COASTGUARD

HMCG Centres routinely broadcast Maritime Safety Information (MSI) every 3 hours from the start times in the Table below, but these can sometimes vary during busy operational times and emergencies. The VHF broadcast channel, either 10, 62, 63 or 64 is announced first on Ch 16. Each broadcast contains one of 3 different Groups of MSI:

Group A, the full broadcast, contains the Shipping forecast, a new Inshore waters forecast for local areas, Gale and strong wind warnings. A Fisherman's 3 day forecast (as appropriate*), Navigational (WZ) warnings and Subfacts & Gunfacts where appropriate ‡. Times of 'A' broadcasts are in bold.

Group B contains a new Inshore waters forecast, new outlook, and Gale warnings. 'B' broadcast times are in plain type.

Group C is a repeat of the Inshore forecast and Gale warnings (as per the previous Group A or B) plus new Strong wind warnings. 'C' broadcast times are italicised.

*A Fisherman's 3 day forecast is broadcast (1 Oct-31 Mar) by Falmouth CG.

‡ Subfacts & Gunfacts are broadcast by Falmouth CGOC.

0.15.1 Actual weather

On request CG Centres and NCI stations may report their actual weather. They stress they are not qualified Met observers; such reports may come from a window view or from passing ships and yachts.

Broadcasts of shipping and inshore waters forecasts by HM Coastguard

Coastguard	Shipping forecast areas	Inshore areas	Broadcast times LT							
			B	C	A	C	B	C	A	C
Dover CGOC	Thames, Dover, Wight	5, 6	0110	*0410*	**0710**	*1010*	1310	*1610*	**1910**	*2210*
Fareham NMOC *(Usually announced as Solent CG)*	Wight, Portland,	6, 7	0130	*0430*	**0730**	*1030*	1330	*1630*	**1930**	*2230*
Falmouth CGOC‡*	Portland, Plymouth, Sole, Lundy, Fastnet	8, 9	0110	*0410*	**0710**	*1010*	1310	*1610*	**1910**	*2210*

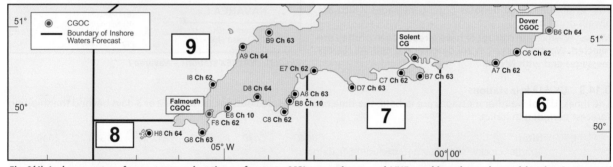

Fig 0(4) Inshore waters forecast areas, locations of remote MSI transmitters and VHF working channels used by the Coastguard
NOTE: Information on this page remains subject to potential change. Readers are therefore advised to check the MCGA Maritime Safety Information leaflet (MCA/064) available at www.mcga.gov.uk, and also the free monthly pdf updates from January 2019 to June 2019 available for download at www.reedsnauticalalmanac.co.uk

Inshore waters forecast areas - remote MSI transmitters and broadcast times

6) North Foreland to Selsey Bill,
DOVER CG, every 3 hours commencing 0110

B6	Langdon (Dover)
C6	Fairlight (Hastings)

7) Selsey Bill to Lyme Regis,
SOLENT CG, every 3 hours commencing 0130

A7	Newhaven (located in Area 6)
B7	Boniface Down (Ventnor IoW)
C7	Needles
D7	Grove Pt (Portland Bill
E7	Beer Head (East Lyme Bay)

8) Lyme Regis to Lands End inc the Isles of Scilly,
FALMOUTH CG, every 3 hours commencing 0110

A8	Berry Head (Torbay)
B8	Dartmouth
C8	East Prawle (Salcombe)
D8	Rame Head
E8	Fowey
F8	Falmouth
G8	Lizard
H8	Isles of Scilly
I8	Trevose Head

9) Land's End to St David's Head, inc the Bristol Channel,
MILFORD HAVEN CG, every 3 hours commencing 0150

A9	Hartland Point
B9	Combe Martin (N Devon)

0.16   OTHER UK/CHANNEL ISLES BROADCASTS

BBC Radio 4 Shipping forecast
BBC Radio 4 broadcasts shipping forecasts at:

0048, 0520 LT[1]	LW, MW, FM
1201 LT	LW only
1754 LT	LW, FM (Sat/Sun)

[1] Includes weather reports from coastal stations

Frequencies

LW	198 kHz
MW	756 kHz (**Redruth**), 774 kHz (**Plymouth**)
FM	92·4–94·6 MHz (**England**), 94·8 MHz (**Channel Is**)

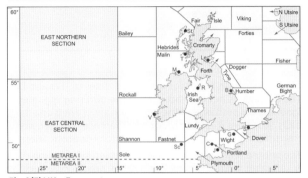

Fig 0(5) UK - Forecast areas

The Shipping Forecast contains:

Time of issue; summary of gale warnings in force at that time; a general synopsis of weather systems and their expected development and movement over the next 24 hrs; sea area forecasts for the same 24 hrs, including wind direction/force, weather and visibility in each; and an outlook for the following 24 hrs.

Gale warnings for all affected areas are broadcast at the earliest break in Radio 4 programmes after receipt, as well as after the next news bulletin.

0.17   FRENCH MSI BROADCASTS

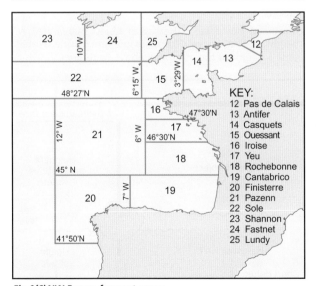

KEY:
12 Pas de Calais
13 Antifer
14 Casquets
15 Ouessant
16 Iroise
17 Yeu
18 Rochebonne
19 Cantabrico
20 Finisterre
21 Pazenn
22 Sole
23 Shannon
24 Fastnet
25 Lundy

Fig 0(6) NW France forecast areas

Weather reports from coastal stations follow the 0048 and 0520 shipping forecasts. They include wind direction and force, present weather, visibility, and sea level pressure and tendency, if available. The stations are shown in Fig 0(5).

0.16.1   BBC Radio 4 Inshore waters forecast
A forecast for UK inshore waters (up to 12M offshore), valid for 24 hrs, is broadcast after the 0048 and 0520 coastal station reports. It includes forecasts of wind direction and force, weather, visibility, sea state and an outlook for a further 24 hrs. It ends with a national inshore outlook for the next 3 days.

Reports of actual weather at St Catherine's Pt (automatic) are only broadcast after the 0048 inshore waters forecast. Scilly also features in the 0520 coastal reports.

0.16.2   Channel Islands broadcasts
BBC Radio Guernsey 93·2, 99·0 MHz; 1116 kHz (☎ 01481 200600), broadcasts weather bulletins for the waters around Guernsey, Herm and Sark at 0630, 0730, 0830 LT Mon-Fri; 0730, 0830 LT Sat/Sun.

BBC Radio Jersey 1026 kHz and 88·8 MHz (☎ 01534 870000), storm warnings on receipt; Shipping forecast for local waters at 0625 & 1625 LT Mon-Fri and 0625 & 0725 LT Sat/Sun.

Jersey Met (☎ +44 1534 445500) provides excellent forecasts on VHF, updated 6 hourly. A recorded, chargeable version can be obtained by calling ☎ 0900 665 0022 if in the CI or UK. From France call ☎ +44 1534 448787; from Guernsey only, call ☎ 12080. For costs visit www.jerseymet.gov.je

Jersey Coastguard (☎ 01534 447705) broadcasts gale warnings, synopsis, 24h forecast, outlook for next 24 hrs and reports from observation stations on VHF Ch 82 (after prior announcement on Ch 16) at 0433 UT, 0646LT*, 0745LT*; 0833, 1245, 1633, 1845, 2033, 2245 UT and on request on VHF Ch's 25 and 82. Gale warnings are also broadcast at 0307, 0907, 1507 and 2107 UT. Note:*broadcast 1hr earlier when DST is being applied.

Jersey Coastguard is part of CG and SAR services at St Helier. www.jersey-harbours.com

CROSS (see 0.12) broadcasts MSI inc weather forecasts on VHF at the times below, in English on request.

CROSS GRIS-NEZ

Ch 79   Belgian border to Baie de la Somme

Dunkerque	0720, 1603, 1920
St Frieux	0710, 1545, 1910

Baie de la Somme to Cap de la Hague

L'Ailly	0703, 1533, 1903

CROSS JOBOURG

Ch 80   Baie de la Somme to Cap de la Hague

Antifer	0803, 1633, 2003
Port-en-Bessin	0745, 1615, 1945
Jobourg	0733, 1603, 1933

Cap de la Hague to Pointe de Penmarc'h

Jobourg	0715, 1545, 1915
Granville	0703, 1533, 1903

CROSS CORSEN

Ch 79   Cap de la Hague to Pointe de Penmarc'h

(Times in **bold** = 1 May to 30 Sep only).

Cap Fréhel	0545, 0803, **1203**, 1633, 2003
Bodic	0533, 0745, **1145**, 1615, 1945
Ile de Batz	0515, 0733, **1133**, 1603, 1933
Le Stiff	0503, 0715, **1115**, 1545, 1915

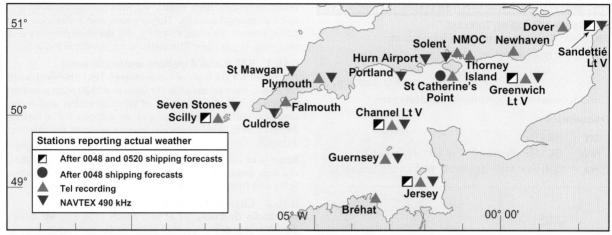

Fig 0(7) Stations reporting actual weather via BBC Radio 4, NAVTEX, telephone recordings

0.18 AUTOMATIC IDENTIFICATION SYSTEM (AIS)

Automatic Identification System (AIS) allows information to be provided to other ships and shore authorities; it is required by SOLAS to be fitted to most vessels over 300 GRT. AIS is widely used to automatically and continuously identify, track and display other vessels and their movements.

Each ship's course and speed vector is shown by a symbol on an AIS screen or overlaid on radar, chart plotter or PC. This data may also appear in a text box as heading, COG & SOG, range, CPA, position, ship's name, and her status – under power or sail, anchored, constrained by draught, restricted in her ability to manoeuvre, not under command, fishing etc.

Many lights, buoys and other aids to navigation (AtoN) are now fitted with AIS, and the number is growing rapidly. On Admiralty charts, these are shown by a magenta circle and the notation 'AIS'. Not all transmitted information is available to all users; it depends on the display system fitted.

AIS Class B transmitters/transceivers for non-SOLAS vessels are available, but are not mandatory. Later sets can receive aids to navigation (A2N) data, showing virtual marks and MMSI numbers from buoys and lighthouses.

Caveats: Many vessels are not fitted with AIS, and some may not have it switched on; some only display 3 lines of text, not a plot; in busy areas only the strongest signals may be shown; AIS may distract a bridge watchkeeper from his visual and radar watch; unlike eyes and radar, AIS does not yet feature in the Colregs; GPS/electronic failures invalidate AIS.

In areas of traffic concentration Class B transmitters may be excluded from the network if no time slots are available.
It is not a radar despite what some advertisements may imply.

0.19 TIDE TABLES AND COMPUTER PREDICTIONS

The Almanac uses UKHO Standard Port tide prediction curves prepared for Admiralty Tide Tables. Standard Port predictions are replacing some established Secondary Ports in ATT to better support offshore activity. The Almanac continues to replicate the existing differences to provide the data in a compact format for planning. These are in italics where alternative information is available. The printed tidal curve is drawn to a mean Spring curve and, in dashed line, a mean

Neap curve, with boxes for associated times of HW and LW. The curve does not take account of the differing arguments (geometrical positions) of the sun, moon and planets on a daily basis. Nor does it take account of differences which owe to the differing volume passing constrictions or obstructions.

The printed tables give condensed planning information. This is refinable from computer generated predictions, such as Easytide, which are accessible on-line and easy to use.

0.20 CALCULATING CLEARANCES BELOW OVERHEAD OBJECTS

A diagram often helps when calculating vertical clearance below bridges, power cables etc. Fig 0(8) shows the relationship to CD. The height of such objects as shown on the chart is usually measured above HAT, so the actual clearance will almost always be more. The height of HAT above CD is given at the foot of each page of the tide tables. Most Admiralty charts now show clearances above HAT, but check the **Heights** block below the chart title.

To calculate clearances, insert the dimensions into the following formula, carefully observing the conventions for brackets:

Masthead clearance = (Height of object above HAT + height of HAT above CD) minus (height of tide at the time + height of the masthead above waterline).

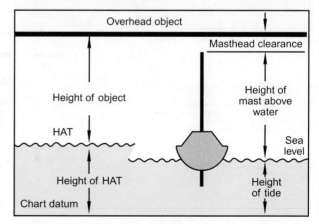

Fig 0(8) Calculating masthead clearance

0.21 IALA BUOYAGE

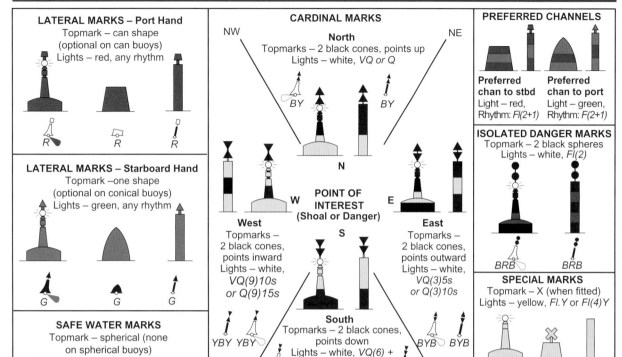

LATERAL MARKS – Port Hand
Topmark – can shape
(optional on can buoys)
Lights – red, any rhythm

R R R

LATERAL MARKS – Starboard Hand
Topmark – one shape
(optional on conical buoys)
Lights – green, any rhythm

G G G

SAFE WATER MARKS
Topmark – spherical (none on spherical buoys)
Lights – white, *Iso, Oc,* or *LFl.10s*

RW RW RW

CARDINAL MARKS

NW   NE

North
Topmarks – 2 black cones, points up
Lights – white, *VQ or Q*

BY BY

N

West
Topmarks –
2 black cones,
points inward
Lights – white,
VQ(9)10s
or *Q(9)15s*

W

POINT OF INTEREST
(Shoal or Danger)

S

E

East
Topmarks –
2 black cones,
points outward
Lights – white,
VQ(3)5s
or *Q(3)10s*

YBY YBY

South
Topmarks – 2 black cones,
points down
Lights – white, *VQ(6) +
LFl.10s*
Q(6) + LFl.15s

BYB BYB

SW YB   YB SE

EMERGENCY WRECK MARKS
Topmark – Upright yellow cross
Lights – *Al BuY 3s 4M*

PREFERRED CHANNELS

Preferred chan to stbd
Light – red,
Rhythm: *Fl(2+1)*

Preferred chan to port
Light – green,
Rhythm: *F(2+1)*

ISOLATED DANGER MARKS
Topmark – 2 black spheres
Lights – white, *Fl(2)*

BRB BRB

SPECIAL MARKS
Topmark – X (when fitted)
Lights – yellow, *Fl.Y* or *Fl(4)Y*

Y Y Y

Y Y

0.22 INTERNATIONAL CODE OF SIGNALS

Code flags, phonetic alphabet (NATO/ITU), Morse code, single-letter signals. INTERNATIONAL PORT TRAFFIC SIGNALS.

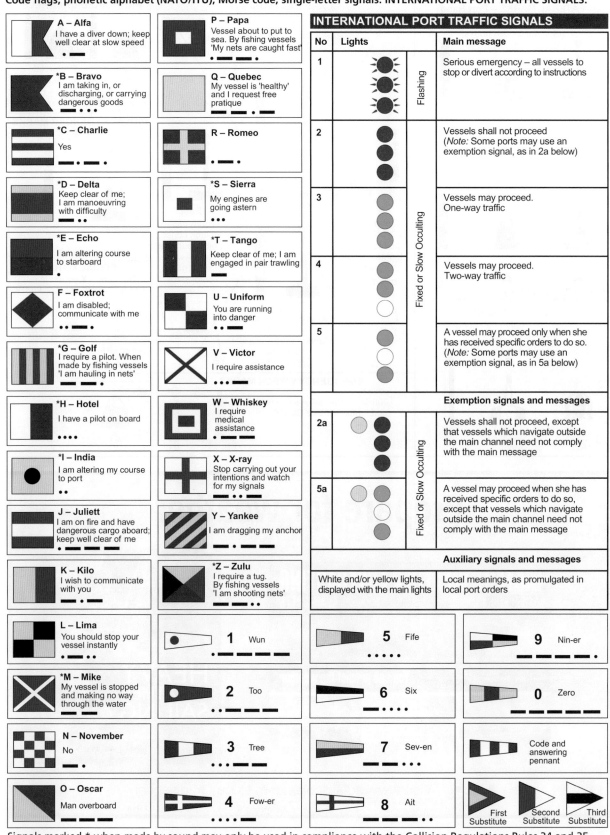

INTERNATIONAL PORT TRAFFIC SIGNALS

No	Lights		Main message
1		Flashing	Serious emergency – all vessels to stop or divert according to instructions
2		Fixed or Slow Occulting	Vessels shall not proceed (*Note:* Some ports may use an exemption signal, as in 2a below)
3			Vessels may proceed. One-way traffic
4			Vessels may proceed. Two-way traffic
5			A vessel may proceed only when she has received specific orders to do so. (*Note:* Some ports may use an exemption signal, as in 5a below)

Exemption signals and messages

No	Lights		Main message
2a		Fixed or Slow Occulting	Vessels shall not proceed, except that vessels which navigate outside the main channel need not comply with the main message
5a			A vessel may proceed when she has received specific orders to do so, except that vessels which navigate outside the main channel need not comply with the main message

Auxiliary signals and messages

White and/or yellow lights, displayed with the main lights	Local meanings, as promulgated in local port orders

Flags and phonetic alphabet

A – Alfa
I have a diver down; keep well clear at slow speed
· —

*B – Bravo
I am taking in, or discharging, or carrying dangerous goods
— · · ·

*C – Charlie
Yes
— · — ·

*D – Delta
Keep clear of me; I am manoeuvring with difficulty
— · ·

*E – Echo
I am altering course to starboard
·

F – Foxtrot
I am disabled; communicate with me
· · — ·

*G – Golf
I require a pilot. When made by fishing vessels 'I am hauling in nets'
— — ·

*H – Hotel
I have a pilot on board
· · · ·

*I – India
I am altering my course to port
· ·

J – Juliett
I am on fire and have dangerous cargo aboard; keep well clear of me
· — — —

K – Kilo
I wish to communicate with you
— · —

L – Lima
You should stop your vessel instantly
· — · ·

*M – Mike
My vessel is stopped and making no way through the water
— —

N – November
No
— ·

O – Oscar
Man overboard
— — —

P – Papa
Vessel about to put to sea. By fishing vessels 'My nets are caught fast'
· — — ·

Q – Quebec
My vessel is 'healthy' and I request free pratique
— — · —

R – Romeo
· — ·

*S – Sierra
My engines are going astern
· · ·

*T – Tango
Keep clear of me; I am engaged in pair trawling
—

U – Uniform
You are running into danger
· · —

V – Victor
I require assistance
· · · —

W – Whiskey
I require medical assistance
· — —

X – X-ray
Stop carrying out your intentions and watch for my signals
— · · —

Y – Yankee
I am dragging my anchor
— · — —

*Z – Zulu
I require a tug. By fishing vessels 'I am shooting nets'
— — · ·

1 Wun
· — — — —

2 Too
· · — — —

3 Tree
· · · — —

4 Fow-er
· · · · —

5 Fife
· · · · ·

6 Six
— · · · ·

7 Sev-en
— — · · ·

8 Ait
— — — · ·

9 Nin-er
— — — — ·

0 Zero
— — — — —

Code and answering pennant

First Substitute   Second Substitute   Third Substitute

Signals marked * when made by sound may only be used in compliance with the Collision Regulations Rules 34 and 35.

0.23 FLAGS AND ENSIGNS

UK WHITE ENSIGN

UK BLUE ENSIGN

UK RED ENSIGN

AUSTRALIA

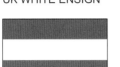

AUSTRIA

BASQUE FLAG

BELGIUM

CANADA

CYPRUS

DENMARK

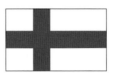

FINLAND

FRANCE

GERMANY

GREECE

GUERNSEY

JERSEY

IRELAND

ISRAEL

ITALY

LIBERIA

MALTA

MONACO

MOROCCO

NETHERLANDS

NEW ZEALAND

NORWAY

PANAMA

POLAND

PORTUGAL

SOUTH AFRICA

SPAIN

SWEDEN

SWITZERLAND

TUNISIA

TURKEY

USA

0.24 LIGHTS AND SHAPES

Vessels being towed and towing

Vessel towed shows sidelights (forward) and sternlight

Tug shows two masthead lights, sidelights, sternlight, yellow towing light

Towing by day — Length of tow more than 200m

Towing vessel and tow display diamond shapes. By night, the towing vessel shows three masthead lights instead of two as for shorter tows

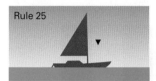

Motor sailing

Cone point down, forward. At night the lights of a power-driven vessel underway

Vessel fishing

All-round red light over all-round white, plus sidelights and sternlight when making way

Fishing/Trawling

A shape consisting of two cones point to point in a vertical line one above the other

Vessel trawling

All-round green light over all-round white, plus sidelights and sternlight when making way

Vessel restricted in her ability to manoeuvre

All-round red, white, red lights vertically, plus normal steaming lights when making way

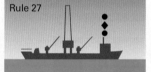

Three shapes in a vertical line: ball, diamond, ball

Not under command

Two all-round red lights, plus sidelights and sternlight when making way

Two balls vertically

Dredger

All round red, white, red lights vertically, plus two all-round red lights (or two balls) on foul side, and two all-round green (or two diamonds) on clear side

Divers down

Letter 'A' International Code

Constrained by draught

Three all-round red lights in a vertical line, plus normal steaming lights. By day — a cylinder

Pilot boat

All-round white light over all-round red, plus sidelights and sternlight when underway, or anchor light

Vessel at anchor

All-round white light; if over 50m, a second light aft and lower

Ball forward

Vessel aground

Anchor light(s), plus two all-round red lights in a vertical line

Three balls in a vertical line

0.25 NAVIGATION LIGHTS

LIGHTS FOR TYPICAL YACHT WITH 3 OPTIONAL VARIANTS

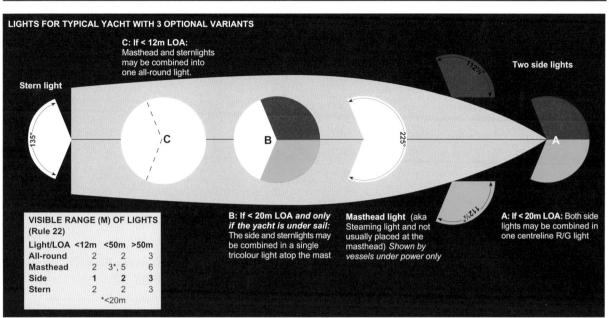

C: If < 12m LOA: Masthead and sternlights may be combined into one all-round light.

Stern light

Two side lights

VISIBLE RANGE (M) OF LIGHTS (Rule 22)			
Light/LOA	<12m	<50m	>50m
All-round	2	2	3
Masthead	2	3*, 5	6
Side	**1**	**2**	**3**
Stern	2	2	3
		*<20m	

B: If < 20m LOA and only if the yacht is under sail: The side and sternlights may be combined in a single tricolour light atop the mast

Masthead light (aka Steaming light and not usually placed at the masthead) *Shown by vessels under power only*

A: If < 20m LOA: Both side lights may be combined in one centreline R/G light

---

PLAN VIEWS OF LIGHTS FOR SAILING VESSELS UNDERWAY AND UNDER SAIL ONLY
Note: If motor-sailing, the lights appropriate for a power-driven vessel must be shown, as below

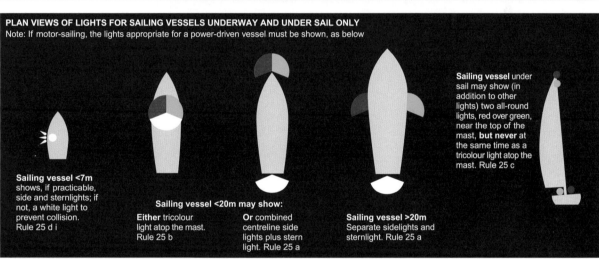

Sailing vessel <7m shows, if practicable, side and sternlights; if not, a white light to prevent collision. Rule 25 d i

Sailing vessel <20m may show:
Either tricolour light atop the mast. Rule 25 b

Or combined centreline side lights plus stern light. Rule 25 a

Sailing vessel >20m Separate sidelights and sternlight. Rule 25 a

Sailing vessel under sail may show (in addition to other lights) two all-round lights, red over green, near the top of the mast, **but never** at the same time as a tricolour light atop the mast. Rule 25 c

---

PLAN VIEWS OF LIGHTS FOR POWER-DRIVEN VESSELS UNDERWAY AND SAILING CRAFT UNDER POWER

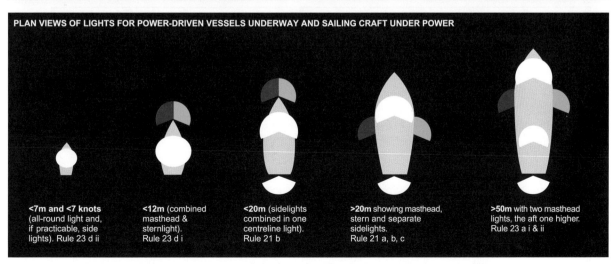

<7m and <7 knots (all-round light and, if practicable, side lights). Rule 23 d ii

<12m (combined masthead & sternlight). Rule 23 d i

<20m (sidelights combined in one centreline light). Rule 21 b

>20m showing masthead, stern and separate sidelights. Rule 21 a, b, c

>50m with two masthead lights, the aft one higher. Rule 23 a i & ii

0.26 TIDAL STREAMS – THE ENGLISH CHANNEL

More UK/Irish tides at www.reedstides.co.uk

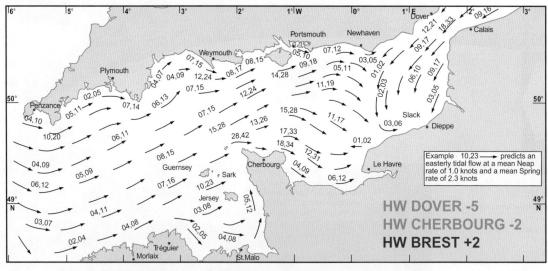

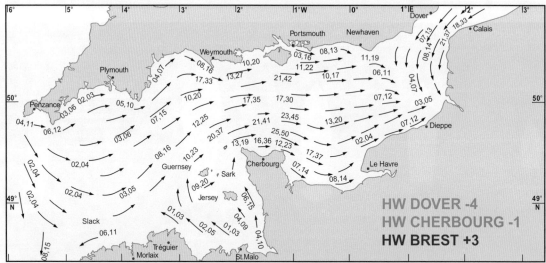

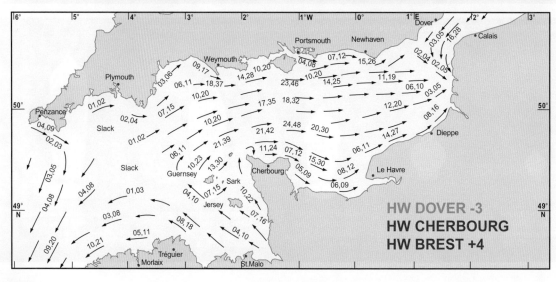

TIDAL STREAMS – THE ENGLISH CHANNEL *contd*

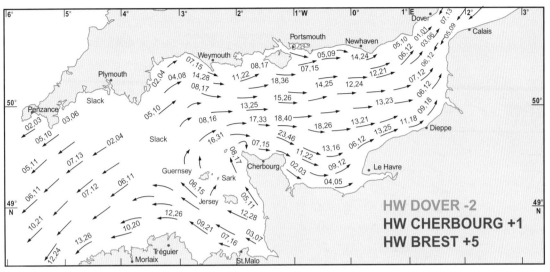

HW DOVER -2
HW CHERBOURG +1
HW BREST +5

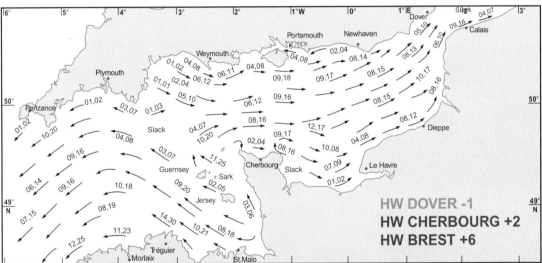

HW DOVER -1
HW CHERBOURG +2
HW BREST +6

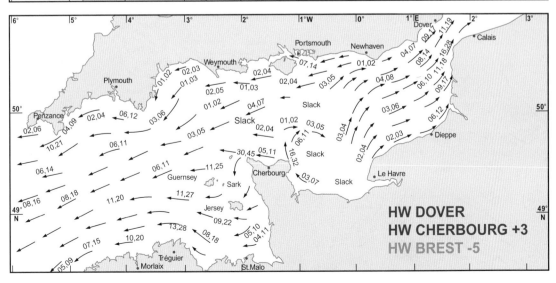

HW DOVER
HW CHERBOURG +3
HW BREST -5

TIDAL STREAMS – THE ENGLISH CHANNEL *contd*

More UK/Irish tides at www.reedstides.co.uk

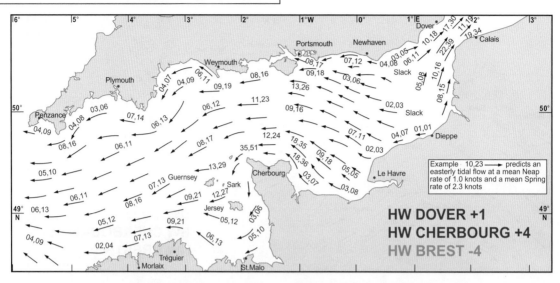

HW DOVER +1
HW CHERBOURG +4
HW BREST -4

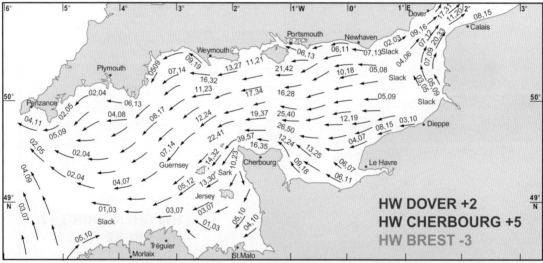

HW DOVER +2
HW CHERBOURG +5
HW BREST -3

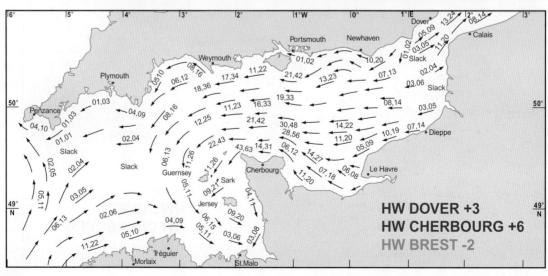

HW DOVER +3
HW CHERBOURG +6
HW BREST -2

TIDAL STREAMS – THE ENGLISH CHANNEL *contd*

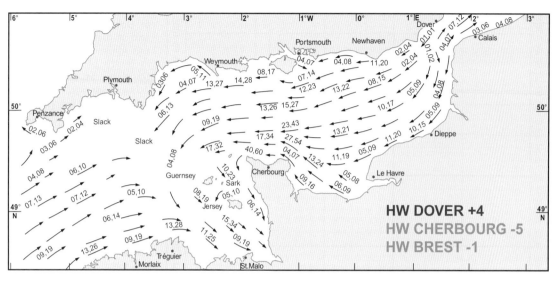

HW DOVER +4
HW CHERBOURG -5
HW BREST -1

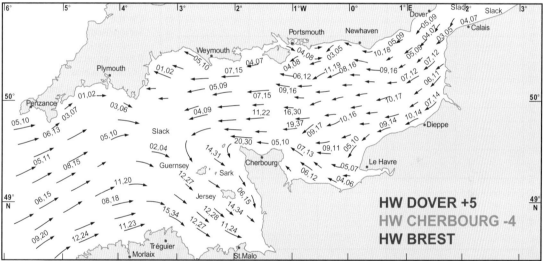

HW DOVER +5
HW CHERBOURG -4
HW BREST

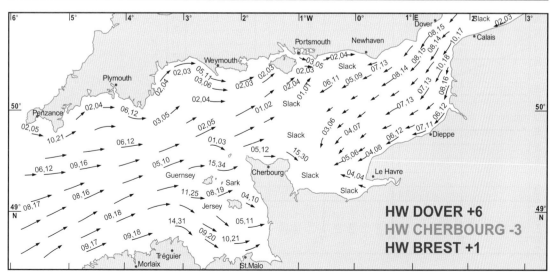

HW DOVER +6
HW CHERBOURG -3
HW BREST +1

0.27 CALCULATING TIDAL STREAM RATES

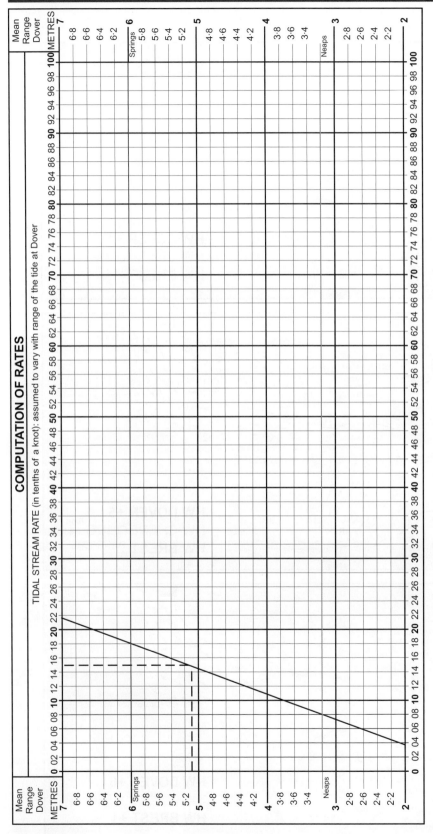

The tidal stream rate at any time may be calculated, assuming that it varies with the range of the tide at Dover. In tidal stream atlases, and on the tidal stream chartlets in this Almanac, the rates are shown in tenths of a knot. Thus '05,27' translates as 0·5k at Neaps and 2·7kn at Springs.

Example: Calculate the tidal stream rate off the north tip of Skye at 0420 UT on a day when the tide at Dover is:

UT	Ht (m)
0328	1·4
0819	6·3
1602	1·1
2054	6·4

The tidal stream chartlet for Skye [Dover HW–4] shows '08,18'. This is a mean Neap rate of 0·8kn and Spring rate of 1·8kn.

On the left hand figure identify the Spring (Red) and Neap (Blue) axes. On these mark the rate of 1.8kn and 0.8kn respectively, using the upper or lower scales.

Draw a diagonal line through these.

Calculate the tidal ranges for Dover:

(6·3 -1·4) = 4·9m   (6·4-1·1)=5·3m

The mean range = (4·9+5·3)/2 = 5·1m

On the vertical axis identify 5·1m and draw a horizontal line cutting the diagonal just drawn. From this intersection read off the required rate on either horizontal axis.

In the example this is 15 (1·5kn).

Times are in UT - add 1 hour in non-shaded areas to convert to BST

0.28 SUNRISE/SET TIMES 2019

The table shows times of Sunrise (SR) and Sunset (SS) for every 3rd day as the times of Sunrise and Sunset never change by more than 8 minutes (and often by only 1–3 minutes) between the given dates.

The table is based on Longitude 0°, so longitude corrections are required, ie add 4 minutes of time for every degree West of Greenwich; subtract if East.

LATITUDE 50°N

	Rise	Set	Rise	Set	Rise	Set	Rise	Set	Rise	Set	Rise	Set
	JANUARY		FEBRUARY		MARCH		APRIL		MAY		JUNE	
1	07 59	16 09	07 34	16 54	06 44	17 42	05 38	18 31	04 37	19 18	03 56	20 00
4	07 58	16 12	07 30	16 59	06 38	17 46	05 31	18 36	04 32	19 23	03 54	20 03
7	07 57	16 15	07 25	17 04	06 32	17 51	05 25	18 41	04 27	19 27	03 53	20 06
10	07 56	16 19	07 20	17 09	06 25	17 56	05 18	18 45	04 22	19 32	03 51	20 08
13	07 54	16 23	07 15	17 14	06 19	18 01	05 12	18 50	04 18	19 36	03 51	20 10
16	07 52	16 28	07 09	17 20	06 12	18 06	05 06	18 55	04 13	19 40	03 50	20 11
19	07 50	16 32	07 04	17 25	06 06	18 11	05 00	19 00	04 09	19 44	03 50	20 12
22	07 47	16 37	06 58	17 30	05 59	18 16	04 54	19 04	04 06	19 48	03 51	20 13
25	07 43	16 42	06 52	17 35	05 53	18 20	04 48	19 09	04 02	19 52	03 52	20 13
28	07 39	16 47	06 46	17 40	05 46	18 25	04 43	19 14	03 59	19 56	03 53	20 13
31	07 35	16 52			05 40	18 30			03 57	19 59		

	JULY		AUGUST		SEPTEMBER		OCTOBER		NOVEMBER		DECEMBER	
1	03 55	20 13	04 29	19 43	05 14	18 45	05 59	17 39	06 49	16 37	07 36	16 01
4	03 57	20 12	04 33	19 38	05 19	18 38	06 04	17 33	06 54	16 32	07 40	16 00
7	03 59	20 10	04 37	19 33	05 23	18 32	06 09	17 26	06 59	16 28	07 44	15 59
10	04 02	20 08	04 42	19 28	05 28	18 25	06 13	17 20	07 04	16 23	07 47	15 58
13	04 05	20 06	04 46	19 23	05 32	18 19	06 18	17 14	07 09	16 19	07 50	15 58
16	04 08	20 03	04 51	19 17	05 37	18 12	06 23	17 08	07 14	16 15	07 52	15 58
19	04 12	20 00	04 55	19 11	05 41	18 06	06 28	17 02	07 18	16 12	07 54	15 59
22	04 15	19 57	04 59	19 05	05 46	17 59	06 32	16 56	07 23	16 08	07 56	16 01
25	04 19	19 53	05 04	18 59	05 50	17 52	06 37	16 50	07 28	16 06	07 57	16 02
28	04 23	19 49	05 08	18 53	05 55	17 46	06 42	16 44	07 32	16 03	07 58	16 05
31	04 27	19 45	05 13	18 47			06 47	16 39			07 59	16 07

0.29 MOONRISE/SET TIMES 2019

The table below gives the times of Moonrise (MR) and Moonset (MS) for every 3rd day; interpolation is necessary for other days. The aim is simply to indicate whether the night in question will be brightly moonlit, partially moonlit

or pitch black – depending, of course, on cloud cover. The table is based on Longitude 0°. To correct for longitude, add 4 minutes of time for every degree West; subtract if East. ** Indicates that the phenomenon does not occur.

LATITUDE 50°N

	Rise	Set	Rise	Set	Rise	Set	Rise	Set	Rise	Set	Rise	Set
	JANUARY		FEBRUARY		MARCH		APRIL		MAY		JUNE	
1	03 06	13 26	05 16	13 50	04 05	12 35	04 34	14 22	03 47	15 23	03 15	17 41
4	06 24	15 07	07 29	16 35	06 03	15 29	05 42	17 36	04 49	18 48	04 54	21 15
7	08 52	17 42	08 49	19 44	07 16	18 41	06 45	21 00	06 16	22 20	07 58	23 48
10	10 22	20 49	09 51	23 00	08 18	22 00	08 18	** **	08 57	00 20	11 52	00 49
13	11 26	** **	11 06	01 18	09 40	00 18	11 07	02 21	12 45	02 16	15 42	01 59
16	12 37	02 23	13 24	04 45	12 11	03 32	15 02	04 12	16 40	03 30	19 19	03 23
19	14 44	06 01	17 15	07 13	16 05	05 42	19 03	05 28	20 27	04 50	22 02	05 39
22	18 26	08 45	21 22	08 41	20 12	07 04	22 46	06 55	23 24	06 56	23 35	08 44
25	22 28	10 17	** **	09 59	23 57	08 27	00 44	09 09	00 40	09 54	00 17	11 58
28	00 56	11 30	03 11	11 47	02 00	10 28	02 38	12 09	01 52	13 08	01 17	15 20
31	04 18	13 07			04 05	13 19			02 52	16 29		

	JULY		AUGUST		SEPTEMBER		OCTOBER		NOVEMBER		DECEMBER	
1	02 46	18 56	04 32	20 18	07 36	20 10	09 13	19 25	11 41	19 55	11 53	20 40
4	05 40	21 45	08 40	21 46	11 37	21 28	12 53	21 16	13 53	22 56	13 11	23 57
7	09 39	23 19	12 36	22 59	15 02	23 23	15 19	** **	15 05	01 06	14 05	02 06
10	13 32	00 05	16 07	00 00	17 16	01 13	16 39	02 13	16 00	04 22	15 11	05 30
13	17 09	01 24	18 39	02 21	18 33	04 23	17 36	05 28	17 11	07 47	17 14	08 57
16	19 59	03 31	20 06	05 27	19 30	07 37	18 40	08 49	19 22	11 06	20 44	11 21
19	21 38	06 32	21 05	08 40	20 37	10 56	20 32	12 10	22 55	13 18	** **	12 42
22	22 41	09 45	22 08	11 57	22 39	14 16	23 51	14 40	01 35	14 36	03 18	13 50
25	23 42	13 02	23 57	15 23	00 54	16 41	02 33	16 09	05 37	15 48	07 05	15 32
28	00 39	16 34	02 03	18 09	05 03	18 09	06 43	17 22	09 24	17 41	09 48	18 22
31	03 16	19 36	06 12	19 45			10 37	19 06			11 14	21 41

0.30   AREA INFORMATION

The 7 geographic Areas are arranged as follows:

An Area map which includes harbours, principal lights, TSS, MRCCs, NCI stations, airports, main ferry routes, magnetic variation and a distance table. Wind farms and other offshore energy installations are not routinely shown.

Tidal stream chartlets showing hourly rates and set.

Lights, buoys and waypoints (LBW) listing positions and characteristics of selected lights and other marks, their daytime appearance, fog signals and Racons. Arcs of visibility and alignment of sector/leading lights are true bearings as seen from seaward. Lights are white unless otherwise stated; any colours are shown between the bearings of the relevant arcs. AIS is widely fitted to navigational marks, but not normally shown in LBW. See the relevant official charts/publications for details.

Passage Information (PI) is at Section 3 in Areas 1,2,3,4 and 7 and at Section 4 in Areas 5 and 6. Further information is geographically arranged between the harbour entries.

Special notes giving data specific to a country or area.

Harbour information (see below).

0.31   HARBOUR INFORMATION

Each harbour entry is arranged as follows:

HARBOUR NAME followed by the County or Unitary Council (or foreign equivalent) and the lat/long of the harbour entrance, or equivalent, for use as the final waypoint.

Harbour ratings (❀ ⚓ ✿), inevitably subjective, which grade a port based on the following criteria:

Ease of access:

❀❀❀   *Can be entered in almost any weather from most directions and at all states of tide, by day or night.*

❀❀   *Accessible in strong winds from most directions; possible tidal or pilotage constraints.*

❀   *Only accessible in calm, settled conditions by day with little or no swell; possible bar and difficult pilotage.*

Facilities available:

⚓⚓⚓   *Good facilities for vessel and crew.*

⚓⚓   *Most domestic needs catered for, but limited boatyard facilities.*

⚓   *Possibly some domestic facilities, but little else.*

Ambience:

✿✿✿   *An attractive place; well worth visiting.*

✿✿   *Average for this part of the coast.*

✿   *Holds no particular attraction.*

CHARTS show Admiralty (AC), Imray, and foreign charts, all smallest scale first. Admiralty Leisure Folios (56XX), which cover most of the UK, Channel Islands and Ireland, and Imray 2000 series folios (2X00) are also shown.

TIDES include a time difference (usually on Dover in the UK), ML, Duration and the harbour's Standard Port. Time and height differences for Secondary Ports are also shown. Standard Port computer prediction is replacing some Secondary Port differences in ATT. The Almanac continues to replicate these differences in italics when alternative information is available on the website.

SHELTER assesses how protected a harbour is from wind, sea, surge and swell. It warns of any access difficulties and advises on safe berths and anchorages.

NAVIGATION gives guidance on the approach and entry, and shows the position of the approach waypoint with its bearing and distance to the harbour entrance or next significant feature. Some waypoints may not be shown on the chartlet. Access times are only stated where a lock, gate, sill or other obstruction restricts entry. Otherwise the minimum charted depth of water in the approaches, where it is less than 2m, is usually shown, but always consult up to date official charts.

Chartlets are based on official charts augmented with local information. Due to their scale, they may not cover the whole area referred to in the text nor do they show every depth, mark, light or feature.

The chartlets are not intended to be used for navigation; positions taken from them should not be used as waypoints in chart plotters. The publisher and editors disclaim any responsibility for resultant accidents or damage if they are so used. The largest scale official chart, properly corrected, should always be used.

Drying areas and an indicative 5m depth contour are shown as: Dries   <5m   >5m

Wrecks around the UK which are of archaeological or historic interest are protected by law. Sites are listed under harbour entries or in Passage Information. Unauthorised interference, including anchoring and diving on these sites, may lead to a substantial fine.

LIGHTS AND MARKS describes, in more detail than is shown on the chartlets, any unusual characteristics of marks, their appearance by day and features not listed elsewhere.

COMMUNICATIONS shows the telephone area code followed by local telephone and VHF contact details for: MRCC/CG, weather, police, doctor/medical, harbourmaster/office, other. Marina contact details are not usually duplicated if they are shown under the marina entry. International telephone calls from/to the UK are described in Special Notes, as are national numbers for emergency services: normally 112 in the EU; 999 in the UK. Radio callsigns, if not obvious, are in *italics*.

FACILITIES describes berthing options and facilities in harbours, marinas and yacht clubs (see the free **Reeds Marina Guide** for detailed marina plans in the UK, Channel Islands and Ireland). Water, electricity, showers and toilets are available in marinas unless otherwise stated. Most yacht clubs welcome visiting crews who belong to a recognised club and arrive by sea. Any rail and air links are also shown.

The overnight cost of a visitor's alongside berth (AB) *is for comparison only and based on information supplied at the time of going to press (Summer 2018).* is the average charge per metre LOA (unless otherwise stated) during high season, usually June to Sept. It includes VAT, harbour dues and, where possible, any tourist taxes (per head). The cost of pile moorings, ⚓s or ⚓s, where these are the norm, may also be given. Shore electricity is usually free abroad but extra in the UK.

The number of ❶ berths is a marina's estimate of how many visitors may be accommodated at any one time. It is always advisable to call the marina beforehand.

TransEurope Marinas (www.transeuropemarinas.com) is an expanding grouping of independent marinas in the UK and abroad. Many hold Blue Flags and 4 or 5 Gold Anchor Awards; it is a condition of membership that they are well-equipped and maintain high standards. They operate a discounted reciprocal berthing scheme and are shown by the symbol ⚓.

ce data

0.32 ENVIRONMENTAL GUIDANCE

- Comply with regulations for navigation and conduct within Marine Nature Reserves, Particularly Sensitive Sea Areas (PSSA) and National Water Parks.
- In principle never ditch rubbish at sea, keep it on board and dispose of it in harbour refuse bins.
- Readily degradable foodstuffs may be ditched at sea when >3M offshore (>12M in the English Channel).
- Foodstuffs and other materials which are not readily degradable should never be ditched at sea.
- Sewage. If you do not have a holding tank, only use the onboard heads when well offshore. A holding tank should be fitted as soon as possible as many countries require them. Pump-out facilities (⚓) are shown in the text.
- Do not pump out holding tanks until >3M offshore.
- Do not discharge foul water into a marina, anchorage or moorings area and minimise on washing-up water.
- Deposit used engine oil and oily waste ashore at a recognised facility. Do not allow an automatic bilge pump to discharge oily bilge water overboard.
- Dispose of toxic waste, (eg some antifoulings, cleaning chemicals, old batteries) at an approved disposal facility.
- Row ashore whenever possible – to minimise noise, wash and disturbance. Land at recognised places.
- Respect wild birds, plants, fish and marine animals. Avoid protected nesting sites and breeding colonies.
- Do not anchor or dry out on vulnerable seabed species, eg soft corals, eel grass.

0.33 DISTANCES (M) ACROSS THE ENGLISH CHANNEL

Approximate distances in nautical miles are by the most direct practicable route and allowing for TSS/shipping lanes.

France/CI \ England	Longships	Falmouth	Fowey	Plymouth bkwtr	Salcombe	Dartmouth	Torbay	Exmouth	Weymouth	Poole Hbr Ent	Needles Lt Ho	Nab Tower	Littlehampton	Shoreham	Brighton	Newhaven	Eastbourne	Folkestone	Dover	Ramsgate
Le Conquet	114	118	134	142	139	152	158	168	181	199	203	236	256	260	264	269	281	313	318	342
L'Aberwrac'h	101	103	114	120	122	133	142	151	171	191	197	215	236	246	247	248	258	297	301	327
Roscoff	110	97	101	97	91	100	107	117	130	144	149	165	184	193	197	200	211	246	252	298
Trébeurden	118	105	107	100	95	103	110	122	128	148	148	168	186	197	201	206	218	257	262	288
Tréguier	131	114	112	102	93	99	104	114	118	132	136	153	170	182	187	192	211	252	253	277
Lézardrieux	139	124	119	107	97	102	105	115	117	129	133	150	167	179	184	189	208	249	250	274
St Quay-Portrieux	151	134	128	120	109	114	115	123	128	133	136	151	174	180	185	188	200	238	242	270
St Cast Le Guildo	164	145	139	127	114	117	120	128	124	129	132	147	170	177	181	184	198	234	239	264
St Malo	172	152	145	132	118	121	124	132	125	130	133	148	170	176	182	185	195	235	240	264
St Helier	156	133	124	110	95	97	100	109	103	108	110	125	145	155	160	163	175	211	218	241
St Peter Port	140	115	105	89	75	71	73	81	81	87	90	105	126	135	139	142	155	190	196	227
Braye (Alderney)	149	129	107	90	74	72	71	74	63	68	71	86	108	115	120	123	135	173	178	202
Cherbourg	173	144	130	111	98	94	93	95	67	64	63	72	89	99	102	107	120	153	158	185
St Vaast	194	165	151	134	119	117	119	127	86	78	72	69	84	91	94	97	104	140	147	178
Ouistreham	231	203	190	173	158	155	159	164	121	111	100	88	93	94	94	93	96	132	133	160
Deauville	238	210	198	181	164	163	160	160	127	114	103	89	90	89	87	87	89	124	126	152
Le Havre	239	210	197	180	164	162	165	159	125	112	98	84	84	83	82	82	83	119	120	146
Fécamp	246	218	206	188	173	169	165	164	127	111	98	81	72	69	66	63	63	98	98	122
Dieppe	273	245	231	214	200	194	191	189	151	133	121	97	86	76	72	67	62	78	80	100
Boulogne	294	264	247	231	216	204	202	199	161	143	132	101	86	77	73	65	54	28	28	51
Calais	307	276	261	244	229	218	214	211	174	155	143	114	99	91	86	78	64	29	24	30

Safety at sea

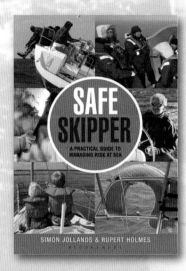

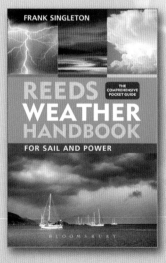

The friendly marina in the heart of
Falmouth, Cornwall

- **Fronting Marina Square, with a choice of restaurants overlooking the harbour**

- Adjacent to the Maritime Museum, in the centre of town

- **Berthing for yachts to over 70m l.o.a & 4m draft**

- Luxury shower facilities

- **Laundry**

- Yachtsmans lounge

- **Free WiFi**

- Tennis court

ort

endenr

" The perfect stopover f
Transatlantic & Bisc
passage

Falmouth, Cornwall, TR11 3
Tel: 01326 211211 | Fax: 01326 3111
Email: marina@portpendennis.c
VHF Channel

www.portpendennis.cor

Southern England

Isles of Scilly to Ramsgate

S England

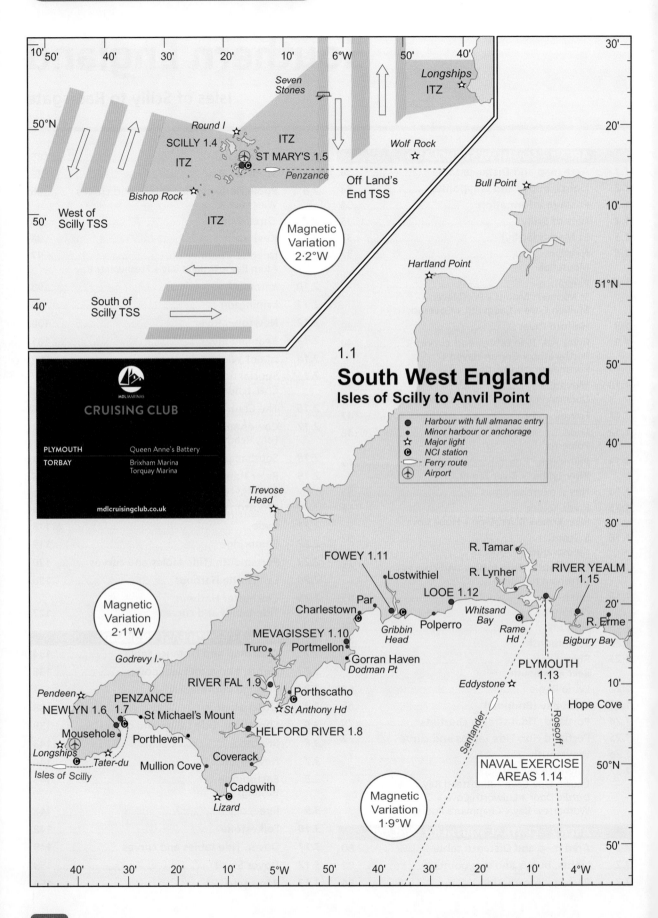

10' 50' 40' 30' 20' 10' 6°W 50' 40' 30'

Seven Stones

Longships ITZ

50°N

Round I

SCILLY 1.4

ITZ

ITZ

ST MARY'S 1.5

Penzance

Wolf Rock

Bishop Rock

Off Land's End TSS

Bull Point

Hartland Point

51°N

West of Scilly TSS

ITZ

Magnetic Variation 2·2°W

40' South of Scilly TSS

1.1
South West England
Isles of Scilly to Anvil Point

	Harbour with full almanac entry
	Minor harbour or anchorage
☆	Major light
Ⓒ	NCI station
	Ferry route
✈	Airport

MDL MARINAS

CRUISING CLUB

PLYMOUTH	Queen Anne's Battery
TORBAY	Brixham Marina
	Torquay Marina

mdlcruisingclub.co.uk

Trevose Head

FOWEY 1.11

Lostwithiel

R. Tamar

R. Lynher

RIVER YEALM 1.15

Par

LOOE 1.12

Charlestown

Magnetic Variation 2·1°W

Godrevy I.

MEVAGISSEY 1.10

Truro

Portmellon

Gorran Haven

Dodman Pt

Gribbin Head

Polperro

Whitsand Bay

Rame Hd

R. Erme

Bigbury Bay

RIVER FAL 1.9

Porthscatho

St Anthony Hd

PLYMOUTH 1.13

Eddystone

Hope Cove

Pendeen

NEWLYN 1.6

PENZANCE 1.7

St Michael's Mount

HELFORD RIVER 1.8

Mousehole

Porthleven

Longships

Tater-du

Mullion Cove

Coverack

Isles of Scilly

Cadgwith

Lizard

Magnetic Variation 1·9°W

Santander

Roscoff

NAVAL EXERCISE AREAS 1.14

50°N

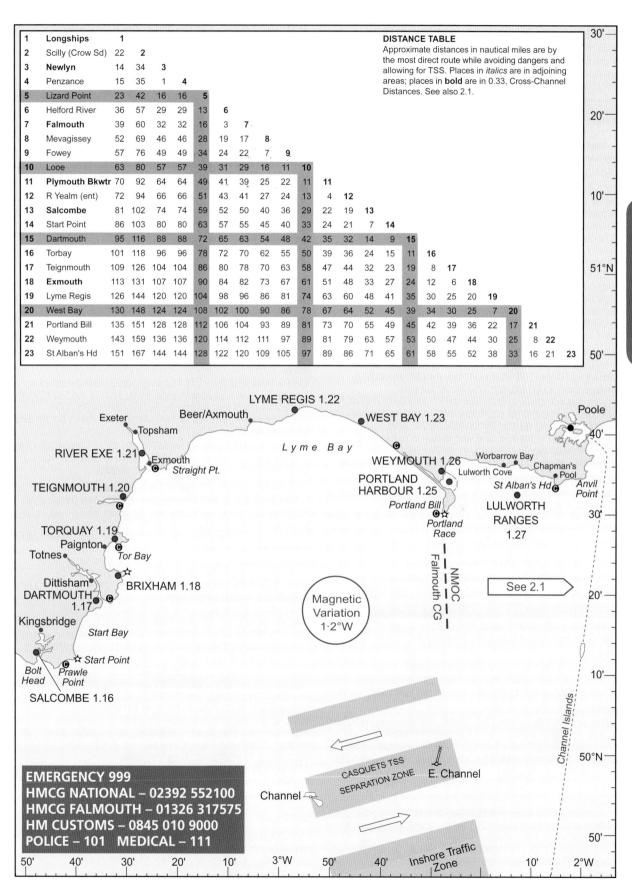

DISTANCE TABLE
Approximate distances in nautical miles are by the most direct route while avoiding dangers and allowing for TSS. Places in *italics* are in adjoining areas; places in **bold** are in 0.33, Cross-Channel Distances. See also 2.1.

#	Place	1	2	3	4	5	6	7	8	9	10	11	12	13	14	15	16	17	18	19	20	21	22	23
1	Longships	1																						
2	Scilly (Crow Sd)	22	2																					
3	Newlyn	14	34	3																				
4	Penzance	15	35	1	4																			
5	Lizard Point	23	42	16	16	5																		
6	Helford River	36	57	29	29	13	6																	
7	Falmouth	39	60	32	32	16	3	7																
8	Mevagissey	52	69	46	46	28	19	17	8															
9	Fowey	57	76	49	49	34	24	22	7	9														
10	Looe	63	80	57	57	39	31	29	16	11	10													
11	Plymouth Bkwtr	70	92	64	64	49	41	39	25	22	11	11												
12	R Yealm (ent)	72	94	66	66	51	43	41	27	24	13	4	12											
13	Salcombe	81	102	74	74	59	52	50	40	36	29	22	19	13										
14	Start Point	86	103	80	80	63	57	55	45	40	33	24	21	7	14									
15	Dartmouth	95	116	88	88	72	65	63	54	48	42	35	32	14	9	15								
16	Torbay	101	118	96	96	78	72	70	62	55	50	39	36	24	15	11	16							
17	Teignmouth	109	126	104	104	86	80	78	70	63	58	47	44	32	23	19	8	17						
18	Exmouth	113	131	107	107	90	84	82	73	67	61	51	48	33	27	24	12	6	18					
19	Lyme Regis	126	144	120	120	104	98	96	86	81	74	63	60	48	41	35	30	25	20	19				
20	West Bay	130	148	124	124	108	102	100	90	86	78	67	64	52	45	39	34	30	25	7	20			
21	Portland Bill	135	151	128	128	112	106	104	93	89	81	73	70	55	49	45	42	39	36	22	17	21		
22	Weymouth	143	159	136	136	120	114	112	111	97	89	81	79	63	57	53	50	47	44	30	25	8	22	
23	St Alban's Hd	151	167	144	144	128	122	120	109	105	97	89	86	71	65	61	58	55	52	38	33	16	21	23

EMERGENCY 999
HMCG NATIONAL – 02392 552100
HMCG FALMOUTH – 01326 317575
HM CUSTOMS – 0845 010 9000
POLICE – 101   MEDICAL – 111

1.2 LIGHTS, BUOYS AND WAYPOINTS

Bold print = light with a nominal range of 15M or more. CAPITALS = place or feature. *CAPITAL ITALICS* = light-vessel, light float or Lanby. *Italics* = Fog signal. ***Bold italics*** = Racon. Many marks/buoys are fitted with AIS, <u>MMSI No</u>; see relevant charts.

ISLES OF SCILLY TO LAND'S END

Bishop Rock ☆ Fl (2) 15s 44m **20M**; part obsc 204°-211°, obsc 211°-233° and 236°-259°; Gy ○ twr with helo platform; ***Racon T, 18M, 254°-215°***; 49°52'·37N 06°26'·74W; <u>992351137</u>.
Round Rk ⚓ 49°53'·10N 06°25'·19W.
Old Wreck ⚓ VQ; 49°54'·26N 06°22'·81W.

ST AGNES and ST MARY'S

Peninnis Hd ⚡ Fl 20s 36m 9M; 231°-117° but part obsc 048°-083° within 5M; W ○ twr on B frame, B cupola; 49°54'·28N 06°18'·21W.
Gugh ⚓ Fl (5) Y 20s; 49°53'·52N 06°18'·73W.
Spanish Ledge ⚓ Q (3) 10s; *Bell;* 49°53'·94N 06°18'·86W.
Woolpack ⚓ Fl G 5s; 49°54'·40N 06°19'·37W.
Bartholomew Ledges ⚓ QR 12m; 49°54'·37N 06°19'·89W.
N Bartholomew ⚓ Fl R 5s; 49°54'·49N 06°19'·99W.
Bacon Ledge ⚓ Fl (4) R 5s; 49°55'·22N 06°19'·26W.
Ldg lts 097·3°: Front, Iso RW (vert) 2s; W △, 49°55'·12N 06°18'·50W. Rear, Oc WR (vert) 10s; Or X on W bcn.
Crow Rock ⚓ Fl (2) 10s; 49°56'·26N 06°18'·49W.
Hats ⚓ VQ (6) + L Fl 10s; 49°56'·21N 06°17'·14W.

AROUND TRESCO, BRYHER and ST MARTIN'S

Tresco Flats, Hulman ⚓ Fl G 4s, 49°56'·29N 06°20'·30W.
Little Rag Ledge ⚓ Fl (2) R 5s, 49°56'·44N 06°20'·43W.
Bryher, Bar ⚓ Q (3) 10s, 49°57'·37N 06°20'·84W.
Bryher, Church Quay ⚓ Q (3) 10s, 49°57'·18N 06°20'·97W.
Spencers Ledge ⚓ Q (6) + L Fl 15s; 49°54'·78N 06°22'·06W.
Steeple Rock ⚓ Q (9) 15s; 49°55'·46N 06°24'·24W.
Round Island ☆ Fl 10s 55m **18M**, also shown in reduced vis; 021°-288°; W ○ twr; *Horn (4) 60s;* 49°58'·74N 06°19'·39W.
St Martin's, Higher Town quay ⚡ Fl R 5s, 49°57'·45N 06°16'·84W.

SCILLY to LAND'S END

Seven Stones Lt V *⚓ Fl (3) 30s 12m* **15M**; R hull; *Horn (3) 60s;* ***Racon O, 15M;*** 50°03'·63N 06°04'·32W; <u>992351023</u>.
Wolf Rock ☆ Fl 15s 34m **16M**; H24; *Horn 30s;* ***Racon T, 10M;*** 49°56'·72N 05°48'·55W; <u>992351128</u>.
Longships ☆ Fl (2) WR 10s 35m **W15M**, R11M; 189°-R-327°-W-189°; also shown in reduced vis; Gy ○ twr with helicopter platform; *Horn 10s;* 50°04'·01N 05°44'·81W.
Carn Base ⚓ Q (9) 15s; 50°01'·48N 05°46'·18W.
Runnel Stone ⚓ Q (6) + L Fl 15s; *Whis;* 50°01'·18N 05°40'·36W.

LAND'S END TO PLYMOUTH

Tater-du ☆ Fl (3) 15s 34m **20M**; 241°-072°; W ○ twr. FR 31m 13M, 060°-072° over Runnel Stone; 50°03'·14N 05°34'·67W.

NEWLYN

Low Lee ⚓ Q (3) 10s; 50°05'·56N 05°31'·38W.
S Pier ⚡ Fl 5s 10m 9M; W ○ twr; 253°-336°; 50°06'·18N 05°32'·57W.
N Pier ⚡ F WG 4m 2M; 238°-G-248°, 50°06'·18N 05°32'·62W.

PENZANCE

S Pier ⚡ Fl WR 5s 11m **W17M**, R12M; 159°-R (unintens)-224°-R-268°-W-344·5°-R-shore; 50°07'·06N 05°31'·68W.
Mountamopus ⚓ Q (6) + L Fl 15s; 50°04'·62N 05°26'·25W.
Lizard ☆ Fl 3s 70m **26M**; 250°-120°, partly visible 235°-250°; W 8-sided twr; *Horn 30s;* 49°57'·61N 05°12'·13W.
Manacle ⚓ Q (3) 10s; *Bell;* 50°02'·81N 05°01'·91W.

FALMOUTH

St Anthony Head ☆ Iso WR 15s 22m, **W16M**, R14M, H24; 295°-W-004°-R (over Manacles)-022°-W-172°; W 8-sided twr; *Horn 30s;* 50°08'·46N 05°00'·96W.
Black Rock Fl (2) 10s 3M; B IDM bcn twr; 50°08'·72N 05°02'·00W.
Black Rock ⚓ Fl R 2·5s; 50°08'·68N 05°01'·74W.
Castle ⚓ Fl G 2·5s; 50°08'·99N 05°01'·62W.

St Mawes ⚓ Q (6) + L Fl 15s; 50°09'·10N 05°01'·42W.
The Governor ⚓ VQ (3) 5s; 50°09'·15N 05°02'·40W.
West Narrows ⚓ Fl (2) R 10s; 50°09'·39N 05°02'·07W.
East Narrows ⚓ Fl (2) G 10s; 50°09'·43N 05°01'·90W.
The Vilt ⚓ Fl (4) G 15s; 50°09'·99N 05°02'·28W.
Northbank ⚓ Fl R 4s; 50°10'·34N 05°02'·26W.
St Just ⚓ QR; 50°10'·44N 05°01'·72W.
Mylor appr chan ⚓ Fl G 6s; 50°10'·79N 05°02'·70W. ⚓ Fl R 5s.
Messack ⚓ Fl G 15s; 50°11'·31N 05°02'·22W.
Carrick ⚓ Fl (2) G 10s; 50°11'·59N 05°02'·74W.
Pill ⚓ Fl (3) G 15s; 50°12'·05N 05°02'·40W.
Turnaware Bar ⚓ Fl G 5s; 50°12'·40N 05°02'·15W.
Dock Basin Dir ⚡ 266°: WRG 2s 15m, 10M; 258°-Iso G-226°-Al WG-264°-Iso W-268°-Al WR-270°-Iso R-274°; 50°09'·38N 05°03'·29W.
N Arm ⚡ QR 5m 3M; 50°09'·42N 05°03'·20W.
Inner Hbr Dir ⚡ 233°: WRG 3s 5m, 9M; 226°-Iso G-231°-Iso W-236°-Iso R-241°; 50°09'·20N 05°03'·94W.
Falmouth Haven Marina ⚡ 2 FR (vert); 50°09'·27N 05°03'·91W.
Falmouth Marina ⚓ VQ (3) 5s; 50°09'·91N 05°04'·99W.

DODMAN POINT and MEVAGISSEY

Naval gunnery targets SSE of Dodman Point:
'A' ⚓ Fl Y 10s; 50°08'·53N 04°46'·37W.
'B' ⚓ Fl Y 5s; 50°10'·30N 04°45'·00W.
'C' ⚓ Fl Y 2s; 50°10'·40N 04°47'·51W.
Gwineas ⚓ Q (3) 10s; *Bell;* 50°14'·48N 04°45'·40W.
Mevagissey, Victoria Pier ⚡ Fl (2) 10s 9m 12M; *Dia 30s;* 50°16'·15N 04°46'·92W.

FOWEY

Cannis Rock ⚓ Q (6) + L Fl 15s; *Bell;* 50°18'·45N 04°39'·88W.
Fowey ⚡ L Fl WR 5s 28m W11M, R9M; 284°-R-295°-W-028°-R-054°; W 8-sided twr, R lantern; 50°19'·63N 04°38'·83W.
St Catherine's Pt ⚡ Fl R 2·5s 15m 2M; vis 150°-295°; 50°19'·69N 04°38'·66W.
Lamp Rock ⚓ Fl G 5s 3m 2M; vis 357°-214°; 50°19'·70N 04°38'·41W.
Whitehouse Pt ⚡ Iso WRG 3s 11m W11M, R/G8M; 017°-G-022°-W-032°-R-037°; R col; 50°19'·98N 04°38'·28W.

POLPERRO, LOOE, EDDYSTONE and WHITSAND BAY

Udder Rock ⚓ VQ (6) + L Fl 10s; *Bell;* 50°18'·93N 04°33'·85W.
POLPERRO, W pier ⚡ FW 4m 4M; FR when hbr closed in bad weather; 50°19'·86N 04°30'·96W.
Spy House Pt ⚡ Iso WR 6s 30m 7M; W288°-060°, R060°-288°; 50°19'·81N 04°30'·69W.
LOOE, Ranneys ⚓ Q (6) + L Fl 15s; 50°19'·85N 04°26'·37W.
Mid Main ⚓ Q (3) 10s 2M; 50°20'·56N 04°26'·94W.
Banjo Pier ☆ Oc WR 3s 8m **W15M**, R12M; 207°-R267°- W-313°-R-332°; 50°21'·06N 04°27'·06W.
White Rock ⚡ Fl R 3s 5m 2M; 50°21'·03N 04°27'·09W.
Eddystone ☆ Fl (2) 10s 41m **17M**. Same twr, Iso R 10s 28m 8M; vis 110°-133° over Hand Deeps; Gy twr, helicopter platform; *Horn 30s;* ***Racon T, 10M;*** 50°10'·84N 04°15'·94W; <u>992351125</u>.
Hand Deeps ⚓ Q (9) 15s; 50°12'·68N 04°21'·10W.

PLYMOUTH

PLYMOUTH SOUND, WESTERN CHANNEL
Draystone ⚓ Fl (2) R 5s; 50°18'·85N 04°11'·07W.
Knap ⚓ Fl G 5s; 50°19'·56N 04°10'·02W.
Plymouth bkwtr W head, ⚡ Fl WR 10s 19m W12M, R9M; 262°-W-208°-R-262°; W ○ twr. Same twr, Iso 4s 12m 10M; vis 033°-037°; *Horn 15s;* 50°20'·07N 04°09'·52W.
Maker ⚡ Fl (2) WRG 10s 29m, W11M, R/G6M; 270°-G330°-W-004°-R-050°; W twr, R stripe; 50°20'·51N 04°10'·87W.
Queens Ground ⚓ Fl (2) R 10s; 50°20'·29N 04°10'·08W.
New Ground ⚓ Fl R 2s; 50°20'·47N 04°09'·43W.
Melampus ⚓ Fl R 4s; 50°21'·15N 04°08'·72W.

PLYMOUTH SOUND, EASTERN CHANNEL
Wembury Pt ⚡ Oc Y 10s 45m; occas; 50°19'·01N 04°06'·63W.
West Tinker ⚓ VQ (9) 10s; 50°19'·25N 04°08'·64W.
East Tinker ⚓ Q (3) 10s; 50°19'·20N 04°08'·30W.

Whidbey ⚓ Oc (2) WRG 10s 29m, W8M, R/G6M; H24; 000°-137·5°-W-139·5°-R-159°; Or and W col; 50°19'·53N 04°07'·27W.
The Breakwater, E head ⚓ L Fl WR 10s 9m W8M, R6M; 190°-R-353°-W-001°-R-018°-W-190°; 50°20'·01N 04°08'·24W.
Staddon Pt ⚓ Oc WRG 10s 15m W8M, R/G5M; H24. 348°-G-038°-W-050°-R-090°; W structure, R bands; 50°20'·17N 04°07'·54W.
Withyhedge Dir ⚡ 070° (for W Chan): WRG 13m W13M, R/G5M; H24; 060°-FG-065°-Al WG (W phase increasing with brg) -069°-FW-071°-Al WR (R phase increasing with brg)-075°-F R-080°; W ▽, orange stripe on col. Same col, Fl (2) Bu 5s; vis 120°-160°; 50°20'·75N 04°07'·44W.

SMEATON PASS (W of Mount Batten and S of The Hoe)
Ldg lts 349°. Front, Mallard Shoal ⚓ Q WRG 5m W10M, R/G3M; W △, Or bands; 233°-G-043°- R-067°- G-087°-W-099°-R-108° (ldg sector); 50°21'·60N 04°08'·33W. Rear, 396m from front, Hoe ⚓ Oc G 1·3s 11m 3M; 310°-040°; W ▽, Or bands; 50°21'·81N 04°08'·39W.
S Mallard ⚓ VQ (6) + L Fl 10s; 50°21'·51N 04°08'·30W.
W Mallard ⚓ QG; 50°21'·57N 04°08'·36W.
S Winter ⚓ Q (6) + L Fl 15s; 50°21'·40N 04°08'·55W.
NE Winter ⚓ QR; 50°21'·54N 04°08'·50W.
NW Winter ⚓ VQ (9) 10s; 50°21'·55N 04°08'·70W.

ENTRANCE TO THE CATTEWATER
QAB (Queen Anne's Battery) ldg lts ⚡ 048·5°. Front, FR; Or/W bcn; 50°21'·84N 04°07'·84W. Rear, Oc R 8s 14m 3M; 139m NE. Fishers Nose ⚡ Fl (3) R 10s 6m 4M; 50°21'·80N 04°08'·01W. Also F Bu ≠ 026·5° with F Bu 50°22'·00N 04°07'·86W, for Cobbler Chan.

DRAKE CHANNEL, THE BRIDGE and THE NARROWS
Ravenness Dir ⚡ 225°: WRG 11m, W13M, R/G5M; vis 217°-FG-221°-Al WG-224° (W phase inc with brg)-FW-226°-Al WR-229° (R phase inc with bearing)-FR-237°; H24; W ▽, O stripe col. In fog, 160°-FW-305°; power failure QY; 50°21'·14N 04°10'·07W.
Asia ⚓ Fl (2) 5s; 50°21'·47N 04°08'·85W.
St Nicholas ⚓ QR; 50°21'·55N 04°09'·20W.
N Drakes Is ⚓ Fl R 4s; 50°21'·52N 04°09'·38W.
E Vanguard ⚓ QG; 50°21'·47N 04°09'·70W.
W Vanguard ⚓ Fl G 3s; 50°21'·49N 04°09'·98W.
Devils Point ⚓ QG 5m 3M; Fl 5s in fog; 50°21'·59N 04°10'·04W.
Battery ⚓ Fl R 2s; 50°21'·52N 04°10'·21W.

The Bridge Channel
No 1, ⚓ QG 4m; 50°21'·03N 04°09'·53W. No 2, ⚓ QR 4m. No 3, ⚓ Fl (3) G 10s 4m. No 4, ⚓ Fl (4) R 10s 4m; 50°21'·09N 04°09'·63W.
Mount Wise, Dir ⚡ 343°: WRG 7m, W13M, R/G5M; H24. 331°-FG-338°-Al WG-342° (W phase increasing with brg)-FW-344°-Al WR-348° (R phase increasing with bearing)-FR-351°. In fog, 341·5°-FW-344·5°; 50°21'·96N 04°10'·33W.
Ocean Court Dir Q WRG 15m, W11M, R/G3M; 010°-G-080°-W-090°-R-100°; 50°21'·85N 04°10'·11W.

PLYMOUTH TO START POINT
RIVER YEALM
Sand bar ⚓ Fl R 5s; 50°18'·59N 04°04'·12W.

SALCOMBE
Sandhill Pt Dir ⚡ 000°: Fl WRG 2s 27m W/R/G 8M; 337·5°-G-357·5°-W-002·5°-R-012·5°; R/W ⚓ on W mast, rear daymark; 50°13'·77N 03°46'·67W. Front daymark, Pound Stone R/W ⚓.
Bass Rk ⚓ Fl R 5s; 50°13'·47N 03°46'·71W.
Wolf Rk ⚓ Fl G 5s; 50°13'·53N 03°46'·58W.
Blackstone Rk ⚓; 50°13'·61N 03°46'·51W.
Ldg lts 042·5°, front Fl 2s 5m 8M, 50°14'·53N 03°45'·31W; rear Fl 5s 45m 8M.

Start Pt ☆ Fl (3) 10s 62m **25M**; 184°-068°. Same twr: FR 55m 9M; 210°-255° over Skerries Bank; *Horn 60s;* 50°13'·34N 03°38'·54W.

START POINT TO PORTLAND BILL
DARTMOUTH
Kingswear Dir ⚡ 328°: Iso WRG 3s 9m 8M; 318°-G-325°-W-331°-R-340°; W ○ twr; 50°20'·81N 03°34'·09W.

Mewstone ⚓ VQ (6) + L Fl 10s; 50°19'·92N 03°31'·89W.
West Rock ⚓ Q (6) + L Fl 15s; 50°19'·86N 03°32'·47W.
Homestone ⚓ QR; 50°19'·61N 03°33'·55W.
Castle Ledge ⚓ Fl G 5s; 50°19'·99N 03°33'·11W.
Checkstone ⚓ Fl (2) R 5s; 50°20'·45N 03°33'·81W.
Dir ⚡ 104·5°: FW 5m 9M; vis 102°-107°; 50°20'·65N 03°33'·80W.

BRIXHAM
Berry Head ☆ Fl (2) 15s 58m **19M**; vis 100°-023°; W twr; 50°23'·98N 03°29'·01W. R lts on radio mast 5·7M NW, inland of Paignton.
Victoria bkwtr ⚡ Oc R 15s 9m 6M; W twr; 50°24'·33N 03°30'·78W.
No 1 ⚓ Fl G; 50°24'·30N 03°30'·89W.
No 2 ⚓ Fl R; 50°24'·32N 03°30'·83W.

PAIGNTON and TORQUAY
⚓ QG (May-Sep); 50°27'·42N 03°31'·80W, 85m off Haldon Pier.
Haldon Pier (E) ⚡ QG 9m 6M; 50°27'·43N 03°31'·73W.
Princess Pier (W) ⚡ QR 9m 6M; 50°27'·46N 03°31'·73W.

TEIGNMOUTH
Outfall ⚓ Fl Y 5s; 50°31'·97N 03°27'·77W, 288°/1·3M to hbr ent.
Bar ⚓ Fl G 2s; 50°32'·44N 03°29'·25W.
Trng wall, middle ⚓ Oc R 6s 4m 3M; 50°32'·33N 03°29'·93W.
The Point ⚓ Oc G 6s 3M + FG (vert); 50°32'·42N 03°30'·05W.

RIVER EXE to SIDMOUTH and AXMOUTH
Exe ⚓ Mo(A) 10s; 50°35'·86N 03°23'·79W.
No 1 ⚓ 50°36'·05N 03°23'·84W.
No 2 ⚓ 50°36'·02N 03°23'·94W.
Exmouth Dir ⚡ 305°: WRG 6m, 6M; 299°-Iso G-304°-Iso W-306°-Iso R-311°; W col, 50°36'·99N 03°25'·34W.
No 10 ⚓ Fl R 3s; 50°36'·73N 03°24'·77W.
No 12 Warren Pt ⚓ 50°36'·91N 03°25'·41W.
Sidmouth ⚡ Fl R 5s 5m 2M; 50°40'·48'N 03°14'·43W.
Axmouth jetty ⚡ Fl G 4s 7m 2M; 50°42'·12N 03°03'·29W.

LYME REGIS
Outfall ⚓ Q (6) + L Fl 15s; 50°43'·17N 02°55'·66W.
Ldg lts 284°: Front, Victoria Pier ⚡ Oc WR 8s 6m, W9M, R7M; 284°-R-104°-W-284°; Bu col; 50°43'·19N 02°56'·17W. Rear, FG 8m 9M.

WEST BAY (BRIDPORT)
W pier root, Dir ⚡ 336°: F WRG 5m 4M; 165°-G-331°-W-341°-R-165°; 50°42'·62N 02°45'·89W.
W pier outer limit ⚡ Iso R 2s 5m 4M; 50°42'·51N 02°45'·83W.
E pier outer limit ⚡ Iso G 2s 5m 4M; 50°42'·53N 02°45'·80W.

PORTLAND BILL TO ANVIL POINT
Portland Bill lt ho ☆ Fl (4) 20s 43m **25M**. vis 221°-244° (gradual change from 1 Fl to 4 Fl); 244°-117° (shows 4 Fl); 117°-141° (gradual change from 4 Fl to 1 Fl). W ○ twr; *Dia 30s;* 50°30'·85N 02°27'·39W. Same twr, FR 19m 13M; 265°-291° over Shambles.
W Shambles ⚓ Q (9) 15s; *Bell;* 50°29'·78N 02°24'·41W.
E Shambles ⚓ Q (3) 10s; *Bell;* 50°31'·26N 02°20'·08W.

PORTLAND HARBOUR
Outer Bkwtr Fort Head (N end) ⚡ QR 14m 5M; 013°-268°; 50°35'·11N 02°24'·87W.
NE Bkwtr (A Hd) ⚡ Fl 2·5s 22m 10M; 50°35'·16N 02°25'·07W.
NE Bkwtr (B Hd) ⚡ Oc R 15s 11m 5M; 50°35'·65N 02°25'·88W.
N Arm (C Hd) ⚡ Oc G 10s 11m 5M; 50°35'·78N 02°25'·95W.

WEYMOUTH
Ldg lts 239·6°: both FR 5/7m 7M; Front 50°36'·46N 02°26'·87W, S Pier hd ⚡ Q 10m 9M; 50°36'·58N 02°26'·49W. IPTS 190m SW.

LULWORTH RANGE TO ANVIL POINT
Targets: DZ 'A' ⚓, Fl Y 2s, 50°33'·34N 02°06'·52W.
off   DZ 'B' ⚓, Fl Y 10s, 50°32'·11N 02°05'·92W.
St Alban's Hd DZ 'C' ⚓, Fl Y 5s, 50°32'·76N 02°04'·56W.
Anvil Pt ⚡ Fl 10s 45m 9M; vis 237°-076° (H24); W ○ twr and dwelling; 50°35'·51N 01°57'·60W. Measured mile close west.

1.3 PASSAGE INFORMATION

More passage information is threaded between the harbours in this area. Notes for crossing the western English Channel are after 1.23; see 0.33 for distances. Admiralty Leisure Folios: 5602 Falmouth to Teignmouth, 5603 Falmouth to Hartland Point, includes the Isles of Scilly. **Bibliography**: *West Country Cruising Companion* (Fernhurst/Fishwick); *Channel Havens* (ACN/Endean); *Shell Channel Pilot* (Imray/Cunliffe); *Channel Pilot* (Admiralty NP27).

ISLES OF SCILLY TO LAND'S END

The 48 islands lie 21-31M WSW of Land's End, with many rocky outcrops and offlying dangers; see AC 34, 883. Care is needed particularly in poor visibility. No one ⚓ gives shelter from all winds and swell, so be ready to move at short notice. Follow the approach transits on AC 34 because the tidal streams are difficult to predict with accuracy. They run harder off Points and over rocks, where overfalls may occur.

The N/S lanes of the Land's End TSS (AC 1148) are roughly defined by Seven Stones rocks to the west and Wolf Rock to the south east. The **Seven Stones** (lt ship, fog sig, Racon, AIS) lie 16M W of Land's End; many of them dry, with ledges in between. **Wolf Rock** (lt ho, fog sig, Racon, AIS) 8M SSW of Land's End, is in contrast steep-to. Beware commercial vessels turning N near Wolf Rock to enter the N-bound lane of the TSS; and those leaving the S-bound lane.

▶*Between Scilly and Land's End (AC 1148) streams are rotatory, clockwise, setting W from HW Dover; N from HW + 2; NE from HW + 4; E from HW –6; SSE from HW – 4; and SW from HW – 2. Sp rates are about 1kn. From Mounts Bay to Scilly leave the Runnel Stone at HW Dover –2; a fair W-going tide lasts for only 3 hrs, with cross tides setting SW then NW to N. Consider arriving at dawn. For the return passage streams are a little less critical.◀*

LAND'S END

The peninsula (AC 1148, 1149, 777 and 2345) is always a critical tidal gate and often a dangerous lee shore. There are many inshore and offlying rocks but no ports of refuge. The main features are: Gwennap Head, with Runnel Stone buoy 1M to the S; Land's End and Longships reef 1M to the W; Cape Cornwall and The Brisons; Pendeen and The Wra.

Passage between **Gwennap Head** and the Runnel Stone (0·5m; 7ca S) is not advised even in calm weather at HW, owing to rocks and uncharted wrecks closer inshore. These dangers are in the R sectors of Longships and Tater-du lts. From Gwennap Hd to Land's End, 2M NW, rocks extend up to 1½ca offshore, and depths are irregular causing a rough seas in strong winds against the tide.

4 cables S of Land's End Armed Knight, a jagged 27m high rock, overlooks **Longships**. This is an extensive and very dangerous reef made up of Carn Bras, on which the lt ho stands, and other rky islets. About 5ca to the E and NE are Kettle's Bottom 5m and Shark's Fin 0·9m, both isolated and very dangerous drying rocks. The ½M wide passage between Land's End and Kettle's Bottom is safe in calm, settled weather, but never at night. To clear Kettle's Bottom and Shark's Fin, keep the Brisons High summit just open W of Low summit brg 001°. In adverse weather navigate in the 3M wide ITZ, well W of Longships.

▶*Local streams exceed 4kn at Sp and are unpredictable.◀*

1M NNE of Land's End, after rounding Cowloe Rks and Little Bo 3·4m and Bo Cowloe 6m, the transit 150° of two bns on the cliffs E of Sennen Cove leads to an ⚓ in about 2m on the S side of **Whitesand Bay**, only safe in fair weather and offshore winds; avoid all underwater cables. Sennen ✠ twr (110m) is conspic, almost on the 150° transit.

Cape Cornwall (conspic ruined chy) is about 3M further N. It overlooks The Brisons, two rocky islets (27 and 22m, High and Low summits) about 5ca to SW. There is no safe passage inside The Brisons. The Vyneck rock 1·8m is 3ca to the NW.

3M NNE is **Pendeen Head** (lt ho) with the Wra (Three Stone Oar), drying rocks, close N. Overfalls and a race extend up to 1½M W and SW of Pendeen; avoid except in calm weather and at slack water. A conspic TV mast is 1.5M S of Pendeen.

TIDAL STRATEGY FOR ROUNDING LAND'S END

The tidal stream chartlets and notes below are reproduced by kind permission of the Royal Cruising Club Pilotage Foundation, *Yachting Monthly* magazine in which they were first published and the author Hugh Davies.

▶*The chartlets referenced to HW Dover, illustrate tidal streams and inshore currents. Streams run hard around Land's End, setting N/S and E/W past it. It is truly a tidal gate, and one which favours a N-bound passage – with careful timing nearly 9½hrs of fair tide can be carried, from HWD –3 to HW +5. Use currents close inshore running counter to the tidal streams.*

Example N-bound: At HWD+1 the N-going flood starts off Gwennap and does not turn NE along the N Cornish coast until HWD+3, but as early as HWD–3 an inshore current is beginning to run N'ly. So, N-bound, use this by arriving off Runnel Stone at HWD–2 and then keep within ¼M of the shore. If abeam Brisons at HWD, the tide and current should serve for the next 6 or 7hrs to make good St Ives, or even Newquay and Padstow.

Example S-bound: If S-bound from St Ives to Newlyn, aim to reach the Runnel Stone by HWD+5, ie with 2hrs of E-going tide in hand for the remaining 9M to Newlyn. This would entail leaving St Ives 5hrs earlier, at HWD, to make the 20M passage; buck a foul tide for the first 3hrs but use an inshore S-going current, keeping as close inshore as prudent, but having to move offshore to clear the Wra and the Brisons. This timing would suit passages from S Wales and the Bristol Channel, going inshore of Longships if the weather suits.

From Ireland, ie Cork or further W, the inshore passage would not benefit. But plan to be off the Runnel Stone at HWD+5 if bound for Newlyn; or at HWD+3 if bound for Helford/Falmouth, with the W-going stream slackening and 5 hrs of fair tide to cover the remaining 20M past the Lizard.◀

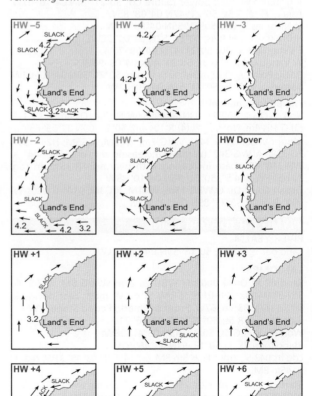

1.4 ISLES OF SCILLY

The Isles of Scilly consist of 48 islands and numerous rocky outcrops, covering an area approx 10M by 7M and lying 21–31M WSW of Land's End. Only St Mary's, St Martin's, Tresco, Bryher, St Agnes and Gugh are inhabited. The islands belong to the Duchy of Cornwall. There is a LB and CG Sector Base at St Mary's.

CHARTS AC 1148, 34, 883, 5603; Imray C10, C7, 2400

TIDES Standard Port is Plymouth. Differences for St Mary's are given in 1.5.

▶*Tidal heights, times, and streams are irregular. See the 12-hourly tidal stream chartlets below and AC 34 for 5 tidal stream diamonds, all referred to HW Plymouth.*◀

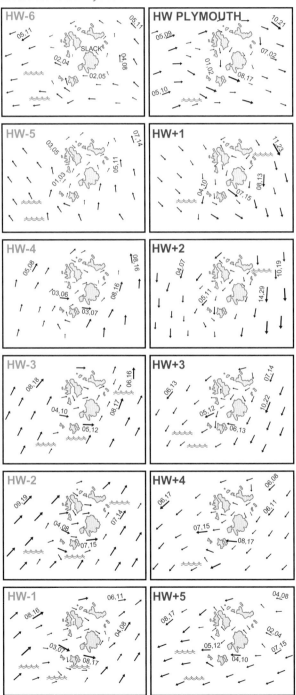

SHELTER The Isles of Scilly are exposed to Atlantic swell and wind. Weather can be unpredictable and fast-changing. Thorough planning and sensible precautions especially with regard to ⚓ and ground tackle are recommended. Boats have been known to drag on fine sand, but holding is mostly good. The islands are most attractive. Listed below are some of the many ⚓s, anti-clockwise:

ST MARY'S Hugh Town Harbour.
Porth Cressa (S of Hugh Town). Beware of dangers on each side of entrance and submarine cables. Good ⚓ (2m) in W/NW'lies, but exposed to swell from SE to SW.
Watermill Cove (NE corner of St Mary's). Excellent shelter in winds S to NW. ⚓ in approx 5m.

ST MARTIN'S, W end. Tean Sound needs careful pilotage, but attractive ⚓ in settled weather. More suitable for shoal draught boats which can ⚓ or take the ground out of main tidal stream in channel. 4 ⚓s £20/night (Star Castle Hotel ☎07730 475559). Hotel re-opened as Karma St Martin's ☎422368.

ST HELEN'S POOL (S of St Helen's Is). Ent via St Helen's Gap to ⚓ in 1·5m–7m. Secure, but may be swell near HW; see chartlet.

OLD GRIMSBY SOUND. Old Grimsby Hbr comprises Green Porth & Raven's Porth, divided by a quay, both dry 2·3m. Beware cable in Green Porth. ⚓s 1½ca NE of quay in 2·5m; access more difficult than New Grimsby. Sheltered in SW'lies but open to swell if wind veers N of W. Facilities: 7 ⚓s (R or Y) £20<18m>£30, ⚓ (quay) ⚓ hotel.

NEW GRIMSBY SOUND Appr (line G) between Tresco and Bryher, or with adequate rise of tide from the S across Tresco Flats. Good shelter except in NW'lies. Popular ⚓ between Hangman Is and the quay in 1·5-4·5m. Beware cables. 22 ⚓s, R or Y, £20<18m>£30; ⚓ £5. Vessels may only beach with Tresco HM permission. ☎423653 mob 07778601237. VHF Ch 08 for ✕ 🛒 ⚓ 🛒 ✉. Ferry to St Mary's.

ST AGNES/GUGH The Cove, well sheltered from W and N winds, except when the sandbar between the islands covers near HWS with a strong NW wind. Beware cables.
Porth Conger, N of the sandbar, sheltered in S'lies except when the bar is covered. Facilities: ⚓ (two quays), ferry to St Mary's; in Middle Town (¾M), 🛒 ✉ ✕ 🍴.

NAVIGATION See 9.1.2 for TSS to the E, S and W. If unable to identify approach ldg lines/marks, approach with caution. Many chans between islands have dangerous shallows, often with rky ledges. Some of the most used ldg lines, as on chartlet, include:

Line A. ⊕ 49°53´·55N 06°18´·00W. *North Carn of Mincarlo ≠ W extremity of Great Minalto 307·1°.* This is the usual approach to St Mary's Road, via St Mary's Sound. From the E or SE avoid Gilstone Rk (dries 4m) 3ca E of Peninnis Hd. Spanish Ledges, off Gugh, are marked by an ECM buoy; thence past Woolpack SCM bn, Bartholomew Ledges lt bn and N Bartholomew PHM buoy Fl R 5s to ent St Mary's Road on 040·5° (see Line B).

Line B. ⊕ 49°54´·69N 06°20´·33W. *St Martin's daymark ≠ the top of Creeb 040·5°.* This clears Woodcock Ledge, breaks in bad weather.

Line C. ⊕ 49°52´·06N 06°21´·24W. *Summit of Castle Bryher ≠ gap between the summits of Great Smith 350·5°.* Via Smith Sound, between the drying rocks off St Agnes and Annet.

Line D. ⊕ 49°52´·40N 06°27´·80W. *Summit of Great Ganilly just open north of Bant's Carn 059°.* From the SW enter Broad Sound between Bishop Rk lt ho and Flemming's Ledge, 7ca to the N; pass between Round Rk NCM and Gunner SCM buoys to Old Wreck NCM By; beware Jeffrey Rk, close to port. Ldg marks are more than 7M off and at first not easy to see. See also Line C, Smith Sound.

Line E. ⊕ 49°54´·34N 06°27´·80W. *St Agnes Old lt ho ≠ Carn Irish 099·7°.* From WNW, passing close N of Crim Rocks and Gunners.

Line F. ⊕ 49°56´·34N 06°26´·46W. *St Agnes Old lt ho ≠ Tins Walbert bcn 127°.* The NW Passage is about 7ca wide with good ldg marks, but beware cross tide. A lit WCM marks Steeple Rk (0·1m). Intercept Line D for St Mary's Road.

Line G. ⊕ 49°59´·00N 06°21´·96W. *Star Castle Hotel ≠ W side of Hangman Is 157°.* From the N, leads into New Grimsby Sound between Bryher and Tresco. Crossing Tresco Flats need adequate rise of tide (plan for max drying ht of 1·7m) and moderate vis.

Line H. ⊕ 49°55´·66N 06°14´·38W. *Summit of Samson Hill ≠ the NE extremity of Innisidgen 284·5°.* From the E or NE: Crow Snd can be rough in strong SE'lies, but, with sufficient rise of tide, is not difficult. From the NE pass close to Menawethan and Biggal Rk, avoiding Trinity Rk and the Ridge, which break in bad weather. Hats SCM buoy marks a shoal with an old boiler, (0·6m) on it. Maintain 284·5° between Bar Pt and Crow Bar (0·7m), pass N of Crow Rk IDM bn (for best water), then turn SSW for St Mary's. **Historic Wrecks** are at 49°52'·2N 06°26'·5W Tearing Ledge, 2ca SE of Bishop Rk Lt; and at 49°54'·30N 06°19'·89W, Bartholomew Ledges, 5ca N of Gugh.

LIGHTS AND MARKS See 1.2 for lts, including Seven Stones lt float, Wolf Rock lt ho and Longships lt ho. A working knowledge of the following daymarks and conspic features will greatly aid pilotage (from NE to SW):

St Martin's E end: Conical bcn twr (56m) with RW bands.

Round Is: roughly conical shaped, with W lt ho 19/55m high.

Tresco: Abbey & FS, best seen from S. Cromwell's Castle and Hangman Is from the NW.

Bryher: Watch Hill, stone bn (43m); rounded Samson Hill.

St Mary's: TV & radio masts (R lts) at N end, are visible from all over Scilly. Crow Rk IDM bn, 11m on rk drying 4·6m, at N.

St Agnes: Old lt ho, ○ W tr, visible from all directions.

Bishop Rk lt ho: Grey ○ tr, 44m; helo pad above lamp.

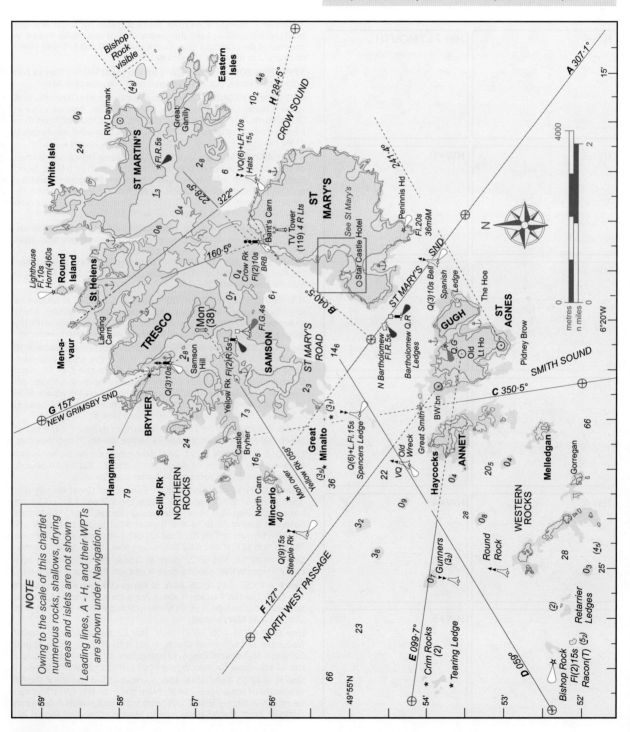

1.5 ST MARY'S

Isles of Scilly 49°55'·14N 06°18'·71W ❄❄❄♒⚓⚓♖♖♖

CHARTS AC 2665, 34, 883, 5603; Imray C10, C7, 2400

TIDES –0630 Dover; ML 3·2; Duration 0600

Standard Port PLYMOUTH (→)

Times				Height (metres)			
High Water		Low Water		MHWS	MHWN	MLWN	MLWS
0000	0600	0000	0600	5·5	4·4	2·2	0·8
1200	1800	1200	1800				
Differences *ST MARY'S*							
–0052	–0103	–0048	–0045	+0·2	–0·1	–0·2	–0·1

SHELTER Good in St Mary's Hbr, except in W/NW gales when ⚓ in Porth Cressa may be more comfortable. Strictly no ⚓ in the approaches and harbour limits (Newman rock to Newford Is).

NAVIGATION WPT 49°53'·96N 06°18'·83W (abeam Spanish Ledge ECM lt buoy) 307°/1·2M via St Mary's Sound to transit line B (040°). The 097·3° transit leads S of Bacon Ledge (0·3m) marked by PHM lt buoy. A charted 151° transit leads into the Pool between Bacon Ledge and the Cow and Calf (drying 0·6m and 1·8m). Pilotage is compulsory for yachts >30m LOA. Hbr speed limit 3kn. NB: Do not impede the ferry *Scillonian III* which arrives about 1200 and sails at 1630 Mon-Fri; Sat times vary with month. Also the blue-hulled cargo ship *Gry Maritha* thrice weekly.

LIGHTS AND MARKS See chartlet and 1.2. 097·3° ldg marks: W bcns; front, white △; rear, orange x; lts as chartlet. The pierhead light has been removed and is awaiting replacement. Buzza Hill twr and power stn chy (48m) are conspic.

COMMUNICATIONS (Code 01720)Ⓗ422392; Dr 422628; Pilot 422078. HM 422768(H24 ansafone), non-HO 07789 273626.
St Mary's Hbr and Pilot Ch **14**, 16 (0800-1700LT).
Falmouth CG covers Scilly and TSS/ITZ off Land's End on Ch 16, 23.

FACILITIES hm@stmarys-harbour.co.uk www.stmarys-harbour.co.uk 38 Y ⚓s lie in 6 trots close E of the ⬧ in 1·5-2·7m; 28 ⚓s are for <12m LOA and 10 for <18m. Fees by LOA: £18<8m<£24·00<12m>£36; multihulls +50%. 3 drying ⚲ against hbr wall with tender landing pontoon; ⚓ at inner drying berth H24. 🛢🔧(Sibleys ☎422431) at middle berth. ⚓ ⛽🔥⚓🔧🔩🏪🖥 @. HM holds mail if addressed c/o HM, St Mary's, Isles of Scilly TR21 0HU.

Hugh Town Essential shops, ⛽ ACA ✉ Ⓑ 🖥🛒 ✕ 🍴.

Porth Cressa ⚓ clear of the marked power cable area, ⚓ 🖥(Sibleys).

Mainland access: Ferry leaves Penzance 0915 Mon-Fri (not Sun) and 1630 St Mary's (Sat varies); booking ☎0845 710 5555, ✈ to St Just, Newquay, Exeter. Booking www.islesofscilly-travel.co.uk ☎(01736) 334220.

1.6 NEWLYN

Cornwall 50°06'·19N 05°32'·58W ❄❄⚓⚓⚓♖♖

CHARTS AC 777, 2345, 5603; Imray C10, C7, 2400

TIDES –0635 Dover; ML 3·2; Duration 0555

Standard Port PLYMOUTH (→)

Times				Height (metres)			
High Water		Low Water		MHWS	MHWN	MLWN	MLWS
0000	0600	0000	0600	5·5	4·4	2·2	0·8
1200	1800	1200	1800				
Differences *NEWLYN*							
–0053	–0108	–0035	–0036	+0·1	0·0	–0·2	0·0

SHELTER Good, but in strong SE'lies heavy swell enters hbr; access H24. FVs take priority. Good ⚓ in Gwavas Lake in offshore winds.

NAVIGATION WPT 50°06'·19N 05°31'·80W, 270°/0·5M to S pier. From NE, beware The Gear and Dog Rk 3½ca NE of hbr ent; from S, avoid Low Lee (1·1m) ECM buoy Q(3)10s and Carn Base (1·8m).

LIGHTS AND MARKS See chartlet and 1.2. Do not confuse S pier hd lt, Fl 5s, with Penzance pierhead lt, Fl WR 5s, 033°/1M. N pierhd lt: G sector (238°-248°) clears The Gear; W sector over hbr.

COMMUNICATIONS (Code 01736) Ⓗ, Dr: as Penzance. HM www.newlynharbour.co.uk 362523 non-HO 361017.
Newlyn Hbr Ch 09 **12** 16 (M-F 0800-1700, Sat 0800-1200LT).

FACILITIES www.newlynharbour.co.uk SW of Mary Williams Pier the first 3-4 finger pontoons are offered to visiting yachts if available. Preference is given to FVs and yachtsmen may be asked to move. Dogs are banned from pontoons/harbour property. NE pontoon: 40⚲<15m (2m on N side, S side shallower); SW pontoon: 40⚲<10m (0·75m below CD) Charges: £12/night<8m>£18/night<12m>£27.00. ⚓ at fish market; key from HM or nightwatchman.

Services ⚓ ⚓ 🖥 ⛽🔧 🔩 🖥 Ⓔ ⚓ 🛒(6t) ⛽.

Town ✉ Ⓑ(am only) 🖥 🍴 ✕ 🍴, bus to Penzance ⇄ ✈.

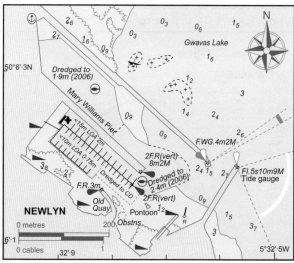

ADJACENT HARBOUR

MOUSEHOLE, Cornwall, **50°04'·97N 05°32'·26W**. ❄❄⚓♖♖♖. AC 2345, 5603. HW +0550 on Dover; Tides as Newlyn; ML 3·2m; Duration 0600. Best appr from S, midway between St Clements Is and bkwtr. N pier lt, 2 FG (vert) = hbr open; 3 FR (vert) = hbr closed. Ent 11m wide; hbr dries approx 2·4m. Shelter good except in NE and S or SE winds; protected by St Clements Is from E'lies. Ent closed with timber baulks from Nov-Apr. HM ☎(01736) 732544 mob 07891 647831. Per night £10, multi hull £12. Facilities limited: ⚓ ⚓ 🍴. Buses to Penzance.

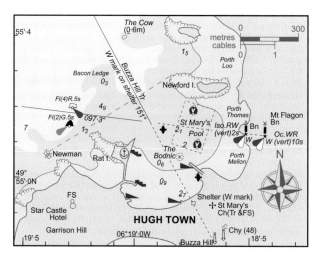

1.7 PENZANCE
Cornwall 50°07'·09N 05°31'·68W ❀❀⊛♨♨♨ ✿✿✿

CHARTS AC 777, 2345, 5603; Imray C10, C7, 2400

TIDES –0635 Dover; ML 3·2; Duration 0550

Standard Port PLYMOUTH (→)

Times				Height (metres)			
High Water		Low Water		MHWS	MHWN	MLWN	MLWS
0000	0600	0000	0600	5·5	4·4	2·2	0·8
1200	1800	1200	1800				

Differences PENZANCE are the same as for Newlyn (1.6)
PORTHLEVEN

–0045	–0105	–0030	–0025	0·0	–0·1	–0·2	0·0

LIZARD POINT

–0045	–0100	–0030	–0030	–0·2	–0·2	–0·3	–0·2

SHELTER Excellent in the wet dock. ⚓/dry out close N of wet dock or ⚓ E of hbr, but Mounts Bay is unsafe ⚓ in S or SE winds, which, if strong, also render the hbr ent dangerous.

NAVIGATION WPT 50°06'·77N 05°31'·06W, 306°/0·5M to S pier hd. Approach dries. Beware The Gear Rk 4·5ca S. Western Cressar (4ca NE of ent) and Ryeman Rks are marked by unlit SCM bns. Hbr speed limit 5kn. Dock gate opens 2hrs before local HW until 1hr after, 12 waiting buoys are sited S of S pier.

LIGHTS AND MARKS Hbr lts often hard to see against shore lts. No ldg lts/marks. IPTS 2 & 3, shown on N knuckle near dock gate, indicates when it is open/shut, but may not be shown for yachts.

8 unlit Y spar buoy racing marks are laid, Apr-Sep, at: 50°07'·30N 05°31'·48W; 50°07'·11N 05°30'·33W; 50°07'·09N 05°32'·00W; 50°06'·63N 05°32'·03W; 50°06'·55N 05°31'·45W; 50°06'·33N 05°32'·48W; 50°06'·25N 05°30'·30W; 50°05'·70N 05°29'·35W.

COMMUNICATIONS (Code 01736) Dr 363866; Ⓗ 362382 HM 366113, non-HO 07967 240660.
VHF Ch 09, **12**, 16 (HW –2 to HW +1, and office hrs).

FACILITIES Wet dock 50🅥<10m £1·80. Registration required. 🏳🛢🛢🛠🛠🏠⛟🅿(3t) 🅿 ⚓ (dry dock).
Penzance SC ☎364989, 🛠🔱✕🅿.

Town 🏠⌂Ⓑ🛒🅿✕🅿@≈✈.

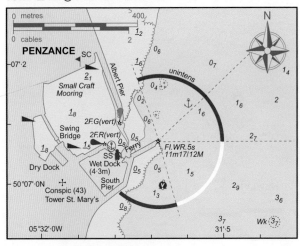

MINOR HARBOURS AROUND MOUNTS BAY

ST MICHAEL'S MOUNT, Cornwall, **50°07'·17N 05°28'·67W**. AC 2345, 777, 5603. HW +0550 on Dover; Tides as Penzance; ML 3·2m; Duration 0550. Beware: Hogus Rks (5·5) 250m NW of hbr; and Outer Penzeath Rk (0·4) 5ca W of hbr; and Maltman Rk (0·9) 1ca SSW of the Mount. Shelter good from N to SE, but only in fair wx. Unlit hbr dries 2·1; approx 3·3m at MHWS and 1·4m at MHWN. Dry out against W pier or ⚓ W of the hbr in 2-3m on sand, space for 6 boats. HM mob ☎07870 400282 or 07917 583284. Facilities: 🔱✕ café. Marazion, 3ca to N: 🖂🛒✕🅿 Mounts Bay SC.

PORTHLEVEN, Cornwall, **50°04'·92N 05°19'·13W**. ❀❀⊛♨♨♨✿. AC 777, 2345, 5603. HW –0635 on Dover; ML 3·1m; Duration 0545. Hbr dries 2m above the old lifeboat house but has approx 2·3m in centre of ent. It is very exposed to W and SW. Beware rks round pier hd and Deazle Rks to W. Lt on S pier, FG 10m 4M, = hbr open. Inside hbr FG, 033°-067°, shown as required to vessels entering. Visitors berth on the quay 2m, E side of inner hbr. HM ☎(01326) 574270. Facilities: Inner Hbr 🗑 £8, 🔱🛠🛠🏠🏠. Village 🖂Ⓑ 🛒🅿✕🅿.

MULLION COVE, Cornwall, **50°00'·90N, 05°15'·54W**. ❀❀⊛♨✿✿✿. AC 777, 2345, 5603. Lizard HW –0630 on Dover, –0050 and –0·2m on HW Plymouth; ML 3·0m; Duration 0545. Porth Mellin hbr dries about 2·4m and is open to SW'lies. Due to lack of space, visiting boats may only stay in hbr briefly to load/unload. No lts. There is a slip on E side of hbr. ⚓ in Mullion Cove where shown on the chart and especially in lee of N end of Mullion Island in approx 4-5m, but best to ask local advice. NT owns the island and the hbr. HM ☎(01326) 240222. Mullion village (1M) 🛒🅿.

Historic Wrecks are located at:
50°03'·44N 05°17'·16W (St Anthony), 2M SE of Porthleven.
50°02'·37N 05°16'·46W (Schiedam), 1·5M N of Mullion Is.
49°58'·54N 05°14'·51W, Rill Cove, 1·4M NW of Lizard Pt.
49°57'·49N 05°12'·98W (Royal Anne) The Stags, Lizard Pt.

LAND'S END TO FALMOUTH

Close E of Gwennap ⚓ off Porthcurno, owing to cables (AC 777). Western approach to Mounts Bay is marked by Tater-du Light. The Bucks (3·3m) are ESE of Tater-du, and Gull Rock (24m) is 9ca NE of close off the E point of Lamorna Cove. Little Heaver (dries) is 100m SW of Gull Rk, and Kemyel Rock (dries) is 1¾ca ENE.

Mousehole is a small drying harbour, sheltered from W and N, but exposed to E or S winds, when entrance may be shut. Approach from SW side of St Clement's Is. In W winds there is good ⚓ off the harbour.

Low Lee, (dangerous rock 1·1m) 4ca NE of Penlee Pt is marked by ECM lit buoy. Carn Base (rk 1·8m) lies 3ca NNW of Low Lee.

Newlyn is the sole harbour in Mount's Bay safe to approach in strong onshore winds, but only near to HW. East towards **Penzance** beware the unmarked Dog Rock (1·1m) and The Gear (1·9m,IDM lt bcn).

From Penzance to St Michael's Mount the head of the bay is shoal, drying 4ca off in places. Dangerous rocks include Cressar, Long, Hogus and Outer Penzeath. Keep south of a limiting line joining Penzance Market Ho Dome (55m) with pierheads of **St Michael's Mount** harbour at 089° to clear these dangers. The tiny hbr dries 2·1m, but is well sheltered, with ⚓ about 1ca W of entrance.

Two dangerous rocks, Guthen and Maltman (0·9m), lie 2ca W and S of St Michael's Mount respectively. 1M SE is The Greeb (7m), with rocks between it and shore. The Bears (dry) lie 1¾ca E of The Greeb. The Stone (dries) is 5ca S of Cudden Pt, while offshore is Mountamopus shoal marked by a SCM lit buoy. Welloe Rk (0·8m) lies 5ca SW of Trewavas Head.

Porthleven is a small tidal harbour, entered between Great and Little Trigg Rocks and the pier on S side. It is closed in bad weather when approach is dangerous. In fair weather there is a good ⚓ off Porth Mellin, about 1½ca NE of Mullion Is; Porth Mellin hbr (dries) is for temp use only. 2·5M W of Lizard Pt is The Boa, a rocky shoal (11m) on which the sea breaks in SW gales.

Lizard Point (Fl 3s70m 26M Horn 30s; AC 2345) is a bold, steep headland. Close inshore reflection of the lt may clearly be seen under some conditions. Dangerous rocks, mostly drying, lie 5-7ca from the lt ho in its SW quadrant. Either side of local LW these rocks are visible. 1M E and 1·5M NE of Lizard lt ho respectively, beware Vrogue (1·8m) and Craggan (1·5m) both dangerous sunken rocks.

A race, hazardous during the strength of either flood or ebb extends 2–3M S; it is worst in W'ly winds against W-going tide. Race conditions with short, heavy seas in westerlies may also exist SE of the Lizard. Pass at least 3M to seaward via 49°55'N 05°13'W. A shorter inner route which is rarely free of rough water may be taken, but go no further N than 49°56'·8N.

▶*In the race slack water occurs at about HW Plymouth –4 and the stream sets NE from HWP –3. The next slack water is at HWP +2½ and the stream sets SW from HWP +3; max springs rates reach 2-3kn. Outside the race the NE-going Channel flood starts at approx HWP –3¾ and the SW-going at HWP +3. Study NP 255 tidal stream atlas, Falmouth to Padstow inc Scilly.*◀

To await slack water, ⚓ (W to E) at: Kynance Cove, Housel Bay (3ca ENE of the lt ho and SSE of a conspic hotel), Parn Voose Cove (49°58'·4N 05°11'·1W), Cadgwith or Coverack.

N of Black Head, rocks extend at least 1ca offshore; a rock drying 1·6m lies off Chynhalls Pt (AC 1267). Coverack is a good ⚓ in W'lies. WSW of Lowland Pt drying rocks extend 2½ca offshore.

The Manacles (dry), 7½ca E and SE of Manacle Pt, are marked by ECM lt buoy and are in R sector of St Anthony Hd lt.

▶*Here the stream sets NE from HW Plymouth –0345, and SW from HW+0200, Sp rates 1·25kn.*◀

There are no offshore dangers on courses NNW to Nare Pt and Helford River or N to Falmouth.

MINOR HARBOURS NORTH EAST OF THE LIZARD

CADGWITH, Cornwall, **49°59'·22N 05°10'·68W**. ⊛⊛⊛♤♧♧♧. AC 154, 2345, 5603. HW –0625 on Dover; –0030 on Plymouth; –0·2m on Devonport; ML 3·0m. Use Differences Lizard Pt under 1.7. Hbr dries; it is divided by The Todden, a rky outcrop. Beware the extension of this, rks called The Mare; also beware Boa Rk to ESE which cover at quarter tide. ⚓ off The Mare in about 2–3m, but not recommended in onshore winds. There are no lts. Many local FVs operate from here and are hauled up on the shingle beach. Facilities: ⊠ ✕ ⌂ shops at Ruan Minor (1M).

COVERACK, Cornwall, **50°01'·44N 05°05'·66W**. ⊛⊛⊛♤♧♧♧. AC 154, 147, 5603. HW –0620 on Dover; ML 3·0m; Duration 0550. From the S beware the Guthens, off Chynhalls Pt; from the N, Davas and other rks off Lowland Pt, and to the NE Manacle Rks (ECM lt buoy). There are no lts. The tiny hbr dries and is full of small FVs. It has 3·3m at MHWS and 2·2m at MHWN. In good weather and offshore winds it is better to ⚓ outside. HM ☎(01326) 280679. Facilities: ⚓ (hotel) ⛽ ☎ from garage (2M uphill) ⊠ ⊞.

1.8 HELFORD RIVER

Cornwall **50°05'·79N 05°06'·06W** (Ent) ⊛⊛⊛♤♧♧♧♧

CHARTS AC 154, 147, 5603; Imray C10, C6, 2400

TIDES –0615 Dover; ML 3·0; Duration 0550

Reference Port PLYMOUTH (⟶)

Times				Height (metres)			
High Water		Low Water		MHWS	MHWN	MLWN	MLWS
0000	0600	0000	0600	5·5	4·4	2·2	0·8
1200	1800	1200	1800				
Differences HELFORD RIVER (Entrance)							
–0030	–0035	–0015	–0010	–0·2	–0·2	–0·3	–0·2
COVERACK							
–0030	–0050	–0020	–0015	–0·2	–0·2	–0·3	–0·2

SHELTER Excellent, except in E'lies when an uncomfortable swell reaches Frenchman's Creek. Yachts may ⚓ in Durgan Bay but not inshore of 3 small buoys marking the eelgrass beds (approx 100m from shore). ⚓/taking the ground is permitted W of Groyne Pt, but not S of Porth Navas Creek owing to oyster beds.

NAVIGATION WPT 50°05'·86N 05°04'·84W, 270°/1·9M to The Pool. From N beware August Rock marked by SHM buoy, Fl G 5s. From SE keep well off Nare Pt and Dennis Hd. At the entrance to Gillan harbour an ECM buoy marks Car Croc, a rock drying 1m. A NCM buoy marks the Voose, a rky reef E of Bosahan Pt.

Beware an extensive drying mud bank N and W of Bar SHM buoy, which is partially obscured by ⚓s. Speed limit 6kn in the river. The channel up to Gweek is buoyed but dries W of 05°10'·35W.

LIGHTS AND MARKS Access by day; no lights. Marks as chartlet; all unlit except seasonal buoys marking August Rk off Rosemullion and Car Croc Q(3) 10s off Dennis Hd.

COMMUNICATIONS (Code 01326) ⊞ 572151; Moorings Officer: ian@helford-river.com ☎250749 mob 07808 071485; water taxi ☎250749, Ch M. Ferry also acts as a water taxi, on request HO. Helford River SC Ch **80** M.

FACILITIES Local info at www.helfordrivermoorings.co.uk. In The Pool 25 G ⚓s, marked 'Visitors' with G pick-up buoys, are randomly numbered (see below); <9m £18, <11m £21, <12m £23,<14m £25; 7th night free. No prior reservations. Rafting essential in high season. Ditch rubbish ashore at the Helford River SC or the Ferryboat Inn (N bank).

N bank: Helford Passage ⚓ ⚓. Porth Navas YC☎340065, limited ⚓ ⚓ ✕ ⌂; www.portnavasyachtclub.co.uk.

S bank: Gillan Creek ⚓ ⚓ ⚓. Helford River SC☎231460 ⚓ ⊡ ✕ ⌂. Helford village ⊠ bus to Falmouth ⇌.

Gweek Quay ☎221657 ⊞221685, info@gweek-quay.com ⚓ ⚓ ☎ ✕ ⚒ ⊞ ⛽⛴(100t) ⚓ ⊠.

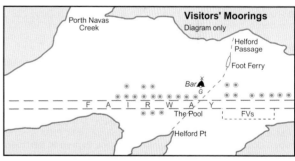

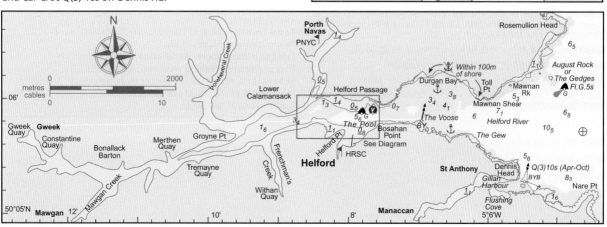

1.9 FALMOUTH & RIVER FAL

Cornwall 50°08'·61N 05°01'·48W (Ent) ✿✿✿✿✿✿✿ ✿✿✿

CHARTS AC 154, 32, 18, 5602; Imray C6, Y58

TIDES –0610 Dover; ML 3·0; Duration 0550

SHELTER Excellent, depending on wind direction; see Facilities. Access H24 in any weather, although the entrance, 1M wide and deep, is rough in fresh on-shore winds over the ebb.

NAVIGATION WPT 50°08'·00N 05°01'·74W, 000°/7ca to Black Rock PHM lt buoy, which marks the edge of the deep water E channel. At night this is advised, although by day small vessels may pass either side of Black Rk (2·1m), IDM Fl(2) 10s, close W of mid-entrance. A Wave Energy Test Area is situated 155° St Anthony Hd 1.5 NM; a Y SPM Fl(5)Y 20s marks its SW corner.

Falmouth is a deep water port with sizeable commercial docks. Do not impede ships up to 90,000 tons nor ⚓ in prohib areas. Outside the buoyed chans large areas are quite shallow; care is needed below half tide.

Speed limits: 8kn in Falmouth inner hbr (ie W of the docks), the Penryn River and, N of Turnaware Bar, in the R Fal/Truro and offlying creeks.

5kn in St Mawes and the Percuil River, Mylor, Restronguet and St Just Creeks and Portscatho (50°10'·83N 04°58'·32W), 2 pages on.

LIGHTS AND MARKS See chartlet and 1.2. R sector of St Anthony Hd lt (H24) covers the Manacles Rocks.

COMMUNICATIONS (Code 01326) Dr Falmouth 212120, Penryn 372502. Ⓗ Falmouth 434700. Ⓗ Truro (A&E) (01872) 250000.

Falmouth Hbr Radio Ch **12** 14 (M-F 0800-1700). HM Truro and launch Ch 12. Falmouth Haven Marina and St Mawes Hbr: Ch 12. Pendennis & Falmouth marinas, Royal Cornwall YC Ch **80/M**. Mylor Yacht Hbr Ch 80/M, 37/M1, Malpas Marine, St Mawes SC: Ch M (HO).

FACILITIES from seaward, clockwise to St Just:

Port Pendennis Marina 50°09'·15N 05°03'·70W. 45+20❶ in 3-4.5m in outer harbour (access H24) www.portpendennis.com ☎211211 £3·50<25m; max LOA 80m, short stay £15. Marina village access HW±3 via ent gate with tfc lts. YC, ✗ ⊡ tennis court.

Falmouth Haven Marina (FHM) 50°09'·22N 05°03'·89W. ☎310990/1; Ch 12. Access H24. Dir bn ▲ WRG 3s on N Quay 233°, min depth 1.4m on ldg line leaving 4 Y buoys to port. 100 ⌒, price standard /Jul-Aug: £2·95/£3·10; short stay (<2 hrs) £1/m. ❶ ⚓ up to 200m E of FHM in 2-5m (see FHC below). 🔋 🚿 ⊡ wi-fi.

Falmouth Harbour Commissioners (FHC) www.falmouthport.co.uk info@falmouthport.co.uk; 21 G ✿s off Prince of Wales pier, prices ✿/⚓: £2·20/£1·45.

Falmouth Marina 50°09'·90N 05°05'·02W up the Penryn River. www.premiermarinas.com ☎316620. Access H24, min depth 2m. In near appr, pass close N of outer pontoon, 2FR (vert), leaving ECM lt bn (hard to see) close to stbd. Ignore unlit PHM & SHM buoys, close NE of ECM bcn, which mark the Penryn R. Berth on J hammerhead/fuel berth, then as directed. A disused pipeline drying 1·8m crosses the marina E/W; adjacent depth gauge shows depth over it. 300+20❶, £3·10<12.5m>£3·30<16m>£3·80, £4·50>20m; £1·54 for 4 hrs (min £12·30).

Services ⬡⊡🔋⚓✕⚒⚓ ⊡△ ⊟(30t) ⬚(25t) 🔋⛟Ⓒ✗ ⊡.

Town ⬤ △ 🔋 ACA ⊠ Ⓑ ⛟ ✗ ⊡ ⇌ ✈ (Newquay).

Fuel Falmouth & Mylor marinas, Falmouth Haven Marina.

Water taxis Falmouth Water Taxi www.falmouthwatertaxi.co.uk ☎07522 446659, VHF Ch 12 (after 0930).

Penryn HM Penryn River ☎373352. Half tide access via drying buoyed channel to Freeman's Wharf 50°10'·00N 05°05'·75W www. freemanswharf.com ☎377509. 45 ½ tide drying ⌒, £2/ft/wk +VAT. ⚓✕✕⊡△🔋🗑🗑 (500m), ACA, ⬚🔋 (20t).

Mylor Yacht Harbour ④ 50°10'·74N 05°03'·14W, NW bank of Carrick Roads. Good shelter in W'lies but exposed to E'lies. PHM and SHM lit buoys mark ent to the fairway to the Yacht Hbr, dredged 2m; access H24. Hbr www.mylor.com enquiries@mylor.com ☎372121. 140 ⌒ + 40 ❶, £3·60; 15 ✿, £2·10. Short stay £3/hr, ✿ £2/hr. ⬛ ⊡🔋🔋⚓✕Ⓔ🔋 ⊟(35t), ⬚(4t) ⛟ ✗ ⊡.

Mylor Pool has many moorings in 1m to 2·2m. Mylor Creek dries or is shoal, <1m. Mylor Creek Boatyard welcomes visitors who can take the ground www.mylorcreekboatyard.com.

St Mawes Harbour 50°09'·49N 05°00'·84W. Keep S of St Mawes SCM buoy marking Lugo Rk, 0·6m. Beware numerous moorings and oyster beds, esp in Percuil River, upper reaches of which dry above village. HM www.stmawesharbour.co.uk ☎270553. 28 ✿s to SW of the hbr toward Carricknath Pt, 6G, call Ch12 for allocation. £17<10m>£22<12m>£33<15m. ✕⚓🔋✕⚒ 🔋. ⚓ above Castle Pt well sheltered from all but SW winds, £0·50/m. Seasonal water taxi 1 May-30 Sep 1000-1600; VHF Ch 12 *St Mawes Hbr water taxi*; £2/person. Outside hrs above taxi available for hire all year mob 07971 846786. 6 RNLI lifejacket lockers on quay.

St Just, good ⚓ off the creek, except in strong W/SW'lies.

FACILITIES UPRIVER TO TRURO See the chartlet overleaf.

Restronguet Creek 50°11'·53N 05°03'·56W, dries completely. Beware Carrick Carlys Rk 0·8m, 3ca E of ent, marked by NCM & SCM posts. Moorings fill the pool (12·4m) so only space to ⚓ is outside. Pandora Inn ☎372678, pontoon dries 0·9m, ⊡ ⛟ ✗ ⊡.

River Fal/Truro River At N end of Carrick Roads beware strong tides and rips off Turnaware Point and Bar. *King Harry chain ferry has right of way over all passing vessels* except those with a pilot aboard. She shows restricted in ability to manoeuvre lights/shapes. Fl Y lts indicate her direction of travel. At Tolverne, ✿ at www. smugglersmoorings.co.uk, ✿ £15<9m>£18<12m>£21 per day, ✕. 2 lit PH buoys mark the shallow water on the large bend where the Fal and Truro rivers divide just below midstream ❶ pontoon.

HM Truro www.portoftruro.co.uk ☎(01872) 272130 administers 4 ❶ pontoons, E of Channals Creek, N of Ruan Creek, off Woodbury Pt and Malpas (max draught 1m): £15/day all LOA; £3 for 2hrs ⚓ £6. 10 visits £120 for pontoon and £50 for anchoring.

A pontoon, W bank just S of King Harry ferry, is for tourist access to Trelissick House (⚓ Fl R 5s). Tenders may berth on the inshore side; yachts may berth outboard briefly for ⚓ only.

Good ⚓s, £6/day, off Tolcarne Creek, Channals Creek, Tolverne, Ruan Creek, Church Creek, Mopus Reach and Malpas.

Upstream, 8ca NW of Malpas a flood barrier gate is usually open, 2 FR/FG (vert). Gate is shut HW ±¼, and 3 Fl R lts shown, if HW >5·6m is forecast due to storm surge.

Malpas Marine ☎ (01872) 271260. ✕ but no ❶ facilities. Local pub.

Truro 50°15'·72N 05°02'·85W. HW Truro, Sp & Nps, is approx HW Falmouth –0022 and –2·0m. Drying ⌒ close SW of HM's office ⊡ ⚓ 🔋. **City** ⇌ all facilities.

YACHT CLUBS
Royal Cornwall YC ☎311105/312126 (Secretary), ⬛ ⚓ ✗ ⊡.
Falmouth Town SC ☎313662.
Falmouth Watersports Association ☎211223.
Port of Falmouth Sailing Association ☎211555.
St Mawes SC ☎270686 ⊡ (visitors most welcome).
Flushing SC ☎374043, **Mylor YC** ☎374391, ⊡.
Restronguet SC ☎374536.

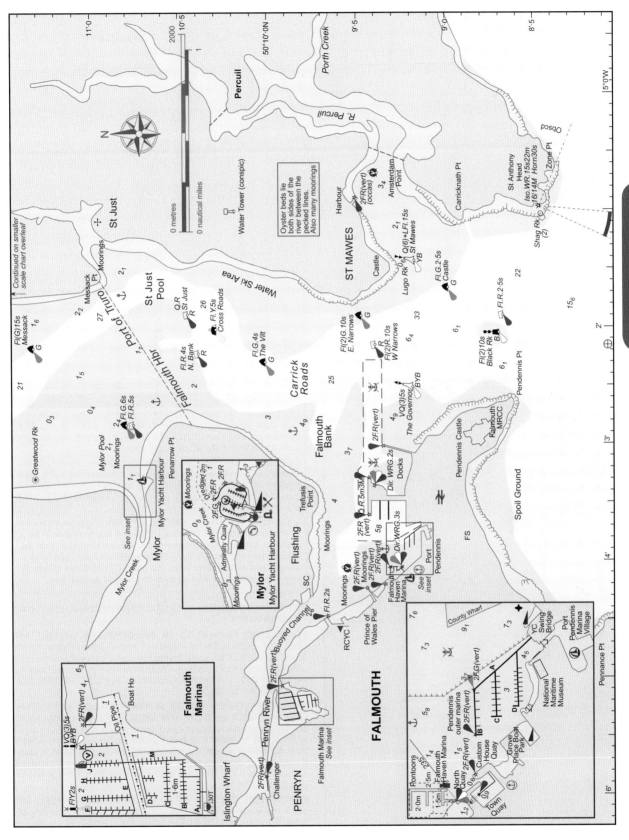

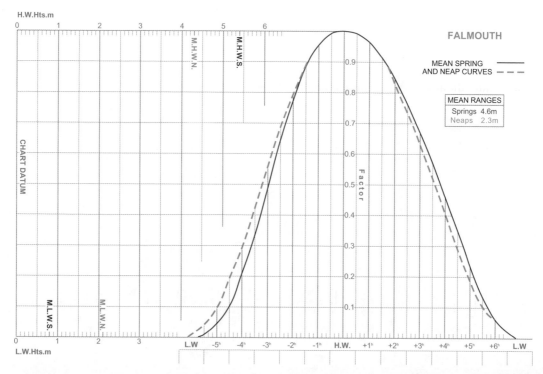

FALMOUTH

MEAN SPRING ──────
AND NEAP CURVES ── ── ──

MEAN RANGES
Springs 4.6m
Neaps 2.3m

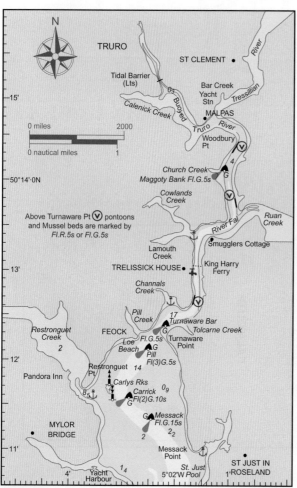

FALMOUTH TO FOWEY

Porthscatho, 3M NNE of St Anthony Hd, is a safe ⚓ in W'lies. Gull Rock (38m high) and The Whelps (dry 4.6m) lie 6ca E/SE of Nare Hd and may be passed inshore. In Veryan Bay beware Lath Rk (2.1m), 1.6M NE of Gull Rock.

On E side of Veryan Bay, Dodman Pt is a 110m rounded shoulder, with a conspic stone cross. Depths are irregular for 1M S, with heavy overfalls in strong winds over sp tide, when it is best to pass 2M off. 3 SPM lt buoys (naval gunnery targets) lie 2.3–4.8M SSE of Dodman Pt.

2.1M NE of Dodman Pt is Gwineas Rk (8m high) and Yaw Rk (0.9m), marked by ECM lt buoy. Passage inside Gwineas Rk is possible, but not advised in strong onshore winds or poor vis. **Gorran Haven**, 1M WSW of Gwineas Rk, a sandy cove with L-shaped pier which dries at sp, is a good ⚓ in offshore winds. So too are **Portmellon**, **Mevagissey Bay** and **St Austell Bay**.

MINOR HARBOURS AND ANCHORAGES FROM ST ANTHONY HEAD TO MEVAGISSEY

PORTSCATHO, Cornwall, **50°10′·84N 04°58′·32W**. AC 1267, 154, 5602. HW −0600 on Dover, HW −0025 and −0.2m on Plymouth; ML 3.0m; Duration 0550. Small drying hbr, but in settled weather and offshore winds pick up a ⚓ or ⚓ outside moorings in good holding. No lts. HM ☎(01872) 580243. Facilities: ⚓ ⚓ ⓟ ⓓ ✉ Ⓑ 1000-1230 Mon, Wed, Fri ▦ ✕ ⌂.

GORRAN HAVEN, Cornwall, **50°14′·49N 04°47′·16W**. AC 1267, 148, 5602. HW −0600 on Dover, HW −0010 and −0.1m on Plymouth. Shelter good with flat sand beach for drying out in offshore wind; good ⚓ 100 to 500m E of harbour. Not suitable ⚓ when wind is in E. Beware Gwineas Rk and Yaw Rk marked by ECM lt buoy. Beware pot markers on appr. Fin keelers without legs should not ⚓ closer than 300m from hbr wall where depth is 1.8m at MLWS. Facilities: ✉ ▦ ✕ ⌂.

PORTMELLON, Cornwall, **50°15′·74N 04°46′·98W**. AC1267, 148, 5602. HW −0600 on Dover, HW −0010 and −0.1m on Plymouth; ML 3.1m; Duration 0600. Shelter good but only suitable as a temp ⚓ in settled weather and offshore winds. Facilities: ✕ ⌂ (summer).

POLKERRIS, Cornwall, **50°20′·15N 04°41′·40W**. AC 148. Tides as Charlestown/Par. Very good ⚓ in NE/SE winds. Facilities: ✕ ⌂.

STANDARD TIME (UT)
For Summer Time add ONE hour in **non-shaded areas**

FALMOUTH   LAT 50°09'N   LONG 5°03'W

TIMES AND HEIGHTS OF HIGH AND LOW WATERS

Dates in red are **SPRINGS**
Dates in blue are NEAPS

YEAR **2019**

JANUARY

Day	Time m	Time m	Day	Time m	Time m
1 TU	0119 4.2 / 0735 1.5	1340 4.3 / 2019 1.3	**16** W	0629 1.8	1224 4.3 / 1911 1.6
2 W	0221 4.3 / 0847 1.5	1441 4.4 / 2120 1.1	**17** TH	0107 4.2 / 0746 1.6	1334 4.4 / 2022 1.4
3 TH	0312 4.5 / 0946 1.1	1531 4.5 / 2212 1.0	**18** F	0215 4.4 / 0855 1.3	1441 4.6 / 2125 1.1
4 F	0356 4.7 / 1036 1.0	1615 4.6 / 2257 0.9	**19** SA	0315 4.7 / 0956 1.0	1541 4.8 / 2223 0.8
5 SA	0436 4.9 / 1120 0.8	1656 4.7 / 2338 0.8	**20** SU	0409 5.0 / 1054 0.7	1636 5.0 / 2318 0.6
6 SU ●	0514 5.0 / 1200 0.8	1735 4.8	**21** M ○	0501 5.2 / 1148 0.4	1729 5.1
7 M	0016 0.8 / 0551 5.0	1237 0.8 / 1812 4.8	**22** TU	0010 0.4 / 0551 5.4	1239 0.2 / 1819 5.2
8 TU	0049 0.8 / 0628 5.0	1309 0.8 / 1848 4.8	**23** W	0059 0.2 / 0640 5.5	1326 0.1 / 1908 5.2
9 W	0119 0.9 / 0703 5.0	1338 0.9 / 1922 4.7	**24** TH	0144 0.2 / 0727 5.5	1411 0.1 / 1953 5.1
10 TH	0146 1.0 / 0736 4.9	1405 1.1 / 1955 4.6	**25** F	0226 0.3 / 0811 5.3	1454 0.3 / 2037 4.9
11 F	0213 1.2 / 0808 4.8	1432 1.2 / 2028 4.4	**26** SA	0308 0.5 / 0855 5.1	1536 0.6 / 2120 4.7
12 SA	0243 1.3 / 0841 4.6	1506 1.3 / 2105 4.3	**27** SU ◑	0351 0.8 / 0941 4.8	1622 0.9 / 2208 4.4
13 SU	0321 1.5 / 0922 4.5	1549 1.5 / 2153 4.1	**28** M	0439 1.2 / 1031 4.5	1713 1.3 / 2306 4.1
14 M ◑	0409 1.6 / 1014 4.3	1643 1.6 / 2253 4.1	**29** TU	0536 1.5 / 1135 4.2	1815 1.5
15 TU	0512 1.8 / 1117 4.3	1754 1.7 / 2359 4.1	**30** W	0024 4.0 / 0643 1.7	1259 4.0 / 1928 1.6
			31 TH	0146 4.0 / 0803 1.6	1415 4.1 / 2047 1.5

FEBRUARY

Day	Time m	Time m	Day	Time m	Time m
1 F	0248 4.2 / 0920 1.4	1512 4.2 / 2149 1.3	**16** SA	0144 4.3 / 0829 1.4	1419 4.4 / 2103 1.3
2 SA	0336 4.5 / 1017 1.2	1558 4.4 / 2238 1.1	**17** SU	0253 4.6 / 0939 1.0	1525 4.6 / 2208 0.9
3 SU	0418 4.7 / 1102 1.0	1640 4.6 / 2320 0.9	**18** M	0352 5.0 / 1041 0.6	1622 4.9 / 2306 0.5
4 M ●	0457 4.9 / 1142 0.8	1719 4.7 / 2357 0.8	**19** TU ○	0444 5.2 / 1136 0.3	1714 5.1 / 2359 0.2
5 TU	0535 5.0 / 1218 0.7	1757 4.8	**20** W	0535 5.5 / 1226 0.0	1804 5.2
6 W	0031 0.8 / 0612 5.1	1251 0.7 / 1833 4.8	**21** TH	0046 0.0 / 0623 5.6	1313 -0.1 / 1850 5.3
7 TH	0100 0.8 / 0647 5.0	1318 0.8 / 1906 4.8	**22** F	0130 0.0 / 0708 5.6	1354 -0.1 / 1932 5.2
8 F	0126 0.8 / 0718 5.0	1342 0.8 / 1936 4.7	**23** SA	0209 0.1 / 0750 5.4	1433 0.1 / 2010 5.1
9 SA	0151 0.9 / 0745 4.9	1409 0.9 / 2001 4.6	**24** SU	0246 0.3 / 0829 5.2	1510 0.4 / 2046 4.8
10 SU	0220 1.0 / 0809 4.8	1439 1.0 / 2020 4.5	**25** M	0324 0.7 / 0906 4.8	1548 0.9 / 2124 4.5
11 M	0253 1.2 / 0832 4.6	1514 1.2 / 2042 4.4	**26** TU ◑	0405 1.1 / 0946 4.4	1631 1.3 / 2210 4.2
12 TU ◑	0332 1.3 / 0907 4.5	1558 1.4 / 2134 4.2	**27** W	0456 1.5 / 1037 4.1	1727 1.7 / 2314 3.9
13 W	0423 1.6 / 1017 4.3	1657 1.6 / 2303 4.1	**28** TH	0600 1.8 / 1200 3.8	1838 1.9
14 TH	0534 1.7 / 1141 4.2	1818 1.7			
15 F	0026 4.1 / 0704 1.7	1302 4.2 / 1946 1.6			

MARCH

Day	Time m	Time m	Day	Time m	Time m
1 F	0102 3.9 / 0719 1.9	1349 3.8 / 2005 1.8	**16** SA	0635 1.7	1242 4.1 / 1920 1.7
2 SA	0222 4.1 / 0853 1.7	1452 4.0 / 2124 1.6	**17** SU	0122 4.2 / 0811 1.4	1405 4.3 / 2047 1.4
3 SU	0314 4.3 / 0956 1.3	1539 4.3 / 2216 1.2	**18** M	0235 4.6 / 0925 1.0	1512 4.6 / 2155 0.9
4 M	0357 4.6 / 1041 1.0	1620 4.5 / 2258 1.0	**19** TU	0335 4.9 / 1027 0.6	1607 4.9 / 2252 0.5
5 TU	0436 4.8 / 1120 0.8	1659 4.7 / 2335 0.8	**20** W	0427 5.3 / 1120 0.2	1657 5.1 / 2343 0.2
6 W ●	0515 5.0 / 1155 0.7	1736 4.8	**21** TH ○	0516 5.5 / 1209 -0.1	1744 5.3
7 TH	0008 0.7 / 0551 5.0	1226 0.6 / 1812 4.9	**22** F	0029 0.0 / 0602 5.6	1253 -0.2 / 1827 5.3
8 F	0037 0.7 / 0625 5.0	1252 0.6 / 1844 4.9	**23** SA	0110 -0.1 / 0646 5.5	1333 -0.1 / 1906 5.3
9 SA	0103 0.7 / 0655 5.0	1318 0.7 / 1912 4.8	**24** SU	0148 0.0 / 0726 5.4	1409 0.1 / 1942 5.2
10 SU	0129 0.7 / 0720 4.9	1345 0.7 / 1932 4.7	**25** M	0222 0.3 / 0801 5.1	1442 0.5 / 2015 4.9
11 M	0157 0.8 / 0738 4.8	1414 0.8 / 1940 4.7	**26** TU	0257 0.7 / 0834 4.8	1515 0.9 / 2049 4.7
12 TU	0229 0.9 / 0753 4.7	1447 1.0 / 2004 4.6	**27** W	0334 1.1 / 0909 4.4	1554 1.4 / 2130 4.3
13 W	0305 1.1 / 0826 4.5	1526 1.3 / 2049 4.4	**28** TH ◑	0420 1.5 / 0954 4.0	1645 1.8 / 2225 4.0
14 TH ◑	0351 1.4 / 0928 4.3	1619 1.6 / 2217 4.1	**29** F	0522 1.9 / 1102 3.7	1755 2.0 / 2351 3.8
15 F	0458 1.6 / 1112 4.1	1739 1.8 / 2357 4.1	**30** SA	0640 2.0 / 1317 3.7	1918 2.0
			31 SU	0147 3.9 / 0809 1.8	1426 3.9 / 2044 1.7

APRIL

Day	Time m	Time m	Day	Time m	Time m
1 M	0245 4.2 / 0921 1.5	1514 4.2 / 2142 1.4	**16** TU	0217 4.6 / 0908 0.9	1456 4.6 / 2136 0.9
2 TU	0330 4.5 / 1008 1.1	1555 4.5 / 2225 1.1	**17** W	0316 4.9 / 1007 0.5	1549 4.9 / 2232 0.5
3 W	0410 4.8 / 1047 0.9	1633 4.7 / 2303 0.9	**18** TH	0407 5.2 / 1059 0.2	1636 5.1 / 2322 0.2
4 TH	0448 4.9 / 1123 0.7	1710 4.8 / 2338 0.7	**19** F ○	0455 5.3 / 1146 0.0	1720 5.2
5 F ●	0524 5.0 / 1155 0.6	1745 4.9	**20** SA	0007 0.0 / 0540 5.4	1230 0.0 / 1801 5.3
6 SA	0009 0.6 / 0558 5.0	1224 0.6 / 1816 4.9	**21** SU	0048 0.0 / 0622 5.3	1309 0.1 / 1839 5.3
7 SU	0038 0.6 / 0629 5.0	1253 0.6 / 1844 4.9	**22** M	0125 0.1 / 0701 5.2	1344 0.3 / 1914 5.1
8 M	0107 0.6 / 0656 4.9	1322 0.6 / 1906 4.9	**23** TU	0159 0.4 / 0736 5.0	1416 0.6 / 1948 5.0
9 TU	0138 0.7 / 0717 4.8	1353 0.8 / 1922 4.8	**24** W	0232 0.7 / 0808 4.7	1448 1.0 / 2022 4.7
10 W	0210 0.8 / 0737 4.7	1427 1.0 / 1951 4.7	**25** TH	0308 1.1 / 0842 4.4	1524 1.4 / 2101 4.4
11 TH	0248 1.0 / 0816 4.5	1507 1.2 / 2042 4.5	**26** F ◑	0351 1.5 / 0925 4.0	1610 1.8 / 2151 4.1
12 F ◑	0335 1.3 / 0927 4.2	1601 1.5 / 2209 4.2	**27** SA	0448 1.8 / 1026 3.8	1716 2.0 / 2258 3.9
13 SA	0444 1.6 / 1102 4.0	1722 1.7 / 2340 4.1	**28** SU	0601 2.0 / 1210 3.7	1833 2.1
14 SU	0621 1.6 / 1230 4.0	1902 1.7	**29** M	0047 3.9 / 0716 1.8	1346 3.9 / 1947 1.9
15 M	0104 4.3 / 0755 1.4	1352 4.3 / 2030 1.3	**30** TU	0204 4.1 / 0826 1.6	1439 4.2 / 2051 1.6

Chart Datum is 2·91 metres below Ordnance Datum (Newlyn). HAT is 5·7 metres above Chart Datum.

》FREE monthly updates. Register at 《
www.reedsnauticalalmanac.co.uk

43

STANDARD TIME (UT)
For Summer Time add ONE hour in **non-shaded areas**

FALMOUTH   LAT 50°09′N   LONG 5°03′W
TIMES AND HEIGHTS OF HIGH AND LOW WATERS

Dates in red are **SPRINGS**
Dates in blue are **NEAPS**

YEAR 2019

MAY

	Time	m		Time	m
1 W	0254 0920 1521 2140	4.4 1.3 4.4 1.2	**16** TH	0255 0942 1527 2207	4.8 0.6 4.8 0.6
2 TH	0336 1004 1600 2223	4.6 1.0 4.6 1.0	**17** F	0346 1034 1613 2257	5.0 0.4 5.0 0.4
3 F	0414 1043 1636 2301	4.8 0.8 4.8 0.8	**18** SA ○	0433 1122 1656 2343	5.1 0.3 5.1 0.3
4 SA ●	0451 1120 1711 2338	4.9 0.7 4.9 0.7	**19** SU	0518 1205 1737	5.1 0.3 5.1
5 SU	0526 1155 1745	5.0 0.6 4.9	**20** M	0025 0600 1245 1815	0.3 5.1 0.4 5.1
6 M	0013 0601 1230 1817	0.6 5.0 0.6 5.0	**21** TU	0103 0639 1321 1851	0.4 5.0 0.5 5.1
7 TU	0048 0635 1304 1849	0.6 4.9 0.6 5.0	**22** W	0138 0715 1354 1926	0.6 4.8 0.8 4.9
8 W	0123 0710 1339 1922	0.6 4.8 0.7 4.9	**23** TH	0213 0748 1427 2001	0.9 4.6 1.1 4.8
9 TH	0200 0746 1417 2002	0.7 4.7 0.9 4.8	**24** F	0248 0822 1501 2038	1.2 4.4 1.4 4.5
10 F	0242 0833 1501 2057	1.0 4.5 1.2 4.6	**25** SA	0326 0904 1541 2123	1.4 4.2 1.7 4.3
11 SA	0334 0938 1600 2208	1.2 4.3 1.4 4.4	**26** SU ◑	0414 0956 1636 2219	1.7 4.0 1.9 4.1
12 SU ◑	0444 1056 1717 2327	1.4 4.2 1.6 4.3	**27** M	0515 1103 1743 2327	1.8 3.8 2.0 4.0
13 M	0609 1218 1842	1.4 4.1 1.5	**28** TU	0622 1227 1851	1.8 3.9 1.9
14 TU	0045 0731 1334 2003	4.4 1.2 4.3 1.3	**29** W	0046 0727 1342 1953	4.1 1.7 4.1 1.7
15 W	0156 0843 1436 2110	4.6 0.9 4.6 0.9	**30** TH	0156 0825 1434 2049	4.3 1.4 4.3 1.4
			31 F	0247 0916 1517 2138	4.5 1.2 4.5 1.2

JUNE

	Time	m		Time	m
1 SA	0331 1002 1556 2224	4.7 1.0 4.7 0.9	**16** SU	0413 1058 1635 2321	4.8 0.6 4.9 0.6
2 SU	0412 1046 1634 2308	4.8 0.8 4.9 0.8	**17** M ○	0458 1143 1716	4.9 0.6 5.0
3 M ●	0454 1128 1714 2351	4.9 0.7 5.0 0.6	**18** TU	0005 0540 1225 1755	0.6 4.9 0.6 5.0
4 TU	0537 1210 1755	5.0 0.6 5.1	**19** W	0045 0620 1303 1832	0.7 4.8 0.8 5.0
5 W	0033 0621 1252 1838	0.6 4.9 0.6 5.1	**20** TH	0122 0657 1337 1909	0.8 4.7 0.9 4.9
6 TH	0115 0705 1333 1922	0.6 4.9 0.7 5.0	**21** F	0156 0732 1409 1943	0.9 4.6 1.1 4.8
7 F	0159 0751 1416 2008	0.6 4.8 0.8 4.9	**22** SA	0229 0807 1439 2019	1.1 4.5 1.3 4.7
8 SA	0245 0840 1504 2100	0.8 4.6 1.0 4.8	**23** SU	0301 0844 1512 2058	1.3 4.3 1.5 4.5
9 SU	0339 0938 1601 2201	1.0 4.4 1.2 4.6	**24** M	0337 0927 1552 2143	1.5 4.2 1.7 4.4
10 M ◑	0441 1044 1706 2309	1.1 4.3 1.3 4.5	**25** TU ◑	0423 1019 1645 2238	1.7 4.1 1.8 4.2
11 TU	0549 1157 1816	1.2 4.2 1.3	**26** W	0521 1119 1748 2338	1.7 4.0 1.9 4.2
12 W	0021 0700 1308 1928	4.5 1.2 4.3 1.2	**27** TH	0625 1222 1855	1.7 4.1 1.8
13 TH	0131 0810 1411 2038	4.5 1.0 4.5 1.1	**28** F	0041 0729 1326 1958	4.2 1.6 4.2 1.6
14 F	0232 0913 1505 2139	4.7 0.9 4.7 0.9	**29** SA	0146 0829 1424 2056	4.4 1.4 4.4 1.3
15 SA	0325 1008 1552 2233	4.8 0.7 4.8 0.7	**30** SU	0244 0924 1516 2150	4.6 1.1 4.7 1.1

JULY

	Time	m		Time	m
1 M	0338 1016 1603 2243	4.7 0.9 4.9 0.9	**16** TU ○	0442 1126 1659 2350	4.7 0.9 4.9 0.8
2 TU ●	0428 1107 1651 2334	4.9 0.8 5.0 0.7	**17** W	0524 1208 1738	4.7 0.8 5.0
3 W	0519 1156 1739	5.0 0.6 5.2	**18** TH	0030 0603 1246 1816	0.8 4.8 0.9 5.0
4 TH	0023 0609 1244 1827	0.5 5.0 0.5 5.3	**19** F	0106 0641 1319 1852	0.8 4.7 0.9 5.0
5 F	0111 0658 1331 1915	0.4 5.0 0.5 5.3	**20** SA	0138 0716 1349 1926	0.9 4.7 1.0 4.9
6 SA	0158 0747 1416 2003	0.5 5.0 0.6 5.2	**21** SU	0206 0749 1415 1958	1.0 4.6 1.2 4.8
7 SU	0245 0835 1502 2052	0.5 4.8 0.7 5.1	**22** M	0232 0822 1441 2030	1.2 4.5 1.3 4.7
8 M	0333 0926 1551 2145	0.7 4.7 0.9 4.9	**23** TU	0301 0855 1513 2106	1.3 4.4 1.5 4.6
9 TU ◑	0425 1022 1645 2244	0.9 4.5 1.1 4.7	**24** W	0337 0935 1553 2151	1.4 4.3 1.6 4.4
10 W	0522 1126 1744 2350	1.1 4.3 1.3 4.4	**25** TH ◑	0423 1026 1647 2247	1.6 4.2 1.8 4.3
11 TH	0624 1236 1850	1.2 4.3 1.4	**26** F	0523 1127 1756 2352	1.7 4.1 1.8 4.2
12 F	0101 0732 1345 2002	4.4 1.3 4.3 1.4	**27** SA	0635 1233 1911	1.7 4.2 1.8
13 SA	0210 0842 1444 2114	4.4 1.2 4.5 1.2	**28** SU	0100 0747 1342 2022	4.3 1.6 4.4 1.5
14 SU	0308 0945 1535 2214	4.5 1.1 4.6 1.1	**29** M	0211 0853 1445 2125	4.5 1.4 4.6 1.2
15 M	0358 1039 1619 2305	4.6 1.0 4.8 0.9	**30** TU	0314 0954 1541 2225	4.7 1.1 4.9 0.9
			31 W	0411 1051 1633 2321	4.9 0.8 5.2 0.6

AUGUST

	Time	m		Time	m
1 TH ●	0504 1146 1724	5.0 0.6 5.3	**16** F	0013 0544 1227 1757	0.8 4.8 0.9 5.1
2 F	0014 0556 1237 1814	0.4 5.1 0.4 5.5	**17** SA	0046 0621 1258 1833	0.8 4.9 0.9 5.1
3 SA	0104 0647 1323 1903	0.2 5.2 0.3 5.5	**18** SU	0115 0656 1324 1906	0.9 4.9 1.0 5.0
4 SU	0150 0734 1407 1950	0.1 5.2 0.3 5.5	**19** M	0139 0728 1347 1934	0.9 4.8 1.0 4.9
5 M	0234 0819 1449 2035	0.2 5.1 0.4 5.3	**20** TU	0202 0756 1411 2000	1.0 4.7 1.1 4.9
6 TU	0316 0903 1532 2120	0.4 4.9 0.7 5.0	**21** W	0228 0820 1440 2026	1.1 4.6 1.3 4.7
7 W ◑	0401 0949 1618 2209	0.8 4.7 1.0 4.7	**22** TH	0300 0845 1514 2057	1.3 4.5 1.5 4.5
8 TH	0449 1042 1711 2309	1.1 4.4 1.3 4.4	**23** F ◑	0338 0925 1559 2154	1.5 4.3 1.7 4.3
9 F	0545 1153 1813	1.4 4.2 1.6	**24** SA	0429 1038 1703 2313	1.7 4.2 1.9 4.2
10 SA	0030 0652 1318 1929	4.1 1.6 4.2 1.7	**25** SU	0544 1156 1832	1.9 4.2 1.9
11 SU	0152 0814 1426 2057	4.1 1.6 4.3 1.6	**26** M	0032 0713 1313 1957	1.8 4.3 1.7
12 M	0254 0929 1519 2202	4.2 1.4 4.5 1.3	**27** TU	0150 0832 1424 2109	4.4 1.5 4.6 1.3
13 TU	0344 1024 1603 2252	4.4 1.2 4.7 1.1	**28** W	0259 0939 1525 2212	4.6 1.2 5.0 0.9
14 W	0426 1110 1642 2335	4.6 1.0 4.9 0.9	**29** TH	0357 1039 1618 2309	4.9 0.8 5.3 0.5
15 TH ○	0506 1151 1720	4.7 0.9 5.0	**30** F ●	0450 1133 1709	5.1 0.5 5.5
			31 SA	0001 0540 1223 1758	0.2 5.3 0.2 5.6

Chart Datum is 2·91 metres below Ordnance Datum (Newlyn). HAT is 5·7 metres above Chart Datum.

》》 **FREE** monthly updates. Register at 《
www.reedsnauticalalmanac.co.uk 《

STANDARD TIME (UT)
For Summer Time add ONE hour in **non-shaded areas**

FALMOUTH   LAT 50°09′N   LONG 5°03′W
TIMES AND HEIGHTS OF HIGH AND LOW WATERS

Dates in red are **SPRINGS**
Dates in blue are NEAPS

YEAR 2019

SW England

SEPTEMBER

#	Time m	#	Time m
1 SU	0049 0.0 / 0629 5.4 / 1309 0.1 / 1845 5.7	**16** M	0045 0.8 / 0631 5.0 / 1256 0.9 / 1841 5.1
2 M	0133 0.0 / 0714 5.4 / 1350 0.1 / 1930 5.6	**17** TU	0109 0.8 / 0702 4.9 / 1319 0.9 / 1908 5.0
3 TU	0214 0.1 / 0755 5.3 / 1429 0.3 / 2011 5.4	**18** W	0133 0.9 / 0727 4.9 / 1344 1.0 / 1931 4.9
4 W	0252 0.4 / 0833 5.1 / 1507 0.6 / 2050 5.0	**19** TH	0159 1.0 / 0745 4.7 / 1413 1.1 / 1948 4.7
5 TH	0331 0.8 / 0911 4.8 / 1548 1.0 / 2131 4.6	**20** F	0229 1.2 / 0758 4.6 / 1445 1.3 / 2009 4.6
6 F	0414 1.3 / 0955 4.4 / 1637 1.5 / ☽ 2221 4.2	**21** SA	0305 1.4 / 0832 4.5 / 1526 1.6 / 2103 4.3
7 SA	0507 1.7 / 1056 4.1 / 1739 1.8 / 2351 3.9	**22** SU	0351 1.7 / 1000 4.3 / 1626 1.8 / ☾ 2248 4.1
8 SU	0614 2.0 / 1247 4.0 / 1859 2.0	**23** M	0504 2.0 / 1130 4.2 / 1804 2.0
9 M	0135 3.9 / 0745 2.0 / 1406 4.2 / 2048 1.8	**24** TU	0014 4.1 / 0649 2.0 / 1252 4.3 / 1941 1.7
10 TU	0239 4.1 / 0914 1.7 / 1500 4.4 / 2149 1.5	**25** W	0137 4.3 / 0817 1.6 / 1407 4.6 / 2056 1.3
11 W	0327 4.4 / 1007 1.4 / 1543 4.7 / 2233 1.2	**26** TH	0248 4.6 / 0926 1.2 / 1509 5.0 / 2158 0.8
12 TH	0407 4.6 / 1049 1.1 / 1621 4.9 / 2312 0.9	**27** F	0344 5.0 / 1024 0.8 / 1601 5.3 / 2253 0.4
13 F	0444 4.8 / 1127 0.9 / 1658 5.1 / 2347 0.8	**28** SA	0433 5.2 / 1116 0.4 / 1650 5.6 / ● 2343 0.1
14 SA	0521 4.9 / 1201 0.8 / 1735 5.2 / ○	**29** SU	0521 5.4 / 1204 0.1 / 1738 5.7
15 SU	0018 0.8 / 0557 5.0 / 1231 0.8 / 1809 5.2	**30** M	0028 0.0 / 0606 5.5 / 1248 0.0 / 1823 5.7

OCTOBER

#	Time m	#	Time m
1 TU	0111 0.0 / 0648 5.5 / 1328 0.1 / 1906 5.6	**16** W	0039 0.8 / 0633 5.0 / 1254 0.8 / 1842 5.0
2 W	0150 0.2 / 0726 5.3 / 1405 0.3 / 1944 5.3	**17** TH	0107 0.8 / 0700 5.0 / 1322 0.9 / 1908 4.9
3 TH	0225 0.5 / 0801 5.1 / 1441 0.7 / 2020 5.0	**18** F	0136 1.0 / 0721 4.9 / 1352 1.0 / 1929 4.7
4 F	0301 0.9 / 0836 4.8 / 1519 1.1 / 2056 4.6	**19** SA	0207 1.1 / 0742 4.7 / 1426 1.2 / 1956 4.5
5 SA	0340 1.4 / 0916 4.5 / 1605 1.6 / ☽ 2141 4.1	**20** SU	0243 1.4 / 0824 4.5 / 1508 1.5 / 2104 4.3
6 SU	0430 1.9 / 1010 4.2 / 1706 2.0 / 2254 3.8	**21** M	0331 1.7 / 0947 4.3 / 1609 1.8 / ☾ 2236 4.1
7 M	0538 2.2 / 1146 4.0 / 1826 2.1	**22** TU	0445 2.0 / 1111 4.2 / 1748 1.9
8 TU	0110 3.8 / 0705 2.2 / 1335 4.1 / 2017 1.9	**23** W	0001 4.1 / 0629 1.9 / 1233 4.3 / 1924 1.6
9 W	0215 4.0 / 0843 1.9 / 1432 4.4 / 2120 1.6	**24** TH	0124 4.3 / 0759 1.6 / 1348 4.6 / 2038 1.2
10 TH	0302 4.3 / 0936 1.5 / 1516 4.7 / 2201 1.2	**25** F	0232 4.6 / 0907 1.2 / 1450 5.0 / 2138 0.8
11 F	0341 4.6 / 1017 1.2 / 1555 4.9 / 2238 1.0	**26** SA	0326 5.0 / 1004 0.7 / 1542 5.3 / 2231 0.4
12 SA	0418 4.8 / 1053 1.0 / 1632 5.1 / 2312 0.8	**27** SU	0413 5.2 / 1055 0.4 / 1630 5.5 / 2320 0.2
13 SU	0454 5.0 / 1128 0.8 / 1708 5.1 / ○ 2344 0.7	**28** M	0458 5.4 / 1142 0.2 / 1716 5.5 / ●
14 M	0530 5.1 / 1159 0.8 / 1742 5.1	**29** TU	0005 0.1 / 0541 5.4 / 1226 0.1 / 1800 5.5
15 TU	0012 0.7 / 0603 5.1 / 1227 0.8 / 1813 5.1	**30** W	0047 0.1 / 0621 5.4 / 1305 0.2 / 1842 5.4
		31 TH	0125 0.3 / 0659 5.3 / 1342 0.4 / 1919 5.1

NOVEMBER

#	Time m	#	Time m
1 F	0200 0.7 / 0733 5.1 / 1418 0.8 / 1954 4.8	**16** SA	0120 0.9 / 0711 5.0 / 1340 0.9 / 1929 4.7
2 SA	0234 1.0 / 0809 4.9 / 1455 1.2 / 2029 4.5	**17** SU	0155 1.1 / 0748 4.8 / 1419 1.1 / 2012 4.5
3 SU	0312 1.5 / 0848 4.6 / 1538 1.6 / 2112 4.1	**18** M	0235 1.3 / 0836 4.7 / 1505 1.3 / 2111 4.3
4 M	0358 1.9 / 0937 4.3 / 1634 1.9 / ☽ 2213 3.8	**19** TU	0326 1.5 / 0940 4.5 / 1608 1.6 / ☾ 2224 4.1
5 TU	0501 2.1 / 1044 4.0 / 1746 2.1	**20** W	0438 1.8 / 1053 4.4 / 1732 1.6 / 2343 4.1
6 W	0015 3.7 / 0617 2.2 / 1237 4.0 / 1907 2.0	**21** TH	0605 1.8 / 1209 4.4 / 1858 1.5
7 TH	0135 3.9 / 0737 2.0 / 1350 4.2 / 2022 1.7	**22** F	0101 4.3 / 0728 1.5 / 1322 4.6 / 2011 1.2
8 F	0226 4.2 / 0843 1.7 / 1440 4.5 / 2114 1.4	**23** SA	0208 4.5 / 0839 1.2 / 1426 4.8 / 2112 0.8
9 SA	0309 4.5 / 0932 1.4 / 1522 4.7 / 2155 1.1	**24** SU	0303 4.8 / 0938 0.8 / 1520 5.1 / 2206 0.6
10 SU	0347 4.7 / 1013 1.1 / 1601 4.9 / 2232 0.9	**25** M	0351 5.0 / 1030 0.6 / 1609 5.2 / 2255 0.4
11 M	0424 4.9 / 1051 0.9 / 1637 5.0 / 2307 0.8	**26** TU	0436 5.2 / 1119 0.4 / 1655 5.2 / ● 2341 0.3
12 TU	0500 5.0 / 1126 0.8 / 1712 5.0 / ○ 2341 0.7	**27** W	0518 5.3 / 1203 0.3 / 1739 5.2
13 W	0534 5.1 / 1159 0.8 / 1747 5.0	**28** TH	0024 0.4 / 0558 5.3 / 1245 0.4 / 1820 5.1
14 TH	0013 0.7 / 0606 5.1 / 1232 0.7 / 1820 5.0	**29** F	0103 0.5 / 0636 5.2 / 1323 0.6 / 1858 4.9
15 F	0046 0.8 / 0638 5.0 / 1306 0.8 / 1854 4.9	**30** SA	0139 0.8 / 0712 5.1 / 1359 0.8 / 1934 4.7

DECEMBER

#	Time m	#	Time m
1 SU	0214 1.1 / 0748 4.9 / 1436 1.1 / 2009 4.5	**16** M	0151 0.9 / 0746 5.0 / 1419 0.8 / 2013 4.6
2 M	0249 1.4 / 0826 4.7 / 1515 1.4 / 2050 4.2	**17** TU	0235 1.0 / 0834 4.9 / 1507 1.0 / 2105 4.5
3 TU	0329 1.7 / 0910 4.5 / 1600 1.7 / 2141 4.0	**18** W	0325 1.2 / 0929 4.7 / 1603 1.2 / 2207 4.3
4 W	0419 1.9 / 1004 4.2 / 1658 1.9 / ☽ 2248 3.9	**19** TH	0425 1.4 / 1031 4.6 / 1709 1.3 / ☾ 2315 4.2
5 TH	0523 2.1 / 1111 4.1 / 1806 1.9	**20** F	0534 1.5 / 1139 4.5 / 1821 1.3
6 F	0014 3.9 / 0633 2.0 / 1231 4.1 / 1914 1.8	**21** SA	0028 4.2 / 0648 1.5 / 1250 4.5 / 1933 1.2
7 SA	0130 4.0 / 0740 1.9 / 1343 4.3 / 2015 1.6	**22** SU	0138 4.4 / 0802 1.3 / 1358 4.6 / 2041 1.1
8 SU	0225 4.3 / 0839 1.6 / 1437 4.5 / 2106 1.3	**23** M	0238 4.6 / 0909 1.1 / 1458 4.7 / 2141 0.9
9 M	0310 4.6 / 0928 1.3 / 1522 4.7 / 2151 1.1	**24** TU	0331 4.8 / 1007 0.9 / 1551 4.8 / 2233 0.7
10 TU	0350 4.7 / 1013 1.1 / 1603 4.8 / 2232 0.9	**25** W	0417 4.9 / 1059 0.7 / 1638 4.9 / 2322 0.6
11 W	0428 4.9 / 1055 0.9 / 1642 4.9 / 2312 0.8	**26** TH	0500 5.0 / 1146 0.6 / 1722 4.9 / ●
12 TH	0505 5.0 / 1135 0.8 / 1722 5.0 / ○ 2352 0.7	**27** F	0006 0.6 / 0540 5.1 / 1229 0.6 / 1803 4.9
13 F	0543 5.1 / 1215 0.7 / 1803 5.0	**28** SA	0047 0.7 / 0619 5.1 / 1308 0.6 / 1842 4.8
14 SA	0031 0.7 / 0622 5.1 / 1256 0.7 / 1845 4.9	**29** SU	0123 0.8 / 0656 5.1 / 1344 0.8 / 1918 4.7
15 SU	0111 0.7 / 0703 5.1 / 1336 0.7 / 1927 4.8	**30** M	0156 1.0 / 0732 5.0 / 1418 1.0 / 1953 4.6
		31 TU	0228 1.2 / 0808 4.8 / 1451 1.2 / 2030 4.4

Chart Datum is 2·91 metres below Ordnance Datum (Newlyn). HAT is 5·7 metres above Chart Datum.

》 FREE monthly updates. Register at 《
www.reedsnauticalalmanac.co.uk

45

1.10 MEVAGISSEY

Cornwall 50°16′·16N 04°46′·93W ✿✿♦♦✿✿✿

CHARTS AC 1267, 148, 147, 5602; Imray C10, C6, 2400

TIDES –0600 Dover; ML 3·1; Duration 0600

Standard Port PLYMOUTH (→)

Times				Height (metres)			
High Water		Low Water		MHWS	MHWN	MLWN	MLWS
0000	0600	0000	0600	5·5	4·4	2·2	0·8
1200	1800	1200	1800				
Differences MEVAGISSEY							
–0015	–0020	–0010	–0005	–0·1	–0·1	–0·2	–0·1

SHELTER Exposed only to E'lies; if >F3 go to Fowey. Appr in strong SE'lies is dangerous. Access all tides.

NAVIGATION WPT 50°16′·16N 04°45′·98W, 270°/6ca to Victoria pier head. Beware rky ledges off N Quay. Speed limit 3kn.

LIGHTS AND MARKS As chartlet and 1.2. S Pier lt ho is conspic.

COMMUNICATIONS (Code 01726) Dr 843701. HM 843305. Ch 16 14 (Summer 0900-2100. Winter 0900-1700); call HM for berth.

FACILITIES www.mevagisseyharbour.co.uk meva.harbour@talk21.com
Outer Hbr Victoria Pier is heavily used by fishing boats ✿ may berth only with prior permission in about 2m, £20 all LOAs, short stay free. Keep clear of 🛢 hose & pleasure boats at inner end. Fishing boats use pier as early/late temporary berth to catch tide for entry/exit.

Bilge keelers/cats can dry out on sandy beach, SE side of W Quay on request to HM. There are 2 trots of 3 cylindrical ✿s off N Pier: moor fore & aft, rafting possible. RNLI lifejacket lockers.

No ⚓ inside the hbr due to overcrowding and many FVs. ⚓ off is only advised in settled weather with no E in the wind.

Inner Hbr (dries 1·5m) reserved for FVs, unless taking on ⚓. ➤ 🗃 🛠 🍴 🛒(1t) 🛢. 🚻 🏴 on W Quay; key held at HM Office.

Village 🛢✉🏪🎦 Ice ✕🛒@, Ⓑ (Jun-Sep1000-1430, Oct-May1000-1300), 🚆 (bus to St Austell), ✈ Newquay.

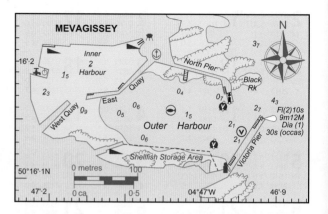

PAR, Cornwall, **50°20′·61N 04°42′·06W**. AC 1267, 148, 31. HW –0555 on Dover. The harbour dries 1·2m and should only be used in an emergency for yachts. There are plans for a marina, but possibly not for several years. Beware Killyvarder Rock (dries 2·4m) 3ca SE of ent, marked by unlit SHM bn. Only enter by day, in calm weather with offshore winds. Facilities in village.

GUNNERY RANGE OFF DODMAN PT AND GRIBBIN HEAD

Naval gunnery practice takes place to seaward of Dodman Point and Gribbin Head, under the control of Flag Officer Sea Training (FOST), HMS Drake, Plymouth PL2 2BG. For info ☎(01752) 557550 (H24) or call *FOST OPS* on VHF Ch 74, which is monitored by all warships in the exercise areas S of Plymouth.

Firing is by day only, approx 1-2 times per week, in a 2hrs block, although actual firing only lasts about 15 mins. Planned firings are broadcast in Gunfacts and Navtex. Advance details are also printed in local newspapers and are available from HMs at Fowey, Looe, Polperro and Mevagissey. Firings are not planned for 2 weeks at Christmas and 4 weeks in August.

Warships, from a position some 2·5 to 9M SSE of Gribbin Hd, fire WSW at 3 target buoys: **A** Fl Y 10s 50°08′·53N 04°46′·37W; **B** Fl Y 5s 50°10′·31N 04°45′·00W; and **C** Fl Y 2s 50°10′·41N 04°47′·51W, about 3·7M SSE of Dodman Pt.

If safety criteria cannot be met, eg owing to vessels in the range area, firing will not take place. A helicopter provides range safety surveillance. A range safety boat will advise other craft of firings and may suggest a course alteration to clear the area. Yachts are legally entitled to transit the range area without undue delay.

MINOR HARBOURS BETWEEN MEVAGISSEY AND FOWEY

CHARLESTOWN, Cornwall, **50°19′·84N 04°45′·35W**. AC 1267, 148, 31. HW –0555 on Dover, –0010 on Plymouth; HW –0·1m on Plymouth; ML 3·1m; Duration 0605. HM ☎01726 70241, VHF Ch 14, 16 (HW –2, only when vessel expected). Still a china clay port, also home to square riggers for filming/publicity.

Enter the inner hbr via entry gate as agreed with HM, but only in W'lies. Ent dries and should only be attempted by day and in offshore winds with calm weather. Gate fee £35, but free if within Square Sail's schedule. Hbr is shut in SE'lies. Waiting buoy 2ca S of hbr. Bkwtrs 2FG & 2FR(vert) 5m 1M. Ent sig: ● (night) = hbr shut. Facilities: www.square-sail.com; 🛢£3.12/m, ⚓🗃🛠 (or pre-arranged tanker) 🛠🔧⚠🛒(8t) ✉✕🛢.

FOWEY TO PLYMOUTH

(AC 1267, 148, 1613) Gribbin Head has a conspic daymark, a 25m high ☐ twr with R & W bands. In bad weather the sea breaks on rocks around the Head. Cannis Rock (4·3m) is 2½ca SE, marked by a lit SCM.

Fowey is a fine natural harbour, accessible in most conditions although exposed to strong SSW'lies. 3M E of Fowey and 5ca offshore at the East end of Lantivet Bay is Udder Rock (0·6m) marked by a lit SCM. 1M ENE of this buoy Larrick Rock (4·3m) is 1½ca off Nealand Pt.

Polperro harbour dries, but the inlet is a good ⚓ in offshore winds. Beware Polca Rock (1m) roughly in the approach. East of Polperro shoals lie 2½ca off Downend Point. A measured Mile lies between Polperro and **Looe**. The channel between Looe Island and Hannafore Point nearly dries; the charted Boat passage is not advised. The Ranneys (dry, marked by SCM lt buoy) are reefs extending 2½ca E and SE of Looe Island where there are overfalls in bad weather. In Whitsand Bay a buoyed artificial reef (the former *HMS Scylla*) lies 1·4M NW of Rame Head.

Eddystone rocks (AC 1613) lie 8M SSW of Rame Head. NE Rock (0·9m) and shoals are 280m NE. Close NW of the light house is the stump of the old lt ho. The sea can break on Hand Deeps (7m), rocks 3·4M NW of Eddystone, marked by a lit WCM.

1.11 FOWEY

Cornwall **50°19'·65N 04°38'·54W** ❄❄❄◊◊◊❀❀❀

CHARTS AC 1267, 148, 31, 5602; Imray C10, C6, 2400

TIDES –0540 Dover; ML 2·9; Duration 0605

Standard Port PLYMOUTH (→)

Times				Height (metres)			
High Water		Low Water		MHWS	MHWN	MLWN	MLWS
0000	0600	0000	0600	5·5	4·4	2·2	0·8
1200	1800	1200	1800				
Differences *FOWEY*							
–0010	–0015	–0010	–0005	–0·1	–0·1	–0·2	–0·2
LOSTWITHIEL							
+0005	–0010	DR	DR	–4·1	–4·1	DR	DR
PAR							
–0010	–0015	–0010	–0005	–0·4	–0·4	–0·4	–0·2

SHELTER Good at Mixtow in all winds. Lower hbr exposed to S-SW, S and SW'ly gales can cause heavy swell and confused seas, especially on the ebb. Entry H24 at any tide in almost any conditions.

NAVIGATION WPT 50°19'·33N 04°38'·80W, 027°/7ca through hbr ent to Whitehouse Pt Dir lt in W sector. Appr in W sector of Fowey lt ho. 3M E of ent beware Udder Rk marked by SCM lt buoy. From SW beware Cannis Rk (4ca SE of Gribbin Hd) marked by SCM lt buoy. Entering hbr, keep well clear of Punch Cross Rks to stbd. Fowey is a busy commercial clay port with ship movements day and night. Speed limit 6kn. Give way to the Bodinnick–Caffa-Mill ferry.

Unmarked, mostly drying chan is navigable up to Golant (1M N of Wiseman's Pt), but moorings restrict ⚓ space. Access on the tide by shoal draught to Lerryn (1·6M) and Lostwithiel (3M) (18m power cables and 5·3m rail bridge).

LIGHTS AND MARKS See chartlet and 1.2. A daymark RW twr 33m on Gribbin Hd (1·3M WSW of hbr ent) is conspic from seaward, as is a white house 3ca E of hbr ent. Fowey light house is conspicuous. The W sector (022°-032°) of Whitehouse Pt dir lt leads through the 200m wide harbour entrance. At the entrance, St Catherine's Point is a lamp box; Lamp Rock is on a G bcn. At Whitehouse Pt a conspic radar and CCTV mast monitors traffic in the entrance.

COMMUNICATIONS (Code 01726) Ⓗ 832241; Dr 829272; HM 832471. *Fowey Hbr Radio* Ch **12** 16 (HO). Hbr Patrol (0900-2000LT) Ch 12. Water taxi (07774 906730) Ch 06. Pilots & Tugs Ch 09, 12.

FACILITIES from seaward:
Polruan Quay (E bank) short stay (2 hrs) pontoon, ◣ ⚓ ◪(3t).
Toms BY ☎870232, ◣◣ ◵◷ ◪ ⬙ ▮♦ ✕ ✎ ⒺⒽ ⛽ ◪(50t) ◻.

FOWEY ◉s (B) are marked 'FHC VISITORS'. 3 x ◉ pontoons are in situ May-Oct. Berth/moor as directed by Hbr Patrol. Average o/night fees on ◉/pontoon: £2·00/m, but £2·50/m on Mixtow Pill inc 30 min free wi-fi. Reductions for 3 or 7 days, except Jul/Aug.

Pont Pill, on the E side, offers double-berth fore and aft ◉s and ◯ in 2m on two 36m floating pontoons; there is also a refuse barge.

Albert Quay HM's Office, ◣◣ ⚓. The 'T' shaped landing pontoon is for short stay (2 hrs). Off the E bank is a trot of blue swinging ◉s (1-50) and another 36m floating pontoon (*Underhills*).

Fuel Pontoon W bank 100m N of Caffa Mill H24 ▮ only, selfservice card sales only.

Berrills BY short stay (2 hrs 0800-1800 then overnight ◉ £2·60/m) landing pontoon, ♿ access, 250m N of Albert Quay, ⚓ ⌗ oil disposal.

Midway between Bodinnick and Mixtow Pill a sheltered double-sided pontoon (*Gridiron*) is off the E bank with 100m of ◉ on its W side.

Mixtow Pill (5ca upriver) 135m walk-ashore pontoon in 2·2m, ◉ on S side. ◵ ⚓ ▮♦ ⒽⓇ ◪(15t) by arrangement. ◉ 30m N of Wiseman's Point.

Royal Fowey YC ☎833573, ⚓⍓ ◵ ✕ ◻ ☎832245 Cafe.
Fowey Gallants SC ☎832335, ◵ ✕ ◻.
Services M, ▮ ⒹⒺ ACA, Ⓔ Ⓡ ◣◣ ✎ ⒺⒽ ✕ (15t).
Town ✕ ◻ ✉ Ⓑ ⇌ (bus to Par), ✈ (Newquay).

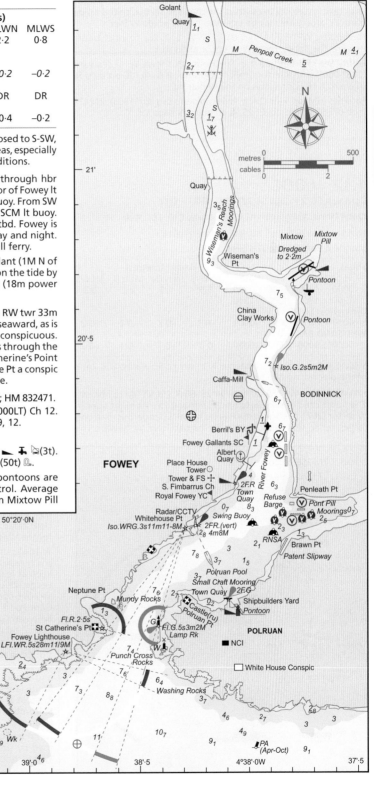

MINOR HARBOUR BETWEEN FOWEY AND LOOE

POLPERRO, Cornwall, **50°19'·78N 04°30'·79W**. AC 1267, 148, 5602. HW −0554 on Dover; HW −0007 and −0·2m on Plymouth; ML 3·1m; Duration 0610. Harbour dries about 2m; depth 2·5m at MHWN.

Shelter Good. Entrance is 9·8m wide (protected by gate in bad weather, but inner basin remains tidal). 6🅥, 3 either side of fairway. Moor fore and aft between 2 of the buoys on one side using short lines taking care not to obstruct the channel. Berths limited inside harbour, alongside may be achieved briefly inside W Pier.

Navigation WPT 50°19'·74N 04°30'·72W, 310°/2ca to inner basin. Beware The Raney, rocks W of ent, and rks to E.

Lights See chartlet and 1.2. Spy House Pt shows W to seaward (288°-060°) with R sectors inshore. Outermost 🅥s lit R/G either side of chan. A FR light shows on the W pier hd.

Facilities HM on Fish Quay, www.polperro.org ☎(01503) 272423 mob 07966 528045; ⌲£1.25. ⚓ on quays. **Village** ✉ ✕ ▱.

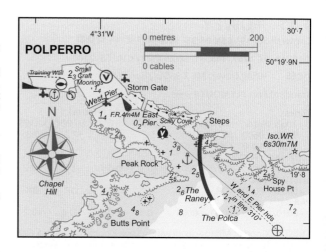

1.12 LOOE

Cornwall **50°21'·04N 04°27'·03W** ❄❄❄🌼🌼🌼🌼🌼

CHARTS AC 1267, 148, 147, 5602; Imray C10, C6, 2400

TIDES −0538 Dover; ML 3·0; Duration 0610

Standard Port PLYMOUTH (→)

Times				Height (metres)			
High Water		Low Water		MHWS	MHWN	MLWN	MLWS
0000	0600	0000	0600	5·5	4·4	2·2	0·8
1200	1800	1200	1800				
Differences LOOE							
−0010	−0010	−0005	−0005	−0·1	−0·2	−0·2	−0·2
WHITSAND BAY							
0000	0000	0000	0000	0·0	+0·1	−0·1	+0·2

SHELTER Good, but uncomfortable in strong SE winds. ⚓ in 2m E of the pier hd; no ⚓ in hbr.

NAVIGATION WPT 50°20'·68N 04°25'·60W, 290°/1·0M to hbr ent. The outer chan has rky outcrops and dries 0·7m. At night appr in W sector (267°-313°) of Banjo Pier hd lt. Ent is dangerous in strong SE'lies, when seas break heavily on the bar. From W, beware The

Ranneys, reef extending 500m SE of Looe Is, marked by SCM buoy Q (6) + L Fl 15s. 3ca NE of hbr ent, avoid the Limmicks, rks extending 1½ca offshore. Do not attempt the rky passage between Looe Is and the mainland except with local knowledge and at HW. At sp, ebb tide runs up to 5kn. The speed limit within the harbour is 5 mph.

LIGHTS AND MARKS See chartlet and 1.2. Looe Island (aka St George's Is) is conspic (45m), 8ca S of the ent. Mid Main ECM lt bcn is off Hannafore Pt, halfway between pier hd and Looe Is. Siren (2) 30s (fishing) at Nailzee Pt. No lts inside hbr.

COMMUNICATIONS (Code 01503) Dr 263195. HM: Ch 16 (occas).

FACILITIES 🅥 berth on W Quay, marked VISITORS, £12.00 all LOA, dries about 3·3m, sloping gently to firm, level sand; rafting is possible and frequently practised.

HM www.looecornwall.com ☎262839, mob 07918 7289550 OH.

W Quay ⚓ ⚓ 🅟 🅟 ⛟ ✕(Wood) ⚒ 🅑 Ⓔ.

E Quay ⚓ For FVs, but access HW ±3 for ⛟ ▱(3t).

Looe SC www.looesailing club.co.uk ☎262559 ✆ ✕ ▱.

Town ✉ Ⓑ 🍺 🅟 ✕ @ ⇌.

PLYMOUTH TO START POINT

Rame Head, on the W side of the entrance to **Plymouth Sound**, is a conspicuous conical headland, with a small chapel on top; rocks extend about 1ca off and wind-over-tide overfalls may be met 1·5M to seaward. Approaching Plymouth from the West, clear Rame Head and Penlee Point by about 8ca, then steer NNE for the Western end of the Breakwater.

At the SE entrance to Plymouth Sound, Great Mewstone (57m) is a conspicuous rocky islet 4ca off Wembury Pt. From the East keep at least 1M offshore until clear of the drying Mewstone Ledge, 2½ca SW of Great Mewstone. The Slimers, which dry, lie 2ca E of Mewstone. E and W Ebb Rocks (awash) lie 2½ca off Gara Point (AC 30). Wembury Bay leads to the **River Yealm**. Between Gara Point and Stoke Point, 2·5M to the E, dangers extend up to 4ca offshore.

In Bigbury Bay beware Wells Rock (1m) and other dangers 5ca S of Erme Head. From Bolt Tail to Bolt Head keep 5ca offshore to clear Greystone Ledge, sunken rocks near Ham Stone (11m), and Gregory Rocks (2m) 5ca SE of Ham Stone. The Little Mew Stone (3m high) and Mew Stone lie directly below Bolt Head. Keep at least 3ca SE of the Mewstones before turning N for **Salcombe**. Overfalls occur off Prawle Point, 2·6M E of Bolt Head.

Naval exercise areas from the Isles of Scilly to Start Point are used by submarines and warships, especially near Plymouth. Yachts should try to stay clear.

HEMISPHERE
RIGGING
SERVICES

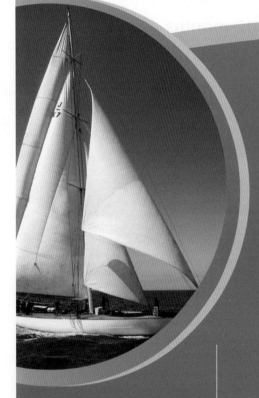

STANDING AND RUNNING RIGGING FOR DINGHIES, RACING AND CRUISING YACHTS

TEL: 07790 225511

MOBILE WORKSHOP
- FOR DOCKSIDE MAINTENANCE AND REPAIRS
- ON THE WATER FITTING AND INSTALLATION

OUR SERVICES
- FULLY INSURED SERVICES
- FIXED PRICE SERVICES
- MAST TUNING
- MAST STEPPING
- WINTER LAY UP
- FREE ESTIMATES
- MAST INSPECTIONS & SAFETY CHECKS
- PROFESSIONAL FRIENDLY ADVICE
- FURLING KITS AND DECK FITTINGS
- RELIABLE FLEXIBLE SERVICE
- DECK GEAR SERVICING & REPLACEMENT

SUPPLIERS OF:

HARKEN®
INNOVATIVE SAILING SOLUTIONS

Southern Ropes

STA-LOK®

 **TALURIT**

WWW.HEMISPHERERIGGINGSERVICES.COM
INFO@HEMISPHERERIGGINGSERVICES.COM

1.13 PLYMOUTH

Devon **50°20'·04N 04°10'·07W** (W Chan) ✿✿✿⚓⚓⚓✿✿✿
50°20'·04N 04°08'·07W (E Chan)

CHARTS AC 1267, 1613, 5602, 1900, 30, 1902, 1901, 1967, 871; Imray C10, C14, C6, 2400

TIDES –0540 Dover; ML 3·3; Duration 0610
Standard Port PLYMOUTH (→)

Times				Height (metres)			
High Water		Low Water		MHWS	MHWN	MLWN	MLWS
0000	0600	0000	0600	5·5	4·4	2·2	0·8
1200	1800	1200	1800				
Differences BOVISAND PIER							
–0010	–0010	–0008	–0009	–0·1	0·0	+0·2	+0·2
TURNCHAPEL (Cattewater)							
0000	0000	+0010	–0015	0·0	+0·1	+0·2	+0·1
JUPITER POINT (R Lynher)							
+0010	+0005	0000	–0005	0·0	0·0	+0·1	0·0
ST GERMANS (R Lynher)							
0000	0000	+0020	+0020	–0·3	–0·1	0·0	+0·2
SALTASH (R Tamar)							
0000	+0010	0000	–0005	+0·1	+0·1	+0·1	+0·1
LOPWELL (R Tavy)							
ND	ND	DR	DR	–2·6	–2·7	DR	DR
CARGREEN (R Tamar)							
0000	+0010	+0020	+0020	0·0	0·0	–0·1	0·0
COTEHELE QUAY (R Tamar)							
0000	+0020	+0045	+0045	–0·9	–0·9	–0·8	–0·4

NOTE: Winds from SE to W increase the flood and retard the ebb; vice versa in winds from the NW to E.

In Plymouth Sound N all traffic N of the Breakwater conforms to IRPC Rule 9 for 'Narrow channels'. There is extensive CCTV coverage of all the enclosed waters.

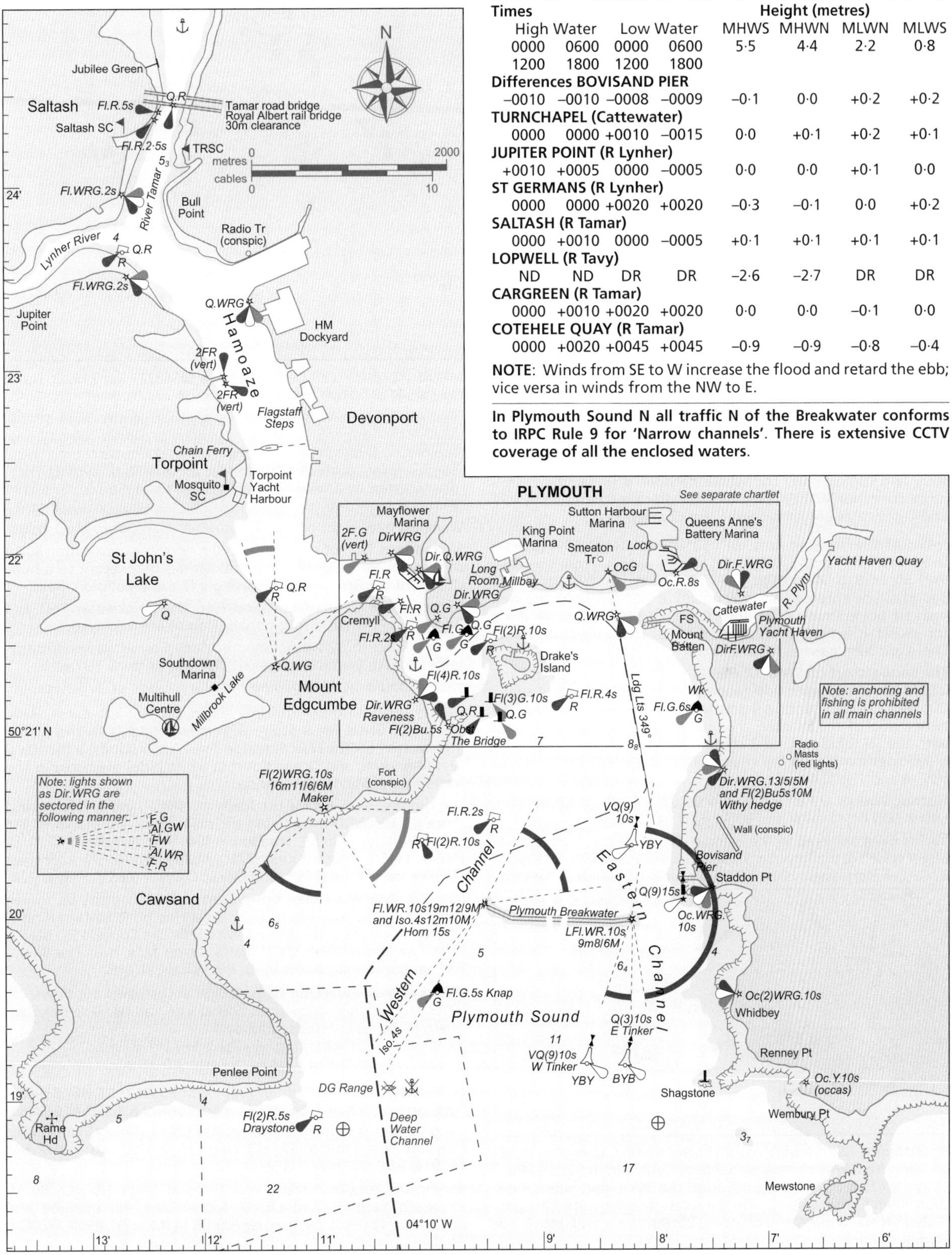

SW England

Chart labels:

Saltash — Fl.R.5s — Saltash SC — Fl.R.2·5s — Jubilee Green — Tamar road bridge — Royal Albert rail bridge 30m clearance — Q.R — TRSC 5₃ — Fl.WRG.2s — River Tamar — Bull Point — Radio Tr (conspic) — Lynher River — 4 — Q.R — R — Fl.WRG.2s — Jupiter Point — Hamoaze — Q.WRG — HM Dockyard — 2FR (vert) — 2FR (vert) — Flagstaff Steps — Devonport — Chain Ferry — Torpoint — Mosquito SC — Torpoint Yacht Harbour — St John's Lake — 2F.G (vert) — DirWRG — Mayflower Marina — Dir.Q.WRG — King Point Marina — Sutton Harbour Marina — Lock — Queens Anne's Battery Marina — See separate chartlet — **PLYMOUTH** — Fl.R — R — Long Room Millbay — Smeaton Tr — OcG — Dir.F.WRG — Yacht Haven Quay — Q.R — R — Cremyll — Fl.R.2S — R — Dir.WRG — Fl.G — G — Q.G — Q.G — G — Fl.G.G — Fl(2)R.10s — R — Q.WRG — Oc.R.8s — Cattewater — FS Mount Batten — Plymouth Yacht Haven — DirF.WRG — R — Drake's Island — Fl(4)R.10s — Fl(3)G.10s — G — Q.R — Q.G — Fl.R.4s — R — Wk — Fl.G.6s — G — Southdown Marina — Q — Q.WG — Mount Edgcumbe — Dir.WRG Raveness — Fl(2)Bu.5s — Obs The Bridge — 7 — 8₈ — Ldg Lts 349° — Note: anchoring and fishing is prohibited in all main channels — Multihull Centre — Millbrook Lake — Radio Masts (red lights) — Dir.WRG.13/5/5M and Fl(2)Bu5s10M Withy hedge — Note: lights shown as Dir.WRG are sectored in the following manner: F.G / Al.GW / FW / Al.WR / F.R — Fl(2)WRG.10s 16m11/6/6M — Maker — Fort (conspic) — Fl.R.2s — R — Fl(2)R.10s — R — Channel — VQ(9) 10s — YBY — Wall (conspic) — Bovisand Pier — Staddon Pt — Q(9)15s — Oc.WRG 10s — Eastern Channel — Cawsand — 6₅ — 4 — Fl.WR.10s19m12/9M and Iso.4s12m10M Horn 15s — Plymouth Breakwater — LFl.WR.10s 9m8/6M — 5 — 6₄ — 4 — Western Channel — Iso.4s — Fl.G.5s Knap — *Plymouth Sound* — Q(3)10s E Tinker — BYB — Oc(2)WRG.10s Whidbey — Renney Pt — Penle Point — DG Range — 11 — VQ(9)10s W Tinker — YBY — Shagstone — Oc.Y.10s (occas) — Wembury Pt — Rame Hd — 5 — 4 — Fl(2)R.5s Draystone — R — Deep Water Channel — Mewstone — 8 — 22 — 04°10' W — 17 — 3₇ — St John's Lake

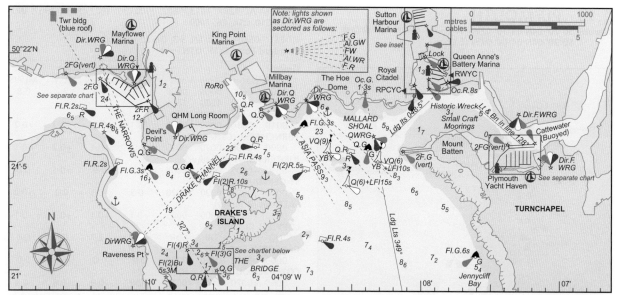

SHELTER

SHELTER Good or better in 4 major and some minor marinas; see Facilities. Around the Sound ⚓s, sheltered according to the wind: Cawsand Bay, Barn Pool (below Mt Edgcumbe), N of Drake's Island, below The Hoe and in Jennycliff Bay. Also good shelter W of Cremyll and off the Hamoaze in the R Lynher and R Tamar above Saltash. Plymouth is a Naval Base (Devonport) and a commercial, ferry and fishing port. The whole harbour is under the QHM's jurisdiction, but Millbay and the Cattewater are controlled by ABP and Cattewater Commissioners (Capt Charlesworth).

NAVIGATION From the west: WPT 50°18'·81N 04°10'·90W (abeam Draystone PHM buoy), 035°/1·5M to W Bkwtr lt. From the east: WPT 50°18'·81N 04°08'·00W, 000°/1·2M to abeam E Bkwtr lt. The Sound can be entered via the W (main) or E Chans which are well marked, but in strong W'lies the E Chan has a hazardous lee shore for yachts. It is vital to keep well clear of the unlit Shag Stone. There are 3·2m patches NE of E Tinker ECM lt buoy.

The Bridge (short cut SW of Drake's Island) is marked by 2 PHM and 2 SHM lt bns, both with tide gauges showing height of tide above CD; least charted depth is 1·3m. There are drying rocky patches either side of the marked channel which is aligned 327°/147° with the LH (blue roof) of 3 conspic high-rise blocks (5ca NW of Mayflower marina). In the chan spring tides reach 346°/1·6kn at HW −4 and 150°/2·4kn at HW +4.

Historic Wrecks are at: 50°21'·73N 04°07'·70W (N of Mt Batten), 50°19'·00N 04°11'·64W and 50°18'·61N 04°12'·05W.

Yachts need not keep to the marked chans but must not impede larger vessels or the Torpoint chain ferries at 50°22'·50N 04°11'·30W. In Cattewater vessels <20m LOA keep clear of vessels >20m LOA.

Speed limits: 10kn* N of Plymouth Bkwtr; 8kn in Cattewater; 4kn N of a line from Fisher's Nose to S side of QAB; 5kn in Sutton Hbr.
*Vessels <15m LOA are exempt from the 10kn limit when more than 400m from the shore or in the access lane for water/jet skiers (Fisher's Nose to the W end of Mount Batten Bkwtr and from Royal Plymouth Corinthian YC to W Mallard buoy).

NAVAL ACTIVITY Call Devonport Ops ☎563777 Ext 2182/3. Naval Ops, ☎501182 (H24) www.qhmplymouth.org.uk gives much info.

Exclusion zones: Vessels must not approach within 50m of any MoD establishment, warship or auxiliary or within 100m of any submarine. Buoys (Y) may be deployed to delineate these zones. Some ships are protected by a zone of 250m while underway. They are escorted by armed vessels; listen to Ch14 (Sound) or Ch13. Failure to respond to warnings will be taken as hostile intent.

Warships have right of way in main/DW chans; obey MoD Police.

Submarines: Do not pass within 200m or cross astern within 800m of any submarine under way. Submarines may secure to a buoy close N of the Breakwater, showing a Fl Y anti-collision light.

Diving: Keep clear of Bovisand Pier, the Breakwater Fort and Ravenness Pt when diving signals (Flag A) are displayed.

Ferries Roscoff: 12/week; 5hrs; Brittany Ferries (www.brittany-ferries.co.uk); Santander: weekly; 19½hrs; Brittany Ferries.

Yachts should avoid impeding the safe passage of larger vessels.

LIGHTS AND MARKS Principal daymarks: conical Rame Head to the W; Great Mew Stone & Staddon Heights to the E; The Breakwater; on The Hoe: Smeaton Tower (R/W bands) & the Naval War Memorial; and Ocean Court (a white bldg) overlooking The Narrows and Mayflower marina.

See chartlets and 1.2 for the many lts, some hard to see against shore lts. Dir WRG lts defining the main chans are shown H24** from: Whidbey (138·5°), Staddon Pt (044°), Withyhedge (070°), W Hoe bn (315°), Western King (271°), Millbay** (048·5°), Ravenness (225°), Mount Wise (343°), and Ocean Court** (085°). ** Not H24.

Notes: In fog W lts may, on request to Port Control, be shown from: Mallard (front ldg lt) Fl 5s; West Hoe bn F; Eastern King Fl 5s; Ravenness Fl (2) 15s; Mount Wise F; Ocean Court Fl 5s. Major lts in The Sound show QY if mains power fails. N of The Bkwtr, four large mooring buoys (C, D, E & F) have Fl Y lts.

Wind strength warning flags (R & W vert stripes) are flown at QAB and Mayflower marinas and at The Camber (HO only) to warn of excessive winds as follows:
| 1 wind flag | = | Force 5–7 (17–27kn) |
| 2 wind flags | = | > Force 7 (>27kn). |

COMMUNICATIONS (Code 01752) QHM 836952; DQHM 836485; Port surveyor 836962; Longroom Port Control 836528; Ⓗ 668080. Longroom Port Control (H24) monitor Ch 13, 14 or 16 underway. Flagstaff Port Control 552413 Flag Ch 13, 16. Cattewater HM 665934; ABP Millbay 662191.

Mayflower & QAB marinas and Torpoint Yacht Hbr: Ch 80 M. Plymouth Yacht Haven: Ch 80. Sutton Lock, for opening and marina: Ch 12 (H24). Cattewater Hbr Ch 14 (Mon-Fri, 0900-1700LT). Millbay Docks Ch 12 14 (only during ferry ops).

FACILITIES Marinas and Rivers from seaward (SE-NW):

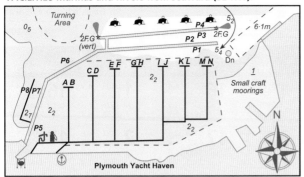

Plymouth Yacht Haven

Plymouth Yacht Haven www.yacht havens.com ☎404231, mobile 07721 498422 (o/night); access H24, channel dredged 2·25m. ➤ 450☐ inc ❶ £3.25 inc ⬡, ⬡⬡⬡⬡(H24) ⬡⬡⬡⬡⬡⬡⬡(75t) mobile ☐(4t) ⬡⬡ ✕ ⬡ wi-fi.

Turnchapel Wharf ryan@turnchapelwharf.com commercial/large yacht. Max draught 5m max LOA50m, ➤ at all states of tide.

Yacht Haven Quay 50°21'·87N 04°06'·61W ☎481190. Access H24 to pontoon in 1·2m, 2 FG (vert); ✕ ⬡⬡ C; dry berths/storage. Ferry: ½hrly from Mount Batten–Barbican, 5 mins: daily 0700-2300, Ch **M** or ☎408590 mob 07930 838614.

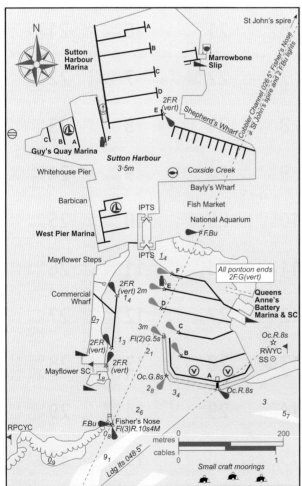

Queen Anne's Battery (QAB) www.queenannesbattery.co.uk ☎671142, mobile 07740 806039. ➤ 235☐+40❶ short stay £8·45, £3·50<12.5m>£4·35<18m>£4·85<18·1m inc ⬡; ⬡⬡⬡⬡(0830-1830, 7/7) ⬡✕✕⬡⬡⬡⬡(25t) ⬡⬡ © ⬡. Extensive facilities.

Sutton Harbour 50°21'·98N 04°07'·96W. West Pier and Guy's Quay marinas are mainly for locals (no shwrs).

Lock operates H24, free. IPTS: sigs 1, 2 & 3. Call *Sutton Lock* Ch 12 for entry and allocation of a berth; secure to floating pontoons either side of lock. The lock retains CD +3·5m in the hbr; when rise of tide >3m, approx HW ±3, free-flow is in operation.

Sutton Hbr Marina www.suttonharbourmarina.com ☎204702, ➤ 467☐ inc ❶ £3.50 inc ⬡, ⬡⬡⬡⬡⬡⬡(25t) ⬡ all maintenance/repairs. Nearby ⬡ ⬡.

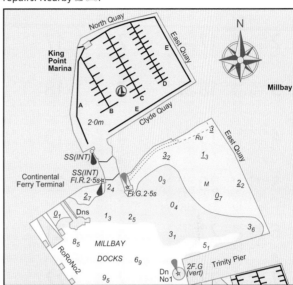

King Point Marina www.kingpointmarina.co.uk ☎424297; 171☐ inc ❶ (development continues), poa approx £3, ⬡⬡⬡⬡⬡ ice, wi-fi(weak).

Mill Bay Village Marina ☎226785, Ch **M**. No ❶.

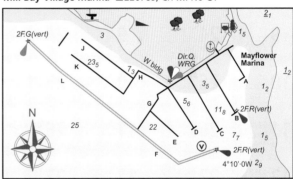

Mayflower Marina ④ www.mayflowermarina.co.uk ☎556633 (HO) mob 07840 116853. 27m max LOA with 3·5m min depth, but 1·5m near ⬡ and fuel berth. Limit speed to maintain steerage. ➤ 396☐ inc 40❶ £3·20 (inc ⬡ wi-fi), £5 then £3/hr<4 hrs. ⬡⬡(H24) ⬡⬡✕ ✕ ⬡⬡⬡⬡(33t) ⬡(1½t) Divers ⬡⬡ ✕ ⬡ ☎500008.

Southdown Marina www.southdownmarina.com ☎823084. Approach dries 2m. ⬡s. ➤ 35☐ in 2m inc❶ £20 all LOA, or at drying quay, ⬡ ⬡ ☐(20t) ⬡.

Torpoint Yacht Hbr www.torpointyachtharbour.co.uk ☎813658. Access H24, 2m. 100 ⬡s 70☐ pre-call for a ❶ berth £2.50, ⬡ ✕ ⬡ ✕ ⬡ ⬡ free water taxi.

YACHT CLUBS
Royal Western Yacht Club of England ☎660077, ⬡ ✕ ⬡.
Royal Plymouth Corinthian YC ☎664327, VHF ChM, ⬡ ✕ ⬡ ➤.
Plym YC ☎404991. **RNSA** ☎567854. **Mayflower SC** ☎ 492566.
Torpoint Mosquito SC ☎812508, ✕ ⬡ visitors welcome.
Saltash SC ☎845988. **Tamar River SC** ☎362741.

STANDARD TIME (UT)
For Summer Time add ONE hour in **non-shaded areas**

PLYMOUTH LAT 50°22'N LONG 4°11'W
TIMES AND HEIGHTS OF HIGH AND LOW WATERS

Dates in red are **SPRINGS**
Dates in blue are **NEAPS**

YEAR 2019

JANUARY

Day	Time	m	Time	m
1 TU	0139 / 0750 / 1404 / 2029	4.6 / 2.0 / 4.8 / 1.8	**16** W 0013 / 0634 / 1242 / 1920	4.4 / 2.2 / 4.5 / 2.0
2 W	0241 / 0859 / 1504 / 2129	4.8 / 1.8 / 4.8 / 1.6	**17** TH 0129 / 0757 / 1401 / 2032	4.5 / 2.0 / 4.7 / 1.8
3 TH	0336 / 0956 / 1558 / 2221	4.9 / 1.6 / 4.9 / 1.5	**18** F 0243 / 0906 / 1515 / 2136	4.8 / 1.7 / 4.9 / 1.5
4 F	0424 / 1045 / 1646 / 2307	5.1 / 1.4 / 5.0 / 1.3	**19** SA 0347 / 1009 / 1617 / 2235	5.1 / 1.4 / 5.1 / 1.2
5 SA	0507 / 1130 / 1728 / 2348	5.3 / 1.2 / 5.1 / 1.2	**20** SU 0443 / 1106 / 1713 / 2330	5.3 / 1.0 / 5.3 / 0.9
6 SU ●	0546 / 1210 / 1805	5.4 / 1.2 / 5.2	**21** M ○ 0536 / 1159 / 1806	5.6 / 0.7 / 5.4
7 M	0025 / 0621 / 1247 / 1840	1.2 / 5.4 / 1.2 / 5.2	**22** TU 0021 / 0626 / 1250 / 1856	0.7 / 5.7 / 0.5 / 5.5
8 TU	0059 / 0655 / 1320 / 1914	1.2 / 5.4 / 1.2 / 5.1	**23** W 0110 / 0715 / 1338 / 1946	0.6 / 5.8 / 0.4 / 5.5
9 W	0129 / 0729 / 1350 / 1949	1.3 / 5.3 / 1.3 / 5.0	**24** TH 0156 / 0803 / 1423 / 2033	0.5 / 5.8 / 0.5 / 5.5
10 TH	0157 / 0803 / 1419 / 2023	1.4 / 5.2 / 1.5 / 4.9	**25** F 0240 / 0849 / 1507 / 2118	0.7 / 5.7 / 0.6 / 5.3
11 F	0224 / 0837 / 1447 / 2057	1.6 / 5.1 / 1.6 / 4.8	**26** SA 0322 / 0934 / 1549 / 2202	0.9 / 5.5 / 0.9 / 5.1
12 SA	0252 / 0910 / 1518 / 2133	1.7 / 4.9 / 1.7 / 4.6	**27** SU ◐ 0404 / 1019 / 1633 / 2248	1.2 / 5.2 / 1.3 / 4.8
13 SU	0326 / 0948 / 1556 / 2215	1.9 / 4.8 / 1.9 / 4.5	**28** M 0450 / 1108 / 1721 / 2342	1.6 / 4.9 / 1.7 / 4.5
14 M ◐	0410 / 1034 / 1647 / 2308	2.0 / 4.7 / 2.0 / 4.4	**29** TU 0544 / 1210 / 1821	1.9 / 4.6 / 2.0
15 TU	0511 / 1132 / 1759	2.2 / 4.6 / 2.1	**30** W 0051 / 0654 / 1325 / 1938	4.4 / 2.1 / 4.4 / 2.1
			31 TH 0203 / 0820 / 1435 / 2059	4.4 / 2.1 / 4.4 / 2.0

FEBRUARY

Day	Time	m	Time	m
1 F	0306 / 0933 / 1535 / 2200	4.6 / 1.9 / 4.6 / 1.8	**16** SA 0209 / 0841 / 1452 / 2116	4.6 / 1.8 / 4.7 / 1.7
2 SA	0400 / 1028 / 1626 / 2250	4.8 / 1.6 / 4.8 / 1.5	**17** SU 0324 / 0953 / 1601 / 2221	4.9 / 1.4 / 4.9 / 1.3
3 SU	0446 / 1114 / 1710 / 2332	5.1 / 1.4 / 4.9 / 1.4	**18** M 0426 / 1054 / 1700 / 2318	5.3 / 1.0 / 5.2 / 0.8
4 M ●	0527 / 1155 / 1749	5.2 / 1.2 / 5.0	**19** TU ○ 0521 / 1148 / 1753	5.6 / 0.6 / 5.4
5 TU	0010 / 0604 / 1232 / 1825	1.1 / 5.3 / 1.1 / 5.1	**20** W 0010 / 0611 / 1238 / 1843	0.5 / 5.8 / 0.3 / 5.6
6 W	0044 / 0640 / 1305 / 1900	1.1 / 5.4 / 1.1 / 5.1	**21** TH 0057 / 0700 / 1324 / 1930	0.2 / 5.9 / 0.1 / 5.6
7 TH	0114 / 0715 / 1333 / 1934	1.1 / 5.3 / 1.1 / 5.1	**22** F 0141 / 0747 / 1406 / 2014	0.2 / 5.9 / 0.2 / 5.6
8 F	0140 / 0748 / 1359 / 2006	1.2 / 5.3 / 1.2 / 5.0	**23** SA 0222 / 0830 / 1446 / 2054	0.3 / 5.8 / 0.4 / 5.4
9 SA	0205 / 0819 / 1425 / 2036	1.3 / 5.2 / 1.3 / 4.9	**24** SU 0300 / 0910 / 1524 / 2131	0.6 / 5.5 / 0.7 / 5.0
10 SU	0231 / 0848 / 1452 / 2105	1.4 / 5.0 / 1.4 / 4.8	**25** M 0338 / 0947 / 1601 / 2205	1.0 / 5.2 / 1.2 / 4.9
11 M	0301 / 0918 / 1524 / 2137	1.5 / 4.9 / 1.5 / 4.7	**26** TU ◐ 0416 / 1024 / 1640 / 2242	1.4 / 4.8 / 1.6 / 4.6
12 TU ◐	0337 / 0956 / 1605 / 2222	1.7 / 4.7 / 1.7 / 4.5	**27** W 0502 / 1109 / 1730 / 2340	1.8 / 4.4 / 2.1 / 4.3
13 W	0425 / 1050 / 1701 / 2325	1.9 / 4.6 / 2.0 / 4.4	**28** TH 0603 / 1237 / 1839	2.2 / 4.2 / 2.3
14 TH	0535 / 1200 / 1825	2.1 / 4.4 / 2.1		
15 F	0043 / 0714 / 1326 / 1959	4.4 / 2.1 / 4.5 / 2.0		

MARCH

Day	Time	m	Time	m
1 F	0121 / 0727 / 1407 / 2018	4.2 / 2.3 / 4.1 / 2.3	**16** SA 0013 / 0646 / 1307 / 1935	4.4 / 2.1 / 4.3 / 2.1
2 SA	0237 / 0910 / 1512 / 2139	4.4 / 2.1 / 4.3 / 2.0	**17** SU 0146 / 0825 / 1439 / 2101	4.5 / 1.8 / 4.6 / 1.7
3 SU	0335 / 1010 / 1605 / 2230	4.6 / 1.7 / 4.6 / 1.6	**18** M 0307 / 0940 / 1549 / 2208	4.9 / 1.3 / 4.9 / 1.2
4 M	0422 / 1055 / 1648 / 2312	4.9 / 1.4 / 4.8 / 1.3	**19** TU 0410 / 1040 / 1646 / 2303	5.3 / 0.8 / 5.2 / 0.8
5 TU	0504 / 1135 / 1727 / 2350	5.1 / 1.1 / 5.0 / 1.1	**20** W 0504 / 1132 / 1737 / 2353	5.6 / 0.4 / 5.5 / 0.4
6 W	0542 / 1211 / 1804	5.3 / 1.0 / 5.1	**21** TH ○ 0554 / 1219 / 1824	5.8 / 0.1 / 5.6
7 TH	0023 / 0619 / 1242 / 1839	1.0 / 5.3 / 0.9 / 5.2	**22** F 0038 / 0641 / 1303 / 1908	0.2 / 5.9 / 0.0 / 5.7
8 F	0053 / 0654 / 1310 / 1913	1.0 / 5.3 / 0.9 / 5.2	**23** SA 0121 / 0725 / 1344 / 1948	0.1 / 5.8 / 0.1 / 5.6
9 SA	0119 / 0728 / 1336 / 1945	1.0 / 5.3 / 1.0 / 5.1	**24** SU 0159 / 0805 / 1421 / 2024	0.2 / 5.7 / 0.4 / 5.5
10 SU	0144 / 0758 / 1402 / 2012	1.0 / 5.2 / 1.0 / 5.0	**25** M 0236 / 0841 / 1456 / 2055	0.5 / 5.4 / 0.7 / 5.2
11 M	0210 / 0826 / 1429 / 2039	1.1 / 5.1 / 1.2 / 5.0	**26** TU 0310 / 0912 / 1529 / 2121	0.9 / 5.1 / 1.2 / 4.9
12 TU	0239 / 0855 / 1459 / 2109	1.2 / 4.9 / 1.4 / 4.8	**27** W 0346 / 0941 / 1605 / 2151	1.4 / 4.7 / 1.7 / 4.6
13 W	0313 / 0931 / 1537 / 2151	1.4 / 4.8 / 1.6 / 4.7	**28** TH ◐ 0428 / 1019 / 1651 / 2237	1.8 / 4.3 / 2.1 / 4.3
14 TH ◐	0358 / 1024 / 1628 / 2253	1.7 / 4.5 / 1.9 / 4.5	**29** F 0525 / 1123 / 1755	2.2 / 4.0 / 2.4
15 F	0503 / 1136 / 1748	2.0 / 4.3 / 2.1	**30** SA 0005 / 0641 / 1334 / 1920	4.1 / 2.4 / 4.0 / 2.4
			31 SU 0201 / 0829 / 1444 / 2102	4.2 / 2.2 / 4.2 / 2.2

APRIL

Day	Time	m	Time	m
1 M	0303 / 0939 / 1536 / 2158	4.5 / 1.8 / 4.5 / 1.8	**16** TU 0250 / 0922 / 1533 / 2148	4.9 / 1.3 / 4.9 / 1.2
2 TU	0352 / 1024 / 1619 / 2241	4.8 / 1.5 / 4.8 / 1.4	**17** W 0351 / 1020 / 1627 / 2243	5.2 / 0.8 / 5.2 / 0.8
3 W	0434 / 1103 / 1659 / 2319	5.0 / 1.2 / 5.0 / 1.2	**18** TH 0443 / 1110 / 1715 / 2331	5.5 / 0.4 / 5.5 / 0.5
4 TH	0514 / 1139 / 1737 / 2353	5.2 / 1.0 / 5.1 / 1.0	**19** F ○ 0532 / 1156 / 1800	5.7 / 0.2 / 5.6
5 F ●	0553 / 1211 / 1814	5.3 / 0.9 / 5.2	**20** SA 0015 / 0617 / 1239 / 1841	0.3 / 5.7 / 0.2 / 5.6
6 SA	0024 / 0629 / 1242 / 1848	0.9 / 5.3 / 0.8 / 5.2	**21** SU 0057 / 0700 / 1319 / 1919	0.4 / 5.7 / 0.3 / 5.6
7 SU	0054 / 0704 / 1311 / 1920	0.9 / 5.3 / 0.9 / 5.2	**22** M 0135 / 0738 / 1355 / 1952	0.4 / 5.5 / 0.5 / 5.4
8 M	0123 / 0736 / 1340 / 1949	0.9 / 5.2 / 0.9 / 5.2	**23** TU 0210 / 0811 / 1428 / 2020	0.7 / 5.3 / 0.9 / 5.2
9 TU	0152 / 0806 / 1409 / 2018	1.0 / 5.1 / 1.1 / 5.1	**24** W 0244 / 0839 / 1501 / 2046	1.0 / 5.0 / 1.3 / 5.0
10 W	0223 / 0839 / 1441 / 2052	1.1 / 5.0 / 1.3 / 5.0	**25** TH 0319 / 0909 / 1535 / 2119	1.4 / 4.7 / 1.7 / 4.7
11 TH	0259 / 0919 / 1521 / 2136	1.3 / 4.8 / 1.5 / 4.8	**26** F ◐ 0400 / 0948 / 1619 / 2203	1.8 / 4.3 / 2.1 / 4.5
12 F ◐	0345 / 1012 / 1614 / 2236	1.6 / 4.5 / 1.9 / 4.6	**27** SA 0453 / 1045 / 1719 / 2306	2.2 / 4.1 / 2.4 / 4.2
13 SA	0453 / 1124 / 1734 / 2354	1.9 / 4.3 / 2.1 / 4.5	**28** SU 0602 / 1236 / 1834	2.3 / 4.0 / 2.4
14 SU	0633 / 1257 / 1919	2.0 / 4.3 / 2.1	**29** M 0104 / 0721 / 1401 / 1953	4.2 / 2.2 / 4.2 / 2.3
15 M	0129 / 0810 / 1427 / 2043	4.6 / 1.7 / 4.6 / 1.7	**30** TU 0219 / 0839 / 1456 / 2103	4.4 / 1.9 / 4.4 / 1.9

Chart Datum is 3·22 metres below Ordnance Datum (Newlyn). HAT is 5·9 metres above Chart Datum.

STANDARD TIME (UT)
For Summer Time add ONE hour in **non-shaded areas**

PLYMOUTH LAT 50°22'N LONG 4°11'W
TIMES AND HEIGHTS OF HIGH AND LOW WATERS

Dates in red are SPRINGS
Dates in blue are NEAPS

YEAR **2019**

SW England

MAY

Day	Time	m	Time	m	Time	m	Time	m
1 W	0312	4.7	0934	1.6	1542	4.7	2154	1.6
2 TH	0358	4.9	1018	1.3	1624	4.9	2237	1.3
3 F	0441	5.1	1058	1.1	1704	5.1	2316	1.1
4 SA ●	0522	5.2	1136	0.9	1743	5.2	2353	1.0
5 SU	0602	5.3	1211	0.9	1820	5.3		
6 M	0028	0.9	0639	5.3	1247	0.8	1855	5.3
7 TU	0103	0.9	0716	5.2	1321	0.9	1928	5.3
8 W	0138	0.9	0752	5.2	1356	1.0	2003	5.2
9 TH	0214	1.1	0831	5.0	1433	1.2	2043	5.1
10 F	0255	1.3	0915	4.8	1517	1.5	2129	4.9
11 SA	0346	1.5	1010	4.6	1614	1.8	2227	4.8
12 SU ◑	0454	1.8	1119	4.4	1729	2.0	2342	4.6
13 M	0621	1.8	1246	4.4	1859	1.9		
14 TU	0111	4.7	0747	1.6	1405	4.6	2018	1.7
15 W	0226	4.9	0856	1.3	1508	4.9	2122	1.3
16 TH	0327	5.2	0954	0.9	1601	5.2	2217	1.0
17 F	0419	5.4	1045	0.7	1649	5.4	2306	0.7
18 SA ○	0508	5.5	1131	0.5	1734	5.5	2351	0.6
19 SU	0553	5.5	1214	0.5	1815	5.5		
20 M	0033	0.5	0634	5.4	1254	0.6	1851	5.5
21 TU	0112	0.7	0712	5.3	1330	0.8	1923	5.4
22 W	0148	0.9	0745	5.1	1404	1.1	1952	5.2
23 TH	0222	1.1	0815	4.9	1437	1.4	2022	5.0
24 F	0257	1.5	0848	4.7	1511	1.7	2057	4.8
25 SA	0335	1.8	0928	4.4	1550	2.0	2139	4.6
26 SU ◑	0422	2.0	1019	4.3	1642	2.2	2232	4.4
27 M	0522	2.2	1127	4.1	1747	2.3	2343	4.3
28 TU	0628	2.2	1253	4.2	1856	2.3		
29 W	0111	4.4	0734	2.0	1401	4.4	2001	2.1
30 TH	0219	4.6	0834	1.8	1455	4.6	2100	1.8
31 F	0314	4.8	0927	1.5	1543	4.8	2151	1.5

JUNE

Day	Time	m	Time	m	Time	m	Time	m
1 SA	0402	4.9	1015	1.3	1628	5.0	2238	1.3
2 SU	0449	5.1	1100	1.1	1711	5.2	2322	1.0
3 M ●	0533	5.2	1143	0.9	1753	5.3		
4 TU	0005	0.9	0616	5.2	1226	0.9	1833	5.4
5 W	0047	0.8	0659	5.3	1307	0.9	1913	5.4
6 TH	0129	0.8	0742	5.2	1349	1.0	1955	5.4
7 F	0213	0.9	0827	5.1	1433	1.1	2039	5.3
8 SA	0259	1.1	0915	5.0	1520	1.3	2127	5.2
9 SU	0351	1.3	1009	4.8	1614	1.5	2223	5.0
10 M ◑	0450	1.5	1112	4.6	1718	1.7	2329	4.8
11 TU	0600	1.6	1224	4.6	1830	1.8		
12 W	0045	4.8	0715	1.6	1335	4.7	1945	1.7
13 TH	0157	4.8	0824	1.4	1437	4.8	2052	1.5
14 F	0259	5.0	0925	1.2	1532	5.0	2150	1.3
15 SA	0354	5.1	1019	1.1	1622	5.2	2242	1.1
16 SU	0444	5.2	1107	0.9	1708	5.3	2329	0.9
17 M ○	0530	5.2	1151	0.9	1750	5.3		
18 TU	0012	0.9	0612	5.2	1232	0.9	1827	5.3
19 W	0052	0.9	0650	5.1	1310	1.0	1901	5.3
20 TH	0129	1.0	0725	5.0	1344	1.2	1933	5.2
21 F	0204	1.2	0758	4.9	1417	1.4	2006	5.1
22 SA	0237	1.4	0833	4.7	1448	1.6	2041	5.0
23 SU	0311	1.6	0911	4.6	1522	1.8	2119	4.8
24 M	0348	1.8	0954	4.4	1601	2.0	2202	4.6
25 TU ◑	0433	2.0	1044	4.3	1652	2.1	2253	4.5
26 W	0531	2.0	1143	4.3	1757	2.2	2355	4.4
27 TH	0636	2.0	1251	4.3	1906	2.1		
28 F	0108	4.5	0740	1.9	1358	4.5	2010	1.9
29 SA	0220	4.6	0841	1.7	1457	4.7	2109	1.7
30 SU	0321	4.8	0937	1.4	1551	4.9	2204	1.4

JULY

Day	Time	m	Time	m	Time	m	Time	m
1 M	0416	5.0	1030	1.2	1641	5.1	2257	1.1
2 TU ●	0508	5.1	1120	1.0	1728	5.3	2347	0.9
3 W	0557	5.2	1209	0.9	1815	5.5		
4 TH	0035	0.8	0646	5.3	1258	0.8	1901	5.6
5 F	0124	0.7	0734	5.3	1344	0.8	1947	5.6
6 SA	0211	0.7	0823	5.3	1430	0.9	2034	5.5
7 SU	0257	0.8	0911	5.2	1517	1.0	2122	5.4
8 M	0345	1.0	1001	5.0	1604	1.2	2212	5.2
9 TU ◑	0435	1.2	1054	4.8	1656	1.5	2308	5.0
10 W	0531	1.4	1154	4.7	1756	1.7		
11 TH	0013	4.8	0635	1.6	1300	4.6	1905	1.8
12 F	0124	4.7	0747	1.7	1404	4.6	2018	1.8
13 SA	0231	4.7	0856	1.6	1504	4.8	2124	1.6
14 SU	0331	4.8	0955	1.5	1558	4.9	2221	1.4
15 M	0424	4.9	1047	1.3	1646	5.1	2311	1.2
16 TU ○	0511	5.0	1133	1.2	1729	5.2	2356	1.1
17 W	0554	5.0	1215	1.1	1807	5.3		
18 TH	0036	1.0	0631	5.0	1253	1.1	1842	5.3
19 F	0113	1.1	0707	5.0	1327	1.2	1917	5.3
20 SA	0146	1.2	0741	5.0	1357	1.3	1950	5.2
21 SU	0216	1.3	0816	4.9	1425	1.4	2024	5.1
22 M	0244	1.4	0851	4.8	1452	1.6	2057	4.9
23 TU	0312	1.6	0926	4.6	1520	1.7	2132	4.8
24 W	0343	1.7	1005	4.5	1556	1.9	2211	4.6
25 TH ◑	0425	1.9	1051	4.4	1646	2.1	2301	4.5
26 F	0526	2.0	1148	4.4	1800	2.2		
27 SA	0005	4.4	0645	2.0	1258	4.4	1923	2.1
28 SU	0124	4.5	0800	1.9	1412	4.6	2035	1.9
29 M	0244	4.6	0906	1.6	1518	4.9	2139	1.5
30 TU	0350	4.9	1007	1.4	1616	5.1	2238	1.2
31 W	0448	5.1	1104	1.1	1708	5.4	2333	0.9

AUGUST

Day	Time	m	Time	m	Time	m	Time	m
1 TH ●	0541	5.3	1157	0.8	1759	5.6		
2 F	0025	0.6	0632	5.4	1247	0.6	1848	5.7
3 SA	0114	0.4	0723	5.5	1335	0.5	1936	5.8
4 SU	0201	0.4	0811	5.5	1419	0.5	2023	5.7
5 M	0245	0.5	0857	5.4	1502	0.7	2108	5.6
6 TU	0327	0.7	0941	5.2	1544	1.0	2153	5.3
7 W ◑	0410	1.1	1026	5.0	1628	1.3	2239	5.0
8 TH	0456	1.4	1117	4.7	1719	1.7	2334	4.7
9 F	0552	1.8	1219	4.5	1822	2.0		
10 SA	0049	4.4	0703	2.0	1332	4.4	1943	2.1
11 SU	0207	4.4	0828	2.0	1440	4.5	2105	1.9
12 M	0313	4.5	0938	1.8	1538	4.8	2207	1.7
13 TU	0408	4.7	1032	1.6	1627	5.0	2257	1.4
14 W	0454	4.9	1118	1.3	1709	5.2	2341	1.2
15 TH ○	0535	5.0	1159	1.2	1747	5.3		
16 F	0020	1.0	0611	5.1	1236	1.1	1823	5.3
17 SA	0055	1.0	0647	5.1	1308	1.1	1857	5.3
18 SU	0125	1.0	0721	5.1	1335	1.2	1931	5.3
19 M	0151	1.1	0754	5.0	1359	1.3	2003	5.2
20 TU	0214	1.3	0826	4.9	1422	1.4	2033	5.1
21 W	0238	1.4	0856	4.8	1447	1.5	2101	4.9
22 TH	0305	1.5	0927	4.7	1518	1.7	2134	4.7
23 F ◑	0340	1.7	1006	4.5	1559	1.9	2219	4.6
24 SA	0428	2.0	1101	4.4	1659	2.2	2324	4.4
25 SU	0544	2.2	1212	4.4	1839	2.2		
26 M	0045	4.4	0726	2.1	1336	4.5	2009	2.0
27 TU	0218	4.5	0844	1.8	1454	4.8	2121	1.6
28 W	0334	4.8	0951	1.4	1557	5.2	2224	1.2
29 TH	0433	5.1	1050	1.0	1652	5.5	2320	0.7
30 F ●	0526	5.4	1143	0.7	1743	5.7		
31 SA	0010	0.4	0616	5.5	1232	0.4	1832	5.9

Chart Datum is 3·22 metres below Ordnance Datum (Newlyn). HAT is 5·9 metres above Chart Datum.

》》 FREE monthly updates. Register at 《
www.reedsnauticalalmanac.co.uk

53

STANDARD TIME (UT)
For Summer Time add ONE hour in **non-shaded areas**

PLYMOUTH LAT 50°22′N LONG 4°11′W
TIMES AND HEIGHTS OF HIGH AND LOW WATERS

Dates in red are **SPRINGS**
Dates in blue are **NEAPS**

YEAR 2019

SEPTEMBER

Time	m		Time	m
1 0058	0.2	**16**	0057	1.0
0705	5.6		0656	5.2
SU 1318	0.3	M	1308	1.1
1919	5.9		1908	5.4
2 0142	0.2	**17**	0121	1.1
0750	5.6		0729	5.2
M 1400	0.3	TU	1331	1.2
2004	5.8		1939	5.3
3 0223	0.3	**18**	0145	1.2
0833	5.5		0759	5.1
TU 1440	0.5	W	1354	1.3
2046	5.6		2008	5.1
4 0302	0.7	**19**	0208	1.3
0913	5.3		0826	5.0
W 1519	0.9	TH	1419	1.4
2125	5.3		2035	5.0
5 0340	1.1	**20**	0235	1.5
0951	5.0		0854	4.9
TH 1558	1.3	F	1450	1.6
2203	4.9		2106	4.8
6 0420	1.6	**21**	0309	1.7
1031	4.7		0932	4.7
F 1643	1.8	SA	1529	1.9
☽ 2246	4.5		2152	4.6
7 0509	2.0	**22**	0354	2.0
1128	4.4		1028	4.5
SA 1742	2.2	SU	1626	2.2
		☽ 2259	4.4	
8 0008	4.2	**23**	0504	2.3
0617	2.3		1141	4.4
SU 1259	4.3	M	1807	2.3
1907	2.4			
9 0146	4.2	**24**	0023	4.3
0802	2.4		0701	2.3
M 1416	4.4	TU	1310	4.5
2050	2.2		1952	2.1
10 0256	4.3	**25**	0205	4.5
0922	2.1		0829	2.0
TU 1517	4.7	W	1436	4.9
2152	1.8		2107	1.6
11 0350	4.6	**26**	0321	4.9
1014	1.7		0937	1.5
W 1605	4.9	TH	1540	5.2
2238	1.5		2209	1.1
12 0434	4.9	**27**	0419	5.2
1057	1.4		1034	1.0
TH 1647	5.2	F	1634	5.6
2319	1.2		2302	0.7
13 0512	5.0	**28**	0509	5.5
1136	1.2		1125	0.6
F 1724	5.3	SA	1724	5.8
2356	1.0		● 2351	0.3
14 0548	5.2	**29**	0557	5.7
1211	1.1		1212	0.4
SA 1800	5.4	SU	1812	5.9
○				
15 0028	1.0	**30**	0036	0.2
0622	5.2		0642	5.8
SU 1242	1.0	M	1256	0.3
1834	5.4		1857	5.9

OCTOBER

Time	m		Time	m
1 0119	0.2	**16**	0051	1.1
0725	5.8		0702	5.3
TU 1337	0.3	W	1305	1.1
1940	5.8		1915	5.3
2 0158	0.4	**17**	0118	1.1
0804	5.6		0732	5.3
W 1415	0.6	TH	1331	1.2
2019	5.6		1945	5.2
3 0235	0.8	**18**	0145	1.3
0840	5.4		0801	5.2
TH 1452	1.0	F	1359	1.4
2054	5.2		2015	5.1
4 0310	1.2	**19**	0214	1.5
0912	5.1		0832	5.0
F 1529	1.4	SA	1432	1.6
2125	4.9		2051	4.9
5 0347	1.7	**20**	0249	1.7
0944	4.8		0913	4.9
SA 1612	1.9	SU	1513	1.8
☽ 2200	4.5		2140	4.6
6 0432	2.2	**21**	0336	2.0
1028	4.5		1009	4.7
SU 1708	2.3	M	1612	2.1
2300	4.1		☽ 2245	4.4
7 0538	2.5	**22**	0448	2.3
1209	4.3		1120	4.6
M 1828	2.5	TU	1752	2.3
8 0120	4.1	**23**	0010	4.4
0717	2.6		0641	2.4
TU 1347	4.4	W	1249	4.6
2023	2.3		1935	2.1
9 0231	4.3	**24**	0151	4.6
0854	2.3		0811	2.0
W 1449	4.6	TH	1416	4.9
2124	1.9		2049	1.6
10 0323	4.6	**25**	0304	4.9
0945	1.9		0917	1.5
TH 1537	4.9	F	1521	5.3
2209	1.6		2148	1.1
11 0406	4.9	**26**	0359	5.3
1027	1.5		1013	1.1
F 1618	5.2	SA	1614	5.6
2248	1.3		2240	0.7
12 0444	5.1	**27**	0448	5.6
1105	1.3		1103	0.7
SA 1656	5.3	SU	1703	5.8
2323	1.1		2328	0.5
13 0520	5.3	**28**	0534	5.7
1139	1.1		1149	0.5
SU 1733	5.4	M	1750	5.9
○ 2355	1.0		●	
14 0555	5.3	**29**	0012	0.4
1211	1.1		0617	5.8
M 1809	5.4	TU	1232	0.4
			1833	5.8
15 0024	1.0	**30**	0054	0.4
0629	5.4		0658	5.8
TU 1238	1.1	W	1313	0.5
1843	5.4		1915	5.7
		31	0132	0.6
			0735	5.6
		TH	1351	0.8
			1952	5.5

NOVEMBER

Time	m		Time	m
1 0208	1.0	**16**	0129	1.3
0809	5.4		0744	5.3
F 1427	1.1	SA	1349	1.3
2025	5.1		2004	5.1
2 0242	1.4	**17**	0204	1.4
0838	5.2		0821	5.2
SA 1504	1.5	SU	1427	1.5
2055	4.8		2045	4.9
3 0318	1.8	**18**	0244	1.7
0910	4.9		0905	5.1
SU 1545	2.0	M	1513	1.7
2131	4.5		2135	4.7
4 0401	2.3	**19**	0334	2.0
0952	4.6		0959	4.9
M 1638	2.3	TU	1614	2.0
☽ 2224	4.2		☽ 2237	4.6
5 0501	2.6	**20**	0444	2.2
1057	4.4		1106	4.8
TU 1747	2.5	W	1738	2.1
			2356	4.5
6 0023	4.1	**21**	0617	2.2
0619	2.7		1227	4.8
W 1255	4.4	TH	1909	2.0
1914	2.4			
7 0149	4.3	**22**	0125	4.6
0752	2.5		0743	2.0
TH 1406	4.6	F	1349	5.0
2034	2.1		2022	1.6
8 0244	4.5	**23**	0237	4.9
0858	2.1		0851	1.6
F 1458	4.8	SA	1455	5.2
2124	1.8		2123	1.3
9 0329	4.8	**24**	0334	5.2
0945	1.8		0948	1.2
SA 1542	5.0	SU	1551	5.4
2205	1.5		2216	1.0
10 0410	5.1	**25**	0424	5.4
1025	1.5		1039	0.9
SU 1624	5.2	M	1641	5.6
2243	1.3		2304	0.8
11 0448	5.2	**26**	0510	5.6
1102	1.3		1126	0.8
M 1703	5.3	TU	1728	5.6
2318	1.2		● 2348	0.7
12 0526	5.4	**27**	0553	5.7
1136	1.2		1210	0.7
TU 1742	5.4	W	1813	5.6
○ 2352	1.1			
13 0603	5.4	**28**	0030	0.7
1210	1.1		0633	5.7
W 1819	5.4	TH	1252	0.8
			1853	5.5
14 0024	1.1	**29**	0110	0.9
0637	5.4		0711	5.6
TH 1243	1.1	F	1331	1.0
1854	5.3		1930	5.3
15 0057	1.1	**30**	0146	1.2
0710	5.4		0744	5.4
F 1315	1.2	SA	1408	1.2
1929	5.2		2003	5.1

DECEMBER

Time	m		Time	m
1 0221	1.5	**16**	0204	1.3
0815	5.2		0816	5.3
SU 1445	1.5	M	1431	1.3
2035	4.8		2045	5.1
2 0256	1.8	**17**	0248	1.4
0849	5.0		0902	5.3
M 1523	1.8	TU	1519	1.4
2112	4.6		2134	4.9
3 0334	2.1	**18**	0337	1.6
0930	4.8		0953	5.1
TU 1608	2.1	W	1613	1.6
2159	4.4		2230	4.8
4 0423	2.4	**19**	0435	1.8
1021	4.6		1052	5.0
W 1704	2.3	TH	1717	1.8
☽ 2302	4.3		☽ 2336	4.7
5 0525	2.5	**20**	0545	2.0
1129	4.5		1201	4.9
TH 1809	2.4	F	1831	1.8
6 0028	4.3	**21**	0050	4.7
0637	2.5		0703	2.0
F 1254	4.5	SA	1316	4.9
1917	2.3		1946	1.7
7 0143	4.4	**22**	0202	4.8
0747	2.3		0817	1.8
SA 1402	4.6	SU	1425	5.0
2020	2.0		2053	1.3
8 0240	4.6	**23**	0304	5.0
0847	2.1		0921	1.5
SU 1457	4.8	M	1526	5.1
2113	1.8		2151	1.3
9 0329	4.9	**24**	0359	5.2
0938	1.8		1017	1.3
M 1546	5.0	TU	1621	5.2
2159	1.5		2243	1.1
10 0414	5.1	**25**	0448	5.4
1023	1.5		1107	1.1
TU 1632	5.1	W	1710	5.3
2242	1.3		2329	1.0
11 0456	5.3	**26**	0533	5.5
1105	1.3		1153	1.0
W 1716	5.2	TH	1756	5.3
2323	1.2		●	
12 0537	5.4	**27**	0013	1.0
1146	1.2		0614	5.5
TH 1758	5.3	F	1236	1.0
○			1837	5.3
13 0003	1.1	**28**	0053	1.1
0616	5.5		0653	5.5
F 1227	1.1	SA	1316	1.1
1838	5.3		1914	5.2
14 0043	1.1	**29**	0130	1.2
0655	5.5		0727	5.4
SA 1307	1.1	SU	1353	1.2
1919	5.3		1948	5.1
15 0123	1.2	**30**	0205	1.4
0734	5.5		0800	5.3
SU 1348	1.2	M	1428	1.4
2000	5.2		2021	4.9
		31	0237	1.6
			0834	5.1
		TU	1502	1.6
			2056	4.8

Chart Datum is 3·22 metres below Ordnance Datum (Newlyn). HAT is 5·9 metres above Chart Datum.

〉〉**FREE** monthly updates. Register at 〈
www.reedsnauticalalmanac.co.uk〈

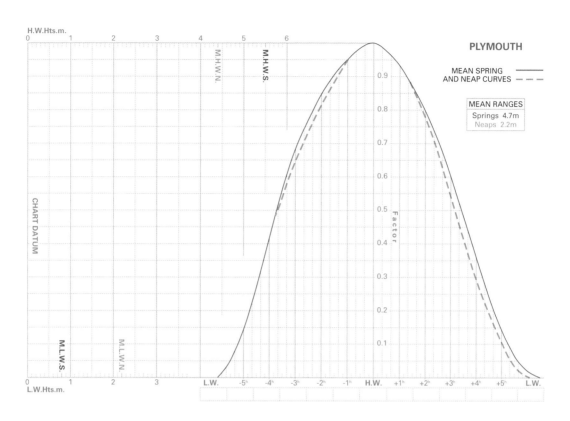

PLYMOUTH

MEAN SPRING AND NEAP CURVES

MEAN RANGES
Springs 4.7m
Neaps 2.2m

1.14 NAVAL EXERCISE AREAS (SUBFACTS AND GUNFACTS)

Submarines and warships frequently exercise south of Plymouth and in the western part of the English Channel. Details of submarine activity and naval gunnery and missile firings are broadcast daily by HM Coastguard. Seek advice when sailing in areas of known submarine activity, and for timings of daily broadcasts. Although the onus for safety is on the relevant naval vessels, if possible it is best to avoid areas where exercises are taking place.

LYNHER (or ST GERMANS) RIVER AC 871. This river flows into The Hamoaze about 0·8M SSW of the Tamar Bridge. On the tide it is navigable for some 4M inland. The channel, entered at Lynher PHM lt buoy 50°23′·78N 04°12′·83W, is marked by 2 more lt buoys in the first mile to Sandacre Pt and carries 2-5m up to Ince Castle. Caution: underwater cables/gaspipe, as charted. Thereafter it carries less than 1m or dries, except at Dandy Hole, a pool with 3-5m, 50°22′·54N 04°16′·29W. Here the navigable chan bends NW then dries completely; it is marked by small R and G posts.

⚓s, amid local moorings, are: off Sand Acre Bay (N bank, beware foul ground); in 2·5-5m at the ent to Forder Lake (N bank opposite Jupiter Pt, where the pontoons and Y moorings are for naval use only); SE of Ince Pt and Castle in about 3m; and at Dandy Hole. St Germans Quay is private, but temp ⌒, M may be pre-arranged with Quay SC ☎(01503) 250370. Facilities: ✉ 🛒 🍺(½M).

RIVER TAMAR AC 871. The river is navigable on the tide for 12M via Calstock and Morwellham to Gunnislake weir. **Jubilee Green** (pontoon in 3m, max LOA 10m) is on W bank close N of the Royal Albert (30m) & Tamar (35m) bridges. Pwr cables (21m) cross 0·4M S of Cargreen (50°26′·55N 04°12′·25W) sited on the W bank 0·7M N of the Tavy with a quay and many local moorings/⚓ in 2·5-5m. Overhead cables 1M N have 16m clearance en-route Weir Quay. **Weir Quay BY** ☎(01822) 840474, ⚓s ⚓ 🛢 🔧 ⚓ 🎣 ▣(20t) 🚤(12t) ⛽. **Weir Quay SC** ☎(01822) 840960, ⚓. Upstream the river S-bends, narrows and partly dries with channel above Halton Quay from 0·1m to >2·5m. N of Cotehele Quay it turns 90° stbd to Calstock, ⚓ in 2m E of viaduct (24m); 🍺 ✉ ≈. **Calstock BY** ☎(01822) 832502. ⚓ 🔧 WC 📻 🚤(8t) ▣(10t).

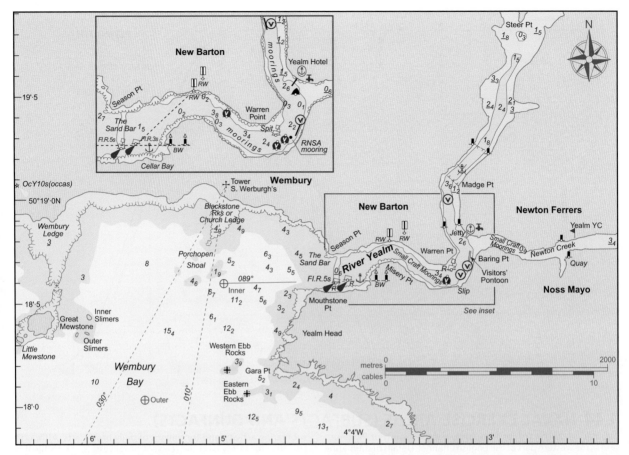

1.15 RIVER YEALM

Devon **50°18'·58N 04°04'·13W** (Ent) ❄❄⚓◊◊✿✿✿

CHARTS AC 1613, 1900, 30, 5602; Imray C10, C6, C14, 2400

TIDES –0522 Dover; ML 3·2; Duration 0615

Standard Port PLYMOUTH (←—)

Times				Height (metres)			
High Water		Low Water		MHWS	MHWN	MLWN	MLWS
0000	0600	0000	0600	5·5	4·4	2·2	0·8
1200	1800	1200	1800				
Differences RIVER YEALM ENTRANCE							
+0006	+0006	+0002	+0002	–0·1	–0·1	–0·1	–0·1

NOTE: Strong SW winds hold up the ebb and raise levels, as does the river if in spate.

SHELTER Very good in river. Entry is easy except in strong SW-W winds; ⚓ in Cellar Bay is open from SW to NW. Space in the hbr is very limited, and in wind-over-tide conditions, when moored yachts lie across the stream, larger vessels may find turning difficult. Vessels >18m LOA may not enter without the HM's permission.

NAVIGATION Outer WPT 50°18'·03N 04°05'·55W, 033°/7ca to the Inner WPT 50°18'·59N 04°04'·98W; then 089°/5½ca to the Sand Bar. Keep St Werburgh's ✠ twr between 010° and 030° to clear the Slimers and the W & E Ebb Rocks, then turn onto 089°; ldg bns (both W △ with B stripe). ≠ clears Mouthstone Ledge, but not the sand bar which extends SW. Two PHM buoys mark the S end of the Sand Bar and must be left to port on entry; the seaward buoy is lit, Fl R 5s. After passing the second (E'ly) buoy turn onto 047° ldg bns (both W with R stripe), on the N shore, then follow line of the river. There is only 1m at MLWS between the Sand Bar and Misery Point. Leave Spit PHM buoy to port and Office SHM to starboard. No ⚓ in river; speed limit 6kn.

LIGHTS AND MARKS Great Mewstone (57m) is conspic 1·5M to W of river ent; no passage between it and the coast.

COMMUNICATIONS (Code 01752) HM 872533; Water taxi ☎07817 132757.

FACILITIES 1⚓ off Misery Pt, 3⚓ off Warren Pt and Ⓥ pontoons in The Pool and 3ca upriver. Pontoon, ⚓/⛽ £1.70. 🚽/♨ at HM.
⚓ ⚓. Yealm YC ☎872291, ⚓ Bistro, ⌂.
Newton Ferrers ⚓ ⚓ ⚓ ✉⚑ ✕⌂.
Noss Mayo ⌂ at Pope's Quay (dries), ⚓ ⚓ ⚓ ✕⌂.
Bridgend (Newton Creek) ⌂ (dries), ⚓ ⊕.
📷 📷 🛢 at Yealmpton (3M).

ADJACENT ANCHORAGES IN BIGBURY BAY

RIVER ERME, Devon, **50°18'·15N 03°57'·67W**. AC1613. HW –0525 on Dover; +0015 and –0·6m on HW Plymouth. Temp day ⚓ in 3m at mouth of drying river, open to SW. Enter only in offshore winds and settled weather. Beware Wells Rk (1m) 1M SE of ent. Appr from SW, clear of Edwards Rk. Ent between Battisborough Is and W. Mary's Rk (dries 1·1m) keeping to the W. No facilities. Two Historic Wrecks are at 50°18'·15N 03°57'·41W and 50°18'·41N 03°57'·19W on W side of the ent.

RIVER AVON, Devon, **50°16'·63N 03°53'·33W**. AC1613. Tides as R Erme, above. Appr close E of conspic Burgh Is (⚓) & Murray's Rk, marked by bcn. Enter drying river HW –1, only in offshore winds and settled weather. Narrow chan hugs cliffy NW shore, then S-bends SE and N off Bantham. Local knowledge or a dinghy recce near LW would assist. Streams run hard, but able to dry out in good shelter clear of moorings. HM ☎(01548) 561196. ✉🛒 ⌂ at Bantham. Aveton Gifford accessible by dinghy, 2·5M.

HOPE COVE, Devon, **50°14'·65N 03°51'·78W**. AC 1613. Tides as R Erme; ML 2·6m; Duration 0615. Popular day ⚓ in centre of cove, but poor holding and only in offshore winds. Appr with old LB ho in SE corner of the cove brg 130° and ⚓ WSW of the pier head, clear of moorings. Basse Rock extends 30m around the pier head. Off the S shore Goody Rock (dries 2·5m) is cleared by brgs of 060° on the pier head and 120° on the old LB ho. No lights. Facilities: very limited in village, good at Salcombe (4M bus).

1.16 SALCOMBE

Devon **50°13'·17N 03°46'·67W (The Bar)** ❀❀⚓⚓⚓❀❀❀

CHARTS AC 1613, 1634, 28, 5602; Imray C10, C6, C5, 2300

TIDES –0523 Dover; ML 3·1; Duration 0615

Standard Port PLYMOUTH (←—)

Times				Height (metres)			
High Water		Low Water		MHWS	MHWN	MLWN	MLWS
0100	0600	0100	0600	5·5	4·4	2·2	0·8
1300	1800	1300	1800				
Differences SALCOMBE							
0000	+0010	+0005	–0005	–0·2	–0·3	–0·1	–0·1
START POINT							
+0015	+0015	+0005	+0010	–0·1	–0·2	+0·1	+0·2

SHELTER Good, but fresh S'lies can cause uncomfortable swell off the town. Better shelter in The Bag at ❶ pontoon. ⚓s at Sunny Cove; off SE shore from YC start line to fuel barge (where poor holding has been reported); 200m W and 300m SSE of Salt Stone SHM bcn 50°15'·19N 03°45'·55W.

NAVIGATION WPT 50°12'·43N 03°46'·67W, 000°/1·3M to Sandhill Pt lt. In normal conditions The Bar (least depth 1m) is not a problem, *but it should not be crossed on an ebb tide with strong onshore winds or swell when dangerous breakers can occur; if in doubt, call*

HM Ch 14 before approaching. The deepest water is close inshore to the west off the leading line. The spring flood reaches 2·5kn. Speed limit 8kn, marked in season by 3 Y SPM between Cadmus and Chapple Rks, but 6kn between Marine Hotel/racing start line and the Saltstone (approx 50°15'·2N 03°45'·5W); minimise wash; radar checks in force. In Jul/Aug within the hbr cruising yachts must motor, not sail. When dinghy racing in progress (Sat/Sun & regatta weeks: Fl Y lt at YC and warning on Ch 14) yachts should use a fairway marked by Y SPM buoys close to the NW bank and straddling the start line. Two Historic Wrecks lie 1M WNW of Prawle Pt at 50°12'·74N 03°44'·40W & 50°12'·73N 03°44'·75W.

LIGHTS AND MARKS Outer 000° ldg marks, both R/W bcns (a conspic gabled house close N is more easily seen): front, Poundstone; rear, Sandhill Pt Dir lt 000°, stay in the W sector (357·5°-002·5°). Close W of ldg line beware Bass Rk (0·8m) marked by PHM lt buoy; and, close E, Wolf Rk (0·6m), marked by SHM lt buoy. When past Wolf Rk, leave Black Stone Rk (5m), G/W bcn, well to stbd and pick up inner ldg lts 042·5°.

COMMUNICATIONS (Code 01548) Dr 842284. HM ☎843791: May to mid-Sep: 0900-1645 Mon-Thu; 0900-1615 Fri-Sun. *Salcombe Hbr Launch* (May to mid-Sep: daily 0715 -2300; rest of year: Mon-Fri 0900-1600) Ch 14. *Harbour taxi* Ch 12. *Fuel Barge* Ch 06. ICC HQ.

FACILITIES www.salcombeharbour.co.uk www.southhams.gov.uk

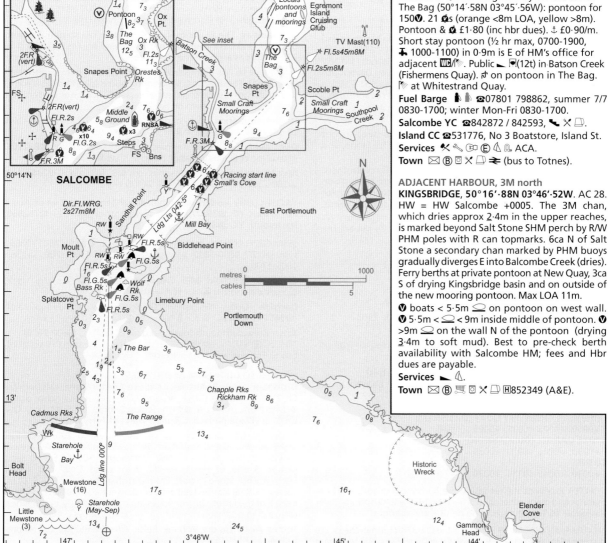

Numerous ⚓s flanking fairway off the town. The Bag (50°14'·58N 03°45'·56W): pontoon for 150❶. 21 ⚓s (orange <8m LOA, yellow >8m). Pontoon & ⚓ £1·80 (inc hbr dues). ⚓ £0·90/m. Short stay pontoon (½ hr max, 0700-1900, ⚓ 1000-1100) in 0·9m is E of HM's office for adjacent ⬚/ꞁ. Public ⬚ꞁ(12t) in Batson Creek (Fishermens Quay). ⚓ on pontoon in The Bag. ꞁ at Whitestrand Quay.

Fuel Barge 🅿🅿 ☎07801 798862, summer 7/7 0830-1700; winter Mon-Fri 0830-1700.
Salcombe YC ☎842872 / 842593, ⚓ ✕ ▭.
Island CC ☎531776, No 3 Boatstore, Island St.
Services ⚒ ⚓ ⬚ Ⓔ ⚓ ⚓ ACA.
Town ✉ Ⓑ ⬚ ✕ ▭ ⇌ (bus to Totnes).

ADJACENT HARBOUR, 3M north

KINGSBRIDGE, 50°16'·88N 03°46'·52W. AC 28. HW = HW Salcombe +0005. The 3M chan, which dries approx 2·4m in the upper reaches, is marked beyond Salt Stone SHM perch by R/W PHM poles with R can topmarks. 6ca N of Salt Stone a secondary chan marked by PHM buoys gradually diverges E into Balcombe Creek (dries). Ferry berths at private pontoon at New Quay, 3ca S of drying Kingsbridge basin and on outside of the new mooring pontoon. Max LOA 11m.
❶ boats < 5·5m ⚓ on pontoon on west wall. ❶ 5·5m < ⚓ < 9m inside middle of pontoon. ❶ >9m ⚓ on the wall N of the pontoon (drying 3·4m to soft mud). Best to pre-check berth availability with Salcombe HM; fees and Hbr dues are payable.
Services ⬚ ⚓.
Town ✉ Ⓑ ⬚ 🛒 ✕ ▭ Ⓗ852349 (A&E).

START POINT TO TEIGNMOUTH

(AC 1613, 1634, 3315) Start Point (Fl(3) 10s 62m25M, *Horn 60s*), 3·3M ENE of Prawle Point, is a long headland with a distinctive cock's comb spine, a W lt ho on the Point, and conspicuous radio masts to the WNW. Black Stone rock (6m high) is visible 2½ca SSE of the lt ho, with Cherrick Rocks (1m) 1ca further S; other drying rocks lie closer inshore. A race may extend 1M to the S and 1·6M to the E. In fair weather any overfalls can be avoided by passing close to seaward of rks; there is no clear-cut inshore passage as such. In bad weather keep at least 2M off.

▶ *The NE-going flood begins at HW Plymouth –2; max (3·1kn Sp) at HWP –1. The SW-going ebb begins at HWP +5; max (2·2kn Sp) at HWP +6. But at HWP +4 it is possible to round Start Point close inshore using a back eddy. On both the flood and the ebb back eddies form between Start Point and Hallsands, 1M NW. Inshore the tide turns 30 mins earlier.◀*

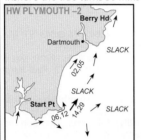
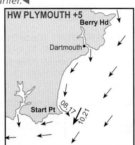

Skerries Bank (least depth 2·1m), on which the sea breaks in bad weather, lies from 9ca to 4·1M NE of Start Pt (AC 1634). In off-shore winds there is a good ⚓ in 3m 1ca off Hallsands (1M NW of Start). A fish farm lies about 3½ ca off. Between **Dartmouth** and **Brixham** rocks extend 5ca offshore. Berry Head (Fl 2 15s), a bold, flat-topped headland (55m), gives a good radar return.

▶ *Here the stream turns N at HW Plymouth –3, and S at HWP +3; max Sp rates 3kn.◀*

In **Torbay** (AC 26) the more obvious dangers are steep-to, but beware the Sunker (awash at CD) 100m SW of Ore Stone, and Morris Rogue (0·8m) 5ca W of Thatcher Rock.

In W'lies Hope Cove, Anstey's Cove (beware the Three Brothers, rks drying 0·6m, S side) and Babbacombe Bay are good ⚓s on sand. There are no offlying dangers from Long Quarry Point for 4M N to **Teignmouth** whose ent is dangerous in onshore winds.

▶ *Off Teignmouth Bar the NNE-going stream begins at HW Plymouth –1½ and the SSW-going at HWP +5¼. In the ent the flood begins at HWP –5½ and the ebb at HWP +¾. The Spr ebb reaches 4–5kn off The Point.◀*

1.17 DARTMOUTH

Devon 50°20'·66N 03°33'·96W ✳✳✳✳⚓⚓⚓⚓✿✿✿

CHARTS AC 1613, 1634, 2253, 5602; Imray C10, C5, 2300

TIDES –0510 Dover; ML 2·8; Duration 0630
DARTMOUTH (→). The differences below are referred to Plymouth as Standard Port, not to Dartmouth.

Times				Height (metres)			
High Water		Low Water		MHWS	MHWN	MLWN	MLWS
0100	0600	0100	0600	5·5	4·4	2·2	0·8
1300	1800	1300	1800				
Differences GREENWAY QUAY (DITTISHAM)							
+0030	+0045	+0025	+0005	–0·6	–0·6	–0·2	–0·2
STOKE GABRIEL (DUNCANNON)							
+0035	+0040	+0020	+0030	–0·9	–0·8	–0·4	–0·3
TOTNES							
+0030	+0040	+0115	+0030	–2·0	–2·1	DR	DR

SHELTER Excellent shelter inside the harbour and upriver. Three marinas and many other �\(for yachts< 14m) and ⚓ options. River navigable to Totnes depending on tide and draught.

NAVIGATION WPT 50°19'·53N 03°32'·83W, 328°/1·5M in the white sector of Kingswear Dir lt. Bayard's Cove Dir lt leads 293° to abeam

Royal Dart YC where the main fairway opens. There is no bar and hbr access is H24, but ent can be difficult in strong SE to SW winds. Speed limit 6kn from Castle Ledge buoy upriver to 1M below Totnes. *Caution: The Lower and Higher car ferries S and N of Dartmouth have right of way; give way early.*

LIGHTS AND MARKS as on the chartlet and/or 1.2. E of the ent, on Inner Froward Pt (167m) is a conspic daymark, obelisk (24·5m). Lateral and cardinal lt buoys mark all dangers to seaward of conspic Dartmouth Castle. Within hbr, all jetty/ pontoon lts to the W are 2FR (vert); and 2FG (vert) to the E.

COMMUNICATIONS (Code 01803) Dr 832212; Ⓗ 832255 HM 832337 835220 (non-HO emergency); *Dartnav* VHF Ch 11 (summer 0730-2100). Darthaven, Dart and Noss marinas Ch 80. Fuel barge Ch 06. Water taxis: *Yacht taxi* (DHNA) Ch 69, 07970 346571 (summer 0800-2300, winter 1000-1600). *Greenway ferry* Ch 10 or 844010 (to Dittisham).

FACILITIES Dart Harbour and Navigation Authority (DHNA), www.dartharbour.org ☎832337, 450⌐+90Ⓥ.
Town jetty (inside) £2·10/m/day (outside 1700-0840)£1·35/m/nt.
S Embankment £1·60; other pontoons, ⚓s and ⚓ £1·55/m; all inc Hbr dues (£0·85) & VAT. ◣ ⚓ WC/⎰ to N of Boat Float.

DHNA Ⓥ pontoons are marked by blue flags and ⚓s are blue with black 'V' or 'Visitors' in yellow and black. Check availability with HM or call DHNA. The most likely Ⓥ berths from S to N are:

W bank: pontoon (max 9m LOA) off Dartmouth YC (May-Sep). Town pontoon W side only, but E side 1700-0845. N'ly of 2 pontoons just S of Dart marina (26'/8m max LOA).

E of fairway: The Ⓥ pontoon by Kingswear rail station is shared by DHNA and Darthaven marina; the former are for 2hrs max. The 3 Ⓥ pontoons are N of the fuel barge.

Only space to ⚓ is E of fairway and Nos 3A to 5 buoys, with HM's agreement. These large unlit mooring buoys (plus Nos 3, 5A and 6) in mid-stream are for commercial vessels/FVs; do not ⚓ over their ground chains, as shown on AC 2253.

Darthaven Marina www.darthaven.co.uk ☎752242, 250⌐+20Ⓥ £2·50+VAT+hbr dues. ⎰ ⛽ ⚓ ✗ ⚒ ⚒ 🖵 Ⓔ ⎇(40t) ⛽ wi-fi.

Dart Marina ⌓ www.dartmarinayachtharbour.com ☎837161. 110⌐ inc Ⓥ £2·50+VAT+hbr dues. ⎰ ✗ wi-fi.

Noss Marina www.premiermarinas.com ☎839087 mob 07920425452. ▲180⌐ £2·98 + dues, ⎰ ⚓ ⚒ 🖵 ⎇(10t) wi-fi.

YACHT CLUBS (visitors welcome): **Royal Regatta**, last week Aug. **Royal Dart YC** ☎752272, short stay pontoon, ⚓ ✗ 🖵 ☎752880. **Dartmouth YC** ☎832305, ◣ ⚓ ⎰ ✗ 🖵.

SERVICES Fuel Barge next to No 6 buoy, Ch 06, ⚓ ⚓; ☎07801 798861 summer 7/7 0800-1800; winter Mon/Wed/Fri 1200-1700, Sat/Sun 1000-1700.

Creekside BY (Old Mill Creek) ☎832649, ▲ ◣ dry dock, ⌐(customer only) ⚓ ⛽ ✗ ⚒ 🖵 ⎇(14t) ⛽.

Dartside Quay (Galmpton Creek): www.dartsidequay.co.uk ☎845445 ▲◣◣◣ ⛽ ⎰ ⚓ ⚓ ✗ ⚒ 🖵 Ⓔ 🍴 ⎇(65t/16t), ⛽ wi-fi.

Town www.dartmouth-tourism.org.uk, 🖊 WC/⎰ ⎰ (0800-2000) 🏨⚓✉Ⓑ⚓✗🖵⇄ ☎555872 (steam train in season to Paignton); bus to Totnes/Paignton, ✈ Exeter.

UPRIVER TO DITTISHAM AND TOTNES

Dittisham ⚓s/⚓s/Berths: For ⚓s above Anchor Stone, call *DartNav* Ch 11. Many WⒶs (black 'V' and LOA) between Anchor Stone & Dittisham; no ⚓. 4 ⚓s off Stoke Gabriel by SHM bcn, QG. WC/⎰ in village park to N (5 mins walk) of Ferry Boat Inn.

R Dart is navigable by day on the flood to Totnes bridge, 5·5M above Dittisham. Use AC 2253 and DHNA annual handbook. HW Totnes = HW Dartmouth +0015. Speed limit 6kn to S end of Home Reach, then 'Dead Slow' (no wash). Leave Anchor Stone to port. From Dittisham brgs of 020° / 310° on successive Boat Houses lead between Lower Back and Flat Owers; or keep E of the latter. Upriver buoys are numbered 1-11 in sequence, rather than odd/even convention for lateral buoys.

Totnes All berths dry. W bank: Baltic Wharf ☎867922, ⌐ (Ⓥ pre-call) ⚓ ⎰ 🍴 🖵(16t) ⎇(35t) ⚒ ⛽.

E bank: (**Ⓥ** call DHNA) for 5⌣ at Steamer Quay, keep clear of ferry berths. Limited ⌣ on soft mud in the W Arm N of Steam Packet Inn ☎863880, ⚓ ▷ ◎ ✕ ⌴. **Totnes** All amenities; mainline ⇌.

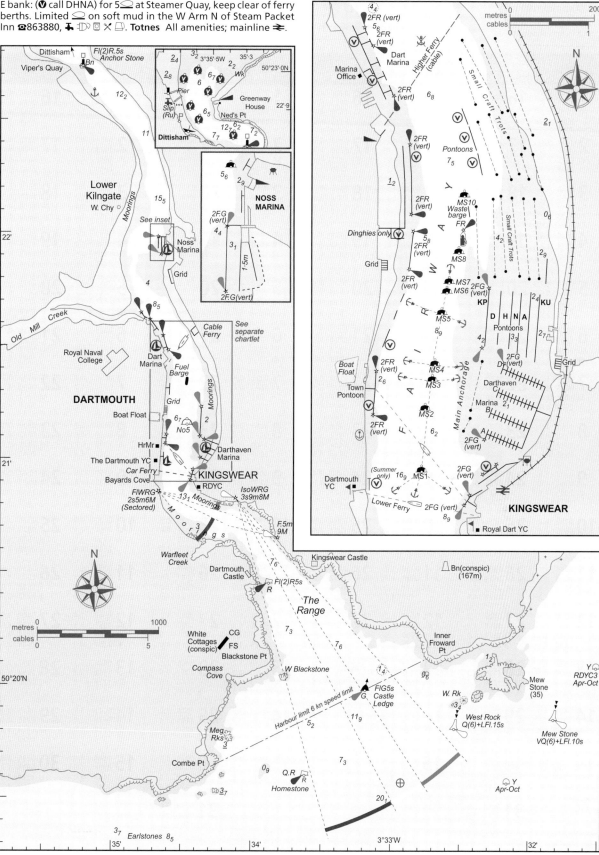

Dartmouth tides

STANDARD TIME (UT)
For Summer Time add ONE hour in **non-shaded areas**

DARTMOUTH LAT 50°21′N LONG 3°35′W
TIMES AND HEIGHTS OF HIGH AND LOW WATERS

Dates in red are SPRINGS
Dates in blue are NEAPS

YEAR **2019**

JANUARY

Day	Time	m	Day	Time	m
1 TU	0155 / 0747 / 1420 / 2025	4.0 / 1.8 / 4.2 / 1.6	16 W	0029 / 0630 / 1257 / 1916	3.8 / 2.0 / 4.0 / 1.8
2 W	0259 / 0856 / 1523 / 2127	4.2 / 1.6 / 4.2 / 1.4	17 TH	0144 / 0753 / 1418 / 2029	3.9 / 1.8 / 4.1 / 1.6
3 TH	0356 / 0953 / 1618 / 2220	4.3 / 1.4 / 4.3 / 1.3	18 F	0301 / 0904 / 1533 / 2134	4.2 / 1.5 / 4.3 / 1.3
4 F	0445 / 1044 / 1707 / 2305	4.5 / 1.2 / 4.4 / 1.1	19 SA	0407 / 1006 / 1638 / 2234	4.5 / 1.2 / 4.5 / 1.0
5 SA	0530 / 1128 / 1750 / 2347	4.7 / 1.0 / 4.5 / 1.0	20 SU	0505 / 1104 / 1736 / 2328	4.7 / 0.8 / 4.7 / 0.7
6 SU ●	0609 / 1209 / 1829	4.8 / 1.0 / 4.6	21 M ○	0559 / 1159 / 1830	5.0 / 0.5 / 4.8
7 M	0024 / 0645 / 1246 / 1904	1.0 / 4.8 / 1.0 / 4.6	22 TU	0020 / 0650 / 1249 / 1920	0.5 / 5.1 / 0.3 / 4.9
8 TU	0057 / 0718 / 1319 / 1938	1.0 / 4.8 / 1.0 / 4.5	23 W	0108 / 0739 / 1336 / 2008	0.4 / 5.2 / 0.2 / 4.9
9 W	0127 / 0752 / 1349 / 2011	1.1 / 4.7 / 1.1 / 4.4	24 TH	0154 / 0826 / 1421 / 2054	0.3 / 5.2 / 0.3 / 4.9
10 TH	0155 / 0825 / 1416 / 2044	1.2 / 4.6 / 1.3 / 4.3	25 F	0237 / 0910 / 1503 / 2138	0.5 / 5.1 / 0.4 / 4.6
11 F	0221 / 0858 / 1445 / 2118	1.4 / 4.5 / 1.4 / 4.2	26 SA	0318 / 0954 / 1546 / 2221	0.7 / 4.9 / 0.7 / 4.5
12 SA	0250 / 0931 / 1515 / 2153	1.5 / 4.3 / 1.5 / 4.0	27 SU ◗	0400 / 1038 / 1628 / 2307	1.0 / 4.6 / 1.1 / 4.2
13 SU	0323 / 1007 / 1553 / 2234	1.7 / 4.2 / 1.7 / 3.9	28 M	0445 / 1126 / 1716 / 2359	1.4 / 4.3 / 1.5 / 4.0
14 M ◗	0405 / 1052 / 1643 / 2326	1.8 / 4.1 / 1.8 / 3.8	29 TU	0539 / 1226 / 1815	1.7 / 4.0 / 1.8
15 TU	0505 / 1149 / 1753	2.0 / 4.0 / 1.9	30 W	0106 / 0649 / 1339 / 1935	3.8 / 1.9 / 3.8 / 1.8
			31 TH	0219 / 0816 / 1452 / 2056	3.8 / 1.9 / 3.8 / 1.8

FEBRUARY

Day	Time	m	Day	Time	m
1 F	0325 / 0930 / 1555 / 2158	4.0 / 1.7 / 4.0 / 1.6	16 SA	0226 / 0838 / 1509 / 2113	4.0 / 1.6 / 4.1 / 1.5
2 SA	0421 / 1027 / 1647 / 2248	4.2 / 1.4 / 4.2 / 1.3	17 SU	0343 / 0950 / 1622 / 2220	4.3 / 1.2 / 4.3 / 1.1
3 SU	0508 / 1113 / 1732 / 2331	4.5 / 1.0 / 4.3 / 1.1	18 M	0446 / 1052 / 1723 / 2317	4.7 / 0.8 / 4.6 / 0.6
4 M ●	0549 / 1155 / 1812	4.6 / 1.0 / 4.4	19 TU ○	0543 / 1147 / 1817	5.0 / 0.4 / 4.8
5 TU	0009 / 0628 / 1231 / 1849	1.0 / 4.7 / 0.9 / 4.5	20 W	0009 / 0636 / 1236 / 1906	0.3 / 5.2 / 0.1 / 5.0
6 W	0043 / 0704 / 1303 / 1923	0.9 / 4.8 / 0.9 / 4.5	21 TH	0056 / 0724 / 1322 / 1953	0.1 / 5.3 / -0.1 / 5.0
7 TH	0112 / 0738 / 1332 / 1957	0.9 / 4.7 / 0.9 / 4.5	22 F	0140 / 0810 / 1404 / 2035	0.0 / 5.3 / 0.0 / 5.0
8 F	0138 / 0811 / 1358 / 2028	1.0 / 4.7 / 1.0 / 4.4	23 SA	0219 / 0852 / 1444 / 2114	0.1 / 5.2 / 0.2 / 4.8
9 SA	0203 / 0841 / 1422 / 2058	1.1 / 4.6 / 1.1 / 4.3	24 SU	0258 / 0931 / 1520 / 2150	0.4 / 4.9 / 0.5 / 4.6
10 SU	0229 / 0909 / 1449 / 2125	1.2 / 4.4 / 1.2 / 4.2	25 M	0334 / 1007 / 1557 / 2224	0.8 / 4.6 / 1.0 / 4.3
11 M	0258 / 0938 / 1520 / 2157	1.3 / 4.3 / 1.3 / 4.1	26 TU ◗	0412 / 1042 / 1636 / 2300	1.2 / 4.2 / 1.4 / 4.0
12 TU ◗	0333 / 1015 / 1600 / 2241	1.5 / 4.1 / 1.5 / 3.9	27 W	0457 / 1127 / 1725 / 2357	1.6 / 3.8 / 1.9 / 3.7
13 W	0421 / 1108 / 1656 / 2342	1.7 / 4.0 / 1.8 / 3.8	28 TH	0558 / 1252 / 1833	2.0 / 3.6 / 2.1
14 TH	0529 / 1217 / 1820	1.9 / 3.8 / 1.9			
15 F	0058 / 0710 / 1341 / 1955	3.8 / 1.9 / 3.9 / 1.8			

MARCH

Day	Time	m	Day	Time	m
1 F	0135 / 0724 / 1423 / 2015	3.6 / 2.1 / 3.6 / 2.1	16 SA	0029 / 0641 / 1321 / 1931	3.8 / 1.9 / 3.7 / 1.9
2 SA	0254 / 0908 / 1531 / 2136	3.8 / 1.9 / 3.7 / 1.8	17 SU	0202 / 0821 / 1456 / 2058	3.9 / 1.6 / 4.0 / 1.5
3 SU	0354 / 1007 / 1625 / 2228	4.0 / 1.5 / 4.0 / 1.4	18 M	0326 / 0937 / 1609 / 2205	4.3 / 1.1 / 4.3 / 1.0
4 M	0443 / 1054 / 1710 / 2311	4.3 / 1.2 / 4.2 / 1.1	19 TU	0430 / 1038 / 1708 / 2302	4.7 / 0.6 / 4.6 / 0.6
5 TU	0526 / 1134 / 1750 / 2349	4.5 / 0.9 / 4.4 / 0.9	20 W	0526 / 1130 / 1800 / 2352	5.0 / 0.2 / 4.9 / 0.2
6 W ●	0606 / 1210 / 1828	4.7 / 0.8 / 4.5	21 TH ○	0617 / 1219 / 1848	5.2 / -0.1 / 5.0
7 TH	0022 / 0643 / 1241 / 1904	0.8 / 4.7 / 0.7 / 4.6	22 F	0037 / 0705 / 1302 / 1931	0.0 / 5.3 / -0.2 / 5.1
8 F	0051 / 0718 / 1309 / 1937	0.7 / 4.7 / 0.7 / 4.6	23 SA	0119 / 0747 / 1342 / 2011	-0.1 / 5.2 / -0.1 / 5.0
9 SA	0118 / 0750 / 1335 / 2007	0.8 / 4.7 / 0.8 / 4.5	24 SU	0158 / 0827 / 1419 / 2046	0.0 / 5.1 / 0.2 / 4.9
10 SU	0143 / 0821 / 1359 / 2034	0.8 / 4.6 / 0.8 / 4.5	25 M	0233 / 0903 / 1453 / 2115	0.3 / 4.8 / 0.5 / 4.6
11 M	0208 / 0847 / 1426 / 2100	0.9 / 4.5 / 1.0 / 4.4	26 TU	0307 / 0932 / 1526 / 2141	0.7 / 4.5 / 1.0 / 4.3
12 TU	0236 / 0915 / 1456 / 2129	1.0 / 4.3 / 1.1 / 4.2	27 W	0342 / 1000 / 1601 / 2211	1.2 / 4.1 / 1.5 / 4.0
13 W	0310 / 0950 / 1533 / 2210	1.2 / 4.2 / 1.4 / 4.1	28 TH ◗	0424 / 1037 / 1646 / 2256	1.6 / 3.7 / 1.9 / 3.7
14 TH ◗	0355 / 1042 / 1624 / 2311	1.5 / 3.9 / 1.7 / 3.9	29 F	0520 / 1141 / 1750	2.0 / 3.4 / 2.2
15 F	0458 / 1152 / 1743	1.8 / 3.7 / 1.9	30 SA	0021 / 0636 / 1350 / 1915	3.5 / 2.2 / 3.4 / 2.2
			31 SU	0218 / 0826 / 1501 / 2100	3.6 / 2.0 / 3.6 / 2.0

APRIL

Day	Time	m	Day	Time	m
1 M	0322 / 0936 / 1555 / 2156	3.9 / 1.6 / 3.9 / 1.6	16 TU	0307 / 0919 / 1552 / 2146	4.3 / 1.1 / 4.3 / 1.0
2 TU	0411 / 1023 / 1640 / 2240	4.2 / 1.3 / 4.2 / 1.2	17 W	0410 / 1018 / 1647 / 2241	4.6 / 0.6 / 4.6 / 0.6
3 W	0456 / 1102 / 1720 / 2317	4.4 / 1.0 / 4.4 / 0.9	18 TH	0505 / 1109 / 1738 / 2329	4.9 / 0.2 / 4.9 / 0.3
4 TH	0537 / 1138 / 1800 / 2353	4.6 / 0.8 / 4.5 / 0.8	19 F ○	0555 / 1156 / 1825	5.1 / 0.0 / 5.0
5 F ●	0616 / 1211 / 1837	4.7 / 0.7 / 4.6	20 SA	0014 / 0641 / 1238 / 1905	0.1 / 5.1 / 0.0 / 5.0
6 SA	0023 / 0654 / 1241 / 1911	0.7 / 4.7 / 0.6 / 4.6	21 SU	0055 / 0723 / 1317 / 1943	0.0 / 5.1 / 0.1 / 5.0
7 SU	0053 / 0727 / 1310 / 1943	0.7 / 4.7 / 0.7 / 4.6	22 M	0133 / 0800 / 1353 / 2015	0.2 / 4.9 / 0.3 / 4.8
8 M	0121 / 0759 / 1338 / 2011	0.7 / 4.6 / 0.7 / 4.6	23 TU	0208 / 0832 / 1426 / 2041	0.5 / 4.7 / 0.7 / 4.6
9 TU	0150 / 0828 / 1407 / 2039	0.8 / 4.5 / 0.9 / 4.5	24 W	0242 / 0901 / 1458 / 2108	0.8 / 4.4 / 1.1 / 4.4
10 W	0220 / 0901 / 1439 / 2112	0.9 / 4.4 / 1.1 / 4.4	25 TH	0316 / 0929 / 1532 / 2139	1.2 / 4.1 / 1.5 / 4.1
11 TH	0256 / 0940 / 1517 / 2155	1.1 / 4.2 / 1.4 / 4.2	26 F ◗	0356 / 1007 / 1614 / 2222	1.6 / 3.7 / 1.9 / 3.9
12 F ◗	0342 / 1031 / 1609 / 2254	1.4 / 3.9 / 1.7 / 4.0	27 SA	0449 / 1103 / 1714 / 2324	2.0 / 3.5 / 2.2 / 3.6
13 SA	0448 / 1141 / 1729	1.7 / 3.7 / 1.9	28 SU	0557 / 1252 / 1828	2.1 / 3.4 / 2.2
14 SU	0011 / 0628 / 1312 / 1913	3.9 / 1.8 / 3.7 / 1.9	29 M	0119 / 0716 / 1418 / 1950	3.6 / 2.0 / 3.6 / 2.1
15 M	0144 / 0806 / 1443 / 2040	4.0 / 1.5 / 4.0 / 1.5	30 TU	0236 / 0835 / 1514 / 2101	3.8 / 1.7 / 3.8 / 1.7

Chart Datum is 2·62 metres below Ordnance Datum (Newlyn). HAT is 5·3 metres above Chart Datum.

STANDARD TIME (UT)
For Summer Time add ONE hour in **non-shaded areas**

DARTMOUTH LAT 50°21'N LONG 3°35W
TIMES AND HEIGHTS OF HIGH AND LOW WATERS

Dates in red are **SPRINGS**
Dates in blue are NEAPS

YEAR 2019

SW England

MAY

Day	Time	m	Time	m	Time	m	Time	m
1 W	0331	4.1	0932	1.4	1602	4.1	2152	1.4
2 TH	0417	4.3	1017	1.1	1645	4.3	2236	1.1
3 F	0502	4.5	1057	0.9	1727	4.5	2315	0.9
4 SA ●	0544	4.6	1135	0.7	1807	4.6	2352	0.8
5 SU	0625	4.7	1211	0.7	1844	4.7		
6 M	0027	0.7	0704	4.7	1245	0.6	1918	4.7
7 TU	0102	0.7	0739	4.6	1319	0.7	1951	4.7
8 W	0136	0.7	0814	4.6	1354	0.8	2026	4.6
9 TH	0212	0.9	0852	4.4	1431	1.0	2104	4.5
10 F	0253	1.1	0936	4.2	1514	1.3	2149	4.4
11 SA	0343	1.3	1029	4.0	1609	1.6	2246	4.2
12 SU ◗	0450	1.6	1137	3.8	1724	1.8	2359	4.0
13 M	0616	1.6	1301	3.8	1853	1.7		
14 TU	0126	4.1	0743	1.6	1422	4.0	2015	1.5
15 W	0243	4.3	0854	1.1	1527	4.3	2120	1.1
16 TH	0345	4.6	0951	0.7	1622	4.6	2216	0.8
17 F	0440	4.8	1043	0.5	1711	4.8	2305	0.5
18 SA	0530	4.9	1129	0.3	1757	4.9	○2350	0.4
19 SU	0616	4.9	1213	0.3	1838	4.9		
20 M	0032	0.3	0659	4.8	1252	0.4	1915	4.9
21 TU	0110	0.5	0735	4.7	1329	0.6	1946	4.8
22 W	0146	0.7	0807	4.5	1402	0.9	2015	4.6
23 TH	0220	0.9	0836	4.3	1434	1.2	2044	4.4
24 F	0254	1.3	0908	4.1	1507	1.5	2118	4.2
25 SA	0332	1.6	0948	3.8	1547	1.8	2159	4.0
26 SU	0418	1.8	1037	3.7	1638	2.0	◗2251	3.8
27 M	0516	2.0	1145	3.5	1742	2.1	2359	3.7
28 TU	0623	2.0	1308	3.6	1851	2.1		
29 W	0126	3.8	0730	1.8	1418	3.8	1958	1.9
30 TH	0236	4.0	0830	1.6	1513	4.0	2057	1.6
31 F	0332	4.2	0924	1.3	1603	4.2	2148	1.3

JUNE

Day	Time	m	Time	m	Time	m	Time	m
1 SA	0423	4.3	1014	1.1	1649	4.4	2237	1.1
2 SU	0510	4.5	1059	0.9	1734	4.6	2321	0.8
3 M ●	0557	4.6	1143	0.7	1816	4.7		
4 TU	0004	0.7	0641	4.6	1224	0.7	1857	4.8
5 W	0046	0.6	0722	4.7	1306	0.7	1936	4.8
6 TH	0128	0.6	0805	4.6	1348	0.8	2017	4.8
7 F	0210	0.7	0848	4.5	1430	0.9	2100	4.7
8 SA	0256	0.9	0936	4.4	1517	1.1	2148	4.6
9 SU	0347	1.1	1029	4.2	1610	1.3	2241	4.4
10 M ◗	0446	1.3	1130	4.0	1712	1.5	2347	4.2
11 TU	0555	1.4	1240	4.0	1825	1.6		
12 W	0100	4.2	0710	1.4	1312	4.1	1941	1.5
13 TH	0212	4.2	0821	1.2	1455	4.2	2049	1.3
14 F	0316	4.4	0923	1.0	1552	4.4	2147	1.1
15 SA	0414	4.5	1017	0.9	1643	4.6	2240	0.9
16 SU	0506	4.6	1105	0.7	1731	4.7	2327	0.7
17 M ○	0554	4.6	1151	0.7	1813	4.7		
18 TU	0011	0.7	0636	4.6	1231	0.7	1851	4.8
19 W	0051	0.7	0713	4.5	1308	0.8	1924	4.7
20 TH	0128	0.8	0747	4.4	1343	1.0	1955	4.6
21 F	0201	1.0	0820	4.3	1414	1.2	2027	4.5
22 SA	0234	1.2	0854	4.1	1446	1.4	2102	4.4
23 SU	0307	1.4	0931	4.0	1518	1.6	2139	4.2
24 M	0344	1.6	1014	3.8	1557	1.8	2222	4.0
25 TU ◑	0429	1.8	1103	3.7	1647	2.0	2311	3.9
26 W	0525	1.8	1200	3.7	1751	2.0		
27 TH	0011	3.8	0631	1.8	1306	3.7	1900	1.9
28 F	0123	3.9	0737	1.7	1413	3.9	2006	1.7
29 SA	0236	4.0	0837	1.5	1515	4.1	2107	1.5
30 SU	0340	4.2	0934	1.2	1610	4.3	2202	1.2

JULY

Day	Time	m	Time	m	Time	m	Time	m
1 M	0437	4.4	1028	1.0	1702	4.5	2255	0.9
2 TU ●	0530	4.5	1119	0.8	1751	4.7	2346	0.7
3 W	0621	4.6	1209	0.7	1838	4.9		
4 TH	0034	0.6	0709	4.7	1256	0.6	1924	5.0
5 F	0122	0.5	0757	4.7	1343	0.6	2010	5.0
6 SA	0208	0.5	0844	4.7	1428	0.7	2056	4.9
7 SU	0255	0.6	0932	4.6	1513	0.8	2143	4.8
8 M	0341	0.8	1020	4.4	1600	1.0	2231	4.6
9 TU ◗	0430	1.0	1112	4.2	1652	1.3	2326	4.4
10 W	0525	1.2	1210	4.1	1750	1.5		
11 TH	0029	4.2	0631	1.4	1314	4.0	1900	1.6
12 F	0139	4.1	0744	1.5	1421	4.0	2015	1.6
13 SA	0249	4.1	0853	1.4	1523	4.2	2122	1.4
14 SU	0351	4.2	0953	1.3	1618	4.3	2220	1.2
15 M	0445	4.3	1046	1.1	1707	4.5	2310	1.0
16 TU ○	0534	4.4	1132	1.0	1751	4.6	2355	0.9
17 W	0617	4.4	1215	0.9	1831	4.7		
18 TH	0035	0.8	0656	4.4	1252	0.9	1906	4.7
19 F	0112	0.9	0731	4.4	1326	1.0	1940	4.7
20 SA	0145	1.0	0804	4.4	1356	1.1	2013	4.6
21 SU	0213	1.1	0837	4.3	1423	1.2	2045	4.5
22 M	0241	1.2	0911	4.2	1449	1.4	2118	4.3
23 TU	0308	1.4	0947	4.0	1517	1.5	2151	4.2
24 W	0340	1.5	1024	3.9	1553	1.7	2230	4.0
25 TH ◑	0421	1.7	1109	3.8	1642	1.9	2318	3.9
26 F	0520	1.8	1204	3.8	1754	2.0		
27 SA	0021	3.8	0641	1.8	1312	3.8	1918	1.9
28 SU	0138	3.9	0756	1.7	1428	4.0	2031	1.7
29 M	0301	4.0	0903	1.4	1536	4.3	2136	1.3
30 TU	0410	4.3	1004	1.2	1636	4.5	2236	1.0
31 W	0510	4.5	1102	0.9	1731	4.8	2331	0.7

AUGUST

Day	Time	m	Time	m	Time	m	Time	m
1 TH ●	0605	4.7	1156	0.6	1822	5.0		
2 F	0024	0.4	0657	4.8	1246	0.4	1912	5.1
3 SA	0113	0.2	0746	4.9	1333	0.3	1959	5.2
4 SU	0159	0.2	0832	4.9	1417	0.3	2045	5.1
5 M	0242	0.3	0917	4.8	1459	0.5	2129	5.0
6 TU	0324	0.5	1001	4.6	1541	0.8	2212	4.7
7 W ◗	0405	0.9	1045	4.4	1624	1.1	2257	4.4
8 TH	0452	1.2	1134	4.1	1713	1.5	2352	4.1
9 F	0546	1.6	1235	3.9	1817	1.8		
10 SA	0104	3.8	0659	1.8	1348	3.8	1940	1.9
11 SU	0223	3.8	0825	1.8	1457	3.9	2102	1.7
12 M	0331	3.9	0936	1.6	1557	4.2	2204	1.5
13 TU	0428	4.1	1031	1.4	1647	4.4	2256	1.2
14 W	0516	4.3	1116	1.1	1732	4.6	2340	1.0
15 TH ○	0558	4.4	1158	1.0	1811	4.7		
16 F	0019	0.8	0636	4.5	1234	0.9	1847	4.7
17 SA	0053	0.8	0710	4.5	1306	0.9	1921	4.7
18 SU	0123	0.9	0744	4.5	1333	1.0	1954	4.7
19 M	0149	0.9	0817	4.4	1357	1.1	2026	4.6
20 TU	0212	1.1	0847	4.3	1419	1.2	2054	4.5
21 W	0235	1.2	0916	4.2	1444	1.3	2122	4.3
22 TH	0301	1.3	0947	4.1	1514	1.5	2153	4.1
23 F ◗	0336	1.5	1025	3.9	1556	1.7	2238	4.0
24 SA	0424	1.8	1118	3.8	1655	2.0	2341	3.8
25 SU	0539	2.0	1228	3.8	1834	2.0		
26 M	0100	3.8	0722	1.9	1351	3.9	2006	1.8
27 TU	0235	3.9	0841	1.6	1511	4.2	2119	1.4
28 W	0353	4.2	0949	1.2	1616	4.6	2222	1.0
29 TH	0455	4.5	1048	0.8	1713	4.9	2318	0.5
30 F ●	0549	4.8	1143	0.5	1806	5.1		
31 SA	0009	0.2	0641	4.9	1231	0.2	1856	5.3

Chart Datum is 2·62 metres below Ordnance Datum (Newlyn). HAT is 5·3 metres above Chart Datum.

» FREE monthly updates. Register at www.reedsnauticalalmanac.co.uk «

61

STANDARD TIME (UT)
For Summer Time add ONE hour in **non-shaded areas**

DARTMOUTH   LAT 50°21'N   LONG 3°35'W
TIMES AND HEIGHTS OF HIGH AND LOW WATERS

Dates in red are **SPRINGS**
Dates in blue are **NEAPS**

YEAR 2019

SEPTEMBER

Day	Time m	Time m	Time m	Time m		Day	Time m	Time m	Time m	Time m
1 SU	0057 0.0	0728 5.0	1316 0.1	1943 5.3		16 M	0055 0.8	0720 4.6	1306 0.9	1932 4.8
2 M	0141 0.0	0813 5.0	1359 0.1	2026 5.2		17 TU	0120 0.9	0751 4.6	1330 1.0	2002 4.7
3 TU	0221 0.1	0855 4.9	1438 0.3	2107 5.0		18 W	0143 1.0	0821 4.5	1353 1.1	2029 4.5
4 W	0259 0.5	0933 4.7	1515 0.7	2146 4.7		19 TH	0206 1.1	0847 4.4	1417 1.2	2056 4.4
5 TH	0337 0.9	1011 4.4	1555 1.1	2222 4.3		20 F	0233 1.3	0915 4.3	1447 1.4	2127 4.2
6 F	0416 1.4	1050 4.1	1639 1.6	◗ 2305 3.9		21 SA	0305 1.5	0952 4.1	1526 1.7	2212 4.0
7 SA	0504 1.8	1145 3.8	1737 2.0			22 SU	0350 1.8	1046 3.9	1621 2.0	◗ 2316 3.8
8 SU	0024 3.6	0613 2.1	1314 3.7	1902 2.2		23 M	0458 2.1	1157 3.8	1802 2.1	
9 M	0202 3.6	0759 2.2	1433 3.8	2048 2.0		24 TU	0038 3.7	0656 2.1	1324 3.9	1949 1.9
10 TU	0313 3.7	0920 1.9	1535 4.1	2149 1.6		25 W	0221 3.9	0826 1.8	1453 4.3	2105 1.4
11 W	0409 4.0	1013 1.5	1626 4.3	2237 1.3		26 TH	0340 4.3	0934 1.3	1600 4.6	2206 0.9
12 TH	0455 4.3	1056 1.2	1708 4.6	2318 1.0		27 F	0439 4.6	1032 0.8	1656 5.0	2301 0.5
13 F	0535 4.5	1136 1.0	1747 4.7	2355 0.8		28 SA	0532 4.9	1123 0.4	1747 5.2	● 2350 0.1
14 SA	0611 4.6	1210 0.9	1823 4.8	○		29 SU	0620 5.1	1211 0.2	1835 5.3	
15 SU	0027 0.8	0647 4.6	1240 0.8	1859 4.8		30 M	0035 0.0	0705 5.2	1255 0.1	1920 5.3

OCTOBER

Day	Time m	Time m	Time m	Time m		Day	Time m	Time m	Time m	Time m
1 TU	0117 0.0	0747 5.2	1335 0.1	2002 5.2		16 W	0050 0.9	0726 4.7	1303 0.9	1939 4.7
2 W	0156 0.2	0827 5.0	1413 0.4	2040 5.0		17 TH	0116 0.9	0755 4.7	1330 1.0	2008 4.6
3 TH	0232 0.6	0902 4.8	1449 0.8	2114 4.6		18 F	0143 1.1	0823 4.6	1358 1.2	2037 4.5
4 F	0306 1.0	0933 4.5	1526 1.2	2145 4.3		19 SA	0211 1.3	0854 4.4	1429 1.4	2111 4.3
5 SA	0343 1.5	1003 4.2	1607 1.7	◗ 2220 3.9		20 SU	0247 1.5	0934 4.3	1510 1.6	2159 4.0
6 SU	0428 2.0	1046 3.9	1703 2.1	2317 3.5		21 M	0332 1.8	1028 4.1	1608 1.9	◗ 2304 3.8
7 M	0532 2.3	1226 3.7	1823 2.3			22 TU	0443 2.1	1138 4.0	1747 2.1	
8 TU	0135 3.5	0713 2.4	1402 3.8	2019 2.1		23 W	0026 3.8	0637 2.2	1304 4.0	1931 1.9
9 W	0248 3.7	0851 2.1	1506 4.0	2122 1.7		24 TH	0207 4.0	0807 1.8	1433 4.3	2047 1.4
10 TH	0342 4.0	0943 1.7	1556 4.3	2206 1.4		25 F	0323 4.3	0915 1.3	1539 4.7	2146 0.9
11 F	0426 4.3	1026 1.3	1639 4.6	2246 1.1		26 SA	0419 4.7	1011 0.9	1635 5.0	2239 0.5
12 SA	0505 4.5	1103 1.1	1718 4.7	2321 0.9		27 SU	0509 5.0	1101 0.5	1726 5.2	2326 0.3
13 SU	0542 4.7	1139 0.9	1756 4.8	○ 2354 0.8		28 M	0557 5.1	1148 0.3	1813 5.3	●
14 M	0619 4.7	1209 0.9	1833 4.8			29 TU	0011 0.2	0641 5.2	1231 0.2	1858 5.2
15 TU	0023 0.8	0654 4.8	1237 0.9	1907 4.8		30 W	0052 0.2	0721 5.2	1311 0.3	1938 5.1
						31 TH	0131 0.5	0758 5.0	1349 0.6	2015 4.9

NOVEMBER

Day	Time m	Time m	Time m	Time m		Day	Time m	Time m	Time m	Time m
1 F	0205 0.8	0830 4.8	1425 0.9	2046 4.5		16 SA	0128 1.1	0806 4.7	1348 1.1	2027 4.5
2 SA	0240 1.2	0900 4.6	1501 1.3	2115 4.2		17 SU	0201 1.2	0842 4.6	1425 1.3	2107 4.3
3 SU	0314 1.6	0930 4.3	1542 1.8	2150 3.9		18 M	0241 1.5	0925 4.5	1510 1.5	2155 4.1
4 M	0357 2.1	1012 4.0	1633 2.1	◗ 2242 3.6		19 TU	0331 1.8	1019 4.3	1610 1.8	◗ 2256 4.0
5 TU	0456 2.4	1114 3.8	1742 2.3			20 W	0439 2.0	1123 4.2	1733 1.9	
6 W	0039 3.5	0614 2.5	1310 3.8	1910 2.2		21 TH	0013 3.9	0611 2.0	1243 4.2	1903 1.8
7 TH	0204 3.7	0749 2.3	1422 4.0	2030 1.9		22 F	0140 4.0	0739 1.8	1405 4.4	2019 1.4
8 F	0302 3.9	0856 1.9	1515 4.2	2122 1.6		23 SA	0254 4.3	0848 1.4	1513 4.6	2120 1.1
9 SA	0348 4.2	0942 1.6	1602 4.4	2203 1.3		24 SU	0353 4.6	0945 1.0	1610 4.8	2214 0.8
10 SU	0430 4.5	1023 1.3	1644 4.6	2241 1.1		25 M	0444 4.8	1037 0.7	1703 5.0	2302 0.6
11 M	0510 4.6	1100 1.1	1726 4.7	2316 1.0		26 TU	0532 5.0	1124 0.6	1751 5.0	● 2348 0.5
12 TU	0549 4.8	1136 1.0	1806 4.8	○ 2351 0.9		27 W	0617 5.1	1209 0.5	1836 5.0	
13 W	0626 4.8	1209 0.9	1843 4.8			28 TH	0029 0.5	0658 5.1	1250 0.6	1917 4.9
14 TH	0023 0.9	0702 4.8	1241 0.9	1918 4.7		29 F	0108 0.7	0734 5.0	1329 0.8	1953 4.7
15 F	0055 0.9	0734 4.8	1314 1.0	1951 4.6		30 SA	0145 1.0	0807 4.8	1405 1.0	2026 4.5

DECEMBER

Day	Time m	Time m	Time m	Time m		Day	Time m	Time m	Time m	Time m
1 SU	0218 1.3	0837 4.6	1442 1.3	2057 4.2		16 M	0201 1.1	0838 4.8	1429 1.1	2106 4.5
2 M	0253 1.6	0910 4.4	1520 1.6	2133 4.0		17 TU	0245 1.2	0922 4.7	1515 1.2	2153 4.3
3 TU	0331 1.9	0949 4.2	1604 1.9	2219 3.8		18 W	0333 1.4	1013 4.5	1608 1.4	2248 4.2
4 W	0418 2.2	1040 4.0	1658 2.1	◗ 2320 3.7		19 TH	0430 1.6	1111 4.4	1711 1.6	◗ 2352 4.1
5 TH	0520 2.3	1147 3.9	1803 2.2			20 F	0539 1.8	1217 4.3	1826 1.6	
6 F	0043 3.7	0632 2.3	1308 3.9	1913 2.1		21 SA	0105 4.1	0659 1.8	1331 4.3	1943 1.5
7 SA	0159 3.8	0743 2.1	1419 4.0	2016 1.8		22 SU	0218 4.2	0814 1.6	1441 4.4	2051 1.3
8 SU	0258 4.2	0845 1.9	1515 4.2	2110 1.6		23 M	0322 4.4	0919 1.3	1545 4.5	2148 1.1
9 M	0347 4.3	0935 1.6	1606 4.4	2156 1.3		24 TU	0418 4.6	1015 1.1	1641 4.6	2241 0.9
10 TU	0434 4.5	1021 1.3	1654 4.5	2240 1.1		25 W	0510 4.8	1105 0.9	1733 4.7	2328 0.8
11 W	0518 4.7	1104 1.1	1738 4.6	2321 1.0		26 TH	0557 4.9	1153 0.8	1819 4.7	●
12 TH	0601 4.8	1146 1.0	1821 4.7	○		27 F	0012 0.8	0639 4.9	1235 0.8	1901 4.7
13 F	0002 0.9	0640 4.9	1225 0.9	1903 4.7		28 SA	0052 0.9	0716 4.9	1315 0.9	1938 4.6
14 SA	0041 0.9	0718 4.9	1305 0.9	1942 4.7		29 SU	0129 1.0	0750 4.8	1352 1.0	2011 4.5
15 SU	0121 1.0	0757 4.9	1347 1.0	2023 4.6		30 M	0202 1.2	0823 4.7	1426 1.2	2043 4.3
						31 TU	0234 1.5	0856 4.5	1459 1.4	2117 4.2

Chart Datum is 2·62 metres below Ordnance Datum (Newlyn). HAT is 5·3 metres above Chart Datum.

>> **FREE** monthly updates. Register at <<
www.reedsnauticalalmanac.co.uk

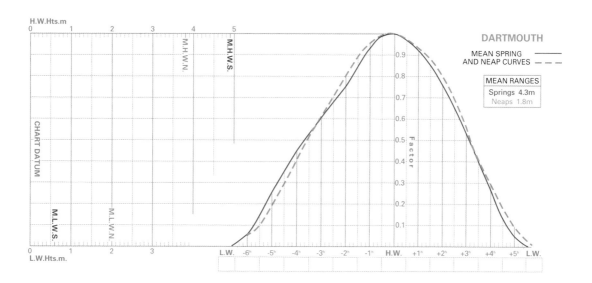

DARTMOUTH

MEAN SPRING ——————
AND NEAP CURVES - - - - -

MEAN RANGES
Springs 4·3m
Neaps 1·8m

1.18 BRIXHAM

Devon 50°24'·31N 03°30'·85W

❈❈❈♒♒♒♒❀❀

CHARTS AC *3315*, 1613, 1634, 5602, 26; Imray C10, C5, 2300

TIDES –0505 Dover; ML 2·9; Duration 0635.

Use **TORQUAY** Differences on **PLYMOUTH**.

SHELTER Very good in marina; good at YC pontoon in SW of hbr, but outer hbr is dangerous in strong NW'lies. ⚓ NW of fairway or in Fishcombe and Elberry Coves W of the hbr; beware water skiers.

NAVIGATION WPT 50°24'·70N 03° 31'·09W, (3ca off chartlet), 159°/0·4M to harbour entrance. Beware fish farm 5Ca WNW of harbour; easy access H24. Inshore around Torbay controlled areas (May-Sep, mainly for swimmers) are marked by unlit Y SPM buoys; boats may enter with caution, speed limit 5kn.

LIGHTS AND MARKS Berry Hd, a conspic headland and good radar return, is 1·25M ESE of entrance. The fairway is marked by two pairs of lit lateral buoys. See chartlet and 1.2 for lt details.

COMMUNICATIONS (Code 01803) Dr 855897; Ⓗ 882153. HM 853321; Pilot 882214; Ch 14 (May-Sep 0800-1800; Oct-Apr 0900-1700, Mon-Fri). Marina Ch 80. Water Taxi: *Shuttle* Ch M.

FACILITIES Brixham Marina
brixhammarina.co.uk ☎882929, ⚓
500⊖ inc ❶ £3·40<12·5m>£4·20<18m, Short stay<4 hrs £8·00.
Events pontoon (no 🗑) £2·32 < 12·5m>£2·90. ⚓(0900-1800, Apr-Oct) ⚙ ◖(50t).
Brixham YC ☎853332, ❶ pontoon, ⚓ ⚓ ⚓ ⊼ ✕ 🗑.
Hbr Office ⚓ each side of fairway, ⚓ ⚓ 🛢 ⊼ 🗑(4t).
Town pontoon ⊖ £1·90, 🗑 no 🚾/🚿.
Town 🏧 ✉ Ⓑ ✎ 🏧 Ⓔ 🏦 ACA ✕ 🗑 bus to Paignton, ⇌ ✈ Exeter.

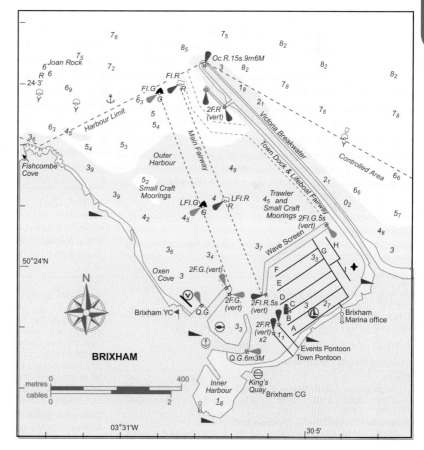

MINOR HARBOUR 2·3M NW OF BRIXHAM

PAIGNTON, Devon, **50°25'·96N 03°33'·36W** 500m S of Paignton pier. AC 1613, 26. HW +0035 and –0·6m on Plymouth; Duration 0640; ML 2·9m. Hbr dries 1·3m, only suitable for max LOA 8·2m. A heavy swell enters in E'lies. Drying rks extend 180m ENE from E pier, ⚡ QR, to Black Rk, ECM twr ⚡ Q (3) 10s 5m 3M. In the approach chan **keep to port**. HM (summer only) ☎(01803) 557812, VHF Ch 14. **Paignton SC** ☎525817. Facilities: ⚓ 🛢 ⚒ ✎ 🏦 ACA.

1.19 TORQUAY

Devon **50°27′·45N 03°31′·73W** ✿✿✿✿◊◊◊✿✿

CHARTS AC 3315, 1613, 5602, 26; Imray C5

TIDES –0500 Dover; ML 2·9; Duration 0640

Standard Port PLYMOUTH (←—)

Times				Height (metres)			
High Water		Low Water		MHWS	MHWN	MLWN	MLWS
0100	0600	0100	0600	5·5	4·4	2·2	0·8
1300	1800	1300	1800				
Differences *TORQUAY*							
+0025	+0045	+0010	0000	–0·6	–0·7	–0·2	–0·1

NOTE: There is often a stand of about 1hr at HW

SHELTER Good, but some swell in hbr with strong SE'lies. ⚓s NW of Hope's Nose at Hope Cove, Anstey's Cove & Babbacombe Bay are well sheltered in W'lies. No ⚓ in hbr.

NAVIGATION WPT 50°27′·03N 03°31′·57W, 339°/0·40M to round SHM buoy 80m WSW of Haldon pier. Semi-blind ent; keep a good lookout. Access at all tides, but in strong SE winds backwash may make the narrow ent difficult. Hbr speed limit 5kn.

Controlled areas (mainly for swimmers, May-Sep) close inshore around Torbay are marked by unlit Y SPM buoys; boats may enter with caution, speed limit 5kn. Watch out for canoeists.

LIGHTS AND MARKS Ore Stone (32m rk) is conspic off Hope's Nose, 2·3M E of hbr ent. Many conspic white bldgs, but none unique. No ldg marks/lts. Lts may be hard to see against town lts.

COMMUNICATIONS (Code 01803) CGOC (01326) 317575; Police 101; Dr 212429; Ⓗ 614567.

Torquay Harbour Ch **14** 16 (May-Sep 0800-1800LT; Oct-Apr 0900-1700, M-F). Marina Ch **80** (H24), M. *Torquay Fuel* Ch M.

FACILITIES Marina www.torquaymarina.co.uk ☎200210, 07764 175611 440◯+60 Ⓥ £3.32<12·5m>£4.09<18m>£4.60<24m; <4hrs £8.00; Ⓞ Ⓑ.

Town Dock HM ☎292429. 4 pontoons and a wavebreak replace moorings in SE part of the hbr. Limited Ⓥ berths on the W side of the wavebreak, £1.94/m/night. Call HM Ch 14.

Haldon Pier ⚓ at root, 96m Ⓥ pontoon in 2m, call HM Ch 14: ◯ raft £1.94, ⚓ ⓓ(long lead). Also used for events.

S Pier S side: ◗ ◮ LPG at fuel pontoon, *Riviera Fuel* VHF Ch M; ☎294509, 07786 370324 (Apr-Sept 0830-1900 W/days, 1000-1900 Sun); ◯(6t). N side: ◯ ⚓.

Inner Hbr A wall with sill (11·6m wide), retains 1·0m–2·8m inside. Above it a footbridge with lifting centre section opens on request

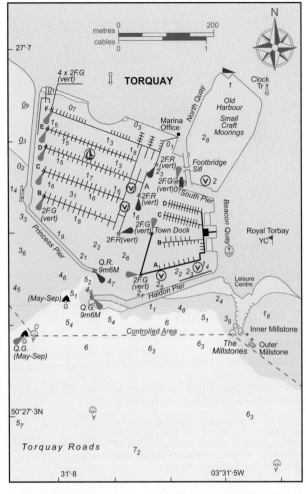

to HM on *Ch 14* when the flapgate is down, approx HW –3½ to +3 (0700-2300 Apr-Sep, 0900-1700 Oct-Mar). IPTS sigs 2 & 4 control ent/exit. Do not pass under the bridge when it is down. ◯ ⚓.

Royal Torbay YC ☎292006, ✕ ◻.

Town ◗ ✕ ✎ ◖Ⓑ◗ Ⓔ ◭ ◭ ACA ✉ Ⓑ ⛴ Ⓞ ✕ ◻ @ ⇌ ✈(Exeter).

1.20 TEIGNMOUTH

Devon **50°32′·37N 03°30′·00W** (Abeam The Point) ✿✿◊◊✿✿

CHARTS AC 3315, 26, 5601, 5602; Imray C10, C5, 2300

TIDES –0450 Dover; ML 2·7; Duration 0625

Standard Port PLYMOUTH (←—)

Times				Height (metres)			
High Water		Low Water		MHWS	MHWN	MLWN	MLWS
0100	0600	0100	0600	5·5	4·4	2·2	0·8
1300	1800	1300	1800				
Differences *TEIGNMOUTH* (Approaches)							
+0020	+0050	+0025	0000	–0·9	–0·8	–0·2	–0·1
TEIGNMOUTH (New Quay)							
+0025	+0055	+0040	+0005	–0·8	–0·8	–0·2	+0·1

Tidal streams. *At the hbr ent the flood starts HW Plymouth –5½ and the ebb at HWP +¾. Be aware that from about HW –3, as The Salty covers, the flood sets NW across it and a small back eddy flows S past the Ⓥ pontoons. HW slack occurs at approx HWP +½.*

SHELTER Good, but entry hazardous in strong onshore winds (NE-S) when surf forms on the bar.

NAVIGATION WPT 50°32′·37N 03°29′·15W, 265°/5ca to trng wall lt (Oc R 6s) ≠ 2 white ◻s on seawall beyond. The E-W chan is dredged as the Bar, Pole and Spratt Sands shift frequently; AC 26 may not immediately show such shifts. Depths reduce to CD over the Bar. Detailed pilotage and hbr notes are in the excellent website below. Appr chan is not well buoyed, contact HM for latest situation. Small buoys laid by Pilots near the appr chan, should not be relied upon. Beware rks off The Ness; and variable extent of The Salty, a bank of hard gravel/sand. Max speed 6kn over the ground; 5kn inside Y buoys off The Ness & Teignmouth beaches.

Clearance under Shaldon bridge is 2·9m at MHWS and approx 6·7m at MLWS; near its N end there is a 9m wide drawbridge section. Avoid a Historic wreck site (50°32′·95N 03°29′·24W; just off chartlet), close inshore ENE of Ch twr.

LIGHTS AND MARKS The Ness, a 50m high red sandstone headland, and church tower are both conspic from afar. Close NE of the latter, just off N edge of chartlet, Teign Corinthian YC bldg (cream colour) is also conspic. Ignore two FR (NNE of The Point); they are not ldg lts. N of The Point, two F Bu lts align 023°, but are not ldg lts. A Y can buoy, Fl Y 5s, at 50°31′·97N 03°27′·78W marks the seaward end of outfall, 103° The Ness 1·3M.

COMMUNICATIONS (Code 01626) Dr 774355; Ⓗ 772161. HM 773165 (or 07796 178456 in emergency or out of hrs).

Hbr Ch **12** 16 (Mon-Fri 0900-1400).

FACILITIES www.teignmouth-harbour. com; info@teignmouth-harbour.com. In suitable weather up to 10 boats can raft on 2 detached 20m ♥ pontoons (Jubilee & Trafalgar) at 50°32′.62N 03°29′.96W in 2·5m; £1.10/m. Third pontoon (Newfoundland) is for local boats only. No ⚓ in hbr due to many moorings and strong tidal streams. In fair weather ⚓ approx 1·5ca SE of The Ness or Teignmouth Pier.

E Quay Polly Steps ⌐ (up to 10m). **Teign Corinthian YC** ☎772734, ⟐£8 ⌂.

Services ⌐ ⚓♦⚒⚔⚓ ⊞ Ⓔ ☕ ⌂(8t) ⛽.

Town ☎ ☏ (1M), ⚓⚓ ⊠ Ⓑ 🖳 ⊡ ✕ ⌂ ⇌ ✈(Exeter).

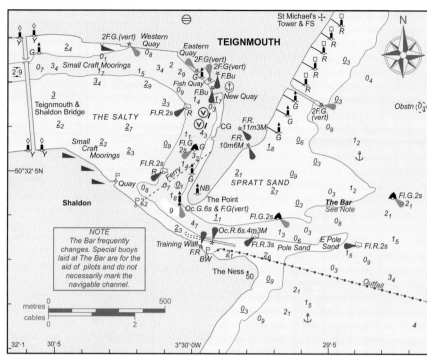

LYME BAY: TEIGNMOUTH TO PORTLAND BILL

Lyme Bay curves 65M in a great arc from Start Pt to Portland Bill (AC 3315).

▶*Tides are weak, rarely more than 0·75kn. From west to east the tidal curve becomes progressively more distorted, especially on the rising tide. The rise is relatively fast for the 1st hr after LW; then slackens noticeably for the next 1½hrs, before resuming the rapid rate of rise. There is often a stand at HW, not very noticeable at Start Pt but lasting about 1½hrs at Lyme Regis.*◀

From Teignmouth to Dawlish rocks extend 1ca offshore. Beware Dawlish Rock (2·1m) about 5ca off North end of town. Pole Sand (dries up to 3·5m) and Warren Sand on the West side of the **River Exe** estuary are liable to shift. A firing range at Straight Point has a danger zone extending 1·5M east marked by 2 Y lit DZ buoys. 3½ M SE of Sidmouth on the 20m contour a mussel farm is marked by 2 pairs of SPM (Fl Y 5s). Another mussel farm, similarly marked, lies 4¾M further S in 25m of water.

Between Torbay and Portland no harbour is accessible in strong onshore winds, and care must be taken not to be caught on a lee shore. Marked by 4 cardinal buoys 2½M E of Hope's Nose is the wk of the *EMSSTROM* (16·5m). In offshore winds there is a good ⚓ NE of Beer Hd, the western-most chalk cliff in England. Golden Cap (186m and conspic) is 3·5M E of **Lyme Regis**. High Ground (3m) and Pollock (4m) are rocky patches 7ca offshore, respectively 1·5M W and 6ca SW of **West Bay** (Bridport). Numerous wks litter the bay, one dangerous to navigation 222°Hardy's Mt 7.8M.

From 6M SE of Bridport Chesil Beach runs SE for about 9M to the N end of the Isle of Portland. From a distance this peninsula looks like an island, with its distinctive wedge-shaped profile sloping down from 145m at the N to sea level at the Bill. Although mostly steep-to, an 18m high, white obelisk on the tip of the Bill warns of a rocky ledge extending about 50m S of the Bill. If the highest window on the lighthouse is seen above the top of the obelisk, you are clear of the outermost rock, but still very close inshore.

If heading up-Channel from Start Point, Dartmouth or Torbay time your passage to pass Portland Bill with a fair tide at all costs, especially at springs. If late, a temporary ⚓ can be found close inshore at Chesil Cove, abeam the highest part of Portland.

1.21 RIVER EXE

Devon **50°36′·94N 03°25′·40W** (Abeam Exmouth) ✿✿⚓⚓✿✿

CHARTS AC 3315, 2290, 5601; Imray C10, C5, 2300

TIDES −0445 Dover; ML 2·1; Duration 0625

Standard Port PLYMOUTH (←)

Times				Height (metres)			
High Water		Low Water		MHWS	MHWN	MLWN	MLWS
0100	0600	0100	0600	5·5	4·4	2·2	0·8
1300	1800	1300	1800				
Differences EXMOUTH (Approaches)							
+0030	+0050	+0015	+0005	−0·9	−1·0	−0·5	−0·3
EXMOUTH DOCK							
+0035	+0055	+0050	+0020	−1·5	−1·6	−0·9	−0·5
STARCROSS							
+0040	+0100	+0055	+0025	−1·4	−1·5	−0·8	−0·1
TURF LOCK							
+0045	+0100	+0034	ND	−1·6	−1·6	−1·2	−0·4
TOPSHAM							
+0045	+0105	ND	ND	−1·5	−1·6	ND	ND

NOTE: In the appr chan the sp ebb reaches 3·3kn; with wind against tide a confused, breaking sea quickly builds. Off Warren Pt the flood stream runs at 3-4kn and the ebb can exceed 4½kn when the banks uncover.

SHELTER Good upstream of No 13 buoy and in Dock.

NAVIGATION WPT 50°35′·92N 03°23′·75W, Exe SWM buoy, 335°/6ca to No 7 SHM buoy. Up-to-date AC 2290 is essential. Leave WPT at approx LW+2 (or when there is sufficient rise of the tide) when hazards can be seen and some shelter obtained, or at HW−1 or HW−2 to reach Exmouth or Topsham, respectively, at slack water.

The approach channel, with a least depth of 0·4m, is well marked with PHM/SHM buoys, some of which are lit. Night entry is not advised, the alignment of the ldg lts is questionable. After No 10 PHM buoy do not cut the corner round Warren Point; turn to the SW after passing No 12 PHM buoy. Beware of the flood stream sweeping NW past the marina entrance into moorings and shallow banks.

The estuary bed is sand/mud, free of rocks. Follow the curve of the channel rather than straight lines between the buoys; some bends are marked on the outside only. *Continued overleaf*

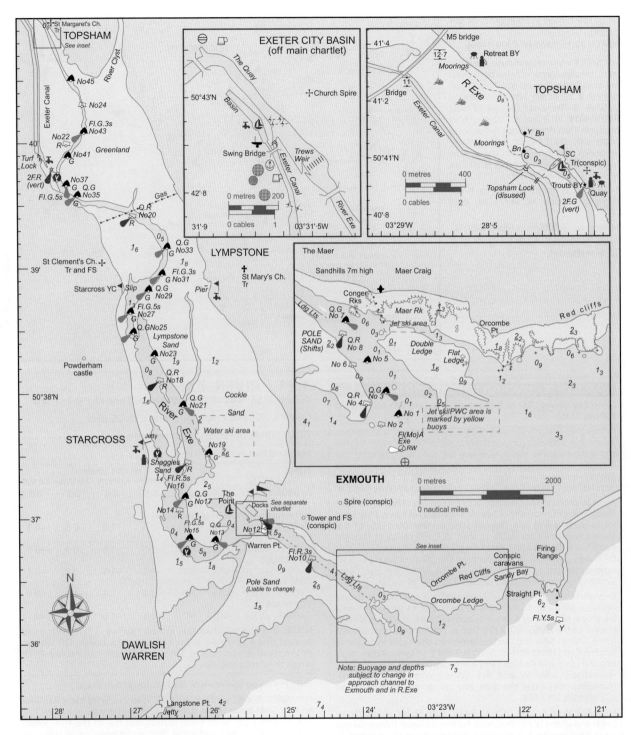

The estuary is an international conservation area.

LIGHTS AND MARKS See chartlet and 1.2. A caravan site is conspic close W of Straight Pt with red cliffs to W and NNE. Exmouth ✠ tr and FS are conspic. At Exmouth follow the 305° directional bn which can be hard to see by day. Following buoys in the upper reaches are lit:

No 22 PHM: Ting Tong   – Fl R 2s.
No 24 PHM: Exe   – Fl R 4s.
No 41 SHM: Turf Green – Fl G 2s.
No 45 SHM: Topsham reach – Fl G 4s.

A jetski/PWC area is marked by yellow buoys to NE of SWM.

Caution: Firing range N of Straight Pt has a danger area to ESE, marked by 2 DZ SPM lt buoys (not the sewer outfall SPM buoy, Fl Y 5s, shown on chartlet). R flags are flown when the range is active (likely times 0800-1600, Mon-Fri); call *Straight Pt Range* VHF Ch 08 16. From the E, check also with safety launch.

COMMUNICATIONS (Code 01395) Dr 273001; Ⓗ 279684. Council River Team (01392) 274306; Port of Exeter Ch 12 16 (Mon-Fri: 0730-1730LT). Exmouth Marina/bridge Ch 14. *Water taxi* Ch M.

FACILITIES **EXMOUTH Marina**, ☎269314, mostly berth holders. Pontoon in ent chan is used by ferries/water taxi and for 🛥. VHF Ch 14 for berth and bridge. Strong cross tides at ent. Footbridge lifts on request 0800-1700; stays open HN. 172� inc at least 4 🚫 in 2m; <6·5m £20, <10m £27, >10m £38, shortstay (4hrs) £7·50 ⚓(15t). **Exe SC** ☎264607, ✕ 🍺.

Town 🛢 🛒 🛠 📮 Ⓔ ⛽ 🏪 ACA ✉ Ⓑ 🛒 🛢 ✕ 🍺 ⚓. 🅿s£12·00, lie near No 15 SHM buoy, off Turf Lock and Topsham SC; call water taxi Ch M ☎07970 918418 for advice.

STARCROSS (01626) **Village** 🛒 🍺 ✉ Ⓑ ⚓. **Starcross Garage** ☎890225, 🅿 🅿 🛠 Gas. **Starcross Fishing & Cruising Club** 2🚫 (☎891996 (evenings) for availability). **Starcross YC** ☎890470.

TURF LOCK, 50°39'·88N 03°28'·06W, operates daily in working hours by prior arrangement with the lock keeper ☎274306. 🚫 £13·90.

Exeter Ship Canal (3m depth). HM ☎01392 265791 or Ch 12 for non-tidal pontoon 🚫 berth in Turf Basin, £20·00 min 2 days (and laying up) 'Turf weekend' £40. Canal transit Mon-Fri 1600-1800 £50(extra at w/e).

TOPSHAM (01392) Here the river carries about 0·4m. Options are � (pontoon in ¾m at LW) at Trouts BY or ⚓ in appr's, dry out at Town Quay in mud on rough wall (☎01392 265791 – £11·00/day) or find a mooring.
Topsham SC ☎877524, 2🚫s 🛠 🛒 ⚓ 🍺.

Trouts BY www.trouts boatyard.co.uk ☎873044. MLWS max depth 0·75m; 🛠 � £18<6m>£25 (+ £20 key deposit); short stay £12. ⚓ 🛢 🛒 🛠 📮 🍺.

Retreat BY www.retreatboatyard.co.uk ☎874720, appr channel dries approx 1m, 🛠 poss � ⚓ 🛢 🛠 🛠 ⚓(10½t) 🍺.
Town 🛢 ⛽ 🍺 ACA ✉ Ⓑ 🛒 🛢 ✕ 🍺 ⚓.

EXETER www.exeter.gov.uk river.canal@exeter.gov.uk Waterways, Civic Centre, Paris St, Exeter EX1 1JN; 0730-1530 M-F, Sat/Sun as req. The HM ☎01392 265791 is responsible for the whole River Exe and Exeter Canal. www.exe-estuary.org makes interesting reading. The City Basin accepts visitors and lay-ups by prior arrangement.
City All facilities; 🛠 🍺 ACA @ ⚓ ✈.

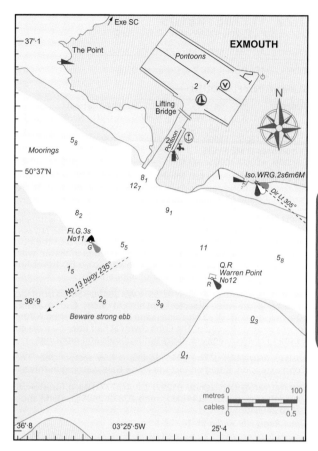

TWO MINOR HARBOURS WEST OF LYME REGIS

BEER, Devon, **50°41'·64N 03°05'·22W**. Beer Roads is an ⚓ in 2-5m on sand, shingle and gravel sheltered by Beer Head from prevailing W'lies, but open to E-SW winds. Beer Head is 142m high, topped by a caravan site. Its chalk cliffs are white, becoming red sandstone to the E towards Seaton. A conspic radio/tel mast is above the centre of the beach. Church tower is conspic. Do not use the historic approach of the beach light ≠ church tower light. All beach lights are purely for illumination, not for navigation. Land on the open beach, as in centuries past; some traditional local boats are still hauled up the beach on skids.

Beer & Seaton are adjacent towns N and NNE of Beer Head.
Beer 🍺 ✉ 🛒 ✕ 🍺. **Seaton** 🅿 🅿 shops ✕ 🍺.

2·5M WSW of Beer Head and 8 cables offshore are the remains of the *MSC Napoli* wreck (50°40'·38N 03°09'·58W), marked by E, S and W cardinal buoys and a No Entry zone, radius 3ca.

AXMOUTH, Devon, **50°42'·13N 03°03'·29W**, 1·85M ENE of Beer Head. AC 3315. HW –0455 on Dover; +0045 and –1·1m on Plymouth; ML 2·3m; Duration 0640. MHWS 4·3m, MHWN 3·2m.

A small drying hbr at the mouth of the R Axe for boats max draught 1·2m, LOA 8·5m (longer by prior arrangement), able to dry out. Appr chan to bar (dries 0·5m) often shifts about a NNE/SSW axis. A dinghy recce (best near LW) to view the access channel and a prior brief by HM/YC are advised. Enter only in settled weather from HW –2½ to +1½ nps; HW –2½ or –½ sp. After 7m wide ent and the pierhead SHM beacon, Fl G 4s 7m 2M, turn 90° to port inside; hug the hbr wall to stbd until into the mooring basin; moor as directed. The low bridge (2m clearance) just N of moorings is the head of navigation.

HM ☎(01297) 22180/07939 044109. **Axe YC** ☎20043: 🛠 🛒 pontoon, 🏗(6t) 🍺. Facilities: 🅿 🛠 🍺. Excellent pilotage notes

at www.axeyachtclub.co.uk are essential reading for first time visitors. **Seaton**, as above for victuals etc.

The chartlet is reproduced by kind permission of the Axe Yacht Club whose Commodore and others revised these notes.

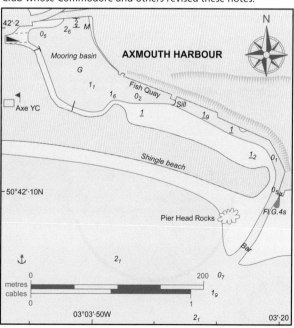

1.22 LYME REGIS

Dorset 50°43'·20N 02°56'·19W ✿✿✿✿✿✿✿✿

CHARTS AC 3315, 5601; Imray C10, C5, 2300

TIDES –0455 Dover; ML 2·4; Duration 0700

Standard Port PLYMOUTH (←—)

Times				Height (metres)			
High Water		Low Water		MHWS	MHWN	MLWN	MLWS
0100	0600	0100	0600	5·5	4·4	2·2	0·8
1300	1800	1300	1800				
Differences LYME REGIS							
+0040	+0100	+0005	–0005	–1·2	–1·3	–0·5	–0·2

NOTE: Rise is relatively fast for the 1st hour after LW, but slackens for the next 1½hrs, after which the rapid rate is resumed. There is often a stand of about 1½hrs at HW.

SHELTER Good in the hbr (dries up to 2.1m), but swell enters in strong E/SE winds when it may be best to dry inside the N Wall. The Cobb, a massive stone pier, protects the W and S of the hbr. A lightweight walk ashore pontoons are provided in season with 2m depth reported at outer ends for boats up to 8T. In settled weather ⚓ as on the chartlet, clear of appr and ⓩs.

NAVIGATION WPT 50°43'·10N 02°55'·64W, 284°/0·35M to front ldg lt; best line just inside W sector until hbr ent opens. Craft drawing >1m may ground LW ±2 on a small sand bar outside the hbr. A 70m rock extension to E end of The Cobb covers at half tide; it is marked by unlit PHM bcn. Beware many fishing floats and moorings.

LIGHTS AND MARKS The 284° ldg line & R/W sectors clear The Cobb extension. R flag on Victoria Pier = Gale warning in force.

COMMUNICATIONS (Code 01297) Dr 445777 HM g.foreshaw@ westdorset-dc.gov.uk ☎442137, mob 07870 240645, DHM mob 07870 240650.
Lyme Regis Hbr Radio Ch **14**, 16.

FACILITIES www.dorsetforyou/lymeregisharbour.com Dry out against Victoria Pier (0.3–1.3m) on clean, hard sand, max LOA 11m. Pontoon � £20 <10m > £25 < 15m > £30 < 20m; 9 Y ⓩs lie ENE of hbr ent in 1·5-2m, £10. ⚓⚓ ⌧⌧⌧ (mobile) ✕✕✕ⒺⒺ.
SC ☎442373, ⚓⚓✕▢.
Power Boat Club ☎07768 725959, ✕▢.
Town ⚓✉Ⓑ♨▢✕▢ bus to Axminster ⇄ ✈ (Exeter).

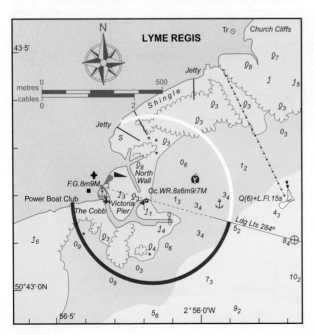

1.23 WEST BAY (BRIDPORT)

Dorset 50°42'·51N 02°45'·81W ✿✿✿✿✿✿✿

CHARTS AC 3315, 5601; Imray C10, C5, 2300

TIDES –0500 Dover; ML 2·3; Duration 0650

Standard Port PLYMOUTH (←—)

Times				Height (metres)			
High Water		Low Water		MHWS	MHWN	MLWN	MLWS
0100	0600	0100	0600	5·5	4·4	2·2	0·8
1300	1800	1300	1800				
Differences BRIDPORT (West Bay)							
+0025	+0040	0000	0000	–1·4	–1·4	–0·6	–0·1
CHESIL BEACH							
+0040	+0055	–0005	+0010	–1·6	–1·5	–0·5	0·0
CHESIL COVE							
+0035	+0050	–0010	+0005	–1·5	–1·6	–0·5	–0·2

NOTE: Rise is relatively fast for first hr after LW; it then slackens for the next 1½hrs, after which the rapid rise is resumed. There is often a stand of about 1½hrs at HW.

SHELTER Access in W'lies and shelter in the outer hbr are improved by the new W pier, but the ent is exposed to SE'lies and swell/surge may affect the outer hbr in any wind direction.

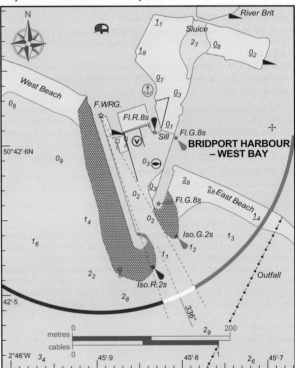

NAVIGATION WPT 50°42'·05N 02°45'·49W, 336°/0·50M to ent in W sector of Dir lt. High Ground shoal 3m is 1·4M W of the ent. 3·7 and 4·6m shoals are SW & SE of hbr. SPM outfall buoy, Fl Y 5s, is 5ca SSW of ent.

LIGHTS AND MARKS See chartlet and 1.2 for lts. High cliffs are conspic close ESE of the hbr.

COMMUNICATIONS (Code 01308) Dr 421109.
Bridport Radio Ch **11** 16. HM 423222, mob 07870 240636

FACILITIES, www. dorsetforyou/westbayharbour.com. **Outer hbr** is silted up to 0·5m below CD. ♥ pontoon is in front of the slipway, close W of ⚡ Fl R 8s. ⌒ £15/night. ⚓–▢▢▢✕✕.

Inner hbr dries, except for a pool scoured 2·7m by sluice water, and is full of FVs and local boats. The entry sill dries 0.1m.

Bridport (1½M N of the hbr) ⚓✉Ⓑ♨▢✕▢⇄ (bus to Axminster) ✈ (Exeter).

CROSSING THE ENGLISH CHANNEL, WESTERN PART

These notes cover cross-Channel passages from harbours between Scilly and Tor Bay to French harbours between Ouessant and Île de Bréhat. See AC 2655, 2656; 2.3 for the Central English Channel; and 0.33 for cross-Channel distances. Passage planning advice and SOLAS V requirements are in Reference Data

The Western Channel differs from the central and eastern parts; it is wider and its French harbours are less easy to approach.

Distance is a major ingredient in any passage plan. The shortest is 88M (Salcombe to Bréhat or Trébeurden); the longest 108M (Fowey to L'Aberwrac'h). Salcombe and Dartmouth are favourite departure points, for the Channel Islands and points West. Falmouth is often preferred for rounding Ushant. Taking an average distance of 100M:

Time at 5kn is 20hrs, or up to 30hrs if the wind is southerly. Therefore it is important to consider:

- **Crew** stamina and experience. A short-handed crew (husband & wife, for example) may cope quite easily with a 20-hr passage, but 30hrs beating to windward might prove too much.

- **Weather**. A window of at least 24hrs (preferably 48hrs) is needed during which the maximum forecast wind is F5 (depending on direction); wind not N'ly (ie onshore along the Brittany coast); with a low chance of fog on that coast or in mid-Channel.

- **ETD and ETA**. Some night sailing is inevitable, but in Jul/Aug there is only about 6hrs of darkness. To avoid leaving, arriving or crossing shipping lanes in the dark, a departure at about 0900 allows you to cross the lanes in daylight and arrive after dawn.

Shipping lanes. Whilst not having to cross a TSS, lines linking the Casquets TSS and Ushant TSS will indicate where the main W-bound and E-bound traffic is likely to be encountered. Be alert for shipping not using this route, and try to avoid close quarters situations. Do not expect big ships to give way to you; they will probably be travelling fast (24kn is not uncommon, even in poor visibility) and may have difficulty detecting you if it is at all rough.

Contingencies. North Brittany is not the easiest of coasts owing to an abundance of offshore rocks which demand precise landfalls and careful pilotage. Fresh onshore winds complicate the task, as does fog. There is no all-weather, all-tide, port of refuge along this coast, so the forecast weather window is crucial. If things are not going as planned, it is no disgrace to turn back or make for an alternative destination. The Channel Islands are a possible diversion, or consider heaving-to for a rest or to await an improvement in the weather.

PORTLAND RACE

South of the Bill lies Portland Race (AC 2255) in which severe and very dangerous sea states occur. Even in settled weather it should be avoided by small craft, although at neaps it may be barely perceptible.

The Race occurs at the confluence of two strong tidal streams which at springs run S down each side of the Isle of Portland for almost 10hrs out of 12hrs. These streams meet the main E-W stream of the Channel, producing large eddies on either side of Portland Bill and a highly confused sea state with heavy overfalls in the Race. The irregular contours of the seabed, which shoals rapidly from depths of about 100m some 2M S of the Bill to as little as 10m on Portland Ledge 1M further N, greatly contribute to the violence of the Race. Portland Ledge strongly deflects the flow of water upwards, so that on the flood the Race lies SE of the Bill and SW on the ebb. Conditions deteriorate with wind-against-tide, especially at springs. In an E'ly gale against the flood stream the Race may spread eastward to The Shambles bank. The Race normally extends about 2M S of the Bill, but further S in bad weather.

The Race can be avoided by passing to seaward of it, ie 3–5M S of the Bill; or by using the inshore passage if conditions suit.

The **inshore passage** offers relatively smooth water between 1ca and 3ca off the Bill (depending on wind). Do not use it in onshore winds >F4/5, nor at night under power due to lobster pot floats and especially not at springs with wind against tide. From W or E, start close inshore at least 2M N of the Bill to utilise the S-going stream; hug the Bill to avoid being set into the Race.

▶ *The tidal stream chartlets at 1.24 which merit careful study show the approx hourly positions of the Race. They are referred to HW Plymouth, to HW Portland for those bound from/to Weymouth, and to HW Dover if passing S of the Race. See also the smaller scale chartlets at 9.1.3 and 9.2.3 and tidal diamonds 'A' on AC 2610 or 'R' on AC 2255.* ◀

Timing is vital to catch 'slackish' water in the inshore passage and to enjoy a fair tide thereafter. The windows are as follows:

▶ **Westbound**: *HW Dover –1 to HW +2; HW Plymouth +5 to HW –5; or HW Portland +4 to HW –6. Timing is easier if starting from Portland Harbour, Weymouth or Lulworth Cove.*

Eastbound: *HW Dover +5 to HW –4; HW Plymouth –2 to HW +2; or HW Portland –3 to HW +1.* ◀

PORTLAND BILL TO ANVIL POINT

The buoyed Shambles bank lies 2–5M E of Portland Bill and is best avoided, as is the gap between it and the Race (AC 2610). In bad weather the sea breaks heavily on it. E of **Weymouth** rocky ledges extend 3ca offshore as far as Lulworth Cove, which provides a reasonable ⚓ in fine, settled weather and offshore winds; as do Worbarrow Bay and Chapman's Pool further E.

A firing range extends 5M offshore between Lulworth Cove and St Alban's Head. Yachts should avoid the range when it is active or pass through as quickly as possible. Beware Kimmeridge Ledges which extend over 5ca seaward. Three rarely used yellow naval target buoys (DZ A, B and C) lie on St Alban's Ledge.

St Alban's Head (107m) is bold and steep-to.

A sometimes vicious race forms over **St Alban's Ledge**, a rocky dorsal ridge (least depth 8·5m) which extends 4M SW from St Alban's Hd.

▶ *1M S of St Alban's Head the E-going flood starts at HW Plymouth –1, and the W-going ebb at HWP +0525; inshore max Sp rates 4¾kn; see also diamond 'U' on AC 2610. Slack water lasts barely 30 mins. The race moves E on the flood and W on the ebb, which is more dangerous since overfalls extend 2½M further SW than on the flood. A back eddy sets almost continuously SSE down the W side of the Head.* ◀

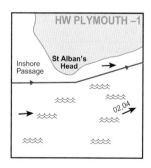
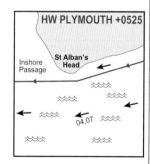

In settled weather and at neaps the race may be barely perceptible and can safely be crossed. Otherwise avoid it by passing to seaward via 50°31'·40N 02°07'·80W (thus clearing Lulworth Range); or by using the inshore passage very close inshore below St Alban's Head. This is at most 5ca wide and avoids the worst of the overfalls, but expect your decks to get wet. In onshore gales stay offshore. The light on St Alban's Hd Iso R 2s is only shown when the ranges are active.

The rugged, cliffy coastline between St Alban's Hd and Anvil Point (Fl 10s 45m 9M) is steep-to quite close inshore. Measured mile beacons stand either side of Anvil Pt lt ho; the track is 083·5°.

1.24 PORTLAND TIDAL STREAMS

More UK/Irish tides at www.reedstides.co.uk

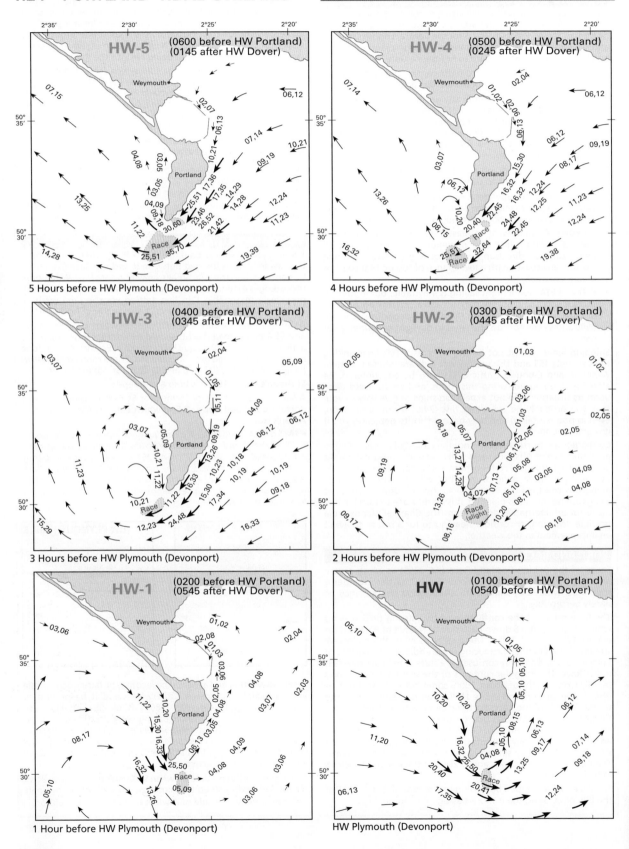

5 Hours before HW Plymouth (Devonport)

4 Hours before HW Plymouth (Devonport)

3 Hours before HW Plymouth (Devonport)

2 Hours before HW Plymouth (Devonport)

1 Hour before HW Plymouth (Devonport)

HW Plymouth (Devonport)

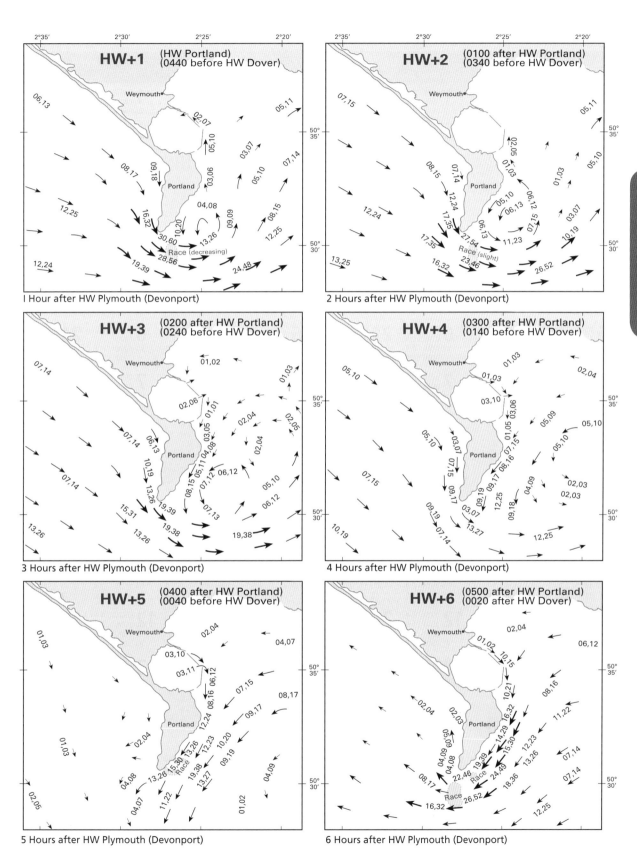

SW England

STANDARD TIME (UT)
For Summer Time add ONE hour in **non-shaded areas**

PORTLAND   LAT 50°34'N   LONG 2°26'W
TIMES AND HEIGHTS OF HIGH AND LOW WATERS

Dates in red are **SPRINGS**
Dates in blue are **NEAPS**

YEAR **2019**

JANUARY

Day	Time	m	Time	m	Time	m	Time	m
1	0242	1.6	0718	0.9	TU 1458	1.6	1956	0.7
16	0106	1.4	0603	0.8	W 1335	1.4	1847	0.6
2	0340	1.7	0827	0.8	W 1559	1.6	2053	0.6
17	0232	1.5	0717	0.7	TH 1500	1.5	2002	0.6
3	0433	1.8	0923	0.7	TH 1656	1.7	2143	0.6
18	0339	1.7	0835	0.7	F 1610	1.6	2114	0.5
4	0523	1.9	1013	0.6	F 1746	1.8	2229	0.5
19	0440	1.9	0944	0.6	SA 1714	1.8	2214	0.4
5	0606	2.0	1059	0.5	SA 1830	1.9	2313	0.4
20	0535	2.1	1044	0.5	SU 1812	1.9	2308	0.3
6	0646	2.1	1141	0.4	SU 1909	1.9	●2353	0.4
21	0625	2.2	1137	0.3	M 1908	2.0	○2358	0.2
7	0721	2.1	1221	0.4	M 1943	1.9		
22	0715	2.3	1225	0.2	TU 2005	2.1		
8	0030	0.4	0755	2.1	TU 1258	0.4	2016	1.9
23	0043	0.2	0807	2.4	W 1310	0.1	2059	2.2
9	0103	0.4	0829	2.0	W 1332	0.4	2049	1.8
24	0127	0.1	0857	2.4	TH 1355	0.1	2140	2.1
10	0133	0.5	0903	1.9	TH 1401	0.4	2121	1.7
25	0210	0.2	0943	2.3	F 1439	0.2	2217	2.0
11	0200	0.5	0933	1.8	F 1430	0.5	2150	1.6
26	0253	0.3	1024	2.1	SA 1524	0.3	2255	1.8
12	0229	0.5	1000	1.7	SA 1503	0.5	2221	1.5
27	0336	0.5	1105	1.9	SU 1610	0.6	◑2338	1.6
13	0303	0.6	1031	1.6	SU 1545	0.5	2259	1.5
28	0422	0.6	1151	1.7	M 1659	0.6		
14	0348	0.6	1115	1.5	M 1637	0.5	◑2352	1.4
29	0036	1.5	0518	0.8	TU 1257	1.5	1759	0.7
15	0452	0.7	1215	1.4	TU 1738	0.6		
30	0155	1.4	0633	0.9	W 1421	1.4	1911	0.8
31	0307	1.5	0752	0.9	TH 1532	1.4	2018	0.7

FEBRUARY

Day	Time	m	Time	m	Time	m	Time	m
1	0408	1.6	0856	0.8	F 1634	1.5	2114	0.7
16	0309	1.5	0820	0.7	SA 1551	1.5	2102	0.5
2	0500	1.7	0951	0.7	SA 1727	1.6	2206	0.6
17	0417	1.7	0941	0.5	SU 1659	1.7	2205	0.4
3	0546	1.9	1041	0.5	SU 1813	1.7	2255	0.5
18	0517	2.0	1037	0.4	M 1802	1.9	2257	0.3
4	0628	2.0	1126	0.4	M 1854	1.8	●2338	0.4
19	0612	2.2	1127	0.2	TU 1902	2.0	○2345	0.1
5	0707	2.1	1208	0.3	TU 1932	1.9		
20	0705	2.3	1213	0.1	W 2003	2.2		
6	0016	0.3	0744	2.1	W 1244	0.3	2006	1.9
21	0029	0.0	0758	2.4	TH 1255	0.0	2050	2.3
7	0050	0.3	0819	2.0	TH 1316	0.3	2037	1.9
22	0111	0.0	0848	2.4	F 1337	0.0	2125	2.2
8	0120	0.3	0850	2.0	F 1344	0.3	2104	1.8
23	0151	0.0	0929	2.3	SA 1416	0.1	2157	2.1
9	0148	0.3	0916	1.8	SA 1411	0.3	2127	1.7
24	0229	0.2	1004	2.1	SU 1455	0.2	2226	1.9
10	0216	0.3	0936	1.7	SU 1441	0.3	2149	1.6
25	0305	0.3	1035	1.9	M 1532	0.4	2253	1.7
11	0245	0.4	1002	1.7	M 1516	0.3	2222	1.6
26	0342	0.5	1108	1.6	TU 1610	0.6	◑2326	1.5
12	0318	0.4	1041	1.6	TU 1559	0.4	◑2308	1.5
27	0422	0.7	1150	1.4	W 1655	0.7		
13	0406	0.5	1134	1.4	W 1656	0.5		
28	0016	1.3	0543	0.9	TH 1318	1.2	1831	0.8
14	0008	1.4	0520	0.6	TH 1245	1.4	1806	0.6
15	0138	1.4	0639	0.7	F 1431	1.4	1928	0.6

MARCH

Day	Time	m	Time	m	Time	m	Time	m
1	0232	1.3	0727	0.9	F 1510	1.2	1948	0.8
16	0059	1.4	0615	0.7	SA 1420	1.3	1913	0.7
2	0346	1.4	0831	0.8	SA 1616	1.3	2047	0.7
17	0251	1.5	1114	0.6	SU 1540	1.4	2056	0.6
3	0440	1.6	0925	0.6	SU 1708	1.5	2141	0.6
18	0401	1.7	0939	0.5	M 1647	1.6	2153	0.4
4	0525	1.8	1015	0.5	M 1753	1.7	2231	0.4
19	0502	1.9	1026	0.3	TU 1749	1.9	2242	0.2
5	0608	2.0	1103	0.3	TU 1835	1.8	2317	0.1
20	0558	2.1	1112	0.1	W 1846	2.1	2329	0.1
6	0648	2.0	1146	0.2	W 1914	1.9	●2357	0.2
21	0651	2.3	1155	0.0	TH 1940	2.2	○	
7	0727	2.1	1224	0.2	TH 1950	2.0		
22	0011	0.0	0743	2.4	F 1235	-0.1	2024	2.3
8	0031	0.2	0803	2.1	F 1254	0.2	2022	1.9
23	0051	0.0	0829	2.4	SA 1313	-0.1	2101	2.3
9	0100	0.2	0835	2.0	SA 1321	0.2	2047	1.9
24	0127	0.0	0907	2.3	SU 1349	0.0	2131	2.1
10	0128	0.2	0856	1.9	SU 1348	0.2	2102	1.8
25	0202	0.1	0936	2.1	M 1423	0.2	2153	1.9
11	0157	0.2	0911	1.8	M 1418	0.2	2122	1.7
26	0235	0.3	1002	1.8	TU 1455	0.4	2215	1.7
12	0226	0.3	0938	1.7	TU 1450	0.2	2154	1.7
27	0308	0.5	1034	1.6	W 1524	0.6	2247	1.5
13	0255	0.3	1017	1.6	W 1526	0.4	2238	1.6
28	0341	0.7	1113	1.4	TH 1549	0.7	◑2328	1.4
14	0334	0.4	1109	1.4	TH 1619	0.5	◑2335	1.4
29	0428	0.8	1207	1.2	F 1625	0.9		
15	0443	0.6	1220	1.3	F 1734	0.6		
30	0031	1.2	0701	0.8	SA 1445	1.1	1918	0.8
31	0318	1.3	0802	0.7	SU 1555	1.3	2018	0.8

APRIL

Day	Time	m	Time	m	Time	m	Time	m
1	0413	1.5	0855	0.6	M 1643	1.5	2111	0.6
16	0344	1.7	0926	0.4	TU 1630	1.7	2134	0.4
2	0458	1.7	0945	0.4	TU 1726	1.6	2201	0.5
17	0444	1.9	1008	0.2	W 1728	1.9	2222	0.3
3	0539	1.8	1033	0.3	W 1807	1.8	2249	0.4
18	0540	2.1	1051	0.1	TH 1821	2.1	2308	0.1
4	0621	2.0	1117	0.2	TH 1847	1.9	2330	0.3
19	0631	2.2	1133	0.0	F 1910	2.2	○2350	0.1
5	0702	2.0	1155	0.2	F 1926	2.0	●	
20	0720	2.3	1212	0.0	SA 1954	2.3		
6	0005	0.2	0741	2.1	SA 1226	0.1	2001	2.0
21	0027	0.0	0804	2.3	SU 1246	0.0	2032	2.2
7	0036	0.2	0813	2.0	SU 1254	0.1	2026	2.0
22	0102	0.1	0838	2.2	M 1320	0.1	2100	2.1
8	0106	0.2	0831	2.0	M 1324	0.2	2038	1.9
23	0135	0.2	0905	2.0	TU 1352	0.3	2122	1.9
9	0137	0.2	0849	1.9	TU 1356	0.2	2101	1.9
24	0209	0.3	0934	1.8	W 1423	0.4	2149	1.8
10	0208	0.2	0921	1.8	W 1430	0.3	2136	1.8
25	0242	0.5	1008	1.6	TH 1450	0.6	2222	1.6
11	0242	0.3	1004	1.6	TH 1507	0.4	2221	1.6
26	0316	0.7	1049	1.4	F 1512	0.7	◑2303	1.4
12	0324	0.4	1059	1.5	F 1600	0.6	◑2320	1.5
27	0359	0.8	1139	1.2	SA 1552	0.9	2356	1.3
13	0434	0.6	1216	1.3	SA 1719	0.7		
28	0630	0.8	1257	1.1	SU 1843	0.9		
14	0045	1.4	0610	0.7	SU 1414	1.3	1909	0.7
29	0131	1.3	0730	0.7	M 1514	1.3	1945	0.8
15	0235	1.5	1053	0.5	M 1529	1.5	2039	0.6
30	0320	1.4	0823	0.6	TU 1603	1.4	2039	0.7

Chart Datum is 0·93 metres below Ordnance Datum (Newlyn). HAT is 2·5 metres above Chart Datum.

》FREE monthly updates. Register at 《
www.reedsnauticalalmanac.co.uk

STANDARD TIME (UT)
For Summer Time add ONE hour in **non-shaded areas**

PORTLAND   LAT 50°34'N   LONG 2°26'W
TIMES AND HEIGHTS OF HIGH AND LOW WATERS

Dates in red are **SPRINGS**
Dates in blue are NEAPS

YEAR **2019**

MAY

Day	Time m				Day	Time m			
1 W	0414 1.6	0912 0.4	1647 1.6	2128 0.5	16 TH	0420 1.8	0943 0.3	1701 1.9	2158 0.4
2 TH	0501 1.7	1000 0.3	1731 1.8	2215 0.4	17 F	0516 2.0	1026 0.2	1753 2.1	2245 0.3
3 F	0546 1.9	1043 0.2	1814 1.9	2258 0.4	18 SA ○	0607 2.1	1107 0.2	1840 2.2	2326 0.2
4 SA ●	0631 2.0	1121 0.2	1855 2.0	2335 0.3	19 SU	0655 2.1	1144 0.1	1923 2.2	
5 SU	0711 2.0	1155 0.2	1932 2.1		20 M	0003 0.2	0736 2.1	1219 0.2	2001 2.2
6 M	0008 0.2	0746 2.0	1227 0.2	2000 2.1	21 TU	0038 0.2	0809 2.0	1253 0.2	2030 2.1
7 TU	0043 0.2	0806 2.0	1302 0.2	2018 2.0	22 W	0114 0.3	0839 1.9	1328 0.3	2058 2.0
8 W	0119 0.2	0833 1.9	1339 0.2	2049 2.0	23 TH	0150 0.4	0912 1.7	1400 0.5	2130 1.8
9 TH	0157 0.3	0913 1.8	1418 0.3	2130 1.9	24 F	0226 0.5	0949 1.6	1428 0.6	2205 1.7
10 F	0239 0.3	1001 1.6	1503 0.6	2219 1.7	25 SA	0301 0.6	1029 1.4	1453 0.7	2245 1.5
11 SA	0331 0.5	1100 1.5	1601 0.6	2318 1.6	26 SU ☽	0342 0.7	1115 1.3	1534 0.8	2330 1.4
12 SU ☽	0439 0.6	1221 1.4	1713 0.7		27 M	0551 0.7	1212 1.2	1638 0.9	
13 M	0038 1.5	0603 0.6	1401 1.4	1838 0.7	28 TU	0030 1.4	0656 0.7	1337 1.3	1902 0.9
14 TU	0214 1.6	1039 0.5	1509 1.5	2006 0.7	29 W	0157 1.4	0750 0.6	1459 1.4	2001 0.8
15 W	0321 1.7	0901 0.4	1607 1.7	2107 0.5	30 TH	0312 1.5	0838 0.5	1557 1.6	2050 0.7
					31 F	0411 1.6	0923 0.4	1648 1.7	2136 0.6

JUNE

Day	Time m				Day	Time m			
1 SA	0505 1.8	1004 0.3	1736 1.9	2220 0.5	16 SU	0543 1.9	1039 0.3	1812 2.1	2305 0.4
2 SU	0554 1.9	1044 0.3	1820 2.0	2302 0.4	17 M ○	0631 1.9	1119 0.3	1856 2.1	2345 0.4
3 M ●	0638 2.0	1124 0.2	1900 2.1	2344 0.3	18 TU	0713 1.9	1157 0.3	1934 2.1	
4 TU	0715 2.0	1205 0.2	1933 2.2		19 W	0022 0.3	0749 1.9	1234 0.3	2008 2.1
5 W	0026 0.3	0749 2.0	1246 0.2	2006 2.1	20 TH	0100 0.3	0822 1.8	1310 0.4	2041 2.0
6 TH	0108 0.3	0827 1.9	1328 0.2	2045 2.1	21 F	0138 0.4	0856 1.7	1345 0.4	2115 1.9
7 F	0152 0.3	0913 1.8	1413 0.3	2130 2.0	22 SA	0214 0.5	0932 1.6	1415 0.5	2149 1.7
8 SA	0240 0.3	1005 1.7	1501 0.4	2220 1.9	23 SU	0248 0.5	1008 1.5	1442 0.6	2225 1.6
9 SU	0333 0.4	1102 1.6	1556 0.5	2315 1.8	24 M	0320 0.6	1047 1.4	1518 0.7	2301 1.5
10 M ☽	0432 0.5	1213 1.5	1657 0.6		25 TU ☽	0401 0.6	1132 1.3	1607 0.7	2344 1.5
11 TU	0022 1.6	0538 0.5	1334 1.5	1803 0.7	26 W	0455 0.6	1229 1.3	1709 0.8	
12 W	0142 1.6	0652 0.6	1441 1.6	1918 0.7	27 TH	0042 1.4	0600 0.6	1347 1.4	1820 0.8
13 TH	0251 1.6	0814 0.5	1538 1.7	2031 0.7	28 F	0159 1.4	0727 0.6	1458 1.5	1938 0.7
14 F	0351 1.7	0909 0.4	1632 1.8	2130 0.6	29 SA	0311 1.5	0830 0.6	1559 1.7	2046 0.6
15 SA	0449 1.8	0955 0.4	1724 2.0	2220 0.5	30 SU	0415 1.6	0920 0.4	1653 1.8	2143 0.5

JULY

Day	Time m				Day	Time m			
1 M	0514 1.8	1011 0.3	1743 2.0	2237 0.5	16 TU ○	0612 1.8	1059 0.4	1833 2.0	2333 0.4
2 TU ●	0606 1.9	1101 0.3	1829 2.1	2328 0.4	17 W	0656 1.8	1141 0.3	1914 2.1	
3 W	0655 2.0	1149 0.2	1913 2.2		18 TH	0012 0.3	0735 1.8	1219 0.3	1951 2.1
4 TH	0015 0.3	0743 2.0	1235 0.2	1959 2.2	19 F	0050 0.3	0809 1.8	1256 0.3	2026 2.0
5 F	0102 0.2	0833 2.0	1320 0.2	2045 2.2	20 SA	0126 0.3	0842 1.8	1330 0.4	2100 1.9
6 SA	0148 0.2	0923 2.0	1406 0.2	2132 2.2	21 SU	0159 0.3	0914 1.7	1400 0.4	2132 1.8
7 SU	0234 0.2	1009 1.9	1452 0.3	2217 2.1	22 M	0228 0.4	0945 1.6	1427 0.5	2201 1.7
8 M	0323 0.3	1056 1.8	1541 0.4	2304 1.9	23 TU	0254 0.4	1016 1.5	1457 0.5	2228 1.6
9 TU ☽	0414 0.4	1150 1.6	1632 0.6	2358 1.7	24 W	0328 0.5	1049 1.4	1537 0.6	2301 1.5
10 W	0508 0.5	1256 1.5	1730 0.7		25 TH ☽	0412 0.5	1134 1.4	1629 0.6	2350 1.4
11 TH	0105 1.6	0609 0.6	1405 1.5	1836 0.8	26 F	0508 0.5	1234 1.4	1734 0.7	
12 F	0218 1.5	0718 0.6	1508 1.6	1951 0.8	27 SA	0054 1.4	0613 0.6	1356 1.4	1844 0.7
13 SA	0324 1.5	0827 0.6	1606 1.7	2102 0.7	28 SU	0218 1.4	0725 0.5	1512 1.6	2002 0.7
14 SU	0426 1.6	0924 0.6	1700 1.8	2200 0.6	29 M	0337 1.5	0844 0.5	1616 1.8	2121 0.6
15 M	0523 1.7	1013 0.5	1749 1.9	2250 0.5	30 TU	0446 1.7	0950 0.4	1714 1.9	2224 0.5
					31 W	0547 1.8	1046 0.3	1807 2.1	2318 0.3

AUGUST

Day	Time m				Day	Time m			
1 TH ●	0642 1.9	1137 0.2	1858 2.3		16 F	0718 1.9	1203 0.3	1933 2.1	
2 F	0007 0.2	0740 2.1	1224 0.1	1949 2.3	17 SA	0034 0.2	0754 1.9	1240 0.3	2009 2.1
3 SA	0052 0.1	0840 2.1	1308 0.1	2039 2.4	18 SU	0108 0.3	0825 1.9	1312 0.3	2042 2.0
4 SU	0136 0.1	0926 2.1	1352 0.1	2126 2.3	19 M	0137 0.3	0854 1.8	1340 0.3	2111 1.9
5 M	0220 0.1	1003 2.0	1435 0.2	2207 2.2	20 TU	0202 0.3	0919 1.7	1406 0.4	2133 1.8
6 TU	0303 0.2	1039 1.9	1518 0.3	2246 2.0	21 W	0227 0.3	0940 1.6	1433 0.4	2152 1.7
7 W	0347 0.3	1123 1.7	1603 0.5	2329 1.8	22 TH	0255 0.4	1007 1.6	1504 0.5	2223 1.6
8 TH	0434 0.5	1212 1.5	1654 0.7		23 F ☽	0332 0.4	1047 1.5	1545 0.6	2309 1.5
9 F	0024 1.7	0529 0.6	1326 1.4	1801 0.8	24 SA	0424 0.5	1142 1.4	1652 0.7	
10 SA	0146 1.4	0639 0.8	1441 1.4	1924 0.9	25 SU	0011 1.4	0533 0.6	1259 1.4	1809 0.7
11 SU	0304 1.4	0755 0.8	1545 1.6	2040 0.8	26 M	0145 1.3	0650 0.6	1441 1.5	1943 0.7
12 M	0411 1.4	0857 0.7	1641 1.7	2138 0.7	27 TU	0321 1.4	0833 0.6	1553 1.7	2120 0.6
13 TU	0507 1.6	0950 0.6	1730 1.9	2229 0.5	28 W	0432 1.6	0939 0.5	1654 1.9	2216 0.4
14 W	0555 1.7	1039 0.5	1814 2.0	2315 0.4	29 TH	0533 1.8	1034 0.3	1750 2.1	2306 0.3
15 TH ○	0639 1.8	1123 0.4	1855 2.1	2356 0.3	30 F ●	0630 2.0	1123 0.2	1842 2.3	2353 0.1
					31 SA	0728 2.2	1209 0.1	1933 2.4	

Chart Datum is 0·93 metres below Ordnance Datum (Newlyn). HAT is 2·5 metres above Chart Datum.

SW England

》》FREE monthly updates. Register at 《
www.reedsnauticalalmanac.co.uk

73

STANDARD TIME (UT)
For Summer Time add ONE hour in **non-shaded areas**

PORTLAND LAT 50°34'N LONG 2°26'W
TIMES AND HEIGHTS OF HIGH AND LOW WATERS

Dates in red are **SPRINGS**
Dates in blue are **NEAPS**

YEAR 2019

SEPTEMBER

Day	DoW	Time	m	Time	m	Time	m	Time	m
1	SU	0036	0.0	0825	2.3	1251	0.0	2025	2.5
2	M	0118	0.0	0907	2.3	1333	0.0	2110	2.4
3	TU	0158	0.0	0942	2.2	1412	0.1	2148	2.2
4	W	0236	0.2	1013	2.0	1450	0.3	2221	2.0
5	TH	0313	0.4	1042	1.8	1529	0.5	2255	1.7
6	F	0351	0.6	1115	1.6	1613	0.7	☽2337	1.5
7	SA	0435	0.8	1209	1.4	1728	0.9		
8	SU	0108	1.3	0604	0.9	1416	1.3	1906	0.9
9	M	0252	1.3	0728	0.9	1529	1.5	2014	0.8
10	TU	0400	1.4	0829	0.8	1624	1.6	2109	0.7
11	W	0451	1.6	0923	0.7	1710	1.8	2159	0.5
12	TH	0536	1.7	1013	0.5	1752	2.0	2247	0.4
13	F	0617	1.9	1059	0.4	1833	2.1	2331	0.2
14	SA	0656	2.0	1141	0.3	1911	2.1	○	
15	SU	0010	0.2	0732	2.0	1218	0.2	1948	2.1
16	M	0043	0.2	0803	2.0	1249	0.3	2020	2.1
17	TU	0109	0.2	0911	2.0	1316	0.3	2046	2.0
18	W	0133	0.3	0849	1.9	1342	0.3	2102	1.8
19	TH	0158	0.3	0904	1.8	1408	0.4	2119	1.7
20	F	0224	0.4	0931	1.7	1435	0.4	2152	1.6
21	SA	0253	0.5	1011	1.6	1507	0.5	2240	1.5
22	SU	0334	0.6	1105	1.5	1611	0.7	☽2345	1.3
23	M	0456	0.7	1222	1.4	1744	0.8		
24	TU	0140	1.3	0629	0.8	1421	1.5	2214	0.7
25	W	0314	1.5	0832	0.7	1536	1.7	2117	0.5
26	TH	0421	1.7	0926	0.5	1637	1.9	2202	0.4
27	F	0520	1.9	1017	0.4	1732	2.1	2249	0.2
28	SA	0614	2.1	1104	0.2	1824	2.3	●2333	0.1
29	SU	0705	2.3	1149	0.1	1914	2.4		
30	M	0015	0.0	0754	2.4	1230	0.1	2002	2.5

OCTOBER

Day	DoW	Time	m	Time	m	Time	m	Time	m
1	TU	0054	0.0	0836	2.4	1309	0.1	2045	2.4
2	W	0130	0.1	0911	2.2	1345	0.2	2120	2.2
3	TH	0204	0.3	0937	2.1	1420	0.4	2149	2.0
4	F	0236	0.5	1000	1.8	1456	0.6	2221	1.7
5	SA	0307	0.7	1032	1.6	1537	0.8	☽2301	1.4
6	SU	0335	0.9	1117	1.4	1711	0.9	2359	1.2
7	M	0353	1.0	1232	1.3	1841	0.9		
8	TU	0244	1.2	0658	1.0	1512	1.4	1942	0.8
9	W	0347	1.4	0759	0.9	1602	1.6	2036	0.7
10	TH	0431	1.6	0852	0.8	1645	1.8	2125	0.5
11	F	0511	1.8	0942	0.6	1726	1.9	2213	0.4
12	SA	0550	1.9	1029	0.5	1805	2.0	2259	0.3
13	SU	0627	2.1	1113	0.4	1844	2.1	○2339	0.2
14	M	0703	2.1	1150	0.3	1921	2.1		
15	TU	0011	0.2	0736	2.1	1221	0.3	1955	2.1
16	W	0038	0.3	0802	2.1	1249	0.3	2018	2.0
17	TH	0104	0.3	0817	2.0	1317	0.3	2032	1.9
18	F	0132	0.4	0836	1.9	1347	0.4	2057	1.8
19	SA	0201	0.4	0904	1.8	1418	0.5	2135	1.7
20	SU	0230	0.5	0949	1.7	1453	0.6	2226	1.5
21	M	0306	0.7	1045	1.6	1559	0.7	☽2337	1.3
22	TU	0434	0.8	1204	1.5	2030	0.8		
23	W	0140	1.3	0617	0.9	1403	1.5	2148	0.7
24	TH	0304	1.5	0814	0.8	1518	1.7	2102	0.5
25	F	0405	1.7	0907	0.6	1617	1.9	2143	0.4
26	SA	0500	2.0	0956	0.5	1712	2.1	2226	0.2
27	SU	0550	2.2	1043	0.3	1803	2.3	2309	0.1
28	M	0638	2.3	1127	0.2	1851	2.4	●2350	0.1
29	TU	0722	2.4	1207	0.2	1936	2.4		
30	W	0026	0.1	0802	2.4	1243	0.2	2015	2.3
31	TH	0100	0.2	0834	2.3	1318	0.3	2048	2.1

NOVEMBER

Day	DoW	Time	m	Time	m	Time	m	Time	m
1	F	0133	0.4	0901	2.1	1354	0.4	2120	1.9
2	SA	0206	0.6	0931	1.9	1431	0.6	2155	1.7
3	SU	0235	0.8	1006	1.7	1514	0.8	2237	1.5
4	M	0255	0.9	1050	1.5	1649	0.9	◐2328	1.3
5	TU	0315	0.9	1146	1.4	1809	0.9		
6	W	0059	1.2	0622	1.1	1419	1.4	1908	0.8
7	TH	0320	1.4	0725	1.0	1524	1.5	2002	0.7
8	F	0359	1.6	0819	0.9	1607	1.7	2051	0.5
9	SA	0436	1.7	0908	0.7	1649	1.8	2138	0.4
10	SU	0514	1.9	0955	0.6	1731	1.9	2223	0.4
11	M	0553	2.1	1039	0.5	1812	2.0	2304	0.3
12	TU	0631	2.2	1117	0.5	1852	2.1	○2336	0.3
13	W	0705	2.2	1151	0.5	1927	2.1		
14	TH	0007	0.3	0732	2.2	1223	0.4	1951	2.0
15	F	0039	0.3	0752	2.1	1258	0.4	2014	1.9
16	SA	0113	0.4	0821	2.1	1334	0.4	2047	1.8
17	SU	0149	0.5	0858	2.0	1414	0.5	2131	1.7
18	M	0227	0.6	0944	1.8	1501	0.6	2226	1.6
19	TU	0317	0.7	1042	1.7	1606	0.7	◑2335	1.4
20	W	0431	0.9	1154	1.6	2013	0.7		
21	TH	0122	1.4	0554	0.9	1334	1.6	2138	0.6
22	F	0242	1.6	0729	0.8	1451	1.7	2035	0.6
23	SA	0340	1.8	0840	0.7	1551	1.8	2115	0.4
24	SU	0433	2.0	0931	0.6	1647	2.0	2159	0.4
25	M	0523	2.1	1019	0.5	1740	2.1	2242	0.3
26	TU	0610	2.3	1103	0.4	1828	2.2	●2323	0.3
27	W	0653	2.3	1143	0.4	1912	2.2		
28	TH	0000	0.3	0732	2.3	1221	0.3	1951	2.1
29	F	0035	0.3	0806	2.2	1258	0.4	2026	2.0
30	SA	0111	0.5	0838	2.1	1336	0.5	2100	1.9

DECEMBER

Day	DoW	Time	m	Time	m	Time	m	Time	m
1	SU	0145	0.6	0913	2.0	1416	0.6	2137	1.7
2	M	0217	0.7	0950	1.8	1500	0.7	2217	1.5
3	TU	0240	0.8	1030	1.6	1600	0.8	2300	1.4
4	W	0307	0.9	1116	1.5	1730	0.8	◐2355	1.3
5	TH	0401	1.0	1215	1.4	1832	0.8		
6	F	0117	1.3	0643	1.0	1342	1.4	1927	0.7
7	SA	0247	1.4	0743	0.9	1459	1.5	2017	0.6
8	SU	0342	1.6	0833	0.8	1557	1.6	2103	0.5
9	M	0429	1.8	0919	0.7	1649	1.8	2146	0.5
10	TU	0514	2.0	1002	0.6	1737	1.9	2226	0.4
11	W	0556	2.1	1043	0.6	1821	2.0	2304	0.4
12	TH	0634	2.2	1124	0.5	1901	2.0	○2343	0.4
13	F	0707	2.2	1205	0.4	1934	2.0		
14	SA	0022	0.3	0739	2.2	1246	0.4	2008	2.0
15	SU	0103	0.4	0816	2.2	1329	0.4	2048	1.9
16	M	0146	0.4	0858	2.1	1415	0.4	2134	1.8
17	TU	0231	0.5	0946	2.0	1505	0.4	2225	1.7
18	W	0321	0.6	1039	1.8	1600	0.5	2325	1.6
19	TH	0419	0.7	1140	1.7	1701	0.7	◑	
20	F	0044	1.5	0524	0.8	1257	1.6	2040	0.7
21	SA	0204	1.5	0636	0.8	1415	1.6	1926	0.6
22	SU	0306	1.7	0757	0.8	1520	1.7	2038	0.6
23	M	0402	1.8	0905	0.7	1620	1.8	2130	0.5
24	TU	0456	2.0	0958	0.6	1718	1.9	2217	0.5
25	W	0545	2.1	1046	0.5	1810	1.9	2301	0.4
26	TH	0631	2.2	1129	0.5	1856	2.0	●2341	0.4
27	F	0712	2.2	1208	0.4	1937	2.0		
28	SA	0019	0.4	0750	2.2	1246	0.4	2013	1.9
29	SU	0056	0.4	0825	2.1	1325	0.4	2047	1.8
30	M	0132	0.5	0900	2.0	1405	0.5	2121	1.7
31	TU	0205	0.6	0935	1.9	1443	0.5	2156	1.6

Chart Datum is 0·93 metres below Ordnance Datum (Newlyn). HAT is 2·5 metres above Chart Datum.

》〉 FREE monthly updates. Register at 〈《
www.reedsnauticalalmanac.co.uk

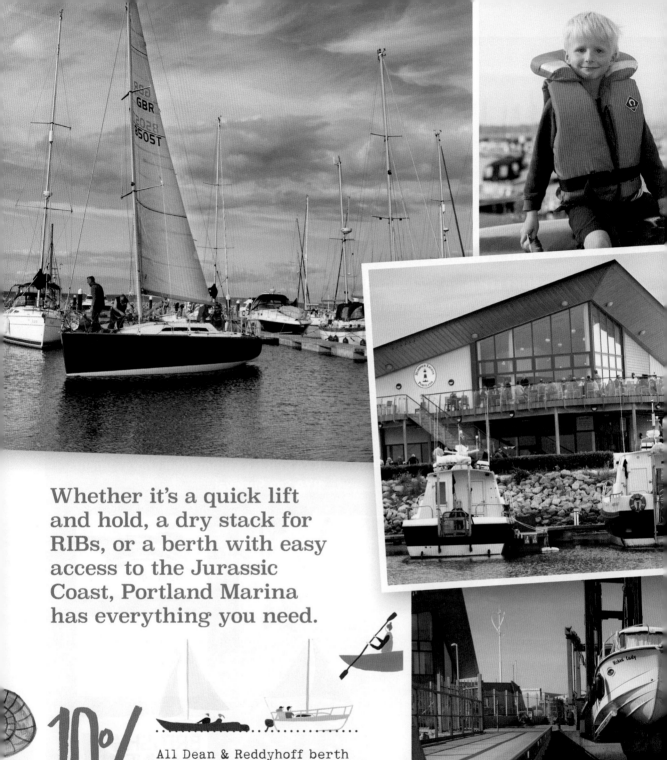

Whether it's a quick lift and hold, a dry stack for RIBs, or a berth with easy access to the Jurassic Coast, Portland Marina has everything you need.

10%

All Dean & Reddyhoff berth holders get a **10%** discount on lifting all year round.

With marinas in **Hampshire**, **Dorset** and the **Isle of Wight**, we're only a short sail away.

03454 30 2012

portland marina

deanreddyhoff.co.uk

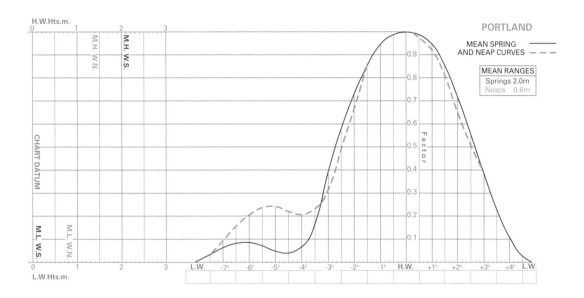

1.25 PORTLAND HARBOUR

Dorset **50°35´·72N 02°25´·91W** (N Ship Chan) ❀❀❀♨♨♨❀

CHARTS AC 2610, 2255, 2268, 5601; Imray C10, C12, C5, C4, 2300

TIDES –0430 Dover; ML 1·0

Standard Port PORTLAND (→)

Times				Height (metres)			
High Water		Low Water		MHWS	MHWN	MLWN	MLWS
0100	0700	0100	0700	2·1	1·4	0·8	0·1
1300	1900	1300	1900				
LULWORTH COVE and MUPE BAY (Worbarrow Bay)							
+0005	+0015	–0005	0000	+0·1	+0·1	+0·2	+0·1

Double LWs occur between Portland and Lulworth Cove. Predictions are for the 1st LW. The 2nd LW occurs 3–4hrs later.

SHELTER The large harbour is exposed to winds from the E and NE, but in westerlies the moorings on the W and S sides are reasonably sheltered by Chesil Beach and the Isle of Portland. The marina is protected by substantial rock breakwaters. ⚓ on W side of hbr in about 3m, as on chartlet, or further N off Castle Cove SC.

NAVIGATION If approaching from the W inshore of Portland Race study the tidal stream diagrams carefully and note the advice in Passage Information.

WPT 50°35´·97N 02°25´·24W, 240°/0·5M to North Ship Channel (NSC). This is the recommended route to the marina via the Marina Access Route, keeping clear of shipping using East Ship Channel (ESC). Small craft (LOA<20m) may use ESC, but must monitor Ch 74 and obey IPTS lights shown on 'C' and Fort Heads. Small craft crossing ESC should remain well clear of the Heads (>0.5M) so that they may be seen by commercial vessels navigating within the channel. Vessels >20m must obtain permission from Portland Harbour Radio before entering or leaving the harbour.

South Ship Channel (SSC) is permanently closed and blocked by overhead wires. 7ca SE of SSC is a noise range, ⚓ marked by 4 Fl Y buoys. Vessel permanently moored is used for military exercises with an exclusion zone of 50m. There is a rocky reef extending about 1·5ca offshore NE of Castle Cove SC, and shoal water E of Small Mouth.

A 12kn speed limit applies throughout the harbour and in port limits, but vessels under 10m LOA are exempt. However, a 6kn speed limit applies to *all* vessels in the areas depicted and within 150m of any breakwater or harbour premises.

Approach to Weymouth & Portland National Sailing Academy (WPNSA) and Portland Marina. The Marina Access Route remains

the recommended route for small craft (<20m LOA) transiting the harbour. Engines must be used (if fitted) when proceeding between the marina/WPNSA and NSC. From NSC head SW to the SWM buoy (see chartlet) then S to the marina entrance, leaving the 2 PHM buoys to port. Obey traffic signals on 'C' Head and on the marina breakwater:

● Flashing = Entrance closed; await instructions (Ch 74 for
● Harbour Control; Ch 80 for Marina)
●

● Fixed = Commercial vessel/sailing fleet departing; do not
● impede
●

● Fixed = Commercial vessel/sailing fleet arriving; do not
● impede
●

LIGHTS AND MARKS The precipitous N face of the Verne, near the N end of the Isle of Portland, is conspic from afar. 'A' Hd lt ho is a conspic W twr, Fl 2·5s 22m 10M.

COMMUNICATIONS (Code 01305) Dorchester 🏥 824055; 🏥 (A&E) 820341. Port Control ☎824044.

Monitor Ch **74** *Portland Hbr Radio* for commercial traffic. *Portland Marina* Ch 80.

FACILITIES **Portland Port**, www.portland-port.co.uk owns the harbour which is a commercial port with an international bunkering station. Port and Harbour Dues (see web site) apply to all vessels in Portland Inner Harbour.

Castle Cove SC ☎783708, ⚓⚓⚓.

WPNSA The 3 pontoons for keel boats are protected by a rock breakwater. admin@wpnsa.org.uk www.wpnsa.org.uk ☎866000. Temporary pontoon berths may be available for 1 or 2 nights by prior arrangement.

RNSA moorings adjacent to marina access route and WPNSA breakwater are private moorings for members occupied Apr-Oct.

Portland Marina The entrance, 50°34´·44N 02°27´·32W, is between the NW end of the marina breakwater and the WPNSA breakwater. www.deanreddyhoff.co.uk ☎03454 302012; 600⚓+❂ berths on pontoons R, S ,T £3·20 inc ⚡ for <14m LOA, short stay <4hrs £8, ⚓ 🅿 ⛽ 🚢(inc bio-diesel) 🛠 🔌 ⚒ ⚓ 🏪 Ⓔ ⚓ 🏴(50t) ⚓ ✕ 🍴.

Castletown is SE of the marina. ✉ Ⓑ ✕ 🍴 501 bus to Weymouth ⇌ ✈ (Hurn).

Chartlet overleaf

PORTLAND HARBOUR *continued*

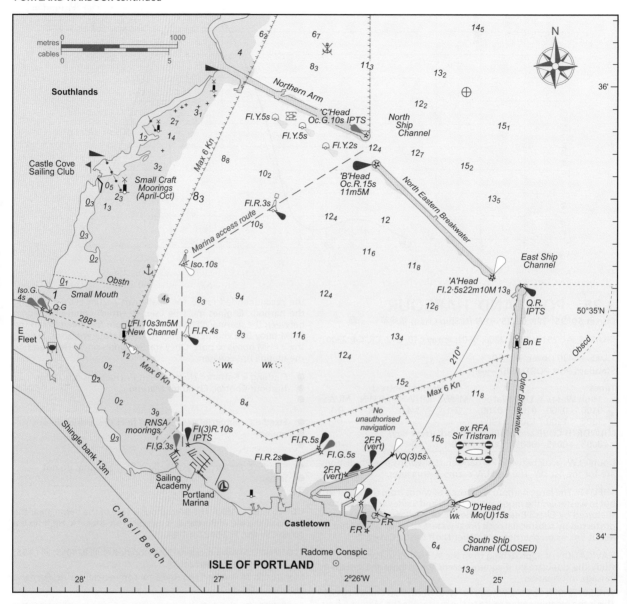

Southlands

Castle Cove
Sailing Club

Small Craft
Moorings
(April-Oct)

Max 6 Kn

Northern Arm

'C'Head
Oc.G.10s IPTS

Fl.Y.5s

Fl.Y.5s

Fl.Y.2s

North
Ship
Channel

'B'Head
Oc.R.15s
11m5M

North Eastern Breakwater

Marina access route

Fl.R.3s

Iso.10s

East Ship
Channel

'A'Head
Fl.2·5s22m10M

Q.R.
IPTS

50°35'N

Bn E

Obscd

Iso.G.
4s

Q.G

Small Mouth

288°

E
Fleet

LFl.10s3m5M
New Channel

Fl.R.4s

Max 6 Kn

Wk

Wk

210°

Max 6 Kn

Outer Breakwater

ex RFA
Sir Tristram

No
unauthorised
navigation

Shingle bank 13m

RNSA
moorings

Fl(3)R.10s
IPTS

Fl.G.3s

Fl.R.5s

Fl.R.2s

Fl.G.5s

2.F.R
(vert)

2.F.R
(vert)

VQ(3)5s

Q

Sailing
Academy
Portland
Marina

Chesil Beach

Castletown

F.R

F.R

'D'Head
Mo(U)15s

Wk

South Ship
Channel (CLOSED)

Radome Conspic

ISLE OF PORTLAND

2°26'W

1.26 WEYMOUTH

Dorset 50°36'·57N 02°26'·58W ✱✱✱❀♦♦❀✿✿✿

CHARTS AC 2610, 2255, 2268, 2172, 5601; Imray C10, C12, C5, C4, 2300

TIDES –0438 Dover; ML 1·1

Standard Port PORTLAND (◄——)

Use Portland predictions and Note. Mean ranges are small: 0·6m at np and 2·0m at sp. NOTE: A LW stand lasts about 4hrs at sp and 1 hr at nps. Due to an eddy, the tidal stream in Weymouth Roads is W-going at all times except for 2hrs, HW –0510 to HW –0310.

SHELTER Good, but swell enters the outer harbour and The Cove in strong E winds. ⚓ feasible in Weymouth Bay NNE of the ent in about 3·5m, 4ca off the beach to seaward of the buoyed bathing area. ❷ berths in The Cove have 1·7m at CD.

NAVIGATION WPT 50°36'·69N 02°26'·23W, 240°/2ca to South Pierhead. The harbour entrance is 1M NW of North Ship Channel into Portland Harbour. Reports indicate a bar may be forming.

When bound to/from the E, check whether Lulworth Firing Ranges are active. See 9.1.29 for further info. A DG Range, marked by 3 SPM buoys (one Fl Y 2s), lies 2·5ca SE of South Pier; ⚓. See ACs 2255 and 2268 for restrictions on ⚓ in Weymouth Bay.

Comply with traffic signals displayed from RW pole near the root of South Pier (see below). Visiting vessels call HM Ch12 before passing pier head for berth info. A rowed ferry has right of way across the fairway E of the LB (see chartlet). Speed limit 'Dead Slow'.

LIGHTS AND MARKS Traffic signals, shown *vertically* on S Pier:

No signals = Proceed with caution;

● ● ● = Large vessel leaving, do not obstruct the pierhead area;
● ● ● = Large vessel is entering, do not leave harbour;
● ● over ● = Harbour closed, do not enter the pierheads area;
● ○ ● = Only proceed on receipt of instructions;
● ● ● (flashing) = All vessels await instructions.

If in any doubt, call *Weymouth Harbour* on Ch 12.

Conspic ✠ spire, 6ca NNW of the harbour ent, is a useful daymark. Pierhead lights may be hard to see against shore lights. Weymouth Eye, a 53m twr on the hard NW of N Pier, is floodlit. Portland 'A' Head lt ho, a conspic W twr 1·7M SE of hbr is readily visible.

Ldg lts 239·6°, 2 FR (H24) and R open ◊s on W poles are 500m inside the pierheads; they are not visible until the S Pier is rounded.

LIFTING BRIDGE allows masted craft access to the marina.

Lifting times (LT):

15 Apr-15 Sep, 0800, 1000, 1200, 1400, 1600, 1800, 2000; **plus Jun, Jul, Aug,** 2100.
16 Sep-14 Apr: 0800, 1000, 1200, 1400, 1600, 1800.
NOTE: Minimum 1hr notice is required for lifts 16 Sep-14 Apr ☎838423 (ansafone) or Ch 12.

Be visible from bridge 5 mins before lift time; listen on Ch 12 for any broadcasts. 3FR or 3FG (vert) on both sides of the bridge are traffic lights and must be obeyed at all times, even when the bridge is closed. Clearance when closed is 2·75m above HAT. A 50m pontoon on the S bank immediately outside bridge is only for craft waiting for an inwards bridge lift and these must take the next available transit. Craft of 1·8m draught or greater should berth as the second boat out on the waiting pontoon at or near LWS or discuss alternative arrangements with the harbour staff – check before remaining o/night. Submerged cable inside bridge has least depth of 1·8m within 10m of pontoons.

COMMUNICATIONS (Code 01305) Dr 774411; 🏥 Dorchester 251150. HM ☎838423.
Weymouth Harbour and *Weymouth Town Bridge* (at opening times): Ch 12 (0730-1700 in winter and up to 0730-2100 in summer). *Weymouth Marina* Ch 80. Ch 60 for ⛽.

FACILITIES (www.weymouth-harbour.co.uk) **Outer Harbour** ⌓ The Cove (S side) on ❷ pontoons <10m LOA> Custom House Quay. Quays from The Cove to lift bridge are for FVs and tourist boats. ⌓ £2·85 inc ⚓, ⚡ (Apr-Sep), £2·20 (Oct-Mar); <4hrs £6·50<6m>£8·50<12m>£10·50. ♿ ▢ (coin operated).

Inner Harbour No ❷ berths on council pontoons W of the bridge.

Weymouth Marina www.deanreddyhoff.co.uk ☎767576, dredged 2·5m, 290⌓ inc ❷ £2·85 inc ⚓ £8 for <14m short stay <4hrs: call Ch 80 for berth.

Fuel ⛽ from jetty W of LB station ☎07747182181 or by bowser (see HM or Harbour Guide for details); no petrol in harbour.

Weymouth SC ☎785481, ⌓. **Royal Dorset YC** ☎786258, ✕ ⌂.
Town ⚓ ▢ ⛽ 📮 🍴 🔧 🔨 Ⓔ 🏤 ⛪ Diver 🏦 ✉ Ⓑ 🚇 ✕ ⌂ @ ⇌ ✈ (Bournemouth).

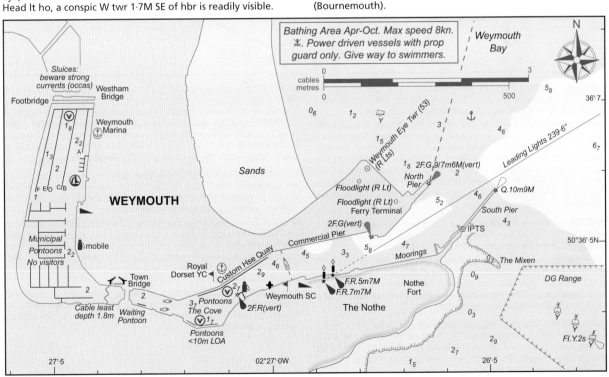

Bathing Area Apr-Oct. Max speed 8kn. ⚓. Power driven vessels with prop guard only. Give way to swimmers.

9.1.29 LULWORTH RANGES

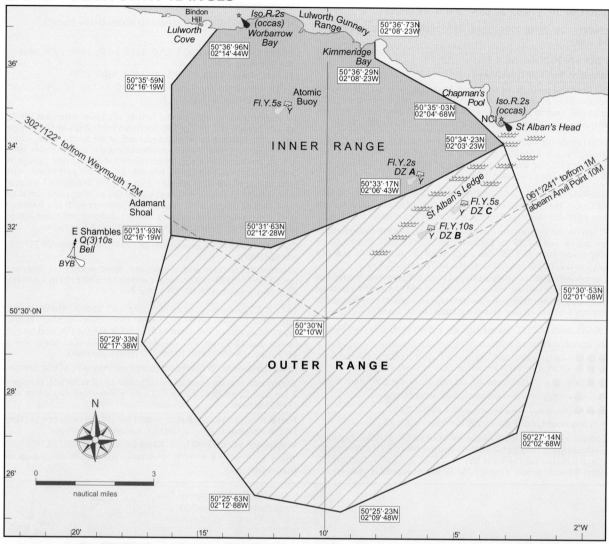

LULWORTH FIRING RANGES comprise an inner (D026) and an outer (D026B) sea danger area; the former is more likely to impact on yachts, so to speak. See chartlet above and AC 2610.

The inner area, shaded pink, extends 5·5M offshore. It runs from just E of Lulworth Cove coastwise to Kimmeridge Bay, thence seaward clockwise and back to just E of Lulworth. If on passage between Weymouth and Anvil Point, a track via 50°30'N 02°10'W just clips the SW corner of the inner range, avoids St Alban's Race and is 3·3M longer than a direct track. See AC 2610 or 5601.4.

The outer area extends 12M offshore and adjoins the seaward boundary of the inner as shown. It is rarely used.

INFORMATION Pre-recorded firing times and range activity are available (H24) from Ansafone ☎(01929) 404819. Clarification may be sought from Range Control (HO) ☎(01929) 404712.

Times can also be obtained from Range Control/Safety Boats Ch 08 Portland CG (Ch 16), the St Alban's Head NCI ☎(01929) 439220; and from local HMs, marinas, YCs and newspapers. Annual firing weekends and No Firing periods are given at www.reedsalmanacs.co.uk in the Almanac's January Update.

NAVAL FIRING Warships may use the inner and outer areas, firing eastward from Adamant Shoal (50°33'N 02°19'W) at the 3 DZ target buoys (up to 3M SW of St Alban's Head), which should be avoided by at least 1M. Warships fly red flags and other vessels/helicopters may patrol the area.

ARMY FIRING takes place on the inner range most weekdays from 0930-1700 (1230 on Fri), often on Tues and Thurs nights for 3–4hrs and for up to six weekends per year. There is NO firing in Aug and on Public Holidays. When firing is in progress red flags (at night Iso R 2s) are flown from St Alban's Head and Bindon Hill. However inshore some red flags are flown whether or not firing is taking place; these mark the boundary of the range's land area.

REGULATIONS Although there is no legal requirement for yachts on passage to keep clear of the ranges, every reasonable effort should be made to do so. Any vessels in the danger area delay the military training programme unnecessarily.

A call to Lulworth Range Control or Lulworth Range Safety Boat on Ch 08 will establish if firings are taking place at the time of your intended transit. If so, Range Control will advise you of the best track to follow, often keeping S of a given latitude (50° 32'.5N). This is shorter than the track shown above, and Range Control will be grateful for your consideration.

Range Safety boats, capable of 30kn, are based in Portland Harbour. When the ranges are active they will intercept yachts in the range and request them (Ch 08) to clear the danger area as quickly as possible.

ANCHORAGES BETWEEN PORTLAND BILL AND ANVIL PT

Helpful to read *Inshore along the Dorset Coast* (Peter Bruce).

CHURCH OPE COVE, Dorset, **50°32´·26N 02°25´·64W**. AC 2255, 2268, 5601.8. Tidal data as for Portland. A small cove on the E side of the Isle of Portland, about midway between the Bill and Portland Hbr. It is used by divers & completely open to the E, but could serve as a temporary ⚓ in about 3m off the shingle beach, to await a fair tide around the Bill.

RINGSTEAD BAY, Dorset, **50°37´·83N 02°20´·48W**. AC 2610, 5601. Tides as for Weymouth, 4M to WSW. Tempy ⚓ in 3-5m towards the E end of the bay. Ringstead Ledges, drying, define the W end of the bay. Rks on the E side restrict the effective width to about 3ca; easiest appr is from SE.

DURDLE DOOR, Dorset, **50°37´·27N 02°16´·58W**. AC 2610, 5601. Tides as for Lulworth Cove, 1M E. Durdle Door is a conspic rock archway. Close E of it Man o' War Cove offers ⚓ for shoal draught in settled weather. To the W, ⚓ may be found – with considerable caution – inside The Bull, Blind Cow, The Cow and The Calf which form part of a rocky reef.

LULWORTH COVE, Dorset, **50°37´·00N 02°14´·82W**. AC 2172, 5601. HW –0449 on Dover. Tides ML 1·2m. Good shelter in fair weather and offshore winds, but ocasionally a squally katabatic wind may blow down from the surrounding cliffs at night. Heavy swell enters the Cove in S/SW winds; if strong the ⚓ becomes untenable. Enter the Cove slightly E of centre. 8kn speed limit. ⚓ in NE quadrant in 2·5m. Holding is poor. Local 🔺 village and 🛥 are on W side. Facilities: 🛥 ⚓ taps in car park and at slip, ✉ ✕ 🏠 museum.

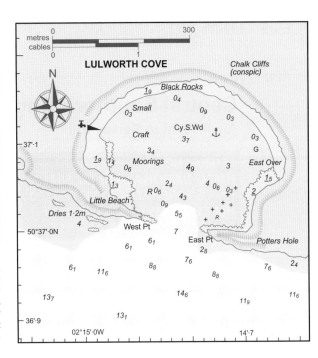

WORBARROW BAY, Dorset, **50°37´·03N 02°12´·08W**. AC 2172, 5601. Tides as Lulworth Cove/Mupe Bay. Worbarrow is a 1½M wide bay, close E of Lulworth Cove. It is easily identified from seaward by the V-shaped gap in the hills at Arish Mell, centre of bay just E of Bindon Hill. Bindon Hill also has a white chalk scar due to cliff falls. Caution: Mupe Rks at W end and other rks 1ca off NW side. ⚓s in about 3m sheltered from W or E winds at appropriate end. The bay lies within Lulworth Ranges; landing prohibited at Arish Mell. No lights/facilities. Tyneham village, deserted since WW2, is close to the E.

CHAPMAN'S POOL, Dorset, **50°35´·53N 02°03´·93W**. AC 2172, 5601. Tidal data: interpolate between Mupe Bay and Swanage. Chapman's Pool, like Brandy Bay and Kimmeridge Bay, is picturesque and comfortable when the wind is offshore. ⚓ in depths of about 3m in centre of bay (to avoid tidal swirl). From here to St Alban's Hd the stream sets SSE almost continuously due to a back eddy. No lts. Facilities: village shop and 'Square & Compass' 🏠 ☎01929 439229, are at Worth Matravers (1.5M walk).

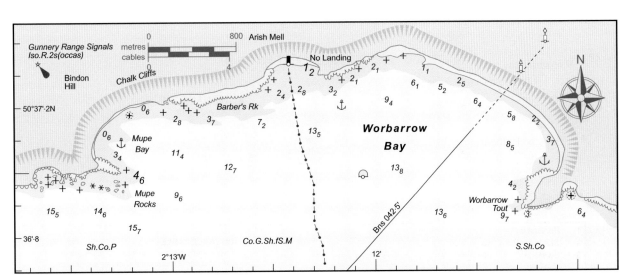

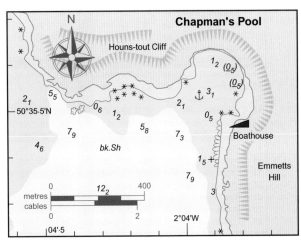

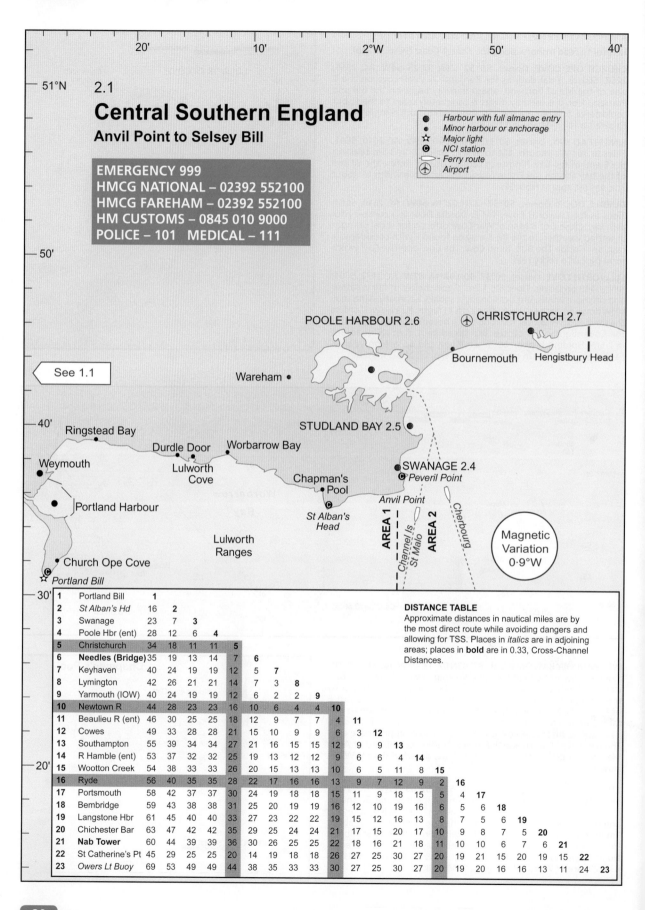

2.1

Central Southern England
Anvil Point to Selsey Bill

EMERGENCY 999
HMCG NATIONAL – 02392 552100
HMCG FAREHAM – 02392 552100
HM CUSTOMS – 0845 010 9000
POLICE – 101   MEDICAL – 111

- Harbour with full almanac entry
- Minor harbour or anchorage
- ☆ Major light
- NCI station
- Ferry route
- ✈ Airport

See 1.1

POOLE HARBOUR 2.6

CHRISTCHURCH 2.7

Bournemouth   Hengistbury Head

Wareham

STUDLAND BAY 2.5

Ringstead Bay

Durdle Door   Worbarrow Bay

Weymouth   Lulworth Cove   SWANAGE 2.4 · Peveril Point

Chapman's Pool

Portland Harbour   Anvil Point

St Alban's Head   AREA 1 | Channel Is St Malo   AREA 2 | Cherbourg

Lulworth Ranges

Church Ope Cove

☆ Portland Bill

Magnetic Variation 0·9°W

DISTANCE TABLE

Approximate distances in nautical miles are by the most direct route while avoiding dangers and allowing for TSS. Places in *italics* are in adjoining areas; places in **bold** are in 0.33, Cross-Channel Distances.

		1	2	3	4	5	6	7	8	9	10	11	12	13	14	15	16	17	18	19	20	21	22	23
1	Portland Bill	1																						
2	*St Alban's Hd*	16	2																					
3	Swanage	23	7	3																				
4	Poole Hbr (ent)	28	12	6	4																			
5	Christchurch	34	18	11	11	5																		
6	**Needles (Bridge)**	35	19	13	14	7	6																	
7	Keyhaven	40	24	19	19	12	5	7																
8	Lymington	42	26	21	21	14	7	3	8															
9	Yarmouth (IOW)	40	24	19	19	12	6	2	2	9														
10	Newtown R	44	28	23	23	16	10	6	4	4	10													
11	Beaulieu R (ent)	46	30	25	25	18	12	9	7	7	4	11												
12	Cowes	49	33	28	28	21	15	10	9	9	6	3	12											
13	Southampton	55	39	34	34	27	21	16	15	15	12	9	9	13										
14	R Hamble (ent)	53	37	32	32	25	19	13	12	12	9	6	6	4	14									
15	Wootton Creek	54	38	33	33	26	20	15	13	13	10	6	5	11	8	15								
16	Ryde	56	40	35	35	28	22	17	16	16	13	9	7	12	9	2	16							
17	Portsmouth	58	42	37	37	30	24	19	18	18	15	11	9	18	15	5	4	17						
18	Bembridge	59	43	38	38	31	25	20	19	19	16	12	10	19	16	6	5	6	18					
19	Langstone Hbr	61	45	40	40	33	27	23	22	22	19	15	12	16	13	8	7	5	6	19				
20	Chichester Bar	63	47	42	42	35	29	25	24	24	21	17	15	20	17	10	9	8	7	5	20			
21	**Nab Tower**	60	44	39	39	36	30	26	25	25	22	18	16	21	18	11	10	10	6	7	6	21		
22	*St Catherine's Pt*	45	29	25	25	20	14	19	18	18	26	27	25	30	27	20	19	21	15	20	19	15	22	
23	*Owers Lt Buoy*	69	53	49	49	44	38	35	33	33	30	27	25	30	27	20	19	20	16	16	13	11	24	23

51°N · 50' · 40' · 30' · 20'

20' · 10' · 2°W · 50' · 40'

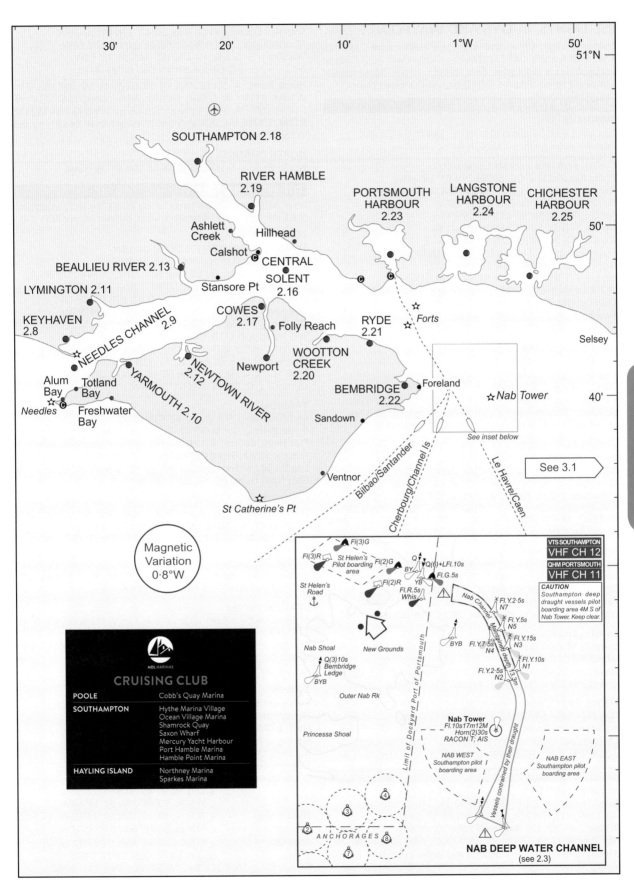

30' 20' 10' 1°W 50' 51°N

SOUTHAMPTON 2.18

RIVER HAMBLE
2.19

Ashlett
Creek   Hillhead
Calshot

PORTSMOUTH
HARBOUR
2.23

LANGSTONE
HARBOUR
2.24

CHICHESTER
HARBOUR
2.25

50'

BEAULIEU RIVER 2.13

CENTRAL
SOLENT
2.16

LYMINGTON 2.11

Stansore Pt

KEYHAVEN
2.8

NEEDLES CHANNEL 2.9

COWES
2.17   Folly Reach

RYDE
2.21

Forts

Selsey

Alum
Bay   Totland
Bay

WOOTTON
CREEK
2.20

Newport

YARMOUTH 2.10

NEWTOWN RIVER
2.12

BEMBRIDGE
2.22

Foreland

Nab Tower

40'

Needles   Freshwater
Bay

Sandown

See inset below

Ventnor

Bilbao/Santander

Cherbourg/Channel Is

Le Havre/Caen

See 3.1

St Catherine's Pt

Magnetic
Variation
0·8°W

Central S England

Fl(3)G

Fl(3)R   St Helen's
Pilot boarding
area

Fl(2)G   Q

Q(6)+LFl.10s

BY   YB

Fl.G.5s

VTS SOUTHAMPTON
VHF CH 12

QHM PORTSMOUTH
VHF CH 11

Fl(2)R
St Helen's
Road

Fl.R.5s
Whis

Nab Channel

Fl.Y.2·5s
N7

CAUTION
Southampton deep
draught vessels pilot
boarding area 4M S of
Nab Tower. Keep clear.

Maintained depth

Fl.Y.5s
N5

Nab Shoal

New Grounds

BYB

Fl.Y.7·5s
N4

Fl.Y.15s
N3

Fl.Y.10s
N1

Q(3)10s
Bembridge
Ledge
BYB

Fl.Y.2·5s
N2

3·3m

Outer Nab Rk

Limit of Dockyard Port of Portsmouth

Princessa Shoal

Nab Tower
Fl.10s17m12M
Horn(2)30s
RACON T; AIS

Vessels constrained by their draught

NAB WEST
Southampton pilot
boarding area

NAB EAST
Southampton pilot
boarding area

4

3

ANCHORAGES 8

2

7

NAB DEEP WATER CHANNEL
(see 2.3)

MDL MARINAS

CRUISING CLUB

POOLE	Cobb's Quay Marina
SOUTHAMPTON	Hythe Marina Village
	Ocean Village Marina
	Shamrock Quay
	Saxon Wharf
	Mercury Yacht Harbour
	Port Hamble Marina
	Hamble Point Marina
HAYLING ISLAND	Northney Marina
	Sparkes Marina

2.2 LIGHTS, BUOYS AND WAYPOINTS

Bold print = light with a nominal range of 15M or more. CAPITALS = place or feature. *CAPITAL ITALICS* = light-vessel, light float or Lanby. *Italics* = Fog signal. ***Bold italics*** = Racon. Some marks/ buoys are fitted with AIS (MMSI No); see relevant charts.

SWANAGE TO ISLE OF WIGHT

SWANAGE
Pier Hd ⚓ 2 FR (vert) 6m 3M; 50°36'·56N 01°56'·95W.
Peveril Ledge ⚓ QR; 50°36'·41N 01°56'·10W.

POOLE BAR and SWASH CHANNEL
Poole Bar (No 1) ⚓ QG; 50°39·29N 01°55'·14W.
(Historic wreck) ⚓ Fl Y 5s; 50°39'·70N 01°54'·86W.
South Hook ⚓ 50°39'·70N 01°55'·20W.
No 2 ⚓ Fl R 2s; 50°39'·23N 01°55'·24W.
No 3 ⚓ Fl G 3s; 50°39'·76N 01°55'·49W.
No 4 ⚓ Fl R 2s; 50°39'·72N 01°55'·60W.
Training Bank ⚓ 2 FR (vert); 50°39'·82N 01°55'·86W.
No 5 ⚓ Fl G 5s; 50°40'·19N 01°55'·81W.
No 6 ⚓ Fl R 4s; 50°40'·14N 01°55'·91W.
Hook Sands (No 7) ⚓ Fl G 3s; 50°40'·50N 01°56'·16W.
Channel (No 8) ⚓ Fl R 2s; 50°40'·45N 01°56'·27W.
Swash (No 9) ⚓ Q (9) 15s; 50°40'·88N 01°56'·70W.
No 10 ⚓ Fl R 4s; 50°40'·84N 01°56'·86W.

EAST LOOE CHANNEL
East Hook ⚓ 50°40'·58N 01°55'·23W.
East Looe 1 ⚓ Fl G 5s; 50°41'·09N 01°55'·82W.
East Looe 2 ⚓ Fl R 4s; 50°41'·07N 01°55'·83W.
East Looe 3 (Limit 10kn) ⚓ Fl G 3s; 50°41'·07N 01°56'·17W.
East Looe 4 (Limit 10kn) ⚓ Fl R 2s; 50°41'·05N 01°56'·17W.
North Hook ⚓ Fl (2) R 5s; 50°40'·97N 01°56'·51W.

BROWNSEA ROADS
No 12 ⚓ Q R; 50°40'·94N 01°57'·17W.
No14 ⚓ Fl R 2s; 50°41'·03N 01°57'·32W.
N Haven ⚓ Q (9) 15s 5m; 50°41'·15N 01°57'·17W.
Brownsea ⚓ Q (3) 10s; 50°41'·14N 01°57'·39W.
Brownsea Island Dir lt 299°F WRG; 296·5°-G-297·8°-AltWG-298·8°-W-299·2°-AltWR-300·2°-R-301·5°; 50°41'·16N 01°57'·67W (only shown for commercial vessels); 2FR(vert); 301·5°-296·5° (H24).

MIDDLE SHIP CHANNEL
Bell (No 15) ⚓ Q (6) + L Fl 15s; 50°41'·36N 01°57'·12W.
No 16 ⚓ VQ R; 50°41'·43N 01°57'·25W.
No 17 ⚓ Fl G 3s; 50°41'·68N 01°57'·02W.
Aunt Betty (No 22) ⚓ Q (3)10s; 50°41'·96N 01°57'·39W.
Diver (No 25) ⚓ Q (9) 15s; 50°42'·29N 01°58'·32W.

NORTH CHANNEL
Salterns Marina Outer Bkwtr Hd ⚓ 2 FR (vert) 2M; Tfc sigs; 50°42'·23N 01°57'·10W.
Parkstone YC platform ⚓ Q 8m 1M; 50°42'·37N 01°58'·08W.
Stakes (No 29) ⚓ Q (6) + L Fl 15s; 50°42'·43N 01°59'·00W.

POOLE BAY
Flag Hd Chine ⚓ Fl (2); 50°41'·79N 01°55'·10W.
Bournemouth Rocks ⚓ 50°42'·32N 01°53'·40W.
Christchurch Ledge ⚓ 50°41'·57N 01°41'·55W (Apr-Oct).
⚓ (x2) Fl(5) Y 20s; 50°37'·98N 01°43'·02W; (265°Needles F'wy).

WESTERN APPROACHES TO THE SOLENT

NEEDLES CHANNEL
Needles Fairway ⚓ L Fl 10s; *Bell*; 50°38'·24N 01°38'·98W.
SW Shingles ⚓ Fl R 2·5s; 50°39'·29N 01°37'·52W.
Bridge ⚓ VQ (9) 10s; ***Racon (T) 10M***; 50°39'·63N 01°36'·88W.
NEEDLES ☆ 50°39'·73N 01°35'·50W; Oc (2) WRG 20s 24m **W17M, R14M, R13M G14M**; ○ Twr, R band and lantern; vis: shore-R-300°-W-083°-R (unintens)-212°-W-217°-G-224° (H24). *Horn (2) 30s.*

Shingles Elbow ⚓ Fl (2) R 5s; 50°40'·37N 01°36'·05W.
Mid Shingles ⚓ Fl (3) R 10s; 50°41'·21N 01°34'·66W.
Warden ⚓ Fl G 2·5s; *Bell*; 50°41'·48N 01°33'·55W.
NE Shingles ⚓ Q (3) 10s; 50°41'·96N 01°33'·41W.

Hurst Point ☆ 50°42'·48N 01°33'·03W; FL (4) WR 15s 23m **W13M, R11M**; W ○ Twr; vis:080°-W(unintens)-104°, 234°-W-244°-R-250°- W-053°. Same structure, Iso WRG 4s 19m **W21M, R18M, G17M**; vis: 038·8°-G-040·8°-W-041·8°-R- 043·8°; By day W7M, R5M, G5M.

NORTH CHANNEL
North Head ⚓ Fl (3) G 10s; 50°42'·69N 01°35'·52W.

THE WESTERN SOLENT

Note: Numerous yellow yacht racing buoys are laid throughout the Solent (seasonal, Mar-Dec). Most, but not all, are lit Fl Y 4s.

SOLENT MARKS
Sconce ⚓ Q; *Bell*; 50°42'·53N 01°31'·43W.
Black Rock ⚓ Fl G 5s; 50°42'·57N 01°30'·59W.
Lymington Bank ⚓ Fl (2) R 5s; *Bell*; 50°43'·10N 01°30'·85W.
Solent Bank ⚓ Fl (3) R 10s; 50°44'·23N 01°27'·37W.
Hamstead Ledge ⚓ Fl (2) G 5s; 50°43'·87N 01°26'18W.
Newtown River ⚓ Q (9) 15s; 50°43'·75N 01°24'·96W.
W Lepe ⚓ Fl R 5s; 50°45'·24N 01°24'·09W.
Salt Mead ⚓ Fl (3) G 10s; 50°44'·51N 01°23'·04W.
Gurnard Ledge ⚓ Fl (4) G 15s; 50°45'·51N 01°20'·59W.
E Lepe ⚓ Fl (2) R 5s; *Bell*; 50°45'·93N 01°21'·07W.
Lepe Spit ⚓ Q (6) + L Fl 15s; 50°46'·78N 01°20'·64W.
Gurnard ⚓ Q; 50°46'·22N 01°18'·84W.

YARMOUTH
East Fairway ⚓ Fl R 2s; 50°42'·62N 01°29'·95W.
Poole Belle ⚓ Fl Y 5s; 50°42'·54N 01°30'·17W.
Pier Head, centre, ⚓ 2 FR (vert) 2M; G col. High intensity FW (occas); 50°42'·51N 01°29'·97W.
Ldg Lts 187·6° Front FG 5m 2M; 50°42'·36N 01°30'·06 W. Rear, 63m from front, FG 9m 2M; both W ◇.

LYMINGTON
Ldg Lts 319·5°, Or posts. Front, FR 12m 8M; 50°45'·19N 01°31'·65W. vis: 309·5°-329·5°. Rear, FR 17m 8M.
Jack in the Basket ⚓ Fl R 2s 9m; 50°44'·27N 01°30'·57W.
No 1 ⚓ Fl G 2s 2m 3M; G △ on pile; 50°44'·41N 01°30'·48W.
No 2 (Cross Boom) ⚓ Fl R 2s 4m 3M; 50°44'·36N 01°30'·58W.
Yacht Haven ldg lts 244°. Front FY 4m; R △; 50°45'·09N 01°31'·53W. Rear, 22m from front, FY 6m; R ▽.

BEAULIEU RIVER
Millennium Dir lt 334°. ⚓ Oc WRG 4s 13m W4M, R3M, G3M; vis: 318°-G-330°-W-337°-R-348°; 50°47'·12N 01°21'·90W.
Beaulieu Spit E ⚓ Fl R 5s 3M; R dolphin; 50°46'·85N 01°21'·76W.
No 1 ⚓ 50°46'·91N 01°21'·70W.
No 2 ⚓ 50°46'·92N 01°21'·78W.

COWES
Prince Consort ⚓ VQ; 50°46'·41N 01°17'·56W.
Trinity House ⚓ Fl Y 5s; 50°46'·31N 01°17'·75W.
No 1 ⚓ Q G; 50°46'·07N 01°18'·03W.
No 2 ⚓ Q R; 50°46'·07N 01°17'·87W.
No 2A ⚓ LFl R 10s; 50°45'·92N 01°17'·70W.
Tide gauge ⚓ L Fl R 5s; 50°45'·78N 01°17'·66W.
E. Cowes Bkwtr Hd ⚓ 2F R; 50°45'·88N 01°17'·52W.

	⚓ Fl G 3s; 50°45'·91N 01°17'·55W.
	⚓ Fl R 3s; 50°45'·89N 01°17'·53W.
Shrape Mud	⚓ Fl R 5s; 50°46'·02N 01°17'·38W.
	⚓ Fl R 5s; 50°46'·00N 01°17'·37W.
Boat Chan Ent	⚓ Fl G 2s; 50°46'·08N 01°17'·05W.
	⚓ Q R; 50°46'·06N 01°17'·04W.

Shrape Bn ⚓ LFl R 10S 3m 3M; 50°46'·09N 01°16'·90W.

CENTRAL SOLENT AND SOUTHAMPTON WATER

Note: Numerous yellow yacht racing buoys are laid throughout the Solent (seasonal, Mar-Dec). Most, but not all, are lit Fl Y 4s.

SOLENT MARKS
Lepe Spit ⚓ Q (6) + L Fl 15s; 50°46'·78N 01°20'·64W.
NE Gurnard ⚓ Fl (3) R 10s; 50°47'·06N 01°19'·42W.
South Bramble ▲ Fl G 2·5s; 50°46'·98N 01°17'·72W.
W Bramble ⚓ VQ (9) 10s; *Bell*; **Racon (T) 3M**; 50°47'·20N 01°18'·65W; AIS.
Thorn Knoll ▲ Fl G 5s; 50°47'·50N 01°18'·44W.
Bourne Gap ⚓ Fl R 3s; 50°47'·83N 01°18'·34W.
West Knoll ⚓ Fl Y 2·5s; 50°47'·43N 01°17'·84W.
North Thorn ▲ QG; 50°47'·92N 01°17'·84W.
East Knoll ▲ Fl (2) G 10s; 50°47'·96N 01°16'·86W.
Stanswood Outfall ⚓ Iso R 10s 6m 5M; 4 FR Lts; 50°48'·26N 01°18'·82W.

CALSHOT REACH
East Knoll ▲ 50°47'·96N 01°16'·83W.
CALSHOT SPIT ⚓ Fl 5s 5m 10M; R hull, Lt Twr amidships; *Horn (2) 60s*; 50°48'·35N 01°17'·64W.
Calshot ⚓ VQ; *Bell*; 50°48'·44N 01°17'·03W.
Castle Point ⚓ IQ R 10s; 50°48'·71N 01°17'·67W.
Reach ▲ Fl (3) G 10s; 50°49'·05N 01°17'·65W .
Black Jack ⚓ Fl (2) R 4s; 50°49'·13N 01°18'·09W.
Hook ⚓ QG; *Horn (1) 15s*; 50°49'·52N 01°18'·30W.
Coronation ⚓ Fl Y 5s; 50°49'·55N 01°17'·62W.
Bald Head ▲ Fl G 2·5s; 50°49'·80N 01°18'·06W.

RIVER HAMBLE
Hamble Pt ⚓ Q (6) + L Fl 15s; 50°50'·15N 01°18'·66W.
Hamble Common ⚓ Dir 351·7°, Oc (2) WRG 12s 5m W4M; R4M; G4M; vis: 348·7°-G-350·7°-W-352·7°-R-354·7°; 50°51'·00N 01°18'·84W.
Sailing Club Dir lt 028·9° ⚓ Iso WRG 6s 5m W4M, R4M, G4M: vis: 025·9°-G-027·9°-W-029·9°-R-031·9°; 50°51'·10N 01°18'·34W.

SOUTHAMPTON WATER
Fawley Terminal SE ⚓ 2 FR(vert) 9m 10M; 50°50'·06N 01°19'·42W.
Fawley Deep ▲ Fl (2) G 4s; 50°50'·42N 01°19'·19W.
Greenland ▲ Iso G 2s; 50°51'·11N 01°20'·38W.
Cadland ⚓ Fl R 3s; 50°51'·02N 01°20'·54W.
After Barn ▲ Fl (2) G 4s; 50°51'·53N 01°20'·81W.
Lains Lake ⚓ Fl (2) R 4s; 50°51'·59N 01°21'·65W.
Hound ▲ Fl (3) G 10s; 50°51'·68N 01°21'·52W.
Netley ▲ Fl G 3s; 50°52'·03N 01°21'·81W.
Deans Elbow ⚓ Oc R 4s; 50°52'·20N 01°22'·85W.
NW Netley ▲ Q G; 50°52'·31N 01°22'·73W.
Moorhead ▲ Q G; 50°52'·55N 01°22'·90W.
Test ⚓ Fl (2) R 4s; 50°52'·59N 01°23'·45W.
Weston Shelf ▲ Fl (2) G 4s; 50°52'·71N 01°23'·26W.

HYTHE
Hythe Pier Hd ⚓ 2 FR (vert) 12m 5M; 50°52'·49N 01°23'·61W.
Hythe Marina Ent ⚓ Q (3) 10s; 50°52'·63N 01°23'·88W.
Hythe Knock ⚓ Fl R 3s; 50°52'·83N 01°23'·81W.

SOUTHAMPTON and RIVER ITCHEN
Swinging Ground No 1 ▲ Oc G 4s; 50°53'·00N 01°23'·44W.
E side. No 1 ⚓ QG; 50°53'·15N 01°23'·40W.
No 2 ⚓ Fl G 5s 2M; 50°53'·29N 01°23'·38W.
No 3 ⚓ Fl G 7s; 50°53'·48N 01°23'·28W.
No 4 ⚓ QG 4m 2M; 50°53'·62N 01°23'·16W.

SOUTHAMPTON and RIVER TEST
Queen Elizabeth II Terminal, S end ⚓ 4 FG (vert) 16m 3M; 50°53'·00N 01°23'·71W.
Gymp ⚓ QR; 50°53'·07N 01°24'·16W.
Town Quay Ldg Lts 329°, both F 12/22m 3/2M.
Gymp Elbow ⚓ Oc R 4s; 50°53'·43N 01°24'·61W.
Dibden Bay ⚓ Q; 50°53'·70N 01°24'·92W.

THE EASTERN SOLENT

Note: Numerous yellow yacht racing buoys are laid throughout the Solent (seasonal, Mar-Dec). Most, but not all, are lit Fl Y 4s.

SOLENT MARKS
West Ryde Middle ⚓ Q (9) 15s; 50°46'·48N 01°15'·79W.
Norris ⚓ Fl (3) R 10s; 50°45'·97N 01°15'·51W.
North Ryde Middle ⚓ Fl (4) R 20s; 50°46'·61N 01°14'·31W.
South Ryde Middle ▲ Fl G 5s; 50°46'·13N 01°14'·16W.
Peel Bank ⚓ Fl (2) R 5s; 50°45'·49N 01°13'·35W.
SE Ryde Middle ⚓ VQ (6)+L Fl 10s; 50°45'·93N 01°12'·10W.
NE Ryde Middle ⚓ Fl (2) R 10s; 50°46'·21N 01°11'·88W.
Mother Bank ⚓ Fl R 3s; 50°45'·49N 01°11'·21W.
Browndown ▲ Fl G 15s; 50°46'·57N 01°10'·95W.
Fort Gilkicker ⚓ Oc G 10s 7M; 50°46'·43N 01°08'·47W.
N Sturbridge ⚓ VQ; 50°45'·33N 01°08'·23W.
Ryde Sands ⚓ Fl R 10s; 50°44'·56N 01°07'·26W.
Ryde Sands ⚓ L Fl R 12s; 50°44'·16N 01°05'·99W.
No Man's Land Fort ⚓ Iso R 2s 21m 8M; 50°44'·40N 01°05'·70W.
Horse Sand Fort ⚓ Iso G 2s 21m 8M; 50°45'·01N 01°04'·34W.
Saddle ▲ VQ (3) G 10s; 50°45'·05N 01°04'·94W.

NORTH CHANNEL and HILLHEAD
Calshot ⚓ VQ; *Bell (1) 30s*; 50°48'·44N 01°17'·03W.
Hillhead ⚓ Fl R 2·5s; 50°48'·07N 01°16'·00W.
E Bramble ⚓ VQ (3) 5s; 50°47'·23N 01°13'·64W.

WOOTTON CREEK
Wootton Beacon ⚓ Q 1M; (NB); 50°44'·53N 01°12'·13W.
Dir lt. Oc WRG 10s vis: 220·8°-G-224·3°-W-225·8°-R-230·8°; 50°44'·03N 01°12'·86W.

RYDE
Ryde Pier ⚓, NW corner, N and E corner marked by 2 FR (vert). In fog FY from N corner, vis: 045°-165°, 200°-320°; 50°44'·34N 01°09'·72W.
Leisure Hbr E side ⚓ 2 FR (vert) 7m 1M. FY 6m shown when depth of water in Hbr greater than 1m; 2 FY 6m when depth exceeds 1·5m; 50°43'·99N 01°09'·29W.

PORTSMOUTH APPROACHES
Horse Sand ▲ Fl G 2·5s; 50°45'·53N 01°05'·27W.
Outer Spit (OSB) ⚓ Q (6) + L Fl 15s; 50°45'·40N 01°05'·54W.
Mary Rose ⚓ Fl Y 5s; 50°45'·80N 01°06'·20W.
Boyne ▲ Fl G 5s; 50°46'·15N 01°05'·26W.
Spit Refuge ⚓ Fl R 5s; 50°46'·15N 01°05'·46W.
Spit Sand Fort ⚓ Fl R 5s; 18m 7M. 50°46'·24N 01°05'·94W.
Castle (NB)▲ Fl (2) G 6s; 50°46'·45N 01°05'·38W.
Southsea S Castle Pile ⚓ Dir 348° WRG 17m 11M day 6M, pile vis: 340°-FG-343°-Al WG (W phase inc with brg)-347°-FW-349°-AlWR(R phase incr with brg)-354°-FR-356°; H24; same structure Fl G 5s; 50°46'·66N 01°05'·65W.
Ridge ⚓ Fl (2) R 6s; 50°46'·44N 01°05'·65W.
No 1 Bar (NB) ▲ Fl (3) G 10s; 50°46'·77N 01°05'·81W.
No 2 ⚓ Fl (3) R 10s; 50°46'·69N 01°05'·97W.
No 3 ▲ QG; 50°47'·08N 01°06'·24W.
No 4 ⚓ QR; 50°47'·01N 01°06'·36W.
BC Outer ⚓ Oc R 15s; 50°47'·32N 01°06'·68W.

PORTSMOUTH HARBOUR
Fort Blockhouse ⚓ Dir 323° WRG 4s 6m WRG 12M; vis: 313°- Oc G-317°-Fl G (phase dec with brg)-321°-Oc W-325°-Fl R (phase inc with brg)-329°-Oc R-333°; 50°47'·37N 01°06'·74W.
Ballast ⚓ Fl R 2·5s; 50°47'·62N 01°06'·83W.
Hbr Ent ⚓ Dir 333·75° WRG 2s 2m 12M; vis: 323·75°-Iso G-329·75°-Fl G (phase dec with brg)-332·75°-Iso (main chan)-334·75°-Fl R (phase inc with brg)-337·75°-Iso R-343·75°; H24; same structure Fuel Jetty ⚓ 2F R (vert) 8m 4M; 50°47'·85N 01°06'·98W.

EASTERN APPROACHES to THE SOLENT
Outer Nab 1 ⚓ VQ (9) 10s; 50°38'·18N 00°56'·88W.

Central S England

Outer Nab 2 ⚓ VQ (3) 5s; 50°38'·43N 00°57'·70W.
Nab Tower ⚓ Fl 10s 17m 12M; *Horn (2) 30s*; RACON T; 50°40'·08N 00°57'·15W; <u>992351136</u>.
N 2 ⚓ Fl Y 2·5s. 6M; 50°41'·03N 00°56'·74W.
N 1 ⚓ Fl Y (4)10s; 50°41'·26N 00°56'·52W.
N 4 ⚓ Fl Y 7·5s; 50°41'·86N 00°57'·24W.
N 3 ⚓ Fl (3) Y 15s; 50°41'·63N 00°56'·74W.
N 5 ⚓ Fl Y 5s; 50°41'·99N 00°56'·97W.
N 7 ⚓ Fl Y 2·5s; 50°42'·35N 00°57'·20W.
New Grounds ⚓ VQ (3) 5s; 50°41'·84N 00°58'·49W.
Nab End ⚓ Fl R 5s; *Whis*; 50°42'·63N 00°59'·49W.
Dean Tail ⚓ Fl G 5s; 50°42'·99N 00°59'·17W.
Dean Tail S ⚓ VQ (6) + L Fl 10s; 50°43'·04N 00°59'·58W.
Dean Tail N ⚓ Q; 50°43'·13N 00°59'·57W.
Horse Tail ⚓ Fl (2) G 10s; 50°43'·23N 01°00'·23W.
Nab East ⚓ Fl (2) R 10s; 50°42'·86N 01°00'·80W.
Dean Elbow ⚓ Fl (3) G 15s; 50°43'·69N 01°01'·88W.
St Helens ⚓ Fl (3) R 15s; 50°43'·36N 01°02'·41W.
Horse Elbow ⚓QG; 50°44'·26N 01°03'·88W.
Cambrian Wreck ⚓ 50°44'·43N 01°03'·43W.
Warner ⚓ QR; *Whis*; 50°43'·87N 01°03'·99W.

BEMBRIDGE
St Helen's Fort ☆ (IOW) Fl (3) 10s 16m 8M; large ○ stone structure; 50°42'·30N 01°05'·05W.

SOUTH EAST COAST of the ISLE OF WIGHT
St Catherine's Point ☆ 50°34'·54N 01°17'·87W; Fl 5s 41m **25M**; vis: 257°-117°; FR 35m 13M (same Twr) vis: 099°-116°.
Ventnor Haven W Bwtr ⚓ 2 FR (vert) 3M; 50°35'·50N 01°12'·30W.
Sandown Pier Hd ⚓ 2 FR (vert) 7m 2M; 50°39'·05N 01°09'·18W.
W Princessa ⚓ Q (9) 15s; 50°40'·16N 01°03'·65W.
Bembridge Ledge ⚓ Q (3) 10s; 50°41'·15N 01°02'·81W

LANGSTONE and APPROACHES
Eastney Pt Fraser Trials Range ⚓ FR, Oc (2) Y 10s, and FY Lts

(occas) when firing taking place; 50°47'·19N 01°02'·22W.
Winner ⚓; 50°45'·10N 01°00'·10W.
Roway Wk ⚓ Fl (2) 5s; 50°46'·11N 01°02'·28W.
Langstone Fairway ⚓ L Fl 10s; 50°46'·32N 01°01'·36W.
Eastney Pt Outfall ⚓ QR 2m 2M; 50°47'·23N 01°01'·68W.
East Milton ⚓ Fl (4) R 10s; 50°48'·16N 01°01'·76W.
NW Sinah ⚓ Fl G 5s; 50°48'·14N 01°01'·58W.

CHICHESTER ENTRANCE
West Pole (tripod) ⚓ Fl R 5s 14m 7M; 50°45'·45N 00°56'·59W.
Bar ⚓ Fl(2) R 10s 10m 4M; 50°45'·92N 00°56'·46W.
Eastoke ⚓ QR; 50°46'·68N 00°56'·11W.
West Winner ⚓ QG; Tide gauge. 50°46'·88N 00°55'·98W.

EMSWORTH CHANNEL
Fishery ⚓ Q (6) + L Fl 15s; 50°47'·38N 00°56'·07W.
NW Pilsey ⚓ Fl G 5s; 50°47'·50N 00°56'·20W.
Verner ⚓ Fl R 10s; 50°48'·20N 00°56'·63W.
Marker Pt ⚓ Fl (2) G 10s 8m; 50°48'·91N 00°56'·72W.
Emsworth ⚓ Q (6) + L Fl 15s; tide gauge; 50°49'·66N 00°56'·76W.

THORNEY CHANNEL
Camber ⚓ Q (6) + L Fl 15s; 50°47'·87N 00°54'·06W.
Pilsey ⚓ Fl (2) R 10s ; 50°47'·98N 00°54'·24W.
Thorney ⚓ Fl G 5s; 50°48'·20N 00°54'·28W.

CHICHESTER CHANNEL
NW Winner ⚓ Fl G 10s; 50°47'·19N 00°55'·92W.
N Winner ⚓ Fl (2) G 10s; 50°47'·31N 00°55'·83W.
Mid Winner ⚓ Fl (3) G 10s; 50°47'·40N 00°55'·72W.
Stocker ⚓ Fl (3) R 10s; 50°47'·45N 00°55'·52W.
Copyhold ⚓ Fl (4) R 10s; 50°47'·50N 00°54'·93W.
East Head Spit ⚓ Fl (4) G 10s; 50°47'·45N 00°54'·82W.
Snowhill ⚓ Fl G 5s; 50°47'·52N 00°54'·34W.
Sandhead ⚓ Fl R 10s; 50°47'·67N 00°54'·25W.
Chalkdock ⚓ Fl (2) G 10s; 50°48'·49N 00°53'·30W.

2.3 PASSAGE INFORMATION

More passage information is threaded between the harbours in this area. See 0.33 for distances across the central part of the English Channel. Admiralty Leisure Folio 5601 covers Exmouth to Christchurch, and 5600 covers The Solent & Approaches. **Bibliography:** *Channel Havens* (ACN/Endean); *Shell Channel Pilot* (Imray/Cunliffe); *Channel Pilot* (Admiralty NP27).

CROSSING THE ENGLISH CHANNEL (CENTRAL PART)

(AC 2656) These notes cover the crossing (about 13hrs) Weymouth/Poole/Solent to Channel Is/Cherbourg/Le Havre and should help in compiling a cross-Channel passage plan. Distances are tabulated in 0.33.

Departure and destination factors to be considered include:

Portland/Weymouth: Good angle on the wind in SW/lies but passage extended by 7/8½M respectively to clear the E end of The Shambles Bank. There is no advantage to closing Portland Bill and Portland Race which should be avoided. Portland Bill Light loom provides a good headmark on night N'ly crossing. Check the tidal streams and monitor course made good.

Poole: Leave on the ebb. At springs with a S/SE wind beware short steep seas in the Swash Channel. Studland Bay or Swanage Roads are useful departure ⚓s. Anvil Pt Light and the Ferries provide useful visual cues on N'ly crossing. Off Handfast, Peveril and Anvil Pts overfalls occur with wind against tide.

Solent ports: Decide whether to leave to the East via Nab Tower, using the tide to gain a lift on the ebb past Bembridge Ledge and across St Catherine's Deep or west via the Needles.

▶*The latter usually requires a fair tide through Hurst Narrows (HW Portsmouth –0100 to +0430) which dictates the ETD.* ◀

The Easterly option has no tidal gate, but is longer if heading to Cherbourg or the Channel Islands and gives a less favourable angle on a SW'ly wind. It can offer better shelter initially in the lee of the IoW, but often has an uncomfortable chop especially on a return passage. Loom of St Catherine's and the Needles are useful guides at night, but can be deceptively far away.

Braye (Alderney) is accessible at all times, but allowance must be made to avoid being swept W to the Casquets at springs. Plan and monitor the track so as to approach from well uptide. Braye Hbr is unsafe in fresh or strong E/NE'lies. From Braye you may make Guernsey, having negotiated the Alderney Race, or the Swinge. A direct passage is achievable off the Casquets or in benign conditions through Ortac Channel.

Cherbourg is accessible H24. Especially at springs, aim well up-tide and approaching the peninsula allow a large offset angle to maintain track. Harbours on the W side of the Cotentin (**Diélette** to **Granville**) are tidally constrained and exposed to the W but are options in E'lies if conditions in the Alderney Race permit on a W-going ebb tide or at slack water.

To the E, **Barfleur** and **St Vaast** are sheltered from the prevailing W'lies, but are tidally constrained. Barfleur dries and yachts off St Vaast until the marina gates open. Further to the E **Le Havre** is a port of refuge although approach requires care.

Thorough planning is a requirement for a safe and efficient crossing. Refer to panel on next page.

- Study the meteorological situation several days before departure so that advantageous wind and weather windows may be predicted and bad weather avoided.

- High pressure may provide quiescent conditions, but is often accompanied by reduced visibility owing to fog. Even if radar equipped and competent in its use, be wary of crossing if fog or poor visibility is forecast.

- For a yacht, consider the forecast wind direction, likely shifts and sea breezes which may affect the angle on the wind. Prevailing winds are SW/W, except in the spring when NE/E winds are equally likely. Check the tidal stream and the effect that this will have on Course to Steer. It is best to place the tide on the lee bow in order to gain the maximum benefit. It is very advantageous to get well to windward and up-tide of the destination. For example, from the Needles it may be beneficial initially to proceed W'wards, working the tides to advantage.

- For motor boats, windows with the wind in the N provide the best down sea conditions, maximising speed and reducing slamming.

- Choose the route, departure points and landfalls so that passage time out of sight of identifiable marks is minimised. This reduces the risk of navigational errors, anxiety and fatigue. A landfall at night or dawn/dusk is frequently easier due to the visibility of navigational marks on clear nights. The loom of a powerful lighthouse is often visible on the clouds at a distance greater than its nominal range.

- Consider tidal constraints at the points of departure and destination, and any tidal gates en route such as Hurst Narrows or the French coast between Cap Barfleur and Cap de la Hague.

▸ *Calculate the hourly direction/rate of tidal streams expected during the crossing so as to lay off the total drift angle required to make good the desired track. Rarely do 6hrs of E-going tide cancel out 6 hrs of W-going (or vice versa); streams off the French coast are usually stronger. Note times and areas of races/overfalls and keep well clear.* ◂

- Consider actions to be taken if fog sets in as the risk of collision is much increased. GPS may provide a position, but navigating 'blind' by radar demands intense attention. Other small boats and yachts may not be detected.

- AIS will only display vessels or marks which transmit.

- Cross on a heading as near 90° as practicable to the Traffic Routeing between the Casquets TSS and the Greenwich Meridian. Consider motoring to expedite such a crossing.

- Keep a very sharp lookout particularly when in the Traffic Routeing area in mid-Channel. Ensure the crew is aware of the directions from which to expect approaching traffic. It is commonsense and good seamanship to alter course early in order to keep well clear of large commercial ships. These may not alter for a small craft despite Rule 18 (a) (iv).

- Make use of additional navigational information such as soundings, noting when crossing distinctive seabed contours; the rising or dipping ranges of major lights; if equipped with radar, use clearly identifiable targets.

- Plan a harbour of refuge in case of fog, bad weather, accident or gear failure. For example, if unable to make Cherbourg in a strong SSW'ly and E-going tide, consider bearing away for St Vaast in the lee of the peninsula. Alternatively, in the event of engine failure identify a sheltered ⚓. In extremis heave to and stay at sea notifying the appropriate CG or safety service.

2.4 SWANAGE

Dorset 50°36'·70N 01°56'·55W ✺✺✺⚓⚓☀☀

CHARTS AC 5601, 2615, 2172; Imray C10, C12, C4, 2300

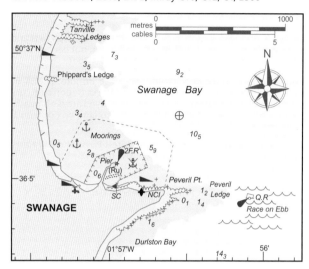

TIDES HW Sp −0235 & +0125, Np −0515 & +0120 on Dover; ML 1·5

Standard Port POOLE HARBOUR (→)

Times				Height (metres)			
High Water		Low Water		MHWS	MHWN	MLWN	MLWS
—	—	0500	1100	2·2	1·7	1·2	0·6
—	—	1700	2300				
Differences SWANAGE							
—	—	−0045	−0050	−0·1	+0·1	+0·2	+0·2

NOTE: From Swanage to Christchurch double HWs occur except at neaps. HW differences refer to the higher HW and are approximate.

SHELTER Large, well sheltered bay with just 1·6m range at springs. 12 ⚓s, good ⚓ in all winds except from NE to SE; Poole is closest port of refuge. ⌓ may be possible on the pier which is owned by a trust and used extensively in the summer by tourist boats, ferries and other commercial concerns. Seek advice at the gate on arrival. Gate closes at dusk Oct to Apr.

NAVIGATION WPT 50°36'·70N 01°56'·55W, 240°/0.3M to Pier Hd. Tidal races off Peveril Point to the south and Handfast Point to the north can be lively in strong onshore winds and an ebb tide. Do not cut inside Peveril Ledge buoy.

Beware of ruins of old pier south of the main pier; Tanville and Phippard's Ledges in west side of bay both dry. At night, main hazard is other boats which may be difficult to see against background lights.

LIGHTS AND MARKS 2 FR (vert) on the Pier, difficult to see due to back scatter of street lights. Peveril Ledge PHM buoy, Q.R, but can be hard to pick out by day.

COMMUNICATIONS (Code 01929) Dr 422231; ⊞ 422202; Pier 427058. Moorings/water taxi mob 07802 480139. No VHF.

FACILITIES 12 ⚓s £10/boat<40'; Water taxi (May-Oct) £2pp return. **Pier** Possible⌓; **Swanage SC** ☎422987 ⬚ ⬛ ⬛ **Boat Park** (Peveril Pt) ⬚ ⬛ ⚓

Services Diving.

Town ⛽ ⛽ (1½M) ⬚ ✉ Ⓑ ✕ ⬚ ⇌ (bus to Wareham and Bournemouth), ✈ (Hurn).

2.5 STUDLAND BAY

Dorset 50°38'·80N 01°55'·50W ❀❀❀❀⚓❀❀❀

CHARTS AC 5601, 2175, 2172; Imray C4

TIDES Approx as for Swanage.

SHELTER Good in winds from S, W and NW, but exposed to N and E. Best ⚓ in about 3m, 4ca WNW of Handfast Pt, but beware poor holding due to weed. Nearest port of refuge is Poole.

NAVIGATION No dangers except very close inshore; N end of bay is shallow. Beware unmarked Redend Rocks (reported to dry 0·5m) ESE of Redend Point. Sp limit 5kn in buoyed areas off beach. Beware of swimmers. PWCs not permitted to land on nor launch from beach.

LIGHTS AND MARKS See chartlet. Handfast Point and Old Harry Rock are conspicuous on approach to the bay.

COMMUNICATIONS None.

FACILITIES Village ⚓ at tap behind beach cafe, 🏪 🏨 ✉ 🛒 ✕ 🍴 hotel. No marine facilities.

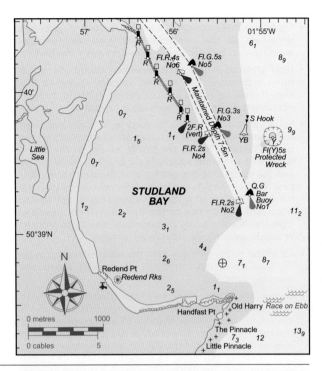

2.6 POOLE HARBOUR

Dorset 50°40'·93N 01°56'·96W (Ent) ❀❀❀❀⚓⚓⚓❀❀❀

CHARTS AC 5601, 2175, 2611; Imray C10, C12, C4, Y23, 2300

TIDES Town Quay ML 1·6. The tide is above ML from about LW+2 to next LW–2.

Standard Port POOLE HARBOUR (→)

Times				Height (metres)			
High Water		Low Water		MHWS	MHWN	MLWN	MLWS
—	—	0500	1100	2·2	1·7	1·2	0·6
		1700	2300				
Differences POOLE HARBOUR ENTRANCE							
—	—	–0025	–0010	0·0	0·0	0·0	0·0
POTTERY PIER							
—	—	+0010	+0010	–0·2	0·0	+0·1	+0·2
CLEAVEL POINT							
—	—	–0005	–0005	–0·1	–0·2	0·0	–0·1
WAREHAM (River Frome)							
—	—	+0130	+0045	0·0	0·0	0·0	+0·3

Daily predictions of the times and heights of HW/LW are for the Standard Port of **POOLE HARBOUR** (near the Ro-Ro terminal). Double HWs occur except at neaps. The height of the 2nd HW is always about 1·8m; the height of the 1st HW varies from springs to neaps. The listed time of HW refers to the highest of the two.

Strong and continuous winds from E to SW may raise sea levels by as much as 0·2m; W to NE winds may lower levels by 0·1m. Barometric pressure can affect the tide by as much as 0·3m. At Wareham the height of LW does not usually fall below 0·7m.

SHELTER An excellent natural harbour with a narrow entrance. Accessible in all conditions, the entrance can become very rough, especially on the ebb in E/SE gales. Yachts may berth at the marinas listed under Facilities. The Town Quay, very exposed in strong E/SE winds, is used by yachts >15m and by arrangement with the Boat Haven. At busy periods yachts may be rafted several deep or allocated a berth in the Port of Poole Marina. Good ⚓s may be found wherever sheltered from the wind and clear of channels, moorings and shellfish beds, especially clear of the buoyed fairways in South Deep and off Pottery Pier (W end of Brownsea Island) or off Shipstal Point, about 1·3M west of Pottery Pier. All are within a Quiet Area (see chartlet and speed limits).

NAVIGATION WPT Poole Bar (No 1 SHM) Buoy, QG, 50°39'·29N 01°55'·14W, 328°/1·95M to Haven Hotel. In very strong SE-S winds the Bar is dangerous, especially on the ebb. Beware cross-Channel high speed ferries which operate through the area. Monitor VHF Ch 14 for information about commercial traffic and MOD activity. In Studland Bay and close to training bank beware lobster pots. From Poole Bar to Shell Bay a **Boat Channel**, suitable for craft <3m draught, parallels the W side of the Swash Channel, close E of the Training Bank, and should be used whenever possible.

East Looe Channel The well lit and buoyed channel, least depth 1.3m, is the recommended approach from the E when height of tide allows. It is liable to shift and the buoys moved accordingly. Check LNTMs from www.phc.co.uk; there may be less water than charted. The groynes to the N are marked by SHM beacons, the two most W'ly are lit, 2 FG (vert).

Within the harbour the two principal channels (Middle Ship and North) lead up to Poole Town Quay. Outside of these there are extensive shoal or drying areas.

Middle Ship Chan is dredged to 7.5m depth and 105m width for ferries and commercial shipping to/from the Continental Freight and Ferry Terminal. Vessels designated 'controlled movements' have moving exclusion zones which must be avoided. Small craft should use the **Boat Channel** which runs parallel to the main channel between the PHM buoys and the unlit beacons further S which mark the edge of the bank. Depth is 2.0m in this channel, but 1.5m closer to the stakes. Caution: When large ferries pass, a temporary but significant reduction in depth may be experienced; it is then prudent to keep close to the PHM buoys.

North Channel is no longer routinely dredged but remains usable for leisure craft; the best water is on the outside of channel bends.

Lulworth gunnery range. Prior to proceeding W all craft should check for activity, as shown in the HM's office and in the Updates to this Almanac.

LIGHTS AND MARKS See chartlet and 2.2 for main buoys, beacons and lights.

Poole Bridge traffic lights, shown from bridge tower:

⬤ (3 vert Fl) = Emergency stop. Do not proceed

⬤ (3 vert) = Do not proceed past the signal

○ + 3 ⬤ vert = Vessels may proceed with caution

⬤ (3 vert) = Vessels may proceed

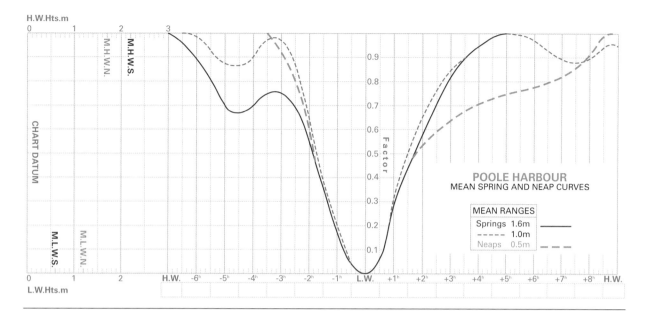

POOLE HARBOUR
MEAN SPRING AND NEAP CURVES

MEAN RANGES
Springs 1.6m
1.0m
Neaps 0.5m

POOLE BRIDGES SUMMER SCHEDULE:

Poole Bridge and the new Twin Sails Bridge co-ordinate openings with *scheduled times* at which vessels are advised to be on station. The timings are biased to outbound traffic am, and inbound pm. The second bridge will open asap after the first, which may open again to release traffic left between the closed bridges. Daily:

Twin Sail: 0530, 0630, 0730, 0930, 1030, 1130, 1230, 1430.

Poole Br: *ASAP thereafter.*

Poole Br: 1530,1630, 1830, 1930, 2130, 2230, 2330.

Twin Sail: *ASAP thereafter*.

+ Sat/Sun: Twin sail: 0830, 1330. Poole Br: 1730, 2030.

Pleasure craft may pass when bridges open for commercial traffic. Timings at www.phc.co.uk. Monitor Ch 12 for more information.

COMMUNICATIONS (Code 01202) ⊞ 665511 HM 440233. *Poole Hbr Control* 440230, Poole Bridge 674115.

VHF Ch **14** 16 (H24), *Poole Bridge* Ch12. Poole Quay Boat Haven, Cobbs Quay (call *CQ Base*) Ch 80; Salterns Marina Ch M 80; Parkstone Haven, Poole YC Haven (*Pike*), Dorset Yacht Co Ch M.

FACILITIES **Marinas** from seaward (harbour dues are usually included, but best to check):

Salterns Marina marina@salterns.co.uk ☎709971, max draught 2·5m, 300⌣ few❶, £3·50/m, short stay £1/m, ▢ ♠ ♠(H24) ♠ ✕ ✎ ⊞ Ⓔ ⚠ ⌣(5t) ⌣(45t) ♠ ☱ ✉ ✕ ⌂. Appr from North Channel near NC 7 SHM buoy, Fl G 3s.

Parkstone YC Haven ☎743610 (Parkstone YC). Access from North Channel near NC 11 SHM buoy, Fl G 3s. Approach chan, dredged 2·5m, is marked by PHM and SHM unlit buoys. Ldg daymarks 006°, both Y ◇s. Some ❶ berths, £4·00; dredged 2m.

Poole Quay Boat Haven ⊕ poolequayboathaven.co.uk ☎649488. max draught 3·0m, 100❶ £4·07 ⊡ £3/day; short stay <4hrs £6·50<10m, £12·50<20m. ⚏ ▢ ⚓ ✎ ⊞. Office hrs 0700-2200 Apr-Sep.

Town Quay (£2·78) **& Port of Poole Marina** ☎660120 by arrangement.

Lake Yard Marina www.lakeyard.com ☎674531. Ent marked by 2FR (vert) and 2FG (vert). ⚓ £1·30, 56⌣+❶ if space £3·20, ⌣(5t) ⌣(50t) Club: ⌂; Water taxi, weekends Apr-Oct.

Sunseeker International Marina ☎685335, 50⌣ ♠ ✕ ✎ ⊞ ▢(30t) ⌣(36t) ♠ ☱ ✕ ⌂.

Cobbs Quay Marina www.cobbsquaymarina.co.uk ☎674299, 850⌣ ❶ £3·25, ▬ ▢ ♠ ♠ LPG ✕ ✎ ⊞ Ⓔ ⚠ ⌣(10t) ♠ ✕ ⌂.

Swinging moorings in Poole Harbour £15, contact PHC through the Boat Haven.

Public Landing Places: Steps at Poole Quay. **Public** ▬: Baiter Harbourside Park, car/trailer parking – machine payment.

Services A complete range of marine services is available; consult marina/HM for exact locations. **Fuel:** Corrals (S side of Poole Quay adjacent bridge) ♠.

Town All domestic facilities, ⇌ ✈ (Bournemouth 12M and Southampton by direct rail link). **Ferries:**

Cherbourg: 2-3/day; 4½hrs (HSS 2¼hrs); Brittany (www. brittany-ferries.co.uk);

St Malo/Jersey/Guernsey: 2/day HSS; 2¾–4½hrs; Condor (www. condorferries.co.uk).

YACHT CLUBS
Royal Motor YC ☎707227, ♠ ✕ ⌂.
Parkstone YC ☎743610 (Parkstone Haven).
Poole YC ☎672687.

REGULATIONS

The following speed limits are strictly enforced:
- **10kn** speed limit applies to the entire harbour.
- **8kn** speed limit applies within 200m of the beach from S Haven Pt NE'wards through E Looe Channel to Flag Head Chine.
- **6kn** speed limit applies from Stakes SCM past Poole Quay and Poole Bridge up to Cobbs Quay in Holes Bay. It is also an advisory speed limit within the S half of the harbour which is designated as a Quiet Area (see chartlet).
- **4kn** speed limit applies within the Poole Quay Boat Haven.

Exemptions The 10kn speed limit does not apply:
- From 1 Oct to 31 Mar to vessels in the North, Middle Ship and Wareham Channels only.
- To water-skiers within the water-ski area between Gold Pt and No 82 PHM buoy (Wareham Channel; see chartlet). Permits must be obtained from the HM.
- Personal Water Craft (PWC) operating in a designated area N of Brownsea Island must not enter the Quiet Area in the S of the harbour, nor linger in the harbour entrance. Permits must be obtained from the HM.

Sandbanks Chain Ferry has right of way over all vessels not carrying a Pilot. It exhibits a B ● above the control cabin when it is about to leave the slipway. A flashing white strobe light is exhibited in the leading direction when the engines are engaged. In fog it sounds 1 long and 2 short blasts every 2 mins. When stationary at night it shows a FW lt; in fog it rings a bell for 5s every 60s.

Central S England

POOLE HARBOUR *continued*

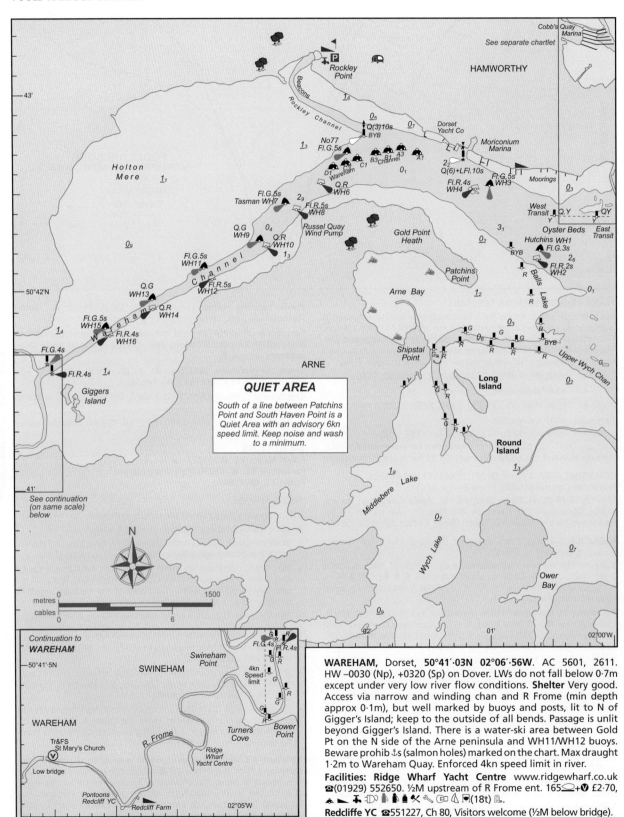

QUIET AREA

South of a line between Patchins Point and South Haven Point is a Quiet Area with an advisory 6kn speed limit. Keep noise and wash to a minimum.

See continuation (on same scale) below

Continuation to
WAREHAM

WAREHAM, Dorset, **50°41′·03N 02°06′·56W**. AC 5601, 2611. HW −0030 (Np), +0320 (Sp) on Dover. LWs do not fall below 0·7m except under very low river flow conditions. **Shelter** Very good. Access via narrow and winding chan and R Frome (min depth approx 0·1m), but well marked by buoys and posts, lit to N of Gigger's Island; keep to the outside of all bends. Passage is unlit beyond Gigger's Island. There is a water-ski area between Gold Pt on the N side of the Arne peninsula and WH11/WH12 buoys. Beware prohib ‡s (salmon holes) marked on the chart. Max draught 1·2m to Wareham Quay. Enforced 4kn speed limit in river.

Facilities: Ridge Wharf Yacht Centre www.ridgewharf.co.uk
☎(01929) 552650. ½M upstream of R Frome ent. 165⌂+🅥 £2·70, ⚓ ⬦ ⚓ 🗲 ⬦ 🅑 🔧 ⚒ ⚙ 🅑 △ 🅟(18t) ⚓.

Redcliffe YC ☎551227, Ch 80, Visitors welcome (½M below bridge).
Wareham Quay ⌂ (free) ⚓ ✕.
Town 🅟 🅟 🅑 ⊠ 🅑 🛒 ✕ 🏨 🚆.

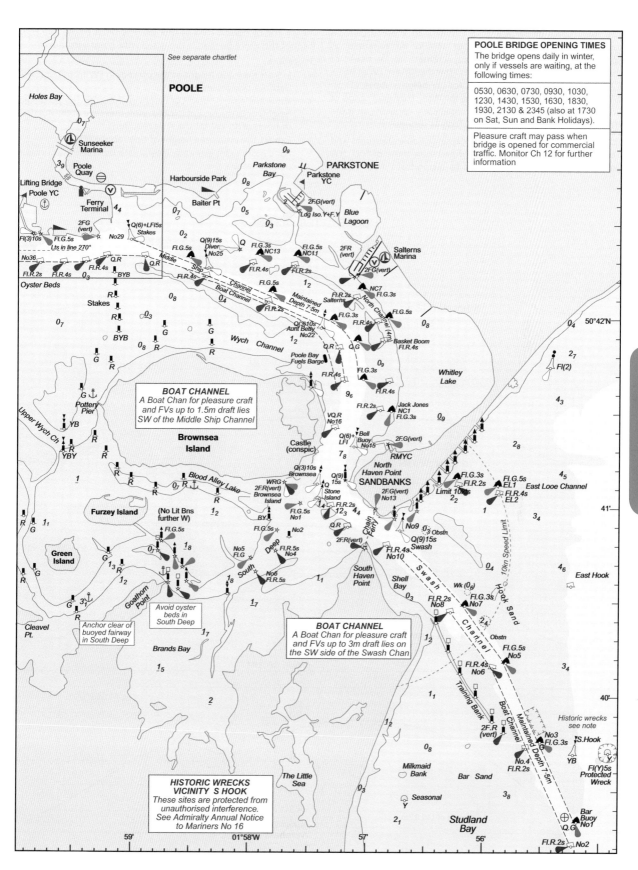

POOLE BRIDGE OPENING TIMES

The bridge opens daily in winter, only if vessels are waiting, at the following times:

0530, 0630, 0730, 0930, 1030, 1230, 1430, 1530, 1630, 1830, 1930, 2130 & 2345 (also at 1730 on Sat, Sun and Bank Holidays).

Pleasure craft may pass when bridge is opened for commercial traffic. Monitor Ch 12 for further information

BOAT CHANNEL
A Boat Chan for pleasure craft and FVs up to 1.5m draft lies SW of the Middle Ship Channel

BOAT CHANNEL
A Boat Chan for pleasure craft and FVs up to 3m draft lies on the SW side of the Swash Chan

HISTORIC WRECKS VICINITY S HOOK
These sites are protected from unauthorised interference. See Admiralty Annual Notice to Mariners No 16

Avoid oyster beds in South Deep

Anchor clear of buoyed fairway in South Deep

Historic wrecks see note

See separate chartlet

POOLE

PARKSTONE

SANDBANKS

HEIGHTS OF TIDE AT POOLE (TOWN QUAY)

The four curves below are an alternative to those shown on page 87. Although their accuracy should be regarded as approximate, they enable a speedy estimate to be made of the Height of Tide at hourly intervals after the time of LW at Poole Town Quay. The curves are drawn for LW heights above CD of: 0·3m, 0·6m (MLWS), 0·9m and 1·2m (MLWN). The small range of tide (neaps 0·4m; springs 1·5m) is immediately apparent, as is the HW stand at neaps and the double HWs at springs.

Note: All references are to LW because at Poole the times and heights of LW are more sharply defined than those of HW. HW times and heights are complicated by a stand of tide at HW at neaps and by double HWs at springs. The times of the two HWs cannot therefore be readily predicted with great accuracy.

Procedure:

- Extract time and height of the preceding LW from the Poole Harbour tide tables (⟶).
- Using the curve whose LW height is closest to the predicted height, note the time of LW and enter the curve at the number of hours after LW for the time required.
- Extract the estimated height of tide (above CD).
- For a more exact estimate, interpolate between the curves if appropriate.

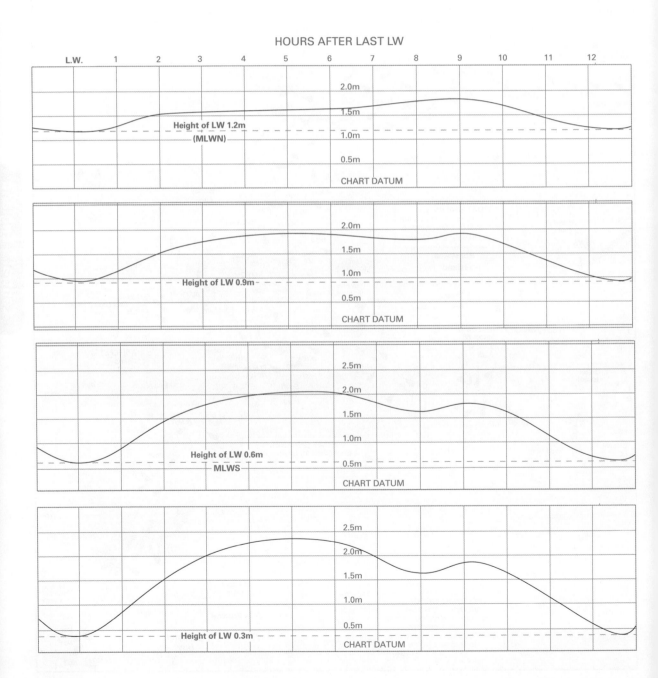

Central S England

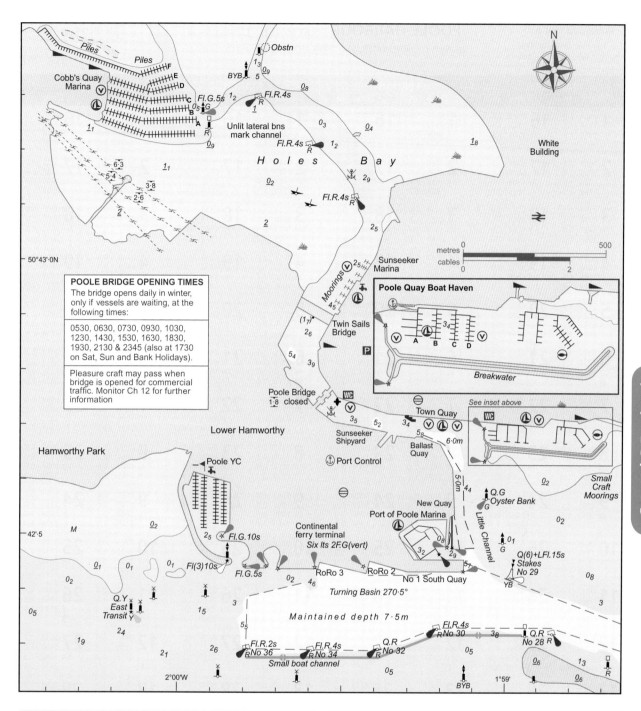

POOLE BRIDGE OPENING TIMES

The bridge opens daily in winter, only if vessels are waiting, at the following times:

0530, 0630, 0730, 0930, 1030, 1230, 1430, 1530, 1630, 1830, 1930, 2130 & 2345 (also at 1730 on Sat, Sun and Bank Holidays).

Pleasure craft may pass when bridge is opened for commercial traffic. Monitor Ch 12 for further information

SAFETY RECOMMENDATIONS

- Check LNTMs (www.phc.co.uk) to keep abreast of changes.
- Look out for large ferries/commercial ships which are limited in manoeuvrability and speed. **Stay out of exclusion zones.** Keep well clear and make your intentions obvious.
- Use designated small boat channels whenever possible.
- Be alert in the harbour entrance where the stream runs hard; it is a particularly hazardous area with great potential for an accident – **the chain ferry has right of way.**
- Keep to the speed limits and be particularly aware of the wash you are creating. This can cause injury to swimmers and other boaters or damage to other craft and property.

- Proceed slowly in the vicinity of Town Quay, marinas and moorings to minimise wash that can damage moored craft.
- Anticipate sailing craft manoeuvres, give them room to tack.
- Poole Harbour Guide is available from the Harbour Master and at www.phc.co.uk. Areas are designated for personal watercraft, waterskiing, wind & kite surfing. 'Quiet Zones' exclude PWC and have speed limits of 6kn.
- Before sailing check the tidal situation and comply with passage planning requirements.
- Bird sensitive areas exist in most bays and lakes in the harbour. Virtually all intertidal mud areas are nominated Sites of Special Scientific Interest (SSSI).

Note – Sea level is above mean tide level from 2hrs after LW to 2hrs before the next LW. HW occurs between 5hrs after LW and 3hrs before the next LW, the time shown is approximate and should be checked for suitability.

STANDARD TIME (UT)
For Summer Time add ONE hour in **non-shaded areas**

POOLE HARBOUR   LAT 50°43′N   LONG 1°59′W
HEIGHTS OF HIGH WATER AND TIMES AND HEIGHTS OF LOW WATERS

Dates in **red** are **SPRINGS**
Dates in blue are **NEAPS**

YEAR 2019

JANUARY

Day	Time	m	Day	Time	m
1 TU	0627 / 1234 / 1558	1.9 / 1.1 / 1.6	16 W	0809 / 1139 / 1524	1.9 / 1.1 / 1.6
2 W	0057 / 0701 / 1330 / 1659	0.9 / 1.9 / 1.0 / 1.7	17 TH	0004 / 0627 / 1244 / 1655	1.0 / 1.8 / 1.0 / 1.7
3 TH	0151 / 0736 / 1421 / 1955	0.9 / 1.9 / 0.9 / 1.8	18 F	0107 / 0654 / 1344 / 1918	0.9 / 1.9 / 0.8 / 1.8
4 F	0241 / 0808 / 1508 / 2029	0.8 / 2.0 / 0.8 / 1.8	19 SA	0205 / 0733 / 1440 / 2006	0.8 / 2.0 / 0.6 / 2.0
5 SA	0328 / 0825 / 1552 / 2053	0.8 / 2.0 / 0.7 / 1.9	20 SU	0300 / 0820 / 1531 / 2054	0.6 / 2.2 / 0.4 / 2.1
6 SU ●	0411 / 0836 / 1633 / 2119	0.8 / 2.1 / 0.6 / 2.0	21 M ○	0351 / 0908 / 1619 / 2139	0.5 / 2.3 / 0.3 / 2.2
7 M	0450 / 0920 / 1710 / 2157	0.8 / 2.2 / 0.6 / 2.0	22 TU	0440 / 0953 / 1706 / 2224	0.4 / 2.4 / 0.2 / 2.3
8 TU	0527 / 1003 / 1744 / 2239	0.9 / 2.1 / 0.7 / 1.9	23 W	0527 / 1036 / 1751 / 2306	0.4 / 2.4 / 0.2 / 2.3
9 W	0601 / 1045 / 1815 / 2323	0.9 / 2.1 / 0.7 / 1.9	24 TH	0614 / 1117 / 1837 / 2348	0.5 / 2.4 / 0.2 / 2.2
10 TH	0635 / 1115 / 1849 / 2359	1.0 / 1.9 / 0.7 / 1.7	25 F	0702 / 1155 / 1925	0.6 / 2.3 / 0.4
11 F	0712 / 1037 / 1927 / 2247	1.0 / 1.9 / 0.8 / 1.7	26 SA	0028 / 0751 / 1229 / 2015	2.0 / 0.7 / 2.1 / 0.6
12 SA	0752 / 1100 / 2009 / 2312	1.0 / 1.8 / 0.8 / 1.6	27 SU ☽	0107 / 0845 / 1249 / 2110	1.8 / 0.9 / 1.9 / 0.8
13 SU	0838 / 1131 / 2058 / 2351	1.1 / 1.8 / 0.9 / 1.6	28 M	0142 / 0946 / 1312 / 2215	1.7 / 1.1 / 1.7 / 1.0
14 M ☽	0932 / 1215 / 2155	1.1 / 1.2 / 1.0	29 TU	0219 / 1055 / 1407 / 2327	1.6 / 1.1 / 1.6 / 1.1
15 TU	0047 / 1033 / 1317 / 2259	1.6 / 1.2 / 1.6 / 1.0	30 W	0313 / 1205 / 1515	1.6 / 1.1 / 1.5
			31 TH	0034 / 0414 / 1308 / 1623	1.1 / 1.6 / 1.1 / 1.5

FEBRUARY

Day	Time	m	Day	Time	m
1 F	0134 / 0509 / 1404 / 1727	1.0 / 1.7 / 0.9 / 1.6	16 SA	0039 / 0641 / 1321 / 1912	1.0 / 1.8 / 0.9 / 1.8
2 SA	0226 / 0559 / 1452 / 1835	1.0 / 1.8 / 0.8 / 1.7	17 SU	0147 / 0724 / 1422 / 1959	0.9 / 1.9 / 0.6 / 1.8
3 SU	0314 / 0653 / 1537 / 2057	0.9 / 1.9 / 0.7 / 1.9	18 M	0246 / 0812 / 1516 / 2045	0.7 / 2.1 / 0.4 / 2.1
4 M ●	0357 / 0802 / 1617 / 2104	0.8 / 2.0 / 0.6 / 1.9	19 TU ○	0339 / 0857 / 1604 / 2127	0.5 / 2.3 / 0.2 / 2.2
5 TU	0435 / 0859 / 1653 / 2137	0.8 / 2.1 / 0.6 / 2.0	20 W	0426 / 0940 / 1649 / 2208	0.4 / 2.4 / 0.0 / 2.3
6 W	0510 / 0945 / 1724 / 2217	0.8 / 2.1 / 0.6 / 2.0	21 TH	0511 / 1021 / 1733 / 2247	0.3 / 2.5 / 0.0 / 2.3
7 TH	0541 / 1027 / 1754 / 2256	0.8 / 2.1 / 0.6 / 2.0	22 F	0555 / 1059 / 1817 / 2325	0.3 / 2.4 / 0.1 / 2.2
8 F	0612 / 1103 / 1825 / 2334	0.8 / 2.0 / 0.6 / 1.9	23 SA	0639 / 1136 / 1900	0.4 / 2.3 / 0.3
9 SA	0645 / 1120 / 1900 / 2237	0.8 / 1.9 / 0.6 / 1.7	24 SU	0001 / 0725 / 1207 / 1946	2.1 / 0.6 / 2.1 / 0.5
10 SU	0722 / 1037 / 1938 / 2240	0.8 / 1.9 / 0.7 / 1.7	25 M	0033 / 0812 / 1215 / 2035	1.9 / 0.8 / 1.9 / 0.8
11 M	0803 / 1059 / 2021 / 2315	0.9 / 1.8 / 0.8 / 1.7	26 TU ☽	0043 / 0905 / 1222 / 2134	1.7 / 1.0 / 1.7 / 1.1
12 TU ☽	0850 / 1140 / 2113	1.0 / 1.7 / 0.9	27 W	0101 / 1013 / 1306 / 2255	1.6 / 1.2 / 1.6 / 1.2
13 W	0003 / 0949 / 1232 / 2216	1.7 / 1.1 / 1.7 / 1.0	28 TH	0213 / 1134 / 1433	1.5 / 1.2 / 1.4
14 TH	0104 / 1058 / 1347 / 2328	1.6 / 1.1 / 1.6 / 1.0			
15 F	0404 / 1211 / 1830	1.6 / 1.0 / 1.7			

MARCH

Day	Time	m	Day	Time	m
1 F	0014 / 0653 / 1246 / 2142	1.2 / 1.7 / 1.1 / 1.8	16 SA	0350 / 1146 / 2011	1.6 / 1.0 / 1.7
2 SA	0118 / 0721 / 1344 / 1959	1.2 / 1.7 / 1.0 / 1.7	17 SU	0022 / 0627 / 1303 / 1908	1.1 / 1.8 / 0.9 / 1.8
3 SU	0211 / 0540 / 1433 / 2029	1.1 / 1.7 / 0.9 / 1.7	18 M	0135 / 0711 / 1406 / 1949	1.0 / 1.9 / 0.6 / 1.9
4 M	0256 / 0630 / 1516 / 2049	1.0 / 1.8 / 0.7 / 1.8	19 TU	0233 / 0758 / 1459 / 2029	0.7 / 2.1 / 0.4 / 2.1
5 TU	0338 / 0735 / 1556 / 2042	0.8 / 1.9 / 0.6 / 1.9	20 W	0323 / 0841 / 1546 / 2109	0.5 / 2.3 / 0.2 / 2.3
6 W ●	0415 / 0838 / 1630 / 2112	0.7 / 2.0 / 0.6 / 2.0	21 TH ○	0409 / 0922 / 1630 / 2147	0.3 / 2.4 / 0.1 / 2.4
7 TH	0447 / 0924 / 1700 / 2151	0.7 / 2.1 / 0.5 / 2.1	22 F	0451 / 1001 / 1711 / 2224	0.3 / 2.5 / 0.1 / 2.4
8 F	0517 / 1005 / 1729 / 2228	0.6 / 2.2 / 0.5 / 2.1	23 SA	0533 / 1038 / 1753 / 2300	0.3 / 2.4 / 0.2 / 2.3
9 SA	0547 / 1042 / 1800 / 2303	0.6 / 2.1 / 0.5 / 2.0	24 SU	0615 / 1114 / 1835 / 2333	0.4 / 2.3 / 0.4 / 2.1
10 SU	0619 / 1114 / 1833 / 2331	0.6 / 2.0 / 0.6 / 1.9	25 M	0657 / 1145 / 1918 / 2359	0.6 / 2.1 / 0.6 / 1.9
11 M	0653 / 1126 / 1908 / 2214	0.7 / 1.9 / 0.6 / 1.8	26 TU	0741 / 1147 / 2004 / 2340	0.6 / 1.8 / 0.9 / 1.8
12 TU	0731 / 1035 / 1949 / 2249	0.7 / 1.9 / 0.8 / 1.8	27 W	0829 / 1145 / 2057	1.0 / 1.7 / 0.9
13 W	0814 / 1115 / 2036 / 2335	0.9 / 1.8 / 0.9 / 1.8	28 TH ☽	0002 / 0926 / 1221 / 2218	1.7 / 1.1 / 1.6 / 1.3
14 TH ☽	0907 / 1205 / 2135	1.0 / 1.7 / 1.1	29 F	0044 / 1129 / 1319 / 2350	1.6 / 1.0 / 1.4 / 1.4
15 F	0031 / 1020 / 1317 / 2258	1.7 / 1.1 / 1.6 / 1.2	30 SA	0621 / 1216 / 2113	1.7 / 1.2 / 1.9
			31 SU	0055 / 0654 / 1316 / 1946	1.3 / 1.7 / 1.1 / 1.7

APRIL

Day	Time	m	Day	Time	m
1 M	0147 / 0541 / 1406 / 2008	1.2 / 1.6 / 0.9 / 1.8	16 TU	0118 / 0651 / 1344 / 1931	1.0 / 1.9 / 0.6 / 2.0
2 TU	0231 / 0614 / 1448 / 2014	1.0 / 1.7 / 0.8 / 1.8	17 W	0213 / 0737 / 1436 / 2009	0.8 / 2.0 / 0.4 / 2.1
3 W	0312 / 0714 / 1527 / 2006	0.9 / 1.9 / 0.7 / 1.9	18 TH	0302 / 0820 / 1523 / 2046	0.6 / 2.2 / 0.3 / 2.3
4 TH	0348 / 0813 / 1601 / 2043	0.8 / 2.0 / 0.6 / 2.1	19 F ○	0347 / 0900 / 1607 / 2123	0.4 / 2.3 / 0.2 / 2.4
5 F ●	0420 / 0859 / 1632 / 2121	0.7 / 2.1 / 0.5 / 2.2	20 SA	0430 / 0938 / 1649 / 2158	0.3 / 2.4 / 0.2 / 2.4
6 SA	0450 / 0940 / 1702 / 2159	0.6 / 2.2 / 0.5 / 2.2	21 SU	0511 / 1016 / 1730 / 2232	0.3 / 2.3 / 0.4 / 2.3
7 SU	0521 / 1018 / 1734 / 2233	0.5 / 2.2 / 0.5 / 2.1	22 M	0552 / 1051 / 1811 / 2304	0.4 / 2.2 / 0.6 / 2.2
8 M	0554 / 1054 / 1808 / 2305	0.6 / 2.1 / 0.6 / 2.0	23 TU	0633 / 1125 / 1854 / 2329	0.6 / 2.0 / 0.8 / 2.0
9 TU	0628 / 1129 / 1844 / 2159	0.6 / 2.0 / 0.7 / 1.9	24 W	0715 / 1143 / 1938 / 2305	0.8 / 1.8 / 1.0 / 1.8
10 W	0705 / 1022 / 1924 / 2232	0.7 / 1.9 / 0.8 / 1.9	25 TH	0759 / 1118 / 2028 / 2331	0.9 / 1.7 / 1.2 / 1.7
11 TH	0748 / 1101 / 2011 / 2318	0.8 / 1.8 / 1.0 / 1.8	26 F ☽	0848 / 1153 / 2133	1.1 / 1.6 / 1.4
12 F ☽	0841 / 1155 / 2113	0.9 / 1.7 / 1.2	27 SA	0010 / 0950 / 1242 / 2312	1.7 / 1.2 / 1.5 / 1.2
13 SA	0017 / 0958 / 1859 / 2246	1.7 / 1.0 / 1.9 / 1.2	28 SU	0552 / 1129 / 2033	1.7 / 1.2 / 1.9
14 SU	0329 / 1129 / 1951	1.7 / 1.0 / 1.8	29 M	0019 / 0622 / 1236 / 2137	1.4 / 1.6 / 1.2 / 1.9
15 M	0011 / 0605 / 1244 / 1854	1.2 / 1.8 / 0.8 / 1.9	30 TU	0112 / 0549 / 1328 / 1943	1.2 / 1.6 / 1.1 / 1.8

Chart Datum: 1·40 metres below Ordnance Datum (Newlyn). HAT is 2·6 metres above Chart Datum.

FREE monthly updates. Register at
www.reedsnauticalalmanac.co.uk

Note – Sea level is above mean tide level from 2hrs after LW to 2hrs before the next LW. HW occurs between 5hrs after LW and 3hrs before the next LW, the time shown is approximate and should be checked for suitability.

STANDARD TIME (UT)
For Summer Time add ONE hour in **non-shaded areas**

POOLE HARBOUR   LAT 50°43'N   LONG 1°59'W
HEIGHTS OF HIGH WATER AND TIMES AND HEIGHTS OF LOW WATERS

Dates in red are **SPRINGS**
Dates in blue are NEAPS

YEAR 2019

MAY

Day	Time m	Time m	Day Time m	Time m
1	0157 1.1	0610 1.7	W 1411 0.9	1914 1.9
16	0148 0.8	0716 2.0	TH 1410 0.6	1947 2.1
2	0237 1.0	0656 1.8	TH 1449 0.8	1935 2.0
17	0238 0.7	0759 2.1	F 1458 0.5	2023 2.2
3	0313 0.8	0747 2.0	F 1525 0.7	2011 2.1
18	0324 0.5	0839 2.2	SA 1543 0.4	○ 2058 2.3
4	0347 0.7	0833 2.1	SA 1600 0.6 ●	2050 2.2
19	0408 0.5	0917 2.2	SU 1627 0.5	2131 2.3
5	0421 0.6	0915 2.2	SU 1635 0.6	2128 2.3
20	0449 0.5	0953 2.2	M 1708 0.6	2204 2.3
6	0456 0.5	0956 2.2	M 1710 0.6	2205 2.2
21	0530 0.5	1029 2.1	TU 1751 0.8	2237 2.2
7	0532 0.5	1035 2.2	TU 1747 0.7	2242 2.2
22	0611 0.6	1106 2.0	W 1833 0.9	2307 2.0
8	0609 0.6	1117 2.1	W 1827 0.8	2317 2.1
23	0652 0.8	1145 1.8	TH 1917 1.1	2244 1.9
9	0650 0.7	1202 2.0	TH 1911 0.9	2234 2.0
24	0734 0.9	1058 1.7	F 2003 1.2	2308 1.8
10	0737 0.8	1255 1.8	F 2004 1.1	2322 1.9
25	0817 1.0	1131 1.6	SA 2053 1.3	2346 1.7
11	0834 0.9	1403 1.8	SA 2114 1.2 ◑	
26	0905 1.1	1215 1.6	SU 2151 1.4	
12	0153 1.8	0951 0.9	SU 1852 2.0	◑ 2238 1.2
27	0033 1.7	1001 1.2	M 1312 1.5	2309 1.4
13	0259 1.8	1110 0.9	M 1951 1.9	2351 1.1
28	0132 1.6	1107 1.2	TU 2029 1.9	
14	0400 1.8	1218 0.8	TU 1832 1.9	
29	0018 1.3	0308 1.6	W 1221 1.1	2134 1.9
15	0053 1.0	0629 1.9	W 1317 0.7	1909 2.0
30	0108 1.2	0604 1.7	TH 1316 1.1	1846 1.9
31	0151 1.1	0641 1.8	F 1402 0.9	1909 2.0

JUNE

Day	Time m	Time m	Day Time m	Time m
1	0231 0.9	0724 1.9	SA 1445 0.8	1938 2.1
16	0302 0.7	0824 2.0	SU 1522 0.7	2037 2.2
2	0312 0.7	0809 2.0	SU 1527 0.7	2016 2.2
17	0348 0.6	0900 2.1	M 1608 0.7	○ 2106 2.2
3	0353 0.6	0854 2.1	M 1609 0.6 ●	2100 2.3
18	0432 0.6	0934 2.1	TU 1652 0.8	2138 2.2
4	0433 0.5	0938 2.2	TU 1650 0.6	2144 2.3
19	0513 0.6	1010 2.1	W 1734 0.9	2214 2.2
5	0514 0.5	1022 2.2	W 1732 0.7	2229 2.3
20	0553 0.7	1048 2.0	TH 1815 1.0	2252 2.1
6	0557 0.5	1107 2.2	TH 1817 0.8	2313 2.2
21	0632 0.8	1130 1.9	F 1855 1.1	2324 2.0
7	0642 0.6	1154 2.1	F 1907 0.9	2358 2.1
22	0709 0.8	1221 1.8	SA 1935 1.2	2252 1.9
8	0733 0.7	1244 2.0	SA 2003 1.0	
23	0747 0.9	1111 1.7	SU 2017 1.2	2324 1.8
9	0044 2.0	0831 0.8	SU 1338 1.9	2109 1.1
24	0830 1.0	1148 1.7	M 2103 1.3	
10	0133 1.9	0936 0.8	M 1441 1.8	◑ 2217 1.1
25	0003 1.7	0918 1.1	TU 1233 1.6	◑ 2156 1.3
11	0224 1.8	1044 0.9	TU 1940 2.0	2324 1.1
26	0052 1.7	1013 1.1	W 1333 1.6	2255 1.3
12	0316 1.8	1149 0.9	W 1813 1.9	
27	0154 1.6	1113 1.1	TH 2026 1.9	2357 1.2
13	0025 1.0	0412 1.8	TH 1248 0.8	1850 2.0
28	0319 1.6	1215 1.1	F 1826 1.8	
14	0121 0.9	0700 1.9	F 1344 0.8	1928 2.0
29	0056 1.7	0439 1.7	SA 1313 1.0	1726 1.9
15	0214 0.8	0744 1.9	SA 1435 0.7	2004 2.1
30	0150 0.9	0545 1.8	SU 1407 0.9	1814 2.0

JULY

Day	Time m	Time m	Day Time m	Time m
1	0240 0.8	0752 2.0	M 1458 0.8	1950 2.1
16	0333 0.7	0854 1.9	TU 1555 0.9	○ 2045 2.1
2	0328 0.6	0840 2.1	TU 1547 0.7 ●	2047 2.3
17	0417 0.7	0923 2.0	W 1638 0.9	2113 2.2
3	0414 0.5	0928 2.2	W 1634 0.7	2136 2.3
18	0458 0.6	0952 2.0	TH 1718 0.9	2152 2.2
4	0500 0.4	1014 2.3	TH 1720 0.7	2223 2.4
19	0536 0.7	1028 2.0	F 1756 0.9	2233 2.2
5	0545 0.4	1058 2.3	F 1808 0.7	2307 2.4
20	0610 0.7	1107 2.0	SA 1830 1.0	2312 2.1
6	0632 0.4	1143 2.2	SA 1857 0.8	2351 2.3
21	0642 0.8	1148 1.9	SU 1904 1.0	2300 1.9
7	0721 0.5	1227 2.1	SU 1950 0.9	
22	0716 0.8	1103 1.8	M 1941 1.1	2303 1.9
8	0032 2.2	0814 0.6	M 1312 2.0	2047 1.0
23	0755 0.9	1121 1.8	TU 2024 1.1	2332 1.8
9	0110 2.0	0911 0.8	TU 1357 1.9	◑ 2149 1.1
24	0840 0.9	1153 1.7	W 2113 1.2	
10	0147 1.9	1014 0.9	W 1443 1.8	2254 1.1
25	0008 1.8	0932 1.0	TH 1236 1.7	◑ 2210 1.2
11	0232 1.8	1120 0.9	TH 1535 1.7	2357 1.1
26	0059 1.7	1030 1.1	F 1339 1.7	2311 1.2
12	0326 1.7	1223 1.0	F 1840 1.9	
27	0214 1.7	1133 1.1	SA 1532 1.7	
13	0057 1.0	0427 1.7	SA 1322 1.0	1917 2.0
28	0014 1.1	0404 1.7	SU 1236 1.1	1648 1.9
14	0154 0.9	0532 1.7	SU 1417 0.9	1952 2.0
29	0116 1.0	0522 1.7	M 1338 1.0	1747 2.0
15	0245 0.8	0819 1.9	M 1508 0.9	2023 2.1
30	0214 0.8	0743 1.9	TU 1436 0.9	1950 2.1
31	0308 0.6	0833 2.1	W 1529 0.7	2042 2.3

AUGUST

Day	Time m	Time m	Day Time m	Time m
1	0358 0.4	0919 2.2	TH 1619 0.6 ●	2129 2.4
16	0440 0.6	0930 2.1	F 1659 0.9	2128 2.3
2	0445 0.3	1004 2.3	F 1706 0.6	2214 2.5
17	0515 0.6	1002 2.1	SA 1733 0.9	2210 2.3
3	0531 0.2	1046 2.4	SA 1753 0.6	2256 2.5
18	0545 0.6	1039 2.1	SU 1803 0.9	2249 2.2
4	0616 0.2	1127 2.3	SU 1839 0.6	2336 2.4
19	0614 0.7	1116 2.0	M 1833 0.9	2322 2.1
5	0702 0.3	1207 2.2	M 1927 0.7	
20	0645 0.7	1147 1.9	TU 1907 0.9	2252 1.9
6	0014 2.3	0750 0.5	TU 1245 2.1	2019 0.9
21	0721 0.8	1054 1.8	W 1946 1.0	2253 1.9
7	0047 2.1	0842 0.7	W 1321 1.9 ●	2117 1.0
22	0801 0.9	1106 1.8	TH 2031 1.1	2322 1.8
8	0111 1.9	0942 0.9	TH 1353 1.8	2223 1.1
23	0849 1.0	1144 1.8	F 2126 1.1	◑
9	0150 1.7	1052 1.1	F 1441 1.7	2332 1.2
24	0007 1.1	0948 1.1	SA 1237 1.8	2232 1.2
10	0250 1.6	1202 1.2	SA 1545 1.7	
25	0112 1.7	1056 1.2	SU 1406 1.7	2342 1.1
11	0038 1.1	0359 1.6	SU 1307 1.2	1909 1.9
26	0615 1.7	1208 1.2	M 1819 1.9	
12	0138 1.0	0506 1.6	M 1405 1.1	1944 1.9
27	0051 1.0	0651 1.8	TU 1318 1.1	1856 2.0
13	0231 0.9	0820 1.8	TU 1455 1.0	2014 2.0
28	0156 0.8	0737 1.9	W 1421 0.9	1943 2.1
14	0318 0.8	0852 1.9	W 1541 0.9	2018 2.1
29	0251 0.6	0824 2.1	TH 1515 0.7	2032 2.3
15	0401 0.7	0913 2.0	TH 1622 0.9	○ 2044 2.2
30	0342 0.3	0907 2.3	F 1604 0.6 ●	2116 2.5
31	0427 0.2	0948 2.4	SA 1649 0.4	2159 2.6

Chart Datum: 1·40 metres below Ordnance Datum (Newlyn). HAT is 2·6 metres above Chart Datum.

Central S England

》》 FREE monthly updates. Register at 《 www.reedsnauticalalmanac.co.uk

93

Note – Sea level is above mean tide level from 2hrs after LW to 2hrs before the next LW. HW occurs between 5hrs after LW and 3hrs before the next LW, the time shown is approximate and should be checked for suitability.

STANDARD TIME (UT)
For Summer Time add ONE hour in **non-shaded areas**

POOLE HARBOUR LAT 50°43'N LONG 1°59'W
HEIGHTS OF HIGH WATER AND TIMES AND HEIGHTS OF LOW WATERS

Dates in red are **SPRINGS**
Dates in blue are **NEAPS**

YEAR **2019**

SEPTEMBER

Time	m	Time	m
1 0511 / 1028 / SU 1733 / 2239	0.1 / 2.5 / 0.4 / 2.6	**16** 0517 / 1009 / M 1734 / 2224	0.6 / 2.2 / 0.8 / 2.3
2 0554 / 1106 / M 1816 / 2318	0.1 / 2.4 / 0.5 / 2.5	**17** 0545 / 1045 / TU 1802 / 2259	0.6 / 2.2 / 0.8 / 2.2
3 0637 / 1144 / TU 1901 / 2354	0.3 / 2.3 / 0.6 / 2.3	**18** 0615 / 1114 / W 1834 / 2324	0.7 / 2.1 / 0.8 / 2.0
4 0722 / 1219 / W 1949	0.5 / 2.1 / 0.8	**19** 0648 / 1027 / TH 1910 / 2216	0.8 / 1.9 / 0.9 / 1.9
5 0025 / 0811 / TH 1245 / 2043	2.1 / 0.8 / 1.9 / 1.0	**20** 0726 / 1029 / F 1951 / 2248	0.9 / 1.9 / 1.0 / 1.9
6 0037 / 0909 / F 1303 / ◑ 2151	1.8 / 1.1 / 1.8 / 1.2	**21** 0809 / 1109 / SA 2042 / 2334	1.0 / 1.9 / 1.1 / 1.8
7 0109 / 1026 / SA 1355 / 2309	1.7 / 1.3 / 1.7 / 1.2	**22** 0903 / 1127 / SU 2152 / ◑	1.2 / 1.8 / 1.2
8 0224 / 1146 / SU 1824	1.6 / 1.3 / 1.9	**23** 0035 / 1024 / M 1314 / 2317	1.6 / 1.3 / 1.7 / 1.2
9 0020 / 0916 / M 1253 / 1852	1.2 / 2.0 / 1.3 / 1.9	**24** 0750 / 1150 / TU 1806	1.8 / 1.3 / 1.9
10 0120 / 0734 / TU 1349 / 1926	1.1 / 1.8 / 1.2 / 1.9	**25** 0034 / 0645 / W 1305 / 1843	1.0 / 1.9 / 1.3 / 2.0
11 0212 / 0807 / W 1437 / 1820	0.9 / 1.8 / 1.1 / 1.9	**26** 0139 / 0725 / TH 1406 / 1929	0.8 / 2.0 / 0.9 / 2.1
12 0257 / 0836 / TH 1521 / 1916	0.8 / 1.9 / 0.9 / 2.0	**27** 0232 / 0807 / F 1457 / 2015	0.5 / 2.2 / 0.7 / 2.3
13 0339 / 0847 / F 1600 / 2014	0.7 / 2.0 / 0.9 / 2.2	**28** 0321 / 0848 / SA 1544 / ● 2058	0.3 / 2.3 / 0.5 / 2.5
14 0416 / 0856 / SA 1636 / ○ 2102	0.6 / 2.1 / 0.8 / 2.3	**29** 0405 / 0927 / SU 1628 / 2139	0.1 / 2.5 / 0.4 / 2.6
15 0448 / 0932 / SU 1706 / 2145	0.6 / 2.2 / 0.8 / 2.3	**30** 0448 / 1005 / M 1710 / 2219	0.1 / 2.5 / 0.4 / 2.6

OCTOBER

Time	m	Time	m
1 0530 / 1042 / TU 1752 / 2257	0.2 / 2.5 / 0.4 / 2.5	**16** 0516 / 1012 / W 1735 / 2235	0.7 / 2.2 / 0.7 / 2.2
2 0612 / 1118 / W 1835 / 2333	0.4 / 2.3 / 0.6 / 2.3	**17** 0548 / 1041 / TH 1807 / 2309	0.7 / 2.2 / 0.7 / 2.1
3 0656 / 1150 / TH 1920	0.7 / 2.1 / 0.8	**18** 0621 / 1052 / F 1842 / 2154	0.8 / 2.0 / 0.8 / 1.9
4 0005 / 0743 / F 1204 / 2011	2.0 / 0.9 / 2.0 / 1.0	**19** 0659 / 1006 / SA 1923 / 2227	0.9 / 2.0 / 0.9 / 1.9
5 0000 / 0839 / SA 1208 / ◑ 2115	1.8 / 1.2 / 1.8 / 1.2	**20** 0742 / 1048 / SU 2011 / 2315	1.1 / 1.9 / 1.0 / 1.7
6 0023 / 1002 / SU 1250 / 2242	1.6 / 1.4 / 1.7 / 1.3	**21** 0836 / 1140 / M 2121 / ◑	1.3 / 1.8 / 1.1
7 0733 / 1127 / M 1749 / 2355	2.0 / 1.4 / 1.8 / 1.2	**22** 0026 / 0630 / TU 1304 / 2258	1.6 / 2.0 / 1.7 / 1.1
8 0853 / 1232 / TU 1822	2.0 / 1.3 / 1.8	**23** 0720 / 1139 / W 1746	1.9 / 1.3 / 1.9
9 0054 / 0708 / W 1325 / 1855	1.1 / 1.9 / 1.3 / 1.8	**24** 0015 / 0630 / TH 1249 / 1825	1.0 / 2.0 / 1.1 / 2.0
10 0144 / 0740 / TH 1412 / 1814	0.9 / 1.9 / 1.1 / 1.9	**25** 0116 / 0706 / F 1346 / 1910	0.7 / 2.1 / 0.9 / 2.1
11 0229 / 0802 / F 1454 / 1856	0.9 / 1.9 / 1.0 / 1.9	**26** 0208 / 0746 / SA 1435 / 1955	0.5 / 2.2 / 0.7 / 2.3
12 0310 / 0750 / SA 1532 / 1948	0.8 / 2.0 / 0.5 / 2.1	**27** 0256 / 0825 / SU 1521 / 2038	0.4 / 2.3 / 0.5 / 2.4
13 0346 / 0819 / SU 1606 / ○ 2036	0.7 / 2.1 / 0.8 / 2.2	**28** 0341 / 0903 / M 1605 / ● 2118	0.3 / 2.4 / 0.4 / 2.5
14 0418 / 0858 / M 1636 / 2118	0.7 / 2.2 / 0.7 / 2.3	**29** 0424 / 0941 / TU 1647 / 2157	0.3 / 2.4 / 0.4 / 2.5
15 0447 / 0937 / TU 1705 / 2158	0.6 / 2.3 / 0.7 / 2.3	**30** 0506 / 1016 / W 1729 / 2236	0.5 / 2.4 / 0.5 / 2.4
		31 0548 / 1051 / TH 1811 / 2313	0.5 / 2.3 / 0.6 / 2.2

NOVEMBER

Time	m	Time	m
1 0632 / 1122 / F 1855 / 2351	0.8 / 2.2 / 0.8 / 2.0	**16** 0604 / 1042 / SA 1826 / 2339	0.8 / 2.1 / 0.7 / 2.0
2 0719 / 1118 / SA 1943 / 2323	1.1 / 2.0 / 1.0 / 1.8	**17** 0645 / 1003 / SU 1909 / 2228	1.0 / 2.0 / 0.8 / 1.8
3 0813 / 1128 / SU 2039 / 2346	1.3 / 1.8 / 1.1 / 1.6	**18** 0732 / 1043 / M 2001 / 2321	1.1 / 2.0 / 0.9 / 1.7
4 0926 / 1204 / M 2157 / ◑	1.4 / 1.7 / 1.2	**19** 0831 / 1140 / TU 2110 / ◑	1.3 / 1.9 / 1.0
5 0036 / 1054 / TU 1257 / 2318	1.5 / 1.5 / 1.6 / 1.2	**20** 0621 / 0956 / W 1436 / 2236	2.0 / 1.3 / 1.8 / 1.0
6 0816 / 1159 / W 1754	2.0 / 1.4 / 1.8	**21** 0718 / 1118 / TH 1545 / 2348	2.0 / 1.3 / 1.8 / 0.9
7 0018 / 0647 / TH 1253 / 1821	1.2 / 1.9 / 1.3 / 1.8	**22** 0611 / 1224 / F 1805	2.0 / 1.1 / 1.9
8 0110 / 0712 / F 1340 / 1818	1.1 / 1.9 / 1.2 / 1.8	**23** 0048 / 0646 / SA 1320 / 1851	0.8 / 2.0 / 0.9 / 2.0
9 0155 / 0720 / SA 1422 / 1844	1.0 / 1.9 / 1.0 / 1.9	**24** 0142 / 0725 / SU 1411 / 1936	0.6 / 2.1 / 0.7 / 2.1
10 0234 / 0718 / SU 1459 / 1927	0.9 / 2.0 / 0.9 / 2.0	**25** 0231 / 0803 / M 1458 / 2018	0.5 / 2.2 / 0.6 / 2.2
11 0310 / 0746 / M 1533 / 2010	0.8 / 2.1 / 0.8 / 2.1	**26** 0318 / 0841 / TU 1543 / ● 2058	0.4 / 2.3 / 0.5 / 2.3
12 0344 / 0823 / TU 1605 / ○ 2053	0.7 / 2.2 / 0.7 / 2.2	**27** 0402 / 0917 / W 1627 / 2137	0.5 / 2.4 / 0.5 / 2.3
13 0417 / 0901 / W 1638 / 2133	0.7 / 2.3 / 0.6 / 2.2	**28** 0446 / 0952 / TH 1709 / 2216	0.5 / 2.4 / 0.5 / 2.2
14 0451 / 0938 / TH 1712 / 2213	0.7 / 2.3 / 0.6 / 2.2	**29** 0529 / 1026 / F 1752 / 2255	0.7 / 2.3 / 0.6 / 2.1
15 0526 / 1012 / F 1748 / 2254	0.8 / 2.2 / 0.7 / 2.1	**30** 0613 / 1059 / SA 1835 / 2336	0.9 / 2.1 / 0.7 / 1.9

DECEMBER

Time	m	Time	m
1 0659 / 1110 / SU 1919 / 2359	1.1 / 2.0 / 0.9 / 1.8	**16** 0639 / 1129 / M 1904	0.9 / 2.1 / 0.7
2 0747 / 1101 / M 2005 / 2319	1.2 / 1.9 / 1.0 / 1.6	**17** 0022 / 0729 / TU 1217 / 1956	1.9 / 1.0 / 2.0 / 0.7
3 0839 / 1134 / TU 2054 / 2359	1.3 / 1.8 / 1.1 / 1.6	**18** 0116 / 0826 / W 1308 / 2056	1.8 / 1.1 / 1.9 / 0.8
4 0943 / 1217 / W 2153 / ◑	1.4 / 1.7 / 1.2	**19** 0219 / 0934 / TH 1401 / ◑ 2206	1.8 / 1.2 / 1.9 / 0.9
5 0055 / 1105 / TH 1312 / 2316	1.5 / 1.4 / 1.6 / 1.2	**20** 0706 / 1047 / F 1457 / 2316	2.0 / 1.2 / 1.8 / 0.9
6 0817 / 1209 / F 1814	1.9 / 1.3 / 1.7	**21** 0803 / 1154 / SA 1557	2.0 / 1.1 / 1.8
7 0020 / 0923 / SA 1259 / 1810	1.2 / 1.9 / 1.2 / 1.7	**22** 0019 / 0632 / SU 1254 / 1839	0.8 / 2.0 / 1.0 / 1.9
8 0108 / 0657 / SU 1341 / 1835	1.1 / 1.9 / 1.1 / 1.8	**23** 0116 / 0710 / M 1348 / 1924	0.7 / 2.0 / 0.8 / 1.9
9 0150 / 0701 / M 1419 / 1909	1.0 / 1.9 / 1.0 / 1.9	**24** 0209 / 0748 / TU 1439 / 2006	0.7 / 2.1 / 0.7 / 2.0
10 0230 / 0718 / TU 1457 / 1948	0.9 / 2.0 / 0.8 / 2.0	**25** 0258 / 0825 / W 1526 / 2046	0.7 / 2.2 / 0.6 / 2.0
11 0310 / 0732 / W 1536 / 2031	0.8 / 2.1 / 0.7 / 2.1	**26** 0346 / 0859 / TH 1611 / ● 2123	0.7 / 2.2 / 0.5 / 2.1
12 0350 / 0819 / TH 1615 / ○ 2115	0.7 / 2.2 / 0.6 / 2.1	**27** 0431 / 0932 / F 1654 / 2200	0.7 / 2.2 / 0.5 / 2.1
13 0430 / 0909 / F 1655 / 2159	0.7 / 2.3 / 0.6 / 2.2	**28** 0514 / 1005 / SA 1736 / 2237	0.8 / 2.2 / 0.6 / 2.0
14 0511 / 0958 / SA 1735 / 2245	0.7 / 2.3 / 0.5 / 2.1	**29** 0557 / 1040 / SU 1816 / 2317	0.9 / 2.2 / 0.6 / 1.9
15 0554 / 1043 / SU 1818 / 2332	0.8 / 2.2 / 0.6 / 2.0	**30** 0639 / 1115 / M 1855 / 2359	1.0 / 2.0 / 0.7 / 1.8
		31 0719 / 1050 / TU 1933 / 2301	1.1 / 1.9 / 0.8 / 1.7

Chart Datum: 1·40 metres below Ordnance Datum (Newlyn). HAT is 2·6 metres above Chart Datum.

SWANAGE TO CHRISTCHURCH BAY

Studland Bay (AC 2172) is a good ⚓ especially in winds from the S and W. Poole Bay offers good sailing in waters sheltered from the W and N, with no offlying dangers.

▶ *1M NE of Durlston Head, Peveril Ledge runs 2½ca seaward, causing a significant race which extends nearly 1M eastwards, particularly on W-going stream against a SW wind. Proceeding towards the excellent shelter of Poole Harbour, overfalls may be met off Ballard Pt and Old Harry on the W-going stream.*◀

(AC 2615) There is deep water quite close inshore between St Alban's Hd and Anvil Pt.

▶ *Tidal streams are weak North of a line between Handfast Point and Hengistbury Head and within Christchurch Bay. The tide runs hard over the ledge at springs, and there may be overfalls.*◀

Hengistbury Hd is a dark reddish headland S of Christchurch Hbr with a groyne extending 1ca S and Beerpan Rks a further 100m E of groyne. Beware lobster pots in this area. Christchurch Ledge extends 2·75M SE from Hengistbury Hd. There is a passage between Christchurch Ledge and Dolphin Bank, some 1½M further SE. A wreck hazardous to surface navigation (PA) has been reported in this chan (2013).

WESTERN APPROACHES TO THE SOLENT

(AC 2035) The Needles are distinctive rocks at the W end of the Isle of Wight. The adjacent chalk cliffs of High Down are conspic from afar, but the Lt Ho may not be seen by day until relatively close. Goose Rk, which dries, is about 50m WNW of the Lt Ho, 100-150m WSW of which is a drying wreck. When rounding the Needles an offing of 1½ca will clear these. The NW side of Needles Chan is defined by the Shingles bank, parts of which dry and on which the sea breaks violently in the least swell. The SE side of the bank is fairly steep-to; the NW side shelves more gradually.

▶ *The ENE-going flood runs from HW Portsmouth +0500 until HW −0130, Sp rates 3.1kn at The Bridge, a reef which runs 8ca W of the Lt ho and is marked by WCM, and 3.9kn at Hurst. The WSW-going ebb runs from HW −0100 until HW +0430, sp rates 4.4kn at Hurst, 3.4kn WSW across the Shingles and at The Bridge. The Needles Chan is well marked and in fair weather presents no significant problems, but in a SW 4 or above over the ebb, breaking seas occur near The Bridge and SW Shingles. In bad weather broken water and overfalls extend along The Bridge. S to W gales against the ebb raise very dangerous breaking seas in the Needles Chan and approaches. The sea state can be at its worst shortly after LW when the flood has just begun. There is then no wind-over-tide situation, but a substantial swell is raised as a result of the recently turned stream.* ◀

In such conditions use the E route to the Solent, S of the IoW and via Nab Tower; or find shelter at Poole or Studland.

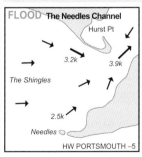

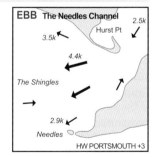

In strong winds the North Channel, N of the Shingles, is preferable to the Needles Channel. The two join S of Hurst Pt, where overfalls and tide rips may be met. Beware The Trap, a very shallow gravel shoal spit, 150m SE of Hurst Castle.

In E winds Alum Bay, with its coloured cliffs close NE of the Needles, is an attractive daytime ⚓ but beware Long Rk (dries) in the middle of the bay, and Five Fingers Rock 1½ca SW of Hatherwood Pt on N side. Totland Bay is good ⚓ in settled weather, but avoid Warden Ledge.

THE SOLENT

(AC 2035, 2036, 2037) There are few dangers in mid-chan. The most significant is Bramble Bank (dries) between Cowes and Calshot. The main fairway (buoyed) passes S and W of the Brambles, but yachts can use the North Chan to the NE of the Brambles at any state of tide.

▶ *Tidal streams are strong at Sp, see 2.14.* ◀

A Precautionary Area between Cowes and Calshot provides priority and safety for large commercial shipping. There are many yacht racing buoys *(seasonal, Mar–Dec)* in the Solent and Southampton Water. Most are fitted with a Fl Y 4s light.

Several inshore spits, banks, rocks and ledges include: Pennington and Lymington Spits on the N shore; Black Rk 4ca W of entrance to Yarmouth; Hamstead Ledge 8ca W of entrance to Newtown River and Saltmead Ledge 1·5M to E; Gurnard Ledge 1·5M W of Cowes; Lepe Middle and Beaulieu Spit, S and W of the ent to Beaulieu R; the shoals off Stone Pt, marked by Lepe Spit SCM buoy; Shrape Mud, which extends N from the breakwater of Cowes Hbr and along to Old Castle Pt; the shoals and isolated rks which fringe the island shore from Old Castle Pt to Ryde, including either side of the ent to Wootton Creek; and Calshot Spit which extends almost into the deep water chan into Southampton Water.

Southampton Water is a commercial waterway with large tankers, container ships, frequent Ro-Ro and High Speed ferries. Yachts should monitor VHF Ch 12 (Southampton VTS) for shipping movements. Between the Fawley Marine Terminal and the BP jetty on the E side the fairway is narrow with little room for yachts to take refuge. To the N there is adequate water for yachts close outboard of the main buoyed channel; the banks are of gently shelving soft mud, apart from foul ground between Hythe and Marchwood. At night, unlit marks and large mooring buoys may be hard to see against the many shore lights. Except in strong N'lies, Southampton Water and the R Test and Itchen provide sheltered sailing. The R Hamble is convenient, but crowded.

Depending on the wind direction, there are many good ⚓s: in W winds on E side of Hurst, as close inshore as depth permits, NE of Hurst lt; in S winds, or in good weather, W of Yarmouth Harbour entrance, as near shore as possible; in winds between W and N in Stanswood Bay, about 1M NE of Stansore Pt. Just N of Calshot Spit there is shelter from SW and W, while Osborne Bay, 2M E of Cowes, is sheltered from winds between S and W; in E winds Gurnard Bay, to the W of Cowes, is preferable. Stokes Bay is well sheltered from the N, and at the E end of the IoW there is good ⚓ in Priory Bay off Bembridge in winds from S, SW or W; but clear out if wind goes into E.

Shoal-draught boats can explore the estuaries at the top of the tide: Ashlett Creek between Fawley and Calshot, Eling up the R Test, and the upper reaches of the R Medina.

ISLE OF WIGHT (SOUTH COAST)

(AC 2045) From the Needles eastward to Freshwater Bay the cliffs can be approached to within 1ca, but beyond the E end of the chalk cliffs there are ledges off Brook and Atherfield which require at least 5ca offing.

▶ *The E-going stream sets towards these dangers. 4M SSW of the Needles the stream turns NNE at HW Portsmouth +0500, and W at HW −0030, Sp rate 2kn. It is safe to pass 2ca off St Catherine's lt ho (conspic), but a race occurs off the point and can be very dangerous at or near Sp against a strong wind, particularly SE of the point on a W-going stream. St Catherine's should then be rounded at least 2M off. 1·25M SE of the point the stream turns E at HW Portsmouth +0520, and W at HW −0055, sp rate 3·75kn.* ◀

Rocks extend about 2½ca either side of Dunnose where a race occurs. In Sandown Bay ⚓ off Shanklin or Sandown where the tidal streams are weak inshore.

▶ *Off the centre of the Bay they turn NE x E at HW Portsmouth +0500, and SW x W at HW −0100, sp rates 2 kn.* ◀

The Yarborough Monument is conspicuous above Culver Cliff. Whitecliff Bay provides ⚓ in winds between W and N. From here to Foreland (Bembridge Pt) the coast is fringed by a ledge of drying rocks extending up to 3ca offshore, and it is advisable to keep to seaward of Bembridge Ledge ECM lt buoy.

2.7 CHRISTCHURCH

Dorset 50°43'·53N 01°44'·33W ❄⚓🌊✿✿✿

CHARTS AC 5601, 2035, 2172; Imray C10, C12, C4, 2300, 2200

TIDES HW Sp −0210, Np, −0140 Dover; ML 1·2

Standard Port PORTSMOUTH (→); Tidal Curve see 2.15 (→)

Times				Height (metres)			
High Water		Low Water		MHWS	MHWN	MLWN	MLWS
0000	0600	0500	1100	4·7	3·8	1·9	0·8
1200	1800	1700	2300				
Differences BOURNEMOUTH							
−0240	+0055	−0050	−0030	−2·6	−2·0	−0·6	−0·2
CHRISTCHURCH (Entrance)							
−0230	+0030	−0035	−0035	−2·9	−2·4	−1·2	−0·2
CHRISTCHURCH (Quay)							
−0210	+0100	+0105	+0055	−2·9	−2·4	−1·0	0·0
CHRISTCHURCH (Tuckton bridge)							
−0205	+0110	+0110	+0105	−3·0	−2·5	−1·0	+0·1

NOTE: Double HWs occur, except near neaps; predictions are for the higher HW. Near neaps there is a stand; predictions are for mid-stand. Tidal levels are for inside the bar; outside, the tide is about 0·6m lower at springs. Water flow in Rivers Avon and Stour cause considerable variations from predicted heights. Tuckton LWs do not fall below 0·7m except under very low river flow conditions.

SHELTER Good in lee of Hengistbury Hd; elsewhere exposed to SW winds. R Stour navigable at HW up to Tuckton, and the R Avon up to the first bridge. Both are well sheltered. Most ⚓s in the hbr dry.

NAVIGATION WPT 50°43'·53N 01°43'·58W, 270°/0·5M to NE end of Mudeford Quay. Outer approach is marked with PHM and SHM buoys; the bar/chan are liable to shift and will not be as shown on the chart. Beware groynes S of Hengistbury Hd, Beerpan, Yarranton/Clarendon Rks; the ebb stream reaches 4-5kn in The Run; best to enter or leave at HW/stand. Night entry is not recommended without local knowledge. Chan inside hbr is narrow and mostly shallow (approx 0·3m) with sand/mud. No ⚓ in chan. No berthing at ferry jetty by Mudeford Quay (beware frequent ferries between Mudeford Quay, The Run and Christchurch). Hbr speed limit 4kn. If fishing from any craft, a local licence is obligatory.

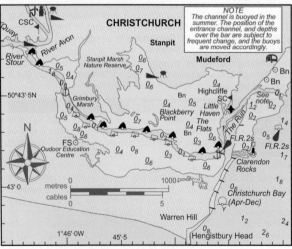

CHRISTCHURCH

NOTE
The channel is buoyed in the summer. The position of the entrance channel, and depths over the bar are subject to frequent change, and the buoys are moved accordingly.

LIGHTS AND MARKS See chartlet and 2.2. Unlit chan buoys in hbr and apps are locally laid Apr-Oct inc; info from ☎483250.

COMMUNICATIONS (Code 01202) Ⓗ 303626; Casualty 704167.

FACILITIES Rossiter Yachts ☎483250, ⚲£30; Little Avon Marina **Riverside Inn Boating Services** ☎477327 max loa 25ft £24/night. **Christchurch SC (CSC)** ☎483150, ⛽£10 limited ⚲ £20, please phone in advance to confirm berth availability.
Services ⚓⚓⚓ 🛠⛽🔧🔨✂🔌 Ⓔ 🏗(10t) ⛽ ACA.
Town ✉ Ⓑ ➔ ✈ (Hurn).

2.8 KEYHAVEN

Hampshire 50°42'·85N 01°33'·26W ❄⚓✿✿

CHARTS AC 5600, 2035, 2021; Imray C4, C3, 2300, 2200

TIDES −0020, +0105 Dover; ML 2·0

Standard Port PORTSMOUTH (→); Tidal Curve see 2.15 (→)

Times				Height (metres)			
High Water		Low Water		MHWS	MHWN	MLWN	MLWS
0000	0600	0500	1100	4·7	3·8	1·9	0·8
1200	1800	1700	2300				
Differences HURST POINT							
−0115	−0005	−0030	−0025	−2·0	−1·5	−0·5	−0·1
TOTLAND BAY							
−0130	−0045	−0035	−0045	−2·2	−1·7	−0·4	−0·1
FRESHWATER BAY							
−0210	+0025	−0040	−0020	−2·1	−1·5	−0·4	0·0

NOTE: Double HWs occur at or near sp; predictions are for the first HW. Off springs there is a stand of about 2hrs; predictions are then for mid-stand.

SHELTER Reasonable, but all moorings and ⚓s are exposed to winds across the marshland. The 'U' bend carries up to 3m but there is only about 0·5m in the entrance. Keyhaven Lake and Mount Lake all but dry. The river gets very congested during the sailing season.

NAVIGATION WPT 50°42'·70N 01°32'·79W is on 308° leading line marked by two B beacons in transit. Follow transit for about 3ca until through the pair of unlit R and G buoys then start turning to port, giving North Point a wide berth. The channel is marked with unlit G buoys. Entry is difficult on the ebb and should not be attempted in strong E winds. Entrance is possible, with local knowledge, via Hawkers Lake (unmarked), approx 0.3m less water than main channel. Beware lobster pots. When approaching from the W beware the Shingles Bank over which seas break and which partly dries. At Hurst Narrows give The Trap a wide berth.

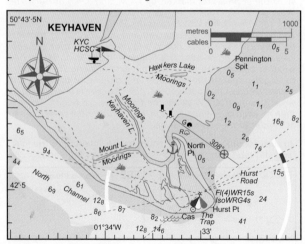

KEYHAVEN

LIGHTS AND MARKS See 2.2, chartlet above and 2.9 for sectors of Hurst Point light.

COMMUNICATIONS (Code 01590) Dr 643022; Ⓗ 677011; *River Warden* VHF Ch ⚓ 645695.

FACILITIES Quay ⚓⚓ Keyhaven YC ☎642165, ⚓⛽£12 🔧 🏗 🛒. **Hurst Castle SC** ☎(01590) 645589, ⚓⚓ 🔧. **New Forest District Council** (R.Warden), ⚓⚓.
W Solent Boat Builders ☎642080, ⚓ ✂🔧 Ⓔ 🏗(9t) ⛽.
Village ✕ 🛒. **Milford-on-Sea** (1M): 🛒🛒⛽🚂✉Ⓑ✕🛒➔ (bus to New Milton), ✈ (Hurn).

2.9 NEEDLES CHANNEL

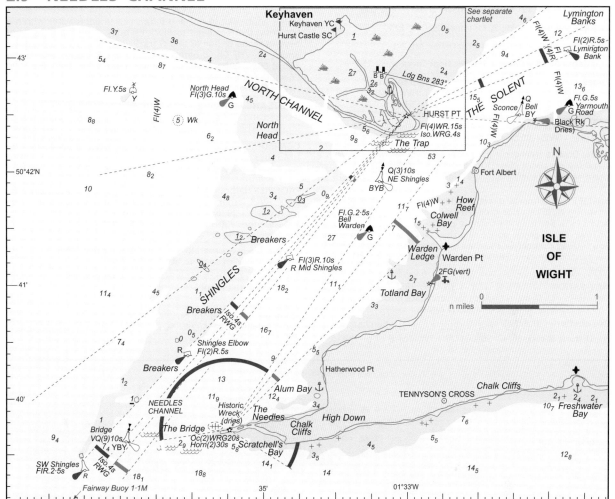

- The NW side of the Needles Channel is defined by the Shingles, a bank parts of which dry, and on which seas breaks in the least swell. The SE side of the bank is fairly steep-to, the NW side is less defined. Dredgers work on the Pot Bank, 4ca S of Bridge WCM.

- On the ebb the stream sets very strongly (3-4kn) WSW across the Shingles. The Needles Channel is well lit and buoyed and in fair weather presents no significant problems. Be aware that even a SW Force 4 against the ebb will raise breaking seas near Bridge and SW Shingles buoys. Contact National Coastwatch Needles on VHF Ch 65 for local conditions.

- In strong winds the North Channel, N of the Shingles, is preferable to the Needles Channel. The two join S off Hurst Point where overfalls and tide rips may be met. Beware The Trap, a shoal spit extending 150m SE of Hurst Castle.

- In bad weather broken water and overfalls extend along The Bridge, a reef which runs 8ca W of the lt ho. Lt's W extremity is marked by Bridge WCM lt buoy, but in calm weather it is feasible to pass E of Bridge buoy to avoid the worst of the ebb stream.

- Gales from S-W raise very dangerous breaking seas against the ebb in the Needles Chan, at points only 250m wide. The sea state can be at its worst shortly after LW when the flood has just begun. Although wind is with tide a substantial swell is raised as a result of turbulence in the recently turned stream. In such conditions consider using North Channel; the E route to the Solent, S of the IoW and via Nab Tower; or find shelter at Poole or Studland.

The Needles are distinctive rocks at the W end of the Isle of Wight (see AC 5600, 2035, 2021). By day the chalk cliffs of High Down are conspic from afar but the lt ho may not be seen by day until relatively close. Goose Rk, dries, is about 50m WNW of the lt ho, 100-150m WSW of which is a drying wreck.

ANCHORAGES BETWEEN THE NEEDLES AND YARMOUTH

ALUM BAY, 50°40'·10N 01°34'·33W. AC 5600, 2021. Tides as for Totland Bay. Very good shelter in E and S winds, but squally in gales. Distinctive white cliffs to S and multi-coloured cliffs and chairlift to E. Appr from due W of chairlift to clear Five Fingers Rk, to the N and Long Rk, a reef drying 0·9m at its E end, to the S. ‡ in about 4m.

TOTLAND BAY, 50°40'·98N 01°32'·86W. AC 5600, 2035. ML 1·9m. Good shelter in E'lies in wide shelving bay between Warden Ledge (rks 4ca offshore) to the N, and Hatherwood Pt to the SW. Appr W of Warden SHM buoy Fl G 2·5s to ‡ out of the stream in 2m between pier (2FG vert) and old LB house; good holding. Colwell Bay, to the N between Warden Pt and Fort Albert, is generally rocky and shallow.

ANCHORAGE EAST OF THE NEEDLES, SOUTH IOW

FRESHWATER BAY, 50°40'·07N 01°30'·61W. AC 5600, 2021. ML 1·6m. Good shelter from the N, open to the S. The bay is 3·2M E of Needles Lt ho and 1·2M E of Tennyson's Cross. Conspic marks: Redoubt Fort on W side; a hotel on N side; Stag and Mermaid Rks to the E. The bay is shallow, with rocky drying ledges ¾ca either side and a rock (dries 0·1m) almost in the centre. Best to ‡ in about 2m just outside the bay. ⊠ ⚏ ✕ ⌂.

2.10 YARMOUTH

Isle of Wight **50°42′·42N 01°30′·05W** ❀❀❀❀⚓⚓⚓ 🏵🏵🏵

CHARTS AC 5600, 2035, 2021; Imray C3, C15, 2200

TIDES Sp −0050, +0150, Np +0020 Dover; ML 2·0

Standard Port PORTSMOUTH (→); Tidal Curve see 2.15 (→)

Times				Height (metres)			
High Water		Low Water		MHWS	MHWN	MLWN	MLWS
0000	0600	0500	1100	4·7	3·8	1·9	0·8
1200	1800	1700	2300				
Differences YARMOUTH							
−0105	+0005	−0025	−0030	−1·7	−1·2	−0·3	0·0

NOTE: Double HWs occur at or near sp; at other times there is a stand lasting about 2hrs. Differences refer to the first HW when there are two; otherwise to the middle of the stand. **See 2.15.**

SHELTER Good from all directions of wind and sea, but swell enters in strong N/NE'lies. Hbr dredged 2m from ent to swing bridge; access H24. Berth on walk ashore pontoons, detached pontoons or Town Quay, giving notice of arrival on Ch 68 for instructions. South Quay commercial use only. In season hbr gets very full and berthing is restricted. 38 Or ⚓s outside hbr (see chartlet). ⚓ well clear.

NAVIGATION WPT 50°42′·58N 01°30′·01W, 188°/2ca to abeam car ferry terminal. Dangers on approach are Black Rock (SHM lt buoy) and shoal water to the N of the E/W bkwtr.

Beware ferries and their wash/turbulence even when berthed.

An **Historic Wreck** 50°42′·55N 01°29′·67W, 2ca E of pierhead, is marked by a Y SPM. Royal Solent moorings lie between the two.

Caution: strong ebb in the ent and parts of the harbour at Sp.

Speed limits: 4kn in hbr and R Yar; 6kn in approaches within a line from the pier head to Black Rk buoy. ⚓ prohib in hbr and beyond swing bridge (R and G traffic lts). All craft must call *Yar Bridge* on VHF Ch 68 before approaching. Bridge opens during BST at 0800, 0900, 1000, 1200, 1400, 1600, 1730, 1830 and 2000 LT; winter months by request only. River is navigable by dinghy at HW to Freshwater.

LIGHTS AND MARKS See 2.2 and chartlet. In emergencies harbour is closed and a R flag is flown from end of Ferry Jetty. In fog a high intensity white light is shown from the Pier Hd and inner E pier.

COMMUNICATIONS (Code 01983 (entire IoW)) Dr 08448 151428. HM 760321.

HM and *Yar Bridge* Ch **68**. Water Taxi Ch 15, 07969 840173.

FACILITIES Harbour www.yarmouth-harbour.co.uk Approx 250🅥, £4·00 (walk ashore pontoons), £2·50 (detached pontoons or ⚓) short stay: £1·39 (walk ashore), £1·00 (other); midweek discounts. 🚹 🚺 (winter 0700-1900, summer 0700-2200 (times vary through the year)) ⛽🔌🛒⚓ 🏗(£4/visit) ⛴🔧🛠(5t) ice.

Hayles 🍴 in SW corner; R Yar Boatyard is located 500m above the bridge, or ½M by road. 🛒🍴🔧🛠🏪⚓🛗🔩.

Town 🏪✉©(PO)🚋 ✕🍴. Ferries to Lymington, ⇌ (Lymington), ✈ (Hurn/Southampton).

YACHT CLUBS
Yarmouth SC ☎760270, 🍴🍺.
Royal Solent YC ☎760256, 🍺🍴✕🍺; visiting crews welcome.

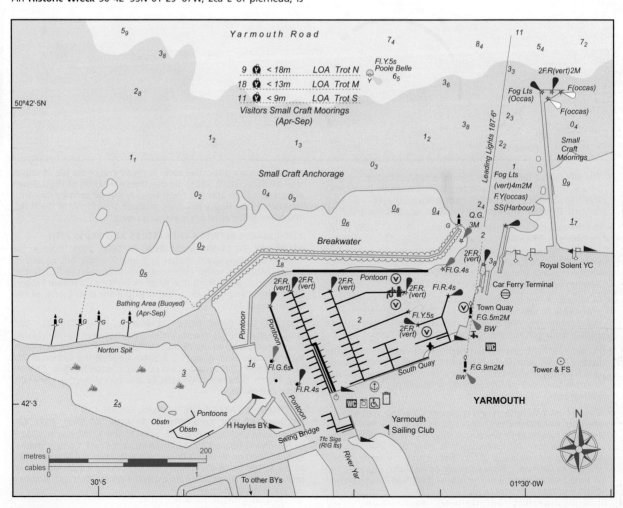

The perfect home for your boat

Lymington Yacht Haven
We look forward to welcoming you in 2019

Your perfect base for exploring the Solent:

- Berths for yachts up to 25m (80ft)
- Full tide access
- 24-hour fuel and service
- Luxury facilities and free Wi-Fi at your berth
- Full marina services, boatyard and brokerage
- On-site chandlery and shop
- The Haven Bar & Restaurant on-site
- An easy 10 minute walk to Lymington High Street
- A range of savings and benefits for Yacht Havens Berth Holders
- Special Winter Berthing Rates (Nov-Feb inclusive)

The Yacht Harbour Association Ltd
UK Coastal Marina of the Year 2016
Over 250 Berths
Lymington Yacht Haven
WINNER
PARTNERED BY
GJW Insurance

Call **01590 677071** or **VHF Ch 80**
or visit **yachthavens.com**

Lymington Yacht Haven

2.11 LYMINGTON

Hampshire 50°45'·13N 01°31'·40W ✿✿✿⚓⚓⚓✿✿✿

CHARTS AC 5600, 2035, 2021; Imray C3, C15, 2200

TIDES Sp −0040, +0100; Np +0020 Dover; ML 2·0

Standard Port PORTSMOUTH (→); Tidal Curve see 2.15 (→)

Times				Height (metres)			
High Water		Low Water		MHWS	MHWN	MLWN	MLWS
0000	0600	0500	1100	4·7	3·8	1·9	0·8
1200	1800	1700	2300				
Differences *LYMINGTON*							
−0110	+0005	−0020	−0020	−1·6	−1·2	−0·4	−0·1

SEE NOTE under Tides at 2.10.

Lymington Harbour Commissioners www.lymington harbour.co.uk
☎672014; HM patrol for berthing advice; wi-fi on river berths.
Dan Bran Pontoon up to 40♥ £15–£35 🔌+£2 ⚓ and use of LTSC.
Long walk ashore pontoon suitable for rallies and other events.
Town Quay public pontoon, many ⚡ ⌷ £18·50-£29/craft (9·5m-
10·5m LOA), £8·20 for short stay; >15m LOA by arrangement with
HM; ⛽ ⌷ (see HM) ⚓ ⚙⚙⚙🚾/🐾.
Scrubbing grids (5) See HM for details.
Berthon Lymington Marina www.berthon.co.uk ☎647405.
250+100♥ £4·05, 🔌⚓⛽⚓⚙⚙⚙✕⚙⚙Ⓔ⚓🚩(75t) 🅿🖳(20t)
🅿 wi-fi.
Town All facilities, ⇌ Ferries to IoW, ✈ (Hurn or Southampton).

YACHT CLUBS **Royal Lymington YC** ☎672677, ⚙✕🖳.
Lymington Town SC ☎674514, ⚙ (Sailability access hoist) ✕🖳.

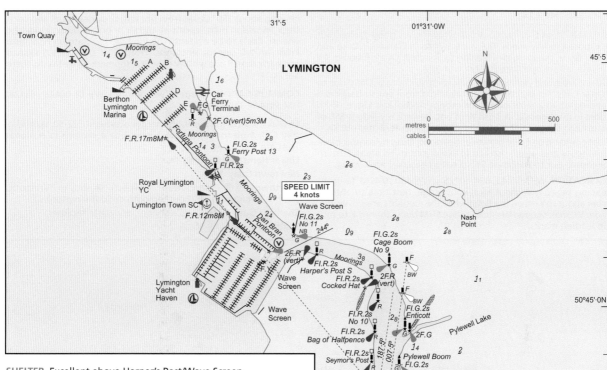

SHELTER Excellent above Harper's Post/Wave Screen.

NAVIGATION WPT 50°44'·0N 01°30'·05W, 320°/5ca to abeam No 1
and Cross Boom bns. Strong cross-tide at river entrance. Min depth
2.2m in mid-chan until above ferry terminal, then 1·8m to Town Quay.
Speed limits (strictly enforced): 6kn N of Jack in the Basket; 4kn
above Wave Screen. ⚓ in the river is prohibited.

> All craft must keep clear of IoW ferries in the river (IRPC Rule 9).
> Do not follow close astern, or overtake as they make the final
> approach to their berth, and be aware of wash. They sound one
> short blast before departure. Ferries may pass each other on N-S
> transit abreast Pylewell Boom.

LIGHTS AND MARKS Conspic YC starting platform is close E of first
(No1) SHM bn, Fl G 2s; Jack-in-the-Basket PHM bn, Fl R 2s, is 1·6ca
SW of platform. Ldg Lts 319·5°, both FR 12/17m 8M on dayglow
Or posts are difficult to see by day. The narrow channel is well
marked by lit bns and one lit PHM buoy, Fl.R.2s to port, Fl.G.2s to
starboard. Two BW posts, Dir FW 007·5°, and two RW posts, Dir
FW 187·5°, are used by ferries passing in the river. Two FY ldg lts
244° mark the ent into Lymington Yacht Haven.

COMMUNICATIONS (Code 01590) Dr 672953; Ⓗ663000; HM 672014.
Marinas VHF Ch **80** M (HO), LHC: *HM* CH 66.

FACILITIES **Lymington Yacht Haven** www.yachthavens.com
☎677071, 2.5m depth, all tides access. 475⌷+100♥, £3·75, ⚙⚙🚽♿
(mobile) 🏴⚓⚓✕⚒⚙⚙🅿🖳🚩(50t) 🖳(4t).

2.12 NEWTOWN RIVER

Isle of Wight **50°43'·45N 01°24'·66W** ❄❄❄❄❄❄❄❄

CHARTS AC 5600, 2035, 2036, 2021; Imray C3, C15, 2200

TIDES Sp −0108, Np +0058, Dover; ML 2·3

Standard Port PORTSMOUTH (→); Tidal Curve see 2.15 (→)

Times				Height (metres)			
High Water		Low Water		MHWS	MHWN	MLWN	MLWS
0000	0600	0500	1100	4·7	3·8	1·9	0·8
1200	1800	1700	2300				
Differences YARMOUTH/NEWTOWN ENTRANCE							
−0105	+0005	−0025	−0030	−1·7	−1·2	−0·3	0·0

NOTE: Double HWs occur at or near springs; at other times there is a stand which lasts about 2hrs. Differences refer to the first HW when there are two. At other times they refer to the middle of the stand. See 2.15.

SHELTER 3½M E of Yarmouth, Newtown gives good shelter, but is exposed to N'ly winds. There are 5 W ⚓s in Clamerkin Lake and 20 W ⚓s in the main arm leading to Shalfleet Quay, R buoys are private; all are numbered; check with HM.

Do not ⚓ beyond yellow buoys indicating 'Anchorage Limit' due to oyster beds. Fin keel boats can stay afloat from ent to Hamstead landing or to yellow buoys indicating 'Anchorage Limit'. If no room in river, good ⚓ in 3-5m W of ent, beware rky ledges SSE of Hamstead Ledge SHM lt buoy, and possible underwater obstns.

Public landing on E side of river at Newtown Quay by conspic black boathouse and where marked to avoid controlled areas. Eastern peninsula out to Fishhouse Pt is a nature reserve; no landing.

NAVIGATION WPT 50°43'·85N 01°25'·21W, 130°/0·42M to front leading beacon. Without local knowledge, best time to enter is from about HW −4, on the flood while the mudflats are still visible. From W, make good Hamstead Ledge SHM buoy, thence E to pick up leading marks.

From E, keep N of Newtown Gravel Banks where fresh W/SW winds over a spring ebb can raise steep breaking seas. Leave WCM buoy, Q(9) 15s, to port.

Min depth to inside the entrance is about 2m. 5kn speed limit in harbour is strictly enforced. Near junction to Causeway Lake depth is only 0·9m and beyond this water quickly shoals.

At ent to Clamerkin Lake (1·2–1·8m) keep to SE to avoid gravel spit off W shore, marked by two SHMs; the rest of chan is marked by occasional buoys. Beware many oyster beds in Western Haven and Clamerkin Lake.

Rifle Range at top of Clamerkin Lake and in Spur Lake; R flags flown during firing. High voltage power line across Clamerkin at 50°42'·81N 01°22'·66W has clearance of only 8·8m and no shore markings.

LIGHTS AND MARKS See 2.2 and chartlet. Conspic TV mast (152m) bearing about 150° (3·3M from hbr ent) provides initial approach track. In season many yachts at anchor or on the buoys inside the hbr are likely to be evident. The ldg bns, 130°, are on Fishhouse Point on NE side of entrance. Front: RW pile with Y-shaped topmark; rear: R pile with W spherical topmark. Once inside, there are no lts.

COMMUNICATIONS (Code 01983 (entire IoW)) Dr 08448 151428; HM 531424; Taxi 884353.

FACILITIES Average mooring charges overnight £22·50, short/day stay £14. ⚓ no formal charge applies but donations to National Trust are welcomed.

Lower Hamstead Landing ⚓.

Newtown Quay ⚓ ⚓.

Newtown ✉ Ⓑ (Yarmouth or Newport), ⇌ (bus to Yarmouth, ferry to Lymington), ✈ (Bournemouth or Southampton).

Shalfleet Quay ⚓⚓ ⚓ ⚓(drying), dry waste facilities.

Shalfleet Village ⚒ ✗ ⌂.

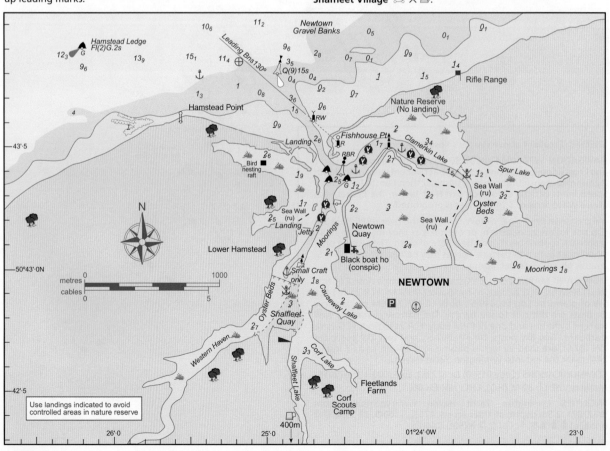

2.13 BEAULIEU RIVER

Hampshire 50°46'·89N 01°21'·72W (Ent) ✿✿✿✿✿✿✿✿

CHARTS AC 5600, 2036, 2021; Imray C3, C15, 2200

TIDES −0100 and +0140 Dover; ML 2·4

Standard Port PORTSMOUTH (→); *Tidal Curve see 2.15 (→)*

Times				Height (metres)			
High Water		Low Water		MHWS	MHWN	MLWN	MLWS
0000	0600	0500	1100	4·7	3·8	1·9	0·8
1200	1800	1700	2300				
BUCKLERS HARD							
−0040	−0010	+0010	−0010	−1·0	−0·8	−0·2	−0·3
STANSORE POINT							
−0030	−0010	−0005	−0015	−0·7	−0·5	−0·3	−0·3

NOTE: Double HWs occur at or near springs; the 2nd HW is approx 1¾hrs after the 1st. On other occasions there is a stand which lasts about 2hrs. Differences refer to the first HW when there are two, or to the middle of the stand. See 2.15.

SHELTER Very good above Needs Ore Pt in all winds. ⚓ possible in reach between Inchmery Ho and Needs Ore Pt, but for facilities proceed to Bucklers Hard Yacht Hbr (3·5M from entrance). No ⚓ E of Inchmery Ho or 1M either side of Bucklers Hard.

NAVIGATION WPT 50°46'·60N 01°21'·40W, close E of seasonal ⚓ (*Mar-Oct*) in 4m, 324°/3·3ca to abeam Beaulieu Spit Dn. Least depth between WPT and No 2 beacon approx 0·9m. Buoys set in from withies to give 1m below CD. If possible proceed under power in the river where speed limit of 5kn applies.

Ldg marks 324° at entrance should be carefully aligned due to shoal water either side of ldg line, although there is possibly more water to NE of line. Beaulieu Spit Dn should be left approx 40m to port. At night keep in W sector of Millennium Bn (Lepe) until No 4 PHM Bn abeam.

Many of the landing stages and slips shown on the chartlet and AC 2021 are privately owned and must not be used without permission. The river is navigable for about 2M above Bucklers Hard, but the uppermost reaches are best explored by dinghy. There are limited alongside drying berths at Beaulieu extending from the weir not more than 20m downstream.

LIGHTS AND MARKS See 2.2 and chartlet. Ldg marks R No 2 Bn, 0·7ca N of Beaulieu Spit Dn aligned W edge of Lepe Ho (≠ 324°). Old CG cottages and Boat Ho are conspic, approx 320m E of Lepe Ho. The river is clearly marked by R and G bns and withies/perches. SHM bns 5, 9, 13, 17, 19, 21; buoys Gins, Fiddlers, Mulberry Fl G 4s; PHM bns 8,12, 20; buoys No 26, 28 Fl R 4s.

COMMUNICATIONS (Code 01590) Dr 612451 or (02380) 845955; Ⓗ 0845 4647.
Beaulieu River Radio Ch 68 (for HM).

FACILITIES Bucklers Hard ⚓ www.beaulieuriver.co.uk ☎616200, 115⚓+40⚓, £3·50/m, short stay £1·50/m. Detached pontoons £2.5/m, ⚓s £2/m, short stay £1.50/m. Below Needs Ore Pt 8 ⚓s £10/boat<40', ⚓ £5.50. ♿ ⚓ 🛢 🚿(0800-1800) 🔋 ⚒ ⚓ 🛒 🚢(35t), scrubbing grid(£25) ⚓ ⛽ 🛒 ✕ 🚻.

Royal Southampton YC (Gins Club) ☎616213, www.rsyc.org.uk; Ch 77(*Sea Echo*); ⚓(temp) 🍴 ⛽ ✕ 🚻.

Village ✉ Ⓑ Ⓒ 🛒 (☎616293), ✕ 🚻(Beaulieu) ⇌ (bus to Brockenhurst), ✈ (Hurn or Southampton).

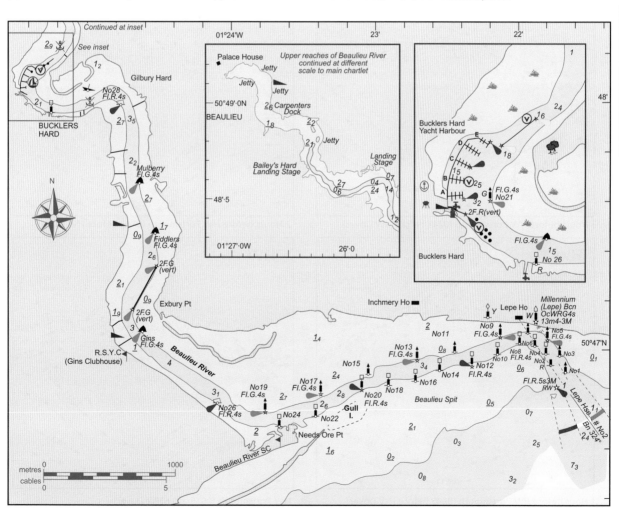

2.14   ISLE OF WIGHT AND SOLENT TIDAL STREAMS

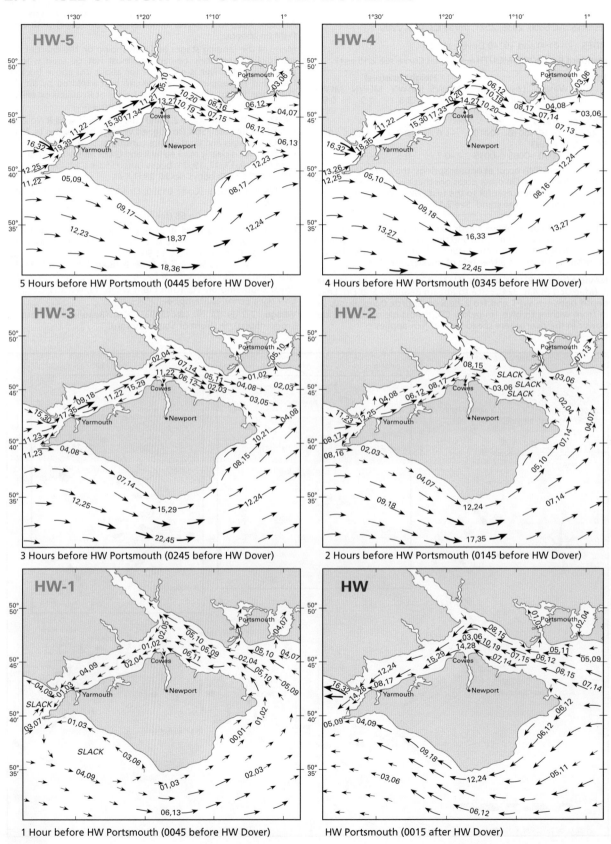

5 Hours before HW Portsmouth (0445 before HW Dover)

4 Hours before HW Portsmouth (0345 before HW Dover)

3 Hours before HW Portsmouth (0245 before HW Dover)

2 Hours before HW Portsmouth (0145 before HW Dover)

1 Hour before HW Portsmouth (0045 before HW Dover)

HW Portsmouth (0015 after HW Dover)

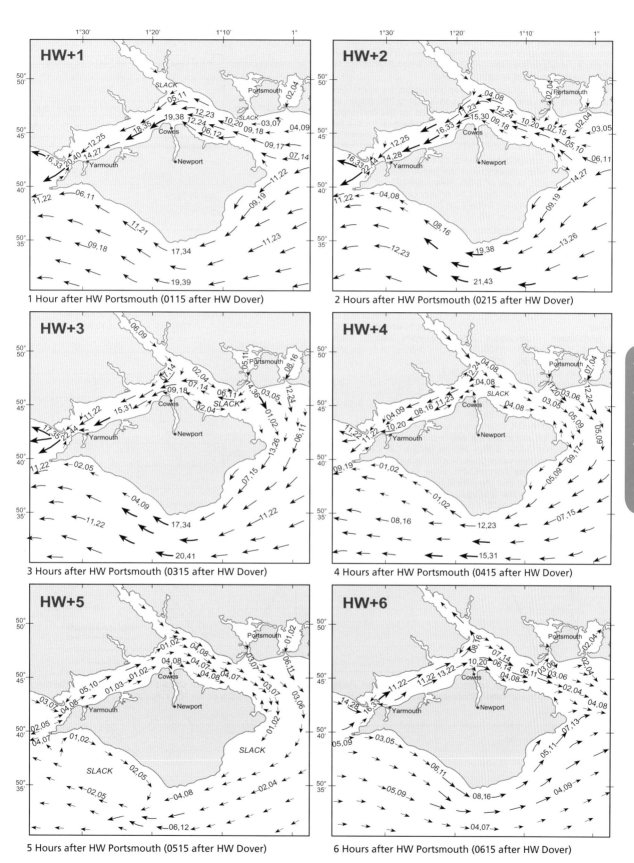

HW+1

1 Hour after HW Portsmouth (0115 after HW Dover)

HW+2

2 Hours after HW Portsmouth (0215 after HW Dover)

HW+3

3 Hours after HW Portsmouth (0315 after HW Dover)

HW+4

4 Hours after HW Portsmouth (0415 after HW Dover)

HW+5

5 Hours after HW Portsmouth (0515 after HW Dover)

HW+6

6 Hours after HW Portsmouth (0615 after HW Dover)

Central S England

2.15 SPECIAL TIDAL CURVES FROM CHRISTCHURCH TO SELSEY BILL

Due to the complex tidal patterns between Christchurch and Selsey Bill, **special curves** for each Secondary Port are used to obtain the most accurate values of times and heights. Because their LW points are more sharply defined than HW, *all times are referenced to LW*. HW times on the tables are approximate.

Critical curve. The spring and neap curves at harbours from Christchurch to Yarmouth differ considerably in shape and duration so a third, 'critical', curve is provided for the Portsmouth range at which the heights of the two HWs are equal for the port concerned. Interpolate between this critical curve and either the spring or neap curve as appropriate.

Note: Whilst the critical curve extends throughout the tidal range, the spring and neap curves stop at the higher HW. For example, with a Portsmouth range of 3·8m (near springs), at 7hrs after LW Lymington the factor should be referenced to the *next* LW. Had the Portsmouth range been 2·0m (near neaps), it should be referenced to the *previous* LW.

Example: Find the height of tide at Christchurch at 1420 on a day when the tidal predictions for Portsmouth are:

18 0110 4·6
SA 0613 1·1
       1318 4·6
       1833 1·0

Standard Port PORTSMOUTH

Times				Height (metres)			
High Water		Low Water		MHWS	MHWN	MLWN	MLWS
0000	0600	0500	1100	4·7	3·8	1·9	0·8
1200	1800	1700	2300				
Differences CHRISTCHURCH ENTRANCE							
−0230	+0030	−0035	−0035	−2·9	−2·4	−1·2	−0·2

- On the tidal prediction form complete fields 2–5, 8–10 (by interpolation) and thus 13–15.
- On the Christchurch tidal graph, plot the HW & LW heights (fields 14 & 15) and join them with a diagonal line.
- Time required (1420) is 3hrs 38mins before LW Christchurch (field 13); from this value go vertically towards the curves.
- Interpolate for the Portsmouth range (3·6m) between the spring curve (3·9m) and the critical curve (2·6m).
- Go horizontally to the diagonal; thence vertically to read off the height of tide at 1420: 1·6m.

STANDARD PORT ...Portsmouth... TIME/HEIGHT REQUIRED...1420...

SECONDARY PORT ...Christchurch... DATE...18 Nov... TIME ZONE... O(UT)

	TIME		HEIGHT		
STANDARD PORT	HW	LW	HW	LW	RANGE
	1	2 1833	3 4·6	4 1·0	5 3·6
Seasonal change	Standard Port		6	6	
DIFFERENCES	7	8 −003S	9 −2·8	10 ~0·3	
Seasonal change	Standard Port		11	11	
SECONDARY PORT	12	13 1758	14 1·8	15 0·7	
Duration	16				

STANDARD PORT ....TIME/HEIGHT REQUIRED................

SECONDARY PORT ....DATE................. TIME ZONE................

	TIME		HEIGHT		
STANDARD PORT	HW	LW	HW	LW	RANGE
	1	2	3	4	5
Seasonal change	Standard Port		6	6	
DIFFERENCES	7	8	9	10	
Seasonal change	Standard Port		11	11	
SECONDARY PORT	12	13	14	15	
	16				

Note* At Tuckton LWs do not fall below 0·7m except under very low river flow conditions.

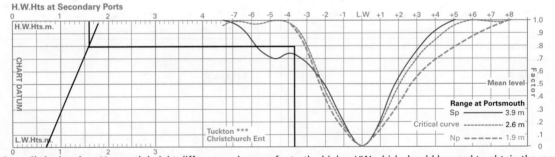

Note: From Christchurch to Yarmouth height differences always refer to the higher HW which should be used to obtain the range at the Secondary Port. HW time differences also refer to the higher HW, but are not required for this calculation.

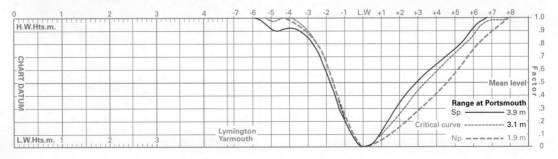

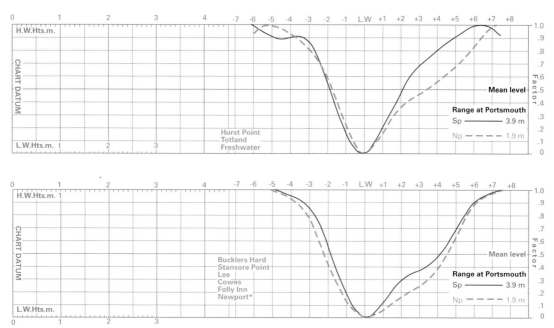

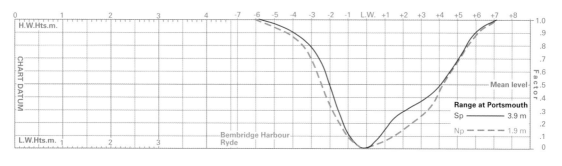

Note* Due to the constriction of the R Medina, Newport requires special treatment since the hbr dries 1·4m. The calculation should be made using the LW time and height differences for Cowes, and the HW height differences for Newport. Any calculated heights which fall below 1·4m should be treated as 1·4m.

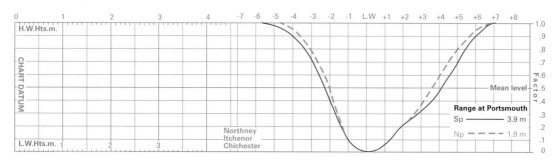

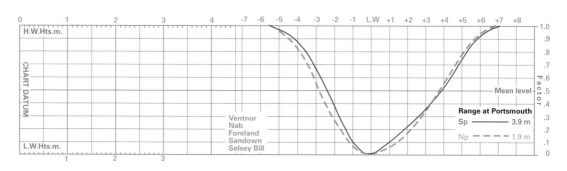

2.16 THE CENTRAL SOLENT

CHARTS AC 5600, 2036.

Reference and websites *Solent Cruising Companion,* Fernhurst; *Solent Hazards,* Peter Bruce; *Solent Year Book* SCRA; *Yachtsman's Guide* ABP Southampton www.southampton vts.co.uk for detailed live port info including: shipping movements, VTS radar displays (Southampton to S of Nab Tr), webcams, tide & weather. See www. bramblemet.co.uk for live weather & tidal info at the Bramble Bk.

Vessel Traffic Service (VTS) **Southampton VTS** operates on VHF Ch **12** 14, advising Solent shipping between the Needles and Nab Tower including Southampton Water. VTS monitors and coordinates the safe passage of commercial ships. It includes a radar service on request. **QHM Portsmouth,** VHF Ch **11,** controls Portsmouth Harbour and its approaches N of a line from Gilkicker Pt to Outer Spit Buoy.

All small craft in the Solent are strongly advised to monitor VHF Ch 12 Southampton, and/or Ch 11 QHM Portsmouth.

Small Craft and Commercial Shipping Always consider the restricted field of vision from large ships at close quarters and their limited ability to manoeuvre at slow speeds with minimal under keel clearance in restricted waters. Smaller ships keeping clear of these vessels will be encountered outside the channel. Maintain a proper lookout, especially astern. If crossing the main fairway, try to do so at 90° to it. Southampton Hbr byelaw 2003/10 refers. Beware the frequent cross-Solent High Speed and Ro-Ro ferries.

The **Precautionary Area** is the potentially dangerous part of the Solent in which large vessels make tight turns into and out of the Thorn Channel leading to Southampton Water. It extends from just NW of Cowes to the Hook buoy N of Calshot Castle (see chartlet and AC 2036). This can be a very busy area. Be particularly alert when between East Lepe and Ryde Middle. Inbound ships pass Prince Consort NCM turning to **port** towards Gurnard NCM, before starting their critical turn to starboard into the Thorn Channel which is, for large ships, very narrow.

Clear Channel Vessels are over 220m LOA and require a clear and unimpeded passage ahead when transiting the Precautionary Area. Vessels may enter the Precautionary Area maintaining a safe distance astern of the 'clear channel vessel'.

Moving Prohibited Zone (MPZ) Any vessel over 150m LOA in the Precautionary Area is enclosed by an MPZ which extends 1000m ahead of the vessel and 100m on either beam. Small craft under 20m LOA must remain outside the MPZ and use seaman-like anticipation of its route and sea-room required when turning.

Escort Towage for VLCCs From S of the Nab Tr VLCCs over 60,000dwt bound to and from Fawley Marine Terminal will have an escort tug secured to the stern by tow wire. Outwards, the escort tug may leave a tanker at Prince Consort. Use extreme caution when passing round the stern of a large tanker and do not pass between it and the escort tug.

Local Signals Clear Channel & MPZ vessels display a black cylinder by day or 3 all-round ● lts (vert) by night. They are normally preceded by a Harbour Patrol Launch (Ch 12 *Callsign SP*) exhibiting a flashing blue light. Southampton patrol launches have 'HARBOUR MASTER' painted on their superstructure. At night a fixed ● all-round lt is shown above the W masthead light.

Coastguard The National Maritime Operations Centre (NMOC) at Fareham, ☎(02392) 552100, co-ordinates SAR activities from Topsham to mid-channel; E to the Greenwich Lt V; thence N to Beachy Hd. Owing to the concentration of small craft in its area the NMOC keeps watch on VHF Ch 67/16. Uniquely, an initial call to *Solent Coastguard* should be made on Ch 67 for routine traffic and radio checks, although radio checks on a working channel with other stations – marinas, yachts etc – are preferred and encouraged.

Solent Yacht Racing Buoys Numerous yellow yacht racing buoys are laid throughout the Solent (*seasonal, Mar-Dec*). Most, but not all, are lit Fl Y 4s. Yellow buoys also mark a seasonal safe swimming and Personal Watercraft area off the beach in Stanswood Bay from Jul-Sep.

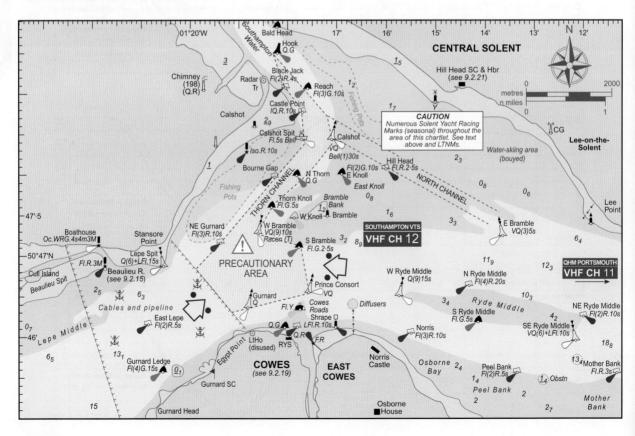

2.17 COWES/RIVER MEDINA

Isle of Wight 50°46'·08N 01°17'·95W ✵✵✵✵♨♨♨ ✿✿✿

CHARTS AC 5600, 2036, 2793; Imray C3, C15, 2200

TIDES +0029 Dover; ML 2·7

Standard Port PORTSMOUTH (→); Tidal Curve see 2.15 (←)

Times				Height (metres)			
High Water		Low Water		MHWS	MHWN	MLWN	MLWS
0000	0600	0500	1100	4·7	3·8	1·9	0·8
1200	1800	1700	2300				
Differences COWES							
–0015	+0015	0000	–0020	–0·5	–0·3	–0·1	0·0
FOLLY INN							
–0015	+0015	0000	–0020	–0·6	–0·4	–0·1	+0·2
NEWPORT							
ND	ND	ND	ND	–0·6	–0·4	+0·1	+0·8

NOTE: Double HWs occur at/near Sp. At other times there is a stand of about 2hrs. Times are for the middle of the stand. See 2.15.

SHELTER Excellent now that outer hbr is protected from N/NE winds by outer bkwtr. ⚓ prohib in hbr. 10 large ⚓s off The Green; pontoons S of chain ferry on both sides of river and in Folly Reach and several marinas. Good ⚓ in Osborne Bay, 2M E, sheltered from SE to W but exposed to wash from large ships.

NAVIGATION WPT 50°46'·35N 01°18'·0W, 173°/2·7ca to mid chan between Nos 1 and 2 buoys. Shrape Mud, on the E side of the hbr, extends to Old Castle Pt. A small craft channel crosses from Shrape Bn to the fairway, passing N of Shrape Bwtr. Depth is 0.2m below CD, but in the approach is close to CD and it should only be used if craft's draught is less than the existing height of tide. R Medina is navigable to Newport, but dries S of Folly Inn.

- Yachts should use engines. Sp limit 6kn enforced in hbr.
- Strong tidal streams run between HW-2½ and HW setting vessels towards W shore in Inner Fairway; see next page
- No sailing through or ⚓ amongst moorings.
- Beware frequent ferry movements, High-Speed to W Cowes, large RoRo ferries to E Cowes, and commercial shipping.
- RoRo ferries use thrust when manoeuvring at E Cowes causing considerable turbulence across the river. Hi-Speed ferries turn short round to stbd when appr Jubilee Pontoon.
- All vessels < 20m must give way to the chain ferry which displays a Y Fl lt when about to depart and under way; daily 0530-0030, Sun 0635-0030. Beware ebb (4·5kn at Sp) in its vicinity.

LIGHTS AND MARKS See 2.2 and chartlet. Outer b'water marked along its length by 5 sync lts Fl Y 2.5s. Small craft chan is marked by Shrape Bn LFl R 10s 3m 3M, and paired PHM and SHM: inbound Fl R 2s / Fl G 2s, Fl R 5s / Fl G 5s (Outer Bwtr) Fl R 3s / Fl G 3s (E Bwtr) with 3 lit Y SPM and R Bn, L Fl R 5s where it joins fairway.

COMMUNICATIONS (Code 01983 (entire IoW)) ♨ (02380) 228844; Ⓗ 524081; Dr 294902; Cowes HM 293952; Monitor *Cowes Hbr Radio* VHF Ch 69 for hbr launches and HM's ⚓s. Yachts >20m LOA should advise arr/dep, and call *Chain Ferry* Ch 69 if passing. Marinas Ch 80. Harbour taxi 07855 767918; *Hbr Taxi* Ch 77 or Sally Water Taxi 07831 331717/299033; *Water Taxi* Ch 06. Casualties: (Ch 16/69) for ambulance at S end of Trinity Landing.

FACILITIES www.cowes.co.uk / www.cowesharbourcommission.co.uk from seaward: **RYS Haven** (strictly members only).

Trinity Landing, on The Parade, short stay on inner side only, pick-up/landing on outer side, call HM Ch 69. The concentration of wash W of the b'water impinges on the pontoon which can be lively.

Cowes Yacht Haven 🅥 www.cowesyachthaven.com ☎299975, 35⌑+ 200 🅥, £3·00, 🅑 🄾 ✕ 🚿 Ⓔ ⚓ 🏴(35t) ⛟(12t) ✕.

Cowes Harbour Shepards Marina, www.shepards.co.uk ☎297821, 🕳 30⌑+100🅥 £2·50, 🅑 🚽 ⛽ 🛒 ✕ ⚓ 🄳 ✕ 🚽 wi-fi.

UKSA ☎294941, info@uksa.org, 5 ⌑ 🅥, £3.30, 🅑 ✕ 🚽.

East Cowes Marina www.eastcowesmarina.co.uk ☎293983, 230 ⌑+150🅥 £3·00, 🅑 🄳 🛒 ✕ 🚽 wi-fi.

Town All marine services and facilities.

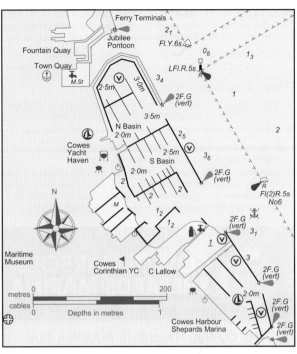

⚓ Town Quay, Trinity Landing, and Whitegates public pontoon.
Fuel Lallows 🍴, Cowes Hbr Fuels pontoon off Souters 🍴.
Scrubbing berths Town Quay, UKSA, Sealift, Folly Inn.

Ferries Red Funnel www.redfunnel.co.uk to/from Southampton: ☎(02380) 334010 Car/pax – E Cowes; foot pax – W Cowes.
Chain Ferry ☎293041, W/E Cowes.

YACHT CLUBS Royal Yacht Squadron ☎292191; **Island SC** ☎296621; **Royal Corinthian YC** ☎293581; **Royal London YC** ☎299727; **Cowes Corinthian YC** ☎296333; **RORC** ☎295144 (Cowes Week); **E Cowes SC** www.eastcowessc.co.uk ☎531687. **Cowes Combined Clubs** ccc@cowesweek.co.uk ☎295744.

RIVER MEDINA, FOLLY REACH TO NEWPORT

FOLLY REACH, 50°44'·03N 01°16'·99W. Above Medham min depth 0·6m to S Folly bn. ⌑ on 🅥 pontoons off W bank (£1·25). *Folly Launch* Ch 72/☎07884 400046; *Folly Waterbus* ☎07974 864627.
Folly Inn ☎297171, 🅑 ♨ 🕳 scrubbing berth ⌑ (pontoon) for 9🅥(max LOA 40') £1·80.

Island Harbour Marina (5ca S of Folly Inn) island-harbour.co.uk ☎539994, 🕳 96⌑+20🅥 £3·20 (£2·80 midweek), 🅑 🄾 ⚓ ✕ 🄳 🏴(50t) ✕ 🄻 ✕ 🄳. Call *Island Harbour Control* on Ch 80 on entering Folly Reach. Excellent shelter; approach via marked channel (dries approx 0·1m) with waiting pontoon to stbd, withies to port. Access HW±3 (draught 1·5m) via lock (width 7·9m) 0800-1730, or by arrangement.

NEWPORT, 50°42'·21N 01°17'·43W. HM ☎525994/ 07970 009589. VHF Ch 69. Tides: Special treatment is required as the hbr dries 1·4m. Use the LW time and height differences for Cowes, and the HW height differences for Newport. Above Island Hbr Marina the chan to Newport (1·2M) dries, but from HW Portsmouth –1½ to +2 it carries 2m or more. S from Folly Inn, the Hbr Authority is IoW Council. Sp limit 6kn to Seaclose, S of Newport Rowing Club (NRC); thence 4kn to Newport. Buoyed and partially lit chan favours W bank. Power lines have 33m clearance. 192° ldg marks/lts are W ◇ bns (lit) on E bank.

Odessa Boatyard ☎524337, odessaboatyard@hotmail.co.uk 75⌑+5🅥s on pontoons downstream on W side £1·50, 🅑 ⚓ 🄳 ⚓ 🍴 slipway (35t) ⛟(6t) 🄻.

Newport Yacht Hbr ☎525994, 50🅥s pontoons on the E/SE sides of the basin have 1·4m at HW ±2 £1·75. Bilge keelers lie alongside pontoons (soft mud); fin keelers against quay wall (firm level bottom). Fender boards can be supplied. ⚓ 🄳 🍴 ⛟(8t) Classic Boat Centre. **Town** 🕳 🄿 🄿 ✕ 🄲.

Chartlets on page 110

TIDAL PREDICTIONS FOR APPRECIATING COWES TIDAL STREAMS
The table on the following page gives appropriate HW predictions for the tidal flows in Cowes Fairway and the small boat channel:

- In the Inner Fairway west of the outer breakwater, a tidal shadow with significantly reduced flow will be experienced immediately west and south of the outer breakwater when the tide is running westwards (HW –2 to HW +5).

- From HW –2·5 to HW, boats following the small boat channel will experience a westerly tidal flow of up to 1.5kn.

- Off the Island Sailing Club pontoon this stream splits into
  A northbound flow towards the harbour entrance
  A southbound flow past the marinas.

- Yachts and small vessels entering or leaving Cowes in the Fairway will be slowed by this approaching the small boat channel and then experience an appreciable lift as well as a set towards the western shore north of the Red Jet jetty.

These HW times mark the start of the stand or 'high water period' as defined by Cowes Commissioners and are useful in appreciating the tidal flows in the Fairway and approaches. The HW times in the table apply throughout the tidal cycle of springs and neaps, simplifying the prediction process.

It should be noted that this table gives the predicted time and height of high water at the beginning of the stand. The normal methodology predicts the first peak high water in ports such as Cowes where there is a double HW or stand. The peak high water will almost invariably be of greater height and later than that shown.

These predictions correlate accurately with those used by Cowes Commissioners and Admiralty Tide Tables. They will deviate from those calculated from the Special Curves (p 219) chart plotters and computer tidal predictions. As with other tables, there will be variations due to meteorological effects.

AREA 2 – Central S England

COWES LAT 50°46′N LONG 1°18′W

2019 TIMES AND HEIGHTS OF HIGH WATER – see note on facing page

JANUARY

Tu	1	0650	3.9	1914	3.6
We	2	0749	3.9	2014	3.7
Th	3	0842	4.0	2108	3.9
Fr	4	0926	4.1	2153	4.0
Sa	5	1009	4.1	2233	4.0
Su	6	1047	4.1	2313	4.1
Mo	7	1124	4.1	2353	4.1
Tu	8	1200	4.1	--	--
We	9	0030	4.1	1231	4.0
Th	10	0101	4.0	1304	4.0
Fr	11	0135	3.9	1337	3.8
Sa	12	0205	3.8	1410	3.7
Su	13	0246	3.8	1453	3.6
Mo	14	0331	3.7	1547	3.5
Tu	15	0434	3.6	1656	3.4
We	16	0540	3.6	1810	3.5
Th	17	0647	3.7	1916	3.6
Fr	18	0746	3.9	2017	3.8
Sa	19	0847	4.1	2114	4.0
Su	20	0940	4.2	2208	4.2
Mo	21	1027	4.4	2316	4.4
Tu	22	1118	4.4	--	--
We	23	0004	4.4	1205	4.4
Th	24	0053	4.4	1307	4.4
Fr	25	0144	4.3	1356	4.3
Sa	26	0240	4.2	1449	4.1
Su	27	0303	4.1	1521	3.9
Mo	28	0357	3.9	1620	3.6
Tu	29	0503	3.8	1729	3.5
We	30	0610	3.7	1844	3.4
Th	31	0718	3.7	1953	3.5

FEBRUARY

Fr	1	0816	3.7	2051	3.7
Sa	2	0907	3.8	2135	3.8
Su	3	0953	3.9	2218	3.9
Mo	4	1028	4.0	2258	4.0
Tu	5	1108	4.1	2334	4.1
We	6	1138	4.1	--	--
Th	7	0006	4.1	1210	4.0
Fr	8	0034	4.0	1240	4.0
Sa	9	0102	4.0	1309	3.9
Su	10	0134	3.9	1343	3.9
Mo	11	0209	3.9	1425	3.8
Tu	12	0252	3.8	1511	3.6
We	13	0347	3.7	1614	3.5
Th	14	0455	3.6	1730	3.4
Fr	15	0606	3.6	1842	3.5
Sa	16	0721	3.7	1954	3.7
Su	17	0828	3.9	2101	4.0
Mo	18	0923	4.1	2153	4.2
Tu	19	1014	4.3	2303	4.4
We	20	1103	4.4	2348	4.4
Th	21	1202	4.5	--	--
Fr	22	0032	4.4	1246	4.4
Sa	23	0118	4.3	1332	4.3
Su	24	0207	4.2	1422	4.1
Mo	25	0302	4.1	1450	3.9
Tu	26	0322	3.9	1543	3.6
We	27	0417	3.6	1648	3.4
Th	28	0525	3.5	1807	3.3

MARCH

Fr	1	0639	3.4	1927	3.4
Sa	2	0753	3.5	2030	3.6
Su	3	0843	3.6	2114	3.7
Mo	4	0927	3.8	2156	3.9
Tu	5	1005	3.9	2233	4.0
We	6	1044	4.0	2306	4.0
Th	7	1116	4.0	2338	4.0
Fr	8	1146	4.1	--	--
Sa	9	0032	4.1	1216	4.1
Su	10	0059	4.1	1245	4.0
Mo	11	0128	4.1	1320	4.0
Tu	12	0138	4.0	1355	3.9
We	13	0222	3.9	1443	3.7
Th	14	0309	3.7	1545	3.5
Fr	15	0421	3.5	1702	3.5
Sa	16	0540	3.5	1820	3.5
Su	17	0659	3.6	1943	3.7
Mo	18	0815	3.8	2050	4.0
Tu	19	0911	4.1	2140	4.2
We	20	0958	4.3	2245	4.4
Th	21	1100	4.4	2327	4.4
Fr	22	1142	4.4	--	--
Sa	23	0009	4.4	1224	4.4
Su	24	0051	4.3	1307	4.3
Mo	25	0135	4.2	1355	4.1
Tu	26	0159	4.0	1417	3.8
We	27	0243	3.8	1511	3.6
Th	28	0334	3.6	1613	3.4
Fr	29	0432	3.3	1728	3.3
Sa	30	0552	3.2	1849	3.3
Su	31	0710	3.2	1957	3.5

APRIL

Mo	1	0812	3.4	2045	3.7
Tu	2	0857	3.6	2126	3.9
We	3	0939	3.8	2200	4.0
Th	4	1011	3.9	2236	4.0
Fr	5	1047	4.0	2330	4.1
Sa	6	1137	4.1	2358	4.1
Su	7	1208	4.1	--	--
Mo	8	0027	4.1	1240	4.1
Tu	9	0100	4.1	1258	4.0
We	10	0118	4.0	1336	3.9
Th	11	0159	3.9	1424	3.7
Fr	12	0247	3.7	1528	3.6
Sa	13	0356	3.5	1645	3.5
Su	14	0515	3.4	1806	3.5
Mo	15	0643	3.5	1932	3.8
Tu	16	0759	3.8	2034	4.0
We	17	0854	4.0	2122	4.2
Th	18	0939	4.2	2225	4.4
Fr	19	1040	4.3	2305	4.4
Sa	20	1121	4.3	2346	4.4
Su	21	1203	4.3	--	--
Mo	22	0025	4.3	1245	4.2
Tu	23	0107	4.2	1333	4.0
We	24	0128	4.0	1351	3.8
Th	25	0206	3.8	1441	3.6
Fr	26	0252	3.5	1537	3.4
Sa	27	0346	3.3	1646	3.3
Su	28	0452	3.1	1759	3.3
Mo	29	0616	3.1	1908	3.5
Tu	30	0726	3.3	2003	3.6

MAY

We	1	0820	3.5	2049	3.8
Th	2	0904	3.7	2125	3.9
Fr	3	0941	3.9	2159	4.0
Sa	4	1037	4.0	2256	4.1
Su	5	1108	4.1	2327	4.2
Mo	6	1143	4.1	--	--
Tu	7	0001	4.2	1221	4.1
We	8	0020	4.1	1304	4.1
Th	9	0102	4.1	1325	3.9
Fr	10	0144	3.9	1417	3.8
Sa	11	0240	3.7	1525	3.7
Su	12	0344	3.6	1636	3.7
Mo	13	0458	3.5	1754	3.7
Tu	14	0619	3.6	1914	3.9
We	15	0734	3.7	2013	4.0
Th	16	0829	3.9	2101	4.2
Fr	17	0920	4.1	2145	4.3
Sa	18	1022	4.2	2244	4.3
Su	19	1103	4.2	2324	4.3
Mo	20	1147	4.1	--	--
Tu	21	0004	4.2	1231	4.1
We	22	0020	4.1	1322	4.0
Th	23	0059	4.0	1333	3.8
Fr	24	0142	3.8	1418	3.7
Sa	25	0225	3.6	1508	3.6
Su	26	0311	3.4	1604	3.5
Mo	27	0403	3.2	1707	3.5
Tu	28	0512	3.2	1813	3.5
We	29	0628	3.3	1912	3.6
Th	30	0731	3.4	2003	3.8
Fr	31	0818	3.6	2044	3.9

JUNE

Sa	1	0903	3.8	2124	4.0
Su	2	0941	3.9	2200	4.1
Mo	3	1043	4.1	2241	4.1
Tu	4	1123	4.1	2322	4.2
We	5	1208	4.1	--	--
Th	6	0006	4.2	1257	4.1
Fr	7	0049	4.1	1357	4.1
Sa	8	0138	4.0	1416	3.9
Su	9	0235	3.8	1520	3.9
Mo	10	0336	3.7	1625	3.8
Tu	11	0441	3.6	1735	3.8
We	12	0554	3.6	1844	3.9
Th	13	0707	3.7	1942	4.0
Fr	14	0804	3.8	2035	4.1
Sa	15	0859	3.9	2122	4.1
Su	16	0943	4.0	2208	4.2
Mo	17	1053	4.0	2248	4.1
Tu	18	1137	4.0	2328	4.1
We	19	1152	4.0	--	--
Th	20	0003	4.0	1234	3.9
Fr	21	0044	4.0	1312	3.9
Sa	22	0122	3.9	1355	3.8
Su	23	0155	3.7	1438	3.7
Mo	24	0237	3.5	1519	3.6
Tu	25	0321	3.4	1609	3.6
We	26	0418	3.3	1707	3.5
Th	27	0527	3.3	1811	3.6
Fr	28	0632	3.4	1907	3.7
Sa	29	0729	3.5	2001	3.8
Su	30	0822	3.7	2049	3.9

JULY

Mo	1	0914	3.9	2134	4.1
Tu	2	1022	4.1	2221	4.2
We	3	1108	4.2	2306	4.2
Th	4	1156	4.2	2353	4.2
Fr	5	1247	4.2	--	--
Sa	6	0041	4.2	1342	4.2
Su	7	0128	4.1	1408	4.1
Mo	8	0219	4.0	1459	4.0
Tu	9	0316	3.8	1603	3.9
We	10	0421	3.7	1704	3.9
Th	11	0526	3.6	1811	3.8
Fr	12	0636	3.6	1914	3.9
Sa	13	0741	3.7	2013	3.9
Su	14	0842	3.8	2103	4.0
Mo	15	0932	3.9	2149	4.0
Tu	16	1014	3.9	2229	4.0
We	17	1059	4.0	2312	4.1
Th	18	1139	4.0	2349	4.0
Fr	19	1214	4.0	--	--
Sa	20	0021	4.0	1251	3.9
Su	21	0058	3.9	1324	3.9
Mo	22	0128	3.8	1356	3.8
Tu	23	0200	3.7	1432	3.7
We	24	0239	3.6	1516	3.7
Th	25	0327	3.4	1608	3.6
Fr	26	0429	3.4	1712	3.6
Sa	27	0538	3.4	1818	3.6
Su	28	0650	3.5	1919	3.7
Mo	29	0753	3.7	2022	3.9
Tu	30	0850	3.9	2114	4.0
We	31	0942	4.0	2205	4.2

AUGUST

Th	1	1054	4.2	2255	4.3
Fr	2	1141	4.3	2342	4.4
Sa	3	1229	4.3	--	--
Su	4	0043	4.4	1318	4.3
Mo	5	0131	4.3	1412	4.2
Tu	6	0223	4.1	1435	4.1
We	7	0252	3.9	1532	4.0
Th	8	0353	3.8	1630	3.9
Fr	9	0454	3.6	1737	3.7
Sa	10	0607	3.5	1843	3.7
Su	11	0722	3.5	1949	3.7
Mo	12	0827	3.7	2045	3.8
Tu	13	0919	3.8	2133	3.9
We	14	0958	3.9	2214	4.0
Th	15	1103	4.1	2249	4.0
Fr	16	1114	4.0	2325	4.1
Sa	17	1150	4.1	2359	4.0
Su	18	1225	4.0	--	--
Mo	19	0026	4.0	1253	4.0
Tu	20	0057	3.9	1321	3.9
We	21	0129	3.9	1353	3.9
Th	22	0206	3.8	1429	3.8
Fr	23	0246	3.6	1518	3.7
Sa	24	0344	3.5	1620	3.6
Su	25	0459	3.4	1736	3.5
Mo	26	0615	3.5	1850	3.6
Tu	27	0729	3.6	2000	3.8
We	28	0836	3.9	2101	4.0
Th	29	0930	4.1	2150	4.3
Fr	30	1018	4.3	2239	4.4
Sa	31	1122	4.4	2338	4.5

SEPTEMBER

Su	1	1206	4.5	--	--
Mo	2	0021	4.5	1252	4.4
Tu	3	0107	4.4	1340	4.3
We	4	0156	4.2	1433	4.2
Th	5	0253	4.0	1541	4.1
Fr	6	0321	3.8	1555	3.8
Sa	7	0423	3.5	1659	3.6
Su	8	0540	3.4	1813	3.5
Mo	9	0701	3.5	1929	3.5
Tu	10	0808	3.6	2026	3.6
We	11	0945	3.9	2109	3.8
Th	12	0938	4.0	2147	3.9
Fr	13	1039	4.1	2225	4.0
Sa	14	1047	4.1	2259	4.1
Su	15	1120	4.1	2331	4.1
Mo	16	1151	4.1	--	--
Tu	17	0000	4.1	1243	4.1
We	18	0029	4.1	1310	4.1
Th	19	0100	4.0	1322	4.0
Fr	20	0138	3.9	1356	3.9
Sa	21	0219	3.8	1438	3.8
Su	22	0310	3.6	1539	3.6
Mo	23	0427	3.5	1659	3.5
Tu	24	0548	3.5	1821	3.5
We	25	0713	3.7	1943	3.8
Th	26	0823	4.0	2043	4.1
Fr	27	0917	4.3	2134	4.3
Sa	28	1001	4.4	2220	4.5
Su	29	1102	4.6	2317	4.6
Mo	30	1143	4.5	2359	4.5

OCTOBER

Tu	1	1226	4.5	--	--
We	2	0043	4.4	1310	4.4
Th	3	0130	4.2	1358	4.2
Fr	4	0225	4.0	1420	4.0
Sa	5	0248	3.8	1512	3.7
Su	6	0353	3.5	1613	3.5
Mo	7	0511	3.4	1731	3.3
Tu	8	0634	3.5	1853	3.4
We	9	0742	3.6	1958	3.5
Th	10	0832	3.8	2045	3.7
Fr	11	0912	4.0	2124	3.9
Sa	12	0944	4.1	2155	4.0
Su	13	1019	4.2	2232	4.1
Mo	14	1050	4.2	2303	4.2
Tu	15	1144	4.2	2332	4.2
We	16	1211	4.2	--	--
Th	17	0022	4.2	1240	4.2
Fr	18	0056	4.1	1254	4.1
Sa	19	0115	4.0	1334	4.0
Su	20	0154	3.8	1417	3.8
Mo	21	0249	3.7	1513	3.6
Tu	22	0408	3.6	1636	3.5
We	23	0533	3.6	1803	3.6
Th	24	0657	3.8	1924	3.8
Fr	25	0804	4.1	2027	4.1
Sa	26	0855	4.3	2115	4.3
Su	27	0941	4.5	2201	4.5
Mo	28	1041	4.6	2242	4.5
Tu	29	1122	4.6	2340	4.5
We	30	1203	4.5	--	--
Th	31	0022	4.4	1244	4.4

NOVEMBER

Fr	1	0109	4.2	1329	4.2
Sa	2	0205	4.0	1350	4.0
Su	3	0223	3.8	1437	3.7
Mo	4	0320	3.6	1530	3.4
Tu	5	0427	3.5	1638	3.3
We	6	0544	3.5	1756	3.2
Th	7	0657	3.6	1912	3.4
Fr	8	0749	3.8	2004	3.6
Sa	9	0835	4.0	2049	3.8
Su	10	0913	4.1	2125	4.0
Mo	11	0948	4.2	2201	4.1
Tu	12	1021	4.2	2232	4.1
We	13	1049	4.2	2329	4.2
Th	14	1144	4.3	--	--
Fr	15	0003	4.2	1159	4.2
Sa	16	0042	4.2	1238	4.2
Su	17	0126	4.1	1318	4.0
Mo	18	0223	4.0	1404	3.9
Tu	19	0244	3.8	1504	3.7
We	20	0400	3.7	1620	3.6
Th	21	0515	3.8	1739	3.6
Fr	22	0637	3.9	1857	3.8
Sa	23	0743	4.1	2002	4.0
Su	24	0836	4.3	2052	4.2
Mo	25	0922	4.4	2139	4.3
Tu	26	1003	4.5	2222	4.3
We	27	1044	4.4	2326	4.3
Th	28	1125	4.4	--	--
Fr	29	0011	4.3	1206	4.3
Sa	30	0058	4.2	1244	4.1

DECEMBER

Su	1	0153	4.1	1326	4.0
Mo	2	0201	3.9	1411	3.8
Tu	3	0251	3.7	1457	3.5
We	4	0344	3.6	1547	3.3
Th	5	0446	3.6	1649	3.2
Fr	6	0552	3.6	1804	3.3
Sa	7	0658	3.7	1915	3.5
Su	8	0750	3.8	2005	3.6
Mo	9	0835	4.0	2051	3.8
Tu	10	0911	4.0	2128	3.9
We	11	0949	4.1	2235	4.1
Th	12	1024	4.2	2310	4.2
Fr	13	1105	4.2	2330	4.2
Sa	14	1146	4.3	--	--
Su	15	0010	4.2	1229	4.2
Mo	16	0056	4.1	1312	4.1
Tu	17	0147	4.0	1402	4.0
We	18	0241	4.0	1500	3.8
Th	19	0347	3.9	1605	3.7
Fr	20	0450	3.9	1714	3.7
Sa	21	0605	3.9	1828	3.7
Su	22	0709	4.0	1934	3.8
Mo	23	0810	4.1	2033	4.0
Tu	24	0857	4.2	2125	4.1
We	25	0945	4.3	2210	4.1
Th	26	1027	4.3	2255	4.2
Fr	27	1112	4.3	2339	4.1
Sa	28	1151	4.2	--	--
Su	29	0017	4.1	1226	4.1
Mo	30	0057	4.0	1307	4.0
Tu	31	0139	3.9	1345	3.9

Central S England

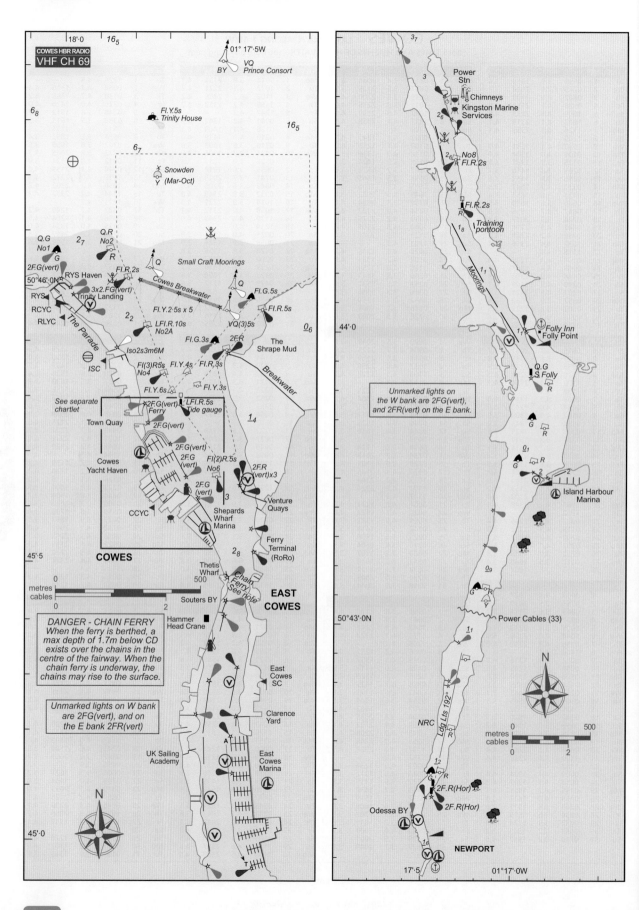

COWES HBR RADIO
VHF CH 69

01° 17'·5W
VQ
BY — Prince Consort

6_8

$18'·0$ 16_5 16_5

Fl.Y.5s
Trinity House

6_7

Snowden
Y (Mar-Oct)

Q.G
No1
G
2F.G(vert)
50°46'·0N
RYS
RCYC
RLYC
The Parade

Q.R
No2
R
2_7
Fl.R.2s
RYS Haven
3x2.FG(vert)
Trinity Landing
V
Iso2s3m6M
ISC

Small Craft Moorings
Q
Cowes Breakwater
Q
Fl.G.5s
Fl.R.5s
VQ(3)5s
Fl.G.3s
2F.R
2_2
Fl.Y.2·5s x 5
LFl.R.10s
No2A

The Shrape Mud

0_6

Breakwater

1_4

Fl(3)R5s
No4
Fl.Y.6s
Fl.Y.4s
Fl.R.3s
Fl.Y.3s

See separate
chartlet
Town Quay
Cowes
Yacht Haven
CCYC

2F.G(vert)
Ferry
2F.G(vert)
2F.G(vert)
2F.G(vert)
LFl.R.5s
Tide gauge
Fl(2)R.5s
No6
V
2F.R
(vert)x3
Venture
Quays
Shepards
Wharf
Marina

3

2_8

Ferry
Terminal
(RoRo)

COWES

metres
cables
0   500
0   2

Thetis
Wharf
Chain
Ferry
See note
Souters BY
Hammer
Head Crane

**EAST
COWES**

East
Cowes
SC

Clarence
Yard

DANGER - CHAIN FERRY
When the ferry is berthed, a
max depth of 1.7m below CD
exists over the chains in the
centre of the fairway. When the
chain ferry is underway, the
chains may rise to the surface.

*Unmarked lights on W bank
are 2FG(vert), and on
the E bank 2FR(vert)*

UK Sailing
Academy

East
Cowes
Marina

V

$45'·5$

$45'·0$

N

A
V
V
T

3_7

Power
Stn
Chimneys
Kingston Marine
Services

3

2_6

2_6 No8
Fl.R.2s

Fl.R.2s
R
Training
pontoon

1_8

Moorings

1_1

Folly Inn
Folly Point
V
1_1

Q.G
S Folly
R

44'·0

*Unmarked lights on
the W bank are 2FG(vert),
and 2FR(vert) on the E bank.*

G
R

0_1
G   R

V
2 2
Island Harbour
Marina

0_9

G   R

50°43'·0N   Power Cables (33)

1_1

Log Lts 192°

NRC
G
R

1_2
G   R
2F.R(Hor)
2F.R(Hor)

Odessa BY
V

1_6
V

NEWPORT

N

metres
cables
0   500
0   2

$17'·5$ 01°17'·0W

Note Double HWs occur at Southampton.
The predictions are for the first HW.

STANDARD TIME (UT)
For Summer Time add ONE hour in **non-shaded areas**

SOUTHAMPTON LAT 50°53'N LONG 1°24'W
TIMES AND HEIGHTS OF HIGH AND LOW WATERS

Dates in red are **SPRINGS**
Dates in blue are NEAPS

YEAR **2019**

Central S England

JANUARY

Time	m		Time	m
1 0012	1.6	**16** 0608	3.9
0630	4.1	1147	1.8
TU 1248	1.7	W 1745	3.6
1854	3.8		
2 0114	1.6	**17** 0019	1.8
0725	4.1	0627	3.9
W 1346	1.5	TH 1256	1.7
2057	4.0	1903	3.8
3 0208	1.5	**18** 0126	1.6
0919	4.3	0740	4.1
TH 1437	1.3	F 1400	1.4
2149	4.1	2014	4.0
4 0257	1.3	**19** 0226	1.3
0959	4.3	0843	4.3
F 1524	1.1	SA 1457	1.0
2240	4.2	2114	4.3
5 0344	1.2	**20** 0321	1.0
0952	4.3	0935	4.5
SA 1609	1.0	SU 1549	0.7
2222	4.2	2204	4.5
6 0429	1.1	**21** 0411	0.7
1035	4.3	1027	4.7
SU 1652	0.9	M 1637	0.4
● 2304	4.2	○ 2312	4.7
7 0511	1.1	**22** 0459	0.5
1117	4.3	1130	4.8
M 1732	0.9	TU 1724	0.2
2342	4.2	2357	4.8
8 0551	1.2	**23** 0545	0.4
1151	4.3	1215	4.8
TU 1808	1.0	W 1809	0.2
9 0016	4.2	**24** 0044	4.7
0625	1.3	0631	0.5
W 1228	4.3	TH 1301	4.7
1838	1.1	1854	0.3
10 0134	4.3	**25** 0133	4.7
0651	1.4	0716	0.6
TH 1300	4.0	F 1351	4.6
1903	1.2	1939	0.5
11 0159	4.3	**26** 0228	4.5
0718	1.4	0802	0.9
F 1333	4.1	SA 1447	4.4
1934	1.3	2026	0.8
12 0232	4.2	**27** 0256	4.3
0754	1.5	0853	1.2
SA 1412	4.0	SU 1557	4.2
2012	1.4	◑ 2119	1.2
13 0313	4.2	**28** 0353	4.1
0837	1.6	0950	1.5
SU 1458	3.9	M 1614	3.9
2059	1.6	2220	1.5
14 0332	4.0	**29** 0455	4.0
0930	1.7	1057	1.7
M 1545	3.8	TU 1718	3.7
◑ 2157	1.7	2332	1.8
15 0422	3.9	**30** 0553	3.9
1036	1.9	1213	1.8
TU 1645	3.7	W 1825	3.7
2308	1.8		
		31 0045	1.8
		0653	3.8
		TH 1321	1.7
		1929	3.7

FEBRUARY

Time	m		Time	m
1 0148	1.7	**16** 0102	1.8
0856	4.0	0708	3.8
F 1419	1.6	SA 1345	1.5
2029	3.8	1957	3.9
2 0241	1.6	**17** 0216	1.5
0944	4.1	0829	4.2
SA 1507	1.3	SU 1448	1.1
2121	3.9	2103	4.2
3 0328	1.3	**18** 0312	1.1
0938	4.1	0923	4.4
SU 1552	1.1	M 1539	0.6
2311	4.2	2153	4.5
4 0413	1.1	**19** 0402	0.7
1018	4.2	1012	4.7
M 1634	0.9	TU 1626	0.3
● 2247	4.2	○ 2258	4.7
5 0454	1.0	**20** 0447	0.4
1100	4.3	1113	4.8
TU 1713	0.8	W 1609	0.1
2325	4.3	2340	4.8
6 0532	1.0	**21** 0531	0.2
1135	4.3	1157	4.8
W 1749	0.8	TH 1753	0.0
2359	4.3		
7 0605	1.1	**22** 0025	4.8
1206	4.3	0614	0.2
TH 1819	0.9	F 1242	4.7
		1835	0.1
8 0102	4.3	**23** 0112	4.7
0630	1.1	0656	0.4
F 1239	4.3	SA 1329	4.6
1842	1.0	1917	0.3
9 0126	4.3	**24** 0203	4.5
0654	1.1	0738	0.6
SA 1334	4.3	SU 1421	4.4
1908	1.0	1959	0.7
10 0157	4.3	**25** 0226	4.3
0726	1.1	0822	1.0
SU 1307	4.2	M 1443	4.1
1944	1.0	2046	1.1
11 0236	4.3	**26** 0315	4.0
0805	1.1	0913	1.4
M 1428	4.1	TU 1541	3.8
2025	1.1	◐ 2141	1.6
12 0321	4.2	**27** 0415	3.8
0738	0.8	1014	1.7
TU 1517	4.0	W 1642	3.6
◑ 2113	1.4	2252	1.9
13 0349	4.0	**28** 0516	3.7
0946	1.6	1133	1.9
W 1609	3.8	TH 1749	3.5
2215	1.7		
14 0444	3.8		
1059	1.8		
TH 1748	3.7		
2336	1.8		
15 0546	3.7		
1224	1.8		
F 1825	3.6		

MARCH

Time	m		Time	m
1 0018	2.0	**16** 0516	3.6
0620	3.6	1206	1.8
F 1255	1.9	SA 1800	3.6
1859	3.5		
2 0129	1.9	**17** 0100	1.9
0816	3.7	0648	3.7
SA 1358	1.7	SU 1340	1.5
2003	3.7	1942	3.9
3 0224	1.7	**18** 0211	1.5
0823	3.7	0814	4.1
SU 1447	1.4	M 1437	1.1
2058	3.9	2047	4.3
4 0310	1.4	**19** 0302	1.0
0953	4.0	0908	4.4
M 1530	1.1	TU 1525	0.6
2141	4.0	2134	4.5
5 0352	1.1	**20** 0347	0.6
0958	4.1	0955	4.6
TU 1611	0.9	W 1609	0.2
2222	4.2	2241	4.7
6 0432	0.9	**21** 0430	0.3
1039	4.2	1056	4.7
W 1649	0.7	TH 1652	0.0
● 2259	4.3	○ 2322	4.8
7 0508	0.9	**22** 0512	0.1
1114	4.3	1138	4.7
TH 1724	0.7	F 1733	0.0
2335	4.3		
8 0540	0.9	**23** 0004	4.7
1145	4.3	0553	0.1
F 1753	0.8	SA 1221	4.7
		1813	0.1
9 0001	4.3	**24** 0048	4.6
0604	0.9	0633	0.3
SA 1237	4.4	SU 1307	4.5
1817	0.8	1852	0.4
10 0056	4.4	**25** 0136	4.5
0629	0.8	0713	0.6
SU 1307	4.4	M 1357	4.4
1844	0.8	1933	0.7
11 0128	4.4	**26** 0236	4.3
0701	0.8	0753	0.9
M 1343	4.4	TU 1414	4.1
1918	0.8	2016	1.2
12 0206	4.4	**27** 0234	4.0
0738	0.8	0839	1.3
TU 1425	4.4	W 1459	3.8
1957	0.9	2108	1.6
13 0230	4.2	**28** 0323	3.7
0821	1.0	0936	1.7
W 1449	4.1	TH 1603	3.6
2043	1.2	◐ 2216	2.0
14 0317	4.0	**29** 0429	3.6
0911	1.4	1052	2.0
TH 1541	3.9	F 1714	3.5
◑ 2139	1.6	2345	2.1
15 0409	3.8	**30** 0539	3.4
1020	1.7	1218	2.0
F 1724	3.8	SA 1915	3.6
2305	1.9		
		31 0101	2.1
		0728	3.5
		SU 1325	1.8
		1925	3.6

APRIL

Time	m		Time	m
1 0158	1.8	**16** 0154	1.5
0746	3.6	0749	4.0
M 1417	1.6	TU 1416	1.1
2026	3.8	2026	4.3
2 0244	1.5	**17** 0242	1.0
0922	3.9	0844	4.3
TU 1501	1.3	W 1503	0.7
2113	4.1	2112	4.5
3 0325	1.2	**18** 0326	0.7
0929	4.0	0931	4.5
W 1542	1.0	TH 1547	0.4
2153	4.2	2223	4.7
4 0403	1.0	**19** 0409	0.4
1011	4.2	1036	4.6
TH 1619	0.8	F 1629	0.2
2231	4.3	○ 2302	4.7
5 0438	0.8	**20** 0450	0.2
1046	4.3	1118	4.6
F 1653	0.7	SA 1709	0.2
● 2303	4.4	2344	4.7
6 0509	0.7	**21** 0530	0.3
1117	4.3	1201	4.5
SA 1723	0.7	SU 1749	0.3
2335	4.4		
7 0537	0.7	**22** 0025	4.6
1211	4.5	0610	0.4
SU 1751	0.7	M 1247	4.4
		1829	0.6
8 0028	4.5	**23** 0109	4.4
0605	0.7	0649	0.6
M 1243	4.5	TU 1339	4.3
1821	0.7	1909	0.9
9 0102	4.5	**24** 0158	4.3
0638	0.6	0728	0.9
TU 1321	4.5	W 1344	4.1
1856	0.8	1951	1.3
10 0141	4.5	**25** 0202	4.0
0716	0.7	0811	1.3
W 1404	4.4	TH 1431	3.9
1936	0.9	2039	1.7
11 0201	4.3	**26** 0243	3.7
0759	1.0	0903	1.7
TH 1429	4.2	F 1523	3.7
2022	1.3	◐ 2143	2.0
12 0249	4.0	**27** 0336	3.5
0849	1.3	1012	1.9
F 1556	4.0	SA 1633	3.6
◑ 2119	1.7	2302	2.2
13 0343	3.7	**28** 0450	3.4
1002	1.7	1128	2.0
SA 1626	3.7	SU 1740	3.6
2300	2.0		
14 0458	3.6	**29** 0018	2.1
1158	1.8	0646	3.5
SU 1754	3.7	M 1239	1.9
		1845	3.6
15 0051	1.8	**30** 0121	1.9
0636	3.7	0749	3.6
M 1321	1.5	TU 1337	1.7
1924	4.0	1947	3.8

Chart Datum: 2·74 metres below Ordnance Datum (Newlyn). HAT is 5·0 metres above Chart Datum.

⟩⟩ **FREE** monthly updates. Register at ⟨ www.reedsnauticalalmanac.co.uk ⟨

111

Southampton tides

Note Double HWs occur at Southampton. The predictions are for the first HW.

STANDARD TIME (UT)
For Summer Time add ONE hour in **non-shaded areas**

SOUTHAMPTON LAT 50°53′N LONG 1°24′W
TIMES AND HEIGHTS OF HIGH AND LOW WATERS

Dates in red are **SPRINGS**
Dates in blue are NEAPS

YEAR 2019

MAY

Day	Time	m	Time	m	Time	m	Time	m
1 W	0210	1.7	0844	3.8	1424	1.4	2120	4.2
16 TH	0214	1.1	0819	4.2	1435	0.9	2048	4.4
2 TH	0251	1.4	0857	4.0	1505	1.2	2119	4.2
17 F	0301	0.8	0908	4.3	1521	0.7	2204	4.6
3 F	0328	1.1	0936	4.1	1543	0.9	2158	4.4
18 SA	0345	0.6	1019	4.4	1604	0.5	○ 2243	4.6
4 SA	0403	0.9	1016	4.3	1619	0.8	● 2231	4.4
19 SU	0427	0.5	1103	4.4	1646	0.5	2324	4.5
5 SU	0436	0.7	1049	4.4	1652	0.7	2306	4.5
20 M	0508	0.5	1148	4.4	1728	0.6		
6 M	0509	0.6	1147	4.5	1726	0.7		
21 TU	0005	4.5	0550	0.6	1239	4.3	1809	0.8
7 TU	0003	4.6	0543	0.6	1223	4.6	1801	0.7
22 W	0047	4.4	0629	0.8	1237	4.2	1850	1.1
8 W	0040	4.6	0621	0.6	1303	4.5	1840	0.8
23 TH	0130	4.2	0708	1.0	1321	4.1	1930	1.4
9 TH	0122	4.5	0701	0.8	1350	4.4	1923	1.0
24 F	0133	4.0	0748	1.3	1406	3.9	2015	1.6
10 F	0145	4.3	0747	1.0	1445	4.3	2013	1.4
25 SA	0213	3.8	0834	1.6	1455	3.8	2107	1.9
11 SA	0233	4.0	0842	1.3	1555	4.1	2120	1.7
26 SU	0259	3.6	0930	1.8	1545	3.7	☾ 2210	2.1
12 SU	0337	3.8	1001	1.6	1624	3.9	☾ 2251	1.8
27 M	0355	3.4	1034	1.9	1645	3.7	2315	2.1
13 M	0452	3.7	1132	1.6	1744	3.9		
28 TU	0500	3.4	1137	1.9	1754	3.7		
14 TU	0015	1.7	0612	3.8	1246	1.4	1859	4.1
29 W	0020	2.0	0614	3.5	1238	1.8	1900	3.8
15 W	0122	1.5	0722	4.0	1346	1.2	2002	4.3
30 TH	0117	1.8	0723	3.7	1333	1.6	1955	4.0
31 F	0204	1.5	0817	3.9	1420	1.4	2041	4.2

JUNE

Day	Time	m	Time	m	Time	m	Time	m
1 SA	0245	1.3	0903	4.1	1503	1.1	2122	4.3
16 SU	0322	0.9	1010	4.3	1542	0.9	2229	4.4
2 SU	0325	1.0	0944	4.3	1544	0.9	2201	4.4
17 M	0407	0.8	1059	4.3	1627	0.9	○ 2311	4.4
3 M	0405	0.8	1023	4.4	1625	0.8	● 2240	4.5
18 TU	0450	0.7	1101	4.2	1711	0.9	2317	4.3
4 TU	0446	0.6	1125	4.5	1705	0.7	2343	4.6
19 W	0533	0.8	1144	4.2	1753	1.0	2355	4.2
5 W	0527	0.6	1206	4.6	1748	0.7		
20 TH	0614	0.9	1224	4.2	1835	1.2		
6 TH	0024	4.6	0609	0.6	1251	4.6	1831	0.8
21 F	0033	4.1	0652	1.1	1303	4.1	1913	1.4
7 F	0109	4.6	0654	0.7	1341	4.5	1919	1.0
22 SA	0114	4.1	0728	1.2	1340	4.0	1949	1.5
8 SA	0200	4.4	0743	0.9	1438	4.4	2012	1.3
23 SU	0151	3.9	0804	1.4	1421	4.0	2028	1.7
9 SU	0258	4.2	0839	1.1	1547	4.3	2114	1.5
24 M	0228	3.8	0845	1.6	1505	3.9	2114	1.8
10 M	0331	4.0	0946	1.3	1611	4.1	☾ 2225	1.6
25 TU	0316	3.7	0937	1.7	1551	3.8	☾ 2209	1.9
11 TU	0434	3.9	1057	1.4	1722	4.1	2336	1.6
26 W	0407	3.5	1036	1.8	1643	3.7	2309	2.0
12 W	0542	3.8	1206	1.4	1827	4.1		
27 TH	0502	3.5	1137	1.8	1740	3.7		
13 TH	0043	1.5	0652	3.9	1310	1.3	1927	4.2
28 F	0009	1.9	0606	3.5	1236	1.7	1847	3.8
14 F	0143	1.3	0751	4.0	1406	1.2	2019	4.3
29 SA	0108	1.7	0718	3.7	1333	1.6	1954	4.0
15 SA	0235	1.1	0925	4.2	1456	1.0	2148	4.4
30 SU	0202	1.4	0821	4.0	1426	1.3	2047	4.2

JULY

Day	Time	m	Time	m	Time	m	Time	m
1 M	0253	1.1	0913	4.2	1516	1.1	2132	4.4
16 TU	0351	1.0	1057	4.2	1612	1.1	○ 2216	4.2
2 TU	0341	0.8	0959	4.4	1604	0.9	● 2220	4.5
17 W	0435	0.9	1147	4.3	1656	1.1	2259	4.2
3 W	0428	0.6	1109	4.6	1651	0.8	2305	4.6
18 TH	0518	0.8	1131	4.2	1739	1.1	2341	4.2
4 TH	0515	0.5	1153	4.6	1737	0.7		
19 F	0558	0.9	1208	4.2	1819	1.1		
5 F	0009	4.7	0601	0.5	1238	4.7	1824	0.7
20 SA	0016	4.2	0635	1.0	1242	4.2	1853	1.3
6 SA	0056	4.6	0647	0.5	1328	4.6	1911	0.8
21 SU	0051	4.1	0705	1.1	1316	4.2	1921	1.4
7 SU	0145	4.5	0735	0.7	1422	4.5	2001	1.0
22 M	0125	4.1	0732	1.2	1429	4.2	1949	1.5
8 M	0240	4.4	0826	0.9	1526	4.4	2055	1.2
23 TU	0201	4.0	0804	1.3	1429	4.1	2026	1.6
9 TU	0343	4.2	0921	1.1	1553	4.2	☾ 2154	1.4
24 W	0242	3.9	0844	1.5	1511	4.0	2112	1.7
10 W	0412	4.0	1023	1.4	1655	4.1	2259	1.6
25 TH	0326	3.8	0935	1.6	1601	3.9	☾ 2210	1.8
11 TH	0517	3.9	1129	1.5	1755	4.1		
26 F	0420	3.7	1040	1.8	1654	3.8	2317	1.8
12 F	0007	1.6	0622	3.8	1236	1.6	1854	4.0
27 SA	0604	3.7	1147	1.8	1752	3.8		
13 SA	0114	1.5	0725	3.9	1339	1.5	1953	4.1
28 SU	0024	1.8	0621	3.6	1254	1.7	1902	3.9
14 SU	0212	1.4	0824	3.9	1435	1.4	2044	4.1
29 M	0129	1.6	0741	3.8	1358	1.5	2017	4.1
15 M	0303	1.2	1009	4.1	1525	1.3	2216	4.3
30 TU	0231	1.2	0850	4.1	1457	1.3	2116	4.4
31 W	0326	0.9	0942	4.4	1550	1.0	2204	4.6

AUGUST

Day	Time	m	Time	m	Time	m	Time	m
1 TH	0417	0.6	1031	4.6	1640	0.7	● 2250	4.7
16 F	0459	0.8	1109	4.2	1720	1.0	2318	4.2
2 F	0504	0.4	1138	4.7	1726	0.5	2353	4.8
17 SA	0537	0.8	1148	4.3	1757	1.0	2352	4.2
3 SA	0549	0.2	1222	4.8	1811	0.5		
18 SU	0612	0.9	1256	4.4	1828	1.2		
4 SU	0038	4.8	0634	0.3	1309	4.7	1856	0.6
19 M	0027	4.2	0639	1.0	1319	4.3	1850	1.2
5 M	0125	4.6	0718	0.4	1400	4.6	1941	0.7
20 TU	0056	4.2	0700	1.1	1346	4.2	1914	1.2
6 TU	0217	4.5	0803	0.7	1501	4.5	2029	1.0
21 W	0154	4.3	0728	1.1	1420	4.3	1947	1.3
7 W	0318	4.3	0853	1.0	1525	4.2	☾ 2123	1.3
22 TH	0208	4.1	0805	1.2	1501	4.3	2029	1.4
8 TH	0347	4.0	0950	1.4	1626	4.1	2225	1.6
23 F	0254	4.0	0850	1.5	1520	4.0	☾ 2120	1.6
9 F	0451	3.8	1056	1.7	1725	4.0	2336	1.8
24 SA	0343	3.8	0947	1.7	1611	3.9	2228	1.8
10 SA	0555	3.7	1209	1.8	1825	3.9		
25 SU	0514	3.8	1103	2.0	1756	3.9	2350	1.9
11 SU	0050	1.8	0700	3.7	1319	1.8	1926	3.9
26 M	0544	3.6	1228	2.0	1822	3.7		
12 M	0154	1.7	0803	3.8	1419	1.7	2026	3.9
27 TU	0111	1.7	0715	3.8	1345	1.7	1956	4.0
13 TU	0247	1.4	0902	3.9	1510	1.5	2200	4.1
28 W	0221	1.3	0838	4.2	1448	1.4	2057	4.3
14 W	0334	1.2	1050	4.2	1556	1.2	2201	4.2
29 TH	0316	0.9	0930	4.5	1540	1.0	2147	4.6
15 TH	0418	0.9	1030	4.2	1639	1.1	○ 2242	4.2
30 F	0404	0.5	1015	4.7	1626	0.6	● 2232	4.8
31 SA	0449	0.2	1120	4.9	1710	0.4	2334	4.9

Chart Datum: 2·74 metres below Ordnance Datum (Newlyn). HAT is 5·0 metres above Chart Datum.

》》 **FREE** monthly updates. Register at 《
www.reedsnauticalalmanac.co.uk 《

Note Double HWs occur at Southampton. The predictions are for the first HW.

STANDARD TIME (UT)
For Summer Time add ONE hour in **non-shaded areas**

SOUTHAMPTON   LAT 50°53'N   LONG 1°24'W
TIMES AND HEIGHTS OF HIGH AND LOW WATERS

Dates in red are **SPRINGS**
Dates in blue are NEAPS

YEAR 2019

SEPTEMBER

Day	Time m	Day	Time m
1 SU	0532 0.1 / 1202 4.9 / 1753 0.3	16 M	0543 0.9 / 1152 4.4 / 1757 1.1 / 2358 4.3
2 M	0017 4.8 / 0614 0.1 / 1248 4.8 / 1835 0.4	17 TU	0608 1.0 / 1244 4.5 / 1818 1.1
3 TU	0103 4.7 / 0656 0.3 / 1336 4.7 / 1918 0.6	18 W	0051 4.4 / 0629 1.0 / 1312 4.5 / 1843 1.1
4 W	0153 4.5 / 0738 0.7 / 1435 4.5 / 2002 1.0	19 TH	0124 4.4 / 0658 1.0 / 1346 4.5 / 1917 1.1
5 TH	0254 4.3 / 0824 1.1 / 1449 4.2 / 2051 1.4	20 F	0203 4.4 / 0734 1.1 / 1402 4.3 / 1957 1.3
6 F	0311 4.0 / 0918 1.5 / 1551 4.0 / ☽ 2152 1.7	21 SA	0248 4.3 / 0816 1.4 / 1450 4.1 / 2043 1.5
7 SA	0418 3.8 / 1026 1.9 / 1652 3.8 / 2308 2.0	22 SU	0312 4.0 / 0908 1.8 / 1538 3.9 / ☽ 2145 1.9
8 SU	0527 3.6 / 1147 2.1 / 1757 3.7	23 M	0447 3.8 / 1027 2.1 / 1729 3.8 / 2330 2.0
9 M	0027 2.0 / 0634 3.6 / 1300 2.1 / 1953 3.8	24 TU	0517 3.6 / 1221 2.1 / 1802 3.7
10 TU	0133 1.9 / 0737 3.7 / 1400 1.9 / 1958 3.8	25 W	0107 1.8 / 0702 3.8 / 1341 1.8 / 1940 4.0
11 W	0226 1.6 / 0839 3.9 / 1450 1.6 / 2052 4.0	26 TH	0210 1.4 / 0823 4.3 / 1436 1.4 / 2040 4.4
12 TH	0312 1.3 / 0927 4.1 / 1534 1.3 / 2211 4.2	27 F	0300 0.9 / 0914 4.6 / 1523 0.9 / 2129 4.6
13 F	0354 1.0 / 1009 4.3 / 1616 1.1 / 2216 4.2	28 SA	0346 0.5 / 0957 4.8 / 1608 0.6 / ● 2232 4.8
14 SA	0434 0.8 / 1046 4.4 / 1655 1.0 / ○ 2257 4.3	29 SU	0429 0.2 / 1059 4.9 / 1650 0.3 / 2313 4.9
15 SU	0511 0.8 / 1118 4.4 / 1729 1.0 / 2330 4.3	30 M	0510 0.1 / 1141 4.9 / 1732 0.3 / 2356 4.9

OCTOBER

Day	Time m	Day	Time m
1 TU	0551 0.2 / 1224 4.8 / 1812 0.4	16 W	0536 1.0 / 1214 4.6 / 1748 1.0
2 W	0040 4.7 / 0631 0.4 / 1310 4.7 / 1852 0.7	17 TH	0025 4.5 / 0602 1.0 / 1243 4.6 / 1818 1.0
3 TH	0129 4.5 / 0712 0.8 / 1405 4.5 / 1934 1.0	18 F	0059 4.6 / 0634 1.0 / 1319 4.6 / 1852 1.0
4 F	0234 4.3 / 0756 1.2 / 1415 4.2 / 2020 1.4	19 SA	0139 4.5 / 0711 1.2 / 1400 4.4 / 1932 1.2
5 SA	0236 3.9 / 0848 1.7 / 1505 3.9 / ☽ 2120 1.8	20 SU	0226 4.3 / 0754 1.5 / 1419 4.1 / 2018 1.5
6 SU	0346 3.8 / 0959 2.1 / 1612 3.7 / 2239 2.1	21 M	0322 4.1 / 0845 1.8 / 1512 3.9 / ☽ 2120 1.9
7 M	0459 3.7 / 1123 2.2 / 1738 3.6 / 2357 2.1	22 TU	0437 3.9 / 1014 2.2 / 1720 3.8 / 2323 2.0
8 TU	0654 3.7 / 1233 2.2 / 1905 3.7	23 W	0510 3.7 / 1215 2.1 / 1749 3.7
9 W	0101 2.0 / 0704 3.7 / 1332 2.0 / 1926 3.7	24 TH	0051 1.8 / 0646 4.0 / 1323 1.8 / 1915 4.0
10 TH	0156 1.7 / 0803 3.9 / 1423 1.7 / 2022 3.9	25 F	0149 1.4 / 0800 4.3 / 1415 1.3 / 2015 4.3
11 F	0242 1.4 / 0852 4.1 / 1507 1.4 / 2143 4.2	26 SA	0237 1.0 / 0851 4.6 / 1501 0.9 / 2104 4.6
12 SA	0324 1.1 / 0936 4.3 / 1547 1.1 / 2148 4.3	27 SU	0322 0.6 / 0932 4.8 / 1544 0.6 / 2212 4.8
13 SU	0403 1.0 / 1013 4.4 / 1624 1.0 / ○ 2229 4.4	28 M	0405 0.4 / 1039 4.9 / 1626 0.4 / ● 2252 4.8
14 M	0439 0.9 / 1050 4.5 / 1657 1.0 / 2302 4.4	29 TU	0446 0.4 / 1119 4.9 / 1708 0.4 / 2336 4.8
15 TU	0510 0.9 / 1119 4.5 / 1724 1.0 / 2331 4.4	30 W	0527 0.4 / 1201 4.8 / 1749 0.5
		31 TH	0020 4.7 / 0608 0.7 / 1245 4.7 / 1829 0.7

NOVEMBER

Day	Time m	Day	Time m
1 F	0109 4.5 / 0649 1.0 / 1334 4.5 / 1910 1.1	16 SA	0041 4.6 / 0617 1.1 / 1259 4.6 / 1836 1.0
2 SA	0124 4.2 / 0732 1.4 / 1339 4.1 / 1955 1.4	17 SU	0124 4.5 / 0657 1.2 / 1343 4.5 / 1918 1.2
3 SU	0209 4.0 / 0823 1.8 / 1430 3.9 / 2049 1.8	18 M	0213 4.4 / 0743 1.5 / 1406 4.2 / 2007 1.5
4 M	0308 3.8 / 0929 2.1 / 1525 3.7 / ☽ 2202 2.0	19 TU	0314 4.2 / 0840 1.8 / 1500 4.0 / ☽ 2115 1.8
5 TU	0420 3.7 / 1048 2.2 / 1640 3.6 / 2317 2.1	20 W	0433 4.1 / 1008 2.0 / 1703 3.9 / 2255 1.9
6 W	0523 3.7 / 1158 2.2 / 1822 3.6	21 TH	0458 3.9 / 1141 2.0 / 1728 3.8
7 TH	0021 2.1 / 0623 3.7 / 1257 2.1 / 1926 3.7	22 F	0016 1.7 / 0616 4.1 / 1251 1.7 / 1845 4.0
8 F	0117 1.9 / 0721 3.9 / 1349 1.8 / 2027 3.9	23 SA	0117 1.4 / 0727 4.3 / 1347 1.4 / 1949 4.2
9 SA	0206 1.6 / 0817 4.1 / 1434 1.5 / 2111 4.1	24 SU	0209 1.1 / 0821 4.5 / 1435 1.1 / 2040 4.4
10 SU	0249 1.4 / 0942 4.4 / 1513 1.3 / 2120 4.2	25 M	0256 0.9 / 0908 4.7 / 1520 0.8 / 2155 4.6
11 M	0328 1.2 / 0940 4.4 / 1549 1.1 / 2157 4.3	26 TU	0340 0.7 / 1021 4.8 / 1604 0.6 / ● 2237 4.6
12 TU	0403 1.1 / 1019 4.5 / 1622 1.0 / ○ 2231 4.4	27 W	0423 0.7 / 1102 4.8 / 1646 0.6 / 2322 4.6
13 W	0436 1.0 / 1049 4.5 / 1652 0.9 / 2332 4.6	28 TH	0506 0.7 / 1144 4.7 / 1728 0.7
14 TH	0508 1.0 / 1148 4.6 / 1724 0.9	29 F	0010 4.5 / 0548 0.9 / 1226 4.6 / 1810 0.8
15 F	0004 4.6 / 0541 1.0 / 1221 4.6 / 1758 0.9	30 SA	0021 4.3 / 0631 1.1 / 1311 4.4 / 1851 1.1

DECEMBER

Day	Time m	Day	Time m
1 SU	0102 4.2 / 0714 1.4 / 1320 4.2 / 1934 1.3	16 M	0114 4.6 / 0652 1.1 / 1332 4.6 / 1914 1.0
2 M	0148 4.1 / 0800 1.7 / 1400 4.0 / 2021 1.6	17 TU	0205 4.5 / 0740 1.3 / 1424 4.4 / 2005 1.2
3 TU	0238 3.9 / 0852 2.0 / 1452 3.8 / 2117 1.9	18 W	0303 4.4 / 0836 1.5 / 1524 4.2 / 2105 1.4
4 W	0332 3.8 / 0956 2.1 / 1546 3.6 / ☽ 2222 2.0	19 TH	0414 4.2 / 0943 1.7 / 1636 4.1 / ☽ 2217 1.6
5 TH	0434 3.8 / 1106 2.2 / 1653 3.6 / 2328 2.1	20 F	0438 4.1 / 1059 1.8 / 1703 3.9 / 2334 1.6
6 F	0535 3.8 / 1212 2.1 / 1759 3.6	21 SA	0547 4.1 / 1213 1.7 / 1814 3.9
7 SA	0028 2.0 / 0726 4.0 / 1307 2.0 / 1946 3.8	22 SU	0042 1.5 / 0654 4.2 / 1317 1.5 / 1922 4.1
8 SU	0121 1.8 / 0823 4.1 / 1353 1.7 / 2041 4.0	23 M	0141 1.4 / 0754 4.3 / 1411 1.3 / 2017 4.2
9 M	0207 1.6 / 0823 4.2 / 1433 1.5 / 2046 4.1	24 TU	0232 1.2 / 0843 4.4 / 1500 1.1 / 2148 4.3
10 TU	0248 1.4 / 0907 4.3 / 1511 1.2 / 2130 4.2	25 W	0320 1.1 / 1009 4.6 / 1546 0.9 / 2234 4.4
11 W	0328 1.2 / 0948 4.4 / 1549 1.0 / 2204 4.3	26 TH	0406 1.0 / 1051 4.6 / 1631 0.8 / ● 2324 4.4
12 TH	0407 1.1 / 1025 4.5 / 1627 0.9 / ○ 2243 4.4	27 F	0451 0.9 / 1133 4.5 / 1714 0.8 / 2326 4.3
13 F	0446 1.0 / 1103 4.6 / 1707 0.8 / 2349 4.6	28 SA	0535 1.0 / 1212 4.5 / 1756 0.8
14 SA	0526 1.0 / 1145 4.6 / 1747 0.8	29 SU	0008 4.3 / 0617 1.1 / 1222 4.3 / 1837 1.0
15 SU	0029 4.6 / 0608 1.0 / 1226 4.6 / 1830 0.8	30 M	0050 4.2 / 0657 1.3 / 1257 4.2 / 1915 1.2
		31 TU	0128 4.1 / 0736 1.5 / 1340 4.1 / 1952 1.4

Chart Datum: 2·74 metres below Ordnance Datum (Newlyn). HAT is 5·0 metres above Chart Datum.

Central S England

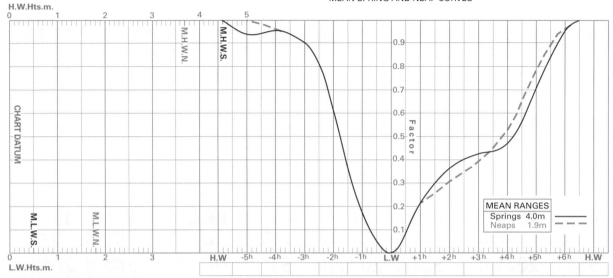

SOUTHAMPTON
MEAN SPRING AND NEAP CURVES

MEAN RANGES
Springs 4.0m
Neaps 1.9m

2.18 SOUTHAMPTON

Hampshire 50°52'·65N 01°23'·37W ✿✿✿♒♒♒♒✿✿

CHARTS AC 5600, 2036, 2041; Imray C3, C15, 2200

TIDES HW (1st) –0001 Dover; ML 2·9

Standard Port SOUTHAMPTON (←—)

Times				Height (metres)			
High Water		Low Water		MHWS	MHWN	MLWN	MLWS
0400	1100	0000	0600	4·5	3·7	1·8	0·5
1600	2300	1200	1800				
Differences REDBRIDGE							
–0020	+0005	0000	–0005	–0·1	–0·1	–0·1	–0·1

NOTE: At springs there are two separate HWs about 2hrs apart; at neaps there is a long stand. Predictions are for the first HW when there are two, otherwise for the middle of the stand. NE gales and a high barometer may lower sea level by 0·6m.

SHELTER Good in most winds, although a choppy sea builds in SE winds >F4, when it may be best to shelter in marinas. Ⓥ berths available in Hythe, Town Quay, Ocean Village, Shamrock Quay and Kemp's Marinas. No specific yacht anchorages but temporary ⚓ is permitted (subject to HM) off club moorings at Netley, Hythe, Weston and Marchwood in about 2m. Keep clear of all channels and Hythe Pier. Public moorings for larger yachts opposite Royal Pier near Gymp Elbow PHM buoy in 4m (contact HM); nearest landing is at Town Quay Marina.

NAVIGATION WPT 50°50'·0N 01°18'·6W, just S of Hamble Point SCM, to Weston Shelf SHM buoy 312°/4M. Main channels are well marked.

- For a Yachtsman's Guide see www.southamptonvts.co.uk.
- Yachts should keep just outside the buoyed lit fairway and when crossing it should do so at 90°, abeam Fawley chy, at Cadland and Greenland buoys, abeam Hythe and abeam Town Quay.

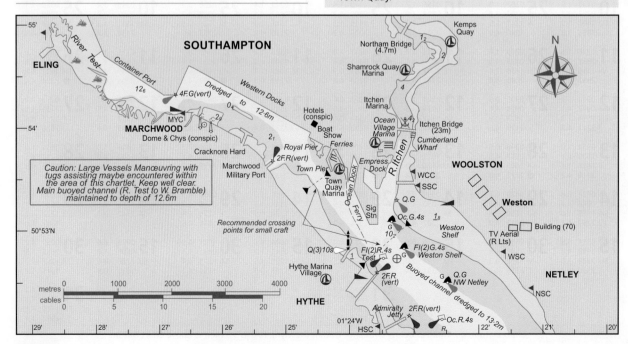

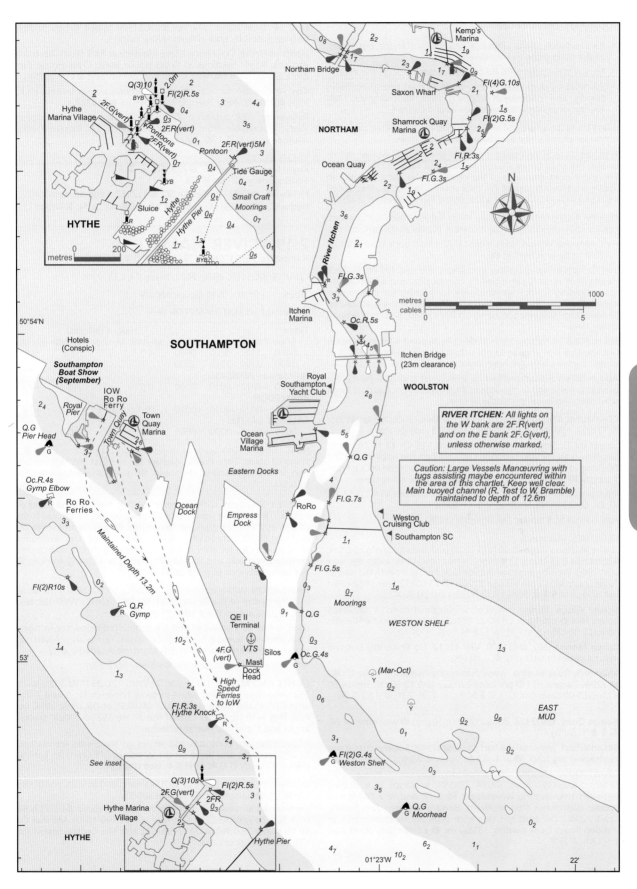

Kemp's Marina

Northam Bridge

Saxon Wharf

Fl(4)G.10s

NORTHAM

Shamrock Quay Marina

Fl(2)G.5s

Ocean Quay

Fl.R.3s

Fl.G.3s

River Itchen

N

Fl.G.3s

Itchen Marina

Oc.R.5s

metres
cables

SOUTHAMPTON

Itchen Bridge (23m clearance)

Hotels (Conspic)

WOOLSTON

50°54'N

Southampton Boat Show (September)

Royal Southampton Yacht Club

RIVER ITCHEN: All lights on the W bank are 2F.R(vert) and on the E bank 2F.G(vert), unless otherwise marked.

IOW Ro Ro Ferry

Royal Pier

Town Quay Marina

Town Quay

Q.G Pier Head

Ocean Village Marina

Q.G

Caution: Large Vessels Manœuvring with tugs assisting maybe encountered within the area of this chartlet. Keep well clear. Main buoyed channel (R. Test to W. Bramble) maintained to depth of 12.6m

Oc.R.4s Gymp Elbow

Eastern Docks

Ro Ro Ferries

Ocean Dock

Fl.G.7s

Weston Cruising Club

Southampton SC

Fl(2)R10s

Empress Dock

RoRo

Maintained Depth 13.2m

Fl.G.5s

WESTON SHELF

Q.R Gymp

QE II Terminal

Q.G

53'

VTS

Silos

Oc.G.4s

Moorings

(Mar-Oct)

4F.G (vert)

Mast Dock Head

High Speed Ferries to IoW

EAST MUD

Fl.R.3s Hythe Knock

Fl(2)G.4s Weston Shelf

See inset

Q(3)10s

2F.G(vert)

Fl(2)R.5s

2FR

Q.G Moorhead

HYTHE

Hythe Marina Village

Hythe Pier

01°23'W

22'

HYTHE inset

Q(3)10

BYB

Fl(2)R.5s

Hythe Marina Village

2FG(vert)

2F.R(vert)

Pontoons

2F.R(vert)

2F.R(vert)5M

Pontoon

Tide Gauge

Small Craft Moorings

Sluice

Hythe

Hythe Pier

BYB

metres 200

HYTHE

SOUTHAMPTON *continued*

- Caution: several large unlit mooring buoys off Hythe, both sides of the main chan, and elsewhere.
- Frequent hi-speed & Ro-Ro ferries operate through the area.
- Essential to keep clear of very large tankers operating from Fawley and commercial shipping from Southampton. NB Precautionary Area between Cowes and Calshot.

R Test There is foul ground at Marchwood and Royal Pier; extensive container port further upstream. Eling Chan dries.

R Itchen Care is necessary, particularly at night. Above Itchen Bridge the chan bends sharply to port and favours the W bank. There are unlit moorings in the centre of the river. Navigation above Northam Bridge (4·7m clearance) is not advisable. Speed limit 6kn in both rivers N of a line Hythe Pier to Weston Shelf.

LIGHTS AND MARKS See 2.2. Main lights shown on chartlet. Fawley chimney (198m, R lts) is conspic by day and night.

Hythe Marina Village, close NW of Hythe Pier: appr chan marked by ECM lit Bn and PHM lit Bn.

Southampton Water divides into Rivers Test and Itchen at Dock Head which is easily identified by conspic silos and a high lattice mast showing traffic sigs which are mandatory for commercial vessels, but may be disregarded by yachts outside the main chans. Beware large ships manoeuvring off Dock Head, and craft leaving either river.

Ent to R Itchen marked by SHM Oc G 4s, beyond which piles with G lts mark E side of chan ldg to Itchen bridge (24·4m); a FW lt at bridge centre marks the main chan.

Above Itchen Bridge, marked by 2 FR (vert) and 2 FG (vert), the principal marks are: Crosshouse lit Bn; Chapel lit Bn. **Caution:** large unlit mooring buoys in middle of river.

COMMUNICATIONS (Code 02380) Hythe Medical Centre ☎845955; Ⓗ 777222. HM ABP & VTS 330022 (339733 outside HO).

Vessel Traffic Services Call: *Southampton VTS* Ch **12** 14 16 (H24). Small craft traffic info broadcast on Ch 12 on the hour 0600-2200 Fri-Sun and Bank Holidays from Easter to last weekend in Oct. From 1 Jun to 30 Sep broadcasts are every day at the same times.

Southampton Hbr Patrol Call: *Southampton Patrol* VHF Ch **12** 16, 01-28, 60-88 (H24). Marinas VHF Ch **80** M.

Fuel barges in R Itchen, call *Wyefuel* Ch 08.

FACILITIES **Marinas:**
Hythe Marina Village ☎ 207073 Ch 80 (call ahead), dredged 2·5m. Access H24 IPTS at lock (2 ● over ○ = free flow) (lock width 8m; max LOA 18m). Waiting pontoon outside lock. Ferries from Hythe pier to Town Quay and Ocean Village. 210⌷+♥ £3·60<12·5m>£4·50<18m (inc ⌁), 🅿⚓🛢🔧×🛒⛽🅿(30t)🛟(12t) ⛽×🅿.

Ocean Village Marina www.oceanvillagemarina.co.uk ☎229385 access H24 ♥ pre-book Ch 80 ⚓ 450⌷ £3·60<12·5m>£4·50<18m 18m>£5·00<24m (inc ⌁), 🅿🛢🔧×⚓🏪×🅿.

Itchen Marina (50) ☎631500, VHF Ch 12. No ♥ berths, but will assist a vessel in difficulty, 🛢🏺🛟(40t).

Shamrock Quay Marina www.shamrockquaymarina.co.uk Ch 80 ☎ 229461 access H24 ♥ pre-book 220⌷+40♥ £3·60<12·5m>£4·50 <18m>£5·00<24m (inc ⌁), 🅿🛢🔧⚓🏪(63t)🛟(12t) ⛽🚭×🅿.

Kemps Quay ☎632323, 260⌷ (all but 50 dry) ♥ welcome £4·26, 🛢🛟⚓🛟(5t).

Saxon Wharf www.saxonwharf.co.uk ☎339490 Superyacht service centre and dry-stack. ♥🅿🛢⌁🔧⚓🏪⛽ wi-fi.

Town Quay Marina ⚓ info@townquay.com access H24 (2·6m), ☎234397, mob 07764 293588; entrance is a dogleg between two floating wavebreaks (☆ 2 FR and ☆ 2 FG) which appear continuous from seaward. Beware adjacent fast ferries. Craft >20m LOA must get clearance from Southampton VTS to ent/dep Town Quay marina. 133⌷ inc ♥ £3·40<13m, short stay £9<13m, 🅿🖥×🅿.

Fuel 🛢 from Itchen Marina ☎80631500 above Itchen Bridge, nearest petrol by hose is from Hythe marina.

Hards at Hythe, Cracknore, Eling, Mayflower Park (Test), Northam (Itchen). Public landings at Cross House hard, Cross House slip, Block House hard (Itchen), Carnation public hard & Cowporters.

City All facilities services ⇄ ✈, Car ferry/passenger catamaran to IoW, ☎019192; ferry Town Quay to Hythe ☎840722.

Reference ABP publish a *Yachtsman's Guide to Southampton Water* obtainable from VTS Centre, Berth 37, Eastern Docks, Southampton SO1 1GG. Please send sae.

YACHT CLUBS
Royal Southampton YC ☎223352, 🚣⚓🛢🖥×🅿.
Southampton SC ☎446575.
Netley SC ☎454272; **Weston SC** ☎452527; **Eling SC** ☎863987.
Hythe SC ☎846563; **Marchwood YC** ☎666141 🚣🔧⚓🛟(10t)🅿.

2.19 RIVER HAMBLE
Hampshire 50°51'·0N 01°18'·50W ❀❀❀❀∴∴∴∴❁❁❁

CHARTS AC 5600, 2036, 2022; Imray C3, C15, 2200

TIDES +0020, −0010 Dover; ML 2·9

Standard Port SOUTHAMPTON (⟵)

Times				Height (metres)			
High Water		Low Water		MHWS	MHWN	MLWN	MLWS
0400	1100	0000	0600	4·5	3·7	1·8	0·5
1600	2300	1200	1800				
Differences WARSASH							
+0020	+0010	+0010	0000	0·0	+0·1	+0·1	+0·3
BURSLEDON							
+0020	+0020	+0010	+0010	+0·1	+0·1	+0·2	+0·2
CALSHOT CASTLE							
0000	+0025	0000	0000	0·0	0·0	+0·2	+0·3

NOTE: Double HWs occur at or near sp; at other times there is a stand of about 2hrs. Predictions are for the first HW if there are two or for the middle of the stand. NE gales can decrease depths by 0·6m.

SHELTER Excellent, with ♥s berths at five main marinas, and at YCs, SCs, BYs and on some Hbr Authority pontoons. Jetty in front of B/W HM's Office Warsash has limited ♥s berths and some ♥s berths at Public Jetty on W bank near Bugle Inn car park. Pontoons at all marinas are lettered A, B et seq from the S end. In mid-stream between Warsash and Hamble Pt Marina is a clearly marked ♥ pontoon, between piles B1 to B6.

NAVIGATION WPT Hamble Point SCM buoy, 50°50'·15N 01°18'·66W, is in the centre of the W sector of Dir 352°, Oc (2) WRG 12s; Nos 1, 3, 5, piles are on the E edge of the G sector.

River can be very crowded and ⚓ is prohibited below the bridges. Unlit piles and buoys are a danger at night. Yachts are advised not to use spinnakers above Warsash Maritime Academy jetty.

Bridge clearances: Road 3·5m; Rly 5·5m; M27 3·8m.

LIGHTS AND MARKS Dir ☆, Oc (2) WRG 12s, 351°-W-353°, leads 352° into river ent. and when midway between No 5 and 7 bns alter 028° in the white sector, 027°-W-029°, of Dir ☆, Iso WRG 6s. Piles Nos 1–10 (to either side of the above 352° and 028° tracks) are lit and fitted with radar reflector.

Above Warsash, pontoons and jetties on the E side are marked by 2FG (vert) lts, and those on the W side by 2FR (vert) lts. Lateral piles are mostly Fl G 4s or Fl R 4s (see chartlet).

COMMUNICATIONS (Code 02380) Seastart 0800 885500. HM 01489 576387, HM patrol mob 07718 146380/81.

Commercial/all vessels >20m call: *Hamble Harbour Radio* Ch **68** (Apr-Sep: daily 0630-2230. Oct-Mar: daily 0700-1830). Marinas Ch **80** M. Water Taxi ☎07827157155. See also The Central Solent for VTS and info broadcasts.

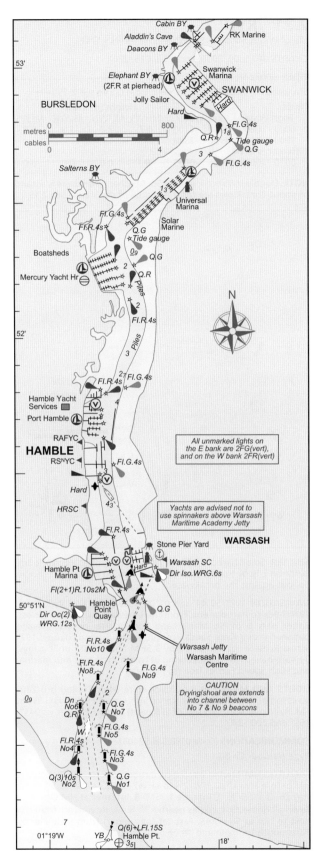

FACILITIES Marinas from seaward:
Hamble Pt Marina : www.hamblepointmarina.co.uk ☎452464, access H24, pre-book Ch 80. 220⌕ £3·70, short stay £9·00; ▯ ✕ ✎ Ⓔ ▯ ▯(65t) ▯(7t) ▯ ▯ ▯.

Stone Pier Yard ☎01489 579589, access H24, 56⌕ £4·00, <4hrs £8; ⚓ ▯ ✕ ✎ △ ♨ ▯(10t) ▯. Nearby: ✉ Ⓒ ▯.

Port Hamble Marina www.porthamblemarina.co.uk ☎452741 access H24 pre-book, Ch 80. ▬ 310⌕+Ⓥ £3·70, short stay £9·00; ▯ ▮ ▮ ▮ Ⓔ ✕ ▯(60t) ▯(7t) ▯ ▯.

Hamble Yacht Services ☎454111 access H24, Ⓥ pre-book, Ch 80. ▬ ▯ ▮ ▮ ✎ ▯ Ⓔ △ ▯.

Mercury Yacht Hbr www.mercuryyachtharbour.co.uk ☎455994, access H24, pre-book Ch 80. 346⌕+Ⓥ £3·70, short stay £9·00; ▬ ▯ ▮ ▮ ✎ ▯ Ⓔ △ ▯(20t) ▯ ▯.

Universal Marina ④ ☎01489 574272, access H24, pre-book Ch 80. 249⌕+Ⓥ £3·40, short stay £8; ▯ ▮ ▯ ✕ ✎ ▯(80t) ▯(7t) ▯/café.

Swanwick Marina ☎01489 884081 (H24), 380⌕+Ⓥ £3·55, £1·60 <4hrs (min £5), ▯ ▮ ▮ ▮ ✕ ✎ Ⓔ △ ▯(60t) ▯(12t) ▯ ✉ ▯ @.

The Hbr Authority jetty in front of the conspic B/W HM's Office at Warsash has limited ⌕ (£2), Ⓥ pontoons in mid-stream (£1.50), ⚓ toilets ashore. Public Jetty on W bank near the Bugle Inn car park (£1·75). See www.hants.gov.uk/hamble harbour.

YACHT CLUBS
Hamble River SC ☎452070
RAFYC ☎452208, ▬ ✕ ▯
Royal Southern YC ☎450300
Warsash SC ☎01489 583575.

SERVICES Wide range of marine services available; consult marina/HM for locations. *Hamble River Guide* available from HM.
Hards At Warsash, Hamble and Swanwick.
Slips at Warsash, Hamble, Bursledon and Lower Swanwick.
Maintenance piles at Warsash by HM's slipway; on W bank upriver of public jetty near Bugle; by slip opposite Swanwick Marina.
✉ (Hamble, Bursledon, Warsash and Lower Swanwick); (Hamble, Bursledon, Sarisbury Green, Swanwick, Warsash).
≈ (Hamble and Bursledon); ✈ (Southampton).

ADJACENT HARBOURS
ASHLETT CREEK, Hants, 50°50'·01N 01°19'·49W. AC 5600, 2038, 2022. Tides approx as Calshot Castle (opposite). Small drying (2·1m) inlet across from R Hamble; best for shoal-draught vessels. Appr at HW close to Fawley Marine Terminal. Unlit chan, marked by 3 PHM buoys, then in sequence a pair of PHM and SHM buoys, a PHM beacon and SHM buoy (No 7). Then follow PHM and SHM beacons. The single dog leg is indicated by an arrow on a PHM. A final SHM buoy leads to the drying sailing club pontoon or the drying quay. Ldg bns are hard to find; local knowledge desirable. Berth at drying quay. Facilities: ▲ ▬ ⚓ Hard ▯. **Ashlett SC.**

HILL HEAD, Hants, 50°49'·08N 01°14'·54W. AC 5600, 2036, 2022. HW +0030 on Dover; use LEE-ON-SOLENT differences. Short term ⚓ for small craft at mouth of R Meon. Bar dries ¼M offshore. Ent dries 1·2m at MLWS. Appr on 030° towards Hill Head SC ho (W, conspic); Small hbr to W inside ent (very narrow due to silting) where small yachts can lie in soft mud alongside wall. Facilities: **Hill Head SC** ☎(01329) 664843. **Hill Head village** ▯ ▯ ▯ ✉ ▯.

2.20 WOOTTON CREEK

Isle of Wight 50°44′·09N 01°12′·77W ✻✻◊✿✿✿

CHARTS AC 5600, 2036, 2022; Imray C3, C15, 2200

TIDES +0023 Dover; ML 2·8. Use RYDE differences.

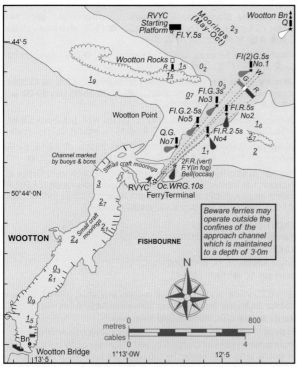

SHELTER Good except in strong N or E winds. Creek dries above the ferry terminal. ABs on RVYC pontoons (dry); No ⚓ in the fairway. Speed limit 5kn.

NAVIGATION WPT Wootton NCM Bn, Q, 50°44′·53N 01°12′·13W, 222°/6·3ca to ferry slip. An obstruction drying 1·4m verges Peel Bank 8 Ca NNE of the Starting Platform. Beware seasonal moorings (May-Oct) laid between the platform & No 1 bn. Large car ferries may operate outside the confines of the approach channel when proceeding to and from Fishbourne Ferry Terminal.

LIGHTS AND MARKS Entrance to approach channel due S of SE Ryde Middle SCM and 1·75M W of Ryde Pier. From the W, pass N of the Starting Platform & No 1 bn. The channel is marked by four SHM bns and two PHMs, all lit. Keep in W sector of Dir lt, Oc WRG 10s, 221°-G-224°-W-225½°-R-230½°. By ferry terminal, turn onto ldg marks on W shore △ ▽, which form a ◊ when in transit 270°.

COMMUNICATIONS (Code 01983) Dr 562955.

FACILITIES Fishbourne, **Royal Victoria YC** ☎882325, ◣ ⬭£1·50/m (shortstay £5·00), ⚓ ⓑ ✕ 🛢. **Village** Wootton Bridge 🏧 🏧 ✉ Ⓑ 🛒 ✕ 🛢 (Ryde), ⇌ (ferry to Portsmouth), ✈ (Southampton).

2.21 RYDE

Isle of Wight 50°43′·98N 01°09′·31W ✻✻◊◊✿✿

CHARTS AC 5600, 2036; Imray C3, C15, 2200

TIDES +0022 Dover; ML 2·8m

Standard Port PORTSMOUTH (→); Tidal Curve see 2.15 (←)

Times				Height (metres)			
High Water		Low Water		MHWS	MHWN	MLWN	MLWS
0000	0600	0500	1100	4·7	3·8	1·9	0·8
1200	1800	1700	2300				
Differences RYDE							
–0010	–0010	–0005	–0005	–0·1	0·0	0·0	0·0

NOTE: Double HWs occur at/near Sp. At other times there is a stand of about 2hrs. Times are for the middle of the stand. See 2.15.

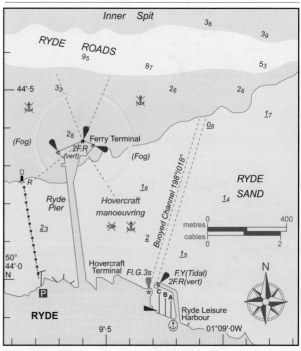

SHELTER Small harbour 300m E of Ryde Pier; dries approx 2·3m. Berth on E'ly of three pontoons; long and fin keel yachts should dry out against the breakwater.

NAVIGATION WPT 50°44′·36N 01°09′·13W (2.7ca E of Ryde Pier), 197°/3·9ca to harbour ent. From the E beware Ryde Sand (dries). Either pass N of No Man's Land Fort or use inshore passage between it and Ryde Sands Bcns (lit PHMs). Drying channel 197° across Ryde Sands is marked by 3 SHM and 3 PHM unlit buoys. Beware hovercraft manoeuvring between Ryde pier and marina, and High Speed Ferries from/to pierhead.

LIGHTS AND MARKS Ryde Church spire (Holy Trinity) brg 200° gives initial appr. Harbour entrance lights are 2 FR and Fl G 3s 7m 1M. Ryde pier is lit by 3 sets of 2FR (vert) and a FY fog lt vis 045°-165° and 200°-320°.

COMMUNICATIONS (Code 01983) Ⓗ 524081.

Ryde Harbour Ch 80.

FACILITIES HM/Marina www.rydeharbour.com ☎613879. Contact HM before arrival. ◣ HW–2½ to +HW2, 100⬭+70♥ £1·50, short stay £6. 🗑 ⓑ ✉ 🏧 🏧 (garage 1·25M) 🛢. Internet at HM Office.
Town all domestic facilities nearby 🛒 ✕ 🛢. Hovercraft from slip next to hbr to Southsea. Fast cat (passenger) from Ryde Pier to Portsmouth for mainland ⇌; ✈ Southampton.

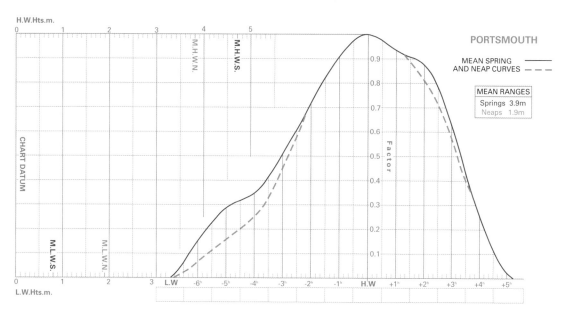

PORTSMOUTH

MEAN SPRING
AND NEAP CURVES - - -

MEAN RANGES
Springs 3.9m
Neaps 1.9m

2.22 BEMBRIDGE

Isle of Wight **50°41'·62N 01°06'·40W** ❀☖☖☖✿✿✿

CHARTS AC 5600, 2037, 2022; Imray C9, C3, C15, 2200

TIDES +0020 Dover

Standard Port PORTSMOUTH (→); Tidal Curve see 2.15 (←)

Times				Height (metres)			
High Water		Low Water		MHWS	MHWN	MLWN	MLWS
0000	0600	0500	1100	4·7	3·8	1·9	0·8
1200	1800	1700	2300				
Differences BEMBRIDGE APPROACHES							
–0010	–0005	0000	+0005	+0·1	+0·1	0·0	+0·1
BEMBRIDGE HARBOUR							
+0020	0000	+0100	+0020	–1·5	–1·4	–1·3	–1·0
VENTNOR							
–0025	–0030	–0025	–0030	–0·8	–0·6	–0·2	+0·2
SANDOWN							
0000	+0005	+0010	+0025	–0·6	–0·5	–0·2	0·0

NOTE: Double HWs occur at/near Sp. At other times there is a stand of about 2hrs. Times are for the middle of the stand. See 2.15.

SHELTER Good, but approach difficult in strong N/NE winds. ▼ berths rafting on long pontoons at Duver Marina dredged to 2m, or dry out, with ⚓s fore and aft, on sandy beach to port just inside harbour ent. No ⚓ in chan and hbr, but Priory Bay is sheltered ⚓ in winds from S to WNW (dries inshore but ½M off gives 1·5m).

NAVIGATION Beware the gravel banks between St Helen's Fort, Nodes Pt and N to Seaview. The bar, between Nos 2A and 3 buoys, almost dries. Tide gauge removed. Depths over the bar updated every 60s are online at www.bembridgeharbour.co.uk. Speed limit 6kn.

LIGHTS AND MARKS St Helen's Fort Fl (3) 10s 16m 8M; no ⚓ within 1ca of Fort. Conspic W daymark on shore where chan turns S. Pontoons and fingers lit at Duver and Fishermans Wharf. Beware many unlit Y racing marks off Bembridge and Seaview (Mar-Oct).

COMMUNICATIONS (Code 01983) Dr 872614; HM 872828; Hbr staff Ch 80; *Water Taxi* Ch 80.

FACILITIES Duver Marina ☎872828, +Ⓥ ⚓ 140⬛Ⓥ from £2.65, ⛽Ⓓ⚓🔌⚒🔧 📷🚿. **Drying out area** £10 flat rate. **St Helen's Quay** ⚓🔒. **Bembridge Marina** permanent berthholders. **Brading Haven YC** ☎872289, ⚓ ✕ 🏠. **Bembridge SC** ☎872686.

Town 🏤🏪🚂✕🛒✉Ⓑ 🚊 (Ryde), ✈ (So'ton).

HARBOUR ON THE SOUTH COAST OF THE ISLE OF WIGHT
VENTNOR HAVEN, 50°35'·53N 01°12'·50W. AC 2045. Bkwtrs to protect FVs. Very exposed to E to SSE. 8Ⓤ (in <F4; seasonal), ☎07976 009260 or VHF Ch 17 for availability. ⚓ (quay) 🏤🏪. **Town** All domestic facilities, some up a steep hill.

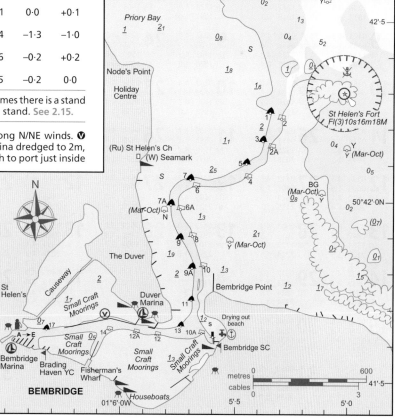

STANDARD TIME (UT) For Summer Time add ONE hour in **non-shaded areas**	**PORTSMOUTH** LAT 50°48'N LONG 1°07'W TIMES AND HEIGHTS OF HIGH AND LOW WATERS	Dates in red are SPRINGS Dates in blue are NEAPS

YEAR 2019

JANUARY

Time	m	Time	m
1 0030 0737 TU 1305 2000	1.6 4.3 1.6 4.1	**16** 0632 1202 W 1901	4.0 1.8 3.8
2 0127 0835 W 1359 2059	1.6 4.4 1.5 4.2	**17** 0031 0740 TH 1312 2011	1.7 4.1 1.6 4.0
3 0219 0925 TH 1448 2153	1.5 4.4 1.3 4.3	**18** 0137 0844 F 1411 2112	1.5 4.3 1.3 4.2
4 0308 1009 F 1535 2239	1.4 4.5 1.2 4.4	**19** 0234 0938 SA 1505 2205	1.3 4.5 1.0 4.5
5 0354 1050 SA 1619 2321	1.3 4.5 1.1 4.4	**20** 0327 1025 SU 1556 2253	1.0 4.7 0.7 4.7
6 0438 1130 SU 1700 ●	1.2 4.5 1.0	**21** 0418 1110 M 1645 ○ 2340	0.8 4.8 0.5 4.8
7 0002 0518 M 1210 1738	4.5 1.2 4.5 1.0	**22** 0506 1155 TU 1731	0.7 4.9 0.4
8 0043 0555 TU 1248 1813	4.5 1.2 4.5 1.0	**23** 0028 0552 W 1242 1816	4.9 0.6 4.9 0.3
9 0121 0629 W 1323 1843	4.5 1.3 4.4 1.1	**24** 0120 0638 TH 1331 1902	4.9 0.6 4.8 0.4
10 0156 0659 TH 1355 1913	4.5 1.4 4.3 1.2	**25** 0217 0724 F 1426 1948	4.8 0.8 4.7 0.7
11 0229 0729 F 1428 1945	4.4 1.4 4.2 1.3	**26** 0317 0813 SA 1525 2038	4.7 1.0 4.5 0.9
12 0304 0805 SA 1508 2023	4.3 1.5 4.1 1.4	**27** 0412 0907 SU 1624 ◑ 2135	4.6 1.3 4.3 1.3
13 0345 0847 SU 1553 2109	4.2 1.6 4.0 1.5	**28** 0503 1011 M 1722 2244	4.4 1.5 4.1 1.5
14 0433 0939 M 1647 ◑ 2204	4.1 1.7 3.9 1.6	**29** 0557 1123 TU 1824 2354	4.2 1.7 3.9 1.7
15 0529 1044 TU 1751 2314	4.0 1.8 3.8 1.7	**30** 0658 1232 W 1931	4.1 1.8 3.9
		31 0059 0802 TH 1335 2038	1.8 4.0 1.7 3.9

FEBRUARY

Time	m	Time	m
1 0159 0901 F 1430 2140	1.7 4.1 1.5 4.0	**16** 0115 0823 SA 1355 2058	1.7 4.1 1.4 4.1
2 0253 0950 SA 1519 2225	1.6 4.2 1.3 4.2	**17** 0222 0922 SU 1453 2153	1.4 4.3 1.1 4.4
3 0340 1032 SU 1603 2305	1.4 4.3 1.1 4.3	**18** 0318 1010 M 1545 2240	1.1 4.6 0.7 4.6
4 0423 1112 M 1644 ● 2343	1.2 4.4 1.0 4.4	**19** 0408 1055 TU 1632 ○ 2326	0.8 4.8 0.4 4.8
5 0503 1151 TU 1721	1.1 4.5 0.9	**20** 0454 1139 W 1717	0.5 4.9 0.2
6 0022 0538 W 1229 1753	4.5 1.1 4.5 0.9	**21** 0012 0538 TH 1225 1800	4.9 0.4 4.9 0.2
7 0059 0609 TH 1303 1823	4.5 1.1 4.4 0.9	**22** 0101 0621 F 1313 1843	4.9 0.4 4.8 0.3
8 0132 0637 F 1333 1851	4.6 1.1 4.4 1.0	**23** 0154 0704 SA 1404 1926	4.9 0.5 4.7 0.5
9 0202 0705 SA 1403 1920	4.4 1.1 4.3 1.0	**24** 0249 0748 SU 1500 2010	4.8 0.6 4.6 0.8
10 0232 0736 SU 1436 1953	4.4 1.1 4.3 1.0	**25** 0338 0834 M 1555 2059	4.6 1.1 4.4 1.2
11 0308 0813 M 1516 2033	4.3 1.2 4.2 1.2	**26** 0424 0929 TU 1649 ◑ 2201	4.4 1.4 4.1 1.6
12 0350 0858 TU 1605 ◑ 2121	4.2 1.3 4.0 1.4	**27** 0515 1041 W 1750 2320	4.1 1.7 3.8 1.8
13 0443 0953 W 1707 2221	4.0 1.6 3.9 1.6	**28** 0619 1159 TH 1902	3.8 1.9 3.7
14 0548 1107 TH 1823 2345	3.9 1.7 3.8 1.7		
15 0705 1240 F 1947	3.9 1.7 3.9		

MARCH

Time	m	Time	m
1 0033 0730 F 1310 2014	2.0 3.7 1.8 3.7	**16** 0642 1227 SA 1933	3.8 1.7 3.9
2 0140 0837 SA 1410 2119	1.9 3.8 1.7 3.9	**17** 0109 0806 SU 1345 2044	1.7 4.0 1.4 4.2
3 0237 0930 SU 1500 2205	1.7 4.0 1.4 4.1	**18** 0215 0905 M 1441 2137	1.4 4.3 1.1 4.4
4 0323 1012 M 1542 2243	1.4 4.2 1.2 4.3	**19** 0307 0952 TU 1530 2223	1.1 4.5 0.7 4.7
5 0403 1051 TU 1621 2320	1.3 4.3 1.0 4.4	**20** 0354 1037 W 1615 2307	0.7 4.7 0.4 4.8
6 0441 1129 W 1656 ● 2356	1.0 4.4 0.9 4.5	**21** 0438 1120 TH 1659 ○ 2352	0.5 4.8 0.2 4.9
7 0514 1205 TH 1728	0.9 4.4 0.8	**22** 0520 1205 F 1741	0.3 4.9 0.2
8 0031 0544 F 1239 1758	4.5 0.9 4.4 0.8	**23** 0037 0601 SA 1252 1821	4.9 0.3 4.8 0.3
9 0103 0612 SA 1308 1826	4.5 0.9 4.4 0.8	**24** 0125 0641 SU 1341 1901	4.9 0.5 4.7 0.6
10 0132 0640 SU 1337 1855	4.5 0.9 4.4 0.8	**25** 0213 0722 M 1433 1942	4.7 0.7 4.6 0.9
11 0200 0710 M 1408 1927	4.5 0.9 4.4 0.9	**26** 0258 0803 TU 1524 2026	4.5 1.0 4.4 1.3
12 0232 0746 TU 1446 2005	4.4 0.9 4.4 1.0	**27** 0341 0849 W 1614 2119	4.3 1.4 4.1 1.7
13 0312 0828 W 1534 2051	4.2 1.1 4.1 1.3	**28** 0428 0953 TH 1712 ◑ 2245	4.0 1.7 3.8 2.0
14 0405 0919 TH 1637 ◑ 2148	4.0 1.4 3.9 1.6	**29** 0528 1123 F 1828	3.7 1.9 3.7
15 0515 1029 F 1800 2315	3.9 1.7 3.8 1.8	**30** 0003 0653 SA 1237 1945	2.1 3.6 2.0 3.8
		31 0113 0808 SU 1340 2051	2.0 3.6 1.8 3.9

APRIL

Time	m	Time	m
1 0212 0905 M 1431 2138	1.8 3.9 1.6 4.1	**16** 0159 0844 TU 1421 2117	1.4 4.3 1.1 4.5
2 0257 0948 TU 1512 2216	1.5 4.1 1.3 4.3	**17** 0248 0932 W 1509 2203	1.1 4.5 0.7 4.7
3 0336 1027 W 1550 2253	1.3 4.2 1.1 4.5	**18** 0334 1016 TH 1554 2246	0.8 4.7 0.5 4.8
4 0411 1104 TH 1625 2328	1.1 4.4 0.9 4.5	**19** 0417 1100 F 1636 ○ 2329	0.5 4.8 0.4 4.9
5 0444 1139 F 1658 ●	0.9 4.4 0.8	**20** 0458 1144 SA 1718	0.4 4.8 0.4
6 0001 0515 SA 1211 1729	4.6 0.8 4.5 0.8	**21** 0012 0538 SU 1231 1758	4.9 0.5 4.7 0.5
7 0032 0545 SU 1242 1800	4.6 0.8 4.5 0.8	**22** 0056 0618 M 1319 1837	4.8 0.6 4.7 0.6
8 0101 0616 M 1313 1831	4.6 0.8 4.5 0.8	**23** 0139 0657 TU 1408 1917	4.7 0.8 4.6 1.0
9 0131 0648 TU 1347 1906	4.5 0.8 4.5 0.9	**24** 0221 0736 W 1456 1958	4.5 1.1 4.4 1.4
10 0205 0724 W 1428 1945	4.5 0.9 4.4 1.0	**25** 0303 0818 TH 1542 2047	4.3 1.4 4.2 1.7
11 0247 0807 TH 1520 2031	4.3 1.1 4.2 1.3	**26** 0347 0911 F 1633 ◑ 2204	4.0 1.7 3.9 2.0
12 0344 0859 F 1628 ◑ 2131	4.1 1.4 4.0 1.7	**27** 0439 1039 SA 1742 2326	3.8 1.9 3.8 2.1
13 0458 1014 SA 1753 2323	3.9 1.7 3.9 1.9	**28** 0555 1154 SU 1905	3.6 2.0 3.7
14 0628 1218 SU 1919	3.8 1.7 4.0	**29** 0033 0727 M 1257 2013	2.1 3.6 1.9 3.9
15 0059 0746 M 1327 2026	1.7 4.0 1.4 4.3	**30** 0133 0830 TU 1349 2103	1.9 3.8 1.7 4.1

Chart Datum: 2·73 metres below Ordnance Datum (Newlyn). HAT is 5·1 metres above Chart Datum.

PORTSMOUTH LAT 50°48′N LONG 1°07′W

TIMES AND HEIGHTS OF HIGH AND LOW WATERS

STANDARD TIME (UT)
For Summer Time add ONE hour in **non-shaded areas**

Dates in red are **SPRINGS**
Dates in blue are NEAPS

YEAR **2019**

MAY

Day	Time	m	Day	Time	m
1 W	0220 / 0917 / 1432 / 2144	1.6 / 4.0 / 1.5 / 4.3	16 TH	0223 / 0908 / 1443 / 2140	1.2 / 4.4 / 0.9 / 4.7
2 TH	0259 / 0958 / 1511 / 2221	1.4 / 4.2 / 1.2 / 4.5	17 F	0309 / 0954 / 1529 / 2223	0.9 / 4.6 / 0.8 / 4.8
3 F	0335 / 1034 / 1548 / 2255	1.2 / 4.3 / 1.0 / 4.5	18 SA	0353 / 1040 / 1612 / ○2305	0.8 / 4.6 / 0.7 / 4.8
4 SA	0410 / 1108 / 1625 / ●2328	1.0 / 4.4 / 0.9 / 4.6	19 SU	0436 / 1126 / 1655 / 2348	0.7 / 4.6 / 0.7 / 4.8
5 SU	0444 / 1141 / 1700	0.9 / 4.5 / 0.8	20 M	0517 / 1213 / 1736	0.7 / 4.6 / 0.8
6 M	0000 / 0518 / 1216 / 1735	4.6 / 0.8 / 4.6 / 0.8	21 TU	0030 / 0557 / 1302 / 1816	4.7 / 0.8 / 4.6 / 1.0
7 TU	0033 / 0554 / 1252 / 1811	4.6 / 0.7 / 4.6 / 0.8	22 W	0112 / 0636 / 1349 / 1856	4.6 / 0.9 / 4.5 / 1.2
8 W	0108 / 0630 / 1333 / 1850	4.6 / 0.8 / 4.6 / 0.9	23 TH	0153 / 0715 / 1434 / 1937	4.5 / 1.1 / 4.4 / 1.4
9 TH	0148 / 0711 / 1420 / 1933	4.5 / 0.9 / 4.5 / 1.1	24 F	0234 / 0755 / 1516 / 2021	4.3 / 1.4 / 4.3 / 1.7
10 F	0236 / 0757 / 1520 / 2024	4.4 / 1.1 / 4.3 / 1.4	25 SA	0316 / 0840 / 1600 / 2116	4.1 / 1.6 / 4.1 / 1.9
11 SA	0338 / 0854 / 1630 / 2131	4.2 / 1.4 / 4.2 / 1.7	26 SU	0403 / 0939 / 1651 / ◑2232	3.9 / 1.8 / 3.9 / 2.0
12 SU	0451 / 1016 / 1744 / ◑2314	4.0 / 1.6 / 4.1 / 1.8	27 M	0459 / 1057 / 1759 / 2341	3.7 / 1.9 / 3.9 / 2.1
13 M	0609 / 1153 / 1858	4.0 / 1.6 / 4.2	28 TU	0618 / 1202 / 1920	3.6 / 1.9 / 3.9
14 TU	0032 / 0721 / 1259 / 2002	1.7 / 4.1 / 1.4 / 4.4	29 W	0039 / 0742 / 1257 / 2019	2.0 / 3.7 / 1.8 / 4.1
15 W	0132 / 0819 / 1354 / 2054	1.4 / 4.3 / 1.2 / 4.6	30 TH	0130 / 0838 / 1344 / 2106	1.8 / 3.9 / 1.6 / 4.2
			31 F	0213 / 0922 / 1428 / 2145	1.5 / 4.1 / 1.4 / 4.4

JUNE

Day	Time	m	Day	Time	m
1 SA	0254 / 1000 / 1510 / 2220	1.3 / 4.3 / 1.2 / 4.5	16 SU	0331 / 1025 / 1551 / 2245	1.0 / 4.4 / 1.1 / 4.6
2 SU	0334 / 1036 / 1552 / 2255	1.1 / 4.4 / 1.0 / 4.6	17 M	0416 / 1113 / 1635 / ○2328	0.9 / 4.5 / 1.0 / 4.6
3 M	0414 / 1114 / 1633 / ●2331	0.9 / 4.5 / 0.9 / 4.7	18 TU	0459 / 1201 / 1719	0.9 / 4.5 / 1.0
4 TU	0455 / 1153 / 1714	0.8 / 4.6 / 0.8	19 W	0010 / 0540 / 1247 / 1800	4.6 / 0.9 / 4.5 / 1.1
5 W	0009 / 0536 / 1236 / 1756	4.7 / 0.7 / 4.7 / 0.9	20 TH	0052 / 0620 / 1331 / 1840	4.5 / 1.0 / 4.5 / 1.2
6 TH	0051 / 0618 / 1323 / 1840	4.7 / 0.7 / 4.7 / 0.9	21 F	0132 / 0657 / 1413 / 1918	4.4 / 1.1 / 4.4 / 1.4
7 F	0136 / 0703 / 1417 / 1927	4.6 / 0.8 / 4.6 / 1.1	22 SA	0211 / 0733 / 1453 / 1955	4.3 / 1.3 / 4.3 / 1.5
8 SA	0229 / 0752 / 1519 / 2021	4.5 / 1.0 / 4.5 / 1.3	23 SU	0249 / 0810 / 1531 / 2035	4.2 / 1.4 / 4.2 / 1.7
9 SU	0331 / 0850 / 1623 / 2126	4.3 / 1.2 / 4.4 / 1.5	24 M	0331 / 0852 / 1612 / 2122	4.0 / 1.6 / 4.1 / 1.8
10 M	0438 / 1002 / 1726 / ◑2243	4.2 / 1.4 / 4.4 / 1.6	25 TU	0418 / 0944 / 1659 / ◑2221	3.9 / 1.7 / 4.0 / 1.9
11 TU	0546 / 1119 / 1831 / 2355	4.1 / 1.4 / 4.4 / 1.6	26 W	0513 / 1047 / 1754 / 2328	3.8 / 1.8 / 4.0 / 1.9
12 W	0652 / 1225 / 1933	4.1 / 1.4 / 4.4	27 TH	0616 / 1154 / 1859	3.7 / 1.8 / 4.0
13 TH	0059 / 0751 / 1323 / 2027	1.5 / 4.2 / 1.3 / 4.5	28 F	0030 / 0728 / 1253 / 2006	1.8 / 3.8 / 1.7 / 4.1
14 F	0154 / 0845 / 1415 / 2116	1.3 / 4.3 / 1.2 / 4.6	29 SA	0124 / 0833 / 1346 / 2100	1.6 / 4.0 / 1.5 / 4.3
15 SA	0244 / 0936 / 1504 / 2201	1.1 / 4.4 / 1.1 / 4.6	30 SU	0214 / 0923 / 1435 / 2144	1.4 / 4.2 / 1.3 / 4.4

JULY

Day	Time	m	Day	Time	m
1 M	0303 / 1008 / 1524 / 2227	1.2 / 4.4 / 1.1 / 4.6	16 TU	0400 / 1104 / 1621 / ○2311	1.1 / 4.4 / 1.2 / 4.5
2 TU	0350 / 1052 / 1612 / ●2309	0.9 / 4.5 / 1.0 / 4.7	17 W	0444 / 1147 / 1705 / 2352	1.0 / 4.4 / 1.1 / 4.5
3 W	0437 / 1137 / 1659 / 2351	0.7 / 4.7 / 0.9 / 4.8	18 TH	0525 / 1228 / 1746	1.0 / 4.5 / 1.1
4 TH	0523 / 1223 / 1745	0.6 / 4.7 / 0.8	19 F	0032 / 0603 / 1310 / 1823	4.5 / 1.0 / 4.5 / 1.2
5 F	0035 / 0608 / 1313 / 1831	4.8 / 0.6 / 4.8 / 0.8	20 SA	0112 / 0638 / 1349 / 1856	4.4 / 1.1 / 4.5 / 1.3
6 SA	0123 / 0654 / 1408 / 1918	4.7 / 0.6 / 4.7 / 0.9	21 SU	0148 / 0709 / 1426 / 1927	4.3 / 1.2 / 4.4 / 1.4
7 SU	0216 / 0743 / 1509 / 2009	4.6 / 0.8 / 4.7 / 1.1	22 M	0222 / 0739 / 1459 / 1958	4.2 / 1.3 / 4.3 / 1.5
8 M	0317 / 0835 / 1607 / 2106	4.5 / 1.0 / 4.6 / 1.3	23 TU	0258 / 0814 / 1534 / 2036	4.2 / 1.4 / 4.3 / 1.5
9 TU	0419 / 0935 / 1703 / ◑2210	4.4 / 1.2 / 4.5 / 1.4	24 W	0339 / 0855 / 1616 / 2122	4.0 / 1.5 / 4.2 / 1.7
10 W	0521 / 1043 / 1800 / 2319	4.2 / 1.4 / 4.4 / 1.6	25 TH	0428 / 0944 / 1704 / ◑2219	3.9 / 1.6 / 4.1 / 1.8
11 TH	0623 / 1150 / 1900	4.1 / 1.5 / 4.3	26 F	0524 / 1047 / 1800 / 2327	3.8 / 1.8 / 4.0 / 1.8
12 F	0025 / 0725 / 1253 / 1958	1.6 / 4.1 / 1.5 / 4.3	27 SA	0630 / 1200 / 1906	3.8 / 1.8 / 4.0
13 SA	0126 / 0826 / 1351 / 2054	1.5 / 4.1 / 1.5 / 4.3	28 SU	0039 / 0743 / 1309 / 2018	1.7 / 3.9 / 1.7 / 4.1
14 SU	0222 / 0925 / 1444 / 2144	1.4 / 4.2 / 1.4 / 4.4	29 M	0143 / 0855 / 1410 / 2118	1.5 / 4.1 / 1.5 / 4.3
15 M	0313 / 1019 / 1534 / 2229	1.2 / 4.3 / 1.3 / 4.4	30 TU	0241 / 0950 / 1506 / 2207	1.2 / 4.3 / 1.3 / 4.5
			31 W	0334 / 1037 / 1558 / 2251	0.9 / 4.6 / 1.0 / 4.7

AUGUST

Day	Time	m	Day	Time	m
1 TH	0423 / 1123 / 1646 / ●2334	0.7 / 4.7 / 0.8 / 4.8	16 F	0507 / 1206 / 1727	0.9 / 4.5 / 1.1
2 F	0510 / 1208 / 1733	0.5 / 4.8 / 0.6	17 SA	0012 / 0542 / 1245 / 1801	4.5 / 0.9 / 4.5 / 1.1
3 SA	0019 / 0555 / 1257 / 1818	4.9 / 0.4 / 4.9 / 0.6	18 SU	0049 / 0614 / 1322 / 1830	4.4 / 1.0 / 4.5 / 1.1
4 SU	0106 / 0640 / 1350 / 1903	4.8 / 0.4 / 4.9 / 0.7	19 M	0123 / 0642 / 1354 / 1857	4.4 / 1.0 / 4.5 / 1.2
5 M	0158 / 0725 / 1449 / 1950	4.7 / 0.6 / 4.8 / 0.8	20 TU	0152 / 0709 / 1424 / 1925	4.3 / 1.1 / 4.4 / 1.2
6 TU	0255 / 0813 / 1545 / 2040	4.6 / 0.8 / 4.7 / 1.1	21 W	0223 / 0739 / 1455 / 1958	4.3 / 1.2 / 4.3 / 1.3
7 W	0356 / 0906 / 1636 / 2138	4.5 / 1.1 / 4.6 / 1.4	22 TH	0300 / 0815 / 1533 / 2039	4.2 / 1.3 / 4.2 / 1.4
8 TH	0455 / 1010 / 1729 / 2246	4.3 / 1.4 / 4.4 / 1.6	23 F	0345 / 0859 / 1620 / ◑2129	4.1 / 1.5 / 4.1 / 1.6
9 F	0555 / 1120 / 1827 / 2356	4.1 / 1.7 / 4.2 / 1.7	24 SA	0441 / 0955 / 1717 / 2236	3.9 / 1.7 / 4.0 / 1.8
10 SA	0701 / 1228 / 1932	4.0 / 1.8 / 4.1	25 SU	0550 / 1112 / 1827	3.8 / 1.9 / 3.9
11 SU	0103 / 0812 / 1332 / 2035	1.7 / 4.0 / 1.8 / 4.1	26 M	0005 / 0715 / 1247 / 1953	1.8 / 3.8 / 1.9 / 4.0
12 M	0204 / 0922 / 1430 / 2129	1.6 / 4.1 / 1.7 / 4.2	27 TU	0126 / 0840 / 1357 / 2101	1.6 / 4.1 / 1.6 / 4.3
13 TU	0258 / 1013 / 1521 / 2213	1.5 / 4.2 / 1.5 / 4.3	28 W	0228 / 0936 / 1455 / 2149	1.3 / 4.4 / 1.3 / 4.5
14 W	0345 / 1050 / 1607 / 2253	1.2 / 4.4 / 1.3 / 4.4	29 TH	0321 / 1021 / 1546 / 2233	0.9 / 4.6 / 1.0 / 4.7
15 TH	0427 / 1128 / 1648 / ○2333	1.0 / 4.5 / 1.1 / 4.4	30 F	0409 / 1105 / 1633 / ●2316	0.6 / 4.8 / 0.7 / 4.9
			31 SA	0455 / 1149 / 1717	0.4 / 5.0 / 0.5

Central S England

Chart Datum: 2·73 metres below Ordnance Datum (Newlyn). HAT is 5·1 metres above Chart Datum.

>> FREE monthly updates. Register at <<
www.reedsnauticalalmanac.co.uk

121

PORTSMOUTH LAT 50°48′N LONG 1°07′W
TIMES AND HEIGHTS OF HIGH AND LOW WATERS

STANDARD TIME (UT)
For Summer Time add ONE hour in **non-shaded areas**

Dates in red are **SPRINGS**
Dates in blue are NEAPS

YEAR 2019

SEPTEMBER

Day	Time	m	Day	Time	m
1 SU	0000 / 0538 / 1236 / 1800	4.9 / 0.3 / 5.0 / 0.5	16 M	0023 / 0558 / 1251 / 1801	4.5 / 0.9 / 4.6 / 1.0
2 M	0046 / 0620 / 1326 / 1843	4.9 / 0.3 / 5.0 / 0.5	17 TU	0055 / 0613 / 1321 / 1828	4.5 / 1.0 / 4.6 / 1.1
3 TU	0135 / 0703 / 1420 / 1926	4.8 / 0.5 / 4.9 / 0.7	18 W	0122 / 0639 / 1347 / 1855	4.5 / 1.1 / 4.5 / 1.1
4 W	0230 / 0747 / 1515 / 2012	4.7 / 0.8 / 4.8 / 1.0	19 TH	0150 / 0708 / 1415 / 1926	4.4 / 1.1 / 4.4 / 1.2
5 TH	0329 / 0835 / 1604 / 2105	4.5 / 1.2 / 4.5 / 1.4	20 F	0225 / 0742 / 1450 / 2004	4.3 / 1.2 / 4.3 / 1.3
6 F	0427 / 0936 / 1654 / 2214	4.3 / 1.6 / 4.3 / 1.7	21 SA	0309 / 0824 / 1538 / 2051	4.2 / 1.5 / 4.1 / 1.6
7 SA	0528 / 1055 / 1753 / 2331	4.0 / 1.9 / 4.0 / 1.9	22 SU	0407 / 0917 / 1641 / 2155	4.0 / 1.8 / 3.9 / 1.8
8 SU	0638 / 1207 / 1905	3.8 / 2.0 / 3.9	23 M	0525 / 1036 / 1802 / 2351	3.8 / 2.1 / 3.8 / 1.9
9 M	0041 / 0755 / 1314 / 2016	1.9 / 3.9 / 2.0 / 3.9	24 TU	0702 / 1243 / 1937	3.9 / 2.0 / 4.0
10 TU	0145 / 0939 / 1414 / 2115	1.8 / 4.0 / 1.8 / 4.1	25 W	0117 / 0824 / 1350 / 2042	1.7 / 4.1 / 1.7 / 4.3
11 W	0238 / 1023 / 1503 / 2156	1.6 / 4.2 / 1.6 / 4.2	26 TH	0215 / 0918 / 1442 / 2130	1.3 / 4.5 / 1.3 / 4.6
12 TH	0324 / 1032 / 1546 / 2234	1.3 / 4.4 / 1.3 / 4.4	27 F	0305 / 1002 / 1530 / 2213	0.9 / 4.8 / 0.9 / 4.8
13 F	0404 / 1105 / 1625 / 2311	1.1 / 4.5 / 1.1 / 4.5	28 SA	0351 / 1045 / 1614 / 2255	0.6 / 5.0 / 0.6 / 4.9
14 SA	0441 / 1141 / 1701 / 2348	1.0 / 4.6 / 1.0 / 4.5	29 SU	0434 / 1147 / 1657 / 2338	0.4 / 5.1 / 0.5 / 5.0
15 SU	0515 / 1217 / 1734	0.9 / 4.6 / 1.0	30 M	0517 / 1211 / 1739	0.3 / 5.1 / 0.4

OCTOBER

Day	Time	m	Day	Time	m
1 TU	0023 / 0558 / 1258 / 1820	5.0 / 0.4 / 5.0 / 0.5	16 W	0026 / 0543 / 1247 / 1800	4.6 / 1.0 / 4.6 / 1.0
2 W	0111 / 0639 / 1346 / 1901	4.9 / 0.6 / 4.9 / 0.8	17 TH	0054 / 0613 / 1314 / 1829	4.6 / 1.1 / 4.6 / 1.0
3 TH	0203 / 0720 / 1435 / 1944	4.7 / 0.9 / 4.7 / 1.1	18 F	0125 / 0644 / 1343 / 1902	4.6 / 1.1 / 4.5 / 1.1
4 F	0300 / 0806 / 1523 / 2032	4.5 / 1.3 / 4.5 / 1.5	19 SA	0201 / 0719 / 1420 / 1940	4.5 / 1.3 / 4.4 / 1.3
5 SA	0357 / 0902 / 1613 / 2140	4.3 / 1.7 / 4.2 / 1.8	20 SU	0248 / 0802 / 1510 / 2028	4.3 / 1.5 / 4.2 / 1.5
6 SU	0457 / 1030 / 1712 / 2305	4.0 / 2.1 / 3.9 / 2.0	21 M	0352 / 0856 / 1619 / 2133	4.1 / 1.9 / 4.0 / 1.9
7 M	0609 / 1143 / 1831	3.8 / 2.2 / 3.7	22 TU	0515 / 1027 / 1746 / 2349	4.0 / 2.1 / 3.9 / 1.9
8 TU	0013 / 0729 / 1249 / 1949	2.1 / 3.8 / 2.1 / 3.8	23 W	0646 / 1234 / 1915	4.0 / 2.0 / 4.0
9 W	0116 / 0923 / 1348 / 2055	1.9 / 4.0 / 2.0 / 4.0	24 TH	0100 / 0802 / 1333 / 2019	1.7 / 4.3 / 1.7 / 4.3
10 TH	0210 / 1001 / 1437 / 2135	1.7 / 4.3 / 1.7 / 4.2	25 F	0154 / 0856 / 1423 / 2107	1.3 / 4.6 / 1.3 / 4.6
11 F	0255 / 1007 / 1519 / 2211	1.1 / 4.4 / 1.1 / 4.4	26 SA	0243 / 0940 / 1508 / 2151	0.6 / 4.8 / 1.0 / 4.8
12 SA	0334 / 1039 / 1556 / 2247	1.2 / 4.6 / 1.2 / 4.5	27 SU	0328 / 1022 / 1552 / 2234	0.7 / 5.0 / 0.7 / 4.9
13 SU	0410 / 1113 / 1630 / 2323	1.1 / 4.7 / 1.1 / 4.6	28 M	0411 / 1104 / 1634 / 2317	0.6 / 5.1 / 0.6 / 5.0
14 M	0443 / 1147 / 1702 / 2356	1.0 / 4.7 / 1.0 / 4.6	29 TU	0453 / 1146 / 1716	0.5 / 5.1 / 0.6
15 TU	0514 / 1219 / 1732	0.9 / 4.7 / 1.0	30 W	0002 / 0534 / 1230 / 1756	4.9 / 0.6 / 5.0 / 0.7
			31 TH	0050 / 0615 / 1314 / 1837	4.8 / 0.8 / 4.9 / 0.9

NOVEMBER

Day	Time	m	Day	Time	m
1 F	0140 / 0656 / 1400 / 1919	4.7 / 1.1 / 4.7 / 1.2	16 SA	0109 / 0627 / 1324 / 1847	4.6 / 1.1 / 4.6 / 1.1
2 SA	0233 / 0740 / 1446 / 2004	4.5 / 1.4 / 4.5 / 1.5	17 SU	0151 / 0706 / 1406 / 1929	4.6 / 1.3 / 4.5 / 1.2
3 SU	0327 / 0833 / 1534 / 2103	4.3 / 1.8 / 4.2 / 1.8	18 M	0243 / 0752 / 1459 / 2019	4.4 / 1.5 / 4.3 / 1.5
4 M	0423 / 0956 / 1628 / 2229	4.1 / 2.0 / 3.9 / 2.0	19 TU	0351 / 0850 / 1610 / 2127	4.3 / 1.8 / 4.1 / 1.7
5 TU	0528 / 1111 / 1741 / 2338	3.9 / 2.2 / 3.7 / 2.1	20 W	0507 / 1026 / 1730 / 2320	4.2 / 2.0 / 4.0 / 1.8
6 W	0646 / 1215 / 1906	3.9 / 2.2 / 3.7	21 TH	0623 / 1205 / 1847	4.2 / 1.9 / 4.1
7 TH	0038 / 0801 / 1314 / 2014	2.0 / 4.0 / 2.0 / 3.9	22 F	0031 / 0732 / 1306 / 1950	1.6 / 4.4 / 1.6 / 4.3
8 F	0132 / 0855 / 1403 / 2103	1.8 / 4.2 / 1.8 / 4.1	23 SA	0127 / 0829 / 1357 / 2042	1.4 / 4.6 / 1.4 / 4.5
9 SA	0217 / 0933 / 1444 / 2143	1.6 / 4.4 / 1.6 / 4.3	24 SU	0216 / 0915 / 1444 / 2128	1.1 / 4.8 / 1.1 / 4.6
10 SU	0256 / 1009 / 1521 / 2221	1.4 / 4.6 / 1.3 / 4.4	25 M	0303 / 0959 / 1528 / 2214	0.9 / 4.9 / 0.9 / 4.7
11 M	0333 / 1044 / 1556 / 2256	1.2 / 4.7 / 1.2 / 4.5	26 TU	0347 / 1041 / 1612 / 2300	0.8 / 4.9 / 0.8 / 4.8
12 TU	0408 / 1117 / 1629 / 2329	1.1 / 4.7 / 1.1 / 4.6	27 W	0431 / 1124 / 1655 / 2347	0.8 / 4.9 / 0.7 / 4.8
13 W	0443 / 1147 / 1703	1.1 / 4.7 / 1.0	28 TH	0514 / 1207 / 1737	0.9 / 4.9 / 0.8
14 TH	0000 / 0516 / 1218 / 1735	4.6 / 1.0 / 4.7 / 1.0	29 F	0035 / 0556 / 1251 / 1818	4.7 / 1.0 / 4.8 / 0.9
15 F	0033 / 0551 / 1249 / 1810	4.7 / 1.1 / 4.7 / 1.0	30 SA	0125 / 0638 / 1334 / 1859	4.7 / 1.2 / 4.6 / 1.2

DECEMBER

Day	Time	m	Day	Time	m
1 SU	0214 / 0721 / 1418 / 1942	4.5 / 1.5 / 4.4 / 1.4	16 M	0146 / 0700 / 1358 / 1924	4.6 / 1.2 / 4.6 / 1.0
2 M	0302 / 0807 / 1502 / 2028	4.3 / 1.7 / 4.2 / 1.7	17 TU	0242 / 0748 / 1452 / 2015	4.6 / 1.3 / 4.4 / 1.2
3 TU	0349 / 0905 / 1550 / 2130	4.2 / 2.0 / 4.0 / 1.9	18 W	0345 / 0845 / 1558 / 2116	4.5 / 1.6 / 4.3 / 1.4
4 W	0441 / 1023 / 1646 / 2247	4.0 / 2.1 / 3.8 / 2.0	19 TH	0450 / 0957 / 1707 / 2236	4.4 / 1.7 / 4.2 / 1.5
5 TH	0545 / 1131 / 1800 / 2350	3.9 / 2.2 / 3.7 / 2.0	20 F	0554 / 1122 / 1815 / 2353	4.4 / 1.7 / 4.1 / 1.5
6 F	0659 / 1230 / 1921	3.9 / 2.1 / 3.7	21 SA	0658 / 1231 / 1919	4.4 / 1.6 / 4.2
7 SA	0044 / 0802 / 1321 / 2022	1.9 / 4.1 / 1.9 / 3.9	22 SU	0055 / 0757 / 1329 / 2016	1.4 / 4.5 / 1.4 / 4.3
8 SU	0131 / 0853 / 1403 / 2111	1.8 / 4.2 / 1.7 / 4.1	23 M	0149 / 0850 / 1420 / 2110	1.3 / 4.6 / 1.3 / 4.4
9 M	0214 / 0935 / 1442 / 2153	1.6 / 4.4 / 1.5 / 4.3	24 TU	0240 / 0939 / 1508 / 2202	1.2 / 4.7 / 1.1 / 4.5
10 TU	0255 / 1013 / 1520 / 2230	1.4 / 4.5 / 1.3 / 4.4	25 W	0328 / 1024 / 1554 / 2252	1.1 / 4.7 / 1.0 / 4.6
11 W	0335 / 1047 / 1559 / 2304	1.3 / 4.6 / 1.1 / 4.5	26 TH	0414 / 1108 / 1639 / 2340	1.1 / 4.7 / 0.9 / 4.6
12 TH	0415 / 1120 / 1638 / 2340	1.1 / 4.7 / 1.0 / 4.6	27 F	0459 / 1151 / 1723	1.1 / 4.7 / 0.9
13 F	0455 / 1154 / 1718	1.1 / 4.7 / 0.9	28 SA	0026 / 0542 / 1233 / 1804	4.6 / 1.1 / 4.6 / 0.9
14 SA	0017 / 0535 / 1232 / 1758	4.7 / 1.0 / 4.7 / 0.8	29 SU	0111 / 0623 / 1315 / 1844	4.6 / 1.2 / 4.5 / 1.0
15 SU	0059 / 0616 / 1312 / 1839	4.7 / 1.1 / 4.7 / 1.0	30 M	0155 / 0703 / 1356 / 1921	4.5 / 1.3 / 4.4 / 1.2
			31 TU	0237 / 0741 / 1436 / 1957	4.4 / 1.5 / 4.3 / 1.4

Chart Datum: 2·73 metres below Ordnance Datum (Newlyn). HAT is 5·1 metres above Chart Datum.

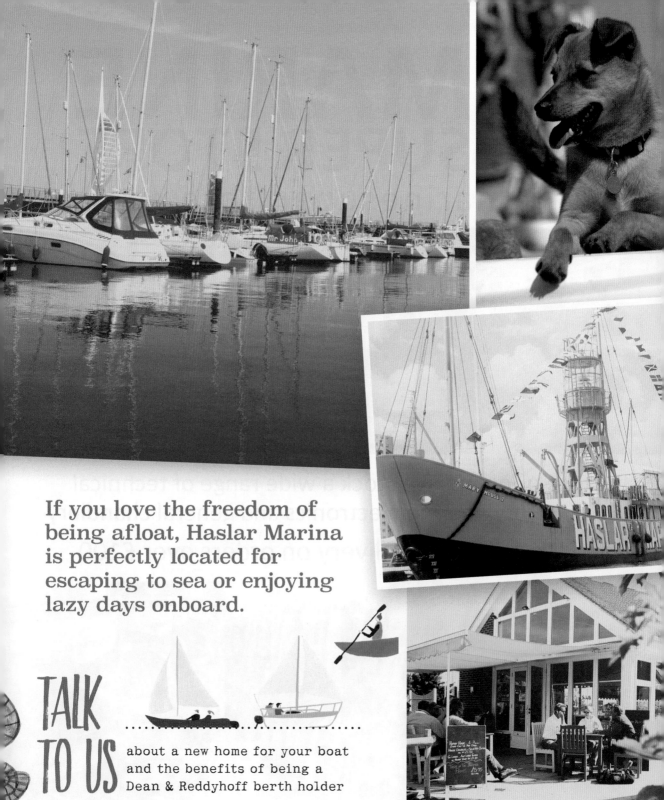

If you love the freedom of being afloat, Haslar Marina is perfectly located for escaping to sea or enjoying lazy days onboard.

TALK TO US

about a new home for your boat and the benefits of being a Dean & Reddyhoff berth holder

With marinas in **Hampshire**, **Dorset** and the **Isle of Wight**, we're only a short sail away.

02392 601201

haslar marina

deanreddyhoff.co.uk

2.23 PORTSMOUTH
Hampshire 50°47'·38N 01°06'·67W (Entrance)

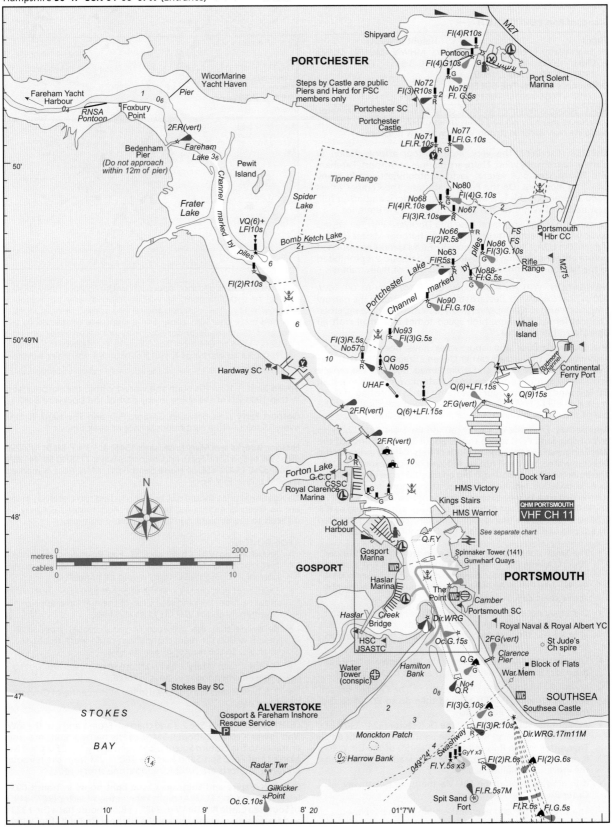

CHARTS AC 5600, 2037, 2625, 2631, 2629, 2628; Imray C9, C3, C15, 2200

TIDES +0020 Dover; ML 2·8

Standard Port PORTSMOUTH (←—)

Times				Height (metres)			
High Water		Low Water		MHWS	MHWN	MLWN	MLWS
0500	1000	0000	0600	4·7	3·8	1·9	0·8
1700	2200	1200	1800				
Differences LEE-ON-THE-SOLENT							
−0005	+0005	−0015	−0010	−0·2	−0·1	+0·1	+0·2

NOTE: Strong winds from NE to SE, coupled with a high barometer, may lower levels by 1m and delay times of HW and LW by 1hr; the opposite may occur in strong W'lies with low pressure.

SHELTER Excellent. This very large harbour affords shelter in some area from any wind. There are marinas at Gosport, Fareham and Portchester, plus several yacht pontoons/jetties and many swinging moorings. On the Portsmouth side, The Camber offers good shelter, but this is a busy commercial dock and often full; beware Isle of Wight car ferries berthing near the entrance.

Portsmouth is a major Naval Base and Dockyard Port; all vessels come under the authority of the Queen's Harbour Master (QHM). If over 20m LOA, QHM's permission (VHF Ch 11) must be obtained before entering, leaving or moving in the harbour. ⚓ is prohibited in the harbour and fishing only allowed clear of main channels.

NAVIGATION WPT No 4 Bar buoy, QR, 50°47'·01N 01°06'·36W, 330°/4·5ca to entrance. Beware strong tidal streams in the narrow entrance which is invariably very busy. Commercial shipping, cross-Channel ferries, warships, high speed ferries and small craft all operate within this area.

The lower harbour and approaches have been dredged to accommodate the aircraft carrier *Queen Elizabeth*. The chan has been widened, deepened and straightened. Buoyage has moved accordingly, but the effect on small craft movements is minimal. On Spit Sand and upharbour new transit marks consisting of three vertical masts lit Fl Y 5s have been positioned specifically for the use of the carrier owing to its size and offset bridge.

Approaches: From the W, yachts can use the Swashway Chan (to NW of Spit Sand Fort) which carries about 2m. The transit of the War Memorial and RH edge of block of flats (049·4°) indicates the deepest water, but need not be followed exactly except at LWS. The Inner Swashway (Round Tr brg approx 035°) carries only 0·3m; local knowledge required. Approaching inshore from the E, a submerged rock barrier extends from Southsea to Horse Sand Ft. The Inshore Boat Passage (0·9m) 1ca off the beach is marked by R & G piles; Main Passage (min depth 1·2m), lies 7ca further S, marked by a G pile and a concrete dolphin lit Q R andQ G respectively.

- **The Small Boat Channel (SBC)** is *mandatory* for craft under 20m LOA entering or leaving harbour. It runs just outside the W edge of the main dredged channel from abeam No 4 Bar buoy to Ballast Bn PHM. A depth gauge, showing height above chart datum, is on BC4.
- Craft may not enter harbour on the E side of the main chan and must not enter the SBC on its E side; craft may enter or leave the SBC anywhere on its W side (beware Hamilton Bank, some of which dries).
- Engines (if fitted) must be used in the SBC between No 4 Bar buoy and Ballast Bn, *which should always be left to port*.
- If crossing to/from Gunwharf Quays or The Camber obtain approval from QHM (Ch 11) and then do so at 90° to the main channel N of Ballast Bcn.
- At night the SBC is covered by the Oc R sector (324°-330°) of the Dir WRG lt on Fort Blockhouse until close to the hbr ent. Thereafter the Iso R 2s sector (337·5°-345°) of the Dir WRG lt on the dolphin E of Gosport Marina leads 341° through the entrance to Ballast Bcn.
- If joining or leaving the SBC via the Inner Swashway, BC Outer Bn *must* be left to port.
- The SBC may be used even if the main channel is closed.

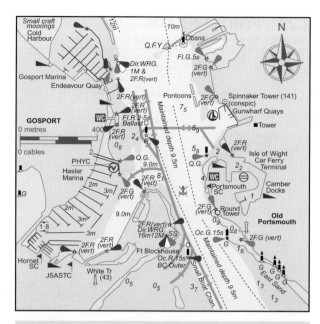

Exclusion zones: Do not approach within 50m of any MoD vessel or establishment, or within 100m of any submarine. Yellow buoys (see chartlet above) are sometimes deployed to delineate the 50m zone. Some ships are protected by a zone of 250m while underway. They are escorted by armed vessels; listen to QHM on VHF 11 or 13 for details. If you fail to respond to warnings it will be assumed you have hostile intentions.

Speed limit is 10kn (through the water) within the harbour and within 1000 yds of the shore in any part of the Dockyard Port.

Gosport Ferry plies between Gosport Pier and The Hard. It runs every 15 mins with a continuous service at peak times.

Historic wrecks are: *Mary Rose* (sank 1545) at 50°45'·8N 01°06'·2W (5ca SSW of Spit Sand Fort), and *Invincible* (sank 1758) *at* 50°44'·36N 01°02'·32W (1·4M ESE of Horse Sand Fort). *Mary Rose* is marked by SPM buoys, Fl Y 5s.

Navigational piles should not be approached too closely as many are on steep-to mudbanks; those lit in Portchester Lake may be difficult to see at night due to background lighting.

FOG ROUTINE
- Broadcast on VHF Ch **11** and **13**, Fog Routine comes into force when QHM considers that visibility is so low that normal shipping movements would be dangerous.
- Small craft may continue at the skipper's discretion, but with great caution, keeping well clear of the main and approach channels. Monitor VHF Ch **11** at all times.
- Be aware that the presence of radar echoes from small vessels within the main channel can cause much doubt and difficulty to the master of a large vessel.
- For their own safety, small craft are strongly advised not to proceed when fog routine is in force.
- 3 W (vert) lights are displayed in fog at the centre of Gosport Ferry pier.

LIGHTS AND MARKS See 2.2 and chartlet. From S and E of the IoW Nab Tower is conspic about 9·5M SE of the harbour entrance. The entrance to the Eastern Solent is between Horse Sand Fort and No Man's Land Fort. Spit Sand Fort lies 160°/1·2M from the harbour entrance. All forts are conspic round stone structures.

St Jude's ✠ spire and Southsea Castle light house in transit 003° leads to Outer Spit SCM. At night keep in the W sector (347°-349°) of Southsea Castle Pile Dir WRG 348°. Ft Blockhouse has a Dir lt 323° and traffic sigs, running toward the Harbour Bn Dir lt (fuel jetty).

FIRING RANGE Tipner Rifle Range, N of Whale Island, danger area extends 2,500m from firing range. R flag or ● lt on Tipner Range FS indicates firing is in progress; yachts should clear the range danger area or transit it as quickly as possible.

COMMUNICATIONS (Code Portsmouth/Gosport 02392) Ⓗ 286000; QHM Hbr Control (H24) 723694; Commercial Docks & Camber Berthing Offices ☎297395.

Yachts should monitor Ch **11** (*QHM*) for traffic and navigational information. *Portsmouth Hbr Radio* (Commercial Port) Ch 14 (H24). *Haslar Marina* and *Port Solent* Ch **80** M (H24). *Gosport Marina* call Ch **80** M (HO). Fareham Marine Ch M (summer 0900-1730). Gunwharf Quay Ch 80. **Naval activities** to the S/SE of Portsmouth and IOW may be advised by Solent CG Ch 67 or ☎552100; or Naval Ops ☎722008. Naval vessels use Ch 13. The positions and times of naval firings and underwater explosions are broadcast daily at 0800 and 1400LT Ch 06; preceded by a Securité call on Ch 16. Warnings of underwater explosions will also be broadcast on Ch 16 at 1hr, at 30 mins and just before the detonation.

FACILITIES
Haslar Marina berths@haslarmarina.co.uk ☎601201. Access H24 580⌇+ 50Ⓥ at L/M pontoons, inboard of green light vessel, £3·25 (max 60m LOA); short stay £10. ⚓ 🅿 ▬ 🔋 ✕ ⚒ ✎ Ⓔ 🖂 🗗 ✕ @ wi-fi. A floating boat lift is at the end of 'F' pontoon. Other Haslar Creek moorings/pontoons are private; Haslar bridge (1·8m clearance).

Gunwharf Quays www.gunwharf-quays.com ☎836732 ⚓ 🏳 🅿 🗗 Ⓔ Ⓥ £3·80 (max 80m LOA), short stay £1·25/m, booking advisable (or try VHF 80 for short notice requirements). £6·00 landing fee/vessel for drop offs to Gunwharf Quays shopping centre. Note Small Boat Channel regulations.

Town Quay (Camber) ☎833166, berth on wall, <12m £18·60, short stay (3hrs) £8·45, pay 'KB Boat Park'; no ⚓/🗗.

Gosport Marina www.premiermarinas.com ☎524811. 471⌇ inc Ⓥ if berths available, £3·10 with £5 🗗 inc, min fee £24·80; short stay £1·48/m, min fee £11·84. ⚓ 🅿 ⚒ 🔋 🛍 ✕ ⚒ Ⓔ 🖂(12t) dry stack storage for motor boats, ⚓ 🗑 ✕ ✕ wi-fi.

Endeavour Quay www.endeavourquay.co.uk ☎584200, fully equipped 🖂 (some ⌇Ⓥ) ✕ ⚒ ⚓ 🧍 🖥(180t) 🖂(35t) ⚓ wi-fi.

Royal Clarence Marina ⚓ www.royalclarencemarina.org. ☎523523, 150⌇, Ⓥ space varies, up to 50m, <12m £3·09 (winter £22·66). 🅿 🗗 🗑 ✎ 🖂 🏳 ✕ wi-fi.

Port Solent portsolent@premiermarinas.com ☎210765. Lock 9m wide. 900⌇ inc Ⓥ, £3·10 (min fee £24·80) with £5 🗗 inc; s/s £1·41/m. 🅿 🗗 🛍 🔋 🧍 ✕ Ⓔ ⚓ 🖂(35t) ⚓ 🗑 ✕ wi-fi.

Portchester Lake is marked by lit and unlit piles. Unusually, PHMs are numbered 57 to 74 (from seaward), and SHMs 95 to 75.

Beware unlit naval buoys at the S end. Do not delay crossing Tipner Range, S limit marked by piles 63/87 and N by 70/78. Portchester Castle is conspic 5ca SSW of marina. Call marina Ch 80 when inbound passing pile 78. Access H24 via chan dredged 1·5m to lock (43m x 9·1m); enter on 3 ● (vert) or on loudspeaker instructions.

WicorMarine Yacht Haven (Portchester) www.wicormarine.co.uk ☎(01329) 237112, some Ⓥ on pontoons, £1·50/m (rallies welcome), ⚓ ⚒ 🗗 🔋 🛍 ✕ ✎ 🖂 Ⓔ 🖂(12t) 🖂(7t) ⚓ 🗑 ✕.

Portsmouth Marine Engineering ☎(01329) 232854, ⌇(drying) Ⓥ £15/night, £40/week, 🏳 🖂(18t).

Upper Quay Marina (dries) upperquaymarine1@gmail.com mob 07712189444, 50⌇+Ⓥ by arrangement, ⚓ 🗗.

TOWN CENTRES
PORTSMOUTH ✕ ✎ 🗗 Ⓔ ⚓ 🗑 ACA 🗛 🖂 Ⓑ ⇌ ✈ (Southampton). **Ferries:**
Brittany Ferries (www.brittany-ferries.co.uk):-
Bilbao: 2/week; 24hrs or 32hrs; Santander: 2/week; 24hrs;
Caen: 3/day; 6hrs; St Malo: daily; 8½hrs;
Cherbourg HSS: up to 3/day; 3hrs.
Le Havre HSS: up to 2/day (summer only); 3¼hrs.
Condor Ferries (www.condorferries.co.uk):
Guernsey/Jersey: 6/week; 6½–12½hrs.
IoW: frequent car and passenger services to Ryde/Fishbourne.
Harbour Tour and Water Taxi ☎(01983) 564602 (Easter – end October 1000-1700) between Historic Dockyard and RN Submarine Museum (Gosport) tickets available inside dockyard Victory Gate.
GOSPORT ✕ ✎ 🗗 Ⓔ ⚓ ACA 🖂 Ⓑ ⇌(Portsmouth) ✈ (S'ton).
FAREHAM (01329) 🖂 Ⓑ 🗑 🗗 ⇌ ✈(Southampton).
Fareham Lake is well marked, partially lit up to Bedenham Pier and unlit thereafter. Chan dries 0·9m in final 5ca to Town Quay.

YACHT CLUBS
Royal Naval Sailing Association ☎521100.
Royal Naval & Royal Albert YC ☎825924, 🗗.
Portsmouth SC ☎ 820596.
Portchester SC ☎376375, detached pontoon⚓ (close No 71 bn).
Hardway SC secretary@hardwaysailingclub.co.uk ⚓ ▬ ⌇ ✕ 🗗.
Gosport CC ☎586838.
Fareham Sailing & Motor Boat Club ☎(01392) 280738.

EASTERN APPROACHES TO THE SOLENT

(AC 2045, 2036) Nab Tower (Fl 10s, *Horn(2) 30s*), a conspic steel and concrete structure 4·5M E of Foreland, marks Nab Shoal for larger vessels but is of no direct significance to yachts. The Nab Channel (maintained depth 13.3m) runs northwards to the East of the tower. It is indicated by Y SPM as far as New Grounds. Keep well clear of shipping manoeuvring in the Pilot Boarding Areas and using the Nab Deep Water Channel.

NW of Nab Tr the E approach to the Solent via Spithead presents few problems and is far safer in SW/W gales than the Needles Channel. A good lookout should be kept for ferry traffic and vessels anchoring in St Helen's Roads.

The main shipping chan is well buoyed and easy to follow, but there is plenty of water for yachts over New Grounds and Warner Shoal when approaching No Man's Land Fort (Iso R 2s) and Horse Sand Fort (Iso G 2s). A submerged barrier lies N of the latter; the barrier to SW of the former has been demolished and therefore yachts may now pass safely SW of No Man's Land Fort; the Ryde Sand Bcns (both PHM and lit) mark the shallows to the SW. Ryde Sand dries extensively and is a trap for the unwary.

Care should be taken to avoid Hamilton Bank to the South of the small boat channel close East of Portsmouth Harbour entrance.

Langstone and Chichester Harbours have offlying sands which are dangerous in strong S'ly winds. E and W Winner flank the entrance to Langstone Hbr. E and W Pole Sands, drying 1m, lie either side of the appr channel to Chichester Hbr. Bracklesham Bay is shallow, with a distinct inshore set at certain states of the tide. SE along a low-lying coast is Selsey Bill with extensive offshore rocks and shoals. Pass these to seaward via Owers SCM lt buoy or the Looe Chan, which in suitable conditions is much used by yachts on passage to/from points E of the Solent. Although lit, the many lobster pot markers demand caution at night; it is dangerous in onshore gales as searoom is limited by extensive shoals on which the sea breaks.

Looe Channel: ▶ *the E-going flood runs from HW Portsmouth +0430 (HW Dover +0500) until HW Portsmouth –0130 (HW Dover –0100), at sp reaching 2.4kn near the Boulder and Street light buoys which mark its narrow western end; they may be hard to see. Max Neap rate is 1.2kn.*

The W-going ebb runs from HW Portsmouth –0130 (HW Dover –0100) until HW Portsmouth +0430 (HW Dover +0500). At Springs reaching 2.6kn near Boulder and Street. Max Neap rate is 1.3kn.◀

Boulder SHM lt buoy is at the W ent to this chan, about 6M SE of Chichester Bar Bn. Medmery Bank, 3·7m, is 1M WNW of Boulder.

2.24 LANGSTONE HARBOUR

Hampshire 50°47'·23N 01°01'·54W (Ent) ✿✿◊◊◊✿✿

CHARTS AC 5600, 2037, 3418; Imray C3, C9, Y33, 2200

TIDES +0022 Dover; ML 2·9

Standard Port PORTSMOUTH (←)

Times				Height (metres)			
High Water		Low Water		MHWS	MHWN	MLWN	MLWS
0500	1000	0000	0600	4·7	3·8	1·9	0·8
1700	2200	1200	1800				
Differences LANGSTONE							
0000	−0015	0000	−0010	+0·1	+0·1	0·0	0·0
NAB TOWER							
+0015	0000	+0015	+0015	−0·2	0·0	+0·2	0·0
SELSEY BILL							
+0010	−0010	+0035	+0020	+0·5	+0·3	−0·1	−0·2

SHELTER Very good in marina (2·4m) to W inside ent, access over tidal flap 1·6m at CD; waiting pontoon. Ent is 7m wide. Buoys on W side of harbour entrance administered by Eastney Cruising Association, max LOA 9m. Or ⚓ out of the fairway in Russell's Lake or Langstone Chan (water ski area); or see HM (E side of ent). Hbr speed limit 10kn; 5kn in Southsea Marina channel.

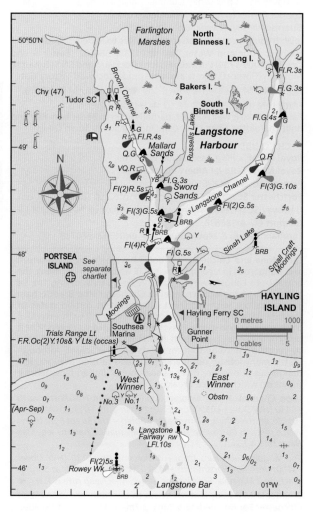

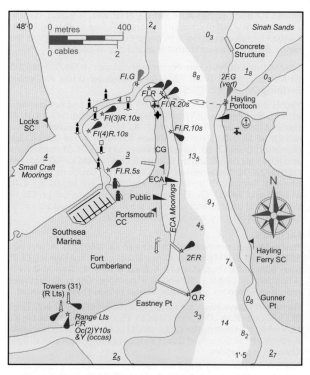

NAVIGATION WPT 50°46'·31N 01°01'·36W, Langstone Fairway SWM Bcn 348°/9·4ca to QR light at ent (Eastney Pt).

Bar has about 1·2m. Entrance channel lies between East and West Winner drying banks, which afford some protection. Appr is straightforward in most conditions, best from HW −3 to +1, but avoid entry against the ebb, esp at sp and in strong onshore winds. In strong S/SE winds do not attempt entry.

LIGHTS AND MARKS Ldg marks (concrete dolphins), or Fairway beacon in line with conspic chy, leads 344° just clear of East Winner. The entrance itself deepens and favours the W side. The narrow approach channel to Southsea Marina is marked by 5 SHM piles, only the first of which is lit, Fl G. There are 9 PHM piles; see chartlet for details.

COMMUNICATIONS (Code 02392) Ⓗ 286000; HM 463419.

Harbour VHF Ch **12** 16 (0900-1700 daily). Marina Ch **80** M.

FACILITIES **Southsea Marina** www.premiermarinas.com ☎822719. Access channel dredged to 0·5m. Marina entrance, width 7m, has R/G lights indicating if the cill is up or down. 320⌂, 8m< £3·10 (min £24·80)<16m< £3·30; inc £5 ⚡ for 2 nights. £1.48/m for <4hrs.
Services: ⚓ ⛵ ⚓ ♨ ⚓ ⛽ ⚒ ⚓ 🅿(25t) ⛴(18t) ⚓ ⚒ ✕ ⌂ wi-fi.

Hayling Pontoon (E side of ent), www.langstoneharbour.org.uk ⌂ (H24 with pre-payment tag available from HO) ⚓. Vacant mooring buoys in harbour may be available, £14·00/night.

Langstone SC ☎9248 4577, ⚓ ⚓ ⚓ ⚓ ⌂.

Eastney Cruising Association (ECA) ☎734103, 6 buoys for craft under 9m by prior arrangement, ⚓ ✕ ⌂. Visitors welcome – access to ECA by intercom at road barrier or ☎827396 (bar).

Tudor SC (Hilsea) ☎ 664948 and 662002, ⚓ ⚓ ⚓ ⌂.

Hayling Ferry SC mob 0780 367571; **Locks SC** mob 07980 856267.

Towns ✉ (Eastney, Hayling), Ⓑ (Havant, Hayling, Emsworth), ⇌ (bus to Havant), ✈ (Southampton).

2.25 CHICHESTER HARBOUR
W. Sussex **50°46'·86N 00°56'·06W** ❀❀❀🔱🔱🔱❀❀❀

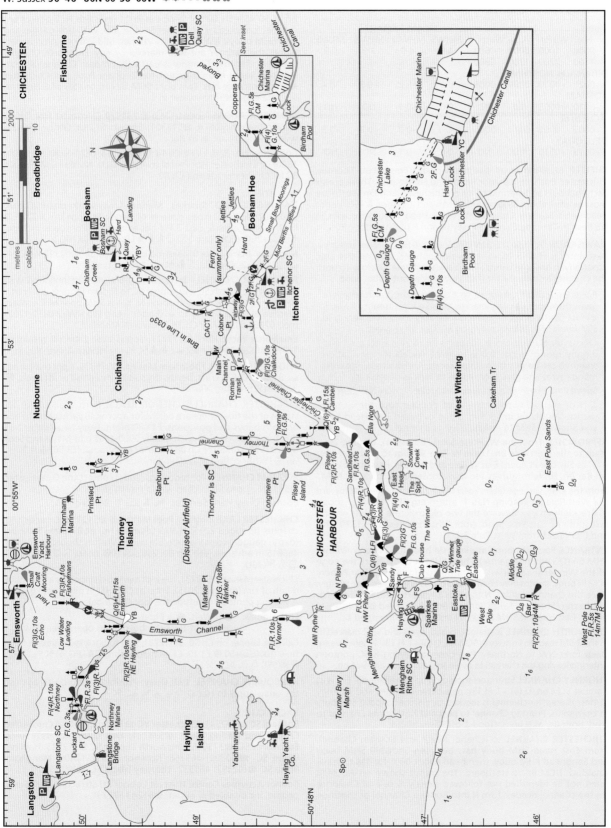

CHARTS AC 5600, 2045, 3418; Imray C9, C3, Y33, 2200

TIDES +0027 Dover; ML 2·8

Standard Port CHICHESTER (→)

Times				Height (metres)			
High Water		Low Water		MHWS	MHWN	MLWN	MLWS
0500	1000	0000	0600	4·9	4·0	1·9	0·9
1700	2200	1200	1800				
Differences NORTHNEY							
+0020	+0010	0000	+0005	0·0	−0·2	−0·2	−0·4
BOSHAM							
+0010	+0005	ND	ND	0·0	−0·1	ND	ND
ITCHENOR							
+0005	0000	−0010	+0005	−0·1	−0·2	−0·2	−0·3
DELL QUAY							
+0015	+0010	ND	ND	0·0	−0·1	ND	ND

SHELTER Excellent in all five main chans: Emsworth, Thorney, Chichester, Bosham, Itchenor Reach and Fishbourne. There are six yacht habours and marinas of which Chichester and Emsworth are most helpful; about 50 ⚓s at Emsworth and Itchenor. ⚓ in Itchenor Reach and Bosham Channel. ⚓s off E Head (uncomfortable in NE'lies); in Thorney Chan off Pilsey Is; and E of Chalkdock Pt.

NAVIGATION WPT 50°45'·32N 00°56'·60W, 013°/1·6M to entrance. Best entry is HW −3 to +1, to avoid confused seas on the ebb, especially in onshore winds >F5. Leave West Pole Bn close to port; the channel Northward is effectively only about 200m wide.

- Bar is periodically dredged to 1.5m below CD. After severe gales the depth can vary markedly and it is prudent to assume a least depth of 0.5m below CD. Be very wary of entering in S'ly winds >F6. Depths may vary ±0·75m after gales.
- Chichester Harbour is very popular for recreational boat and dinghy racing. With 10,000 resident vessels the navigation channels can be very busy at weekends. Anticipate sailing craft manoeuvres and give them room to tack as necessary.
- Speed limit 8kn in all areas N of East Stoke Pt; max fine £2,500.
- ⚓ N of E Head remains the most popular spot for visiting yachts. Vessels should ⚓ S of a line between East Head Spit and Snowhill SHM to remain clear of channel.

APPROACHES: From the W, Horse Sand Fort astern, bearing 265°, leads to the West Pole Bn. Leave W Pole 30-80m to port and alter 013° as Eastoke Pt opens E of W Pole and Bar bns. Pass approx 50m off Bar Bn. For latest survey see www.conservancy.co.uk or LNTM.

From the E/SE, via Looe Chan, keep W for 2M, then alter NW toward the West Pole Bn keeping Nab Tr astern brg 184° to clear shoals in Bracklesham Bay. Beware the two old concrete targets on and just south of East Pole Sand. Note: Historic Wrecks lie 137°/1·2M and 095°/3·2M from West Pole Bn.

ENTRANCE Pass between Eastoke PHM buoy (QR) and W Winner SHM bn (QG and tide gauge). Three SHM lit buoys mark the edge of the Winner shoal (dries), to starboard of the ent. Near Fishery SCM lit buoy, depths may change and buoys are moved accordingly. Here the channel divides: N towards Emsworth and ENE towards Chichester. Stocker's Sands (dries 1·9m), is marked by 2 lit PHM buoys. East Head lit SHM marks W end of recommended ⚓.

EMSWORTH CHANNEL is straight, broad, deep and well marked/lit in the 2·5M reach to Emsworth SCM bn, Q (6) + L Fl 15s, where Sweare Deep forks NW to Northney. Pass close to Hayling Is SC pontoon entering/leaving this channel to avoid shallow patch just S of bn.

THORNEY CHANNEL Strangers should go up at half-flood. Ent is at Camber SCM bn, Q (6) + L Fl 15s; pass between Pilsey and Thorney Lt bns, thereafter channel is marked by perches. Above Stanbury Pt chan splits, Prinsted Channel to port and Nutbourne Channel to stbd; both dry at N ends. Good ⚓ off Pilsey Island.

CHICHESTER CHANNEL to Itchenor Reach and Bosham Channel. From East Head SHM buoy pass between Snowhill SHM buoy and Sandhead PHM buoy, then head about 030° for 1M to leave Chaldock NCM Bn to starboard. The transit shown on AC 3418 need not be identified nor followed exactly. 6 ca E of Chaldock Bn the channel divides: turn N into Bosham Channel or continue

ESE into Itchenor Reach for Birdham Pool and Chichester Marina, beyond which is the drying Fishbourne Channel to Dell Quay.

LIGHTS AND MARKS Approach marked by West Pole PHM tripod bn at 50°45'·45N 00°56'·59W; Bar PHM bn is 6ca N. E side of ent channel is marked by W Winner SHM lit pile, with tide gauge; and by 3 SHM buoys: NW Winner, N Winner, Mid Winner. All channels within the harbour are well marked by day. Emsworth Channel is well lit; Thorney Channel is partly lit; Bosham Channel is unlit. Itchenor Reach is unlit except for Birdham Pool and Chichester Marina entrance bns. Two unlit SHM lie inshore NE of E Head to indicate edge of shoal water in ⚓.

COMMUNICATIONS (Code 01243) 🏥 787970 Chichester Hbr Office www.conservancy.co.uk ☎512301; Emsworth Hbr Office 376422. www.chimet.co.uk ⚓ at Bar.
Chichester Hbr Radio VHF Ch **14** 16 (Apr-Sep: 0830-1700; Sat 0900-1300) or Chichester Hbr Patrol (w/e Apr-Oct). Northney Marina Ch **80** M. Chichester Marina Ch 80.

FACILITIES
HAYLING ISLAND Sparkes Marina www.sparkesmarina.co.uk ☎(02392) 463572. From close N of Sandy Pt appr on ≠ 277° (2 bns); thence alter S, past 3 PH bns to marina. Pontoons in 1·6m. 150⛵+20♥ £3·80 (inc 🚿) 🅿 🛢 🚽 ⚓ LPG ✕ ⚒ 🛒(16t) ✕.

Northney Marina northney@mdlmarinas.co.uk ☎02392 466321. ⚓ 228⛵+20♥ £3·80. 🅿 🛢 ✕ ⚒ 🖭 🔧(35t) ✕.

EMSWORTH CHANNEL Jetty (dries 1·6m) abreast millpond wall between Emsworth SC and Slipper SC to N, 50m long, ✭ 2FR (vert). Free for <2hrs stay, ⚓. ⚓ at South St, Kings St, Slipper Mill; contact Warden ☎376422. Ferry to ⚓ (Easter – end Sep HW±2 0830-1700) mob 07864 915247 or Ch 14 *Emsworth Mobile*.

Emsworth Yacht Hbr ☎377727, access over 2·4m sill retaining 1·5m. 200⛵+20♥ £2·95, ⚓ 🛢 🚽 ✕ ⚒ 🖭 🔧(50t) 🛒(40t) ✕.

THORNEY CHANNEL Thornham Marina, ☎375335. Appr chan dries, ⚓ 81⛵+2♥ £23/craft, berths dry suiting multihull or lift keels. ✕ ⚒ ☕ 🔧(12t) 🛒(10t) ✕ 🚽.

CHICHESTER CHANNEL/ITCHENOR REACH Hard available at all stages of the tide. There are berths for approx 40 yachts at Itchenor on both buoys and pontoons. £11·80/night. Ferry to ⚓/water taxi (BH & w/e Apr – mid-May/Oct; daily 0900-1800 mid-May – end Sep) mob 07970 378350 or Ch 08 *Itchenor Ferry*. ⚓ ⚓ (£3·60) ⚓ ⚒ ✕ ⚒ 🖭 🇪 🏴(hbr office) ✕ 🚽(Ship Inn).

BOSHAM CHANNEL For moorings (200+) contact the Quaymaster ☎573336. ⚓ prohib in chan which mostly dries.
Bosham Quay Hard Services ⚓ ⚓ ⚓ ⚓. EC Wed.

CHICHESTER LAKE Birdham Pool Marina ⚓ ☎512310. Enter drying channel (dredged 2015) HW±3 (depth >1m) at Birdham SHM bn (depth gauge) access via lock (5m wide). Visitors should call 07831 466815 in advance. VHF Ch 80. 250⛵+10♥ £2·60, 🅿 🅿 🛢 ✕ ⚒ 🖭 🇪 🔧(30t).

Chichester Marina ☎512731, the well marked channel has a minimum depth of ½m (dredged 2015). Enter at CM SHM lit pile, depth gauge. A waiting pontoon is outside the lock (7m wide).
Lock signals: ● = Wait; ● = Enter.
Q (top of tower) = both gates open (free flow).
Call lock keeper on Ch 80, ☎512731, H24 to book exit outwith free-flow. 1000⛵+50 ♥ £3·10, £1·48/m for <4hrs, 🅿 🛢 LPG, 🅿 ✕ ⚒ 🖭 🇪 ☕ 🔧(65t) 🚽 🛒 ACA ✕ 🚽.

FISHBOURNE CHANNEL Dell Quay: Possible drying berth against the quay, apply to Hbr Office. ⚓ ⚓ ⚓ ✕ 🖭.

YACHT CLUBS
Bosham SC ☎572341; **Chichester YC** ☎512918, 🅿 ✕ 🚽; **Chichester Cruiser and Racing Club** ☎371731; **Dell Quay SC** ☎785080; **Emsworth SC** ☎373065; **Emsworth Slipper SC** ☎372523; **Hayling Island SC** ☎(02392) 463768; **Itchenor SC** ☎512400; **Mengham Rithe SC** ☎(02392) 463337; **Thorney Island SC** ☎371731.

Cobnor Activities Centre Trust (at Cobnor Pt) attracts many young people, inc disabled 🅿 afloat. ☎01243 572791.

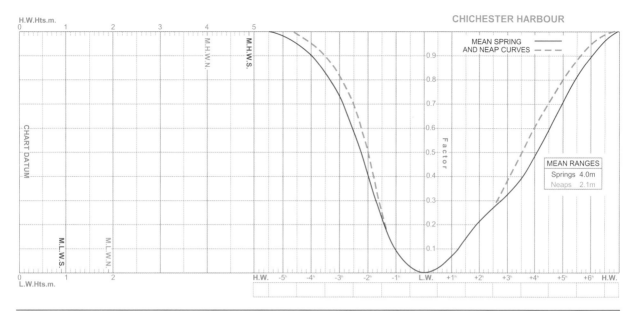

CHICHESTER HARBOUR

Central S England

STANDARD TIME (UT)
For Summer Time add ONE hour in **non-shaded areas**

CHICHESTER HARBOUR   LAT 50°47'N   LONG 0°56'W
TIMES AND HEIGHTS OF HIGH AND LOW WATERS

Dates in red are **SPRINGS**
Dates in blue are NEAPS

YEAR 2019

JANUARY

Time	m		Time	m
1 0033	1.6	**16**	0644	4.1
0736	4.3		1215	1.9
TU 1310	1.6	W	1916	4.0
2015	4.2			
2 0136	1.6	**17**	0039	1.8
0846	4.4		0751	4.3
W 1410	1.5	TH	1321	1.7
2124	4.3		2024	4.2
3 0232	1.5	**18**	0143	1.6
0939	4.5		0849	4.5
TH 1501	1.4	F	1419	1.4
2212	4.4		2119	4.4
4 0320	1.4	**19**	0240	1.4
1019	4.6		0938	4.7
F 1545	1.3	SA	1510	1.1
2251	4.5		2208	4.7
5 0402	1.3	**20**	0331	1.1
1055	4.6		1023	4.9
SA 1626	1.2	SU	1558	0.9
2328	4.5		2257	4.8
6 0442	1.3	**21**	0420	0.9
1133	4.6		1110	5.0
SU 1704	1.1	M	1646	0.7
●		○	2347	4.9
7 0006	4.6	**22**	0508	0.8
0519	1.3		1159	5.0
M 1211	4.6	TU	1735	0.6
1741	1.1			
8 0044	4.6	**23**	0040	5.0
0555	1.3		0557	0.7
TU 1248	4.6	W	1250	5.0
1817	1.1		1823	0.5
9 0120	4.6	**24**	0133	5.0
0631	1.3		0646	0.7
W 1322	4.5	TH	1341	5.0
1852	1.2		1912	0.6
10 0155	4.5	**25**	0223	5.0
0706	1.4		0735	0.8
TH 1431	4.5	F	1433	4.9
1927	1.3		2002	0.7
11 0229	4.5	**26**	0312	4.9
0742	1.5		0826	1.0
F 1431	4.4	SA	1526	4.7
2003	1.4		2054	1.0
12 0306	4.4	**27**	0402	4.7
0820	1.7		0921	1.3
SA 1513	4.4	SU	1620	4.5
2042	1.6	◑	2152	1.3
13 0350	4.2	**28**	0453	4.5
0903	1.8		1025	1.5
SU 1603	4.1	M	1716	4.2
2128	1.7		2257	1.6
14 0441	4.1	**29**	0548	4.3
0959	1.9		1135	1.7
M 1700	4.0	TU	1818	4.0
◑ 2226	1.8			
15 0538	4.1	**30**	0007	1.8
1106	2.0		0650	4.1
TU 1804	3.9	W	1248	1.8
2333	1.9		1931	3.9
		31	0117	1.8
			0804	4.1
		TH	1355	1.7
			2100	4.0

FEBRUARY

Time	m		Time	m
1 0219	1.7	**16**	0130	1.7
0916	4.2		0825	4.3
F 1449	1.5	SA	1406	1.4
2201	4.2		2103	4.3
2 0307	1.6	**17**	0231	1.4
1007	4.3		0921	4.6
SA 1532	1.4	SU	1459	1.1
2245	4.3		2157	4.6
3 0348	1.4	**18**	0322	1.1
1045	4.4		1010	4.8
SU 1610	1.2	M	1547	0.7
2321	4.5		2246	4.9
4 0426	1.3	**19**	0410	0.8
1121	4.5		1058	5.0
M 1647	1.0	TU	1634	0.5
● 2355	4.6	○	2335	5.0
5 0503	1.2	**20**	0456	0.6
1157	4.6		1145	5.1
TU 1723	1.0	W	1720	0.3
6 0028	4.6	**21**	0024	5.1
0538	1.1		0542	0.5
W 1231	4.6	TH	1234	5.1
1758	0.9		1806	0.3
7 0100	4.7	**22**	0111	5.2
0612	1.1		0628	0.5
TH 1303	4.6	F	1322	5.1
1832	0.9		1852	0.4
8 0131	4.6	**23**	0156	5.1
0646	1.1		0713	0.6
F 1334	4.6	SA	1409	4.9
1905	1.0		1937	0.6
9 0201	4.6	**24**	0240	5.0
0718	1.2		0758	0.8
SA 1406	4.5	SU	1457	4.7
1936	1.1		2023	0.9
10 0234	4.5	**25**	0326	4.7
0750	1.3		0845	1.1
SU 1441	4.4	M	1547	4.5
2008	1.3		2112	1.2
11 0310	4.3	**26**	0415	4.4
0825	1.4		0940	1.5
M 1523	4.2	TU	1640	4.2
2045	1.4	◑	2214	1.6
12 0356	4.2	**27**	0508	4.1
0908	1.6		1055	1.8
TU 1617	4.0	W	1740	3.9
◑ 2133	1.6		2334	1.9
13 0453	4.1	**28**	0609	3.9
1009	1.8		1219	1.9
W 1723	3.9	TH	1851	3.7
2242	1.8			
14 0601	4.0			
1134	1.8			
TH 1839	3.9			
15 0010	1.9			
0715	4.1			
F 1259	1.7			
1958	4.0			

MARCH

Time	m		Time	m
1 0053	2.0	**16**	0644	3.9
0722	3.8		1239	1.6
F 1332	1.8	SA	1934	4.0
2027	3.8			
2 0158	1.9	**17**	0116	1.7
0852	3.9		0801	4.1
SA 1426	1.6	SU	1348	1.4
2147	4.0		2047	4.3
3 0247	1.7	**18**	0218	1.3
0956	4.1		0905	4.4
SU 1509	1.4	M	1442	1.0
2231	4.3		2144	4.6
4 0327	1.5	**19**	0308	1.0
1035	4.3		0958	4.7
M 1547	1.2	TU	1530	0.7
2305	4.4		2234	4.9
5 0404	1.2	**20**	0354	0.6
1107	4.4		1045	4.9
TU 1623	1.0	W	1616	0.4
2336	4.6		2320	5.1
6 0439	1.1	**21**	0438	0.4
1139	4.6		1131	5.1
W 1658	0.8	TH	1700	0.3
●		○		
7 0006	4.7	**22**	0004	5.2
0514	0.9		0522	0.3
TH 1211	4.6	F	1218	5.1
1732	0.8		1744	0.3
8 0035	4.7	**23**	0047	5.2
0548	0.9		0605	0.4
F 1241	4.6	SA	1302	5.1
1806	0.8		1827	0.4
9 0104	4.7	**24**	0129	5.1
0621	0.9		0647	0.5
SA 1312	4.6	SU	1347	4.9
1837	0.8		1909	0.6
10 0133	4.7	**25**	0210	4.9
0652	0.9		0728	0.7
SU 1343	4.6	M	1431	4.7
1907	0.9		1949	0.9
11 0204	4.6	**26**	0252	4.7
0722	1.0		0808	1.0
M 1416	4.5	TU	1517	4.5
1938	1.0		2030	1.2
12 0236	4.5	**27**	0338	4.4
0755	1.1		0850	1.4
TU 1453	4.2	W	1607	4.2
2013	1.2		2117	1.6
13 0316	4.3	**28**	0427	4.0
0835	1.3		0947	1.7
W 1542	4.1	TH	1703	3.9
2058	1.5	◑	2239	2.0
14 0413	4.1	**29**	0525	3.7
0929	1.5		1133	1.9
TH 1652	3.9	F	1810	3.7
◑ 2205	1.7			
15 0526	3.9	**30**	0017	2.1
1102	1.7		0638	3.6
F 1810	3.9	SA	1252	1.9
2354	1.9		1936	3.7
		31	0125	2.0
			0810	3.6
		SU	1350	1.7
			2114	3.9

APRIL

Time	m		Time	m
1 0217	1.8	**16**	0155	1.3
0929	3.9		0849	4.3
M 1436	1.5	TU	1418	1.0
2201	4.2		2128	4.6
2 0259	1.5	**17**	0246	1.0
1010	4.1		0945	4.6
TU 1516	1.2	W	1508	0.7
2235	4.4		2217	4.9
3 0336	1.2	**18**	0333	0.7
1042	4.3		1032	4.8
W 1553	1.0	TH	1553	0.5
2305	4.6		2300	5.0
4 0412	1.0	**19**	0417	0.5
1113	4.5		1116	5.0
TH 1628	0.9	F	1637	0.4
2335	4.7	○	2342	5.1
5 0446	0.9	**20**	0459	0.4
1144	4.6		1200	5.0
F 1702	0.8	SA	1720	0.5
●				
6 0004	4.7	**21**	0022	5.1
0519	0.8		0541	0.5
SA 1216	4.7	SU	1244	5.0
1735	0.8		1801	0.6
7 0035	4.7	**22**	0103	5.0
0551	0.8		0621	0.6
SU 1248	4.7	M	1326	4.9
1807	0.8		1841	0.8
8 0106	4.7	**23**	0143	4.8
0623	0.8		0659	0.8
M 1322	4.6	TU	1409	4.7
1839	0.8		1919	1.0
9 0138	4.7	**24**	0223	4.6
0655	0.8		0737	1.0
TU 1357	4.6	W	1452	4.5
1913	0.9		1957	1.3
10 0212	4.5	**25**	0305	4.3
0732	0.9		0816	1.3
W 1437	4.4	TH	1538	4.2
1952	1.1		2041	1.6
11 0253	4.3	**26**	0350	4.0
0814	1.1		0905	1.6
TH 1529	4.2	F	1629	4.0
2040	1.4	◐	2142	1.9
12 0351	4.1	**27**	0442	3.8
0910	1.4		1026	1.9
F 1636	4.1	SA	1727	3.8
◐ 2151	1.7		2325	2.1
13 0503	4.0	**28**	0546	3.6
1044	1.6		1200	1.9
SA 1749	4.0	SU	1841	3.7
2338	1.8			
14 0618	3.9	**29**	0041	2.0
1214	1.5		0711	3.6
SU 1908	4.1	M	1305	1.8
			2007	3.8
15 0054	1.6	**30**	0139	1.8
0737	4.1		0834	3.8
M 1322	1.3	TU	1357	1.6
2027	4.3		2109	4.1

Chart Datum: 2·74 metres below Ordnance Datum (Newlyn). HAT is 5·3 metres above Chart Datum.

》》 **FREE** monthly updates. Register at 《
www.reedsnauticalalmanac.co.uk

STANDARD TIME (UT)
For Summer Time add ONE hour in **non-shaded areas**

CHICHESTER HARBOUR LAT 50°47′N LONG 0°56′W
TIMES AND HEIGHTS OF HIGH AND LOW WATERS

Dates in red are **SPRINGS**
Dates in blue are NEAPS

YEAR 2019

MAY

Time	m		Time	m
1 0225	1.6	**16** 0221	1.1
0926	4.0	0928	4.5
W 1440	1.4	TH 1443	1.0
2151	4.3	2156	4.8
2 0305	1.3	**17** 0310	0.9
1005	4.2	1017	4.7
TH 1518	1.2	F 1530	0.8
2225	4.5	2238	4.9
3 0340	1.1	**18** 0355	0.7
1039	4.4	1100	4.8
F 1553	1.1	SA 1615	0.8
2257	4.6	○ 2317	4.9
4 0413	1.0	**19** 0437	0.7
1113	4.5	1143	4.8
SA 1627	1.0	SU 1657	0.8
● 2330	4.7	2357	4.9
5 0446	0.9	**20** 0518	0.7
1148	4.6	1225	4.8
SU 1701	0.9	M 1738	0.9
6 0003	4.7	**21** 0037	4.8
0520	0.8	0558	0.8
M 1225	4.7	TU 1308	4.7
1736	0.9	1817	1.1
7 0039	4.7	**22** 0118	4.6
0555	0.8	0636	1.0
TU 1304	4.7	W 1349	4.6
1813	0.9	1856	1.2
8 0116	4.7	**23** 0157	4.5
0634	0.8	0715	1.1
W 1345	4.6	TH 1431	4.5
1854	1.0	1935	1.4
9 0156	4.6	**24** 0237	4.3
0716	0.9	0755	1.3
TH 1431	4.5	F 1513	4.3
1939	1.1	2018	1.6
10 0244	4.4	**25** 0318	4.1
0804	1.0	0841	1.6
F 1526	4.4	SA 1557	4.1
2034	1.4	2111	1.9
11 0343	4.2	**26** 0405	3.9
0906	1.3	0942	1.8
SA 1627	4.3	SU 1646	4.0
2150	1.6	◑ 2225	2.0
12 0448	4.1	**27** 0458	3.8
1029	1.4	1059	1.9
SU 1732	4.2	M 1742	3.9
◑ 2315	1.6	2344	2.0
13 0557	4.0	**28** 0601	3.7
1146	1.4	1207	1.9
M 1843	4.2	TU 1853	3.9
14 0025	1.5	**29** 0049	1.9
0711	4.1	0723	3.7
TU 1252	1.3	W 1305	1.8
2000	4.4	2007	4.0
15 0127	1.3	**30** 0142	1.7
0828	4.3	0833	3.9
W 1351	1.1	TH 1353	1.6
2106	4.6	2100	4.2
		31 0224	1.5
		0920	4.1
		F 1434	1.5
		2140	4.4

JUNE

Time	m		Time	m
1 0300	1.3	**16** 0336	1.0
1000	4.3	1044	4.5
SA 1511	1.3	SU 1555	1.1
2217	4.5	2253	4.7
2 0335	1.1	**17** 0419	1.0
1038	4.5	1126	4.6
SU 1549	1.2	M 1638	1.1
2253	4.6	○ 2333	4.6
3 0412	1.0	**18** 0500	1.0
1119	4.6	1208	4.6
M 1628	1.1	TU 1719	1.2
● 2332	4.7		
4 0451	0.9	**19** 0014	4.6
1202	4.6	0540	1.0
TU 1710	1.0	W 1249	4.5
		1758	1.3
5 0013	4.7	**20** 0054	4.5
0533	0.8	0619	1.1
W 1248	4.7	TH 1330	4.5
1754	1.0	1837	1.3
6 0058	4.7	**21** 0133	4.4
0619	0.8	0658	1.2
TH 1335	4.7	F 1409	4.5
1842	1.0	1915	1.5
7 0145	4.6	**22** 0211	4.3
0707	0.8	0737	1.3
F 1426	4.6	SA 1447	4.4
1933	1.1	1956	1.6
8 0237	4.5	**23** 0250	4.2
0801	1.0	0819	1.5
SA 1520	4.6	SU 1527	4.3
2031	1.3	2041	1.7
9 0334	4.4	**24** 0332	4.1
0902	1.1	0906	1.6
SU 1617	4.5	M 1609	4.2
2139	1.4	2134	1.9
10 0434	4.3	**25** 0420	3.9
1012	1.3	1002	1.8
M 1719	4.4	TU 1656	4.1
◑ 2250	1.5	◐ 2237	2.0
11 0536	4.2	**26** 0512	3.9
1119	1.3	1101	1.8
TU 1817	4.4	W 1749	4.0
2356	1.5	2339	2.0
12 0643	4.1	**27** 0611	3.8
1222	1.3	1158	1.8
W 1926	4.4	TH 1852	4.0
13 0058	1.4	**28** 0036	1.8
0758	4.2	0722	3.9
TH 1323	1.3	F 1252	1.8
2035	4.5	2000	4.2
14 0157	1.3	**29** 0129	1.7
0907	4.3	0831	4.1
F 1420	1.2	SA 1344	1.6
2130	4.6	2055	4.3
15 0249	1.1	**30** 0216	1.4
0959	4.5	0922	4.3
SA 1510	1.2	SU 1432	1.4
2214	4.6	2139	4.5

JULY

Time	m		Time	m
1 0301	1.2	**16** 0404	1.1
1007	4.4	1113	4.4
M 1518	1.3	TU 1622	1.3
2221	4.6	○ 2314	4.5
2 0345	1.0	**17** 0444	1.1
1052	4.6	1152	4.5
TU 1605	1.1	W 1701	1.3
● 2304	4.7	2354	4.5
3 0431	0.9	**18** 0523	1.0
1140	4.7	1230	4.5
W 1653	1.0	TH 1740	1.3
2351	4.7		
4 0519	0.8	**19** 0033	4.5
1231	4.7	0601	1.0
TH 1742	1.0	F 1308	4.5
		1817	1.3
5 0040	4.7	**20** 0111	4.4
0608	0.7	0638	1.1
F 1323	4.8	SA 1344	4.5
1833	0.9	1854	1.3
6 0131	4.7	**21** 0146	4.3
0659	0.7	0715	1.2
SA 1414	4.8	SU 1419	4.4
1925	1.0	1930	1.4
7 0223	4.7	**22** 0220	4.3
0752	0.8	0751	1.3
SU 1506	4.7	M 1453	4.4
2020	1.1	2007	1.6
8 0318	4.6	**23** 0258	4.2
0849	0.9	0828	1.4
M 1558	4.7	TU 1531	4.2
2119	1.2	2047	1.7
9 0414	4.4	**24** 0341	4.1
0950	1.1	0909	1.6
TU 1651	4.5	W 1615	4.1
◑ 2224	1.4	2135	1.8
10 0511	4.3	**25** 0431	4.0
1053	1.3	1001	1.7
W 1747	4.4	TH 1705	4.0
2328	1.5	◐ 2235	1.9
11 0611	4.1	**26** 0527	3.9
1156	1.4	1101	1.8
TH 1848	4.3	F 1801	4.0
		2340	1.9
12 0033	1.5	**27** 0630	3.9
0722	4.1	1204	1.8
F 1300	1.5	SA 1907	4.1
1958	4.3		
13 0136	1.5	**28** 0045	1.7
0842	4.1	0745	4.0
SA 1401	1.5	SU 1309	1.7
2104	4.3	2015	4.2
14 0233	1.4	**29** 0146	1.5
0945	4.2	0852	4.2
SU 1454	1.4	M 1408	1.5
2154	4.4	2110	4.4
15 0321	1.2	**30** 0240	1.2
1032	4.3	0944	4.4
M 1540	1.3	TU 1502	1.3
2235	4.4	2157	4.6
		31 0329	0.9
		1032	4.6
		W 1552	1.0
		2243	4.8

AUGUST

Time	m		Time	m
1 0418	0.7	**16** 0501	0.9
1121	4.8	1210	4.5
TH 1641	0.9	F 1718	1.1
● 2331	4.8		
2 0506	0.6	**17** 0013	4.5
1211	4.9	0538	0.9
F 1730	0.7	SA 1244	4.6
		1754	1.1
3 0020	4.9	**18** 0047	4.5
0555	0.5	0613	0.9
SA 1303	4.9	SU 1316	4.6
1819	0.7	1828	1.1
4 0111	4.9	**19** 0119	4.5
0645	0.5	0647	1.0
SU 1352	4.9	M 1347	4.5
1908	0.7	1901	1.2
5 0202	4.8	**20** 0150	4.4
0734	0.6	0718	1.1
M 1441	4.9	TU 1417	4.4
1959	0.9	1932	1.3
6 0254	4.7	**21** 0223	4.3
0826	0.8	0748	1.2
TU 1531	4.7	W 1451	4.3
2052	1.1	2004	1.4
7 0347	4.5	**22** 0301	4.2
0922	1.1	0822	1.4
W 1621	4.6	TH 1531	4.2
◑ 2152	1.3	2042	1.6
8 0442	4.3	**23** 0349	4.0
1024	1.4	0905	1.6
TH 1714	4.3	F 1621	4.0
2300	1.6	◐ 2136	1.8
9 0540	4.1	**24** 0448	3.9
1131	1.6	1007	1.8
F 1812	4.1	SA 1722	3.9
		2254	1.9
10 0009	1.7	**25** 0556	3.8
0647	3.9	1132	1.9
SA 1240	1.7	SU 1831	3.9
1920	4.0		
11 0118	1.7	**26** 0019	1.8
0815	3.9	0712	3.9
SU 1345	1.7	M 1252	1.8
2041	4.0	1944	4.1
12 0217	1.5	**27** 0129	1.5
0938	4.1	0828	4.2
M 1439	1.6	TU 1357	1.5
2145	4.2	2048	4.4
13 0304	1.4	**28** 0226	1.2
1027	4.2	0926	4.5
TU 1523	1.5	W 1451	1.2
2227	4.3	2139	4.6
14 0345	1.2	**29** 0316	0.8
1105	4.4	1015	4.7
W 1603	1.3	TH 1540	0.9
2302	4.4	2226	4.8
15 0424	1.0	**30** 0403	0.5
1138	4.5	1102	4.9
TH 1641	1.2	F 1627	0.6
○ 2337	4.5	● 2313	5.0
		31 0450	0.4
		1150	5.1
		SA 1713	0.5

Central S England

Chart Datum: 2·74 metres below Ordnance Datum (Newlyn). HAT is 5·3 metres above Chart Datum.

》FREE monthly updates. Register at 《
www.reedsnauticalalmanac.co.uk

131

STANDARD TIME (UT)
For Summer Time add ONE hour in **non-shaded areas**

CHICHESTER HARBOUR   LAT 50°47'N   LONG 0°56'W
TIMES AND HEIGHTS OF HIGH AND LOW WATERS

Dates in red are **SPRINGS**
Dates in blue are **NEAPS**

YEAR 2019

SEPTEMBER

Day	Time m	Time m	Time m	Time m	Day	Time m	Time m	Time m	Time m
1 SU	0001 5.0	0536 0.3	1238 5.1	1759 0.5	16 M	0021 4.6	0544 0.8	1246 4.7	1759 1.0
2 M	0050 5.0	0623 0.4	1325 5.1	1845 0.6	17 TU	0052 4.6	0616 0.9	1314 4.6	1830 1.0
3 TU	0139 4.9	0709 0.5	1412 5.0	1931 0.8	18 W	0122 4.5	0645 1.0	1342 4.5	1858 1.1
4 W	0228 4.8	0756 0.8	1459 4.8	2018 1.0	19 TH	0152 4.4	0714 1.1	1412 4.4	1929 1.2
5 TH	0319 4.5	0845 1.1	1548 4.5	2111 1.4	20 F	0227 4.3	0746 1.3	1448 4.3	2006 1.4
6 F ◑	0413 4.3	0944 1.5	1640 4.2	2221 1.7	21 SA	0311 4.1	0828 1.5	1539 4.1	2055 1.6
7 SA	0510 4.0	1100 1.8	1738 4.0	2342 1.8	22 SU ◐	0417 3.9	0927 1.8	1650 3.9	2215 1.8
8 SU	0616 3.8	1216 1.9	1846 3.8		23 M	0532 3.9	1111 1.9	1804 3.9	
9 M	0054 1.8	0742 3.8	1323 1.9	2017 3.8	24 TU	0000 1.8	0649 3.9	1239 1.8	1919 4.0
10 TU	0153 1.7	0929 4.0	1417 1.7	2142 4.0	25 W	0112 1.5	0807 4.2	1343 1.5	2028 4.3
11 W	0240 1.5	1016 4.2	1500 1.5	2222 4.2	26 TH	0208 1.1	0909 4.5	1435 1.1	2123 4.6
12 TH	0320 1.2	1051 4.4	1539 1.3	2249 4.4	27 F	0258 0.8	0959 4.9	1523 0.8	2210 4.9
13 F	0358 1.0	1118 4.5	1616 1.1	2319 4.5	28 SA ●	0345 0.5	1044 5.1	1608 0.5	2255 5.1
14 SA ○	0435 0.9	1146 4.6	1652 1.0	2350 4.6	29 SU	0430 0.3	1128 5.2	1653 0.4	2341 5.1
15 SU	0511 0.8	1216 4.7	1727 1.0		30 M	0514 0.3	1213 5.2	1736 0.4	

OCTOBER

Day	Time m	Time m	Time m	Time m	Day	Time m	Time m	Time m	Time m
1 TU	0028 5.1	0558 0.4	1258 5.2	1819 0.5	16 W	0024 4.6	0543 0.9	1242 4.7	1759 1.0
2 W	0116 5.0	0641 0.6	1343 5.0	1901 0.8	17 TH	0056 4.6	0614 1.0	1312 4.6	1830 1.0
3 TH	0204 4.8	0723 0.9	1428 4.8	1943 1.1	18 F	0129 4.5	0646 1.1	1343 4.5	1905 1.1
4 F	0253 4.6	0805 1.2	1516 4.5	2027 1.4	19 SA	0206 4.4	0723 1.2	1419 4.4	1945 1.3
5 SA	0346 4.3	0854 1.6	1608 4.2	2125 1.7	20 SU ●	0253 4.2	0808 1.5	1513 4.2	2035 1.5
6 SU ◑	0442 4.0	1013 1.9	1705 3.9	2303 1.9	21 M	0403 4.1	0909 1.8	1630 4.0	2156 1.7
7 M	0545 3.8	1144 2.1	1811 3.7		22 TU ◐	0516 4.0	1056 1.9	1744 4.0	2338 1.7
8 TU	0019 2.0	0703 3.7	1251 2.0	1936 3.7	23 W	0630 4.1	1218 1.8	1857 4.1	
9 W	0118 1.8	0901 3.9	1346 1.9	2119 3.9	24 TH	0047 1.5	0745 4.3	1321 1.5	2008 4.3
10 TH	0208 1.6	0949 4.2	1432 1.6	2157 4.2	25 F	0145 1.2	0850 4.6	1414 1.1	2106 4.6
11 F	0251 1.3	1021 4.4	1512 1.3	2225 4.4	26 SA	0236 0.9	0940 4.9	1502 0.8	2155 4.9
12 SA	0330 1.1	1047 4.6	1550 1.1	2253 4.5	27 SU	0323 0.6	1025 5.1	1548 0.6	2239 5.0
13 SU ○	0406 1.0	1115 4.7	1625 1.0	2323 4.6	28 M ●	0408 0.5	1107 5.2	1631 0.5	2324 5.1
14 M	0441 0.9	1144 4.7	1658 0.9	2354 4.6	29 TU	0451 0.5	1149 5.2	1714 0.6	
15 TU	0513 0.9	1213 4.7	1730 0.9		30 W	0010 5.0	0533 0.6	1233 5.1	1755 0.7
					31 TH	0057 4.9	0615 0.8	1317 4.9	1835 0.9

NOVEMBER

Day	Time m	Time m	Time m	Time m	Day	Time m	Time m	Time m	Time m
1 F	0144 4.8	0655 1.1	1401 4.7	1915 1.1	16 SA	0115 4.6	0628 1.1	1325 4.6	1850 1.1
2 SA	0232 4.6	0736 1.4	1446 4.4	1958 1.4	17 SU	0159 4.5	0711 1.3	1409 4.5	1936 1.2
3 SU	0322 4.3	0822 1.7	1536 4.2	2048 1.7	18 M	0252 4.4	0801 1.5	1507 4.3	2031 1.4
4 M ◑	0416 4.1	0924 2.0	1631 3.9	2210 1.9	19 TU ◐	0357 4.3	0907 1.7	1618 4.2	2149 1.6
5 TU	0514 3.9	1100 2.1	1733 3.8	2336 2.0	20 W	0504 4.2	1036 1.8	1727 4.1	2312 1.6
6 W	0621 3.9	1212 2.1	1847 3.7		21 TH	0611 4.3	1151 1.7	1836 4.2	
7 TH	0038 1.9	0740 3.9	1312 2.0	2009 3.8	22 F	0019 1.5	0720 4.4	1254 1.5	1946 4.4
8 F	0133 1.7	0851 4.1	1402 1.7	2108 4.1	23 SA	0119 1.3	0827 4.7	1351 1.3	2049 4.6
9 SA	0219 1.5	0934 4.4	1445 1.5	2147 4.3	24 SU	0213 1.1	0921 4.9	1442 1.0	2140 4.8
10 SU	0300 1.4	1008 4.6	1523 1.3	2221 4.5	25 M	0303 0.9	1005 5.0	1529 0.9	2226 4.9
11 M	0336 1.2	1040 4.7	1556 1.2	2253 4.6	26 TU ●	0349 0.9	1047 5.1	1613 0.8	2310 4.9
12 TU ○	0409 1.1	1110 4.7	1628 1.1	2325 4.6	27 W	0432 0.9	1128 5.0	1655 0.8	2356 4.9
13 W	0441 1.1	1141 4.8	1700 1.0		28 TH	0515 1.0	1210 4.9	1737 0.9	
14 TH	0000 4.6	0513 1.1	1214 4.7	1733 1.0	29 F	0042 4.8	0556 1.1	1254 4.8	1817 1.0
15 F	0036 4.6	0549 1.1	1248 4.7	1810 1.0	30 SA	0128 4.7	0637 1.3	1337 4.6	1858 1.2

DECEMBER

Day	Time m	Time m	Time m	Time m	Day	Time m	Time m	Time m	Time m
1 SU	0214 4.6	0718 1.5	1420 4.4	1940 1.4	16 M	0156 4.7	0707 1.2	1404 4.6	1934 1.1
2 M	0300 4.4	0802 1.7	1506 4.2	2028 1.6	17 TU	0249 4.6	0800 1.3	1501 4.5	2030 1.2
3 TU	0348 4.2	0854 1.9	1557 4.1	2128 1.8	18 W	0348 4.5	0902 1.4	1604 4.4	2136 1.3
4 W ◑	0439 4.1	1005 2.1	1652 3.9	2244 2.0	19 TH ◐	0448 4.5	1013 1.6	1707 4.3	2246 1.4
5 TH	0536 4.0	1125 2.1	1754 3.8	2353 2.0	20 F	0548 4.4	1122 1.6	1812 4.3	2351 1.5
6 F	0640 4.0	1232 2.1	1907 3.8		21 SA	0653 4.4	1227 1.5	1921 4.3	
7 SA	0051 1.9	0748 4.1	1330 1.9	2014 4.0	22 SU	0054 1.4	0759 4.6	1330 1.4	2030 4.4
8 SU	0144 1.8	0844 4.3	1417 1.7	2105 4.2	23 M	0154 1.4	0859 4.7	1426 1.3	2128 4.5
9 M	0226 1.7	0927 4.5	1454 1.6	2146 4.4	24 TU	0248 1.3	0949 4.9	1516 1.1	2218 4.6
10 TU	0302 1.5	1004 4.6	1526 1.4	2223 4.5	25 W	0336 1.2	1031 4.8	1601 1.0	2304 4.5
11 W	0336 1.4	1038 4.7	1559 1.3	2300 4.6	26 TH ●	0420 1.2	1112 4.8	1644 1.0	2349 4.7
12 TH ○	0412 1.3	1113 4.7	1635 1.1	2340 4.6	27 F	0502 1.2	1154 4.8	1725 1.0	
13 F	0452 1.2	1151 4.8	1715 1.0		28 SA	0032 4.7	0543 1.2	1236 4.7	1806 1.1
14 SA	0022 4.7	0533 1.2	1232 4.8	1758 1.0	29 SU	0114 4.6	0623 1.2	1317 4.6	1846 1.2
15 SU	0107 4.7	0619 1.1	1316 4.7	1844 1.0	30 M	0155 4.6	0703 1.4	1358 4.5	1926 1.4
					31 TU	0235 4.5	0743 1.5	1438 4.4	2007 1.4

Chart Datum: 2·74 metres below Ordnance Datum (Newlyn). HAT is 5·3 metres above Chart Datum.

〉〉 FREE monthly updates. Register at 〈
www.reedsnauticalalmanac.co.uk 〈

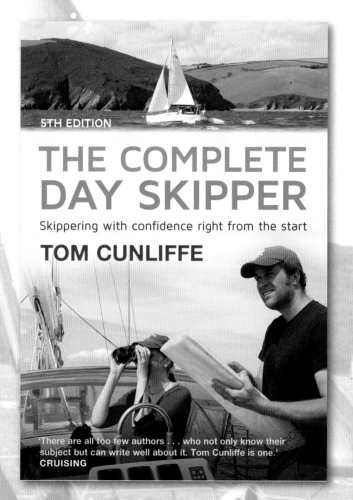
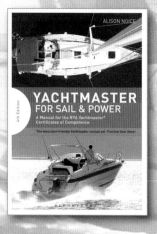
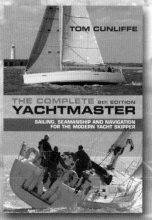
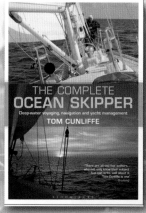
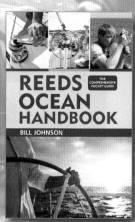

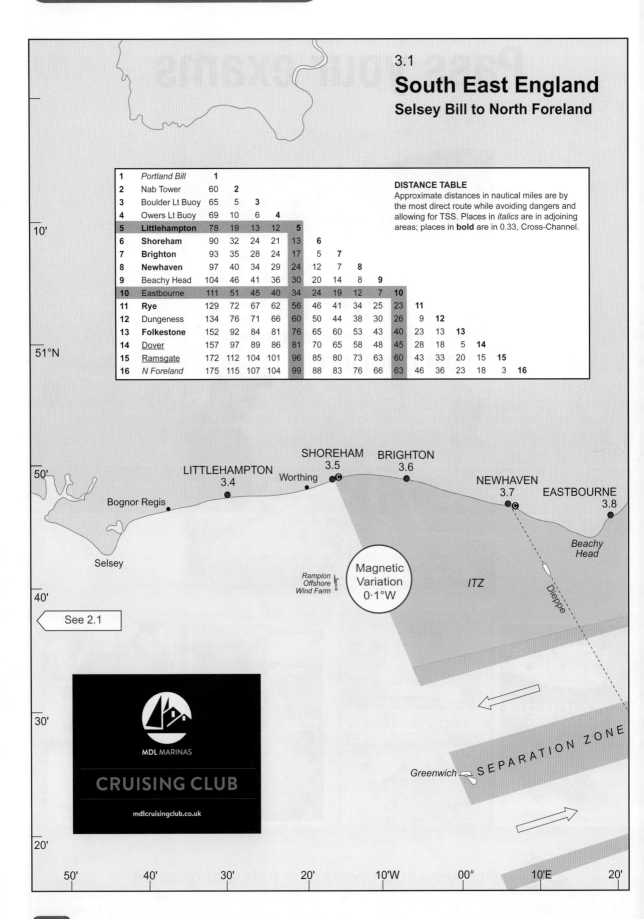

3.1
South East England
Selsey Bill to North Foreland

		1	2	3	4	5	6	7	8	9	10	11	12	13	14	15	16
1	*Portland Bill*	**1**															
2	Nab Tower	60	**2**														
3	Boulder Lt Buoy	65	5	**3**													
4	Owers Lt Buoy	69	10	6	**4**												
5	**Littlehampton**	78	19	13	12	**5**											
6	**Shoreham**	90	32	24	21	13	**6**										
7	**Brighton**	93	35	28	24	17	5	**7**									
8	**Newhaven**	97	40	34	29	24	12	7	**8**								
9	Beachy Head	104	46	41	36	30	20	14	8	**9**							
10	Eastbourne	111	51	45	40	34	24	19	12	7	**10**						
11	**Rye**	129	72	67	62	56	46	41	34	25	23	**11**					
12	Dungeness	134	76	71	66	60	50	44	38	30	26	9	**12**				
13	**Folkestone**	152	92	84	81	76	65	60	53	43	40	23	13	**13**			
14	*Dover*	157	97	89	86	81	70	65	58	48	45	28	18	5	**14**		
15	*Ramsgate*	172	112	104	101	96	85	80	73	63	60	43	33	20	15	**15**	
16	*N Foreland*	175	115	107	104	99	88	83	76	66	63	46	36	23	18	3	**16**

DISTANCE TABLE
Approximate distances in nautical miles are by the most direct route while avoiding dangers and allowing for TSS. Places in *italics* are in adjoining areas; places in **bold** are in 0.33, Cross-Channel.

SHOREHAM
3.5

BRIGHTON
3.6

LITTLEHAMPTON
3.4

Worthing

NEWHAVEN
3.7

EASTBOURNE
3.8

Bognor Regis

Selsey

Beachy
Head

Rampion
Offshore
Wind Farm

Magnetic
Variation
0·1°W

ITZ

Dieppe

See 2.1

MDL MARINAS
CRUISING CLUB
mdlcruisingclub.co.uk

Greenwich

SEPARATION ZONE

50' 40' 30' 20' 10'W 00° 10'E 20'

10'

51°N

50'

40'

30'

20'

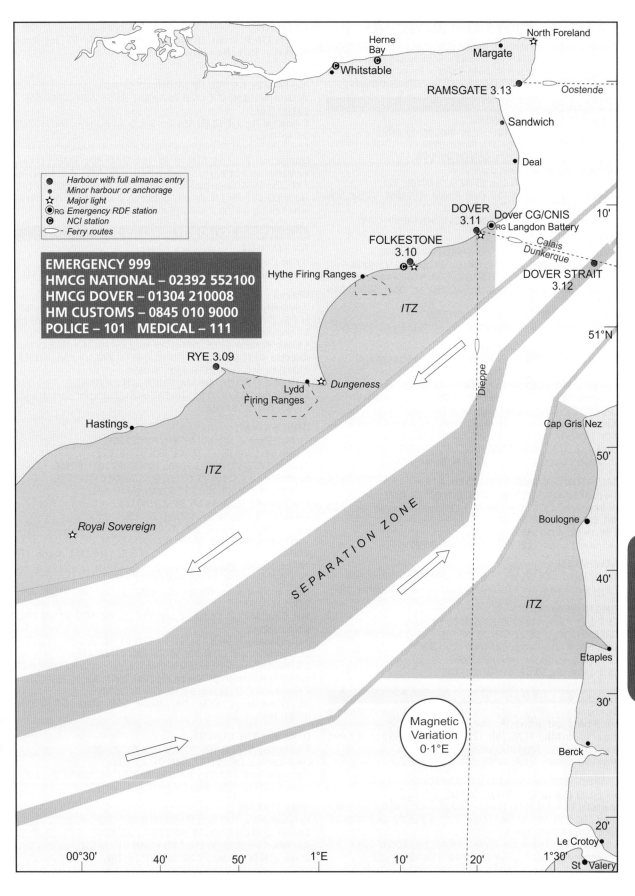

North Foreland

Herne
Bay

Margate

Whitstable

RAMSGATE 3.13

Oostende

Sandwich

Deal

● Harbour with full almanac entry
● Minor harbour or anchorage
☆ Major light
◉RG Emergency RDF station
Ⓒ NCI station
⊂⊐ Ferry routes

DOVER
3.11

Dover CG/CNIS

RG Langdon Battery

10'

FOLKESTONE
3.10

Calais
Dunkerque

DOVER STRAIT
3.12

EMERGENCY 999
HMCG NATIONAL – 02392 552100
HMCG DOVER – 01304 210008
HM CUSTOMS – 0845 010 9000
POLICE – 101   MEDICAL – 111

Hythe Firing Ranges

ITZ

51°N

RYE 3.09

Lydd
Firing Ranges

Dungeness

Dieppe

Cap Gris Nez

Hastings

50'

ITZ

S E P A R A T I O N   Z O N E

Royal Sovereign

Boulogne

40'

ITZ

Etaples

30'

Magnetic
Variation
0·1°E

Berck

20'

Le Crotoy

00°30'   40'   50'   1°E   10'   20'   1°30'

St Valery

SE England

3.2 LIGHTS, BUOYS AND WAYPOINTS

Bold print = light with a nominal range of 15M or more. CAPITALS = place or feature. *CAPITAL ITALICS* = light-vessel, light float or Lanby. *Italics* = Fog signal. ***Bold italics*** = Racon. Some marks/buoys are fitted with AIS (<u>MMSI No</u>); see relevant charts.

OWERS TO BEACHY HEAD

SELSEY BILL and THE OWERS
S Pullar ⚓ VQ (6) + L Fl 10s; 50°38'·84N 00°49'·29W.
Pullar ⚓ Q 15s; 50°40'·47N 00°50'·09W.
Boulder ▲ Fl G 2·5s; 50°41'·56N 00°49'·09W.
Street ⚓ QR; 50°41'·69N 00°48'·89W.
Mixon Bn ⚓ Q(6) + L Fl 15s; 50°42'·37N 00°46'·32E.
Owers ⚓ Q (6) + L Fl 15s; *Bell*; ***Racon (O) 10M***; 50°38'·59N 00°41'·09W.
E Borough Hd ⚓ Q (3) 10s *Bell*; 50°41'·54N 00°39'·09W.

LITTLEHAMPTON
West Pier Hd ⚓ QR 7m 6M; 50°47'·88N 00°32'·46W.
Training Wall Hd ⚓ QG 10m 2M; 50°47'·87N 00°32'·38W.
Ldg lts 346°. Front, E Pier Hd ⚓ FG 6m 7M; B col. Rear, 64m from front, Oc W 7·5s 9m10M; W twr; vis: 290°-W- 042°; 50°48'·09N 00°32'·51W.
Outfall ⚓ Fl Y 5s; 50°46'·27N 00°30'·53W.
Littlehampton ⚓ Fl (5) Y 20s 9m 5M; 50°46'·19N 00°29'·54W.

WORTHING and SHOREHAM
Met Mast ⚓ Fl Y 2·5s 12m 2M; 50°41'·29N 00°20'·59W.
Outfall ⚓ Fl R 2·5s 3m; 50°48'·38N 00°20'·34W.
Express ⚓ Fl Y 5s; (Apr-Oct); 50°47'·28N 00°17'·09W.
W Bkwtr Head ⚓ Fl R 5s 7m 7M; 50°49'·49N 00°14'·89W.
Ldg lts 355°. Middle Pier Front, Oc 5s 8m 10M; W watch-house, R base; tidal Lts, tfc sigs; *Horn 20s*. Rear, 192m from front, Fl 10s 12m 10M; Gy twr vis: 283°-103°; 50°49'·85N 00°14'·89W.
Outfall ⚓ 50°49'·47N 00°14'·39W.
Shoreham Outfall ⚓ Q (6) + L Fl 15s; 50°47'·88N 00°13'·72W.

BRIGHTON and BRIGHTON MARINA
Black Rk Ledge ⚓ Fl Y 4s; 50°48'·07N 00°06'·46W.
W Bkwtr Hd ⚓ QR 10m 7M; W ○ structure, R bands; *Horn (2) 30s*; 50°48'·50N 00°06'·38W.
E Bkwtr Hd ⚓ QG 8m 7M.
Saltdean Outfall ⚓ Fl Y 5s; 50°46'·72N 00°02'·13W.

NEWHAVEN
Bkwtr Head ⚓ Oc (2) 10s 17m 12M; 50°46'·56N 00°03'·50E.
E Pier Hd ⚓ Iso G 10s 12m 6M; W twr; 50°46'·81N 00°03'·59E.

OFFSHORE MARKS
CS 1 ⚓ Fl Y 2·5s; *Whis*; 50°33'·69N 00°03'·92W.
GREENWICH ⚓ 50°24'·54N 00°00'·10E; Fl 5s 12m **15M**; Riding light FW; R hull; ***Racon (M) 10M***; *Horn 30s*.
CS 2 ⚓ Fl Y 5s; 50°39'·14N 00°32'·60E.
CS 3 ⚓ Fl Y 10s; 50°52'·04N 01°02'·18E.

BEACHY HEAD TO DUNGENESS

Beachy Head ⚓ 50°44'·03N 00°14'·49E; Fl (2) 20s 31m 8M; W round twr, R band and lantern; vis: 248°-101°; (H24).
Royal Sovereign ⚓ Fl 20s 28m 12M; W ○ twr, R band on W cabin on col; *Horn (2) 30s*; 50°43'·45N 00°26'·09E.
Royal Sovereign ⚓ QR; 50°44'·23N 00°25'·84E.

EASTBOURNE and SOVEREIGN HARBOUR
SH ⚓ L Fl 10s; 50°47'·40N 00°20'·71E.
Martello Tower ⚓ Fl (3) 15s 12m 7M.; 50°47'·24N 00°19'·83E.
Dir lt 260·5° Fl WRG 5s 4m 1M; vis: 255°-G-259°-W-262°-R-265°; 50°47'·28N 00°19'·71E.
S Bkwtr Hd ⚓ Fl (4) R 12s 3m 6M; 50°47'·30N 00°20'·03E.
St Leonard's Outfall ⚓ Fl Y 5s; 50°49'·31N 00°31'·95E.

HASTINGS

Ldg Lts 356·3°. Front, FR 14m 4M; 50°51'·29N 00°35'·38E. Rear, West Hill, 357m from front, FR 55m 4M; W twr.

RYE

Rye Fairway, L Fl 10s; 50°54'·04N 00°48'·04E.
W Groyne Hd No 2 ⚓ LFl R 7s 7m 6M; 50°55'·58N 00°46'·55E; <u>992351156</u>.
E Arm Hd No 1 ⚓ Q (9) 15s 7m 5M; G △; *Horn 7s*; 50°55'·73N 00°46'·46E.
Dungeness Outfall ⚓ Q (6) + L Fl 15s; 50°54'·45N 00°58'·21E.
Dungeness ☆ 50°54'·81N 00°58'·56E; Fl 10s 40m **21M**; B ○ twr, W bands and lantern, floodlit; Part obsc 078°-shore; (H24).
F RG 37m 10M (same twr); vis: 057°-R-073°-G-078°-196°-R-216°; *Horn (3) 60s*; FR Lts shown between 2·4M and 5·2M WNW when firing taking place. QR on radio mast 1·2M NW.

DUNGENESS TO NORTH FORELAND

FOLKESTONE
Hythe Flats Outfall ⚓ Fl Y 5s; 51°02'·52N 01°05'·32E.
Breakwater Head ☆ Fl (2) 10s 14m **22M**; 51°04'·56N 01°11'·69E.

DOVER
Admiralty Pier Extension Head ☆ 51°06'·69N 01°19'·66E; Fl 7·5s 21m **20M**; W twr; vis: 096°-090°, obsc in The Downs by S Foreland inshore of 226°; *Horn 10s*; Int Port Tfc sigs.
S Bkwtr W Hd ☆ 51°06'·78N 01°19'·80E;Oc R 30s 21m **18M**; W twr.
Knuckle ☆ 51°07'·04N 01°20'·49E; Fl (4) WR 10s 15m **W15M**, R13M; W twr; vis: 059°-R-239°-W-059°.
N Head ⚓ Fl R 2·5s 11m 5M; 51°07'·20N 01°20'·61E.
Eastern Arm Hd ⚓ Fl G 5s 12m 5M; *Horn (2) 30s*; Int port tfc sigs; 51°07'·31N 01°20'·59E.

DOVER STRAIT
Bullock Bank ⚓ VQ; 50°46'·94N 01°07'·60E.
Ridens SE ⚓ VQ (3) 5s; 50°43'·47N 01°18'·87E.
Colbart SW ⚓ VQ (6) + L Fl 10s; 50°48'·86N 01°16'·30E.
South Varne ⚓ Q (6) + L Fl 15s; 50°55'·64N 01°17'·30E.
Mid Varne ⚓ VQ(9)10s; 50°58'·94N 01°19'·88E.
East Varne ⚓ VQ(3)5s; 50°58'·22N 01°20'·90E.
Colbart N ⚓ VQ; 50°57'·45N 01°23'·29E.
Varne NW ⚓ Q; 51°00'·80N 01°22'·70E.
Varne NE ⚓ Q (3) 10s; 50°59'·80N 01°22'·70E.
VARNE ⚓ 51°01'·29N 01°23'·90E; Fl R 5s 12m **15M**; ***Racon (T) 10M***; *Horn 30s*; <u>992351038</u>.
CS 4 ⚓ Fl (4) Y 15s; 51°08'·62N 01°33'·92E.
MPC ⚓ Fl Y 2·5s; ***Racon (O) 10M***; 51°06'·12N 01°38'·20E; <u>992351122</u>.
SW Goodwin ⚓ Q (6) + LFl 15s; 51°08'·50N 01°28'·88E; <u>992351036</u>.
S Goodwin ⚓ Fl (4) R 15s; 51°10'·60N 01°32'·26E.
SE Goodwin ⚓ Fl (3) R 10s; 51°12'·99N 01°34'·45E.
E GOODWIN ⚓ 51°13'·26N 01°36'·37E; Fl 15s 12m **23M**; R hull with lt twr amidships; ***Racon (T) 10M***; *Horn 30s*; <u>992351035</u>.
E Goodwin ⚓ Q (3) 10s; 51°15'·67N 01°35'·69E.
NE Goodwin ⚓ Q(3) 10s; ***Racon (M) 10M***. 51°20'·31N 01°34'·16E.

DEAL and THE DOWNS
Trinity Bay VQ (9) 10s; 51°11'·60N 01°29'·00E.
Deal Bank ⚓ QR; 51°12'·92N 01°25'·57E.
Goodwin Fork ⚓ Q (6) + L Fl 15s; *Bell*; 51°14'·38N 01°26'·70E.
Downs ⚓ Fl (2) R 5s; *Bell*; 51°14'·50N 01°26'·22E.

GULL STREAM
W Goodwin ▲ Fl G 5s; 51°15'·61N 01°27'·38E.
S Brake ⚓ Fl (3) R 10s; 51°15'·77N 01°26'·82E.
NW Goodwin ⚓ Q (9) 15s; 51°16'·74N 01°28'·55E.
Brake ⚓ Fl (4) R 15s; 51°16'·98N 01°28'·19E.

N Goodwin ▲ Fl G 2·5s; 51°18'·12N 01°30'·35E.
Gull Stream ⚓ QR; 51°18'·26N 01°29'·69E.
Gull ⚐ VQ (3) 5s; 51°19'·57N 01°31'·30E.
Goodwin Knoll ▲ Fl (2) G 5s; 51°19'·57N 01°32'·20E.

RAMSGATE CHANNEL
B2 ▲ Fl (2) G 5s; 51°18'·26N 01°23'·93E.
W Quern ⚐ Q (9) 15s; 51°18'·98N 01°25'·39E.

RAMSGATE
RA ⚐ Q(6) + L Fl 15s; 51°19'·60N 01°30'·13E.
E Brake ⚓ Fl R 5s; 51°19'·47N 01°29'·20E.
No 1 ⚐ QG; 51°19'·56N 01°27'·29E.
No 2 ⚐ Fl (4) R 10s; 51°19'·46N 01°27'·28E.
No 3 ⚐ Fl G 2·5s; 51°19'·56N 01°26'·61E.
No 4 ⚐ QR; 51°19'·46N 01°26'·60E.
N Quern ⚐ Q; 51°19'·41N 01°26'·11E.
No 5 ⚐ Q (6) + L Fl 15s; 51°19'·56N 01°25'·91E.
No 6 ⚓ Fl (2) R 5s; 51°19'·46N 01°25'·91E.
South Bkwtr Hd ⚐ VQ R 10m 5M; 51°19'·46N 01°25'·41E.
N Bkwtr Hd ⚐ QG 10m 5M; 51°19'·56N 01°25'·47E.
Western Marine terminal Dir lt 270° ⚐, Oc WRG 10s 10m 5M;

B △, Or stripe; vis: 259°-G-269°-W- 271°-R-281°; 51°19'·51N 01°24'·85E. Rear 493m from front Oc 5s 17m 5M; B ▽, Or stripe; vis: 263°-278°.

BROADSTAIRS and NORTH FORELAND
Broadstairs Knoll ⚓ Fl R 2·5s; 51°20'·88N 01°29'·48E.
Pier SE End ⚐ 2 FR (vert) 7m 4M; 51°21'·50N 01°26'·74E.
Elbow ⚐ Q; 51°23'·23N 01°31'·59E.

North Foreland ☆ 51°22'·49N 01°26'·70E; Fl (5) WR 20s 57m **W19M, R16M, R15M**; W 8-sided twr; vis: shore-W-150°-R(**16M**)-181°-R(**15M**)-200°-W-011°; H24; 992351020.

OFFSHORE MARKS
SANDETTIE ⚓ 51°09'·36N 01°47'·12E; Fl 5s 12m **15M**; R hull with lt twr amidships; *Racon; Horn 30s;* 992351029.

F1 ⚐ Fl (4) Y 15s; 51°11'·21N 01°44'·91E.
South Falls ⚐ Q (6) + L Fl 15s; 51°13'·84N 01°43'·93E.
Sandettie W ▲ Fl (3) G 12s; 51°15'·09N 01°54'·47E.
Mid Falls ⚐ Fl (3) R 10s; 51°18'·63N 01°46'·99E.
Inter Bank ⚐ Fl Y 5s; *Bell; Racon (M) 10M;* 51°16'·47N 01°52'·23E.
F2 ⚐ Fl (4) Y 15s; 51°20'·41N 01°56'·19E.

3.3 PASSAGE INFORMATION
More passage information is threaded between the harbours in this area. See 9.0.5 for distances across the eastern part of the English Channel. Admiralty Leisure Folio 5605 covers Chichester to Ramsgate including the Dover Strait. **Bibliography:** *Shell Channel Pilot* (Imray/Cunliffe); *Channel Pilot and Dover Strait Pilot* (Admiralty NP27 & 28).

THE EASTERN CHANNEL
One of the greatest densities of commercial shipping occurs in this area. In such waters the greatest danger is a collision with a larger vessel, especially in poor visibility. Even when coastal cruising it is essential to know the limits of the TSS and ITZ. For example, the SW-bound TSS lane from the Dover Strait passes only 3.5M off Dungeness. Radar surveillance of the Dover Strait is maintained at all times by the Channel Navigation Information Service (CNIS). In addition to the many large ships using the traffic lanes, ferries cross between English and continental ports; fishing vessels operate both inshore and offshore; in the summer months many yachts may be encountered; and static dangers such as lobster pots and fishing nets are concentrated in certain places.

▶ *Although the rates of tidal streams vary with locality, they are greatest in the narrower parts of the Channel and off major headlands. In the Dover Strait spring rates can reach 4kn, but elsewhere in open water they seldom exceed 2kn.* ◀

N'ly winds, which may give smooth water and pleasant sailing off the English coast, can cause rough seas on the French coast. In strong S'lies the English coast between the Isle of Wight and Dover is very exposed and shelter is hard to find. The Dover Strait has a funnelling effect, which can increase the wind speed and cause very rough seas.

ON PASSAGE UP CHANNEL
▶ *There are three tidal gates; the Looe, Beachy Head and Dungeness. These form a tidal sequence which can enable a reasonably fast yacht to carry a fair tide from Selsey Bill to Dover. Based on a mean speed over the ground of 7kn, transit the Looe at about slack water, HW Dover +0500 (HW Portsmouth +0430); Beachy Head will be passed at HW D –0100; Dungeness at HW D +0300 and Dover at HW +0530, only stemming the first of the ebb in the last hour. The down-Channel passage is less efficient and a stop at Sovereign Harbour or Brighton may be considered.* ◀

SELSEY BILL AND THE OWERS
(AC 1652) Selsey Bill is a low headland off which lie several drying rocks and shoals. Just W and SW of Selsey, The Streets extend 1·25M seaward. 1·25M SSW of the Bill are The Grounds (or Malt Owers) and The Dries. Inshore 1M East of The Dries 1·25M South of the lifeboat house on E side of Selsey Bill is The Mixon, a group of rocks with a red lit bn at the E end. The Owers extend 3M to the S and 5M to the SE of the Bill. Looe Chan is immediately S of these dangers marked by buoys at the W end, where it is narrowest, between Brake (or Cross) Ledge to the N and Boulder Bk to the South. It is navigable in good visibility and suitable weather. Beware of lobster pots.

▶ *The E-going stream begins at HW Portsmouth +0430, and the W-going at HW Portsmouth –0135, spring rate 2·5kn.* ◀

In poor visibility, bad weather or at night keep S of the Owers SCM lt buoy, 7M SE of Selsey Bill, marking SE end of the Outer Owers.

▶ *Much of the Outer Owers is less than 3m, and large parts virtually dry. A combination of tidal streams and strong winds produces heavy breaking seas and overfalls over a large area.* ◀

THE OWERS TO BEACHY HEAD
(AC 1652) The coast from Selsey Bill to Brighton is low, faced by a shingle beach and with few offlying dangers. Bognor Rks (dry in places) extend 1·75M ESE from a point 1M W of the pier, and Bognor Spit extends another 0·5M further E and S. Middleton ledge are rocks running 8ca offshore, about 2M E of Bognor pier, with depths of less than 1m. Shelley Rks lie 5ca S of Middleton ledge, with depths of less than 1m.

Winter Knoll, 2·5M SSW of Littlehampton, has depths of 2·1m. Kingston Rks, depth 2m, lie about 3·25M ESE of Littlehampton. An unlit outfall bn is 3ca off Goring-by-Sea (2M W of Worthing). Grass Banks, an extensive shoal with a wreck swept to 1·2m at W end, lie 1M S of Worthing pier with Elbow shoal, depth 2·5m, at their E extremity. A met mast 103m high (Mo(U)W 15s, *Horn Mo(U)*) is 7M S of Worthing together with Rampion Wind Farm. Early decision to pass N or S is recommended.

Off Shoreham, Church Rks, with depth of 0·3m, lie 1·5M W of the harbour entrance and 2½ca offshore. Jenny Rks, with depth 0·9m, are 1·25M E of the entrance, 3ca offshore. A diffuser is marked by a SCM and a wreck swept to 3.3m by a Y SPM.

High chalk cliffs are conspic from Brighton to Beachy Head. With the exception of outfalls, the only danger more than 3ca offshore is a Wk (10m). From Birling Gap to Head Ledge a shelf extends 4ca S of the cliff, and Beachy Hd lt ho (FL(2) 20s 31m 8M) is built between these. Beware of fishing gear markers.

3.4 LITTLEHAMPTON

W. Sussex 50°47′·87N 00°32′·43W ⚓🌊🌊🏴🏴

CHARTS AC 5605, 1652, 1991; Imray C12, C9, 2100

TIDES +0015 Dover; ML 2·8

Standard Port SHOREHAM (→)

Times				Height (metres)			
High Water		Low Water		MHWS	MHWN	MLWN	MLWS
0500	1000	0000	0600	6·3	4·8	1·9	0·6
1700	2200	1200	1800				
Differences LITTLEHAMPTON (ENT)							
+0010	0000	−0005	−0010	−0·4	−0·4	−0·2	−0·2
ARUNDEL							
ND	+0120	ND	ND	−3·1	−2·8	ND	ND
PAGHAM							
+0015	0000	−0015	−0025	−0·7	−0·5	−0·1	−0·1
BOGNOR REGIS							
+0010	−0005	−0005	−0020	−0·6	−0·5	−0·2	−0·1

NOTE: Tidal hts in hbr are affected by flow down R Arun. Tide seldom falls lower than 0·9m above CD.

SHELTER Good, but ent is hazardous in strong SE winds which cause swell up the hbr. The bar is rough in SW'lies. Visitors can berth at Town Quay, where marked, in front of HM's office.

NAVIGATION WPT 50°47′·53N 00°32′·30W, 346°/0·6M to front ldg lt. Bar 5ca offshore.

- Approach and entrance dry up to 0·9m.
- Shoal bank on West side of channel South of East Pier.
- The ebb runs fast (3–5kn) at springs.
- From HW−1 to HW+4 a strong W-going tidal stream sets across the ent; keep to E side. Speed limit 6½kn from 15m S of W Pier – Old Town Bridge in Arundel. NO WASH passing moored craft.

On E side of ent chan a training wall which covers at half-tide is marked by 7 poles and lit bn at S end. The W pier is a long, prominent structure of wood piles; beware shoal ground within its arm. A tide gauge on end shows height of tide above CD. To obtain depth on the bar subtract 0·8m from indicated depth.

River Arun. ⚓ Trawling is prohibited where submarine pipelines run across the channel from a position close N of the Lt ho to the first yacht berths, and SE of a retractable footbridge (3·6m clearance MHWS; 9·4m above CD). Sited 3ca above Town Quay, this gives access for masted craft to Littlehampton Marina. It is opened by request to HM before 1630 previous day. The River Arun is navigable on the tide by small, unmasted craft for 24M.

LIGHTS AND MARKS High-rise bldg (38m) is conspic 0·4M NNE of hbr ent. A pile with small platform and ☆, Fl Y (5) 20s 5M, is 2·5M SE of hbr ent at 50°46′·1N 00°29′·5W.

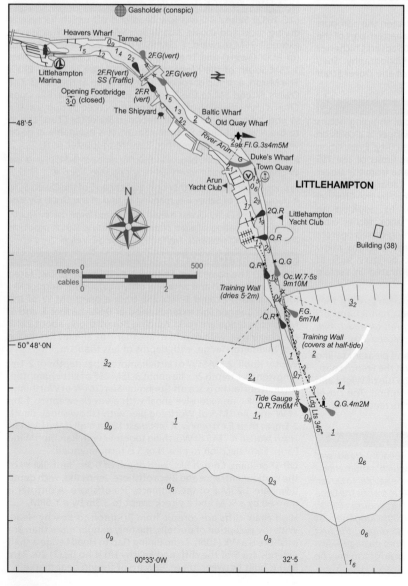

Ldg lts 346°: Front FG on B column; Rear, lt ho Oc W 7·5s at root of E breakwater, 290°-W-042°. Craft approaching from the W should keep to seaward of the 2m contour until W Pier is positively identified. The Fl G 3s lt at Norfolk Wharf leads craft upstream, once inside hbr ent. The outer ends of some Yacht Club mooring pontoons on the W bank are lit QR.

When Pilot boat with P1 at the bow displays the Pilot flag 'H' or ○ over ● lts, all boats keep clear of entrance; large ship moving.

Footbridge sigs, from high mast to port:
Fl ● = open
Fl ● = bridge moving
○ = closed

Bridge's retractable centre section (22m wide) has 2 FR (vert) to port and 2 FG (vert) to stbd at both upstream and downstream ends.

COMMUNICATIONS (Code 01903) Dr 714113. HM www.littlehampton.org.uk 721215.
VHF Ch 71 16 (0900-1700LT); Pilots Ch 71 16 when vessel due. Bridge Ch 71. Marinas Ch 80 M (HO).

FACILITIES Town Quay pontoon connected to shore £20 <8m>£23<11m>£32, ⚓ ⊕ ▣ 🚻 ▶ in harbour office. **Services** ▬ (pay hbr dues).

Littlehampton Yacht Club ☎713990, ⚓⚓▭ www.littlehamptonyachtclub.co.uk.

The Shipyard: www.shipyardlittlehampton. co.uk ☎713327, ▬ 6◯⚓ £1·00, ⚓ ⊕ ⚒(Wood) ♺ ▣(50t).

Arun YC ☎716016, ⚓ ▬ 92◯+10⦿ £14·00/yacht (dries), & ⚓ ⊕ × ▭.

Marina www.littlehamptonmarina.co.uk ☎713553, ▬ 120◯ £2·00, & ⚓ ⚓ ⚒ ⚒ ▣(12t) 🛢 × ▭.

Ship and Anchor Marina ☎(01243) 551262, 2M upriver at Ford, ▬ 50◯+⦿⚓⚒⚒⚓🛢 × ▭.

Town 🏪🏪 ACA ✉ Ⓑ 🛢 × ▭ ⇌ ✈ (Gatwick).

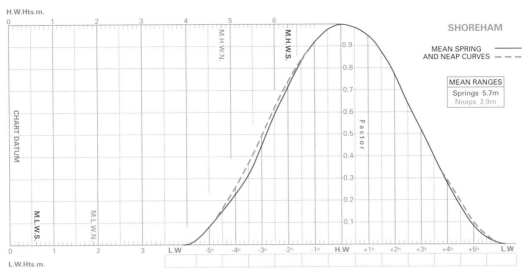

SHOREHAM

MEAN SPRING ———
AND NEAP CURVES − − −

MEAN RANGES
Springs 5.7m
Neaps 2.9m

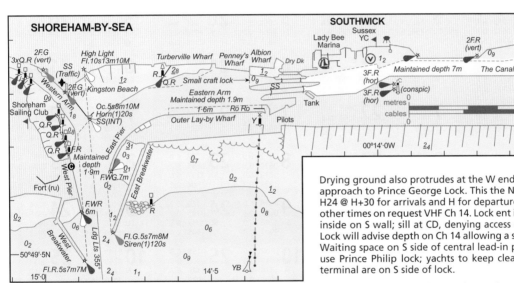

Drying ground also protrudes at the W end of Albion Wharf, the approach to Prince George Lock. This the N'ly of two locks opens H24 @ H+30 for arrivals and H for departures; also, if not busy, at other times on request VHF Ch 14. Lock ent is 5·5m wide, pontoon inside on S wall; sill at CD, denying access only at about LWS±1. Lock will advise depth on Ch 14 allowing a safety margin of 0·2m. Waiting space on S side of central lead-in pier. Commercial ships use Prince Philip lock; yachts to keep clear. A Ro-Ro ramp and terminal are on S side of lock.

LIGHTS AND MARKS Radio mast is conspic 170m N of High lt. SCM lit buoy, marks outfall diffusers, 157°/1·8M from hbr ent. Chimney 100m high, silver with blk top. 3F.R(hor) at top and 3F.R(hor) mid height. SYC yacht race signal mast on Shoreham Beach 00°16'·06W.

Ldg lts 355°: front Oc 5s 8m 10M; rear High lt, Fl 10s 13m 10M.

Traffic Sigs IPTS (Sigs 2 and 5, Oc) are shown from Middle Pier. Note: ⬤ Fl lt exempts small craft.
Oc R 3s (from LB ho, directed at E or W Arms) = No exit.

Lock Sigs (Comply strictly to avoid turbulence):
3 ⬤ (vert) = do not approach lock.
⬤ ○ ⬤ (vert) = clear to approach lock.

COMMUNICATIONS (Code 01273) HM 598100; Locks 592366; Dr 461101; Ⓗ (01903) 205111.
HM and lock call *Shoreham Hbr Radio* VHF Ch **14** 16 (H24). Lock advises Marina of arrivals, 0830-1800 Mon-Sat; 1000-1400 Sun.

FACILITIES 🅱 👤 on N side of W Arm HW±3; ACA.

Lady Bee Marina ☎596880 (0900-1700), mob 07802 848915 (best to book), access as lock times, ⚓ 110⌑ +10 Ⓥ £25/craft<12m inc ⟨Đ⟩ 📻 🅿 🔧 🔩 🅱 ⛽ 🔺 ⚒. FV at Nicholson's Marina ☎598100.

Sussex YC ☎464868, welcomes visitors, on a drying ½ tide pontoon in the Western Arm (limited Ⓥ) but with only 1⌑ in the canal, so prior notice advised; also, ♿ ✕ ⬜.

Town ✉ Ⓑ 🛒 🗑 ⛽ ⇌ ✈ (Gatwick).

3.5 SHOREHAM

W. Sussex **50°49'·53N 00°14'·85W** ❀ ✴ ◊ ◊ ✿ ✿

CHARTS AC 5605, 1652, 2044; Imray C12, C9, 2100

TIDES +0009 Dover; ML 3·3; Duration 0605
Standard Port SHOREHAM (→)

Times				Height (metres)			
High Water		Low Water		MHWS	MHWN	MLWN	MLWS
0500	1000	0000	0600	6·3	4·8	1·9	0·6
1700	2200	1200	1800				
Differences WORTHING							
+0010	0000	−0005	−0010	−0·1	−0·2	0·0	0·0

SHELTER Excellent, once through the lock and into The Canal. The shallow water (dredged 1·9m) at the ent can be very rough in strong onshore winds and dangerous in onshore gales. Lady Bee and Aldrington (E end of The Canal) marinas, least depth 2m, both welcome visitors. They are managed by the Hbr Authority. Visitors are advised not to use the drying Western Arm if possible. Hbr speed limit = 4kn.

NAVIGATION WPT 50°49'·23N 00°14'·81W, 355°/0·52M to front ldg lt. From E beware Jenny Rks (0·9m) and, from the W, Church Rks (0·3m). In the Eastern Arm the North shore off Turberville Wharf has drying patches within magenta dashed line - avoid.

STANDARD TIME (UT)
For Summer Time add ONE
hour in **non-shaded areas**

SHOREHAM   LAT 50°50'N   LONG 0°15'W
TIMES AND HEIGHTS OF HIGH AND LOW WATERS

Dates in red are SPRINGS
Dates in blue are NEAPS

YEAR **2019**

JANUARY

Time	m		Time	m
1 0108	1.5	**16** 0611	5.1
0722	5.4	1235	1.9
TU 1342	1.5	W 1853	5.0
1956	5.2		
2 0209	1.5	**17** 0105	1.9
0825	5.2	0725	5.3
W 1439	1.4	TH 1345	1.7
2057	5.4	2000	5.2
3 0304	1.4	**18** 0214	1.6
0919	5.7	0826	5.6
TH 1530	1.2	F 1447	1.3
2149	5.6	2058	5.6
4 0352	1.3	**19** 0313	1.3
1006	5.9	0921	5.9
F 1616	1.1	SA 1541	1.0
2234	5.8	2152	6.0
5 0436	1.1	**20** 0406	1.0
1049	6.0	1012	6.2
SA 1658	1.0	SU 1632	0.7
2315	5.9	2244	6.3
6 0516	1.1	**21** 0455	0.7
1127	6.0	1102	6.5
SU 1736	0.9	M 1721	0.5
● 2352	6.0	○ 2335	6.5
7 0553	1.1	**22** 0544	0.6
1201	6.0	1151	6.6
M 1813	0.9	TU 1810	0.3
8 0026	6.0	**23** 0024	6.6
0627	1.1	0632	0.5
TU 1233	6.0	W 1240	6.6
1847	0.9	1858	0.3
9 0057	6.0	**24** 0113	6.6
0701	1.1	0721	0.6
W 1302	5.9	TH 1328	6.5
1921	1.0	1947	0.4
10 0126	5.9	**25** 0201	6.5
0736	1.2	0810	0.6
TH 1331	5.7	F 1416	6.3
1955	1.1	2036	0.5
11 0154	5.7	**26** 0248	6.4
0810	1.4	0900	0.8
F 1403	5.6	SA 1504	6.1
2029	1.3	2126	0.8
12 0226	5.6	**27** 0335	6.1
0846	1.5	0952	1.0
SA 1439	5.4	SU 1554	5.7
2104	1.5	◗ 2218	1.1
13 0304	5.4	**28** 0426	5.7
0927	1.7	1049	1.4
SU 1521	5.2	M 1650	5.3
2146	1.7	2318	1.5
14 0351	5.1	**29** 0524	5.3
1018	1.8	1156	1.6
M 1614	5.0	TU 1754	5.0
◗ 2240	1.8		
15 0451	5.1	**30** 0029	1.7
1123	1.9	0631	5.1
TU 1727	4.9	W 1310	1.8
2350	1.9	1913	4.9
		31 0141	1.8
		0749	5.1
		TH 1415	1.7
		2032	5.0

FEBRUARY

Time	m		Time	m
1 0243	1.7	**16** 0147	1.7
0857	5.2	0803	5.3
F 1512	1.5	SA 1425	1.4
2132	5.3	2041	5.4
2 0335	1.5	**17** 0257	1.4
0950	5.5	0906	5.7
SA 1559	1.2	SU 1526	1.0
2219	5.5	2141	5.9
3 0420	1.3	**18** 0353	1.0
1033	5.7	1001	6.1
SU 1641	1.0	M 1619	0.6
2300	5.8	2235	6.3
4 0459	1.1	**19** 0443	0.6
1111	5.9	1053	6.4
M 1719	0.9	TU 1708	0.3
● 2336	5.9	○ 2325	6.6
5 0535	1.0	**20** 0531	0.4
1145	5.9	1142	6.6
TU 1754	0.8	W 1755	0.2
6 0009	6.0	**21** 0013	6.7
0609	1.0	0617	0.3
W 1215	5.9	TH 1230	6.7
1828	0.8	1842	0.1
7 0038	6.0	**22** 0059	6.8
0643	1.0	0704	0.3
TH 1242	5.9	F 1315	6.6
1901	0.9	1928	0.2
8 0103	6.0	**23** 0143	6.7
0715	1.0	0750	0.4
F 1310	5.9	SA 1358	6.5
1933	0.9	2013	0.3
9 0128	5.9	**24** 0224	6.5
0745	1.1	0834	0.6
SA 1339	5.8	SU 1440	6.2
2000	1.0	2056	0.6
10 0158	5.8	**25** 0306	6.1
0813	1.2	0918	0.9
SU 1411	5.7	M 1524	5.8
2028	1.2	2140	1.0
11 0232	5.7	**26** 0350	5.7
0847	1.3	1005	1.3
M 1448	5.5	TU 1613	5.3
2104	1.3	◗ 2229	1.5
12 0312	5.5	**27** 0441	5.2
0929	1.5	1102	1.7
TU 1533	5.3	W 1711	4.9
◗ 2152	1.6	2336	1.9
13 0402	5.3	**28** 0544	4.8
1026	1.7	1230	2.0
W 1630	5.0	TH 1825	4.6
2254	1.8		
14 0508	5.0		
1141	1.9		
TH 1756	4.8		
15 0019	1.9		
0644	5.0		
F 1309	1.8		
1931	5.0		

MARCH

Time	m		Time	m
1 0111	2.1	**16** 0614	4.8
0703	4.7	1243	1.8
F 1350	1.9	SA 1913	4.9
2005	4.7		
2 0221	1.9	**17** 0130	1.8
0834	4.8	0747	5.1
SA 1450	1.7	SU 1409	1.5
2114	5.0	2028	5.4
3 0316	1.7	**18** 0244	1.4
0932	5.2	0854	5.6
SU 1539	1.4	M 1512	1.0
2201	5.4	2129	5.9
4 0400	1.4	**19** 0339	0.9
1015	5.5	0950	6.0
M 1620	1.1	TU 1603	0.6
2240	5.7	2222	6.3
5 0439	1.0	**20** 0427	0.5
1052	5.7	1041	6.4
TU 1656	0.9	W 1650	0.3
2315	5.9	2310	6.6
6 0514	1.0	**21** 0513	0.3
1125	5.9	1128	6.6
W 1731	0.8	TH 1736	0.1
● 2345	6.0	○ 2356	6.8
7 0547	0.9	**22** 0559	0.2
1153	5.9	1214	6.7
TH 1805	0.7	F 1821	0.1
8 0012	6.0	**23** 0039	6.8
0620	0.8	0643	0.2
F 1219	6.0	SA 1257	6.6
1837	0.7	1904	0.2
9 0036	6.0	**24** 0120	6.7
0651	0.8	0725	0.3
SA 1246	6.0	SU 1337	6.5
1907	0.8	1945	0.4
10 0102	6.0	**25** 0158	6.5
0719	0.9	0806	0.5
SU 1316	6.0	M 1416	6.2
1932	0.8	2025	0.7
11 0132	6.0	**26** 0235	6.1
0745	0.9	0844	0.9
M 1347	5.9	TU 1455	5.8
2000	1.0	2103	1.1
12 0205	5.9	**27** 0314	5.6
0818	1.0	0924	1.3
TU 1423	5.8	W 1540	5.3
2037	1.1	2147	1.6
13 0243	5.7	**28** 0402	5.1
0900	1.3	1012	1.7
W 1505	5.5	TH 1636	4.9
2123	1.4	◗ 2243	2.0
14 0330	5.4	**29** 0503	4.7
0954	1.7	1121	2.0
TH 1600	5.1	F 1745	4.6
◗ 2224	1.7		
15 0433	5.0	**30** 0027	2.2
1106	1.8	0618	4.5
F 1723	4.8	SA 1317	2.1
2352	1.9	1915	4.6
		31 0154	2.1
		0757	4.6
		SU 1421	1.8
		2043	4.9

APRIL

Time	m		Time	m
1 0250	1.8	**16** 0226	1.3
0904	5.0	0839	5.5
M 1510	1.5	TU 1453	1.0
2132	5.3	2112	5.9
2 0334	1.4	**17** 0320	0.8
0948	5.3	0934	6.0
TU 1551	1.2	W 1543	0.6
2210	5.6	2203	6.3
3 0411	1.1	**18** 0407	0.5
1023	5.6	1024	6.3
W 1627	1.0	TH 1629	0.4
2243	5.8	2250	6.6
4 0446	1.0	**19** 0452	0.3
1054	5.8	1110	6.5
TH 1702	0.8	F 1714	0.3
2313	6.0	○ 2334	6.7
5 0520	0.8	**20** 0537	0.2
1123	5.9	1154	6.5
F 1737	0.6	SA 1757	0.3
● 2339	6.1		
6 0553	0.8	**21** 0015	6.7
1152	6.0	0619	0.3
SA 1810	0.7	SU 1236	6.5
		1839	0.4
7 0007	6.1	**22** 0055	6.5
0624	0.8	0700	0.4
SU 1223	6.1	M 1315	6.3
1839	0.7	1918	0.6
8 0037	6.1	**23** 0131	6.3
0653	0.8	0738	0.6
M 1254	6.1	TU 1352	6.1
1907	0.8	1956	0.8
9 0109	6.1	**24** 0206	6.0
0723	0.8	0815	0.9
TU 1328	6.0	W 1430	5.8
1940	0.9	2033	1.2
10 0143	6.0	**25** 0243	5.6
0759	0.9	0853	1.3
W 1405	5.9	TH 1512	5.4
2019	1.1	2115	1.6
11 0223	5.8	**26** 0327	5.1
0843	1.1	0938	1.6
TH 1449	5.6	F 1605	5.0
2108	1.4	◗ 2207	2.0
12 0311	5.4	**27** 0427	4.7
0938	1.4	1037	1.9
F 1547	5.2	SA 1709	4.7
◗ 2212	1.7	2319	2.2
13 0416	5.0	**28** 0537	4.5
1052	1.7	1203	2.1
SA 1715	4.9	SU 1820	4.6
2341	1.9		
14 0600	4.8	**29** 0108	2.2
1227	1.7	0652	4.5
SU 1858	5.0	M 1337	1.9
		1940	4.8
15 0115	1.7	**30** 0212	1.9
0731	5.1	0810	4.8
M 1352	1.4	TU 1431	1.6
2013	5.4	2043	5.2

Chart Datum: 3·27 metres below Ordnance Datum (Newlyn). HAT is 6·9 metres above Chart Datum.

STANDARD TIME (UT)
For Summer Time add ONE hour in **non-shaded areas**

SHOREHAM   LAT 50°50′N   LONG 0°15′W
TIMES AND HEIGHTS OF HIGH AND LOW WATERS

Dates in red are **SPRINGS**
Dates in blue are NEAPS

YEAR **2019**

MAY

Day	Time	m	Day	Time	m
1 W	0258 / 0901 / 1514 / 2125	1.5 / 5.1 / 1.4 / 5.5	16 TH	0256 / 0912 / 1520 / 2140	0.9 / 5.8 / 0.7 / 6.2
2 TH	0337 / 0939 / 1552 / 2159	1.3 / 5.5 / 1.1 / 5.8	17 F	0345 / 1003 / 1607 / 2226	0.6 / 6.1 / 0.6 / 6.4
3 F	0413 / 1014 / 1629 / 2232	1.0 / 5.7 / 1.0 / 6.0	18 SA	0430 / 1049 / 1651 / 2310	0.5 / 6.2 / 0.5 / 6.5
4 SA ●	0448 / 1049 / 1705 / 2305	0.9 / 5.9 / 0.9 / 6.1	19 SU	0514 / 1133 / 1735 / 2351	0.5 / 6.3 / 0.5 / 6.4
5 SU	0523 / 1124 / 1740 / 2339	0.8 / 6.0 / 0.8 / 6.2	20 M	0557 / 1215 / 1816	0.5 / 6.3 / 0.6
6 M	0557 / 1159 / 1814	0.7 / 6.1 / 0.8	21 TU	0030 / 0637 / 1254 / 1855	6.3 / 0.6 / 6.2 / 0.8
7 TU	0013 / 0631 / 1236 / 1848	6.2 / 0.7 / 6.1 / 0.8	22 W	0106 / 0715 / 1331 / 1932	6.1 / 0.8 / 6.0 / 1.0
8 W	0049 / 0708 / 1314 / 1927	6.2 / 0.7 / 6.1 / 0.9	23 TH	0141 / 0751 / 1408 / 2009	5.9 / 1.0 / 5.8 / 1.2
9 TH	0128 / 0749 / 1356 / 2012	6.0 / 0.8 / 5.9 / 1.0	24 F	0217 / 0829 / 1448 / 2050	5.5 / 1.2 / 5.5 / 1.5
10 F	0212 / 0837 / 1445 / 2104	5.8 / 1.0 / 5.4 / 1.3	25 SA	0258 / 0912 / 1535 / 2139	5.2 / 1.5 / 5.2 / 1.8
11 SA	0304 / 0934 / 1547 / 2210	5.5 / 1.3 / 5.4 / 1.6	26 SU ◑	0351 / 1004 / 1632 / 2238	4.9 / 1.8 / 4.9 / 2.0
12 SU ◑	0413 / 1046 / 1710 / 2332	5.1 / 1.5 / 5.2 / 1.7	27 M	0456 / 1107 / 1734 / 2348	4.6 / 1.9 / 4.8 / 2.1
13 M	0545 / 1211 / 1837	5.0 / 1.5 / 5.2	28 TU	0602 / 1219 / 1837	4.6 / 2.0 / 4.9
14 TU	0055 / 0708 / 1328 / 1949	1.5 / 5.2 / 1.3 / 5.5	29 W	0106 / 0704 / 1331 / 1936	2.0 / 4.7 / 1.8 / 5.1
15 W	0202 / 0816 / 1428 / 2048	1.2 / 5.5 / 1.0 / 5.9	30 TH	0207 / 0801 / 1426 / 2028	1.7 / 5.0 / 1.6 / 5.4
			31 F	0253 / 0851 / 1511 / 2113	1.4 / 5.3 / 1.3 / 5.7

JUNE

Day	Time	m	Day	Time	m
1 SA	0334 / 0935 / 1552 / 2154	1.2 / 5.6 / 1.1 / 5.9	16 SU	0410 / 1030 / 1632 / 2249	0.8 / 5.9 / 0.8 / 6.1
2 SU	0414 / 1017 / 1632 / 2234	1.0 / 5.8 / 0.9 / 6.1	17 M ○	0455 / 1115 / 1716 / 2331	0.7 / 6.0 / 0.8 / 6.1
3 M ●	0453 / 1058 / 1712 / 2314	0.8 / 6.0 / 0.9 / 6.2	18 TU	0538 / 1157 / 1757	0.7 / 6.1 / 0.9
4 TU	0533 / 1140 / 1753 / 2354	0.7 / 6.1 / 0.8 / 6.2	19 W	0010 / 0618 / 1236 / 1835	6.1 / 0.8 / 6.0 / 1.0
5 W	0614 / 1222 / 1835	0.7 / 6.2 / 0.8	20 TH	0046 / 0655 / 1301 / 1912	6.0 / 0.9 / 5.9 / 1.1
6 TH	0036 / 0658 / 1306 / 1920	6.2 / 0.7 / 6.2 / 0.8	21 F	0121 / 0731 / 1348 / 1949	5.8 / 1.0 / 5.8 / 1.2
7 F	0120 / 0744 / 1353 / 2008	6.1 / 0.7 / 6.1 / 0.9	22 SA	0155 / 0808 / 1424 / 2027	5.6 / 1.2 / 5.6 / 1.4
8 SA	0208 / 0835 / 1445 / 2103	5.9 / 0.9 / 5.9 / 1.1	23 SU	0230 / 0848 / 1502 / 2110	5.4 / 1.3 / 5.4 / 1.6
9 SU	0303 / 0932 / 1546 / 2205	5.7 / 1.1 / 5.7 / 1.3	24 M	0311 / 0932 / 1545 / 2159	5.1 / 1.5 / 5.2 / 1.8
10 M ◑	0408 / 1038 / 1654 / 2316	5.4 / 1.2 / 5.6 / 1.4	25 TU ◐	0401 / 1022 / 1639 / 2255	4.9 / 1.7 / 5.0 / 1.9
11 TU	0522 / 1149 / 1807	5.3 / 1.3 / 5.5	26 W	0505 / 1120 / 1741 / 2357	4.8 / 1.8 / 5.0 / 1.9
12 W	0028 / 0637 / 1259 / 1917	1.4 / 5.3 / 1.3 / 5.6	27 TH	0612 / 1223 / 1843	4.8 / 1.9 / 5.0
13 TH	0134 / 0747 / 1401 / 2020	1.2 / 5.4 / 1.1 / 5.7	28 F	0101 / 0713 / 1327 / 1941	1.8 / 4.9 / 1.7 / 5.2
14 F	0231 / 0848 / 1456 / 2114	1.1 / 5.6 / 1.0 / 5.9	29 SA	0202 / 0809 / 1426 / 2034	1.6 / 5.2 / 1.5 / 5.5
15 SA	0323 / 0941 / 1545 / 2203	0.9 / 5.8 / 0.9 / 6.1	30 SU	0255 / 0901 / 1517 / 2122	1.3 / 5.5 / 1.3 / 5.8

JULY

Day	Time	m	Day	Time	m
1 M	0343 / 0950 / 1604 / 2209	1.1 / 5.8 / 1.1 / 6.0	16 TU ○	0440 / 1100 / 1700 / 2315	1.0 / 5.8 / 1.1 / 5.9
2 TU ●	0428 / 1037 / 1650 / 2254	0.9 / 6.0 / 0.9 / 6.2	17 W	0521 / 1141 / 1740 / 2353	0.9 / 5.9 / 1.0 / 5.9
3 W	0514 / 1125 / 1736 / 2340	0.7 / 6.2 / 0.8 / 6.3	18 TH	0600 / 1219 / 1818	0.9 / 6.0 / 1.0
4 TH	0600 / 1212 / 1824	0.6 / 6.3 / 0.7	19 F	0029 / 0636 / 1254 / 1853	5.9 / 0.9 / 5.9 / 1.1
5 F	0027 / 0648 / 1301 / 1911	6.3 / 0.5 / 6.3 / 0.7	20 SA	0101 / 0711 / 1326 / 1928	5.8 / 1.0 / 5.9 / 1.1
6 SA	0115 / 0737 / 1350 / 2002	6.3 / 0.5 / 6.3 / 0.7	21 SU	0131 / 0746 / 1355 / 2003	5.7 / 1.0 / 5.8 / 1.2
7 SU	0204 / 0828 / 1440 / 2054	6.1 / 0.6 / 6.2 / 0.8	22 M	0200 / 0821 / 1424 / 2039	5.6 / 1.2 / 5.6 / 1.4
8 M	0256 / 0922 / 1533 / 2151	5.9 / 0.8 / 6.0 / 1.0	23 TU	0232 / 0857 / 1456 / 2118	5.4 / 1.3 / 5.5 / 1.5
9 TU ◑	0352 / 1019 / 1630 / 2252	5.7 / 1.0 / 5.8 / 1.2	24 W	0310 / 0935 / 1536 / 2202	5.2 / 1.5 / 5.3 / 1.7
10 W	0453 / 1121 / 1731 / 2358	5.5 / 1.2 / 5.6 / 1.4	25 TH ◐	0356 / 1021 / 1627 / 2258	5.0 / 1.7 / 5.1 / 1.9
11 TH	0600 / 1227 / 1838	5.3 / 1.4 / 5.4	26 F	0457 / 1121 / 1737	4.9 / 1.9 / 5.0
12 F	0104 / 0713 / 1334 / 1948	1.4 / 5.3 / 1.4 / 5.4	27 SA	0005 / 0621 / 1234 / 1855	1.9 / 4.8 / 1.9 / 5.1
13 SA	0207 / 0824 / 1434 / 2051	1.4 / 5.3 / 1.4 / 5.5	28 SU	0116 / 0733 / 1346 / 2001	1.8 / 5.0 / 1.8 / 5.3
14 SU	0304 / 0923 / 1528 / 2145	1.3 / 5.5 / 1.3 / 5.7	29 M	0222 / 0835 / 1450 / 2058	1.6 / 5.4 / 1.5 / 5.7
15 M	0354 / 1015 / 1616 / 2232	1.1 / 5.7 / 1.1 / 5.8	30 TU	0319 / 0930 / 1545 / 2151	1.2 / 5.7 / 1.2 / 6.0
			31 W	0410 / 1022 / 1634 / 2240	0.9 / 6.1 / 0.9 / 6.2

AUGUST

Day	Time	m	Day	Time	m
1 TH ●	0459 / 1113 / 1722 / 2329	0.6 / 6.3 / 0.7 / 6.4	16 F	0540 / 1200 / 1757	0.9 / 6.0 / 1.0
2 F	0547 / 1202 / 1811	0.4 / 6.5 / 0.5	17 SA	0009 / 0614 / 1232 / 1831	5.9 / 0.9 / 6.0 / 1.0
3 SA	0018 / 0635 / 1251 / 1859	6.5 / 0.3 / 6.6 / 0.5	18 SU	0038 / 0648 / 1259 / 1904	5.9 / 0.9 / 6.0 / 1.0
4 SU	0106 / 0723 / 1339 / 1948	6.3 / 0.3 / 6.6 / 0.5	19 M	0103 / 0721 / 1323 / 1936	5.8 / 0.9 / 5.9 / 1.1
5 M	0153 / 0812 / 1425 / 2037	6.4 / 0.4 / 6.5 / 0.6	20 TU	0130 / 0751 / 1348 / 2006	5.8 / 1.1 / 5.8 / 1.2
6 TU	0241 / 0901 / 1512 / 2128	6.2 / 0.6 / 6.3 / 0.8	21 W	0159 / 0818 / 1419 / 2035	5.7 / 1.2 / 5.7 / 1.4
7 W	0329 / 0952 / 1600 / 2222	5.9 / 0.9 / 5.9 / 1.1	22 TH	0233 / 0850 / 1455 / 2112	5.5 / 1.4 / 5.6 / 1.5
8 TH	0422 / 1047 / 1654 / 2323	5.5 / 1.3 / 5.6 / 1.5	23 F ◐	0313 / 0932 / 1539 / 2202	5.3 / 1.6 / 5.3 / 1.8
9 F	0523 / 1153 / 1757	5.2 / 1.6 / 5.2	24 SA	0404 / 1028 / 1638 / 2310	5.0 / 1.9 / 5.0 / 2.0
10 SA	0035 / 0636 / 1308 / 1914	1.7 / 4.9 / 1.8 / 5.1	25 SU	0521 / 1146 / 1811	4.8 / 2.0 / 4.9
11 SU	0147 / 0803 / 1416 / 2032	1.7 / 5.0 / 1.7 / 5.2	26 M	0036 / 0704 / 1317 / 1937	2.0 / 4.9 / 1.9 / 5.1
12 M	0248 / 0910 / 1513 / 2131	1.5 / 5.2 / 1.6 / 5.4	27 TU	0159 / 0816 / 1432 / 2042	1.7 / 5.3 / 1.6 / 5.5
13 TU	0340 / 1002 / 1602 / 2219	1.3 / 5.5 / 1.3 / 5.7	28 W	0304 / 0916 / 1530 / 2137	1.3 / 5.7 / 1.2 / 6.0
14 W	0424 / 1046 / 1644 / 2300	1.1 / 5.8 / 1.2 / 5.8	29 TH	0356 / 1010 / 1620 / 2228	0.8 / 6.2 / 0.8 / 6.3
15 TH ○	0503 / 1125 / 1722 / 2337	1.0 / 5.9 / 1.1 / 5.9	30 F ●	0443 / 1100 / 1707 / 2317	0.5 / 6.5 / 0.5 / 6.6
			31 SA	0530 / 1148 / 1754	0.3 / 6.7 / 0.4

Chart Datum: 3·27 metres below Ordnance Datum (Newlyn). HAT is 6·9 metres above Chart Datum.

FREE monthly updates. Register at
www.reedsnauticalalmanac.co.uk

SHOREHAM LAT 50°50'N LONG 0°15'W
TIMES AND HEIGHTS OF HIGH AND LOW WATERS

STANDARD TIME (UT)
For Summer Time add ONE hour in **non-shaded areas**

Dates in red are **SPRINGS**
Dates in blue are **NEAPS**

YEAR **2019**

SEPTEMBER

Day	Time	m	Day	Time	m
1 SU	0004 / 0616 / 1235 / 1840	6.7 / 0.2 / 6.8 / 0.3	16 M	0010 / 0621 / 1227 / 1837	6.0 / 0.9 / 6.1 / 1.0
2 M	0051 / 0703 / 1319 / 1926	6.7 / 0.2 / 6.8 / 0.4	17 TU	0035 / 0653 / 1250 / 1907	6.0 / 0.9 / 6.1 / 1.0
3 TU	0135 / 0749 / 1402 / 2012	6.6 / 0.4 / 6.6 / 0.5	18 W	0101 / 0719 / 1317 / 1933	6.0 / 1.0 / 6.0 / 1.1
4 W	0219 / 0834 / 1445 / 2058	6.3 / 0.6 / 6.3 / 0.8	19 TH	0130 / 0745 / 1347 / 2001	5.9 / 1.1 / 5.9 / 1.2
5 TH	0303 / 0919 / 1529 / 2146	6.0 / 1.0 / 5.9 / 1.2	20 F	0203 / 0817 / 1422 / 2039	5.8 / 1.3 / 5.7 / 1.4
6 F	0351 / 1009 / 1619 / 2242	5.5 / 1.4 / 5.4 / 1.6	21 SA	0241 / 0900 / 1504 / 2128	5.5 / 1.6 / 5.4 / 1.7
7 SA	0449 / 1112 / 1721	5.1 / 1.9 / 5.0	22 SU	0331 / 0957 / 1601 / 2235	5.2 / 1.9 / 5.0 / 2.0
8 SU	0003 / 0601 / 1244 / 1839	2.0 / 4.7 / 2.1 / 4.8	23 M	0445 / 1116 / 1738	4.8 / 2.1 / 4.8
9 M	0126 / 0742 / 1358 / 2015	2.0 / 4.7 / 2.0 / 4.9	24 TU	0007 / 0644 / 1258 / 1919	2.1 / 4.9 / 2.0 / 5.0
10 TU	0230 / 0856 / 1457 / 2117	1.8 / 5.1 / 1.7 / 5.2	25 W	0142 / 0801 / 1418 / 2028	1.8 / 5.3 / 1.6 / 5.5
11 W	0321 / 0946 / 1543 / 2202	1.5 / 5.5 / 1.4 / 5.6	26 TH	0248 / 0902 / 1515 / 2124	1.3 / 5.8 / 1.1 / 6.0
12 TH	0404 / 1026 / 1623 / 2240	1.2 / 5.8 / 1.2 / 5.8	27 F	0339 / 0955 / 1602 / 2214	0.8 / 6.3 / 0.7 / 6.4
13 F	0441 / 1102 / 1659 / 2315	1.0 / 6.0 / 1.0 / 5.9	28 SA	0425 / 1042 / 1648 / 2301	0.5 / 6.7 / 0.4 / 6.7
14 SA	0515 / 1135 / 1732 / 2344	0.9 / 6.1 / 1.0 / 6.0	29 SU	0510 / 1128 / 1733 / 2346	0.3 / 6.9 / 0.3 / 6.8
15 SU	0549 / 1203 / 1805	0.9 / 6.1 / 0.9	30 M	0555 / 1212 / 1818	0.2 / 6.9 / 0.3

OCTOBER

Day	Time	m	Day	Time	m
1 TU	0030 / 0638 / 1255 / 1901	6.7 / 0.3 / 6.8 / 0.4	16 W	0006 / 0624 / 1220 / 1838	6.1 / 1.0 / 6.2 / 1.0
2 W	0113 / 0722 / 1335 / 1944	6.6 / 0.5 / 6.6 / 0.6	17 TH	0036 / 0651 / 1250 / 1906	6.1 / 1.0 / 6.1 / 1.1
3 TH	0154 / 0804 / 1415 / 2026	6.3 / 0.8 / 6.3 / 0.9	18 F	0107 / 0720 / 1322 / 1938	6.0 / 1.1 / 6.0 / 1.2
4 F	0235 / 0846 / 1457 / 2109	6.0 / 1.1 / 5.8 / 1.3	19 SA	0141 / 0756 / 1358 / 2018	5.9 / 1.3 / 5.8 / 1.4
5 SA	0321 / 0931 / 1545 / 2159	5.5 / 1.6 / 5.3 / 1.8	20 SU	0221 / 0841 / 1442 / 2109	5.6 / 1.6 / 5.4 / 1.7
6 SU	0418 / 1028 / 1647 / 2312	5.0 / 2.1 / 4.9 / 2.1	21 M	0313 / 0941 / 1541 / 2217	5.3 / 1.9 / 5.1 / 2.0
7 M	0528 / 1209 / 1803	4.7 / 2.3 / 4.6	22 TU	0432 / 1102 / 1721 / 2348	4.9 / 2.1 / 4.8 / 2.0
8 TU	0057 / 0705 / 1334 / 1947	2.2 / 4.7 / 2.2 / 4.7	23 W	0626 / 1239 / 1901	5.0 / 2.0 / 5.1
9 W	0204 / 0828 / 1433 / 2052	2.0 / 5.0 / 1.9 / 5.1	24 TH	0120 / 0742 / 1357 / 2009	1.8 / 5.4 / 1.6 / 5.5
10 TH	0255 / 0918 / 1518 / 2136	1.6 / 5.4 / 1.5 / 5.5	25 F	0226 / 0843 / 1453 / 2105	1.3 / 5.9 / 1.1 / 6.0
11 F	0337 / 0957 / 1556 / 2213	1.3 / 5.8 / 1.2 / 5.8	26 SA	0318 / 0934 / 1541 / 2155	0.9 / 6.4 / 0.7 / 6.4
12 SA	0413 / 1031 / 1631 / 2244	1.1 / 6.0 / 1.1 / 5.9	27 SU	0403 / 1020 / 1626 / 2241	0.6 / 6.7 / 0.5 / 6.6
13 SU	0447 / 1101 / 1704 / 2312	1.0 / 6.1 / 0.9 / 6.0	28 M	0448 / 1105 / 1710 / 2326	0.4 / 6.8 / 0.4 / 6.7
14 M	0520 / 1127 / 1737 / 2339	0.9 / 6.2 / 0.9 / 6.1	29 TU	0531 / 1148 / 1754	0.4 / 6.8 / 0.4
15 TU	0553 / 1152 / 1809	0.9 / 6.2 / 0.9	30 W	0009 / 0614 / 1229 / 1837	6.7 / 0.5 / 6.7 / 0.5
			31 TH	0051 / 0656 / 1308 / 1918	6.5 / 0.7 / 6.5 / 0.7

NOVEMBER

Day	Time	m	Day	Time	m
1 F	0131 / 0737 / 1347 / 1959	6.3 / 0.9 / 6.2 / 1.0	16 SA	0050 / 0704 / 1304 / 1925	6.1 / 1.1 / 6.1 / 1.1
2 SA	0211 / 0817 / 1427 / 2039	6.0 / 1.3 / 5.7 / 1.4	17 SU	0129 / 0745 / 1344 / 2009	6.0 / 1.3 / 5.8 / 1.3
3 SU	0255 / 0900 / 1513 / 2125	5.6 / 1.7 / 5.3 / 1.8	18 M	0213 / 0833 / 1432 / 2101	5.7 / 1.5 / 5.5 / 1.5
4 M	0349 / 0953 / 1613 / 2225	5.2 / 2.1 / 4.9 / 2.1	19 TU	0309 / 0933 / 1533 / 2207	5.4 / 1.7 / 5.2 / 1.7
5 TU	0453 / 1106 / 1723 / 2359	4.8 / 2.3 / 4.6 / 2.3	20 W	0426 / 1050 / 1702 / 2330	5.2 / 1.9 / 5.0 / 1.8
6 W	0606 / 1252 / 1843	4.8 / 2.3 / 4.6	21 TH	0559 / 1215 / 1833	5.2 / 1.8 / 5.2
7 TH	0123 / 0734 / 1356 / 2007	2.1 / 4.9 / 2.0 / 4.9	22 F	0052 / 0714 / 1328 / 1943	1.6 / 5.5 / 1.5 / 5.5
8 F	0218 / 0834 / 1444 / 2056	1.8 / 5.3 / 1.7 / 5.3	23 SA	0158 / 0816 / 1427 / 2041	1.3 / 5.9 / 1.1 / 5.8
9 SA	0302 / 0915 / 1524 / 2133	1.5 / 5.6 / 1.4 / 5.6	24 SU	0252 / 0909 / 1518 / 2133	1.0 / 6.2 / 0.8 / 6.2
10 SU	0340 / 0949 / 1600 / 2205	1.3 / 5.9 / 1.2 / 5.8	25 M	0341 / 0957 / 1605 / 2221	0.8 / 6.5 / 0.6 / 6.4
11 M	0416 / 1020 / 1634 / 2236	1.1 / 6.0 / 1.0 / 6.0	26 TU	0427 / 1042 / 1650 / 2306	0.7 / 6.6 / 0.5 / 6.5
12 TU	0450 / 1050 / 1708 / 2309	1.0 / 6.1 / 1.0 / 6.1	27 W	0511 / 1125 / 1734 / 2350	0.6 / 6.6 / 0.6 / 6.5
13 W	0525 / 1122 / 1742 / 2341	1.0 / 6.2 / 0.9 / 6.2	28 TH	0554 / 1206 / 1817	0.7 / 6.5 / 0.7
14 TH	0557 / 1154 / 1815	1.0 / 6.2 / 0.9	29 F	0031 / 0635 / 1246 / 1858	6.4 / 0.9 / 6.3 / 0.8
15 F	0015 / 0629 / 1228 / 1848	6.2 / 1.0 / 6.2 / 1.0	30 SA	0112 / 0715 / 1324 / 1937	6.2 / 1.1 / 6.1 / 1.0

DECEMBER

Day	Time	m	Day	Time	m
1 SU	0151 / 0754 / 1403 / 2016	6.0 / 1.3 / 5.7 / 1.3	16 M	0124 / 0739 / 1339 / 2005	6.1 / 1.1 / 6.0 / 1.0
2 M	0233 / 0835 / 1445 / 2059	5.7 / 1.6 / 5.4 / 1.6	17 TU	0211 / 0829 / 1428 / 2057	6.0 / 1.2 / 5.8 / 1.2
3 TU	0320 / 0923 / 1537 / 2149	5.4 / 1.9 / 5.0 / 1.9	18 W	0306 / 0926 / 1526 / 2157	5.8 / 1.4 / 5.5 / 1.4
4 W	0415 / 1019 / 1639 / 2249	5.1 / 2.1 / 4.8 / 2.1	19 TH	0410 / 1033 / 1636 / 2306	5.6 / 1.5 / 5.3 / 1.5
5 TH	0516 / 1128 / 1744	4.9 / 2.2 / 4.7	20 F	0524 / 1146 / 1754	5.5 / 1.6 / 5.3
6 F	0002 / 0619 / 1251 / 1849	2.1 / 4.9 / 2.1 / 4.8	21 SA	0018 / 0637 / 1256 / 1908	1.5 / 5.5 / 1.5 / 5.3
7 SA	0118 / 0722 / 1355 / 1949	2.0 / 5.1 / 1.9 / 5.0	22 SU	0126 / 0744 / 1359 / 2014	1.4 / 5.7 / 1.3 / 5.5
8 SU	0215 / 0816 / 1443 / 2039	1.8 / 5.4 / 1.6 / 5.3	23 M	0227 / 0843 / 1456 / 2112	1.2 / 5.9 / 1.1 / 5.8
9 M	0301 / 0900 / 1524 / 2123	1.6 / 5.6 / 1.4 / 5.6	24 TU	0320 / 0936 / 1546 / 2204	1.1 / 6.1 / 0.9 / 6.0
10 TU	0342 / 0940 / 1603 / 2203	1.4 / 5.9 / 1.2 / 5.8	25 W	0409 / 1024 / 1634 / 2252	1.0 / 6.2 / 0.8 / 6.1
11 W	0420 / 1018 / 1641 / 2242	1.2 / 6.1 / 1.0 / 6.0	26 TH	0455 / 1108 / 1719 / 2336	0.9 / 6.3 / 0.7 / 6.2
12 TH	0458 / 1056 / 1718 / 2321	1.1 / 6.2 / 0.9 / 6.1	27 F	0539 / 1150 / 1802	0.9 / 6.3 / 0.8
13 F	0536 / 1134 / 1756	1.0 / 6.2 / 0.9	28 SA	0018 / 0619 / 1230 / 1841	6.2 / 1.0 / 6.2 / 0.9
14 SA	0000 / 0614 / 1213 / 1836	6.2 / 1.0 / 6.2 / 0.9	29 SU	0057 / 0658 / 1307 / 1919	6.1 / 1.0 / 6.0 / 0.9
15 SU	0041 / 0655 / 1254 / 1918	6.2 / 1.0 / 6.2 / 0.9	30 M	0134 / 0735 / 1343 / 1956	6.0 / 1.2 / 5.8 / 1.1
			31 TU	0211 / 0812 / 1419 / 2034	5.8 / 1.4 / 5.5 / 1.3

Chart Datum: 3·27 metres below Ordnance Datum (Newlyn). HAT is 6·9 metres above Chart Datum.

>> **FREE** monthly updates. Register at <<
www.reedsnauticalalmanac.co.uk

3.6 BRIGHTON
E. Sussex 50°48'·53N 00°06'·38W 🏵🏵🏵♨♨♨✿✿✿

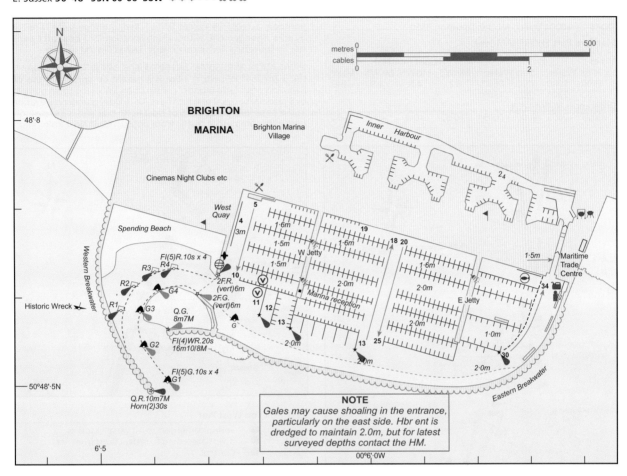

NOTE

Gales may cause shoaling in the entrance, particularly on the east side. Hbr ent is dredged to maintain 2.0m, but for latest surveyed depths contact the HM.

CHARTS AC 5605, 1652, 1991; Imray C12, C31, C9, 2100

TIDES +0004 Dover; ML 3·5; Duration 0605

Standard Port SHOREHAM (◄—)

Times				Height (metres)			
High Water		Low Water		MHWS	MHWN	MLWN	MLWS
0500	1000	0000	0600	6·3	4·8	1·9	0·6
1700	2200	1200	1800				
Differences BRIGHTON MARINA							
−0002	−0002	0000	0000	+0·2	+0·2	+0·1	+0·1

SHELTER Good in the marina under all conditions, but in strong S'ly winds confused seas can make the final appr very rough. Speed limit 5kn. Depths reported are less than shown, HM recommends that yachts drawing 2m or more do not to enter hbr at LW ± 2hrs.

NAVIGATION WPT 50°48'·23N 00°06'·39W, 000°/0·26M to W bkwtr lt.

- Ent chan dredged 1·7m, but after gales shoaling occurs especially on E side; craft drawing >1·5m should keep to the W side of chan until past the second SHM buoy.
- In heavy weather, best appr is from SSE to avoid worst of the backlash from bkwtrs; beware shallow water E of ent in R sector of lt Fl (4) WR 20s.

W-going stream starts at Brighton HW−1½ and E-going at HW+4½. Inshore the sp rate reaches approx 1·5kn. A Historic Wreck is at 50°48'·6N 00°06'·49W, immediately W of the marina's W bkwtr.

LIGHTS AND MARKS The marina is at the E end of the town, where white cliffs extend eastward. Daymark: conspic white hospital block, brg 334° leads to ent. Six Y spar lt buoys (up to 2M offshore) used as racing buoys:

1. 50°48'·06N 00°06'·41W
2. 50°47'·61N 00°08'·43W
3. 50°46'·61N 00°07'·00W
4. 50°47'·00N 00°15'·33W
5. 50°48'·40N 00°19'·40W
6. 50°47'·40N 00°05'·00W

A sewer outfall can buoy, Fl Y 5s, is 1·1M off the coast. Navigational lts may be hard to see against shore glare:

E bkwtr hd QG 8m 7M. W bkwtr hd, tr R/W bands, QR 10m 7M; Horn (2) 30s. Inner Hbr lock controlled by normal R/G lts, 0800-1800LT.

COMMUNICATIONS (Code 01273) Ⓗ 696955; Dr 686863; HM 819919; Call: *Brighton Control* VHF Ch **M** 80 16 (H24).

FACILITIES **Brighton Marina** www. premiermarinas.com ☎819919; Inner Hbr has least depth of 2·4m. 1600◯ inc 🅥 £2·41, £1·00 <4hrs; ♿🅿♠🔌🛠🔧 🕕 🅔 ♂ 🚻 🏳(50t) ⛴(35t) Divers ⚓ ACA ✕ 🛢. Fuel pontoon ♠♠♠ (H24).

Brighton Marina YC ☎818711, ✕ 🛢.
Bus service from marina; timetable info ☎674881. Electric railway runs from marina to Palace Pier, Mar-Oct.

Town ✉ Ⓑ 🚊 ✕ 🛢 ⇌ ✈ (Gatwick).

3.7 NEWHAVEN ❀❀❀♨♨♨♧♧

E. Sussex **50°46´·84N 00°03´·53E**

CHARTS AC 5605, 1652, 2154; Imray C12, C31, C9, 2100

TIDES +0004 Dover; ML 3·6; Duration 0550

Standard Port SHOREHAM (←—)

Times				Height (metres)			
High Water		Low Water		MHWS	MHWN	MLWN	MLWS
0500	1000	0000	0600	6·3	4·8	1·9	0·6
1700	2200	1200	1800				
Differences NEWHAVEN							
–0003	–0005	0000	+0005	+0·5	+0·4	+0·2	+0·2

SHELTER Good in all weathers, but in strong onshore winds there is often a dangerous sea at the ent. Appr from the SW, to pass 50m off bkwtr hd to avoid heavy breaking seas on E side of dredged chan. At marina (mostly dredged to 1m), berth on inside of ❷ pontoon, access H24 except LWS±1. Swing bridge 0.4M N opens by prior arrangement, with waiting pontoon for 1 boat off E bank run by University Technical College. Lewes lies 7M upriver.

NAVIGATION WPT 50°46´·24N 00°03´·60E, 348°/0·32M to W bwtr lt. Wk 3·7m 254° W bwtr lt 1·8ca is a hazard on appr from the W. Rampion wind farm is under construction SW of the town marked by 8 cardinal buoys, 4 of which have AIS. E Quay is under renovation as the base for operations S of No 2 RoRo pier and is increasingly busy with commercial traffic from pontoon S of No 1 RoRo pier.

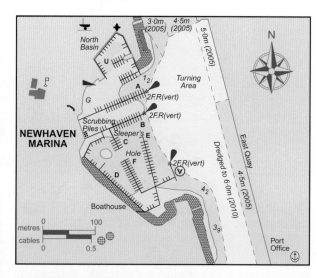

SS (Traffic) on West Pier

Fl	●	Serious emergency. All vessels stop or
Fl	●	divert according to instructions from Port
Fl	●	Control
F	●	No vessel to proceed contrary to this
F	●	signal
F	●	
F	●	
F	●	Small vessels may proceed. Two way traffic
F	○	
F	●	Proceed only when instructed by Port
F	○	Control. All other vessels keep clear
F	●	

Caution: Hbr silts and dredging is continuous. Do not leave marina when 3FR(vert) are lit at NE ent. Do not proceed S of RoRo terminal when 3FR(vert) are lit at NW corner of pontoon. Beware ferries; check on VHF Ch 12. Speed limit 5kn.

LIGHTS AND MARKS Lt ho on west breakwater is conspic.

COMMUNICATIONS (Code 01273) Dr 515076; ℍ696955; HM/Port Control vts@newhavenport authority.co.uk ☎612926 (H24). 'Newhaven Radio' Ch 12 (H24). Swing bridge opening Ch 12. Marina Ch **80** M (0800-1700).

FACILITIES Marina www.newhavenmarina.co.uk ☎513881; 280⌂+20❷ £2·30, ♿▯🅗(HO,1ca N of marina entrance) ♨❌⚓ ▦🛒(18t) 🛟(10t) ⛽🚮❌🗑; ⚓☎612612 (£17·50).

Newhaven and Seaford SC ☎(01323) 890077, ⚓⛴.

Town: 🏧Ⓔ⛽ ACA ✉Ⓑ🛒❌🗑➔✈ (Gatwick); **Ferries:** Dieppe; 2-3/day; 4hrs; Transmanche Ferries (transmancheferries.co.uk).

3.8 EASTBOURNE

E. Sussex 50°47'·34N 00°19'·90E ※❀♨♨♨❀❀

CHARTS AC 536, 5605; Imray 12, C31, C8, 2100

TIDES –0005 Dover; ML 3·8; Duration 0540

Standard Port SHOREHAM (←—)

Times				Height (metres)			
High Water		Low Water		MHWS	MHWN	MLWN	MLWS
0500	1000	0000	0600	6·3	4·8	1·9	0·6
1700	2200	1200	1800				
Differences EASTBOURNE							
–0010	–0005	+0015	+0020	+1·1	+0·6	+0·2	+0·1

SHELTER Very good, but in strong SE'lies the entrance may be hazardous and should be avoided at HW±1½. Access via buoyed channel (2m) and twin locks into inner basin (4m). The channel is prone to shoaling after gales and is dredged regularly. If uncertain of up-to-date situation, contact HM before entry.

NAVIGATION WPT 50°47'·37N 00°20'·81E, SWM buoy 'SH', L Fl 10s, 259°/0·45M to hbr ent. There are shoals to the NE in Pevensey Bay and from 2·5M SE toward Royal Sovereign lt (tide rips). From Beachy Hd, keep 0·75M offshore to clear Holywell Bank. At LW keep to ldg line/W sector in 3m. 5kn speed limit in approach chan and Sovereign Harbour.

LIGHTS AND MARKS See 3.2 and chartlet. From E, by day R roofs are conspic. Dir lt, Fl WRG 5s leads 258° through appr channel. Large wreck on N side marked by 2 lit SHM. Eastbourne pier, 2 FR, is 2M S of hbr ent; unlit PHM buoy is approx 5ca S.

COMMUNICATIONS (Code 01323) Dr470370 Ⓗ 417400. HM 470099. **VHF Ch 17 call** *Sovereign Hbr* **at SWM.**

FACILITIES Sovereign Harbour www.premiermarinas.com ☎470099, 860◻ £3·10<8m, £3·30<12.5m, ⚒🔲⛟🅿🛒(H24) ⬛🏴(50t) 🛠🔧⛽. **Sovereign Harbour YC** ☎470888, ✗ 🍴 visitors welcome. **Retail Park** with S/market and all domestic facilities. **Town** (2½M) all needs, ⇌ ✈ (Gatwick).

ADJACENT ANCHORAGE

HASTINGS, E Sussex, **50°50'·88N 00°35'·50E**. AC 536. Tides, see RYE; ML 3·8m; Duration 0530. Strictly a settled weather ⚓ or emergency shelter. The stone breakwater is in disrepair and only protects FVs. Beware dangerous wreck 3ca SE of pier hd. Ldg lts 356°, both FR 14/55m 4M: front on W metal column; rear 357m from front, on 5-sided W tr on West Hill. Pier under refit, now lit Q R 1s. W bkwtr hd Fl R 2·5s 5m 4M; Fl G 5s 2m, 30m from head of No 3 Groyne (E bkwtr). A historic wreck (*Amsterdam*) is about 2M W of pier, close inshore at 50°50'·7N 00°31'·65E. Facilities: ACA (St Leonards). Few marine services, but domestic facilities at Hastings and St Leonards. YC ☎(01424) 420656.

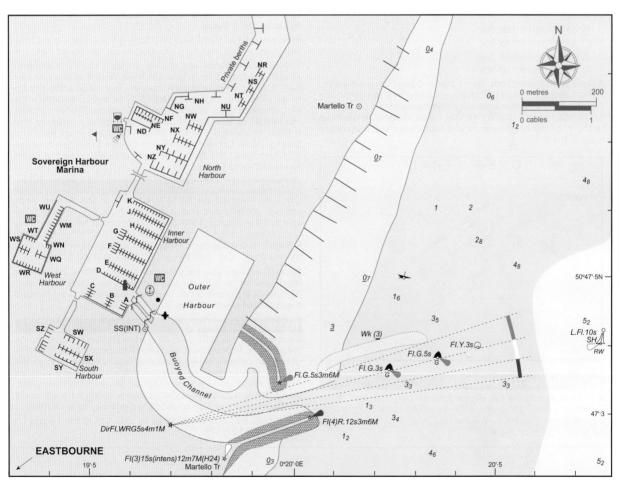

EASTBOURNE

CROSSING THE EASTERN ENGLISH CHANNEL

This section applies to crossings, ranging from the short (4–5hrs) Dover Strait route to Boulogne or Calais to the moderate (10–15hrs) ones from Brighton, Newhaven or Eastbourne to Dieppe and adjacent French ports. Distance tables are at 9.0.5. Routes cross the Dover Strait TSS where the SW-bound lane lies only 7M S of Beachy Head and 4.5M SE of Dover.

- Dover to Calais or Boulogne is only about 25M but crosses the most congested part of Dover Strait TSS.

- Plan to use the tidal streams to minimise crossing time and to be up-tide and to windward of the destination port when clear of the TSS.

- Always cross the TSS on a heading of 90° to the traffic flow, and minimise the time spent in the TSS. Be prepared to motor.

- When crossing the TSS, listen to the VHF broadcasts of navigational and traffic information made by CNIS on Ch 11. These include details of vessels which appear to be contravening Rule 10.

- Keep a very sharp lookout for ships in the traffic lanes and cross-Channel ferries. Ensure that the crew is aware of the directions from which to expect traffic in the TSS.

- Unless equipped with radar, and you are competent in its operation, do not attempt the crossing if fog or poor visibility is forecast. Even with radar, navigate with extreme caution.

- A Voluntary Separation Scheme (VSS) is in use to help ferries safely cross the Dover TSS. It is not a formal separation scheme and is not shown on official charts, but small craft should be aware of its existence. Most cross-Channel ferries can be expected to be using the VSS but high speed ferries are not included in the scheme. The VSS is bounded by the following points:

  1) 51°05'·35N 01°28'·00E
  2) 51°00'·10N 01°40'·00E
  3) 50°59'·30N 01°39'·10E
  4) 51°04'·70N 01°26'·80E

A separation line extends from a point midway between points 2 and 3 above to 1M W of CA6 buoy. Parallel to and 1·5M to the E of this zone a further separation line extends from 51°06'·40N 01°29'·90E to 51°01'·20N 01°41'·60E. Calais to Dover ferries keep to the W of this line; ferries to and from Dover and Dunkerque keep to the E.

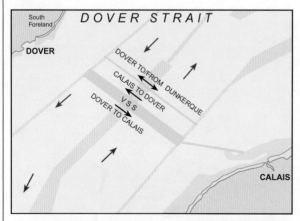

BEACHY HEAD TO DUNGENESS

▶ *2M S of Beachy Hd the W-going stream begins at HW Dover +0030, and the E-going at HW Dover –0520, sp rates 2·25kn. In bad weather there are overfalls off the Head, which should then be given a berth of 2M.◀*

(AC *536*) Royal Sovereign lt tr is 7·4M E of Beachy Head. The extensive Royal Sovereign shoals lie from 3M NW of the tr to 1·5M N of it, and have a minimum depth of 3·5m.

▶ *There are strong eddies over the shoals at Sp, and the sea breaks on them in bad weather.* ◀

On the direct course from the Royal Sovereign to Dungeness there are no dangers. Along the coast in Pevensey and Rye Bays there are drying rocky ledges or shoals extending 5ca offshore in places. These include Boulder Bank near Wish Tr, S of Eastbourne; Oyster Reef off Cooden; Bexhill Reef off Bexhill-on-Sea; Bopeep Rocks off St Leonards; and the shoals at the mouth of R Rother, at the entrance to Rye. There are also banks 2–3M offshore, on which the sea builds in bad weather.

Avoid the firing range danger area between Rye and Dungeness (lt, fog sig, RC). The nuclear power station is conspic at S extremity of the low-lying spit. The point is steep-to on SE side.

▶ *2M SE of Dungeness the NE-going flood starts at HW Dover –0100, max Sp rate 1.9kn. The SW-going ebb starts at HW Dover +0430, max Sp rate 2.1kn.* ◀

DUNGENESS TO NORTH FORELAND

(AC 1892, 1828) From Dungeness to Folkestone the coast forms a bay. Beware Roar bank, depth 2·7m, E of New Romney: otherwise there are no offlying dangers apart from Hythe firing range. Good ⚓ off Sandgate in offshore winds.

▶ *Off Folkestone the E-going stream starts at HW Dover –0155, Sp rate 2kn; the W-going at HW Dover +0320, sp rate 1·5kn.* ◀

Passing Dover and S Foreland keep 1M offshore. Do not pass too close to Dover as the many ferries leave at speed and there can be considerable backwash. It is advisable to call Dover Port Control (Ch 74) to identify yourself before passing the harbour entrances. 8M S of Dover in the TSS is the Varne, a shoal 7M long, with least depth 3·3m and heavy seas in bad weather, marked by Lanby and 5 buoys.

▶ *Between S and N Foreland the N-going stream begins at about HW Dover –0150, and the S-going at about HW Dover +0415.* ◀

The Goodwin Sands are drying, shifting shoals, extending about 10M from S to N, and 5M from W to E at their widest part. The E side is relatively steep-to, but large areas dry up to 3m. The sands are well marked by lit buoys. Kellett Gut is an unmarked chan about 5ca wide, running SW/NE through the middle of the sands, but it is not regularly surveyed and is liable to change. The Gull Stream (buoyed) leads from The Downs, inside Goodwin Sands and outside Brake Sands to the S of Ramsgate. The Ramsgate channel leads inside the Brake Sands and Cross Ledge.

LYDD FIRING RANGES

Off Lydd, centred on 50°54'N 00°53'E, a Sea Danger Area extends 3M offshore and stretches E from Rye Fairway buoy to a N/S line approx 1·5M W of Dungeness lt ho. When firing takes places, about 300 days p.a. 0830–1630LT (often to 2300), R flags/R lts are displayed ashore and a Range Safety Craft may be on station. Call Lydd Ranges Ch **73** or ☎01303 225518.

Radar fixes may also be obtained by VHF. While vessels may legally transit through the Sea Danger Area, Masters of vessels of all kinds are requested to pass S of Stephenson Shoal to avoid disruption of military firings.

HYTHE FIRING RANGES

(Centred on 51°02'N 01°03'E) have a Sea Danger Area extending 2M offshore, from Hythe to Dymchurch (approx 5M and 8M WSW of Folkestone hbr). While vessels may legally transit through the Sea Danger Area, they are requested to keep clear, and should not enter or remain in it for other purposes. When firing takes place, about 300 days pa 0830–1630 LT (often to 2300), R flags/R lts are displayed ashore and a Range Safety Craft may be on station.

Radar fixes may also be obtained by VHF. Call *Hythe Ranges* Ch **73** or ☎01303 225879.

3.9 RYE

E. Sussex **50°55'·60N 00°46'·58E** ❀❀⚓⚓☆☆☆

CHARTS AC 5605, 536, 1991; Imray C12, C31, C8, 2100

TIDES ML 2·0; Duration 3·25hrs sp, 5hrs nps

Standard Port DOVER (⟶)

Times				Height (metres)			
High Water		Low Water		MHWS	MHWN	MLWN	MLWS
0000	0600	0100	0700	6·8	5·3	2·1	0·8
1200	1800	1300	1900				
Differences RYE (approaches)							
+0005	–0010	ND	ND	+1·0	+0·7	ND	ND
RYE HARBOUR							
+0005	–0010	DR	DR	–1·4	–1·7	DR	DR
HASTINGS							
0000	–0010	–0030	–0030	+0·8	+0·5	+0·1	–0·1

SHELTER Very good in R Rother which dries completely to soft mud. Rye Bay is exposed to prevailing SW'lies with little shelter, when there is good ⚓ in lee of Dungeness (6M to E). In N'lies ⚓ 5ca N of the Rye Fairway buoy.

Rye Hbr is a small village, ¾M inside ent on W bank, used by commercial shipping. Berth initially on Admiralty Jetty (E bank) and see HM for � or M. No ⚓. Max speed 6kn.

Rye Town (a Cinque Port) is 2M upriver. Enter via Rock Channel for ❶ ⌚ along NE side of Strand Quay.

NAVIGATION WPT Rye Fairway SWM lt buoy, 50°54'·04N 00°48'·02E, 329°/1·8M to W Arm tripod lt, 50°55'·55N 00°46'·65E. For details of Lydd and Hythe Firing Ranges see Passage Information (⟵).

- Bar dries 2·75m about 2ca offshore and needs care when wind >F6 from SE to SW. Also shoals E and W of ent with ground swell or surf; narrow ent (42m) and chan (30m).
- Enter HW –2 to HW +2; flood runs 4·5kn (max HW –2 to HW –1).
- Depth of water over the bar can be judged by day from horizontal timbers at base of West Arm tripod structure (approx 2ca N of the bar) these are set at 1·5, 3 and 4·5m above CD.

Note: A Historic Wreck (*Anne*) is about 4M WSW of Rye, close inshore at 50°53'·42N 00°41'·91E.

LIGHTS AND MARKS W Arm lt LFl R 7s 7m 6M, wooden tripod, radar reflector. E Arm hd, Q (9) 15s 7m 5M; Horn 7s, G △. On E Pier a floodlit 'Welcome to Rye' sign may be helpful. Rock Chan entrance marked by a QR and QG lt buoy.

IPTS (Sigs 2 & 5 only) are shown to seaward (3M) from HM's office and upriver (1M) from HM's office.

COMMUNICATIONS (Code 01797) Dr 222031; Ⓗ 224499. HM 225225 Ch 14 (0900-1700LT, HW±2 or when vessel due). Call HM and monitor Ch 14 before arrival/departure.

FACILITIES (from seaward) **Admiralty Jetty** ⚓⚓⚓⚓
Rye Hbr ⚓ HW±3 (26t launch £10) ⚒⚒ ⒺⒺⒺ ⚓(3t/15t) ⚓ ACA
Rye Hbr SC (Sec'y) ☎223376.

Rye Town, Strand Quay ⌚, £14.75<8m>£17.74<10m>£23.57 wood fender posts against steel piled wall with numbered ladders. ❶ must stop at timber staging on E bank and report to HM Office. ⚓⚓⚓⚓⚓⚓

Town ✉ Ⓑ ⚓ ⇌ ✈ (Lydd).

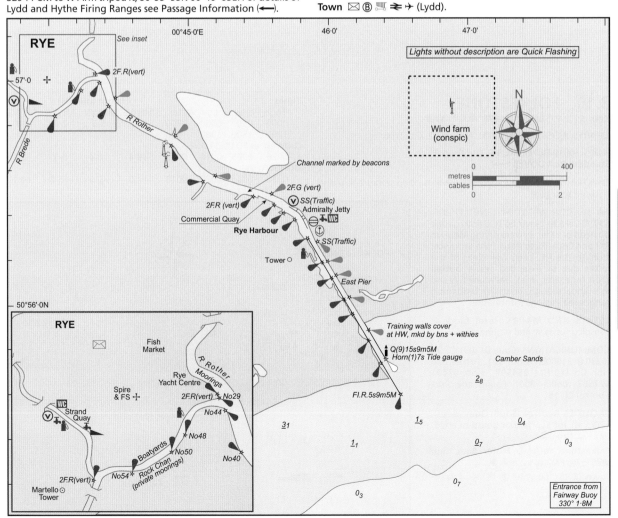

SE England

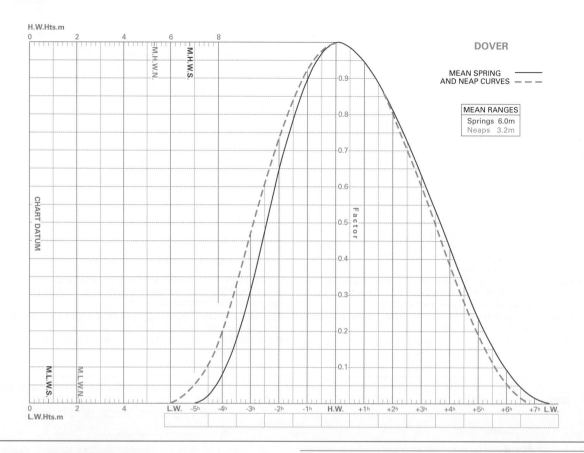

Dover tidal curve — Mean Spring and Neap curves

DOVER		
MEAN SPRING		————
AND NEAP CURVES		- - - -

MEAN RANGES	
Springs	6.0m
Neaps	3.2m

3.10 FOLKESTONE

Kent **51°04'·59N 01°11'·67E** ⚓❄❄❄❄❄

CHARTS AC 5605, 1892, 1991; Imray C12, C8, 2100

TIDES –0010 Dover; ML 3·9; Duration 0500
Standard Port DOVER (⟶)

SHELTER Good except in strong E-S winds when seas break at the harbour entrance.

NAVIGATION WPT 51°04'·33N 01°11'·89E, 330°/0·26M to bkwtr hd lt. Depth gauge on head of E Pier. Beware drying Mole Hd Rks and Copt Rks to stbd of the entrance; from/to the NE, keep well clear of the latter due to extended sewer outfall pipe. Inner Hbr, dries, has many FVs and local shoal draught boats. For Lydd firing ranges see Passage Information.

LIGHTS AND MARKS Hotel block is conspicuous at W end of Inner Hbr.

COMMUNICATIONS (Code 01303) CGOC (01304) 210008; Police 101. Folkestone Hbr Co Ltd/HM 254597, mob 07401 627563 (0600–1800); *Folkestone Port Control* Ch 15.

FACILITIES Berth on **South Quay** (dries); £20/craft for ⌣ (fender board needed) or ⚓. ▂ (free) ⚓.

Folkestone Yacht & Motor Boat Club ☎251574, F&A ⚓▂◣ 🍴⚓📕🛒.

Town 🏧 📞(100m) ✉ Ⓑ 🛒 ✗ 🍴 ⇌ ✈ (Lydd).

Times				Height (metres)			
High Water		Low Water		MHWS	MHWN	MLWN	MLWS
0000	0600	0100	0700	6·8	5·3	2·1	0·8
1200	1800	1300	1900				
Differences FOLKESTONE							
–0020	–0005	–0010	–0010	+0·4	+0·4	0·0	–0·1
DUNGENESS							
–0010	–0015	–0020	–0010	+1·0	+0·6	+0·4	+0·1

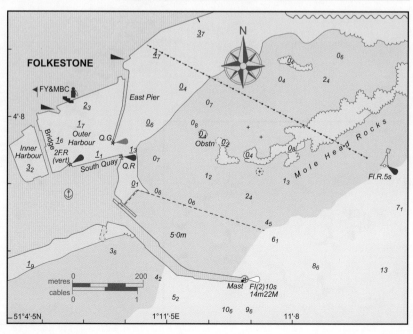

3.11 DOVER

Kent **51°06'·74N 01°19'·73E** (W ent) 🏴🏴🏴♨♨♨🌸🌸
       51°07'·25N 01°20'·61E (E ent)

CHARTS AC 5605,1892, 1828, 1698; Imray C30, C12, C8, 2100

TIDES 0000 Dover; ML 3·7; Duration 0505

Standard Port DOVER (→)

Times				Height (metres)			
High Water		Low Water		MHWS	MHWN	MLWN	MLWS
0000	0600	0100	0700	6·8	5·3	2·1	0·8
1200	1800	1300	1900				
Differences DEAL							
+0012	+0010	+0004	+0002	−0·5	−0·3	0·0	+0·1

SHELTER Very good in marina. ⚓ only when marinas full as directed by Port Control, Craft will be escorted and sufficient cable must be deployed to prevent dragging. Keep clear of extensive seasonal (Apr-Oct) swim zone along N shore.

NAVIGATION WPT from SW, 51°06'·18N 01°19'·67E, 000°/0·5M to Admiralty Pier lt ho. WPT from NE, 51°07'·30N 01°21'·41E, 270°/0·5M to S end Eastern Arm. Beware overfalls outside the bkwtrs and the frequent ferries and high speed craft using both ents. Strong tides across ents and high walls make ent under sail slow and difficult; use of engine is very strongly recommended.

- **Specific permission** to ent/leave the hbr via E or W ent must first be obtained from Port Control on Ch **74**; Call *Port Control* when 2M off and comply with instructions. Clearance for small/slow craft is not normally given until within 200m of ent. ♥ are welcomed by hbr launch and usually escorted to the marina.
- Vessels with a draught >2.5m bound for the marina must inform Port Control, Ch **74**, prior to entry as must yachts without engine.
- **If no VHF:** Stay safe distance clear of either entrance, call Port Control +44(0)1304 206063 or make visual contact (lamp) with Port Control (E Arm) who may direct launch to assist.
- Extensive redevelopment of the W docks and Prince of Wales Pier is underway. Avoid buoyed exclusion zone marking new marina.

Narrow ent between new jetty and lit bcn leads to Granville Dock. Note: A historic wreck is adjacent to the Eastern Arm bkwtr at 51°07'·6N 01°20'·7E (see chartlet). There are 4 more historic wrecks on the Goodwin Sands.

LIGHTS AND MARKS As per 3.2 and chartlet. **IPTS** (for all vessels) are shown for the E ent by Port Control (conspic twr) at S end of E Arm and for W entrance, from the head of Admiralty Pier. Q W lt from Port Control twr or patrol launch = keep clear of ent you are approaching. IPTS controls Wick Channel.

COMMUNICATIONS (Code 01304) Ⓗ (01233) 633331. Port Control 240400 ext 5530 HM 240400 ext 4522. Duty mob 07836 262713. Call *Dover Port Control* (also Hbr launch) Ch **74**, 12, 16 for clearance to enter. Request clearance to leave hbr as soon as clear of the marina. *Dover Marina* Ch **80**, only within marina. *Dover Coastguard* Ch 16, maintains TSS surveillance. Chan Nav Info Service (CNIS) broadcasts tfc/ nav/weather/tidal info Ch 11 at H+40; also, if vis <2M, at H+55.

FACILITIES Dover Harbour Marina ⚓ ☎241663; on arrival berth at reception pontoon to visit marina office or await gate opening call 241669. 3 berthing options are available in the marina:
- Tidal hbr, E of waiting pontoon (1·5m), access H24. 107🛟 on 3 pontoons (C, B, A) in 2·5m, £2·70. Mostly for single night stop.
- Granville Dock, gate open approx HW−3½ to +4½, 133🛟, £2·40.
- Wellington Dock. Gate and swing bridge open HW ±2 nps and approx HW−1½ to HW+1½ or 2½ sp, depending on range. In the final appr, especially near LW, stay in deep water as defined by the W sector (324°-333°) of the F WR lt, 2 unlit SHM poles and a G conical buoy. IPTS are shown, plus a small Fl 🔴 lt 5 mins before Wellington bridge is swung open. 160🛟 £2·20.

Services www.doverport.co.uk ⚓ 370🛟 inc ♥ ⛽ 🔧 🛠(H24) 🔩(0500–2100), LPG 🔥 ⚒ ⚒ 🖥 Ⓔ ⚓ 🏗(50t) ⛴ 🔒 ACA, wi-fi.
Royal Cinque Ports YC www.rcpyc.co.uk ☎206262, ⚓ 🍺.
White Cliffs M and YC www.wcmyc.co.uk ☎211666.
Town 🏦 🏧(H24) ✉ Ⓑ 🛒 ✕ 🍽 ≠.
Ferries Services to Calais (passenger/vehicle), Dunkerque (vehicle).

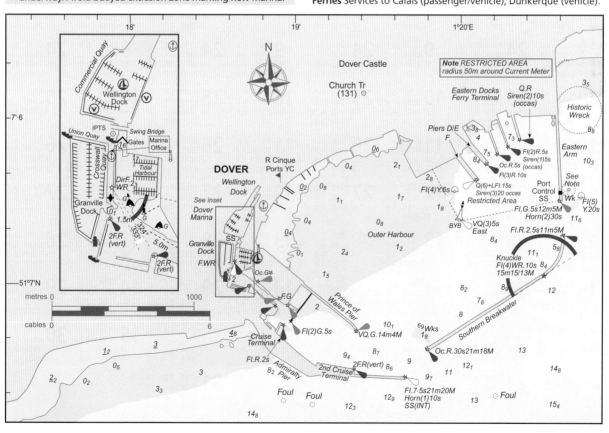

STANDARD TIME (UT)
For Summer Time add ONE hour in **non-shaded areas**

DOVER   LAT 51°07'N   LONG 1°19'E
TIMES AND HEIGHTS OF HIGH AND LOW WATERS

Dates in red are SPRINGS
Dates in blue are NEAPS

YEAR 2019

JANUARY

Day	Time	m	Day	Time	m
1 TU	0142 / 0710 / 1419 / 1955	1.9 / 5.8 / 1.7 / 5.7	16 W	0032 / 0558 / 1322 / 1851	2.2 / 5.6 / 2.0 / 5.4
2 W	0251 / 0815 / 1526 / 2052	1.8 / 5.9 / 1.6 / 5.9	17 TH	0154 / 0712 / 1433 / 1954	2.1 / 5.7 / 1.8 / 5.7
3 TH	0357 / 0912 / 1629 / 2140	1.7 / 6.1 / 1.4 / 6.0	18 F	0304 / 0813 / 1537 / 2048	1.8 / 6.0 / 1.5 / 6.0
4 F	0454 / 1000 / 1719 / 2223	1.5 / 6.2 / 1.3 / 6.2	19 SA	0405 / 0908 / 1636 / 2139	1.5 / 6.3 / 1.2 / 6.3
5 SA	0540 / 1042 / 1800 / 2301	1.3 / 6.3 / 1.3 / 6.4	20 SU	0502 / 0959 / 1734 / 2228	1.2 / 6.6 / 1.0 / 6.5
6 SU	0619 / 1120 / 1835 / 2338	1.2 / 6.4 / 1.3 / 6.5	21 M	0558 / 1049 / 1830 / 2316	0.9 / 6.8 / 0.8 / 6.9
7 M	0654 / 1156 / 1906	1.2 / 6.4 / 1.3	22 TU	0654 / 1137 / 1924	0.7 / 6.9 / 0.7
8 TU	0014 / 0725 / 1230 / 1934	6.5 / 1.2 / 6.3 / 1.3	23 W	0002 / 0746 / 1224 / 2012	7.0 / 0.6 / 6.9 / 0.7
9 W	0048 / 0754 / 1302 / 2004	6.5 / 1.3 / 6.2 / 1.4	24 TH	0049 / 0835 / 1312 / 2056	7.0 / 0.5 / 6.8 / 0.7
10 TH	0118 / 0826 / 1331 / 2036	6.4 / 1.4 / 6.1 / 1.4	25 F	0136 / 0920 / 1400 / 2137	7.0 / 0.5 / 6.7 / 0.8
11 F	0144 / 0859 / 1358 / 2111	6.3 / 1.4 / 5.9 / 1.5	26 SA	0225 / 1003 / 1451 / 2218	6.8 / 0.7 / 6.4 / 1.1
12 SA	0213 / 0936 / 1430 / 2148	6.1 / 1.6 / 5.8 / 1.7	27 SU	0315 / 1047 / 1545 / 2302	6.6 / 1.0 / 6.1 / 1.4
13 SU	0250 / 1016 / 1512 / 2230	6.0 / 1.7 / 5.6 / 1.9	28 M	0409 / 1135 / 1644 / 2355	6.2 / 1.4 / 5.7 / 1.8
14 M	0337 / 1103 / 1607 / 2321	5.8 / 1.9 / 5.5 / 2.1	29 TU	0510 / 1232 / 1754	5.8 / 1.7 / 5.4
15 TU	0439 / 1204 / 1728	5.6 / 2.0 / 5.3	30 W	0057 / 0622 / 1336 / 1911	2.0 / 5.6 / 1.9 / 5.4
			31 TH	0208 / 0743 / 1444 / 2023	2.1 / 5.5 / 1.9 / 5.5

FEBRUARY

Day	Time	m	Day	Time	m
1 F	0320 / 0854 / 1554 / 2122	2.0 / 5.7 / 1.8 / 5.7	16 SA	0230 / 0757 / 1510 / 2035	2.0 / 5.8 / 1.7 / 5.8
2 SA	0427 / 0949 / 1653 / 2208	1.7 / 5.9 / 1.6 / 6.0	17 SU	0341 / 0900 / 1618 / 2131	1.6 / 6.1 / 1.4 / 6.2
3 SU	0520 / 1032 / 1739 / 2247	1.5 / 6.1 / 1.4 / 6.2	18 M	0447 / 0955 / 1724 / 2221	1.2 / 6.5 / 1.0 / 6.6
4 M	0602 / 1107 / 1816 / 2322	1.3 / 6.2 / 1.3 / 6.4	19 TU	0550 / 1045 / 1825 / 2308	0.8 / 6.7 / 0.8 / 6.9
5 TU	0637 / 1139 / 1848 / 2355	1.2 / 6.3 / 1.2 / 6.5	20 W	0648 / 1131 / 1917 / 2353	0.5 / 6.9 / 0.6 / 7.1
6 W	0708 / 1210 / 1918	1.2 / 6.3 / 1.2	21 TH	0739 / 1215 / 2003	0.3 / 7.0 / 0.4
7 TH	0026 / 0738 / 1239 / 1948	6.5 / 1.1 / 6.3 / 1.2	22 F	0036 / 0824 / 1258 / 2042	7.2 / 0.2 / 6.9 / 0.5
8 F	0053 / 0809 / 1304 / 2019	6.5 / 1.1 / 6.2 / 1.2	23 SA	0119 / 0904 / 1341 / 2117	7.1 / 0.3 / 6.8 / 0.6
9 SA	0116 / 0841 / 1328 / 2051	6.4 / 1.2 / 6.2 / 1.3	24 SU	0203 / 0941 / 1425 / 2153	7.0 / 0.5 / 6.5 / 0.9
10 SU	0142 / 0913 / 1357 / 2123	6.4 / 1.3 / 6.2 / 1.4	25 M	0248 / 1018 / 1512 / 2230	6.7 / 0.9 / 6.2 / 1.3
11 M	0216 / 0946 / 1433 / 2159	6.3 / 1.4 / 6.1 / 1.6	26 TU	0336 / 1059 / 1605 / 2314	6.3 / 1.4 / 5.8 / 1.7
12 TU	0257 / 1024 / 1519 / 2241	6.2 / 1.6 / 5.8 / 1.8	27 W	0431 / 1150 / 1708	5.8 / 1.8 / 5.4
13 W	0348 / 1113 / 1618 / 2338	5.9 / 1.9 / 5.5 / 2.1	28 TH	0015 / 0538 / 1256 / 1823	2.1 / 5.4 / 2.2 / 5.2
14 TH	0456 / 1223 / 1752	5.6 / 2.0 / 5.3			
15 F	0102 / 0637 / 1354 / 1930	2.2 / 5.5 / 2.0 / 5.5			

MARCH

Day	Time	m	Day	Time	m
1 F	0130 / 0704 / 1408 / 1949	2.3 / 5.2 / 2.2 / 5.2	16 SA	0028 / 0620 / 1325 / 1913	2.2 / 5.4 / 2.1 / 5.4
2 SA	0247 / 0839 / 1520 / 2100	2.2 / 5.4 / 2.0 / 5.5	17 SU	0205 / 0748 / 1450 / 2023	2.0 / 5.6 / 1.8 / 5.8
3 SU	0400 / 0938 / 1625 / 2149	1.9 / 5.7 / 1.8 / 5.9	18 M	0322 / 0853 / 1604 / 2120	1.6 / 6.0 / 1.4 / 6.2
4 M	0458 / 1019 / 1716 / 2227	1.5 / 6.0 / 1.5 / 6.2	19 TU	0433 / 0948 / 1713 / 2210	1.1 / 6.4 / 1.0 / 6.6
5 TU	0541 / 1050 / 1755 / 2301	1.3 / 6.2 / 1.3 / 6.4	20 W	0539 / 1037 / 1811 / 2256	0.7 / 6.7 / 0.7 / 6.9
6 W	0616 / 1117 / 1827 / 2331	1.2 / 6.3 / 1.2 / 6.5	21 TH	0635 / 1120 / 1900 / 2338	0.4 / 6.9 / 0.5 / 7.1
7 TH	0647 / 1145 / 1857 / 2359	1.1 / 6.3 / 1.1 / 6.5	22 F	0723 / 1201 / 1943	0.2 / 7.0 / 0.4
8 F	0717 / 1212 / 1928	1.0 / 6.4 / 1.1	23 SA	0019 / 0804 / 1240 / 2019	7.2 / 0.2 / 7.0 / 0.4
9 SA	0024 / 0748 / 1237 / 1958	6.5 / 1.0 / 6.4 / 1.0	24 SU	0100 / 0841 / 1319 / 2053	7.1 / 0.3 / 6.8 / 0.6
10 SU	0049 / 0819 / 1301 / 2029	6.5 / 1.0 / 6.4 / 1.1	25 M	0140 / 0915 / 1400 / 2125	6.9 / 0.6 / 6.6 / 0.9
11 M	0115 / 0849 / 1330 / 2100	6.6 / 1.1 / 6.4 / 1.2	26 TU	0221 / 0948 / 1443 / 2159	6.6 / 1.0 / 6.3 / 1.3
12 TU	0148 / 0920 / 1406 / 2133	6.5 / 1.2 / 6.3 / 1.4	27 W	0306 / 1023 / 1533 / 2236	6.2 / 1.5 / 5.9 / 1.8
13 W	0228 / 0955 / 1450 / 2214	6.4 / 1.5 / 6.1 / 1.6	28 TH	0359 / 1107 / 1632 / 2331	5.7 / 2.0 / 5.4 / 2.2
14 TH	0316 / 1041 / 1546 / 2308	6.0 / 1.8 / 5.7 / 2.0	29 F	0503 / 1214 / 1742	5.3 / 2.3 / 5.2
15 F	0421 / 1146 / 1716	5.6 / 2.1 / 5.3	30 SA	0054 / 0621 / 1334 / 1905	2.4 / 5.1 / 2.4 / 5.1
			31 SU	0214 / 0809 / 1446 / 2026	2.3 / 5.2 / 2.2 / 5.4

APRIL

Day	Time	m	Day	Time	m
1 M	0325 / 0912 / 1549 / 2119	2.0 / 5.5 / 1.9 / 5.8	16 TU	0305 / 0840 / 1546 / 2103	1.5 / 6.0 / 1.4 / 6.3
2 TU	0423 / 0952 / 1641 / 2158	1.6 / 5.9 / 1.6 / 6.1	17 W	0417 / 0934 / 1653 / 2153	1.1 / 6.4 / 1.1 / 6.6
3 W	0508 / 1020 / 1723 / 2230	1.3 / 6.1 / 1.3 / 6.3	18 TH	0521 / 1021 / 1749 / 2237	0.7 / 6.7 / 0.8 / 6.9
4 TH	0545 / 1047 / 1758 / 2300	1.2 / 6.3 / 1.2 / 6.4	19 F	0615 / 1103 / 1836 / 2319	0.5 / 6.8 / 0.6 / 7.0
5 F	0618 / 1114 / 1830 / 2328	1.0 / 6.4 / 1.1 / 6.5	20 SA	0701 / 1141 / 1917 / 2359	0.3 / 6.9 / 0.5 / 7.1
6 SA	0651 / 1142 / 1903 / 2354	1.0 / 6.4 / 1.0 / 6.6	21 SU	0741 / 1219 / 1954	0.4 / 6.9 / 0.6
7 SU	0724 / 1208 / 1936	0.9 / 6.5 / 1.0	22 M	0038 / 0816 / 1258 / 2027	7.0 / 0.5 / 6.8 / 0.7
8 M	0021 / 0756 / 1236 / 2008	6.6 / 0.9 / 6.5 / 1.0	23 TU	0118 / 0848 / 1337 / 2059	6.8 / 0.8 / 6.6 / 1.0
9 TU	0051 / 0827 / 1308 / 2040	6.6 / 1.0 / 6.6 / 1.1	24 W	0158 / 0918 / 1419 / 2130	6.5 / 1.1 / 6.4 / 1.4
10 W	0126 / 0859 / 1346 / 2115	6.6 / 1.2 / 6.4 / 1.3	25 TH	0241 / 0948 / 1506 / 2203	6.1 / 1.6 / 6.0 / 1.7
11 TH	0207 / 0936 / 1433 / 2157	6.4 / 1.4 / 6.1 / 1.6	26 F	0331 / 1023 / 1601 / 2249	5.7 / 2.0 / 5.6 / 2.1
12 F	0258 / 1023 / 1533 / 2253	6.0 / 1.7 / 5.7 / 1.9	27 SA	0432 / 1122 / 1705	5.3 / 2.3 / 5.3
13 SA	0410 / 1128 / 1713	5.6 / 2.1 / 5.4	28 SU	0011 / 0542 / 1252 / 1817	2.4 / 5.1 / 2.5 / 5.2
14 SU	0014 / 0613 / 1308 / 1854	2.1 / 5.4 / 2.1 / 5.5	29 M	0133 / 0704 / 1405 / 1933	2.3 / 5.1 / 2.3 / 5.4
15 M	0148 / 0736 / 1433 / 2005	1.9 / 5.6 / 1.8 / 5.8	30 TU	0240 / 0820 / 1506 / 2032	2.0 / 5.4 / 2.0 / 5.7

Chart Datum: 3·67 metres below Ordnance Datum (Newlyn). HAT is 7·4 metres above Chart Datum.

DOVER LAT 51°07'N LONG 1°19'E
TIMES AND HEIGHTS OF HIGH AND LOW WATERS

STANDARD TIME (UT)
For Summer Time add ONE hour in **non-shaded areas**

Dates in red are SPRINGS
Dates in blue are NEAPS

YEAR 2019

MAY

Day	Time	m	Day	Time	m
1 W	0335 / 0904 / 1557 / 2115	1.7 / 5.7 / 1.7 / 6.0	16 TH	0352 / 0914 / 1624 / 2130	1.1 / 6.3 / 1.2 / 6.5
2 TH	0422 / 0937 / 1642 / 2149	1.4 / 6.0 / 1.4 / 6.2	17 F	0456 / 1000 / 1721 / 2216	0.8 / 6.5 / 0.9 / 6.7
3 F	0504 / 1008 / 1722 / 2221	1.2 / 6.2 / 1.2 / 6.4	18 SA	0550 / 1042 / 1809 / ○2259	0.7 / 6.6 / 0.8 / 6.8
4 SA	0544 / 1040 / 1800 / ●2253	1.1 / 6.3 / 1.1 / 6.5	19 SU	0636 / 1121 / 1851 / 2340	0.6 / 6.7 / 0.7 / 6.8
5 SU	0622 / 1111 / 1837 / 2324	1.0 / 6.5 / 1.0 / 6.6	20 M	0715 / 1159 / 1929	0.7 / 6.7 / 0.8
6 M	0659 / 1142 / 1914 / 2357	0.9 / 6.6 / 0.9 / 6.6	21 TU	0019 / 0750 / 1238 / 2004	6.7 / 0.8 / 6.6 / 0.9
7 TU	0735 / 1216 / 1950	0.9 / 6.6 / 0.9	22 W	0059 / 0823 / 1318 / 2036	6.6 / 1.0 / 6.5 / 1.1
8 W	0032 / 0809 / 1254 / 2025	6.6 / 1.0 / 6.6 / 1.0	23 TH	0138 / 0852 / 1359 / 2106	6.3 / 1.3 / 6.3 / 1.4
9 TH	0112 / 0845 / 1337 / 2105	6.6 / 1.1 / 6.4 / 1.2	24 F	0220 / 0919 / 1442 / 2138	6.0 / 1.6 / 6.1 / 1.7
10 F	0159 / 0926 / 1429 / 2151	6.3 / 1.4 / 6.2 / 1.4	25 SA	0306 / 0951 / 1531 / 2219	5.7 / 1.9 / 5.8 / 1.9
11 SA	0256 / 1016 / 1536 / 2250	6.0 / 1.7 / 5.8 / 1.7	26 SU	0401 / 1037 / 1626 / ◑2319	5.4 / 2.1 / 5.5 / 2.2
12 SU	0417 / 1124 / 1701 / ◑	5.7 / 1.9 / 5.6	27 M	0503 / 1147 / 1729	5.2 / 2.3 / 5.4
13 M	0009 / 0556 / 1253 / 1827	1.8 / 5.6 / 2.0 / 5.7	28 TU	0037 / 0610 / 1309 / 1834	2.2 / 5.2 / 2.3 / 5.4
14 TU	0132 / 0716 / 1411 / 1940	1.7 / 5.7 / 1.7 / 5.9	29 W	0146 / 0714 / 1414 / 1934	2.1 / 5.3 / 2.1 / 5.6
15 W	0243 / 0820 / 1520 / 2039	1.4 / 6.0 / 1.4 / 6.2	30 TH	0243 / 0807 / 1509 / 2022	1.8 / 5.6 / 1.8 / 5.8
			31 F	0334 / 0850 / 1558 / 2103	1.5 / 5.8 / 1.6 / 6.1

JUNE

Day	Time	m	Day	Time	m
1 SA	0423 / 0928 / 1645 / 2142	1.3 / 6.1 / 1.3 / 6.3	16 SU	0523 / 1022 / 1742 / 2242	1.0 / 6.3 / 1.1 / 6.5
2 SU	0510 / 1006 / 1730 / 2220	1.1 / 6.3 / 1.2 / 6.5	17 M	0611 / 1103 / 1827 / ○2324	1.0 / 6.5 / 1.0 / 6.5
3 M	0554 / 1043 / 1813 / ●2258	1.0 / 6.5 / 1.0 / 6.6	18 TU	0651 / 1142 / 1907	1.3 / 6.5 / 1.0
4 TU	0637 / 1121 / 1855 / 2338	0.9 / 6.6 / 0.9 / 6.6	19 W	0003 / 0727 / 1221 / 1943	6.5 / 1.1 / 6.5 / 1.1
5 W	0718 / 1202 / 1936	0.9 / 6.6 / 0.9	20 TH	0042 / 0800 / 1300 / 2016	6.4 / 1.2 / 6.5 / 1.2
6 TH	0020 / 0759 / 1247 / 2019	6.6 / 1.0 / 6.6 / 0.9	21 F	0120 / 0828 / 1339 / 2046	6.2 / 1.4 / 6.4 / 1.4
7 F	0107 / 0840 / 1335 / 2104	6.5 / 1.1 / 6.5 / 1.1	22 SA	0158 / 0856 / 1417 / 2117	6.0 / 1.5 / 6.2 / 1.5
8 SA	0158 / 0926 / 1430 / 2154	6.4 / 1.3 / 6.4 / 1.2	23 SU	0238 / 0928 / 1456 / 2154	5.8 / 1.7 / 6.0 / 1.7
9 SU	0259 / 1018 / 1532 / 2252	6.1 / 1.4 / 6.2 / 1.4	24 M	0322 / 1007 / 1539 / 2238	5.6 / 1.9 / 5.8 / 1.9
10 M	0410 / 1121 / 1639 / ◑2359	5.9 / 1.6 / 6.0 / 1.5	25 TU	0415 / 1055 / 1631 / ◑2335	5.4 / 2.0 / 5.6 / 2.0
11 TU	0530 / 1231 / 1753	5.8 / 1.7 / 5.9	26 W	0515 / 1157 / 1732	5.3 / 2.2 / 5.5
12 W	0107 / 0646 / 1340 / 1906	1.5 / 5.8 / 1.7 / 6.0	27 TH	0043 / 0618 / 1312 / 1834	2.0 / 5.3 / 2.2 / 5.5
13 TH	0214 / 0752 / 1446 / 2011	1.4 / 5.9 / 1.5 / 6.1	28 F	0150 / 0716 / 1419 / 1931	1.9 / 5.4 / 2.0 / 5.7
14 F	0320 / 0857 / 1551 / 2106	1.3 / 6.0 / 1.4 / 6.3	29 SA	0250 / 0807 / 1517 / 2021	1.7 / 5.7 / 1.8 / 5.9
15 SA	0426 / 0937 / 1651 / 2156	1.1 / 6.2 / 1.2 / 6.4	30 SU	0346 / 0853 / 1612 / 2108	1.5 / 5.9 / 1.5 / 6.2

JULY

Day	Time	m	Day	Time	m
1 M	0439 / 0938 / 1703 / 2154	1.3 / 6.2 / 1.2 / 6.4	16 TU	0549 / 1048 / 1808 / ○2312	1.3 / 6.3 / 1.2 / 6.3
2 TU	0530 / 1022 / 1752 / ●2240	1.1 / 6.4 / 1.1 / 6.6	17 W	0631 / 1127 / 1849 / 2349	1.3 / 6.4 / 1.2 / 6.3
3 W	0619 / 1107 / 1840 / 2326	1.0 / 6.6 / 0.9 / 6.7	18 TH	0707 / 1204 / 1925	1.3 / 6.5 / 1.2
4 TH	0708 / 1153 / 1929	0.9 / 6.7 / 0.8	19 F	0025 / 0738 / 1241 / 1956	6.3 / 1.3 / 6.5 / 1.2
5 F	0013 / 0756 / 1240 / 2017	6.7 / 0.9 / 6.8 / 0.8	20 SA	0100 / 0806 / 1316 / 2026	6.2 / 1.4 / 6.4 / 1.3
6 SA	0102 / 0842 / 1329 / 2106	6.6 / 0.9 / 6.8 / 0.8	21 SU	0133 / 0833 / 1348 / 2055	6.1 / 1.4 / 6.3 / 1.4
7 SU	0153 / 0928 / 1421 / 2154	6.5 / 1.0 / 6.7 / 0.9	22 M	0203 / 0904 / 1415 / 2128	6.0 / 1.5 / 6.2 / 1.5
8 M	0248 / 1014 / 1515 / 2244	6.3 / 1.2 / 6.5 / 1.0	23 TU	0232 / 0938 / 1446 / 2204	5.8 / 1.6 / 6.0 / 1.6
9 TU	0348 / 0919 / 1612 / ◑2338	6.1 / 1.4 / 6.3 / 1.2	24 W	0306 / 1017 / 1526 / 2246	5.7 / 1.8 / 5.9 / 1.8
10 W	0453 / 1201 / 1715	5.9 / 1.6 / 6.0	25 TH	0352 / 1102 / 1618 / ◑2339	5.5 / 2.0 / 5.7 / 2.0
11 TH	0037 / 0605 / 1304 / 1826	1.4 / 5.7 / 1.7 / 5.9	26 F	0501 / 1202 / 1728	5.3 / 2.2 / 5.5
12 F	0140 / 0717 / 1410 / 1939	1.5 / 5.7 / 1.8 / 5.9	27 SA	0051 / 0625 / 1325 / 1844	2.1 / 5.3 / 2.2 / 5.6
13 SA	0246 / 0821 / 1518 / 2045	1.6 / 5.7 / 1.7 / 6.0	28 SU	0208 / 0732 / 1439 / 1950	1.9 / 5.5 / 2.0 / 5.8
14 SU	0355 / 0917 / 1624 / 2142	1.5 / 5.9 / 1.5 / 6.1	29 M	0314 / 0829 / 1542 / 2047	1.7 / 5.8 / 1.7 / 6.1
15 M	0458 / 1005 / 1721 / 2230	1.4 / 6.1 / 1.3 / 6.2	30 TU	0414 / 0920 / 1640 / 2140	1.4 / 6.1 / 1.3 / 6.4
			31 W	0510 / 1009 / 1735 / 2229	1.2 / 6.5 / 1.1 / 6.6

AUGUST

Day	Time	m	Day	Time	m
1 TH	0606 / 1057 / 1830 / ●2317	1.0 / 6.7 / 0.8 / 6.8	16 F	0647 / 1145 / 1905	1.3 / 6.5 / 1.2
2 F	0701 / 1143 / 1923	0.8 / 6.9 / 0.6	17 SA	0004 / 0715 / 1219 / 1933	6.3 / 1.3 / 6.6 / 1.2
3 SA	0004 / 0751 / 1229 / 2013	6.8 / 0.7 / 7.0 / 0.5	18 SU	0035 / 0741 / 1249 / 2000	6.3 / 1.3 / 6.5 / 1.2
4 SU	0051 / 0836 / 1315 / 2059	6.8 / 0.7 / 7.0 / 0.5	19 M	0102 / 0808 / 1314 / 2029	6.2 / 1.3 / 6.4 / 1.2
5 M	0139 / 0917 / 1403 / 2141	6.7 / 0.8 / 6.9 / 0.6	20 TU	0125 / 0838 / 1336 / 2100	6.2 / 1.3 / 6.4 / 1.3
6 TU	0228 / 0957 / 1452 / 2224	6.6 / 0.9 / 6.8 / 0.8	21 W	0147 / 0910 / 1404 / 2131	6.1 / 1.4 / 6.3 / 1.5
7 W	0320 / 1039 / 1544 / ◑2309	6.3 / 1.2 / 6.4 / 1.2	22 TH	0218 / 0943 / 1440 / 2206	6.0 / 1.6 / 6.1 / 1.7
8 TH	0416 / 1127 / 1641	6.0 / 1.6 / 6.1	23 F	0258 / 1022 / 1525 / ◗2249	5.8 / 1.9 / 5.9 / 1.9
9 F	0002 / 0521 / 1227 / 1747	1.6 / 5.7 / 1.9 / 5.7	24 SA	0351 / 1111 / 1627 / 2349	5.5 / 2.1 / 5.6 / 2.2
10 SA	0106 / 0637 / 1336 / 1907	1.9 / 5.5 / 2.1 / 5.6	25 SU	0518 / 1226 / 1807	5.3 / 2.3 / 5.4
11 SU	0215 / 0755 / 1450 / 2031	2.0 / 5.5 / 2.0 / 5.6	26 M	0125 / 0707 / 1405 / 1932	2.2 / 5.4 / 2.2 / 5.6
12 M	0329 / 0901 / 1605 / 2135	1.9 / 5.7 / 1.8 / 5.9	27 TU	0248 / 0812 / 1518 / 2036	2.0 / 5.7 / 1.8 / 6.2
13 TU	0439 / 0952 / 1705 / 2223	1.7 / 6.0 / 1.5 / 6.1	28 W	0354 / 0908 / 1621 / 2131	1.6 / 6.1 / 1.4 / 6.4
14 W	0531 / 1033 / 1753 / 2301	1.5 / 6.2 / 1.3 / 6.2	29 TH	0455 / 0958 / 1720 / 2221	1.2 / 6.6 / 1.0 / 6.7
15 TH	0613 / 1110 / 1833 / 2333	1.4 / 6.4 / 1.2 / 6.3	30 F	0554 / 1044 / 1818 / ●2307	1.0 / 6.9 / 0.7 / 6.9
			31 SA	0649 / 1129 / 1912 / 2351	0.8 / 7.1 / 0.5 / 7.0

Chart Datum: 3·67 metres below Ordnance Datum (Newlyn). HAT is 7·4 metres above Chart Datum.

SE England

»» FREE monthly updates. Register at ««
www.reedsnauticalalmanac.co.uk

151

Dover tides – Dover Strait

STANDARD TIME (UT)
For Summer Time add ONE hour in **non-shaded areas**

DOVER LAT 51°07′N LONG 1°19′E
TIMES AND HEIGHTS OF HIGH AND LOW WATERS

Dates in red are **SPRINGS**
Dates in blue are **NEAPS**

YEAR 2019

SEPTEMBER

Day	Time	m		Day	Time	m
1 SU	0737 / 1212 / 1959	0.6 / 7.2 / 0.4		16 M	0004 / 0713 / 1217 / 1932	6.4 / 1.2 / 6.6 / 1.1
2 M	0034 / 0818 / 1256 / 2041	7.0 / 0.6 / 7.2 / 0.4		17 TU	0028 / 0741 / 1239 / 2001	6.4 / 1.2 / 6.5 / 1.2
3 TU	0117 / 0855 / 1340 / 2119	6.9 / 0.7 / 7.1 / 0.5		18 W	0049 / 0811 / 1302 / 2031	6.4 / 1.3 / 6.5 / 1.3
4 W	0202 / 0931 / 1426 / 2157	6.7 / 0.9 / 6.9 / 0.9		19 TH	0113 / 0842 / 1330 / 2101	6.4 / 1.4 / 6.5 / 1.4
5 TH	0250 / 1009 / 1514 / 2237	6.4 / 1.2 / 6.5 / 1.3		20 F	0144 / 0914 / 1405 / 2134	6.3 / 1.5 / 6.3 / 1.6
6 F	0343 / 1052 / 1609 / 2326	6.0 / 1.7 / 6.0 / 1.8		21 SA	0224 / 0951 / 1449 / 2215	6.1 / 1.8 / 6.0 / 1.9
7 SA	0444 / 1149 / 1713	5.6 / 2.1 / 5.6		22 SU	0315 / 1039 / 1548 / 2311	5.7 / 2.1 / 5.6 / 2.3
8 SU	0030 / 0557 / 1306 / 1835	2.2 / 5.3 / 2.4 / 5.3		23 M	0433 / 1148 / 1750	5.3 / 2.4 / 5.4
9 M	0147 / 0725 / 1428 / 2021	2.3 / 5.3 / 2.3 / 5.4		24 TU	0044 / 0647 / 1336 / 1921	2.4 / 5.4 / 2.3 / 5.6
10 TU	0306 / 0842 / 1548 / 2125	2.2 / 5.6 / 2.0 / 5.7		25 W	0226 / 0756 / 1457 / 2025	2.1 / 5.7 / 1.9 / 6.0
11 W	0419 / 0934 / 1649 / 2209	1.9 / 6.0 / 1.6 / 6.0		26 TH	0337 / 0853 / 1603 / 2120	1.7 / 6.2 / 1.4 / 6.4
12 TH	0510 / 1013 / 1734 / 2243	1.6 / 6.3 / 1.4 / 6.3		27 F	0439 / 0942 / 1704 / 2208	1.3 / 6.7 / 1.0 / 6.8
13 F	0549 / 1048 / 1810 / 2311	1.4 / 6.5 / 1.2 / 6.4		28 SA	0537 / 1027 / 1801 / 2252	0.9 / 7.0 / 0.6 / 7.0
14 SA	0621 / 1120 / 1839 / 2337	1.3 / 6.6 / 1.2 / 6.4		29 SU	0628 / 1116 / 1852 / 2333	0.7 / 7.2 / 0.4 / 7.1
15 SU	0647 / 1150 / 1905	1.3 / 6.6 / 1.2		30 M	0714 / 1152 / 1937	0.6 / 7.3 / 0.4

OCTOBER

Day	Time	m		Day	Time	m
1 TU	0013 / 0753 / 1234 / 2016	7.1 / 0.6 / 7.3 / 0.4		16 W	0716 / 1206 / 1935	1.2 / 6.6 / 1.2
2 W	0053 / 0829 / 1316 / 2052	7.0 / 0.7 / 7.1 / 0.7		17 TH	0019 / 0747 / 1233 / 2005	6.5 / 1.2 / 6.6 / 1.2
3 TH	0136 / 0904 / 1359 / 2127	6.8 / 1.0 / 6.8 / 1.0		18 F	0047 / 0819 / 1303 / 2036	6.5 / 1.3 / 6.6 / 1.4
4 F	0221 / 0939 / 1446 / 2203	6.4 / 1.3 / 6.4 / 1.5		19 SA	0121 / 0852 / 1340 / 2110	6.5 / 1.5 / 6.4 / 1.6
5 SA	0312 / 1019 / 1540 / 2247	6.0 / 1.8 / 5.9 / 2.0		20 SU	0203 / 0931 / 1426 / 2153	6.2 / 1.7 / 6.1 / 1.9
6 SU	0412 / 1113 / 1643 / 2353	5.6 / 2.3 / 5.5 / 2.5		21 M	0256 / 1020 / 1529 / 2249	5.8 / 2.0 / 5.6 / 2.3
7 M	0521 / 1234 / 1801	5.3 / 2.5 / 5.2		22 TU	0424 / 1130 / 1741	5.4 / 2.3 / 5.4
8 TU	0116 / 0644 / 1402 / 1956	2.6 / 5.3 / 2.4 / 5.3		23 W	0019 / 0623 / 1314 / 1905	2.4 / 5.5 / 2.2 / 5.6
9 W	0237 / 0809 / 1520 / 2101	2.4 / 5.5 / 2.1 / 5.7		24 TH	0203 / 0734 / 1435 / 2009	2.2 / 5.8 / 1.8 / 6.0
10 TH	0345 / 0904 / 1618 / 2142	2.0 / 5.9 / 1.7 / 6.0		25 F	0315 / 0832 / 1542 / 2103	1.7 / 6.3 / 1.3 / 6.4
11 F	0436 / 0944 / 1701 / 2213	1.7 / 6.2 / 1.4 / 6.2		26 SA	0417 / 0921 / 1643 / 2150	1.3 / 6.7 / 0.9 / 6.7
12 SA	0515 / 1017 / 1735 / 2238	1.5 / 6.5 / 1.3 / 6.4		27 SU	0513 / 1007 / 1739 / 2233	1.0 / 7.0 / 0.7 / 6.9
13 SU	0547 / 1048 / 1804 / 2304	1.3 / 6.6 / 1.2 / 6.5		28 M	0603 / 1049 / 1828 / 2312	0.8 / 7.2 / 0.5 / 7.0
14 M	0615 / 1116 / 1833 / 2330	1.3 / 6.6 / 1.1 / 6.5		29 TU	0647 / 1131 / 1912 / 2351	0.7 / 7.2 / 0.5 / 7.0
15 TU	0644 / 1142 / 1903 / 2355	1.2 / 6.6 / 1.1 / 6.5		30 W	0727 / 1211 / 1950	0.7 / 7.2 / 0.6
				31 TH	0031 / 0803 / 1253 / 2025	6.9 / 0.8 / 7.0 / 0.9

NOVEMBER

Day	Time	m		Day	Time	m
1 F	0113 / 0839 / 1335 / 2059	6.8 / 1.1 / 6.7 / 1.2		16 SA	0029 / 0802 / 1247 / 2019	6.6 / 1.3 / 6.6 / 1.4
2 SA	0157 / 0914 / 1421 / 2132	6.5 / 1.4 / 6.3 / 1.7		17 SU	0109 / 0839 / 1328 / 2056	6.5 / 1.4 / 6.4 / 1.6
3 SU	0246 / 0950 / 1514 / 2209	6.1 / 1.8 / 5.9 / 2.1		18 M	0155 / 0921 / 1419 / 2142	6.3 / 1.6 / 6.1 / 1.8
4 M	0342 / 1038 / 1614 / 2304	5.8 / 2.2 / 5.4 / 2.5		19 TU	0254 / 1013 / 1529 / 2239	6.0 / 1.8 / 5.7 / 2.1
5 TU	0446 / 1154 / 1724	5.5 / 2.5 / 5.2		20 W	0417 / 1123 / 1717	5.7 / 2.0 / 5.6
6 W	0033 / 0557 / 1320 / 1851	2.7 / 5.3 / 2.5 / 5.2		21 TH	0002 / 0550 / 1251 / 1840	2.2 / 5.7 / 2.0 / 5.7
7 TH	0153 / 0716 / 1431 / 2012	2.5 / 5.5 / 2.2 / 5.5		22 F	0132 / 0704 / 1407 / 1945	2.1 / 5.9 / 1.7 / 6.0
8 F	0258 / 0818 / 1528 / 2058	2.2 / 5.8 / 1.9 / 5.8		23 SA	0243 / 0805 / 1514 / 2040	1.7 / 6.2 / 1.3 / 6.3
9 SA	0350 / 0902 / 1613 / 2130	1.9 / 6.1 / 1.6 / 6.1		24 SU	0347 / 0857 / 1617 / 2129	1.4 / 6.6 / 1.0 / 6.5
10 SU	0432 / 0938 / 1651 / 2159	1.6 / 6.3 / 1.4 / 6.3		25 M	0445 / 0945 / 1714 / 2213	1.1 / 6.8 / 0.8 / 6.7
11 M	0508 / 1009 / 1726 / 2228	1.4 / 6.4 / 1.2 / 6.4		26 TU	0537 / 1029 / 1804 / 2254	1.0 / 6.9 / 0.7 / 6.8
12 TU	0543 / 1039 / 1802 / 2257	1.3 / 6.5 / 1.2 / 6.5		27 W	0622 / 1112 / 1848 / 2334	0.9 / 7.0 / 0.8 / 6.8
13 W	0617 / 1108 / 1837 / 2326	1.2 / 6.6 / 1.1 / 6.6		28 TH	0704 / 1153 / 1927	0.9 / 6.9 / 0.9
14 TH	0652 / 1138 / 1912 / 2356	1.2 / 6.6 / 1.1 / 6.6		29 F	0014 / 0742 / 1235 / 2002	6.8 / 1.0 / 6.7 / 1.1
15 F	0727 / 1210 / 1945	1.2 / 6.6 / 1.2		30 SA	0055 / 0819 / 1316 / 2035	6.7 / 1.2 / 6.5 / 1.4

DECEMBER

Day	Time	m		Day	Time	m
1 SU	0137 / 0854 / 1400 / 2106	6.5 / 1.4 / 6.2 / 1.7		16 M	0106 / 0836 / 1326 / 2053	6.6 / 1.2 / 6.4 / 1.4
2 M	0222 / 0928 / 1448 / 2138	6.2 / 1.7 / 5.9 / 2.0		17 TU	0155 / 0922 / 1419 / 2140	6.5 / 1.3 / 6.2 / 1.5
3 TU	0311 / 1008 / 1542 / 2220	5.9 / 2.0 / 5.5 / 2.3		18 W	0252 / 1014 / 1523 / 2235	6.3 / 1.5 / 6.0 / 1.7
4 W	0406 / 1102 / 1644 / 2320	5.7 / 2.2 / 5.3 / 2.5		19 TH	0357 / 1115 / 1640 / 2341	6.1 / 1.6 / 5.8 / 1.9
5 TH	0508 / 1215 / 1751	5.5 / 2.3 / 5.2		20 F	0510 / 1224 / 1803	5.9 / 1.7 / 5.7
6 F	0044 / 0615 / 1327 / 1900	2.5 / 5.4 / 2.3 / 5.3		21 SA	0054 / 0625 / 1333 / 1913	1.9 / 5.9 / 1.5 / 5.8
7 SA	0155 / 0719 / 1426 / 1956	2.4 / 5.6 / 2.0 / 5.5		22 SU	0204 / 0733 / 1439 / 2014	1.8 / 6.1 / 1.5 / 6.0
8 SU	0254 / 0810 / 1519 / 2040	2.1 / 5.8 / 1.8 / 5.8		23 M	0311 / 0833 / 1546 / 2107	1.6 / 6.2 / 1.3 / 6.2
9 M	0344 / 0851 / 1606 / 2118	1.8 / 6.0 / 1.5 / 6.0		24 TU	0416 / 0926 / 1650 / 2156	1.4 / 6.4 / 1.2 / 6.3
10 TU	0430 / 0928 / 1651 / 2153	1.6 / 6.2 / 1.3 / 6.2		25 W	0514 / 1015 / 1743 / 2240	1.2 / 6.5 / 1.1 / 6.5
11 W	0512 / 1005 / 1733 / 2228	1.4 / 6.4 / 1.2 / 6.4		26 TH	0603 / 1100 / 1829 / 2321	1.1 / 6.6 / 1.0 / 6.6
12 TH	0553 / 1041 / 1814 / 2304	1.2 / 6.5 / 1.1 / 6.6		27 F	0647 / 1141 / 1909	1.0 / 6.6 / 1.1
13 F	0633 / 1119 / 1854 / 2341	1.1 / 6.6 / 1.1 / 6.6		28 SA	0001 / 0727 / 1221 / 1945	6.6 / 1.1 / 6.5 / 1.2
14 SA	0713 / 1157 / 1933	1.1 / 6.6 / 1.1		29 SU	0041 / 0804 / 1301 / 2017	6.6 / 1.2 / 6.4 / 1.4
15 SU	0021 / 0753 / 1240 / 2012	6.7 / 1.1 / 6.6 / 1.2		30 M	0120 / 0838 / 1340 / 2046	6.5 / 1.3 / 6.2 / 1.5
				31 TU	0159 / 0909 / 1420 / 2114	6.3 / 1.5 / 6.0 / 1.7

Chart Datum: 3·67 metres below Ordnance Datum (Newlyn). HAT is 7·4 metres above Chart Datum.

»»**FREE** monthly updates. Register at
www.reedsnauticalalmanac.co.uk««

3.12 DOVER STRAIT

For orientation only – due to scale not all lights and buoys are shown

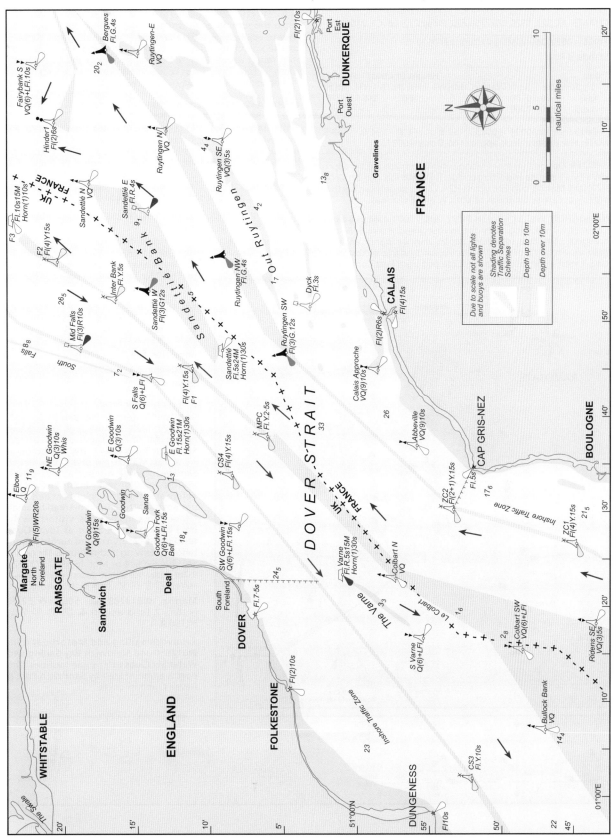

3.13 RAMSGATE

Kent 51°19'·51N 01°25'·50E ✿✿✿✿♦♦♦♦✿✿✿

CHARTS AC 323, 1828, 1827, 5605/6; Imray C30, C8, C1, 2100

TIDES +0030 Dover; ML 2·7; Duration 0530

Standard Port

Times				Height (metres)			
High Water		Low Water		MHWS	MHWN	MLWN	MLWS
0000	0600	0100	0700	5·7	4·0	1·4	0·6
1200	1800	1300	1900				
RICHBOROUGH							
–0015	–0015	+0014	+0023	–1·8	–1·3	–1·0	–0·5

NOTE: HW Broadstairs is approx HW Dover +0037.

SHELTER Options: (a) Inner Marina, min depth 2m. Access approx HW ±2 via flap gate and lifting bridge; (b) W Marina, min 2m, access H24; (c) E Marina, min 2m, access H24. Larger vessels can berth on outer wavebreak pontoons of both W and E marinas.

NAVIGATION WPT 51°19'·43N 01°27'·70E, 270°/1·45M to S bkwtr. Commercial shipping uses the well-marked main E-W chan dredged 7·5m. *Due to silting, depths may be significantly less than shown; parts of Eastern Marina, particularly, almost dry at LWS. Latest information may be obtained from Port Control.*

For ent/dep yachts must use the Recommended Yacht Track on the S side of the main buoyed chan. Ent/dep under power, or advise Port Control if unable to motor. Ent/dep Royal Hbr directly; cross the turning basin without delay **keeping close to the W Pier to avoid shoal patch alongside E Pier.** Holding area to the S of the S bkwtr must be used by yachts to keep the hbr ent clear for freight vessels. Beware Dike Bank to the N and Quern Bank close S of the chan. Cross Ledge and Brake shoals are further S. Speed limit 5kn. See www.rma.eu.com.

LIGHTS AND MARKS Ldg lts 270°: front Dir Oc WRG 10s 10m 5M; rear, Oc 5s 17m 5M. N bkwtr hd = QG 10m 5M; S bkwtr hd = VQ R 10m 5M. At E Pier, **IPTS** (Sigs 2 and 3) visible from seaward and from within Royal Hbr, control appr into hbr limits (abeam Nos 1 & 2 buoys) and ent/exit to/from Royal Hbr. In addition a Fl Orange lt = ferry is under way; no other vessel may enter Hbr limits from seaward or leave Royal Hbr. Ent to inner marina controlled by separate IPTS to stbd of ent. Siren sounded approx 10 mins before gate closes; non-opening indicated by red ball or light.

COMMUNICATIONS (Code 01843) Dr 852853 ⊞ 225544. Hbr Office 572100; Broadstairs HM 861879.

Listen and contact *Ramsgate Port Control* on Ch 14 when intending to enter or leave Royal Hbr. Only when in Royal Hbr call *Ramsgate Marina* Ch 80 for a berth. Ramsgate Dock Office must be called on Ch 14 for information on Inner Marina Lock.

FACILITIES Marina www.portoframsgate.co.uk ☎572100, ⬩ £32/craft<5t; 510 ⬩+300♥ £2.96; ⬩ ♣ ⬩ ⬩ ⚒ ⬩ ⬩ ⬩ ⓔ ⬩ ⬩(40t) ⬩ ACA.

Royal Hbr ⬩ ⬩(0600-2200).

Royal Temple YC ☎591766, ⬩.

Town ⬩ ✉ ⑧ ⬩ ✕ ⬩ ⇌.

ADJACENT HARBOUR

SANDWICH, Kent, 51°16'·83N 01°21'·20E. AC 1827 1828. Richborough differences above; ML 1·4m; Duration 0520. HW Sandwich Town quay is HW Richborough +1. Access via narrow drying channel to Sandwich; arrive off ent at HW Dover. Visitors should seek local knowledge before arriving by day; night ent definitely not advised. The chan is marked by small lateral buoys and beacons. Visitors' berths on S bank of the R Stour at Town Quay ☎(01304) 612162. Limited turning room before the swing bridge (opens 1hr notice ☎01304 620644 or Mobile 0860 378792), 1·7m clearance when closed. Facilities: ⬩ HW±2 ✉ ⑧ ⬩ ✕ ⬩ ⇌ ✈ (Manston).

Marina ☎613783, ⬩ 50⬩+♥ £2.20 (<18m, 2·1m draught) ♣ ⬩ ⬩ ⚒ ⬩ ⬩(15t) ⬩ ⬩.

Sandwich Sailing and Motorboat Club ☎617650 and **Sandwich Bay Sailing and Water Ski Clubs** offer some facilities. The port is administered by Sandwich Port & Haven Commissioners.

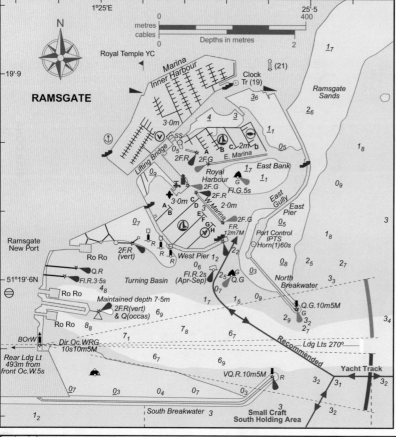

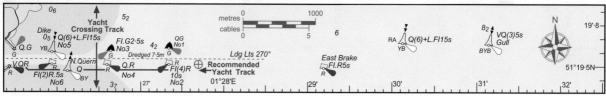

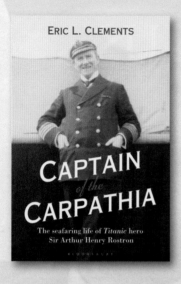
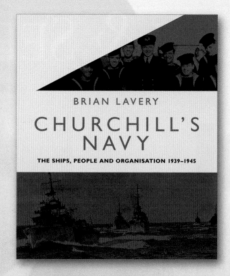
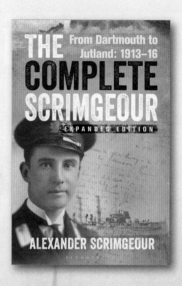
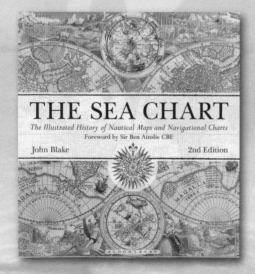
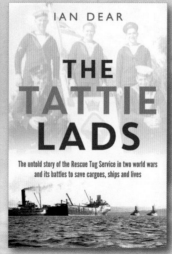
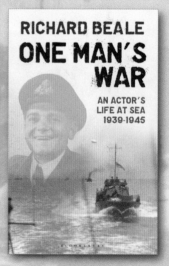

Northern France

Dunkerque to the L'Aber-Ildut, including the Channel Islands

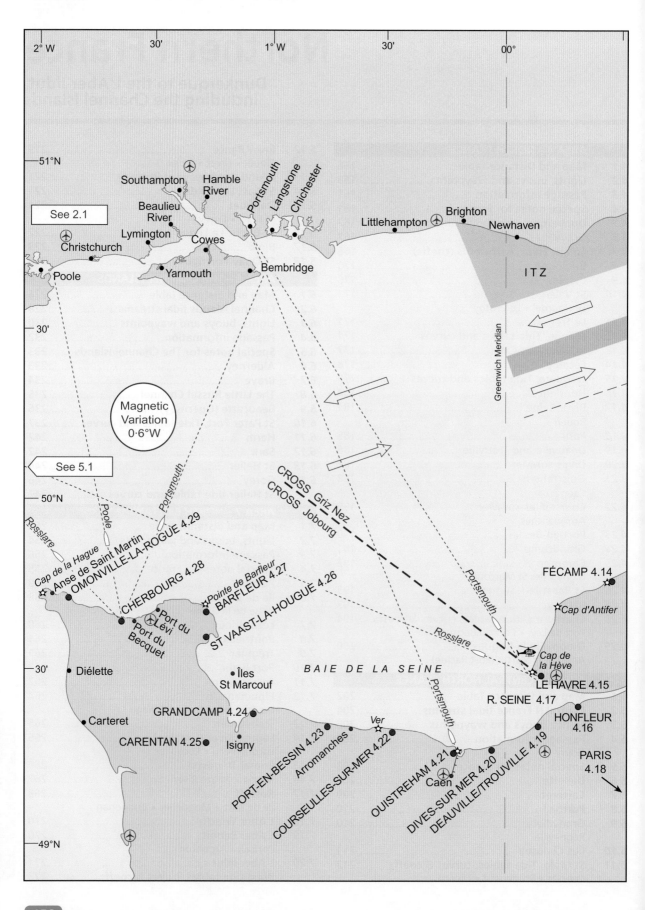

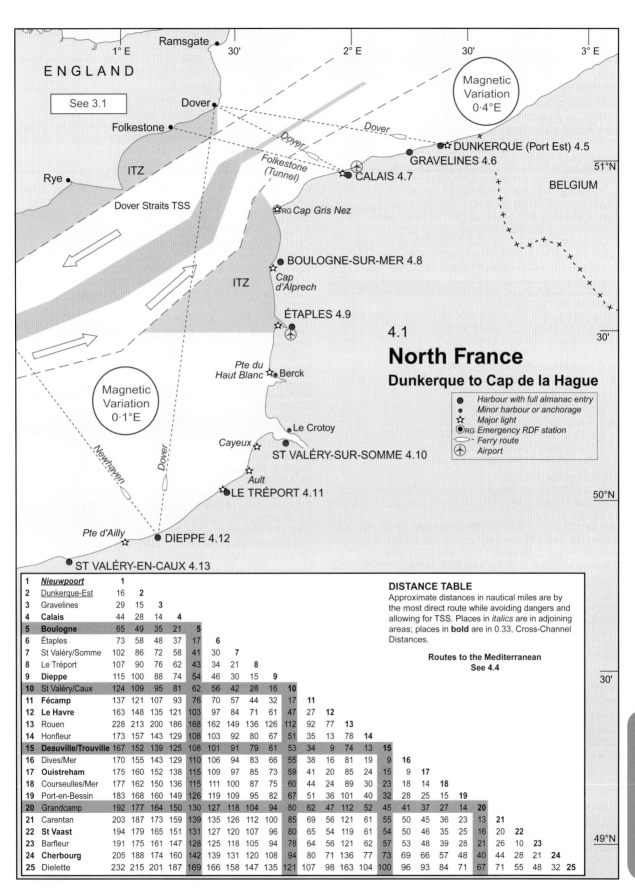

ENGLAND

See 3.1

Ramsgate
Dover
Folkestone
Rye
ITZ
Dover Straits TSS
Newhaven
Dover

Magnetic Variation 0·4°E

Dover
Folkestone (Tunnel)
DUNKERQUE (Port Est) 4.5
GRAVELINES 4.6
CALAIS 4.7
RG Cap Gris Nez
BELGIUM
51°N
30'

BOULOGNE-SUR-MER 4.8
Cap d'Alprech
ITZ
ÉTAPLES 4.9

Magnetic Variation 0·1°E

Pte du Haut Blanc
Berck
Le Crotoy
Cayeux
ST VALÉRY-SUR-SOMME 4.10
Ault
LE TRÉPORT 4.11
Pte d'Ailly
DIEPPE 4.12
ST VALÉRY-EN-CAUX 4.13
50°N

4.1
North France
Dunkerque to Cap de la Hague

- ● Harbour with full almanac entry
- ● Minor harbour or anchorage
- ☆ Major light
- ◉RG Emergency RDF station
- ⌐⌐ Ferry route
- ✈ Airport

DISTANCE TABLE

Approximate distances in nautical miles are by the most direct route while avoiding dangers and allowing for TSS. Places in *italics* are in adjoining areas; places in **bold** are in 0.33, Cross-Channel Distances.

Routes to the Mediterranean
See 4.4

30'

#	Place	1	2	3	4	5	6	7	8	9	10	11	12	13	14	15	16	17	18	19	20	21	22	23	24	25
1	*Nieuwpoort*	1																								
2	Dunkerque-Est	16	2																							
3	Gravelines	29	15	3																						
4	**Calais**	44	28	14	4																					
5	**Boulogne**	65	49	35	21	5																				
6	Étaples	73	58	48	37	17	6																			
7	St Valéry/Somme	102	86	72	58	41	30	7																		
8	Le Tréport	107	90	76	62	43	34	21	8																	
9	**Dieppe**	115	100	88	74	54	46	30	15	9																
10	St Valéry/Caux	124	109	95	81	62	56	42	28	16	10															
11	**Fécamp**	137	121	107	93	76	70	57	44	32	17	11														
12	**Le Havre**	163	148	135	121	103	97	84	71	61	47	27	12													
13	Rouen	228	213	200	186	168	162	149	136	126	112	92	77	13												
14	Honfleur	173	157	143	129	108	103	92	80	67	51	35	13	78	14											
15	**Deauville/Trouville**	167	152	139	125	108	101	91	79	61	53	34	9	74	13	15										
16	Dives/Mer	170	155	143	129	110	106	94	83	66	55	38	16	81	19	9	16									
17	**Ouistreham**	175	160	152	138	115	109	97	85	73	59	41	20	85	24	15	9	17								
18	Courseulles/Mer	177	162	150	136	115	111	100	87	75	60	44	24	89	30	23	18	14	18							
19	Port-en-Bessin	183	168	160	149	126	119	109	95	82	67	51	36	101	40	32	28	25	15	19						
20	**Grandcamp**	192	177	164	150	130	127	118	104	94	80	62	47	112	52	45	41	37	27	14	20					
21	Carentan	203	187	173	159	139	135	126	112	100	85	69	56	121	61	50	50	45	36	23	13	21				
22	**St Vaast**	194	179	165	151	131	127	120	107	96	80	65	54	119	61	54	50	46	35	25	16	20	22			
23	Barfleur	191	175	161	147	128	125	118	105	94	78	64	56	121	62	57	53	48	39	28	21	26	10	23		
24	**Cherbourg**	205	188	174	160	142	139	131	120	108	94	80	71	136	77	73	69	66	57	48	40	44	28	21	24	
25	Dielette	232	215	201	187	169	166	158	147	135	121	107	98	163	104	100	96	93	84	71	67	71	55	48	32	25

49°N

N France

4.2 LIGHTS, BUOYS AND WAYPOINTS

Bold print = light with a nominal range of 15M or more. CAPITALS = place or feature. *CAPITAL ITALICS* = light-vessel, light float or Lanby. *Italics* = Fog signal. ***Bold italics*** = Racon. Many marks/buoys are fitted with AIS (MMSI No); see relevant charts.

OFFSHORE MARKS: W Hinder to Dover Strait

Fairy South ₹ VQ (6) + L Fl 10s; 51°21'·20N 02°17'·31E.
Fairy West ₹ VQ (9) 10s 6M; 51°23'·89N 02°09'·27E. 992271127.
Hinder 1 ₹ Fl (2) 6s; 51°20'·80N 02°10'·93E.
Bergues ₹ Fl G 4s 7m 4M; 51°17'·14N 02°18'·63E.
Bergues S ₹ Q (6) + L Fl 15s; 51°15'·09N 02°19'·42E.
Ruytingen E ₹ VQ; 51°14'·55N 02°17'·93E.
Ruytingen N ₹ VQ 4M; 51°13'·16N 02°10'·28E.
Ruytingen SE ₹ VQ (3) 15s; 51°09'·20N 02°08'·94E.992271126.
Ruytingen NW ₹ Fl G 4s 3M; 51°09'·11N 01°57'·30E.
Ruytingen W ₹ VQ 4M; 51°06'·93N 01°50'·45E.
Ruytingen SW ₹ Fl (3) G 12s 3M; 51°04'·33N 01°45'·84E.
Sandettié N ₹ VQ 6M; 51°18'·35N 02°04'·73E.
Sandettié E*Fl R 4s 7m 3M;51°14'·88N 02°02'·65E.992271119.

DUNKERQUE TO BOULOGNE

PASSE DE ZUYDCOOTE

E12 ₹ VQ (6) + L Fl 10s; 51°07'·92N 02°30'·61E. (Belgium)
CME ₹ Q (3) 10s; 51°07'·30N 02°30'·00E.
E11 ₹ Fl G 4s; 51°06'·85N 02°30'·83E.
E10 ₹ Fl (2) R 6s; 51°06'·26N 02°30'·40E.
E9 ₹ Fl (2) G 6s; 51°05'·61N 02°29'·60E.
E8 ₹ Fl (3) R 12s; 51°05'·12N 02°28'·60E. (E7 does not exist)

PASSE DE L'EST

E6 ₹ QR; 51°04'·86N 02°27'·08E.
E4 ₹ Fl R 4s; 51°04'·58N 02°24'·52E.
E1 ₹ Fl (2) G 6s; 51°04'·12N 02°23'·04E.
E2 ₹ Fl (2) R 6s; 51°04'·35N 02°22'·31E.
₹ Q (6) + L Fl 15s; 51°04'·28N 02°21'·73E.

DUNKERQUE PORT EST

E jetty ☆ Fl (2) R 6s 12m 10M; R □, W pylon; 51°03'·59N 02°21'·20E.
W jetty ☆ Fl (2) G 6s 35m 11M; W twr, brown top; 51°03'·63N 02°20'·95E.
Ldg lts 137·5°, front Q7m11M 51°02'·98N 02°22'·05E, rear, Q10m11M, 114m from front, both W cols, R tops, synched.
Inner W jetty ₹ Q 11m 9M; 51°03'·33N 02°21'·43E.
Dunkerque lt ho ☆ Fl (2) 10s 59m **26M**; 51°02'·93N 02°21'·86E.

DUNKERQUE INTERMEDIATE CHANNEL

DW30 ₹ QR; 51°04'·14N 02°20'·16E.
DW29 ₹ QG; 51°03'·85N 02°20'·21E.
DW16 ₹ Fl (2) R 6s; 51°03'·52N 02°08'·62E.
DKB ₹ VQ (9) 10s; 51°02'·95N 02°09'·26E.
DW12 ₹ Fl (3) R 12s; 51°03'·38N 02°05'·36E.
DW11 ₹ Fl (3) G 12s; 51°02'·79N 02°05'·53E. (DW10-7 omitted)

GRAVELINES

W jetty ₹ Fl (2) WG 6s 9m W/G4M; 085°-W-224°-G-085°; Y ○ twr, G top; 51°00'·94N 02°05'·49E.

PASSE DE L'OUEST

DW6 ₹ VQ R; 51°02'· 81N 02°00'· 80E.
DW5 ₹ QG; 51°02'·20N 02°00'·92E.
DKA ₹ L Fl 10s; 51°02'·55N 01°56'·96E.
RCE (Ridens de Calais East) ₹ Fl G 4s; 51°02'·29N 01°52'·98E.
Dyck ₹ Fl R 4s; ***Racon B***; 51° 02'·90N 01°51'·80E.
RCA (Ridens de Calais Approach) ₹ Q; 51°01'·00N 01°48'·53E. 992271125.

CALAIS

E jetty ☆ Fl (2) R 6s 12m **17M**; (in fog two Fl (2) 6s (vert) on request); Gy twr, R top; *Horn (2) 40s;* 50°58'·39N 01°50'·45E.
E Jetty root, Dir ₹ WG 12m 1M; 089·4°-FG-093·7°-Al WG-094°·3-FW-098·6°; metal structure; 50°58'·25N 01°51'·20E.

W jetty ₹ Iso G 3s 12m 9M; (in fog Fl 5s on request); W twr, G top; *Bell 5s;* 50°58'·24N 01°50'·40E.
Calais ☆ Fl (4) 15s 59m **22M**; vis 073°-260°; W 8-sided twr, B top; 50°57'·68N 01°51'·21E (440m E of marina entry gate).

CALAIS, WESTERN APPROACH

CA6 ₹ Fl (3) R 12s; 50°58'·63N 01°49'·92E.
CA4 ₹ Fl (2)R 6s; 50°58'·38N 01°48'·65E.
CA1 ₹ Fl G 4s; 50°57'·64N 01°46'·14E (0·5M NNW of Sangatte).
Sangatte ₹ Oc WG 4s 13m W8M, G5M; 065°-G-089°-W-152°-G-245°; W pylon, B top; 50°57'·19N 01°46'·50E.
Les Quénocs ₹ VQ; 50°56'·85N 01°41'·12E.
Abbeville wreck ₹ VQ (9) 10s; 50°56'·08N 01°37'·58E.
Cap Gris-Nez ☆ Fl 5s 72m **29M**; 005°-232°; W twr, B top; 50°52'·09N 01°34'·96E.

OFFSHORE MARKS: DOVER STRAIT TSS, French side

Colbart N ₹ Q 6M; 50°57'·50N 01°23'·30E. 992271121.
Colbart SW₹ VQ (6) + L Fl 10s 8m; 50°48'·87N 01°16'·32E.
ZC2 (Zone Cotière) ₹ Fl (2+1) Y 15s 5M; 50°53'·54N 01°30'·89E.
ZC1 ₹ Fl (4) Y 15s 4MN; 50°44'·99N 01°27'·21E.
Ridens SE ₹ VQ (3) 5s 6M; 50°43'·48N 01°18'·87E.
Bassurelle ₹ Fl (4) R 15s 6M; ***Racon B, 5-8M***; 50°32'·74N 00°57'·69E.
Vergoyer N ₹ VQ 5M; 50°39'·67N 01°22'·21E.
Vergoyer NW ₹ Fl (2) G 6s 4M; 50°37'·16N 01°17'·85E.
Vergoyer E ₹ VQ (3) 5s 6M; 50°35'·76N 01°19'·70E.
Vergoyer W ₹ Fl G 4s 4M; 50°34'·66N 01°13'·57E.
Vergoyer SW ₹ VQ (9) 10s 6M; 50°27'·01N 01°00'·03E.

BOULOGNE TO DIEPPE

BOULOGNE

Bassure de Baas ₹ VQ; 50°48'·53N 01°33'·05E.
Approches Boulogne ₹ VQ (6) + L Fl 10s 8m 6M; 50°45'·31N 01°31'·07E.
Digue N (detached) ₹ Fl (2) R 6s 10m 7M; 50°44'·71N 01°34'·18E.
Digue S (Carnot) ☆ Fl (2+1) 15s 25m **19M**; W twr, G top; 50°44'·44N 01°34'·05E.
Clearing brg 122·4° : Front, FG in a neon ▽ 4m 5M; 50°43'·71N 01°35'·66E. Rear, FR 44m 11M; intens 113°-133°; 560m from front.
Inner NE jetty ₹ FR 11m 7M; 50°43'·91N 01°35'·24E. Inner SW jetty ₹ FG 17m 5M; W col, G top; *Horn 30s;* 50°43'·90N 01°35'·11E.
Cap d'Alprech ☆ Fl (3) 15s 62m **23M**; W twr, B top; 50°41'·91N 01°33'·75E, 2·5M S of hbr ent.

LE TOUQUET/ÉTAPLES

Pointe de Lornel ₹ VQ (9) 10s 6m 3M; 50°33'·24N 01°35'·12E.
Mérida wreck ₹ VQ (3) 5s 50°32'·85N 01°33'·44E.
Camiers lt ho ₹ Oc (2) WRG 6s 17m W10M, R/G7M; 015°-G-090°-W-105°-R-141°; R pylon; 50°32'·86N 01°36'·28E.
Canche Est groyne ₹ Fl R 4s 8m; 50°32'·57N 01°35'·66E.
Le Touquet ☆ Fl (2) 10s 54m **25M**; Or twr, brown band, W&G top; 50°31'·43N 01°35'·52E.
Pointe du Haut-Blanc ☆ Fl 5s 44m **23M**; W twr, R bands, G top; 50°23'·90N 01°33'·67E (Berck).

BAIE DE LA SOMME

ATSO ₹ Mo (A) 12s; 50°14'·00N 01°28'·08E (shifts frequently).
Pte du Hourdel ₹ Oc (3) WG 12s 19m, W12M, G9M; 053°-W-248°-G-323°; tidal sigs; *Horn (3) 30s;* W twr, G top; 50°12'·90N 01°33'·98E.
Cayeux-sur-Mer ☆ Fl R 5s 32m **22M**; W twr, R top; 50°11'·65N 01°30'·72E.
Le Crotoy ₹ Oc (2) R 6s 19m 8M; 285°-135°; W pylon; 50°12'·91N 01°37'·40E. Marina ₹ Fl R & Fl G 2s 4m 2M; 50°12'·98N 01°38'·20E.

ST VALÉRY-SUR-SOMME

Trng wall head, ₹ Fl G 2.5s 2m 1M; 50°12'·25N 01°35'·85E,.
Embankment head ₹ Iso G 4s 9m 9M; 347°-222°; W pylon, G top; 50°12'·25N 01°36'·02E.
La Ferté môle ₹ Fl R 4s 9m 9M; 000°-250°; W pylon, R top; 50°11'·18N 01°38'·14E (ent to marina inlet).

LE TRÉPORT
Ault ☆ Oc (3) WR 12s 95m **W15M**, R11M; 040°-W-175°-R-220°; W twr, R top; 50°06'·28N 01°27'·23E (4M NE of Le Tréport).
W jetty ☆ Fl (2) G 10s 15m **20M**; W twr, G top; 50°03'·88N 01°22'·14E.

DIEPPE
W jetty ⚓ Iso G 4s 11m 8M; W twr, G top; *Horn 30s;* 49°56'·27N 01°04'·97E.
Quai de la Marne ⚓ QR 12m 3M; 49°55'·93N 01°05'·20E, E quay.
Pointe d'Ailly ☆ Fl (3) 20s 95m **31M**; W □ twr, G top; *Horn (3) 60s;* 49°54'·96N 00°57'·50E.

DIEPPE TO LE HAVRE
SAINT VALÉRY-EN-CAUX
W jetty ⚓ Fl (2) G 6s 13m 11M; W twr, G top; 49°52'·40N 00°42'·54E.
Paluel power station, restricted area, Paluel 2 ⚓ Fl Y 4s; 49°52'·33N 00°38'·45E; Wave recorder ⚓ Fl(5) Y 20s; 49°52'·30N 00°37'·87E; Paluel 1 ⚓ Fl Y 4s; 49°52'·33N 00°37'·53E.

FÉCAMP
N jetty ☆ Fl (2) 10s 15m **16M**; Gy twr, R top; 49°45'·94N 00°21'·80E.
Fécamp NW, Met Mast FECO1-YO2 ⚓ Mo(U) W 15s 13m 10M, Aero F R 36m, Aero UQ R 60m, Aero UQ W 60m(day) Gy pylon, Y platform; Horn Mo(U) 30s; 49°50'·85N 00°13'·14E. <u>992271122</u>.

PORT DU HAVRE-ANTIFER
Cap d'Antifer ☆ Fl 20s 128m **29M**; 021°-222°; Gy 8-sided twr, G top, on 90m cliffs; 49°41'·01N 00°09'·93E.
A17 ⚓ Iso G 4s; 49°41'·53N 00°01'·75E.
A18 ⚓ QR; 49°42'·02N 00°02'·18E. Cross the chan W of A17/18.
Ldg lts 127·5°, both Dir Oc 4s 113/135m **22M**; 127°-128°; by day F **33M** 126·5°-128·5° occas. **Front** ☆, 49°38'·31N 00°09'·12E.

LE HAVRE, APPROACH CHANNEL
Cap de la Hève ☆ Fl 5s 123m **24M**; 225°-196°; W 8-sided twr, R top; 49°30'·74N 00°04'·16E.
LHA ⬭ Mo (A) 12s 10m 6M; R&W; *Racon, 8-10M* (a series of 8 dots, or 8 groups of dots; distance between each dot or group represents 0·3M); 49°31'·38N 00°09'·86E. Reserve lt Mo (A).
Ldg lts 106·8°, both Dir F 36/78m **25M** (H24); intens 106°-108°; Gy twrs, G tops. Front, 49°28'·91N 00°06'·50E; rear, 0·73M from front.
LH3 ⚓ QG; 49°30'·84N 00°04'·02W. (LH1 & 2 buoys do not exist)
LH4 ⚓ QR; 49°31'·11N 00°03'·90W.
LH7 ⚓ Iso G 4s; 49°30'·25N 00°00'·82W.
LH8 ⚓ Fl (2) R 6s; 49°30'·44N 00°00'·70W.
Note the W-E longitude change. (LH9 buoy does not exist)
LH13 ⚓ Fl G 4s; 49°29'·32N 00°03'·62E (Ent to Port 2000).
LH14 ⚓ Fl R 4s; 49°29'·67N 00°03'·43E; 1m shoal depth close W.
LH16 ⚓ Fl (2) R 6s; 49°29'·45N 00°04'·28E.
LH 2000 ⚓ VQ (9) 10s; 49°29'·14N 00°04'·78E (Ent to Port 2000).

LE HAVRE
Digue N ☆ Fl R 5s 15m **21M**; IPTS; W ○ twr, R top; 49°29'·19N 00°05'·44E.
Digue S ⚓ VQ (3) G 2s 15m 11M; W twr, G top; 49°29'·05N 00°05'·38E.
Marina ent, W spur ⚓ Fl (2) R 6s 3M; 49°29'·22N 00°05'·53E.

THE SEINE ESTUARY UP TO HONFLEUR
CHENAL DE ROUEN
Nord du Mouillage ⚓ Fl (4) Y 15s; 49°28'·80N 00°00'·22E.
No 2 ⚓ QR; *Racon T*; 49°27'·70N 00°00'·60E.
No 4 ⚓ Fl R 2·5s; 49°27'·19N 00°01'·97E. Yachts keep N of chan.
Amfard SW ⚓ QR; 49°26'·30N 00°04'·82E.
No 10 ⚓ Fl R 4s; 49°26'·10N 00°06'·39E.
Digue du Ratier ⚓ VQ 10m 4M; 49°25'·94N 00°06'·59E.
Falaise des Fonds ☆ Fl (3) WRG 12s 15m, **W17M**, R/G13M; 040°-

G-080°-R-084°-G-100°- W-109°-R-162°-G-260°; W twr, G top; 49°25'·47N 00°12'·85E.
No 20 ⚓ Fl (2) R 6s; 49°25'·85N 00°13'·71E. (over to Honfleur)

HONFLEUR
Digue Ouest ⚓ QG 10m 6M; 49°25'·67N 00°13'·83E.
Digue Est ⚓ Q 9m 8M; *Horn (5) 40s;* 49°25'·67N 00°13'·95E.
Inner E jetty head, ⚓ Fl R 2M; 49°25'·40N 00°14'·10E.
No 22 ⚓ Fl R 4s; 49°25'·85N 00°15'·37E.

TROUVILLE TO COURSEULLES
CHENAL DE ROUEN TO DEAUVILLE and TROUVILLE
Ratelets ⚓ Q (9) 15s; 49°25'·29N 00°01'·71E.
Semoy ⚓ VQ (3) 5s; 49°24'·15N 00°02'·35E, close to 148° ldg line.
Trouville SW ⚓ VQ (9) 10s; 49°22'·54N 00°02'·56E.

DEAUVILLE and TROUVILLE
Ldg lts 145°, both Oc R 4s 11/17m 12/10M: Front, East inner jetty (*estacade*); 330°-150°; W twr, R top; 49°22'·03N 00°04'·48E. Rear, Pte de la Cahotte; synch; 120°-170°; 49°21'·93N 00°04'·58E.
W trng wall ⚓ Fl WG 4s 10m W9M, G6M; 005°-W-176°-G-005°; B pylon, G top; 49°22'·37N 00°04'·11E. Also 4 unlit SHM bcns.
E trng wall ⚓ Fl (4) WR 12s 8m W7M, R4M; 131°-W-175°-R-131°; W pylon, R top; 49°22'·22N 00°04'·33E. Also 3 unlit PHM bcns.
W outer bkwtr ⚓ Iso G 4s 9m 5M; 49°22'·11N 00°04'·33E.
West inner jetty (*estacade*) ⚓ QG 11m 9M; 49°22'·03N 00°04'·43E.

DIVES-SUR-MER *Note the E-W longitude change.*
DI ⚓ Iso 4s; 49°19'·17N 00°05'·86W.
No 1 ⚓ VQ G; 49°18'·50N 00°05'·67W. Ch bys freq moved.
No 2 ⚓ VQ R; 49°18'·51N 00°05'·56W.
No 3 ⚓ QG 7m 4M; W pylon, G top; 49°18'·30N 00°05'·55W.
No 5 ⚓ Fl G 4s 8m 4M; W pylon, G top; 49°18'·09N 00°05'·50W.
Bcns 3 & 5, if damaged, may be temporarily replaced by buoys.
No 7 ⚓ Fl G 4s; 49°17'·65N 00°05'·31W.

OUISTREHAM and CAEN
Merville ⚓ VQ; 49°19'·65N 00°13'·39W; spoil ground buoy.
Ouistreham ⚓ VQ (3) 5s; wreck buoy; 49°20'·42N 00°14'·81W.
Ldg lts 185°, both Dir Oc (3+1) R 12s 10/30m **17M**; intens 183·5°-186·5°, synch. **Front** ☆, E jetty, W mast, R top, 49°16'·99N 00°14'·81W. **Rear** ☆, 610m from front, tripod, R top.
No 1 ⚓ QG; 49°19'·19N 00°14'·67W.
No 2 ⚓ QR; 49°19'·17N 00°14'·43W.
Barnabé ⚓ QG 7m 5M; W pylon, G top; 49°18'·02N 00°14'·76W.
St-Médard ⚓ QR 7m 5M; W pylon, G top; 49°18'·02N 00°14'·62W.
Riva ⚓ Fl G 4s 9m 3M; W pylon, G top; 49°17'·73N 00°14'·79W.
Quilbé ⚓ Fl R 4s 9m 3M; W pylon, R top; 49°17'·72N 00°14'·67W.
Ouistreham lt ho ☆ Oc WR 4s 37m **W17M**, R13M; 115°-R-151°-W-115°; W twr, R top; 49°16'·79N 00°14'·87W.

COURSEULLES-SUR-MER
Courseulles ⚓ Iso 4s 00°27'·68W.
W jetty ⚓ Iso WG 4s 7m; W9M, G6M; 135°-W-235°-G-135°; brown pylon on dolphin, G top; 49°20'·41N 00°27'·37W.
E jetty ⚓ Oc (2) R 6s 9m 7M; 49°20'·26N 00°27'·39W.

COURSEULLES TO ST VAAST
Ver ☆ Fl (3)15s 42m **26M**; obsc'd by cliffs of St Aubin when brg >275°; conspic lt ho, W twr, Gy top; 49°20'·41N 00°31'·13W.

ARROMANCHES
Ent buoys: ⚓ 49°21'·35N 00°37'·26W; ⚓ 49°21'·25N 00°37'·30W.
Bombardons ⚓ wreck buoys; 49°21'·66N 00°38'·97W.

PORT-EN-BESSIN
Ldg lts 204°, both Oc (3) 12s 25/42m 10/11M; synch. Front, 069°-339°, W pylon, G top; 49°20'·96N 00°45'·53W. Rear; 114°-294°, W and Gy ho; 93m from front.
E mole ⚓ Oc R 4s 14m 7M, R pylon; 49°21'·12N 00°45'·38W.
W mole ⚓ Fl WG 4s 14m, W10M, G7M; G065°-114·5°, W114·5°-065°; G pylon; 49°21'·17N 00°45'·43W.

COASTAL MARKS

Omaha Beach, 1M off : ⌁ 49°22'·66N 00°50'·28W; ⌁ 49°23'·17N 00°51'·93W; ⌁ 49°23'·66N 00°53'·74W.
Broadsword ⌁ Q (3) 10s, wreck buoy; 49°25'·34N 00°52'·96W.
Est du Cardonnet ⌁ VQ (3) 5s; 49°26'·83N 01°01'·10W.

GRANDCAMP

Les Roches de Grandcamp: No 1 ⌁ 49°24'·72N 01°01'·75W;
No 3 ⌁ 49°24'·92N 01°03'·70W; No 5 ⌁ 49°24'·78N 01°04'·98W.
Ldg lts 146°, both Dir Q 9/12m **15M**, 144·5°-147·5°. **Front** ☆, 49°23'·42N 01°02'·90W. **Rear** ☆,102m from front.
Jetée Est ⌁ Oc (2) R 6s 9m 9M; *Horn Mo(N) 30s;* 49°23'·53N 01°02'·96W.
Jetée Ouest ⌁ Fl G 4s 9m 6M; 49°23'·47N 01°02'·96W.

ISIGNY-SUR-MER

IS, small B/Y ⌁, no topmark (⌁ on AC2135); 49°24'·28N 01°06'·37W.
No 1 ⌁, Fl R 4s; 49°23'·60N 01°07·27W.
Dir lts 173°, both Dir Q WRG 7m 9M; 49°19'·57N 01°06·78W.
Training wall heads ⌁ Fl G 4s; 49°21'·40N 01°07'·20W; ⌁ Fl R 4s, 49°21'·40N 01°07'·10W, off Pte du Grouin.

CARENTAN

C-I ⌁ Iso 4s; 49°25'·44N 01°07'·08W; 210°/1·76M to 1 & 2 bys.
No 1 ⌁ Fl G 2·5s; 49°23'·93N 01°08'·52W.
No 2 ⌁ Fl R 2·5s; 49°23'·88N 01°08'·37W.
Trng wall ⌁ Fl (4) G 15s; G △ on G bcn;49°21'·96N 01°09'·95W.
Trng wall ⌁ Fl (4) R 15s; R ☐ on R bcn;49°21'·93N 01°09'·878W.

ÎLES SAINT-MARCOUF

Iles St-Marcouf ⌁ VQ (3) 5s 18m 8M; ☐ Gy twr, G top; 49°29'·86N 01°08'·81W.
Ouest-Saint-Marcouf ⌁ Q (9) 15s; 49°29'·73N 01°11'·97W.
Saint Floxel ⌁ 49°30'·64N 01°13'·94W.
Quineville ⌁ Q (9) 10s, wreck buoy; 49°31'·79N 01°12'·38W.

ST VAAST TO POINTE DE BARFLEUR

ST VAAST-LA-HOUGUE

Ldg lts 267°: Front, La Hougue Oc 4s 9m 10M; W pylon, G top; 49°34'·25N 01°16'·37W. Rear, Morsalines Oc (4) WRG 12s 90m, W11M, R/G8M; 171°-W-316°-G-321°-R-342°-W-355°; W 8-sided twr, G top; 49°34'·16N 01°19'·10W, 1·8M from front.
Le Manquet ⌁ 49°34'·26N 01°15'·56W.
Le Bout du Roc ⌁ 49°34'·68N 01°15'·27W.
La Dent ⌁ 49°34'·57N 01°14'·20W.
Le Gavendest ⌁ Q (6) + L Fl 15s; 49°34'·36N 01°13'·89W.
Jetty ⌁ Dir Oc (2) WRG 6s 12m W10M, R/G7M; 219°-R-237°-G-310°-W-350°-R-040°; W 8-sided twr, R top; 49°35'·17N 01°15'·41W.
Pte de Saire ⌁ Oc (2+1) 10s 11m 10M; squat W twr, G top; 49°36'·36N 01°13'·78W.

BARFLEUR

Ldg lts 219·5°, both Oc (3) 12s 7/13m 10M; synch. Front, W ☐ twr;49°40'·18N 01°15'·61W. Rear, 085°-355°; Gy and W ☐ twr, G top; 288m from front.
La Grotte ⌁ 49°41'·06N 01°14'·86W.
Roche-à-l'Anglais ⌁ 49°40'·78N 01°14'·93W.
La Vimberge ⌁ 49°40'·54N 01°15'·25W.
W jetty ⌁ Fl G 4s 8m 6M; 49°40'·32N 01°15'·57W.
E jetty ⌁ Oc R 4s 5m 6M; 49°40'·31N 01°15'·47W.
La Jamette ⌁ 49°41'·87N 01°15'·59W.
Pte de Barfleur ☆ Fl (2) 10s 72m **25M**; obsc when brg less than 088°; Gy twr, B top; 49°41'·78N 01°15'·96W. 992271217.

POINTE DE BARFLEUR TO CAP DE LA HAGUE

Les Équets ⌁ Q 8m 3M; 49°43'·62N 01°18'·36W.
Basse du Rénier ⌁ VQ 8m 4M; 49°44'·84N 01°22'·09W.
Les Trois Pierres ⌁ 49°42'·90N 01°21'·80W.

Anse de Vicq, 158° ldg lts; both Iso R 4s 8/14m 6M; front 49°42'·20N 01°23'·95W.
La Pierre Noire ⌁ Q (9) 15s 8m 4M;49°43'·54N 01°29'·07W.

PORT DU LÉVI

Cap Lévi ☆ Fl R 5s 36m **17M**; Gy ☐ twr; 49°41'·75N 01°28'·38W. 992271213.
Port Lévi ⌁ Oc (2) WRG 6s 7m 7M; 055°-G-083°-W-105°-R-163°; W & Gy hut, W lantern; 49°41'·24N 01°28'·34W.

PORT DU BECQUET

Ldg lts 186·3°, both intens 183°-189·3°; synch. Front, Dir Oc (3) 12s 8m 10M; W 8-sided twr; 49°39'·23N 01°32'·84W. Rear, Dir Oc (3) 12s 13m 7M. W 8-sided twr, R top; 49m from front.

CHERBOURG, EASTERN ENTRANCES

Passe Collignon ⌁ Fl (2) R 6s 5m 4M; 49°39'·59N 01°34'·24W.
Passe de l'Est, Jetée des Flamands ldg lts 189°, both Q 9/16m 13M. Front, 49°39'·33N 01°35'·94W . Rear, 516m from front.
Roches du Nord-Ouest ⌁ Fl R 2·5s; 49°40'·64N 01°35'·28W.
La Truite ⌁ Fl R 15s; 49°40'·33N 01°35'·49W.
Fort d'Île Pelée ⌁ Oc (2) WR 6s 19m; W10M, R7M; 055°-W-120°-R-055°; W & R pedestal; 49°40'·21N 01°35'·08W.
Fort de l'Est ⌁ 49°40'·28N 01°35'·92W, Iso G 4s 19m 9M.
Fort Central ⌁ VQ (6) + L Fl 10s 5m 4M; 322°-032°; 49°40'·40N 01°37'·04W.

CHERBOURG, PASSE DE L'OUEST

CH1 ⊙; V-AIS; 49°43'·24N 01°42'·09W.
Passe de l'Ouest, Dir Q , lights in line140·3°; Front, Jetée du Homet 5m **17M**; W △; 49° 39'·54N 01°37'·97W. Rear, Gare Maritime, 0.99M from front; 35m **17M**; intens 138·8° - 141·8°, shows 24H, sync with front; W △ on Gy pylon; 49°38'·78N 01°37'·00W.
Fort de l'Ouest ☆ Fl (3) WR 15s 19m **W24M, R20M**; 122°-W-355°-R-122°; Gy tr, R top; 49°40'·45N 01°38'·87W. 992271214.
Fort de l'Ouest ⌁ Fl R 4s; 49°40'·39N 01°38'·89W.
⌁ Q (6) + L Fl 15s; 49°40'·34N 01°38'·65W.
Digue de Querqueville ⌁ Fl (2) G 6s 8m 6M; W col, G top; 49°40'·30N 01°39'·80W.
Inner ldg lts 124·3°; both intens 114·3°-134·3°: Front, Digue du Homet head, QG 10m 6M; 49°39'·48N 01°36'·96W. Rear, Dir Iso G 4s 16m 10M; W col, B bands, 397m from front.
La Ténarde ⌁ VQ 8m 4M; 49°39'·74N 01°37'·75W.

CHERBOURG, PETITE RADE and MARINA

Entrance, W side, Digue du Homet ⌁ QG 10m 8M; intens 114·3°-134·3°; W pylon, G top; 49°39'·48N 01°36'·96W.
E side, ⌁ Q R, off Jetée des Flamands; 49°39'·44N 01°36'·60W.
Marina ent, E side, ⌁ Fl (3) R 12s 6m 6M; W col, R lantern; 49°38'·91N 01°37'·08W.
W mole ⌁ Fl (3) G 12s 7m 6M; G pylon; 49°38'·87N 01°37'·15W.
E quay⌁ Fl (4) R 15s 3m 3M; R bcn; 49°38'·79N 01°37'·12W.
Wavescreen pontoon, N end ⌁ Fl (4) G 15s 4m 2M; W post, G top.

CHERBOURG TO CAP DE LA HAGUE

Raz de Bannes ⌁ 49°41'·32N 01°44'·53W.
Omonville Dir lt 257°: Iso WRG 4s13m; W10M, R/G7M; 180°-G-252°-W-262°-R-287°; W pylon, G top; 49°42'·24N 01°50'·15W.
L'Étonnard ⌁ 49°42'·33N 01°49'·84W.
Basse Bréfort ⌁ VQ 8m 4M; 49°43'·90N 01°51'·15W.
Jobourg Nuclear plant chimney, R lts; 49°40'·80N 01°52'·91W.
La Plate ⌁ Fl (2+1) WR 10s 11m; W9M, R6M; 115°-W-272°-R-115°; Y 8-sided twr, with B top; 49°43'·97N 01°55'·74W.
Cap de la Hague (Gros du Raz) ☆ Fl 5s 48m **23M**; Gy twr, W top; *Horn 30s;* 49°43'·31N 01° 57'·26W. 992271218.
La Foraine ⌁ VQ (9) 10s, 12m 6M; 49°42'·90N 01°58'·31W.

4.3 PASSAGE INFORMATION

More passage information is threaded between harbours in this Area. **Bibliography**: *Shell Channel Pilot* (Imray/Cunliffe). *Channel Havens* (ACN/Endean). *Channel Islands, Cherbourg Peninsula and North Brittany* (Imray/RCC/Carnegie); *Dover Strait NP28 and Channel NP27 Pilots*.

NORTH EAST/CENTRAL FRANCE

The coasts of Picardy and Normandy are hardly more than an overnight passage from harbours on the UK's S coast – eg, Brighton to Fécamp is only 65M. However, as some of the harbours dry, a boat which can take the ground is an advantage, see also **784, Pl Deauville to Grandcamp**. See Area 3 for a chart of the Dover Strait TSS. Notes on the English Channel and cross-Channel passages appear below and also in Area 3, Introduction for cross-Channel distances.

DUNKERQUE TO BOULOGNE

Offshore, a series of banks lies roughly parallel with the coast (AC 323, 1892, 2451): Sandettié bank, Outer Ruytingen midway between Sandettié and the coast, and the Dyck banks which extend NE'wards for 30M from a point 5M NE of Calais. There are well-buoyed channels between some of these banks, but great care is needed in poor visibility. In general the banks are steep-to on the inshore side, and slope to seaward. In bad weather the sea breaks on the shallower parts.

Dunkerque Port Est, the old port, has several good yacht marinas and all facilities. Dunkerque Port Ouest is a commercial/ferry port which yachts should not enter. About 3M SW is the drying hbr of **Gravelines**, which should not be entered in strong onshore winds. If E-bound, the Passe de Zuydcoote (4·4m) is a popular inshore link to West Diep and the Belgian ports.

The sea breaks heavily in bad weather on Ridens de Calais, 3M N of **Calais** (AC 1351) and also on Ridens de la Rade immediately N and NE of the harbour where the least charted depth is 0·1m.

The NE-bound traffic lane of the Dover Strait TSS is only 3M off C Gris Nez. Keep a sharp lookout not only for coastal traffic in the ITZ, but also cross-Channel ferries. ▶*1M NW of C Gris Nez the NE-going stream begins at HW Dieppe –0150, and the SW-going at HW Dieppe +0355, sp rates 4kn.*◀ Between C Gris Nez (lt, fog sig) and Boulogne the coastal bank dries about 4ca offshore.

4.4 SPECIAL NOTES FOR FRANCE

Instead of 'County' the 'Département' is given. For details of documentation apply to the French Tourist Office, 300 High Holborn, London, WC1V 7JH; ☎020 7061 6631; info.uk@franceguide.com www.franceguide.com.

STANDARD TIME is –0100, to which DST should be added.

AFFAIRES MARITIMES is a part of the central government Ministry of Transport. It oversees all maritime activities (commercial, fishing, pleasure) and their harmonious development. Information on navigation and other maritime issues can be supplied by a representative whose ☎ is given under each port; often the nearest commercial or fishing port. The Head Office is: Ministère des Transports et de la Mer, Bureau de la Navigation de Plaisance, 3 Place de Fontenoy, 75700 Paris, ☎01·44·49·80·00. www.ecologique-solidaire.gouv.fr/politiques/plaisance-et-loisirs-nautiques.

CHARTS Official charts are issued by SHOM (*Service Hydrographique et Oceanographique de la Marine*), the Navy's Hydrographer. SHOM charts are excellent for the smaller French ports not covered by UKHO. A free, annual chart catalogue *Le petit catalogue* is downloadable from www.shom.fr. It includes chart agents and mail order specialists. In this Almanac under 'Facilities', SHOM means a chart agent. Any new edition of a SHOM chart receives a new, different chart number.

TIDAL LEVELS SHOM charts refer elevations of lights to MHWS. Heights of other features are referred to Mean Sea Level (*Niveau moyen*) or as indicated on the chart; it is important to check the chart's title block. Clearance under bridges etc is referred to MSL, MHWS or HAT depending on the recency of the chart. Useful French tidal terms:

BOULOGNE TO DIEPPE

In the approaches to Pas de Calais a number of shoals lie offshore (AC 2451): The Ridge (or Le Colbart) Les Ridens, Bassurelle, Vergoyer, Bassure de Baas and Battur. In bad weather, and particularly with wind against tide, the sea breaks heavily on all these shoals.

From **Boulogne** (AC 438) to Pte de Lornel the coast dries up to 5ca offshore. ▶*Off Boulogne's Digue Carnot the N-going stream begins HW Dieppe –0130, and the S-going at HW Dieppe +0350, sp rates 1·75kn.*◀

Le Touquet and **Étaples** lie in the Embouchure de la Canche, entered between Pte de Lornel and Pte du Touquet, and with a drying bank extending 1M seaward of a line joining these two points. Le Touquet-Paris-Plage lt is shown from a conspic twr, 1M S of Pte du Touquet.

▶*Off the estuary entrance the N-going stream begins about HW Dieppe –0335, sp rate 1·75kn; and the S-going stream begins about HW Dieppe +0240, sp rate 1·75kn.*◀

From Pte du Touquet the coast runs 17M S to Pte de St Quentin, with a shallow coastal bank which dries to about 5ca offshore, except in the approaches to the dangerous and constantly changing Embouchure de l'Authie (Pte du Haut-Blanc, about 10M S) where it dries up to 2M offshore.

Baie de la Somme, between Pte de St Quentin and Pte du Hourdel (lt) is a shallow, drying area of shifting sands. The access channel to **St Valéry-sur-Somme** and **Le Crotoy**, which runs close to Pte du Hourdel, is well buoyed, but the whole estuary dries out 3M to seaward, and should not be approached in strong W/NW winds.

Offshore there are two shoals, Bassurelle de la Somme and Quémer, on parts of which the sea breaks in bad weather. ▶*4·5M NW of Cayeux-sur-Mer (lt) the stream is rotatory anti-clockwise. The E-going stream begins about HW Dieppe –0200, and sets 070° 2·5kn at sp: the W-going stream begins about HW Dieppe +0600, and sets 240° 1·5kn at sp.*◀

S of Ault (lt) the coast changes from low sand dunes to medium-height cliffs. Between **Le Tréport** and Dieppe, 14M SW, rky banks, drying in places, extend 5ca offshore. Banc Franc-Marqué (3·6m) lies 2M offshore, and some 3M N of Le Tréport. About 3M NW of Le Tréport, Ridens du Tréport (5·1m) should be avoided in bad weather. A prohibited area extends 6ca off Penly nuclear power station and is marked by 2 SPM lt buoys.

LAT	Plus basse mer astronomique (PBMA)
MHWS	Pleine mer moyenne de VE (PMVE)
MHWN	Pleine mer moyenne de ME (PMME)
MLWN	Basse mer moyenne de ME (BMME)
MLWS	Basse mer moyenne de VE (BMVE)
ML/MSL	Niveau moyen (NM)
HW	Vive-eau (VE)
LW	Morte-eau (ME)

TIDAL COEFFICIENTS Popular and much used in France, these are included for a selection of strategic ports, Calais, Cherbourg, St Malo, Brest and Port Bloc, together with explanatory notes.

PUBLIC HOLIDAYS New Year's Day, Easter Sunday and Monday, Labour Day (1 May), Ascension Day, Armistice Day 1945 (8 May), Whit Sunday and Monday, National (Bastille) Day (14 July), Feast of the Assumption (15 Aug), All Saints' Day (1 Nov), Remembrance Day (11 Nov), Christmas Day.

REGULATIONS Flares must be in-date. Carry original ship's registration papers. Yachts built since 2007 must have black water tank(s) or a treatment system. Mandatory boat equipment in French waters is listed at www.developpement-durable.gouv.fr/spip.php?page=article&id_article=5476.

FACILITIES The cost of a visitor's overnight berth is based on high season rates, mostly quoted as €/metre LOA, although sometimes beam may also be taken into account. Low season rates and concessions can be a big saving. The fee usually includes electricity, but rarely showers which range between €1·20 and €2. VAT (TVA) is 20%. Port à Sec offers savings for longer stays if not living aboard. **Fuel** can be obtained H24 in many marinas, self-service, using credit cards to activate/pay at the pumps. Lacking a suitable card, payments can always be made during office hours at the Capitainerie.

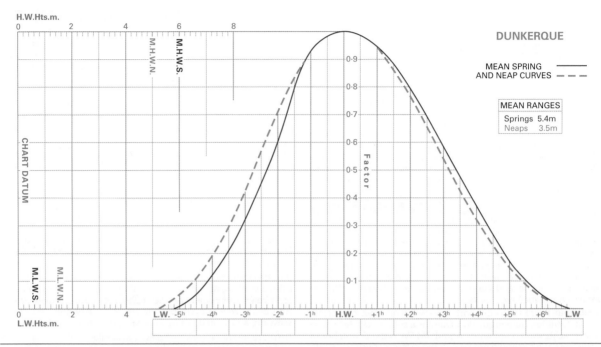

DUNKERQUE

MEAN SPRING
AND NEAP CURVES — — —

MEAN RANGES
Springs 5.4m
Neaps 3.5m

TELEPHONES ☎ numbers contain 10 digits, the first 2 digits being Zone codes (01 to 05) as appropriate to location. Mobile numbers are prefixed 06. Info Nos, eg the recorded weather (Auto) are prefixed 08. The ringing tone is long, equal on/off tones (slower than UK engaged tone). Rapid pips mean the call is being connected. Engaged tone is like that in UK.

To call France from UK, dial + (or 00) 33, then the 9 digit number, omitting the 0 from the Zone code prefix. To call UK from France dial + (or 00) 44 followed by the Area Code (omitting the first 0) and the number. Phonecards for public phones may be bought at the PTT, many Hbr offices, cafés and tabacs. Cheap rates are 2130-0800 Mon-Fri 1330-0800 Sat–Mon.

⚠ **Emergencies, dial:**
15   Ambulance (SAMU), better than a doctor;
17   Police, better than a gendarmerie;
18   Fire (also divers);
112   General emergencies;
196   CROSS, use for fast response.

BRITISH CONSULS IN FRANCE
The British Consular service assists UK nationals who are travelling abroad and have encountered difficulties or distress ranging from loss of passport or money to the death of or serious injury to a crew member or fellow traveller.

British Consuls at harbours on the Channel and Atlantic coasts are *the first port of call when help is needed*. The relevant Tel No is given under each harbour.

SIGNALS International Port Traffic Signals (0.20) and Storm warning signals are used in many French ports. Tidal signals are now rarely seen in French ports.

INLAND WATERWAYS

TOLLS are due on waterways managed by Voies Navigable de France (VNF). Licence discs (*vignettes*) must be visibly displayed stbd side forward. They are obtainable:

• Online from www.vnf.fr/vignettesVNF/ in English. Pay online by Visa, Mastercard or e-Carte Bleue.
• By credit card, phone: 03·21·63·24·30.
• By cash only at 35 listed VNF offices.

Licence fees, 2018, were calculated with a set charge per metre depending on length of stay and LOA added to a fixed fee depending on four classes of size. Validity was for either 1 year,

(Jan-Dec) 1 month (30 days) or 7 days. Note: All time periods are consecutive. 1 day licences are also available on request.

Duration	LOA x €/m	<8m	8–11m	11–14m	>14m
1 year	€8·40/m	+€86·50	+€198·50	+€375·10	+€493·70
30 days	€7·50/m	+€27·50	+€40·00	+€52·40	+€66·90
7 days	€3·90/m	+€15·20	+€23·00	+€30·60	+€38·10

e.g. 10m yacht for 30 days = (€7·50 x 10) + €40·00 = €115·00.

See: www.french-waterways.com for additional information.

CLOSURES
Dates of closures (*chomages*), plus lock hours, water levels and other practicalities, can be obtained from www.vnf.fr or the French National Tourist Office in London (see overleaf).

VNF Head Office, 175 Rue Ludovic Boutleux, BP 820, 62400 Bethune. ☎03·21·63·24·30. www.vnf.fr

QUALIFICATIONS
Helmsmen of craft <15m LOA and not capable of >20kph (11kn) must have an International Certificate of Competence validated for inland waterways, plus a copy of the CEVNI rules. For larger, faster craft the requirements are under review. Copies of police regs for inland waterways must be on board (electronic if easy to access). Download general regs (RGPNI) at: www.development-durable.gouv.fr/Police-de-la-navigation and special regs (RPP) at: www. vnf.fr/vnf/content.vnf?action=content&occ_id=37793.

INLAND WATERWAYS TO THE MEDITERRANEAN
The quickest route is R Seine - Paris (see 4.17), Paris - St Mammès – Canal du Loing – Canal de Briare – Canal Latéral à la Loire – Canal du Centre – Saône – Rhône. Approx 1318km (824M) 182 locks; allow 4 weeks. Max dimensions: LOA 38·5m, beam 5m, draft 1·8m, air draft 3·5m.

Alternatives above St Mammès: R Yonne-Laroche, then either Canal de Bourgogne – Saône; or Canal du Nivernais – Canal du Centre – Saône. Both are slowed by many locks.

There are other routes via R Oise/Aisne or R Marne, both continuing via Canal de la Marne à la Saône.

The Saône can also be accessed via the canals of N France, entered at Calais, Dunkerque or Belgian and Dutch ports.

CANALS IN WESTERN FRANCE
See (5.13) Inland Waterways, especially the Ille et Rance Canal/ Vilaine River from St Malo to Arzal/Camoel in S Brittany.

STANDARD TIME UT –01
Subtract 1 hour for UT
For French Summer Time add ONE hour in **non-shaded areas**

DUNKERQUE LAT 51°03'N LONG 2°22'E
TIMES AND HEIGHTS OF HIGH AND LOW WATERS

Dates in red are **SPRINGS**
Dates in blue are **NEAPS**

YEAR 2019

JANUARY

Day	Time m	Time m	Time m	Time m
1 TU	0323 1.5	0902 5.2	1558 1.2	2139 5.3
2 W	0432 1.4	1002 5.3	1704 1.1	2235 5.4
3 TH	0533 1.3	1055 5.4	1758 1.1	2324 5.5
4 F	0622 1.1	1142 5.6	1843 1.0	
5 SA	0007 5.6	0704 1.0	1225 5.7	1922 1.0
6 SU ●	0046 5.7	0742 0.9	1304 5.8	1958 1.0
7 M	0122 5.7	0818 0.8	1341 5.8	2032 1.0
8 TU	0155 5.8	0853 0.8	1416 5.8	2106 1.0
9 W	0227 5.7	0927 0.8	1448 5.7	2139 1.1
10 TH	0300 5.7	1000 0.9	1523 5.6	2212 1.1
11 F	0335 5.6	1032 1.0	1600 5.5	2245 1.3
12 SA	0412 5.4	1106 1.1	1641 5.3	2322 1.4
13 SU	0452 5.2	1145 1.2	1726 5.1	
14 M ◑	0005 1.5	0542 5.1	1233 1.4	1822 5.0
15 TU	0057 1.6	0643 5.0	1333 1.5	1924 4.9
16 W	0205 1.7	0748 4.9	1450 1.5	2031 5.0
17 TH	0326 1.6	0856 5.0	1604 1.4	2140 5.1
18 F	0435 1.4	1003 5.3	1708 1.1	2241 5.4
19 SA	0535 1.1	1101 5.6	1805 0.9	2334 5.7
20 SU	0630 0.9	1152 5.9	1857 0.7	
21 M ○	0021 5.9	0721 0.6	1240 6.1	1946 0.6
22 TU	0107 6.1	0809 0.4	1328 6.3	2033 0.5
23 W	0153 6.2	0857 0.3	1416 6.3	2120 0.5
24 TH	0239 6.1	0945 0.2	1504 6.3	2207 0.6
25 F	0325 6.1	1032 0.3	1553 6.1	2254 0.7
26 SA	0412 5.9	1120 0.5	1642 5.9	2341 1.0
27 SU ◐	0501 5.7	1210 0.7	1736 5.6	
28 M	0033 1.2	0556 5.5	1305 1.0	1838 5.3
29 TU	0131 1.4	0705 5.2	1407 1.2	1951 5.0
30 W	0238 1.6	0822 5.0	1517 1.4	2103 4.9
31 TH	0355 1.6	0933 5.0	1635 1.4	2210 5.0

FEBRUARY

Day	Time m	Time m	Time m	Time m
1 F	0510 1.4	1036 5.1	1740 1.3	2309 5.1
2 SA	0606 1.2	1131 5.3	1828 1.2	2356 5.3
3 SU	0650 1.0	1215 5.5	1908 1.0	
4 M ●	0035 5.5	0728 0.9	1253 5.7	1942 1.0
5 TU	0109 5.7	0802 0.8	1328 5.8	2014 0.9
6 W	0140 5.7	0834 0.7	1359 5.8	2046 0.9
7 TH	0208 5.8	0907 0.7	1428 5.8	2119 0.9
8 F	0237 5.8	0938 0.7	1459 5.8	2150 0.9
9 SA	0309 5.7	1008 0.8	1531 5.7	2220 1.0
10 SU	0339 5.6	1039 0.8	1602 5.5	2252 1.1
11 M	0409 5.5	1113 1.0	1634 5.4	2329 1.2
12 TU ◐	0446 5.4	1153 1.1	1720 5.2	
13 W	0014 1.4	0540 5.2	1244 1.3	1828 5.0
14 TH	0112 1.6	0655 5.0	1356 1.5	1944 4.9
15 F	0235 1.6	0814 4.9	1523 1.5	2106 4.9
16 SA	0400 1.5	0937 5.1	1641 1.2	2224 5.2
17 SU	0514 1.2	1049 5.5	1749 1.0	2323 5.6
18 M	0616 0.8	1144 5.8	1845 0.7	
19 TU ○	0011 5.9	0709 0.5	1231 6.1	1934 0.5
20 W	0055 6.1	0757 0.2	1316 6.3	2019 0.4
21 TH	0138 6.2	0842 0.1	1401 6.4	2104 0.4
22 F	0220 6.2	0927 0.1	1446 6.3	2147 0.4
23 SA	0303 6.2	1011 0.1	1530 6.1	2230 0.6
24 SU	0346 6.0	1055 0.3	1615 5.9	2312 0.8
25 M	0431 5.9	1138 0.6	1701 5.6	2356 1.1
26 TU ◐	0520 5.6	1225 1.0	1755 5.2	
27 W	0047 1.4	0621 5.2	1322 1.3	1903 4.9
28 TH	0152 1.6	0740 4.8	1432 1.6	2025 4.6

MARCH

Day	Time m	Time m	Time m	Time m
1 F	0313 1.7	0903 4.8	1600 1.7	2143 4.7
2 SA	0442 1.6	1015 4.9	1716 1.5	2248 4.9
3 SU	0545 1.3	1114 5.2	1808 1.3	2337 5.2
4 M	0631 1.0	1158 5.4	1848 1.1	
5 TU	0016 5.4	0707 0.8	1235 5.6	1920 0.9
6 W ●	0049 5.6	0739 0.7	1308 5.8	1951 0.8
7 TH	0118 5.7	0810 0.6	1337 5.8	2022 0.7
8 F	0145 5.8	0842 0.6	1404 5.9	2054 0.7
9 SA	0212 5.9	0913 0.5	1432 5.9	2125 0.7
10 SU	0241 5.8	0943 0.6	1500 5.8	2155 0.8
11 M	0308 5.8	1013 0.7	1527 5.7	2226 0.9
12 TU	0337 5.7	1045 0.8	1558 5.5	2301 1.0
13 W	0413 5.6	1123 1.0	1641 5.3	2343 1.2
14 TH ◐	0502 5.3	1212 1.2	1746 5.0	
15 F	0040 1.4	0619 5.0	1321 1.5	1912 4.8
16 SA	0201 1.6	0748 4.9	1455 1.5	2045 4.8
17 SU	0334 1.5	0923 5.1	1622 1.3	2210 5.1
18 M	0456 1.1	1037 5.5	1734 1.0	2309 5.5
19 TU	0601 0.7	1131 5.9	1830 0.7	2356 5.8
20 W	0653 0.4	1217 6.1	1917 0.5	
21 TH ○	0037 6.1	0739 0.2	1259 6.3	2000 0.4
22 F	0116 6.2	0823 0.1	1340 6.3	2042 0.4
23 SA	0157 6.2	0905 0.1	1422 6.2	2123 0.4
24 SU	0238 6.2	0947 0.2	1505 6.1	2203 0.5
25 M	0320 6.1	1026 0.4	1546 5.9	2242 0.7
26 TU	0402 5.9	1106 0.7	1628 5.6	2323 1.0
27 W	0449 5.6	1148 1.1	1717 5.2	
28 TH ◑	0009 1.3	0544 5.2	1238 1.4	1818 4.8
29 F	0108 1.6	0659 4.8	1346 1.8	1941 4.5
30 SA	0228 1.8	0829 4.6	1516 1.8	2107 4.5
31 SU	0403 1.7	0947 4.8	1640 1.6	2216 4.8

APRIL

Day	Time m	Time m	Time m	Time m
1 M	0512 1.4	1046 5.1	1737 1.3	2307 5.1
2 TU	0600 1.1	1131 5.4	1818 1.1	2347 5.4
3 W	0637 0.9	1208 5.6	1851 0.9	
4 TH	0021 5.6	0709 0.7	1241 5.7	1921 0.8
5 F ●	0051 5.7	0740 0.6	1310 5.8	1953 0.7
6 SA	0117 5.8	0813 0.5	1336 5.9	2027 0.6
7 SU	0144 5.9	0846 0.5	1403 5.9	2100 0.6
8 M	0212 5.9	0918 0.5	1432 5.8	2132 0.6
9 TU	0242 5.9	0950 0.6	1502 5.7	2205 0.7
10 W	0315 5.8	1024 0.8	1538 5.6	2242 0.9
11 TH	0355 5.6	1104 1.0	1624 5.3	2326 1.1
12 F ◐	0450 5.4	1155 1.2	1734 5.0	
13 SA	0025 1.3	0610 5.1	1307 1.5	1858 4.8
14 SU	0146 1.4	0736 5.0	1440 1.5	2030 4.8
15 M	0317 1.3	0911 5.1	1605 1.3	2153 5.1
16 TU	0437 1.0	1021 5.5	1715 1.0	2250 5.5
17 W	0541 0.6	1114 5.9	1809 0.7	2335 5.8
18 TH	0632 0.4	1158 6.1	1855 0.6	
19 F ○	0015 6.0	0718 0.2	1238 6.2	1938 0.5
20 SA	0053 6.1	0800 0.2	1317 6.2	2019 0.5
21 SU	0133 6.2	0841 0.2	1358 6.1	2059 0.5
22 M	0214 6.1	0921 0.3	1439 6.0	2138 0.5
23 TU	0256 6.0	0959 0.6	1520 5.8	2216 0.7
24 W	0339 5.8	1037 0.9	1601 5.5	2255 0.9
25 TH	0423 5.5	1117 1.1	1646 5.2	2338 1.2
26 F ◐	0515 5.2	1203 1.5	1740 4.8	
27 SA	0031 1.5	0619 4.8	1303 1.8	1851 4.5
28 SU	0141 1.7	0741 4.6	1425 1.9	2017 4.5
29 M	0309 1.7	0904 4.7	1549 1.7	2131 4.7
30 TU	0423 1.4	1006 5.0	1651 1.4	2226 5.0

Chart Datum is 2·69 metres below IGN Datum. HAT is 6·4 metres above Chart Datum.

N France

>> FREE monthly updates. Register at <<
www.reedsnauticalalmanac.co.uk

163

STANDARD TIME UT –01
Subtract 1 hour for UT
For French Summer Time add ONE hour in **non-shaded areas**

DUNKERQUE   LAT 51°03'N   LONG 2°22'E
TIMES AND HEIGHTS OF HIGH AND LOW WATERS

Dates in red are **SPRINGS**
Dates in blue are NEAPS

YEAR 2019

MAY

Date	Time	m	Date	Time	m
1 W	0516 / 1053 / 1737 / 2310	1.2 / 5.3 / 1.2 / 5.3	16 TH	0517 / 1052 / 1745 / 2311	0.6 / 5.8 / 0.8 / 5.7
2 TH	0558 / 1133 / 1815 / 2347	0.9 / 5.5 / 1.0 / 5.5	17 F	0609 / 1137 / 1833 / 2352	0.5 / 5.9 / 0.7 / 5.9
3 F	0633 / 1208 / 1849	0.8 / 5.7 / 0.8	18 SA	0655 / 1217 / 1916 ○	0.4 / 6.0 / 0.6
4 SA ●	0018 / 0707 / 1237 / 1923	5.6 / 0.7 / 5.8 / 0.7	19 SU	0031 / 0738 / 1256 / 1957	6.0 / 0.5 / 6.0 / 0.6
5 SU	0046 / 0742 / 1305 / 1959	5.8 / 0.6 / 5.8 / 0.6	20 M	0112 / 0818 / 1337 / 2037	6.0 / 0.5 / 5.9 / 0.6
6 M	0116 / 0819 / 1336 / 2036	5.9 / 0.6 / 5.9 / 0.6	21 TU	0155 / 0857 / 1418 / 2116	6.0 / 0.6 / 5.9 / 0.6
7 TU	0149 / 0856 / 1411 / 2113	5.9 / 0.6 / 5.9 / 0.6	22 W	0238 / 0935 / 1459 / 2155	5.9 / 0.8 / 5.7 / 0.7
8 W	0225 / 0933 / 1448 / 2152	5.9 / 0.6 / 5.8 / 0.7	23 TH	0321 / 1013 / 1540 / 2234	5.7 / 1.0 / 5.5 / 0.9
9 TH	0306 / 1012 / 1533 / 2233	5.8 / 0.8 / 5.6 / 0.8	24 F	0404 / 1051 / 1622 / 2314	5.5 / 1.2 / 5.3 / 1.1
10 F	0355 / 1056 / 1628 / 2322	5.7 / 1.0 / 5.4 / 1.0	25 SA	0450 / 1133 / 1710 / 2359	5.2 / 1.4 / 5.0 / 1.3
11 SA	0458 / 1151 / 1736	5.4 / 1.2 / 5.1	26 SU ◐	0543 / 1224 / 1806	4.9 / 1.6 / 4.8
12 SU ◑	0023 / 0608 / 1303 / 1846	1.1 / 5.2 / 1.4 / 5.0	27 M	0055 / 0825 / 1328 / 1912	1.5 / 4.7 / 1.8 / 4.6
13 M	0140 / 0725 / 1427 / 2011	1.2 / 5.2 / 1.4 / 5.0	28 TU	0208 / 0759 / 1448 / 2026	1.6 / 4.7 / 1.7 / 4.7
14 TU	0300 / 0852 / 1543 / 2128	1.1 / 5.3 / 1.2 / 5.2	29 W	0322 / 0910 / 1555 / 2131	1.5 / 4.9 / 1.5 / 4.9
15 W	0414 / 0959 / 1650 / 2224	0.9 / 5.5 / 1.0 / 5.5	30 TH	0422 / 1005 / 1649 / 2223	1.3 / 5.1 / 1.3 / 5.1
			31 F	0512 / 1050 / 1734 / 2305	1.1 / 5.3 / 1.1 / 5.4

JUNE

Date	Time	m	Date	Time	m
1 SA	0554 / 1129 / 1814 / 2342	0.9 / 5.5 / 0.9 / 5.6	16 SU	0635 / 1200 / 1858	0.7 / 5.7 / 0.8
2 SU	0634 / 1204 / 1854	0.8 / 5.7 / 0.9	17 M	0016 / 0719 / 1242 / 1940 ○	5.8 / 0.7 / 5.7 / 0.7
3 M ●	0016 / 0714 / 1238 / 1934	5.7 / 0.7 / 5.8 / 0.7	18 TU	0059 / 0759 / 1323 / 2020	5.8 / 0.8 / 5.8 / 0.7
4 TU	0052 / 0755 / 1315 / 2016	5.6 / 0.6 / 5.9 / 0.6	19 W	0143 / 0838 / 1404 / 2100	5.9 / 0.8 / 5.7 / 0.7
5 W	0132 / 0838 / 1358 / 2059	5.9 / 0.6 / 5.7 / 0.5	20 TH	0225 / 0916 / 1443 / 2137	5.8 / 0.9 / 5.7 / 0.7
6 TH	0216 / 0921 / 1444 / 2144	6.0 / 0.7 / 5.8 / 0.6	21 F	0305 / 0952 / 1520 / 2214	5.7 / 1.0 / 5.5 / 0.9
7 F	0305 / 1007 / 1535 / 2231	5.8 / 0.8 / 5.7 / 0.6	22 SA	0344 / 1029 / 1558 / 2251	5.5 / 1.1 / 5.4 / 1.0
8 SA	0359 / 1056 / 1629 / 2322	5.8 / 0.9 / 5.5 / 0.7	23 SU	0424 / 1106 / 1640 / 2330	5.4 / 1.3 / 5.2 / 1.2
9 SU	0456 / 1151 / 1726	5.6 / 1.1 / 5.4	24 M	0509 / 1148 / 1727	5.2 / 1.4 / 5.1
10 M ◑	0021 / 0557 / 1256 / 1828	0.9 / 5.5 / 1.2 / 5.2	25 TU ◑	0013 / 0559 / 1237 / 1820	1.3 / 5.0 / 1.6 / 4.9
11 TU	0128 / 0706 / 1407 / 1943	0.9 / 5.3 / 1.3 / 5.2	26 W	0107 / 0656 / 1338 / 1920	1.4 / 4.9 / 1.7 / 4.8
12 W	0239 / 0825 / 1516 / 2057	0.9 / 5.3 / 1.2 / 5.3	27 TH	0214 / 0758 / 1451 / 2024	1.5 / 4.9 / 1.6 / 4.9
13 TH	0347 / 0932 / 1622 / 2156	0.8 / 5.4 / 1.1 / 5.4	28 F	0324 / 0903 / 1556 / 2126	1.4 / 5.0 / 1.5 / 5.0
14 F	0451 / 1028 / 1721 / 2247	0.7 / 5.6 / 1.0 / 5.6	29 SA	0424 / 1001 / 1652 / 2221	1.2 / 5.2 / 1.3 / 5.2
15 SA	0547 / 1116 / 1812 / 2333	0.7 / 5.6 / 0.9 / 5.7	30 SU	0516 / 1051 / 1742 / 2309	1.1 / 5.4 / 1.1 / 5.4

JULY

Date	Time	m	Date	Time	m
1 M	0605 / 1136 / 1829 / 2352	0.9 / 5.6 / 0.9 / 5.7	16 TU	0009 / 0705 / 1234 / 1927 ○	5.6 / 1.0 / 5.6 / 0.8
2 TU ●	0651 / 1218 / 1915	0.8 / 5.7 / 0.7	17 W	0052 / 0745 / 1313 / 2007	5.7 / 0.9 / 5.7 / 0.7
3 W	0035 / 0737 / 1301 / 2001	5.9 / 0.7 / 5.9 / 0.6	18 TH	0133 / 0822 / 1351 / 2044	5.8 / 0.9 / 5.7 / 0.7
4 TH	0120 / 0824 / 1348 / 2048	6.0 / 0.6 / 5.9 / 0.4	19 F	0211 / 0857 / 1425 / 2119	5.8 / 0.9 / 5.7 / 0.7
5 F	0208 / 0911 / 1436 / 2136	6.1 / 0.6 / 5.9 / 0.4	20 SA	0246 / 0932 / 1458 / 2153	5.7 / 1.0 / 5.7 / 0.8
6 SA	0258 / 0959 / 1525 / 2224	6.1 / 0.7 / 5.9 / 0.4	21 SU	0320 / 1005 / 1531 / 2226	5.6 / 1.1 / 5.6 / 0.9
7 SU	0350 / 1048 / 1615 / 2314	6.0 / 0.8 / 5.8 / 0.5	22 M	0355 / 1038 / 1607 / 2259	5.5 / 1.1 / 5.4 / 1.0
8 M	0442 / 1139 / 1706	5.8 / 0.9 / 5.6	23 TU	0433 / 1113 / 1647 / 2335	5.4 / 1.3 / 5.3 / 1.1
9 TU ◑	0008 / 0537 / 1236 / 1803	0.6 / 5.7 / 1.1 / 5.5	24 W	0516 / 1152 / 1732	5.2 / 1.4 / 5.1
10 W	0107 / 0640 / 1338 / 1909	0.8 / 5.5 / 1.2 / 5.4	25 TH ◑	0017 / 0605 / 1239 / 1826	1.3 / 5.1 / 1.5 / 5.0
11 TH	0211 / 0752 / 1444 / 2022	0.9 / 5.3 / 1.3 / 5.3	26 F	0110 / 0702 / 1340 / 1926	1.4 / 4.9 / 1.7 / 4.9
12 F	0317 / 0902 / 1552 / 2128	1.0 / 5.3 / 1.3 / 5.3	27 SA	0219 / 0805 / 1458 / 2033	1.5 / 4.9 / 1.7 / 4.9
13 SA	0425 / 1004 / 1659 / 2227	1.0 / 5.3 / 1.2 / 5.4	28 SU	0335 / 0914 / 1610 / 2141	1.4 / 5.0 / 1.5 / 5.1
14 SU	0529 / 1101 / 1757 / 2321	1.0 / 5.4 / 1.1 / 5.5	29 M	0442 / 1020 / 1713 / 2243	1.3 / 5.2 / 1.3 / 5.4
15 M	0621 / 1150 / 1845	1.0 / 5.5 / 0.9	30 TU	0541 / 1116 / 1809 / 2336	1.1 / 5.5 / 1.0 / 5.7
			31 W	0635 / 1204 / 1900	0.9 / 5.7 / 0.7

AUGUST

Date	Time	m	Date	Time	m
1 TH ●	0023 / 0724 / 1249 / 1948	5.9 / 0.7 / 5.9 / 0.5	16 F	0117 / 0803 / 1331 / 2024	5.8 / 1.0 / 5.8 / 0.7
2 F	0109 / 0811 / 1334 / 2035	6.1 / 0.6 / 6.0 / 0.3	17 SA	0151 / 0835 / 1402 / 2056	5.8 / 0.9 / 5.8 / 0.7
3 SA	0155 / 0858 / 1419 / 2122	6.2 / 0.5 / 6.1 / 0.2	18 SU	0222 / 0908 / 1430 / 2128	5.8 / 0.9 / 5.8 / 0.7
4 SU	0243 / 0944 / 1505 / 2209	6.2 / 0.6 / 6.1 / 0.2	19 M	0251 / 0939 / 1500 / 2158	5.8 / 1.0 / 5.7 / 0.8
5 M	0331 / 1031 / 1551 / 2256	6.2 / 0.6 / 6.1 / 0.3	20 TU	0321 / 1009 / 1530 / 2227	5.7 / 1.0 / 5.6 / 1.0
6 TU	0420 / 1117 / 1639 / 2345	6.1 / 0.8 / 5.9 / 0.5	21 W	0353 / 1039 / 1601 / 2259	5.6 / 1.1 / 5.5 / 1.0
7 W ◑	0511 / 1207 / 1732	5.8 / 1.0 / 5.7	22 TH	0423 / 1113 / 1633 / 2335	5.4 / 1.3 / 5.4 / 1.2
8 TH	0038 / 0608 / 1303 / 1833	0.8 / 5.5 / 1.2 / 5.4	23 F ◑	0501 / 1154 / 1718	5.2 / 1.4 / 5.2
9 F	0137 / 0715 / 1408 / 1947	1.0 / 5.2 / 1.4 / 5.2	24 SA	0021 / 0602 / 1247 / 1831	1.4 / 5.0 / 1.6 / 5.0
10 SA	0244 / 0830 / 1522 / 2103	1.3 / 5.0 / 1.5 / 5.1	25 SU	0124 / 0717 / 1402 / 1950	1.6 / 4.8 / 1.8 / 4.9
11 SU	0401 / 0943 / 1641 / 2213	1.4 / 5.0 / 1.4 / 5.2	26 M	0250 / 0837 / 1532 / 2112	1.6 / 4.8 / 1.7 / 5.0
12 M	0514 / 1048 / 1744 / 2313	1.3 / 5.1 / 1.2 / 5.4	27 TU	0412 / 0958 / 1648 / 2226	1.5 / 5.1 / 1.5 / 5.3
13 TU	0609 / 1142 / 1832	1.2 / 5.3 / 1.0	28 W	0523 / 1101 / 1753 / 2323	1.2 / 5.4 / 1.0 / 5.7
14 W	0001 / 0652 / 1223 / 1913	5.6 / 1.1 / 5.5 / 0.8	29 TH	0621 / 1150 / 1846	0.9 / 5.5 / 0.6
15 TH	0041 / 0729 / 1259 / 1950 ○	5.7 / 1.0 / 5.7 / 0.7	30 F ●	0010 / 0710 / 1233 / 1933	6.1 / 0.7 / 5.8 / 0.4
			31 SA	0053 / 0755 / 1314 / 2019	6.3 / 0.6 / 6.2 / 0.2

Chart Datum is 2·69 metres below IGN Datum. HAT is 6·4 metres above Chart Datum.

FREE monthly updates. Register at
www.reedsnauticalalmanac.co.uk

STANDARD TIME UT –01
Subtract 1 hour for UT
For French Summer Time add
ONE hour in **non-shaded areas**

DUNKERQUE LAT 51°03′N LONG 2°22′E

TIMES AND HEIGHTS OF HIGH AND LOW WATERS

Dates in red are **SPRINGS**
Dates in blue are NEAPS

YEAR **2019**

SEPTEMBER

Time	m	Time	m
1 0137	6.4	**16** 0153	5.9
0839	0.5	0840	0.9
SU 1356	6.2	M 1359	5.9
2103	0.1	2059	0.7
2 0221	6.4	**17** 0218	5.9
0923	0.5	0910	0.9
M 1439	6.2	TU 1426	5.9
2148	0.2	2129	0.8
3 0306	6.2	**18** 0246	5.8
1007	0.6	0939	1.0
TU 1523	6.2	W 1454	5.8
2232	0.3	2157	0.9
4 0351	6.0	**19** 0312	5.7
1050	0.8	1009	1.0
W 1608	6.0	TH 1520	5.7
2316	0.6	2227	1.0
5 0439	5.8	**20** 0339	5.6
1135	1.0	1041	1.2
TH 1659	5.7	F 1551	5.6
		2301	1.2
6 0003	0.9	**21** 0414	5.4
0532	5.4	1120	1.4
F 1226	1.3	SA 1633	5.3
◗ 1758	5.4	2344	1.4
7 0059	1.3	**22** 0506	5.1
0637	5.1	1211	1.6
SA 1330	1.6	SU 1740	5.0
1914	5.1	◗	
8 0209	1.6	**23** 0047	1.7
0757	4.8	0639	4.8
SU 1451	1.7	M 1325	1.8
2039	4.9	1920	4.9
9 0336	1.7	**24** 0216	1.8
0920	4.8	0810	4.8
M 1620	1.6	TU 1502	1.7
2157	5.0	2052	5.0
10 0455	1.6	**25** 0350	1.6
1032	5.0	0940	5.0
TU 1726	1.3	W 1627	1.4
2259	5.3	2211	5.4
11 0551	1.3	**26** 0506	1.2
1124	5.3	1044	5.5
W 1814	1.0	TH 1735	0.9
2345	5.5	2308	5.8
12 0633	1.1	**27** 0604	0.9
1203	5.5	1132	5.8
TH 1854	0.8	F 1828	0.5
		2353	6.2
13 0022	5.7	**28** 0651	0.7
0708	1.0	1212	6.1
F 1236	5.7	SA 1914	0.3
1927	0.7	●	
14 0056	5.8	**29** 0034	6.4
0739	1.0	0735	0.6
SA 1307	5.8	SU 1251	6.3
○ 1958	0.7	1958	0.2
15 0126	5.9	**30** 0115	6.4
0809	0.9	0817	0.5
SU 1334	5.9	M 1331	6.3
2028	0.7	2041	0.2

OCTOBER

Time	m	Time	m
1 0156	6.4	**16** 0146	6.0
0859	0.6	0842	0.9
TU 1412	6.3	W 1355	6.0
2123	0.3	2100	0.8
2 0238	6.2	**17** 0213	5.9
0940	0.7	0907	1.0
W 1455	6.2	TH 1424	5.9
2205	0.5	2130	0.9
3 0322	6.0	**18** 0242	5.8
1022	0.8	0944	1.0
TH 1540	6.0	F 1454	5.8
2246	0.8	2202	1.0
4 0407	5.7	**19** 0313	5.7
1104	1.1	1018	1.1
F 1629	5.7	SA 1529	5.7
2329	1.2	2238	1.2
5 0457	5.4	**20** 0352	5.4
1151	1.4	1059	1.3
SA 1726	5.3	SU 1614	5.4
◗		2324	1.5
6 0020	1.6	**21** 0447	5.1
0558	5.0	1152	1.5
SU 1251	1.7	M 1728	5.1
1840	4.9	◗	
7 0129	1.9	**22** 0027	1.7
0720	4.7	0620	4.8
M 1413	1.9	TU 1306	1.7
2010	4.8	1902	5.0
8 0301	2.0	**23** 0156	1.8
0849	4.6	0748	4.8
TU 1548	1.8	W 1440	1.6
2132	4.9	2033	5.1
9 0425	1.8	**24** 0329	1.6
1002	4.9	0918	5.1
W 1657	1.4	TH 1605	1.3
2233	5.2	2152	5.5
10 0523	1.5	**25** 0443	1.3
1054	5.2	1022	5.5
TH 1747	1.1	F 1712	0.9
2318	5.5	2248	5.9
11 0606	1.2	**26** 0541	0.9
1133	5.5	1109	5.8
F 1826	0.9	SA 1806	0.5
2355	5.7	2333	6.1
12 0640	1.1	**27** 0629	0.8
1208	5.7	1148	6.1
SA 1858	0.8	SU 1852	0.4
13 0028	5.9	**28** 0012	6.3
0709	1.0	0713	0.7
SU 1238	5.8	M 1227	6.3
○ 1927	0.8	● 1936	0.3
14 0056	5.9	**29** 0051	6.3
0739	0.9	0754	0.6
M 1304	5.9	TU 1306	6.3
1957	0.7	2017	0.4
15 0121	5.9	**30** 0131	6.3
0810	0.9	0835	0.6
TU 1328	5.9	W 1349	6.3
2029	0.7	2058	0.5
		31 0214	6.2
		0916	0.7
		TH 1432	6.2
		2139	0.7

NOVEMBER

Time	m	Time	m
1 0256	6.0	**16** 0225	5.9
0957	0.9	0928	0.9
F 1517	6.0	SA 1441	5.9
2219	1.0	2147	1.0
2 0340	5.7	**17** 0303	5.7
1038	1.1	1007	1.0
SA 1604	5.7	SU 1524	5.7
2300	1.3	2228	1.2
3 0427	5.4	**18** 0349	5.5
1122	1.4	1052	1.2
SU 1658	5.3	M 1618	5.5
2346	1.7	2317	1.4
4 0522	5.0	**19** 0451	5.2
1215	1.7	1146	1.4
M 1803	4.9	TU 1730	5.3
◑		◑	
5 0045	1.9	**20** 0020	1.6
0632	4.7	0605	5.0
TU 1326	1.9	W 1257	1.5
1927	4.7	1845	5.2
6 0208	2.1	**21** 0141	1.7
0800	4.6	0723	5.0
W 1457	1.8	TH 1420	1.4
2051	4.8	2008	5.2
7 0334	1.9	**22** 0303	1.6
0916	4.8	0848	5.1
TH 1611	1.6	F 1538	1.2
2152	5.0	2126	5.5
8 0438	1.7	**23** 0415	1.3
1012	5.1	0953	5.5
F 1705	1.3	SA 1645	0.9
2240	5.3	2223	5.8
9 0526	1.4	**24** 0515	1.1
1056	5.4	1043	5.7
SA 1748	1.1	SU 1742	0.7
2320	5.6	2311	6.0
10 0603	1.2	**25** 0606	0.9
1133	5.6	1126	6.0
SU 1822	0.9	M 1831	0.6
2355	5.7	2353	6.1
11 0636	1.1	**26** 0652	0.8
1205	5.7	1207	6.1
M 1854	0.9	TU 1915	0.5
		●	
12 0024	5.8	**27** 0033	6.1
0708	1.0	0735	0.7
TU 1233	5.8	W 1249	6.2
○ 1926	0.8	1957	0.6
13 0050	5.9	**28** 0113	6.1
0742	0.9	0816	0.7
W 1300	5.9	TH 1332	6.2
2001	0.8	2038	0.7
14 0118	5.9	**29** 0155	6.0
0817	0.9	0857	0.7
TH 1331	6.0	F 1417	6.1
2036	0.8	2118	0.9
15 0150	5.9	**30** 0238	5.9
0852	0.9	0938	0.8
F 1405	6.0	SA 1501	5.9
2111	0.9	2157	1.1

DECEMBER

Time	m	Time	m
1 0320	5.7	**16** 0302	5.8
1018	1.0	1004	0.8
SU 1546	5.7	M 1526	5.9
2236	1.3	2226	1.1
2 0402	5.4	**17** 0351	5.6
1059	1.2	1051	0.9
M 1633	5.4	TU 1619	5.7
2317	1.6	2315	1.2
3 0449	5.1	**18** 0445	5.4
1144	1.5	1144	1.0
TU 1724	5.1	W 1718	5.5
4 0005	1.8	**19** 0012	1.4
0542	4.9	0545	5.3
W 1237	1.7	TH 1245	1.1
◑ 1826	4.8	◑ 1822	5.4
5 0106	2.0	**20** 0119	1.5
0647	4.7	0652	5.2
TH 1348	1.8	F 1356	1.2
1942	4.7	1937	5.3
6 0224	2.0	**21** 0232	1.5
0805	4.7	0809	5.2
F 1504	1.7	SA 1508	1.1
2055	4.9	2054	5.4
7 0335	1.8	**22** 0342	1.4
0914	4.9	0920	5.4
SA 1607	1.5	SU 1616	1.0
2151	5.1	2157	5.5
8 0433	1.6	**23** 0448	1.2
1008	5.1	1018	5.5
SU 1658	1.3	M 1718	0.9
2237	5.3	2251	5.6
9 0520	1.4	**24** 0546	1.1
1052	5.4	1109	5.7
M 1742	1.1	TU 1813	0.8
2318	5.5	2339	5.7
10 0600	1.2	**25** 0636	0.9
1131	5.5	1155	5.9
TU 1821	1.0	W 1900	0.8
2353	5.7		
11 0639	1.1	**26** 0022	5.8
1204	5.7	0721	0.8
W 1859	0.9	TH 1240	5.9
		● 1943	0.8
12 0025	5.8	**27** 0104	5.9
0717	0.9	0804	0.7
TH 1238	5.8	F 1324	6.0
○ 1938	0.9	2024	0.8
13 0059	5.9	**28** 0146	5.9
0757	0.8	0844	0.7
F 1314	5.9	SA 1407	6.0
2018	0.8	2102	0.9
14 0137	5.8	**29** 0225	5.8
0838	0.8	0923	0.8
SA 1354	5.8	SU 1449	5.8
2059	0.9	2139	1.0
15 0218	5.9	**30** 0303	5.7
0920	0.8	1001	0.9
SU 1438	6.0	M 1528	5.7
2141	0.9	2215	1.2
		31 0340	5.5
		1037	1.0
		TU 1606	5.5
		2251	1.3

Chart Datum is 2·69 metres below IGN Datum. HAT is 6·4 metres above Chart Datum.

》》 FREE monthly updates. Register at 《《
www.reedsnauticalalmanac.co.uk

165

N France

4.5 DUNKERQUE (PORT EST)

Nord **51°03'·62N 02°21'·09E** ✵✵✵♤♤♤✿✿

CHARTS AC 323, 1872, 1350, 5605.11; SHOM 6651, 7057; Navi 1010; Imray C30

TIDES Dunkerque is a Standard Port (⟵). +0050 Dover; ML 3·2; Duration 0530. See Gravelines for Dunkerque Port Ouest.

SHELTER Good; access at all tides/weather, but fresh NW–NE winds cause heavy seas at ent and scend in hbr.

NAVIGATION Yachts must not use the W Port. From the east, WPT 51°04'·29N 02°21'·72E (SCM lt buoy) 210°/0·8M to hbr ent. From Nieuwpoort Bank WCM lt buoy, the E1-12 buoys (Lights, buoys and waypoints) lead S of Banc Hills but shore lights make navlights and shipping hard to see. Avoid very large ships manoeuvring into/out of Charles de Gaulle lock.

From the west, WPT 51°03'·83N 02°20'·25E (DW29 lt buoy) 111°/0·6M to hbr ent. Fetch Dyck PHM buoy, 5M N of Calais, thence to DKA SWM buoy, and via DW channel lt buoys past W Port to WPT. Streams reach about 3½kn.

LIGHTS AND MARKS A chimney (116m) 1M WSW of hbr ent and other industrial structures are conspic. The W jetty lt ho is a distinctive, brown brick twr, green top. The E jetty lt pylon is small, slender and less obvious; the jetty covers at HW but is lit by 9 bright sodiums. The ldg lts, W cols, R tops, and the tall lt ho, W tr with B top, lead 137°/1M towards the marinas. (The F Vi(olet) 179° and 185° ldg lts define big ship turning circles.)

IPTS (full) E jetty head and seaward of Port du Grand Large.

Lock Enter Trystram lock (5m) for marinas in Bassin du Commerce and Bassin de la Marine via 3 bridges: (a) at S end of the lock; (b) at SE side of Darse No 1; and (c) at ent to Bassin du Commerce.

The daily schedule (1 Apr–3 Sep, 7/7, LT) for lock and bridges is:

Outward
Bridges (c) and (b) both open 0815, 1045, 1415, 1615, 1830. Bridge (a) and lock open 0850, 1105, 1435, 1635, 1900.

Inward
Lock opens 1020, 1530, 1830, 2100. Bridge (a) opens 30 mins later. Bridges (b) and (c) open 20 mins after (a); the 1110 opening is only on Sat/Sun.

Lock/bridge, light and sound signals: ● ● = No entry
2 long blasts = Request open lock or bridge
● + Fl ● = Standby to secure on the side of the Fl ●
● + Fl ● = Enter lock and secure on side of the Fl ●

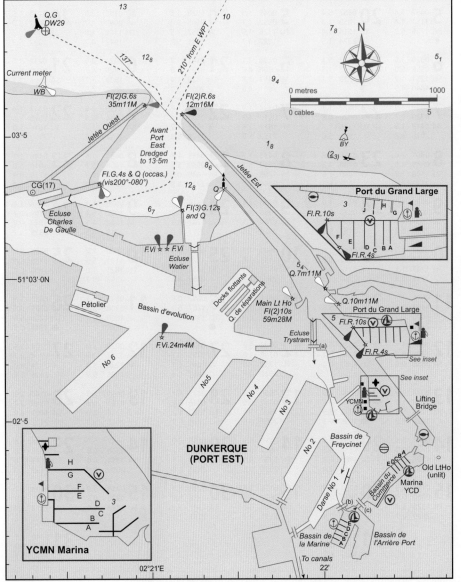

COMMUNICATIONS Monitor *Dunkerque VTS* Ch **73** (H24) for traffic info (English spoken);
CROSS 03·21·87·21·87, VHF Ch 16; ⚓ 03·28·66·45·25; Auto 08·92·68·08·59; Police 03·28·59·13·22; ⊖ 03·28·29·25·50; Brit Consul 01·44·51·31·00; Ⓗ 03·28·66·70·01; HM 03·28·28·75·96. www.portdedunkerque.fr Port Control 03·28·28·78·78; Aff Mar 03·28·63·13·93; Locks VHF Ch 14. Marinas VHF Ch 09.

FACILITIES from seaward: The 3 *marinas are municipal and ❶ berths are at different rates; all on the same website www. dunkerque-marina.com

*Port du Grand Large ⚓ H24 access (3m). ☎03.28.63.23.00. 315+25❶ €2·50. ♠ ♠ (0800-1200, 1400-1830 daily) ◥(30t) ✕ ☐ ▢ ── YC.

YC Mer du Nord (YCMN) ☎ 03·28·66·79·90. 250+50❶ €2·48 plus €0·20 pp/day tax, ⬔ €1·00. H24 access (3m). secreteriat@ ycmn.com www.ycmn.com ⬚ ✎ ♠ ♠ 0800-1200, 1400-1800 daily, ✕ ☐ ▢ wi-fi ◥(7t).

*Bassin de la Marine 170 inc❶ €1·90. ☎03·28·24·58·80. Access to/from canals via lock at SW end of Darse No 1.

*Bassin du Commerce 140+30 ❶ €1·90. Quai des Hollandais (SE bank pontoon) €1·15. YC de Dunkerque (YCD) marina (5m) ☎03.28.21.13.77. ♠ ⚓ ⬚.

Town ♠⬔✕☐▤🍺♠SHOM ✎ 🖂 Ⓔ✕🏧⬚✉Ⓑ⇌✈(Lille). Vehicle ferry (no foot passengers) to Dover.

4.6 GRAVELINES

Nord **51°00'·94N 02°05'·55E** ✷✷⚓⚓✿✿

CHARTS AC 323, 5605.11, 1350; SHOM 6651, 7057; Navi 1010

TIDES HW +0045 on Dover (UT); ML 3·3m; Duration 0520

Standard Port DUNKERQUE (⟵)

Times				Height (metres)			
High Water		Low Water		MHWS	MHWN	MLWN	MLWS
0200	0800	0200	0900	6·0	5·0	1·5	0·6
1400	2000	1400	2100				
Differences GRAVELINES (3M SW of Dunkerque Ouest)							
−0010	−0015	−0010	−0005	+0·5	+0·3	+1·0	0·0
DUNKERQUE PORT OUEST (3M ENE of Gravelines)							
−0010	−0015	−0010	−0005	+0·4	+0·3	0·0	0·0

SHELTER Good, but do not enter in strong winds. Appr chan has only 0·5m at LWS. Access approx HW ±3 for shoal draft, ±2 if draft >1·5m; max draft 2·5m. For a first visit HW is recommended.

In strong NE or SW winds the ancient, manually operated bridge over the entry gate cannot be swung, stopping entry and exit to/from the basin.

NAVIGATION WPT 51°01'·30N 02°05'·13E, 142°/0·43M to bkwtr heads; old lt ho (unlit, conspic B/W spiral) in transit with wine-glass shaped water twr 142°. Beware strong E-going stream across ent near HW.

There is often a dredger working at entrance. Do not cut between the E bkwtr head and a NCM lt bcn close N of it.

The approach chan dries to soft mud, all but a trickle; it is about 15m wide and marked by lateral dolphins. Keep to the W on entry and to the E when inside. 650m before the ent to Bassin Vauban the chan turns 60° stbd.

Use pontoon (waiting only) on port hand side of approach to ent, the entry gate into the marina stays open approx HW ±3 in season, HW ±1½ out of season. A digital gauge shows depth over the sill which dries 0·6m. ❤ berth to stbd.

The River Aa, pronounced Ar-Ar, flows through the Bassin via sluice gates upstream which are usually shut.

LIGHTS AND MARKS Lts as chartlet and Lights, buoys and waypoints. A nuclear power stn with 6 towers (63m) is 1M NE. 1·8M further NE is the ent to Dunkerque Ouest, commercial & ferry port.

The disused lt ho, with continuous black and white spiral band, is very distinctive. The W bkwtr lt shows an approach lt across a sector of arc 139° (085°-224°) G landwards.

COMMUNICATIONS ☎ 08·92·68·02·59; Police 03·28·23·15·17/ 03·28·24·56·45; Brit Consul 01·44·51·31·00; Aff Mar 03·28·26·73·00. Entry gate 03·28·23·13·42, VHF Ch 09 (HO); Marina VHF Ch 09 (HO); Tourist Office 03·28·51·94·00.

FACILITIES

Bassin Vauban General Council Harbour Office, ☎03·28·65·45·24. port. de.plaisance.gravelines@wanadoo.fr www.portvaubangravelines. com M-F 0800-1200, 1345-1715, 430+20❤⌂, €1·90.Max LOA 15m, max beam for entry gate 9·8m. ✗⌂⚓⊕↻🚽🔲🛒 free wi-fi, bike/car hire. Large well equipped ☇▷(10t, 3t)🚩(7t)⚓⊞🔧🛠.

Berth on ❤ pontoon No 1, first to stbd, or as directed. The marina almost dries, but fin keelers sink upright into soft, glorious mud. It may be advisable to close the engine cooling water seacock after shutting down and to clean the water strainer thereafter.

Town 🏤🏬 (H24 Super-U 🍴 600m) ✗⌂🔲✉️Ⓑ⇌.

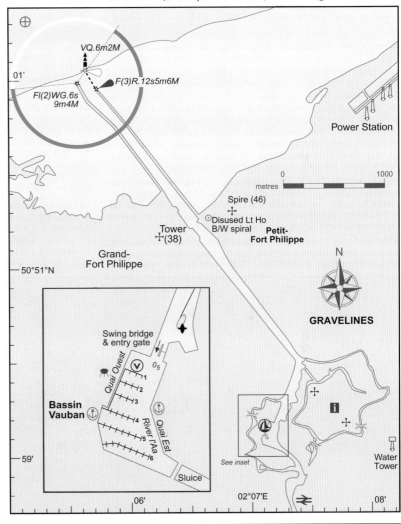

4.7 CALAIS

Pas de Calais **50°58'·33N 01°50'·42E** ✿❀⚓⚓✿✿

CHARTS AC 1892, 323, 1351, 5605.10; SHOM 7323, 6651, 7258; Navi 1010; Imray C8

TIDES +0048 Dover; ML 4·0; Duration 0525

Standard Port DUNKERQUE (←—)

Times				Height (metres)			
High Water		Low Water		MHWS	MHWN	MLWN	MLWS
0200	0800	0200	0900	6·0	5·0	1·5	0·6
1400	2000	1400	2100				
Differences CALAIS							
−0020	−0030	−0015	−0005	+1·3	+1·0	+0·5	+0·2
SANDETTIE BANK (11M N of Calais)							
−0010	−0015	−0015	−0010	−0·6	−0·6	−0·2	−0·1
WISSANT (8M WSW of Calais)							
−0035	−0050	−0030	−0010	+2·0	+1·5	+0·9	+0·4

SHELTER Very good in the marina (Bassin de l'Ouest, 3-6m). In strong NW-NE winds hbr ent is very rough with heavy swell; access H24. 22 W waiting buoys outside tidal gate (times below).

NAVIGATION WPT 50°57'·64N 01°46'·14E (CA 1 buoy) 077°/2·7M to Jetée Ouest hd, keeping clear to the S of ferry track 083°/263°.

Beware Ridens de la Rade, about 4ca N of ent, a sandbank (0·1m) on which seas break. From the E it is better to keep N of this bank until able to round CA4 PHM lt buoy and appr from the W. But, with sufficient rise of tide, Ridens can be crossed on a SW'ly track.

Byelaws require yachts to have engine running (even if sailing) and to keep clear of ferries/commercial vessels. Speed limits: Avant-Port & Bassin Est 10kn; Arrière-Port 8kn; elsewhere 5kn.

LIGHTS AND MARKS Conspic daymarks: Cap Blanc-Nez and Dover Patrol monument, 5·5M WSW of hbr ent; the lt ho(W, B top) in town centre; two silos on N side of Bassin Est; R/W chimney (78m) 700m SE of Bassin Est; Hbr Control in pyramidal bldg at ent to Arrière Port. Lts as chartlet and Lights, buoys and waypoints.

IPTS (full code) shown from Ⓐ (Jetée Est) Ⓑ and Ⓒ as on chartlet, must be strictly obeyed; no VHF chatter should be necessary. On arrival, first comply with Ⓐ, then Ⓑ to enter Arrière Port.

If no sigs are shown (ie no ferries under way) yachts may enter/exit. Or follow a ferry entering, keeping to the stbd side of the fairway and not impeding said ferry. Keep a very sharp lookout, especially for ferries closing fast from astern.

One ● (alongside lowest IPTS lt) = dredger working.

On leaving, first comply with Ⓒ, then Ⓑ, Ⓐ for seaward. Yachts are often cleared outward as a convoy; monitor Ch 17.

Bassin de l'Ouest (marina). 17m wide entry gate (sill 2m above CD) stays open HW −2 to +2½. Gate signals:

●	=	All movements prohib.
◐	=	10 mins before gate opens.
●	=	Movement authorised.
4 blasts	=	Request permission to enter.

The co-located Hénon swing bridge opens at HW −2, −1, HW, +1¼, +2½. Times are usually strictly observed. The monthly schedule is posted on www.plaisance-opale.com. The lock gates to Bassin de l'Ouest, giving access to the marina, are reported as having been removed until further notice. A gauge shows depth of water over the sill.

Bassin Carnot: If bound for the canals, lock gates open HW −1½ to HW +¾. Lock sigs are as per the CEVNI code. Call VHF Ch 17 or sound 2 blasts to request permission to enter.

COMMUNICATIONS Whilst underway in apprs and hbr, monitor VHF Ch **17** 16 (H24) *Calais Port Control* for VTS, marina and Carnot lock (occas); Channel Navigation Info Service (CNIS): *Gris Nez Traffic* Ch **13** 79 16 (H24). Info broadcasts in English & French Ch 79, at H + 10, also at H + 25 when vis is < 2M; CROSS 03·21·87·21·87, *CROSS Gris Nez* VHF Ch 15, 67, **68**, 73; Ⓑ 03·21·96·31·20; ♨ 03·21·33·24·25; Auto 08·92·68·02·62; Police 03·21·34·86·00; ⊖ 03·21·34·75·40; Brit Consul 01·44·51·31·00; Ⓗ 03·21·46·33·33; HM (Port) 03·21·96·31·20; Aff Mar 03·28·59·19·33.

FACILITIES **Marina** ☎03·21·34·55·23 calais-marina@calais-port. fr www.plaisance-opale.com; 250 + 30 ♥ €2·36 Berth/raft on pontoon A, 1st to stbd, or on fingers S side of B pontoon. ⚓♂↯ ⬛ ⬛(9t). **YC** ☎03·21·82·28·05, www.calais-voile.fr, ⬛. For mast unstepping call Calais Nautic, ☎03·21·96·07·57.

Town ✕⬛🍴🛒⬛🔵⚒⚠⬛✉Ⓑ⇌✈ ferries to Dover.

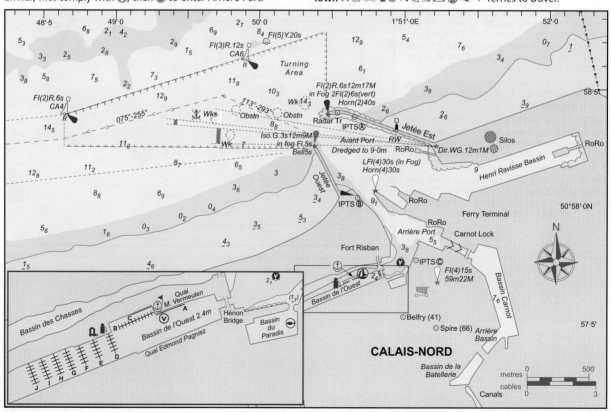

CALAIS TIDAL COEFFICIENTS 2019

Date	Jan am	Jan pm	Feb am	Feb pm	Mar am	Mar pm	Apr am	Apr pm	May am	May pm	June am	June pm	July am	July pm	Aug am	Aug pm	Sept am	Sept pm	Oct am	Oct pm	Nov am	Nov pm	Dec am	Dec pm
1	56	57	53	57	37	40	50	56	56	61	69			72	86	92	112	113	115	112	90	82	74	68
2	60	63	62	66	45	50		61		66	73	77	76	81	96	100	113	111	107	102	75	67	62	56
3	66	69	69		55	60	67	71	71	75	81	84	85	88	103	104	107	102	94	86	59	51	50	45
4	71	74	73	76	65	70	76	80	79	83	86	88	91	93	104	103	96	88	77	69	44	39	41	38
5	76		78	80		74	83	86	86	88	89	89	94	94	100	96	80	71	59	51	34	33	36	35
6	78	79	82	83	78	81	88	90	89	90	89	87	93	92	91	85	62	54	43	37	34	36	37	39
7	80	80	83	83	83	86	90	90	90	89	85	82	89	86	79	72	47	42	34	33	41	46	43	47
8	80	79	82	81	87	88	90	88	87	84	78	74	82	77	65	58	38	38	36	40	51	56	52	57
9	78	76	79	77	88	88	86	83	81	77	70	66	72	68	53	49	40	44	45	51	61	66	61	66
10	74	72	74	70	86	85	78	74	71	66	62	60	64	60	46	46	49	55	57	62	71	75	70	74
11	69	66	66	62	82	79	68	63	61	56	58	58	58	56	48	51	60		67		78	81	78	81
12	62	59	58	54	74	70	57	51	53	50	59	61	56	57	54	59	65	69	72	76		84		83
13	55	52	49	46	65	59	47	44	50	52	63	67	58	61		63	74	77	80	82	85	87	85	86
14	48	45	43	42	54	49	44	46	56	62	70	73	63	66	66	70	80	82	85	86	87	87	87	87
15	43	42	43	47	44	42	52	59	67	73	76			69	73	76	84	85	87	88	86	84	86	85
16	43	45	52	59	41	44	67	76	79	84	78	80	71	73	78	79	85	85	87	86	82	79	82	79
17	48	53	67	76	50	58	84			88	82	82	75	76	80	81	85	83	85	82	75	70	76	72
18	59	65	84	92	66	76	91	97	92	94	83	82	77	78	81	80	82	79	79	75	66	61	69	65
19	72	79		99	85	93	102	105	95	95	81	80	77	77	79	78	76	72	71	66	56	52	62	59
20	86	92	105	110	100		107	108	95	93	78	76	76	74	75	73	68	63	61	55	49	48	58	57
21		97	113	115	107	111	107	105	91	87	73	70	72	70	70	66	58	52	50	45	49	53	58	61
22	102	105	115	113	114	115	101	97	84	79	66	62	67	64	62	58	47	42	41	40	57	63	64	67
23	107	108	110	104	115	113	91	85	74	68	59	55	61	57	53	49	39	37	42	47	70	76	71	75
24	108	106	98	91	109	104	78	70	63	57	51	47	54	50	45	41	39	44	54	62	83	88	79	82
25	102	98	83	74	97	90	63	55	52	47	44	42	47	44	39	39	51	60	71	80	92	96	84	
26	92	86	65	57	82	73	48	41	42	38	40	39	42	41	41	45	69	78	88		98		86	87
27	79	71	49	43	64	55	36	32	36	35	40	41	41	42	51	59		87	95	101	99	99	88	87
																				106				
28	64	58	38	36	47	40	31	32	35	37	44	48	45	50	67	75	96	103		109	98	96	87	85
29	52	48			35	32	35	39	40	45	52	57	55	61		83	108	113	111	111	93	89	83	80
30	46	45			31	34	45	50	49	54	62	67	67	74	91	98	115	116	109	106	85	80	77	73
31	47	50			39	44			59	64				80	104	108			102	96			69	65

Tidal coefficients indicate the magnitude of the tide on any particular day, without having to look up and calculate the range, and thus determine whether it is springs, neaps or somewhere in between. This table is valid for all areas covered by this Almanac. Typical values are:

120	Very big spring tide
95	**Mean spring tide**
70	Average tide
45	**Mean neap tide**
20	Very small neap tide

See also coefficients for Cherbourg (p 196), St Malo (p 215) and Brest (p 277).

N France

4.8 BOULOGNE-SUR-MER

Pas de Calais 50°44'·58N 01°34'·10E (hbr ent) ✦✦✦♒♒♒♒

CHARTS AC 1892, 2451, 438; SHOM 6824, 7247, 7416,7323; Navi 1010, 1011; Imray C8, C31

TIDES 0000 Dover; ML 4·9; Duration 0515
Standard Port DUNKERQUE (←)

Times				Height (metres)			
High Water		Low Water		MHWS	MHWN	MLWN	MLWS
0200	0800	0200	0900	6·0	5·0	1·5	0·6
1400	2000	1400	2100				
Differences BOULOGNE							
−0045	−0100	−0045	−0025	+2·8	+2·2	+1·1	+0·5

SHELTER Good, except in strong NW'lies. Entry H24 at all tides and in most weather. Exposed ⚓ in N part of outer hbr; no ⚓ SW of pecked magenta line.

NAVIGATION WPT 50°44'·71N 01°33'·00E, 100°/0·71M to hbr ent. No navigational dangers, but keep W and S of the free-standing head of Digue Nord, as outer part of bkwtr covers at HW. After rounding Digue Carnot, head S for 325m to a white □ mark (often faded) on bkwtr, thence track 123° to inner ent; caution drying banks on either side. Obey IPTS; keep clear of FVs. Monitor Port Control, VHF Ch 12, on entry and departure.

LIGHTS AND MARKS 'Approches Boulogne' SCM buoy is 295°/2·1M from Digue Carnot lt. Cap d'Alprech lt ho is 2·5M S of hbr ent. Monument is conspic 2M E of hbr ent. Cathedral dome is conspic, 0·62M E of marina. There are no ldg lts/marks.

IPTS are shown from SW jetty hd (FG ☆) and from Quai Gambetta, opposite ☆ 3FG ▽, visible from marina. A Y lt ● by top IPTS = dredger working, but does not prohibit movements.

COMMUNICATIONS CROSS 03·21·87·21·87; ☎ 08·92·68·02·62; Auto 08·92·65·08·08; Forecasts by CROSS Gris Nez VHF Ch 79 at H+10 (078-1910LT); Maritime Police 03·21·30·87·09; Emergency 17; ⊜ 03·21·80·89·90; Brit Consul 01·44·51·31·00; ℍ 03·21·99·33·33; Aff Mar 03·21·30·53·23; Dr 03·21·71·33·33; Control tower 03·21·10·35·47, VHF Ch 12 (H24); Bassin Napoléon Lock VHF Ch 12 (H24); Marina Ch 09.

FACILITIES Visitors berth in marina (2·7m) on W side of river. FVs berth opposite at Quai Gambetta. If R Liane is sluicing beware strong current through the moorings, especially at LW.

Marina ☎06·76·98·74·98; boulogne-marina@littoralhautsdefrance. cci.ft www.portboulogne.com 130 inc 70🅥 €2·30 plus €0·20pp/ day tax. ⚓ HW±3, 🆆🅒(card access) Ⓔ ⚒ 🛢 ⚓ (20t). Flagler ⚓ ☎03·21·33·90·75, also for sail repairs and ⚓. YC Boulonnais ☎03·21·31·80·67, ✕ ⏢ ⚓.

Bassin Napoléon pontoons are for locals and long-stay 🅥 (week or month; 12-25m LOA) by arrangement. Sanson lock (Ch 12) works HW−3¼ to HW+4½ with ent/exits approx every 35 mins, and freeflow HW −2 to +¼. Daily schedules from HM or website.

Town ✕ ⏢ 🛒 🏧 SHOM ⊠ Ⓑ ⇌ ✈ Le Touquet.

4.9 ÉTAPLES

Pas de Calais 50°31'·00N 01°38'·00E ✦✦♒♒♒♒

CHARTS AC 2451; SHOM 7416; Navi 1011; Imray C31

TIDES −0010 Dover; ML 5·3; Duration 0520
Standard Port DIEPPE (→)

Times				Height (metres)			
High Water		Low Water		MHWS	MHWN	MLWN	MLWS
0100	0600	0100	0700	9·3	7·4	2·5	0·8
1300	1800	1300	1900				
Differences LE TOUQUET (ÉTAPLES)							
+0010	+0015	+0030	+0030	+0·2	+0·2	+0·3	+0·4
BERCK							
+0005	+0015	+0025	+0030	+0·4	+0·5	+0·3	+0·4

SHELTER Good, except in strong W/NW'lies.

NAVIGATION WPT 50°32'·88N 01°32'·36E is on the boundary between W and G sectors of Camiers lt; 090°/0·7M to Mérida unlit

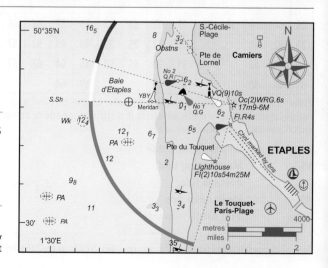

ÉTAPLES *continued*

WCM buoy (50°32'·88N 01°33'·47E) marking a drying wreck 2M NW of Le Touquet lt ho. Estuary dries 1M offshore; do not attempt entry in even moderate SW/W winds, when seas break heavily a long way out. Careful identification of marks is essential.

The outer part of the buoyed chan constantly shifts, such that the chan ent may lie N or S of the ⊕. Nos 1 & 2 buoys (QG and QR) should be visible from the WPT. Bcn, Fl R 4s, 4ca SW of Camiers lt, marks NE end of covering training wall (also marked by posts).

LIGHTS AND MARKS See Lights, buoys and waypoints. Le Touquet is at the S end of 175m high hills, visible for 25M. Le Touquet lt ho, orange with brown band, is conspic; Camiers' R pylon is hard to see by day.

4.10 ST VALÉRY-SUR-SOMME

Somme **50°11'·34N 01°38'·90E** ❀❀⚓⚓✿✿✿

CHARTS AC 2451; SHOM 7416; Navi 1011; Imray C31

TIDES LE HOURDEL –0010 Dover; ML —.

Standard Port DIEPPE (➞)

Times				Height (metres)			
High Water		Low Water		MHWS	MHWN	MLWN	MLWS
0100	0600	0100	0700	9·3	7·4	2·5	0·8
1300	1800	1300	1900				
Differences ST VALÉRY-SUR-SOMME							
+0035	+0035	ND	ND	+0·9	+0·7	ND	ND
LE HOURDEL							
+0020	+0020	ND	ND	+0·8	+0·6	ND	ND
CAYEUX							
+0000	+0005	+0015	+0010	+0·5	+0·6	+0·4	+0·4

SHELTER The 3 hbrs (Le Hourdel, Le Crotoy and St Valéry-sur-Somme) are well sheltered.

NAVIGATION WPT 50°14'·00N 01°28'·08E ('ATSO' SWM buoy, Mo (A) 12s). The estuary dries up to 3M seaward of Le Hourdel. It is open to the W and can be dangerous in onshore winds >F6. Silting is a major problem. The sands build up in ridges offshore, but inside Pte du Hourdel are mostly flat, except where R Somme and minor streams scour their way to the sea. If Dover tidal range >4·4m, there is enough water at HW ±1 for vessels <1·5m draft.

Leave 'ATSO' buoy at HW St Valéry –2, then head SE twds R/W lt ho on dunes near Cayeux-sur-Mer to beginning of buoyed chan (S1) Follow chan buoys S1 to S50 (some lit) in strict order (no corner-cutting!) to small unlit WCM bifurcation buoy, E of Pte du Hourdel. Here fork stbd to St Valéry and its covering training wall with four light beacons; see chartlet and Lights, buoys and waypoints. Continue between the tree-lined promenade to stbd and a covering dyke marked by PHM bns. Bias your track towards the town side, but comply with any buoys. At W lt bcn on head of Digue du Nord turn stbd past town quay into marina, dredged

2m. Enter St Valéry Marina (max draft 2·0m) at HW ±1; usually met and given a berth on pontoons 4–6. If N-bound leave St Valéry HW–2 for a fair tide.

Abbeville Canal (max draft 2·0m) is navigable, by day only, to Abbeville. Lock 03·22·60·80·23; pre-call Amiens ☎06·74·83·60·69.

LIGHTS AND MARKS Cayeux-sur-Mer lt ho (W + R top, Fl R 5s) and Le Crotoy water twr are conspic. Pte du Hourdel lt ho is partly obscured by dunes until close to. See also Lights, buoys and waypoints.

COMMUNICATIONS CROSS 03·21·87·21·87; Auto 08·92·68·08·80; Police 17 or 03·22·60·82·08; ⊖ 03·22·24·04·75; British Consul 01·44·51·31·00; Dr 03·22·26·92·25; Aff Mar 03·21·30·87·19 (Boulogne); HM VHF Ch 09 (St Valéry HW ±2); Le Crotoy YC (Jul/ Aug only).

FACILITIES Marina/YC Sport Nautique Valéricain ❀⚓☎03·22·60·24·80. snval@wanadoo.fr www.portsaintvalery.com Access HW ±2 for 1·5m draft. 250 + 30**Ⓥ** ⬡ €2·35, multihulls €2·80. 🖥 🅿(€5·00) ⚡ free wi-fi, 🅿 🛢 ⬡ Ⓔ ⚓(10t).

Town 🏨 🍴 (500m) ✕ 🛒 🛍 🛢 🏧 ✉ Ⓑ ⇌ and ✈ (Abbeville). **Ferry** see Dieppe, Boulogne or Calais.

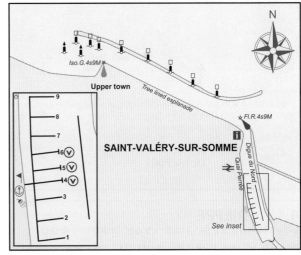

See inset

SAINT-VALÉRY-SUR-SOMME

ADJACENT HARBOURS

LE HOURDEL A basic FV hbr. Access HW±1½ for 1·5m draft. After lt ho, head SW via unmarked side chan to hug the end of shingle spit. Dry out on hard sand. Impossible to leave before HW –2.

LE CROTOY Somme 50°13'·07N 01°37'·99E ❀❀⚓✿✿ From the bifurcation buoy follow the chan E, marked by buoys C1-C10. Enter hbr very close to FV stages (portside). The badly silted marina can be entered by max draft 0·8m when tidal range at Dover >4·4m (approx Coeff 85). Secure at last FV stage and ask YC for berth on the two drying pontoons. Departure impossible before HW –2. **Marina** (280) ☎03·22·27·83·11 (office in YC bldg), ⚓(6t) 🅿. **YC Nautique de la Baie de Somme** ☎03·22·27·83·11, 🖥. Town ✕ 🛒 🛍 🏨 🍴 ✉ Ⓑ ⇌.

COMMUNICATIONS CROSS 03·21·87·21·87; Auto 08·36·68·08·62; ⊖ 03·21·05·01·72; Brit Consul 01·44·51·31·00; Dr 03·21·05·14·42; Aff Mar Étaples 03·21·94·61·50; HM VHF Ch 09.

FACILITIES **Marina** ☎03·21·84·54·33. cn.canche@wanadoo.fr 182 + 18**Ⓥ** €2·04. (1·2m) close downriver of low bridge; beware up to 5kn current. Best to berth on first pontoon hammerhead near HW slack. 🅿 🖥 ⬡ 🛠 ⚒ ⚓(8t); Quay 🖥(130t). **YC** ☎03.21.94.74.26, 🖥 🅿.

Town 🏨 🍴 ✕ 🛒 🛍 ✉ Ⓑ ⇌ ✈ Le Touquet.

Le Touquet Drying moorings to stbd off Le Touquet YC. VHF Ch 09, 77. **Cercle Nautique du Touquet** ☎03·21·05·12·77, ✕ 🖥 ⚓ ⚓ ⚓. **Services** 🖥 🅿 🛢 🛠 ⚒ ⚓. **Town** 🏨 🍴 ✕ 🛏 🛒 🛢 ✉ Ⓑ ⇌ ✈.

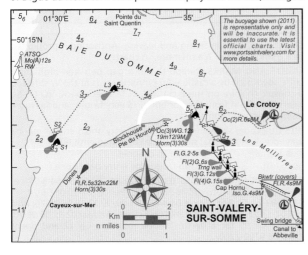

SAINT-VALÉRY-SUR-SOMME

N France

4.11 LE TRÉPORT

Seine Maritime **50°03'·89N 01°22'·20E** ❀❀♨♨♨❁❁

CHARTS AC 2451, 1354; SHOM 7207, 7416, 7417; Navi 1011; Imray C31

TIDES –0025 Dover; ML 5·0; Duration 0530

Standard Port DIEPPE (→)

Times				Height (metres)			
High Water		Low Water		MHWS	MHWN	MLWN	MLWS
0100	0600	0100	0700	9·3	7·4	2·5	0·8
1300	1800	1300	1900				
Differences LE TRÉPORT							
+0005	+0000	+0005	+0015	+0·3	+0·2	+0·1	+0·1

SHELTER Good in marina, E of FVs at S side of S Basin (3·7m).

NAVIGATION WPT 50°04'·27N 01°21'·90E, 146°/0·4M to ent. Coast dries to approx 300m off the pier hds. Shingle spit extends NW from E pier hd; keep well to the W. Entry difficult in strong on-shore winds which cause scend in Avant Port. Strong cross-current at entrance; 5kn speed limit. Ent chan and most of Avant Port dry; S chan is dredged 1·5m, but prone to silting; avoid grounding on the triangle between the two channels. Lock opens HW±4; yachts can wait against Quai Francois I with sufficient rise of tide. Berth, if possible, on smooth S wall of lock; vert wires rigged for warps. N wall is corrugated steel shuttering on which fenders can snag. Beware current when leaving.

LIGHTS AND MARKS Le Tréport lies between high chalk cliffs, with crucifix (floodlit) 101m above town and conspic church S of hbr. IPTS at the root of the E Jetty. There are 150° ldg marks, but they are hard to see and not strictly necessary.

Lock sigs shown from head of piled ent chan: 3 ● (vert) = enter; 3 ● (vert) = no entry.

COMMUNICATIONS [LB] 02·35·86·30·27; CROSS 03·21·87·21·87; Auto 06·70·63·43·40; Police 02·35·86·12·11; ⊖ 02·35·86·15·34; Brit Consul 01·44·51·31·00; Dr 02·35·86·16·23; Port HM 06·87·70·30·90; *Capitainerie Le Tréport* VHF Ch 12 (HW ±4) for lock/marina; Aff Mar 02·35·06·96·70.

FACILITIES **Marina/lock** ☎02·35·50·63·06. www.treport.cci.fr accueilcci@treport.cci.fr 110+15ⓥ €2·13, ⚓⛽🛢⚓🅿️🔧 Ⓔ. **Port de Commerce** is an overflow; half-price rates for yachts. 🏗(10t). **YC de la Bresle** ☎02·35·50·17·60, 🍴. **Town** 🏦🏤✕🛒🔧⚓✉Ⓑ ⚡, ✈ Dieppe.

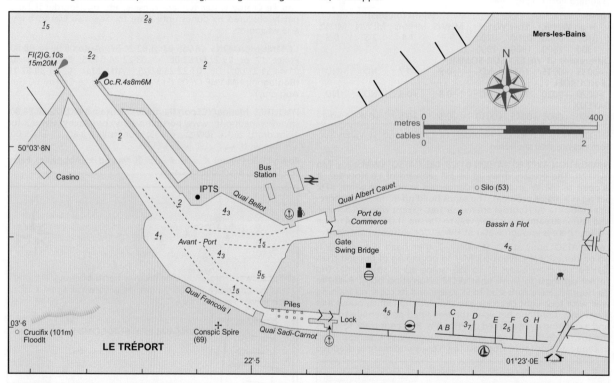

4.12 DIEPPE

Seine Maritime **49°56'·31N 01°05'·00E** ❀❀❀♨♨❁❁❁

CHARTS AC 2451, 2148, 1355; SHOM 7417, 7317; Navi 1011, 1012; Imray C31

TIDES –0011 Dover; ML 4·9; Duration 0535; Dieppe is a Standard Port (→)

SHELTER Very good. Access all tides, but ent exposed to NW-NE winds, which cause a heavy scend. A fixed wavebreak at the E end of the marina is effective. 3kn max speed in marina.

NAVIGATION WPT 49°56'·66N 01°04'·83E (off chartlet) 167°/4ca to W jetty head. Ent to hbr is simple, but beware strong stream across ent. Ent chan and Avant Port are dredged 4·5m. Be alert for ferry at the outer hbr, and small coasters or FVs in the narrow chan between outer hbr and marina. Due to restricted visibility, monitor Ch 12 during arr/dep. It is vital to comply precisely with IPTS (below) both entering and leaving. Anchoring is prohibited in a triangular area 7ca offshore, as charted.

LIGHTS AND MARKS Dieppe lies in a gap between high chalk cliffs, identified by a castle with pinnacle roofs at the W side and a conspic ch spire on the E cliff. DI, ECM wreck buoy, VQ (3) 5s, bears 288°/2·5M from the hbr ent. IPTS very bright, and far more visible than jetty lights.

IPTS (full code) shown from root of W jetty and S end of marina wavebreak. Additional sigs may be combined with IPTS:

● ● to right = Dredger in channel.
ⓦ to left = Bassin Duquesne entry gate open.

COMMUNICATIONS CROSS 03·21·87·21·87; [LB] 02·35·84·8·55; 🚁 03·21·31·52·23; Auto 08·36·68·08·76; Police 02·35·06·96·74; ⊖ 02·35·82·24·47; Brit Consul 01·44·51·31·00; Ⓗ 02·35·06·76·76; Port HM 02·35·84·10·55, *Dieppe Port* Ch **12** (HO); Aff Mar 02·35·06·96·70; Marina Ch 09 for a berth. Marina monitors Ch 12 and opens 1/7-24/8 0700-2200.

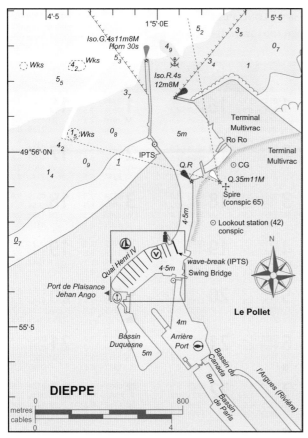

DIEPPE

metres
cables

0 ——— 800 (metres)
0 ——— 4 (cables)

FACILITIES

Marina Jehan Ango ☎02·35·84·10·55. smpd-plaisance@portde dieppe.fr www.portdedieppe.fr 450 ⌲ inc 50🅥 €3·04, on pontoons 10, 9, 8 in about 4·5m but not on hammerheads. If 10 & 9 are full, late arrivals ⌲/raft on E side of pontoon 8; Pontoon 1 W side finger is for boats >14m LOA. Use gangway intercom for shore access code number (if arriving out of hrs, for code call Ch 12). 🛁🚿 cr/card H24, call Ch 09 if unmanned. 🔧🖭 Ⓔ ⚒ ⛽ ⛴(12t).

Cercle de la Voile de Dieppe ☎02·35·84·32·99, ⚔ 🖵 🏴 WC(code access).

Bassin Duquesne (70 ⌲) access via lifting bridge/entry gate, and **Bassin de Paris** (70 ⌲) are only for local yachts.

Town ⚔🛒🚊🗑🏤🖻@✉Ⓑ⇌✈. Ferry Dieppe-Newhaven; www. transmancheferries.co.uk

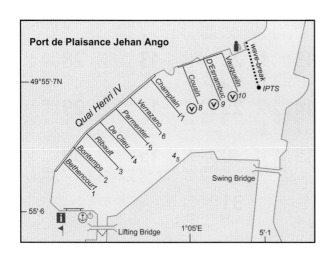

Port de Plaisance Jehan Ango

DIEPPE TO LE HAVRE

From **Dieppe** to Pte d'Ailly lighthouse the coast (AC 2451) is fringed by a bank, drying in places, up to 4ca offshore.

▶*Off Dieppe the ENE-going stream begins about HW Dieppe –0505, and the WSW-going at about HW Dieppe +0030, sp rates 2kn. E of Pte d'Ailly an eddy runs W close inshore on the first half of the E-going stream.*◀

Drying rks extend 5ca off Pte d'Ailly, including La Galère, a rk which dries 6·8m, about 3ca N of the Pointe (AC 2147). Dangerous wrecks lie between about 1·5M NNW and 1·2M WNW of the lighthouse, the former marked by a NCM light buoy. About 6M N of Pte d'Ailly, Grands and Petits Écamias are banks with depths of 11m, dangerous in a heavy sea.

Between Pte d'Ailly and **St Valéry-en-Caux** (lighthouse) there are drying rks 4ca offshore in places. About 1·5M E of Pte de Sotteville a rocky bank (4·2m) extends about 1M NNW; a strong eddy causes a race, Raz de Saint-Michel, over this bank.

There are rky ledges, extending 4ca offshore in places. Immediately E of St Valéry-en-Caux shallow sandbanks, Les Ridens, with a least depth of 0·6m, extend about 6ca offshore.

▶*At St Valéry-en-Caux entrance the E-going stream begins about HW Dieppe –0550, and the W-going stream begins about HW Dieppe –0015, sp rates 2·75kn. E of the ent a small eddy runs W on the E-going stream.*◀

From St Valéry-en-Caux to **Fécamp**, 15M WSW, the coast consists of high, sheer chalk cliffs broken by valleys. 3M W of St Valéry-en-Caux, Paluel nuclear power station is conspicuous with the NE/NW limits of its seaward prohibited area marked by two SPM lt buoys. Close NE of Fécamp off Pte Fagnet, Les Charpentiers are unmarked drying rocks extending almost 2ca offshore.

▶*At Fécamp pierheads the NNE-going stream begins about HW Le Havre –0340, up to 2·9kn sp; the SSW-going about HW Le Havre +0220, much weaker. Diamond A on AC 1354 gives tidal stream data about 9ca WNW of the entrance.*◀

From Fécamp to Cap d'Antifer lighthouse drying rocks extend up to 2½ca offshore. L'Aiguille, a prominent (51m) pinnacle rock is just off Étretat,1·7M NNE of Cap d'Antifer. A large offshore wind farm (WIP 2018/19) is being built 7·5M NW of Fécamp.

Port du Havre-Antifer, a VLCC harbour lies, 1·5M S of Cap d'Antifer (AC 2146, 2990). A 2M long breakwater extends about 1·5M seaward, and should be given a berth of about 1M, and considerably more in heavy weather with wind against tide when there may be a race.

The buoyed approach channel starts 11M NW of the harbour. Commercial vessels have priority but generally traffic is not heavy; yachts should maintain a good lookout and cross it at 90° to its axis as quickly as possible just W of A23/A24 lt bys. If a ship is underway in the channel, crossing vessels should advise *Le Havre d'Antifer* VHF Ch 22 of their position and intentions.

▶*Off Cap d'Antifer the NE-going stream begins about HW Le Havre –0430, and the SW-going at about HW Le Havre +0140. There are eddies close inshore E of Cap d'Antifer on both streams.*◀

Le Havre is a large commercial port, as well as a yachting centre. From the NW, the most useful mark is the Le Havre Lanby 'LHA' 9M W of C de la Hève (lt). The well-buoyed/lit access channel runs 6M ESE from LH 3/4 buoys to the hbr ent.

Strong W'lies cause rough water over shoal patches either side of the channel. Coming from the N or NE, there is deep water close off C de la Hève, but from here steer S to join the main ent chan. Beware Banc de l'Éclat (1·0m) which lies on N side of main channel near LH 14 buoy about 1·5M from harbour entrance.

W and S of Digue Sud yachts should not impede vessels entering/ leaving Port 2000, a container terminal. If inbound to the marina it is best to cross early to the N side of the channel.

STANDARD TIME UT –01
Subtract 1 hour for UT
For French Summer Time add ONE hour in **non-shaded areas**

DIEPPE LAT 49°56'N LONG 1°05'E
TIMES AND HEIGHTS OF HIGH AND LOW WATERS

Dates in red are **SPRINGS**
Dates in blue are **NEAPS**

YEAR 2019

JANUARY

Day	Time m	Time m		Day	Time m	Time m
1 TU	0206 2.2	0749 7.9		**16** W	0042 2.6	0632 7.3
	1443 2.1	2021 7.8			1325 2.7	1913 7.3
2 W	0314 2.1	0855 8.1		**17** TH	0156 2.5	0751 7.5
	1548 1.9	2124 8.0			1443 2.4	2028 7.6
3 TH	0415 1.9	0950 8.3		**18** F	0313 2.2	0858 8.0
	1645 1.7	2216 8.3			1553 1.9	2130 8.1
4 F	0506 1.7	1037 8.6		**19** SA	0420 1.7	0956 8.5
	1732 1.5	2301 8.5			1654 1.4	2226 8.7
5 SA	0549 1.6	1118 8.7		**20** SU	0519 1.3	1049 9.0
	1812 1.4	2340 8.6			1752 1.0	2318 9.1
6 SU ●	0628 1.5	1154 8.8		**21** M ○	0614 1.0	1140 9.4
	1850 1.3				1847 0.6	
7 M	0015 8.7	0705 1.4		**22** TU	0009 9.5	0708 0.7
	1229 8.9	1926 1.2			1229 9.7	1939 0.4
8 TU	0050 8.8	0740 1.4		**23** W	0058 9.7	0800 0.6
	1304 8.9	2001 1.3			1318 9.8	2028 0.3
9 W	0124 8.7	0814 1.5		**24** TH	0145 9.7	0847 0.6
	1338 8.7	2033 1.4			1405 9.7	2114 0.4
10 TH	0158 8.6	0844 1.7		**25** F	0232 9.5	0932 0.7
	1412 8.5	2102 1.6			1451 9.4	2156 0.7
11 F	0231 8.4	0915 1.9		**26** SA	0317 9.2	1015 1.1
	1445 8.3	2133 1.8			1537 9.0	2238 1.1
12 SA	0304 8.1	0948 2.0		**27** SU ◐	0403 8.7	1058 1.5
	1520 8.0	2208 2.0			1625 8.5	2321 1.6
13 SU	0341 7.9	1028 2.4		**28** M	0453 8.2	1146 1.9
	1600 7.7	2250 2.3			1719 7.9	
14 M ◐	0425 7.6	1116 2.6		**29** TU	0012 2.1	0552 7.7
	1650 7.4	2341 2.5			1245 2.3	1825 7.4
15 TU	0520 7.4	1214 2.7		**30** W	0116 2.5	0704 7.4
	1754 7.2				1357 2.5	1944 7.2
				31 TH	0231 2.6	0823 7.4
					1513 2.4	2100 7.4

FEBRUARY

Day	Time m	Time m		Day	Time m	Time m
1 F	0344 2.4	0930 7.7		**16** SA	0236 2.4	0827 7.7
	1620 2.1	2200 7.8			1524 2.1	2107 7.9
2 SA	0444 2.1	1021 8.1		**17** SU	0355 1.9	0936 8.3
	1713 1.8	2246 8.1			1634 1.5	2210 8.5
3 SU	0532 1.8	1103 8.4		**18** M	0501 1.4	1035 8.9
	1756 1.5	2325 8.4			1737 0.9	2306 9.1
4 M ●	0613 1.6	1140 8.7		**19** TU ○	0602 0.9	1128 9.4
	1834 1.3				1836 0.5	2357 9.6
5 TU	0000 8.7	0650 1.4		**20** W	0658 0.5	1218 9.8
	1213 8.8	1910 1.1			1928 0.1	
6 W	0033 8.8	0725 1.3		**21** TH	0044 9.9	0749 0.3
	1247 8.9	1944 1.1			1304 10.0	2015 0.0
7 TH	0105 8.9	0757 1.3		**22** F	0129 9.9	0834 0.3
	1319 8.9	2015 1.1			1348 9.9	2058 0.1
8 F	0137 8.9	0827 1.3		**23** SA	0211 9.7	0915 0.5
	1351 8.9	2044 1.2			1431 9.7	2135 0.5
9 SA	0208 8.8	0856 1.4		**24** SU	0252 9.4	0952 0.8
	1422 8.7	2112 1.3			1512 9.2	2210 1.0
10 SU	0238 8.6	0926 1.6		**25** M	0332 8.9	1027 1.3
	1453 8.4	2143 1.6			1553 8.6	2245 1.6
11 M	0311 8.3	0959 1.9		**26** TU ◐	0414 8.3	1107 1.9
	1528 8.1	2217 1.9			1640 7.9	2324 2.2
12 TU ◐	0347 8.0	1039 2.2		**27** W	0505 7.6	1158 2.4
	1609 7.8	2259 2.2			1739 7.2	
13 W	0432 7.7	1129 2.4		**28** TH	0024 2.8	0614 7.0
	1703 7.4	2354 2.5			1308 2.8	1902 6.8
14 TH	0533 7.4	1234 2.6				
	1819 7.2					
15 F	0107 2.7	0701 7.3				
	1359 2.5	1952 7.3				

MARCH

Day	Time m	Time m		Day	Time m	Time m
1 F	0146 3.0	0747 6.9		**16** SA	0035 2.7	0627 7.2
	1434 2.8	2036 6.9			1329 2.6	1928 7.2
2 SA	0313 2.8	0908 7.3		**17** SU	0211 2.5	0806 7.5
	1553 2.4	2141 7.4			1502 2.1	2051 7.8
3 SU	0422 2.4	1002 7.8		**18** M	0337 2.0	0920 8.2
	1651 1.9	2227 8.0			1617 1.4	2157 8.5
4 M	0513 1.9	1044 8.2		**19** TU	0446 1.3	1021 8.9
	1736 1.5	2305 8.4			1723 0.8	2252 9.2
5 TU	0554 1.5	1120 8.6		**20** W	0548 0.8	1113 9.4
	1815 1.3	2338 8.7			1820 0.4	2340 9.6
6 W	0631 1.3	1153 8.8		**21** TH ○	0643 0.5	1201 9.8
	1850 1.1				1911 0.1	
7 TH	0010 8.9	0705 1.2		**22** F	0025 9.9	0731 0.3
	1225 9.0	1923 0.9			1245 9.9	1955 0.0
8 F	0042 9.0	0736 1.1		**23** SA	0107 9.9	0813 0.3
	1257 9.1	1953 0.9			1327 9.9	2034 0.2
9 SA	0112 9.0	0806 1.1		**24** SU	0147 9.7	0851 0.4
	1327 9.1	2022 0.9			1406 9.6	2108 0.5
10 SU	0142 9.0	0835 1.1		**25** M	0224 9.4	0925 0.8
	1358 9.0	2051 1.1			1444 9.2	2139 1.0
11 M	0212 8.9	0906 1.3		**26** TU	0301 8.9	0956 1.3
	1429 8.8	2120 1.3			1523 8.6	2207 1.6
12 TU	0244 8.7	0936 1.5		**27** W	0339 8.3	1031 1.9
	1503 8.4	2150 1.6			1604 7.8	2247 2.3
13 W	0319 8.3	1012 1.8		**28** TH ◐	0423 7.5	1117 2.5
	1542 8.0	2229 2.0			1657 7.1	2340 2.9
14 TH ◐	0401 7.9	1058 2.2		**29** F	0526 6.9	1222 2.9
	1632 7.5	2323 2.4			1816 6.6	
15 F	0458 7.4	1201 2.5		**30** SA	0102 3.2	0701 6.6
	1747 7.1				1350 3.0	1958 6.7
				31 SU	0234 3.1	0834 6.9
					1513 2.6	2109 7.2

APRIL

Day	Time m	Time m		Day	Time m	Time m
1 M	0348 2.5	0932 7.5		**16** TU	0320 1.9	0903 8.2
	1616 2.1	2156 7.8			1559 1.4	2139 8.6
2 TU	0441 2.0	1014 8.0		**17** W	0428 1.3	1003 8.8
	1704 1.6	2234 8.3			1703 0.9	2232 9.2
3 W	0524 1.6	1051 8.5		**18** TH	0529 0.9	1054 9.3
	1744 1.3	2309 8.7			1759 0.5	2318 9.5
4 TH	0602 1.3	1125 8.8		**19** F ○	0621 0.6	1140 9.6
	1821 1.1	2342 8.9			1847 0.4	
5 F ●	0637 1.1	1158 9.0		**20** SA	0001 9.7	0707 0.4
	1855 0.9				1222 9.7	1929 0.3
6 SA	0014 9.2	0710 1.0		**21** SU	0042 9.7	0748 0.5
	1231 9.1	1927 0.8			1302 9.6	2006 0.5
7 SU	0045 9.1	0742 0.9		**22** M	0120 9.5	0824 0.6
	1302 9.2	1958 0.8			1341 9.4	2039 0.8
8 M	0116 9.1	0813 0.9		**23** TU	0157 9.3	0857 0.9
	1334 9.1	2028 0.9			1418 9.0	2109 1.2
9 TU	0149 9.1	0846 1.1		**24** W	0232 9.0	0928 1.3
	1408 8.9	2059 1.2			1456 8.5	2138 1.7
10 W	0222 8.8	0919 1.3		**25** TH	0309 8.2	1002 1.8
	1445 8.6	2133 1.5			1535 7.9	2216 2.3
11 TH	0259 8.5	0956 1.6		**26** F ◐	0351 7.6	1044 2.4
	1526 8.1	2213 1.9			1624 7.2	2304 2.9
12 F ◐	0343 8.0	1042 2.0		**27** SA	0446 7.0	1141 2.8
	1618 7.6	2307 2.4			1731 6.7	
13 SA	0443 7.5	1145 2.4		**28** SU	0016 3.2	0607 6.6
	1736 7.2				1300 3.0	1859 6.7
14 SU	0021 2.6	0614 7.2		**29** M	0143 3.1	0737 6.7
	1315 2.4	1914 7.4			1420 2.8	2017 7.1
15 M	0156 2.5	0749 7.5		**30** TU	0257 2.7	0844 7.2
	1446 2.0	2035 7.9			1526 2.3	2111 7.6

Chart Datum is 4·43 metres below IGN Datum. HAT is 10·1 metres above Chart Datum.

》》 FREE monthly updates. Register at **《《**
www.reedsnauticalalmanac.co.uk

STANDARD TIME UT –01
Subtract 1 hour for UT
For French Summer Time add
ONE hour in **non-shaded areas**

DIEPPE LAT 49°56'N LONG 1°05'E
TIMES AND HEIGHTS OF HIGH AND LOW WATERS

Dates in red are SPRINGS
Dates in blue are NEAPS

YEAR 2019

MAY

Day	Time	m		Day	Time	m
1 W	0354	2.2		16 TH	0405	1.3
	0933	7.8			0940	8.7
	1618	1.8			1638	1.1
	2155	8.1			2207	9.0
2 TH	0442	1.7		17 F	0504	1.0
	1014	8.3			1031	9.0
	1703	1.5			1732	0.9
	2233	8.5			2254	9.2
3 F	0525	1.4		18 SA	0555	0.9
	1052	8.6			1117	9.2
	1744	1.2			1818	0.8
	2309	8.8		○	2337	9.3
4 SA	0604	1.2		19 SU	0640	0.8
	1128	8.9			1159	9.3
	1822	1.0			1859	0.8
●	2344	9.0				
5 SU	0641	1.0		20 M	0017	9.3
	1203	9.0			0720	0.8
	1858	0.9			1239	9.2
					1936	0.9
6 M	0018	9.1		21 TU	0055	9.2
	0718	0.9			0757	0.9
	1238	9.1			1318	9.1
	1933	0.9			2011	1.1
7 TU	0052	9.2		22 W	0132	9.0
	0753	0.9			0839	1.1
	1314	9.1			1355	8.8
	2008	1.0			2045	1.4
8 W	0128	9.1		23 TH	0208	8.7
	0829	0.9			0905	1.4
	1352	9.0			1433	8.4
	2044	1.1			2116	1.8
9 TH	0207	8.9		24 F	0246	8.2
	0907	1.2			0939	1.8
	1433	8.7			1512	8.0
	2123	1.5			2152	2.2
10 F	0249	8.6		25 SA	0326	7.7
	0948	1.5			1017	2.2
	1519	8.3			1556	7.5
	2208	1.8			2234	2.6
11 SA	0337	8.1		26 SU	0414	7.2
	1038	1.8			1105	2.6
	1617	7.8			1650	7.1
	2303	2.2		◗	2333	2.9
12 SU	0441	7.7		27 M	0516	6.9
	1142	2.1			1206	2.8
	1733	7.6			1758	6.9
◖						
13 M	0016	2.4		28 TU	0043	3.0
	0605	7.5			0631	6.8
	1304	2.1			1317	2.7
	1857	7.7			1911	7.0
14 TU	0142	2.2		29 W	0154	2.8
	0728	7.7			0742	7.1
	1426	1.8			1425	2.5
	2011	8.1			2015	7.4
15 W	0259	1.8		30 TH	0257	2.4
	0840	8.2			0842	7.5
	1536	1.4			1524	2.1
	2114	8.6			2107	7.9
				31 F	0352	2.0
					0932	8.0
					1617	1.7
					2153	8.3

JUNE

Day	Time	m		Day	Time	m
1 SA	0442	1.6		16 SU	0528	1.2
	1016	8.4			1057	8.7
	1704	1.4			1750	1.2
	2235	8.6			2316	8.9
2 SU	0528	1.3		17 M	0614	1.1
	1057	8.6			1140	8.8
	1748	1.2			1832	1.0
	2314	8.9		○	2356	9.0
3 M	0612	1.1		18 TU	0655	1.1
	1137	8.9			1220	8.9
	1830	1.1			1911	1.2
●	2353	9.0				
4 TU	0654	1.0		19 W	0034	8.9
	1217	9.0			0734	1.1
	1912	1.0			1259	8.8
					1949	1.3
5 W	0033	9.2		20 TH	0112	8.8
	0736	0.9			0811	1.2
	1259	9.1			1336	8.7
	1953	1.0			2025	1.5
6 TH	0114	9.2		21 F	0149	8.6
	0818	0.9			0846	1.4
	1342	9.1			1414	8.5
	2035	1.1			2059	1.7
7 F	0158	9.0		22 SA	0226	8.4
	0900	1.0			0919	1.6
	1429	8.9			1451	8.2
	2119	1.3			2131	2.0
8 SA	0245	8.8		23 SU	0303	8.0
	0946	1.2			0953	1.9
	1519	8.6			1529	7.9
	2207	1.6			2207	2.3
9 SU	0338	8.4		24 M	0343	7.7
	1038	1.5			1031	2.2
	1616	8.2			1611	7.6
	2303	1.8			2251	2.5
10 M	0439	8.1		25 TU	0429	7.3
	1139	1.7			1118	2.4
	1721	8.0			1700	7.3
◖				◗	2344	2.7
11 TU	0008	2.0		26 W	0526	7.1
	0548	7.9			1214	2.6
	1248	1.8			1801	7.2
	1831	8.0				
12 W	0121	2.0		27 TH	0047	2.7
	0739	7.8			0633	7.1
	1359	1.7			1318	2.6
	1940	8.1			1910	7.3
13 TH	0231	1.8		28 F	0154	2.6
	0811	8.1			0743	7.3
	1506	1.6			1425	2.4
	2045	8.3			2015	7.6
14 F	0337	1.6		29 SA	0259	2.3
	0914	8.3			0845	7.6
	1608	1.4			1528	2.1
	2142	8.6			2111	8.0
15 SA	0436	1.4		30 SU	0359	1.9
	1009	8.6			0939	8.0
	1703	1.3			1625	1.7
	2231	8.8			2201	8.4

JULY

Day	Time	m		Day	Time	m
1 M	0453	1.5		16 TU	0554	1.4
	1028	8.4			1127	8.5
	1717	1.4			1812	1.5
	2247	8.7		○	2341	8.7
2 TU	0545	1.2		17 W	0636	1.3
	1115	8.8			1206	8.7
	1807	1.2			1853	1.4
●	2332	9.0				
3 W	0635	1.0		18 TH	0019	8.8
	1201	9.0			0715	1.2
	1855	1.0			1243	8.7
					1931	1.4
4 TH	0018	9.2		19 F	0054	8.8
	0724	0.8			0753	1.2
	1248	9.2			1318	8.7
	1943	0.9			2007	1.4
5 F	0105	9.3		20 SA	0130	8.7
	0812	0.7			0828	1.3
	1335	9.3			1353	8.6
	2031	0.9			2040	1.6
6 SA	0153	9.3		21 SU	0205	8.6
	0900	0.7			0859	1.4
	1424	9.2			1427	8.5
	2118	1.0			2109	1.7
7 SU	0241	9.1		22 M	0238	8.4
	0947	0.8			0928	1.6
	1513	9.0			1459	8.3
	2205	1.2			2140	1.9
8 M	0331	8.8		23 TU	0311	8.1
	1035	1.0			1000	1.8
	1603	8.7			1533	8.0
	2255	1.4			2215	2.2
9 TU	0424	8.5		24 W	0347	7.8
	1125	1.4			1037	2.1
	1658	8.4			1611	7.7
◖	2349	1.7			2258	2.4
10 W	0522	8.1		25 TH	0431	7.5
	1221	1.7			1122	2.4
	1758	8.1			1658	7.4
				◗	2349	2.6
11 TH	0051	1.9		26 F	0526	7.2
	0628	7.9			1218	2.6
	1325	1.9			1759	7.2
	1906	7.9				
12 F	0159	2.0		27 SA	0052	2.7
	0739	7.8			0638	7.1
	1433	2.0			1326	2.7
	2015	7.9			1917	7.3
13 SA	0307	1.9		28 SU	0208	2.6
	0849	7.9			0758	7.3
	1539	1.9			1442	2.4
	2119	8.1			2030	7.6
14 SU	0411	1.8		29 M	0321	2.2
	0951	8.1			0906	7.8
	1638	1.8			1552	2.0
	2214	8.4			2131	8.1
15 M	0507	1.6		30 TU	0425	1.7
	1042	8.3			1004	8.3
	1729	1.6			1652	1.6
	2300	8.5			2226	8.4
				31 W	0523	1.3
					1057	8.8
					1748	1.2
					2317	9.0

AUGUST

Day	Time	m		Day	Time	m
1 TH	0620	0.9		16 F	0003	8.8
	1148	9.2			0658	1.2
	1843	0.9			1224	8.8
●					1913	1.3
2 F	0006	9.4		17 SA	0036	8.9
	0715	0.6			0733	1.1
	1237	9.4			1256	8.9
	1936	0.7			1947	1.3
3 SA	0055	9.6		18 SU	0108	8.9
	0806	0.4			0805	1.2
	1324	9.6			1328	8.8
	2025	0.6			2017	1.4
4 SU	0142	9.6		19 M	0140	8.8
	0853	0.3			0834	1.2
	1410	9.6			1359	8.8
	2111	0.6			2045	1.5
5 M	0228	9.5		20 TU	0210	8.7
	0936	0.5			0901	1.4
	1456	9.4			1428	8.6
	2154	0.8			2113	1.7
6 TU	0314	9.0		21 W	0240	8.4
	1018	0.8			0929	1.6
	1541	9.0			1458	8.3
	2236	1.2			2143	1.9
7 W	0400	8.8		22 TH	0312	8.1
	1101	1.3			1001	1.9
	1628	8.8			1531	8.0
◖	2322	1.6			2219	2.2
8 TH	0451	8.2		23 F	0349	7.8
	1148	1.8			1039	2.3
	1722	8.0			1612	7.7
				◖	2304	2.5
9 F	0016	2.1		24 SA	0437	7.4
	0552	7.7			1130	2.6
	1247	2.2			1705	7.3
	1828	7.6				
10 SA	0123	2.4		25 SU	0003	2.8
	0708	7.4			0544	7.0
	1359	2.5			1237	2.9
	1947	7.5			1823	7.1
11 SU	0239	2.1		26 M	0124	2.8
	0830	7.4			0717	7.1
	1514	2.4			1405	2.8
	2102	7.7			1956	7.3
12 M	0351	2.1		27 TU	0251	2.4
	0939	7.7			0841	7.6
	1621	2.1			1527	2.2
	2202	8.0			2109	7.9
13 TU	0451	1.8		28 W	0403	1.8
	1031	8.1			0946	8.2
	1715	1.8			1633	1.6
	2249	8.4			2209	8.6
14 W	0540	1.5		29 TH	0507	1.2
	1114	8.4			1042	8.9
	1759	1.6			1733	1.1
	2328	8.6			2303	9.2
15 TH	0621	1.3		30 F	0607	0.7
	1151	8.7			1133	9.4
	1838	1.4			1831	0.8
○				●	2352	9.6
				31 SA	0702	0.4
					1221	9.7
					1923	0.5

Chart Datum is 4·43 metres below IGN Datum. HAT is 10·1 metres above Chart Datum.

》》 FREE monthly updates. Register at 《《
www.reedsnauticalalmanac.co.uk

N France

175

STANDARD TIME UT –01
Subtract 1 hour for UT
For French Summer Time add
ONE hour in **non-shaded areas**

DIEPPE LAT 49°56′N LONG 1°05′E
TIMES AND HEIGHTS OF HIGH AND LOW WATERS

Dates in red are **SPRINGS**
Dates in blue are NEAPS

YEAR 2019

SEPTEMBER

Time	m		Time	m
1 0039	9.8	**16**	0042	9.0
0751	0.2		0739	1.1
SU 1307	9.9	M	1300	9.0
2011	0.4		1952	1.2
2 0124	9.9	**17**	0112	9.0
0836	0.2		0807	1.1
M 1350	9.8	TU	1329	9.0
2053	0.5		2019	1.3
3 0208	9.7	**18**	0142	8.9
0916	0.4		0834	1.3
TU 1432	9.6	W	1357	8.8
2133	0.7		2048	1.5
4 0250	9.4	**19**	0211	8.7
0953	0.8		0901	1.5
W 1513	9.1	TH	1427	8.6
2210	1.2		2116	1.7
5 0332	8.8	**20**	0243	8.4
1029	1.4		0930	1.9
TH 1555	8.6	F	1500	8.3
2249	1.7		2148	2.0
6 0418	8.1	**21**	0319	8.0
1111	2.1		1005	2.3
F 1644	7.9	SA	1538	7.9
◗ 2338	2.3		2230	2.4
7 0516	7.4	**22**	0405	7.5
1207	2.6		1054	2.7
SA 1750	7.3	SU	1630	7.4
		◗	2328	2.8
8 0045	2.7	**23**	0510	7.1
0635	7.0		1204	3.0
SU 1326	3.0	M	1748	7.0
1919	7.0			
9 0211	2.8	**24**	0050	2.9
0812	7.0		0651	7.0
M 1453	2.8	TU	1338	2.9
2046	7.3		1932	7.2
10 0332	2.4	**25**	0229	2.5
0924	7.5		0822	7.6
TU 1604	2.4	W	1508	2.4
2146	7.8		2051	7.9
11 0434	2.0	**26**	0346	1.8
1014	8.1		0929	8.3
W 1659	1.9	TH	1617	1.6
2230	8.3		2152	8.7
12 0522	1.6	**27**	0451	1.1
1054	8.5		1025	9.1
TH 1741	1.6	F	1717	1.0
2307	8.6		2245	9.3
13 0601	1.3	**28**	0550	0.6
1128	8.8		1114	9.6
F 1817	1.4	SA	1813	0.6
2340	8.8	●	2334	9.7
14 0636	1.2	**29**	0643	0.3
1159	8.9		1200	9.9
SA 1851	1.3	SU	1904	0.4
○				
15 0012	9.0	**30**	0019	9.9
0708	1.1		0730	0.2
SU 1230	9.0	M	1244	10.0
1922	1.2		1949	0.4

OCTOBER

Time	m		Time	m
1 0102	9.9	**16**	0045	9.1
0812	0.3		0739	1.1
TU 1325	9.9	W	1300	9.1
2030	0.5		1954	1.2
2 0144	9.7	**17**	0115	9.0
0836	0.2		0808	1.3
W 1405	9.6	TH	1330	8.9
2106	0.8		2025	1.4
3 0224	9.3	**18**	0147	8.8
0923	1.1		0837	1.5
TH 1444	9.1	F	1402	8.8
2141	1.3		2055	1.6
4 0304	8.8	**19**	0221	8.5
0957	1.7		0909	1.8
F 1524	8.5	SA	1436	8.4
2217	1.9		2129	1.9
5 0347	8.0	**20**	0259	8.1
1033	2.3		0946	2.2
SA 1609	7.8	SU	1517	8.0
◗ 2302	2.5		2212	2.3
6 0441	7.3	**21**	0347	7.6
1127	2.9		1036	2.7
SU 1712	7.1	M	1611	7.5
		◗	2310	2.7
7 0007	3.0	**22**	0455	7.2
0559	6.8		1146	3.0
M 1250	3.3	TU	1732	7.1
1843	6.7			
8 0135	3.1	**23**	0032	2.8
0741	6.8		0635	7.2
TU 1422	3.1	W	1321	2.9
2017	7.0		1913	7.3
9 0259	2.7	**24**	0210	2.4
0856	7.3		0802	7.7
W 1535	2.6	TH	1450	2.2
2118	7.6		2030	8.0
10 0402	2.2	**25**	0327	1.7
0945	7.9		0908	8.5
TH 1628	2.0	F	1558	1.6
2201	8.1		2131	8.7
11 0449	1.7	**26**	0430	1.1
1023	8.4		1003	9.1
F 1711	1.6	SA	1657	1.0
2238	8.5		2224	9.3
12 0529	1.4	**27**	0527	0.7
1057	8.8		1051	9.6
SA 1748	1.4	SU	1751	0.7
2311	8.8		2312	9.6
13 0605	1.2	**28**	0618	0.5
1129	9.0		1136	9.8
SU 1822	1.2	M	1840	0.5
○ 2343	9.0	●	2356	9.8
14 0638	1.1	**29**	0703	0.5
1200	9.1		1219	9.9
M 1854	1.2	TU	1924	0.5
15 0014	9.1	**30**	0039	9.8
0709	1.1		0744	0.6
TU 1231	9.1	W	1259	9.7
1924	1.2		2003	0.7
		31	0119	9.6
			0821	0.9
		TH	1338	9.5
			2040	1.0

NOVEMBER

Time	m		Time	m
1 0159	9.2	**16**	0129	8.9
0855	1.3		0820	1.5
F 1417	9.0	SA	1344	8.9
2114	1.4		2042	1.4
2 0239	8.7	**17**	0207	8.7
0928	1.8		0857	1.7
SA 1456	8.4	SU	1424	8.6
2149	1.9		2121	1.7
3 0320	8.0	**18**	0250	8.3
1004	2.4		0939	2.1
SU 1540	7.8	M	1509	8.1
2231	2.5		2206	2.1
4 0410	7.4	**19**	0342	7.9
1054	3.0		1031	2.4
M 1636	7.1	TU	1606	7.7
◗ 2328	3.0	◗	2303	2.4
5 0517	6.9	**20**	0451	7.5
1208	3.3		1138	2.7
TU 1753	6.7	W	1723	7.4
6 0046	3.2	**21**	0020	2.5
0644	6.8		0616	7.5
W 1333	3.2	TH	1303	2.6
1921	6.8		1849	7.6
7 0207	2.9	**22**	0147	2.2
0804	7.1		0735	7.9
TH 1446	2.8	F	1425	2.1
2030	7.3		2003	8.0
8 0312	2.5	**23**	0301	1.8
0859	7.7		0841	8.4
F 1543	2.3	SA	1533	1.6
2120	7.8		2106	8.6
9 0404	2.0	**24**	0404	1.5
0942	8.2		0937	8.9
SA 1630	1.9	SU	1633	1.2
2200	8.3		2201	9.0
10 0448	1.6	**25**	0501	1.0
1020	8.6		1027	9.3
SU 1711	1.5	M	1726	0.9
2237	8.6		2250	9.3
11 0528	1.4	**26**	0551	0.9
1055	8.8		1113	9.5
M 1749	1.4	TU	1815	0.8
2312	8.9	●	2335	9.4
12 0605	1.3	**27**	0636	0.9
1129	9.0		1156	9.5
TU 1824	1.2	W	1858	0.8
○ 2346	9.0			
13 0639	1.2	**28**	0018	9.4
1201	9.1		0717	0.9
W 1858	1.2	TH	1236	9.5
			1938	0.9
14 0019	9.0	**29**	0058	9.3
0712	1.2		0755	1.1
TH 1234	9.1	F	1315	9.3
1932	1.2		2016	1.1
15 0053	9.0	**30**	0138	9.0
0746	1.3		0831	1.4
F 1308	9.1	SA	1354	8.9
2006	1.4		2052	1.4

DECEMBER

Time	m		Time	m
1 0218	8.6	**16**	0202	8.9
0906	1.8		0853	1.5
SU 1434	8.5	M	1419	8.8
2127	1.8		2119	1.3
2 0258	8.2	**17**	0249	8.6
0940	2.3		0938	1.7
M 1515	7.9	TU	1508	8.5
2204	2.2		2207	1.6
3 0342	7.7	**18**	0341	8.3
1024	2.7		1030	2.0
TU 1603	7.4	W	1603	8.2
2249	2.7		2300	1.8
4 0434	7.2	**19**	0440	8.0
1118	3.0		1129	2.2
W 1701	7.0	TH	1707	7.9
◗ 2347	2.9	◗		
5 0538	7.0	**20**	0003	2.0
1227	3.2		0549	7.9
TH 1811	6.9	F	1238	2.2
			1818	7.8
6 0057	3.0	**21**	0115	2.1
0650	7.0		0700	8.0
F 1339	3.0	SA	1353	2.1
1923	7.0		1931	7.9
7 0206	2.8	**22**	0228	1.9
0758	7.3		0809	8.2
SA 1444	2.7	SU	1504	1.8
2025	7.4		2039	8.2
8 0308	2.4	**23**	0335	1.7
0853	7.8		0911	8.5
SU 1541	2.2	M	1607	1.5
2116	7.8		2139	8.5
9 0401	2.0	**24**	0435	1.5
0939	8.2		1006	8.8
M 1629	1.9	TU	1703	1.3
2201	8.2		2232	8.8
10 0448	1.7	**25**	0527	1.3
1020	8.5		1054	9.0
TU 1714	1.6	W	1753	1.1
2241	8.5		2319	8.9
11 0531	1.5	**26**	0613	1.3
1059	8.8		1138	9.1
W 1755	1.4	TH	1838	1.0
2320	8.8	●		
12 0611	1.4	**27**	0002	9.0
1136	9.0		0655	1.2
TH 1835	1.2	F	1219	9.2
○ 2358	9.0		1919	1.0
13 0650	1.3	**28**	0043	9.0
1214	9.1		0735	1.3
F 1914	1.1	SA	1259	9.1
			1957	1.1
14 0037	9.1	**29**	0121	8.9
0730	1.2		0813	1.4
SA 1253	9.1	SU	1337	8.9
1954	1.1		2034	1.3
15 0118	9.0	**30**	0159	8.7
0811	1.3		0848	1.7
SU 1335	9.1	M	1415	8.6
2035	1.2		2108	1.5
		31	0237	8.4
			0921	2.0
		TU	1452	8.3
			2140	1.9

Chart Datum is 4·43 metres below IGN Datum. HAT is 10·1 metres above Chart Datum.

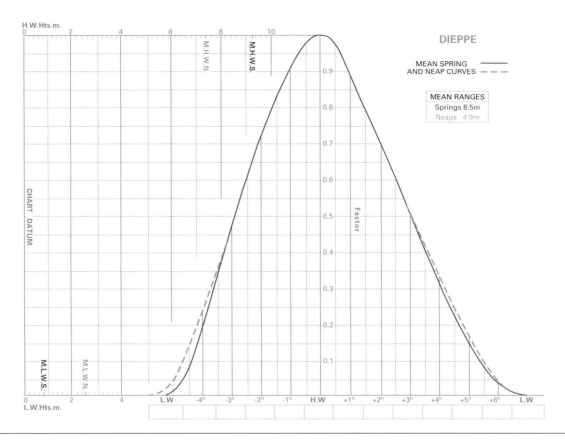

DIEPPE

MEAN SPRING ———
AND NEAP CURVES – – –

MEAN RANGES
Springs 8.5m
Neaps 4.9m

4.13 ST VALÉRY-EN-CAUX

Seine Maritime **49°52'·48N 00°42'·64E** ❀❀❀⚓⚓⚓❀❀

CHARTS AC 2451; SHOM 7417; Navi 1012; Imray C31

TIDES –0044 Dover; ML 4·6; Duration 0530

Standard Port DIEPPE (◄——)

Times				Height (metres)			
High Water		Low Water		MHWS	MHWN	MLWN	MLWS
0100	0600	0100	0700	9·3	7·4	2·5	0·8
1300	1800	1300	1900				
Differences ST VALÉRY-EN-CAUX							
–0005	–0005	–0015	–0015	–0·5	–0·4	–0·1	–0·1

SHELTER Good, but confused seas break at entrance in fresh W-NE'lies.

NAVIGATION WPT 49°52'·97N 00°42'·14E, 155°/6ca to W pier. Ent is easy, but tidal streams across ent are strong. Shingle builds up against W wall; hug the E side. Inside the pier hds, wave-breaks (marked by posts) reduce the surge. Avant Port dries 3m. 5 W waiting buoys are in the pool NE of gate.

Gate is open HW ±2½ if coeff >79; if coeff < 41 gate is open HW –2½ to HW+1¾ (bridge opens H & H+30 in these periods). When entry gate first opens, strong current may complicate berthing on Ⓥ pontoon W side; best to wait >15 mins (i.e. second bridge opening). Ⓥ also berth on 'A' pontoon. Traffic sigs at the bridge/gate are coordinated to seaward/inland: ● = ent/exit; ● = no ent/exit; ● + ● = no movements.

LIGHTS AND MARKS From N or W Paluel nuclear power stn is conspic 3M W of ent. From E a TV mast is 1M E of ent. Hbr is hard to see between high chalk cliffs, but lt ho (W twr, G top) on W pierhead is fairly conspic. Lts as on chartlet and LBW.

COMMUNICATIONS CROSS 03·21·8728·76; Police 02·35·97·00·17; British Consul 01·44·51·31·00; Ⓗ 02·35·97·00·11; Doctor 02·35·97·05·99; Aff Mar 02·35·10·34·34; HM VHF Ch 09 (French only).

FACILITIES **Marina** ☎02·35·97·01·30, mob 06·07·31·56·95. www.ville-saint-valery-en-caux.fr 575 + 20 Ⓥ €2·20; max LOA 14m. ⚓ 🛢 Ⓔ 🔧 ⚒ 🚁(5/10t). **CN Valeriquais** ☎02.35.97.25.49, 🗐 🏴. **Town** ╳ 🛎 🛒 🏧 🏨 🛢 ✉ Ⓑ bus to ⇌ at Yvetot; ✈ and ferry Dieppe.

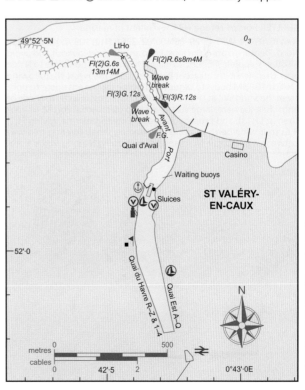

N France

4.14 FÉCAMP

Seine Maritime 49°45'·92N 00°21'·80E ❀❀⊛◊◊◊✿✿

CHARTS AC 2451, 2148, 1354; SHOM 6857, 7417, 7207; Navi 1012; Imray C31

TIDES –0044 Dover; ML 4·9; Duration 0550

Standard Port DIEPPE (←)

Times				Height (metres)			
High Water		Low Water		MHWS	MHWN	MLWN	MLWS
0100	0600	0100	0700	9·3	7·4	2·5	0·8
1300	1800	1300	1900				
Differences FÉCAMP							
–0015	–0010	–0030	–0040	–0·8	–0·5	+0·3	+0·4
ETRETAT 49°42'·6N 00°12'·0E							
–0020	–0015	–0050	–0050	–1·1	–0·8	+0·3	+0·4

SHELTER Excellent in basins, but in even moderate W/NW winds a considerable surf runs off the ent and scend can make the Avant Port uncomfortable. Access difficult in onshore winds >F5; do not attempt >F7. Bassin Bérigny for long stay; see HM.

NAVIGATION WPT 49°46'·01N 00°21'·36E, 107°/3ca to hbr ent. Beware the Charpentier Rks off Pte Fagnet and strong cross currents, depending on tides. Ent chan (1·5m) is prone to silting, best water is close to N jetty.

LIGHTS AND MARKS Ent lies SW of conspic ⌖, sig stn, TV mast and 5 wind turbines (2 Q) on Pte Fagnet cliff, 142m high. See chartlet and Lights, buoys and waypoints for lts. Pilots use the QR aligned with QG as a 082° ldg line.

IPTS on column at root of S jetty; a W lt is shown next to top lt when Bassin Bérigny gate is open.

COMMUNICATIONS CROSS 03·21·87·21·87; ▣02·35·28·00·91; Police 02·35·10·45·00; ⊜ 02·35·28·19·40; British Consul 01·44·51·31·00; ⊞ 02·35·28·05·13; HM 02·35·28·25·53, VHF Ch 12 16 (HW –3 to +1); Marina VHF Ch 09 & Écluse Bérigny.

4.15 LE HAVRE

Seine Maritime 49°29'·12N 00°05'·43E ❀❀⊛⊛◊◊◊✿✿

CHARTS AC 2613, 2146, 2990; SHOM 7418, 6683; Navi 526, 1012; Imray C31

TIDES –0103 Dover; ML 4·9; Duration 0543. Le Havre is a Standard Port (←). There is a stand of about 3hrs at HW.

SHELTER Port of refuge H24. Excellent in marina.

NAVIGATION WPT 49°29'·86N 00°02'·33E (LH12) 108°/2·1M to Digue Nord. Pleasure craft <20m must keep an all-round lookout and avoid all other shipping. Inbound to marina from N or W, hold close to N side appr chan, (caution: adjacent Banc de l'Éclat, 1m, is just N), following LH10 to 16 PHM buoys to Digue Nord where marina ent opens to port. From the S and SW, at LH11 or 13 SHM buoys cross directly to N side of appr chan (beware: large vessels ent/dep commercial Port SE of appr chan at LH13), and proceed as above. Entering marina, giving Fl(2) R 6s spur wide berth to avoid surrounding shoal, steering mid-way between it and Digue Normand before turning N along Digue Nord.

FACILITIES Marina ④ ☎02·35·28·13·58 plaisance@portdefecamp.fr www.portdefecamp.fr. 725 + 75 ♥ fingers at pontoon C, €2·40. ⚓ ♪ (H24 with any cr/card) ▯ ⚑ ⚓ ⚓ ◺ ⚓ ☢ Ⓔ ⚒ ⌕(mob 16t). Société des Régates de Fécamp ☎02·35·28·08·44; ⌧.

Bassin Bérigny is entered via a gate & swing bridge, HW –2½ to HW +¾; max LOA for visitors is 45m. ☎02.35.28.23.76, ⚓ ◺ ⌕(30t).

Town ✗ ▯ ▤ ⚓ ⊠ Ⓑ ⇌. ✈ Le Havre.

LIGHTS AND MARKS 2 chys, RW, on 106·8° ldg line, and twr of St Joseph's ⌖ all conspic. IPTS at Digue Nord head. If safe, Port Control may permit small craft to ent/dep against signal on request.

COMMUNICATIONS CROSS 02·33·52·16·16; ⚓ 02·35·42·21·06; Auto 08·92·68·08·76; ⊜ 02·35·41·33·51; Brit Consul 01·44·51·31·00; Dr 02·35·41·23·61; HM 02·35·21·23·95; Port Control & Radar VHF Ch 12, 20; Aff Mar 02·35·19·29·99; Marina VHF Ch 09; Sas Quinette (lock) VHF Ch 88.

FACILITIES Marina ☎02·35·21·23·95. www.lehavreplaisance. com capitainerie@lehavre-plaisance.fr 973 + 90♥ pontoons P in 4·5m (up to 40m LOA by prior arrangement), O or A 3·5m, (best shelter, close to office), ⚓, €3·70, ⚓ ♪ ✗ ▤ ▯ ⚓ ⚓ ◺ Ⓔ ⚒ ⚒ SHOM ACA ⚓ ⛴(16t) de-mast ⌕(6t). Port Vauban, arrange with HM, €3·20, 1 month min. stay in centre of main shopping centre/ facilities. Website has bridge/lock times (twice p/day).

Sport Nautique et Plaisance du Havre (SNPH) ☎02·35·21·01·41, www.snph.org/ ⌧. Sté Régates du Havre (SRH) ☎02·35·42·41·21, ✗ ⌧(shut Aug). City All facilities.

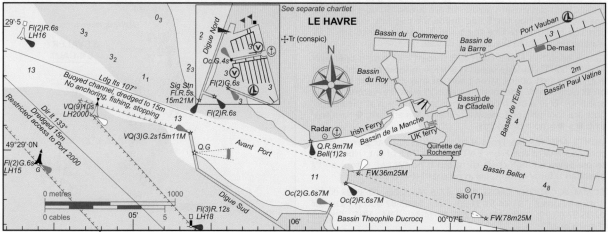

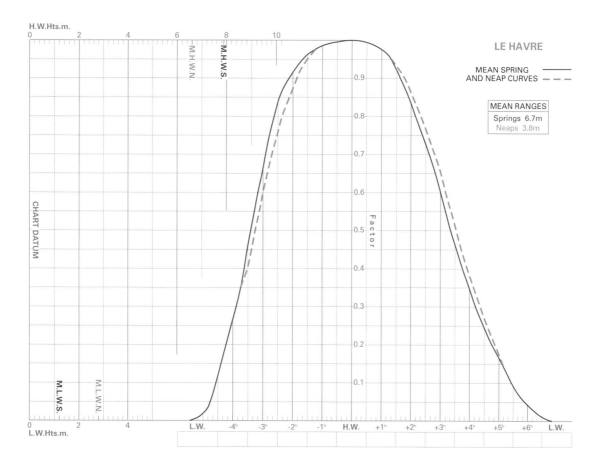

LE HAVRE

MEAN SPRING ———
AND NEAP CURVES – – –

MEAN RANGES
Springs 6.7m
Neaps 3.8m

H.W.Hts.m.
0 2 4 6 8 10

M.H.W.N.
M.H.W.S.

CHART DATUM

Factor
0·9
0·8
0·7
0·6
0·5
0·4
0·3
0·2
0·1

M.L.W.S.
M.L.W.N.

0 2 4
L.W.Hts.m.

L.W. -4ʰ -3ʰ -2ʰ -1ʰ H.W. +1ʰ +2ʰ +3ʰ +4ʰ +5ʰ +6ʰ L.W.

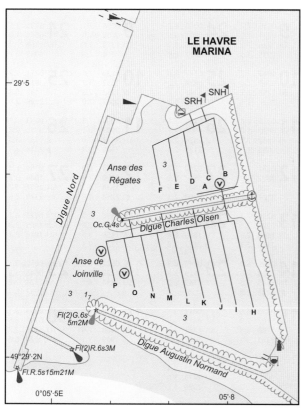

LE HAVRE MARINA

— 29'·5

SRH SNH

Anse des Régates

3

F E D A C B
V

Digue Nord

3
Oc.G.4s Digue Charles Olsen

Anse de Joinville
V
V
P O N M L K J I H

3 1·7
Fl(2)G.6s 5m2M
3

49°29'·2N
✳ Fl(2)R.6s3M

☆ Fl.R.5s15m21M

Digue Augustin Normand

0°05'·5E 05'·8

THE SEINE ESTUARY

The Seine estuary is entered between Le Havre and Deauville, and is encumbered by shallow and shifting banks which extend seawards to Banc de Seine, 15M W of Le Havre. With wind against tide there is a heavy sea on this bank. N and E of it there are 4 big ship waiting anchorages. The estuary is carefully regulated by the port authorities at Le Havre and Rouen. Given the large amount of commercial traffic to/from both these ports, it is sensible for yachtsmen to study AC 2146, 2990 carefully prior to entering the area, so as to gain a sound working knowledge of their routes and the general layout. **Note that the Greenwich Meridian lies about 2·7M W of Cap de la Hève and Deauville; extra caution is necessary in this area to ensure you are using the correct longitude sign (E/W) for your navigation.**

Yachts leaving **Le Havre** Marina near LW should give the inner spur, Fl (2) R 6s, a wide berth due to shoal ground. If bound for **Honfleur, Rouen** or **Paris**, do not try a short cut east of PHM No 2 (Q R pillar buoy at 49°27'·70N 00°00'·60E); to do so risks grounding on Spoil Ground (0·4m). ▶ *The SW-going stream begins at HW Le Havre +0400, and the NE-going at HW Le Havre –0300, sp rates 1·5kn.* ◀ Between Deauville and Le Havre the sea can be rough in W winds. The inshore **Passage des Pêcheurs**, over the Digue du Ratier, between Honfleur and Deauville is only good HW±2 and requires local knowledge.

Chenal du Rouen is the main chan into R Seine, and carries much commercial tfc. The S side of the chan is defined by Digue du Ratier, a training wall which extends E to Honfleur. Yachts should stay on the N side of the channel, just outside the PHM buoys, whether east or west bound. This rule applies between No 2 buoy and Tancarville bridge.

With 24hrs access/exit, **Honfleur** is a useful starting port when bound upriver. See routes to the Mediterranean (Special notes), R Seine, Rouen and Paris.

STANDARD TIME UT –01
Subtract 1 hour for UT
For French Summer Time add
ONE hour in **non-shaded areas**

LE HAVRE   LAT 49°29'N   LONG 0°07'E
TIMES AND HEIGHTS OF HIGH AND LOW WATERS

Dates in red are SPRINGS
Dates in blue are NEAPS

YEAR **2019**

JANUARY

Time	m	Time	m
1 0113 2.5 / 0656 7.1 / TU 1348 2.6 / 1925 7.0		**16** 0544 6.7 / 1227 3.1 / W 1819 6.6	
2 0219 2.4 / 0758 7.2 / W 1455 2.4 / 2027 7.1		**17** 0100 2.9 / 0657 6.8 / TH 1349 2.8 / 1933 6.8	
3 0321 2.3 / 0851 7.4 / TH 1553 2.2 / 2119 7.3		**18** 0224 2.6 / 0803 7.1 / F 1508 2.3 / 2038 7.2	
4 0414 2.1 / 0937 7.5 / F 1641 1.9 / 2204 7.5		**19** 0332 2.1 / 0901 7.5 / SA 1607 1.8 / 2134 7.6	
5 0458 2.0 / 1018 7.7 / SA 1722 1.8 / 2243 7.6		**20** 0430 1.7 / 0953 7.8 / SU 1702 1.4 / 2225 7.9	
6 0537 1.9 / 1054 7.7 / SU 1800 1.7 / ● 2319 7.6		**21** 0525 1.4 / 1042 8.1 / M 1757 1.0 / ○ 2313 8.1	
7 0615 1.8 / 1128 7.8 / M 1836 1.6 / 2353 7.6		**22** 0619 1.1 / 1130 8.3 / TU 1850 0.7	
8 0652 1.8 / 1201 7.7 / TU 1911 1.6		**23** 0001 8.3 / 0711 1.0 / W 1217 8.3 / 1940 0.6	
9 0026 7.5 / 0727 1.9 / W 1234 7.7 / 1944 1.8		**24** 0048 8.3 / 0800 1.0 / TH 1304 8.3 / 2026 0.7	
10 0100 7.4 / 0757 2.1 / TH 1307 7.5 / 2015 1.9		**25** 0135 8.1 / 0844 1.2 / F 1350 8.1 / 2110 1.0	
11 0132 7.3 / 0829 2.3 / F 1341 7.4 / 2046 2.2		**26** 0221 7.9 / 0927 1.5 / SA 1437 7.8 / 2151 1.5	
12 0207 7.1 / 0902 2.5 / SA 1417 7.1 / 2119 2.4		**27** 0308 7.6 / 1009 1.9 / SU 1526 7.4 / ◐ 2233 2.0	
13 0246 7.0 / 0939 2.7 / SU 1500 6.9 / 2159 2.6		**28** 0400 7.2 / 1056 2.4 / M 1622 7.0 / 2322 2.5	
14 0332 6.8 / 1025 3.0 / M 1553 6.7 / ◐ 2249 2.9		**29** 0502 6.9 / 1153 2.7 / TU 1733 6.7	
15 0432 6.7 / 1121 3.1 / TU 1701 6.6 / 2349 3.0		**30** 0025 2.8 / 0615 6.8 / W 1303 2.9 / 1854 6.6	
		31 0137 2.9 / 0730 6.8 / TH 1418 2.8 / 2010 6.7	

FEBRUARY

Time	m	Time	m
1 0249 2.8 / 0834 7.0 / F 1526 2.5 / 2108 7.0		**16** 0150 2.8 / 0737 6.9 / SA 1441 2.5 / 2021 7.1	
2 0351 2.5 / 0924 7.2 / SA 1621 2.2 / 2153 7.2		**17** 0310 2.3 / 0844 7.3 / SU 1546 1.9 / 2121 7.5	
3 0441 2.2 / 1004 7.4 / SU 1706 1.9 / 2231 7.4		**18** 0413 1.8 / 0939 7.7 / M 1649 1.3 / 2212 7.9	
4 0524 2.0 / 1039 7.6 / M 1746 1.7 / ● 2304 7.6		**19** 0515 1.3 / 1029 8.1 / TU 1749 0.8 / ○ 2259 8.2	
5 0602 1.8 / 1112 7.7 / TU 1822 1.5 / 2336 7.6		**20** 0613 0.9 / 1116 8.3 / W 1843 0.5 / 2345 8.3	
6 0637 1.7 / 1144 7.7 / W 1856 1.5		**21** 0703 0.7 / 1202 8.4 / TH 1929 0.4	
7 0007 7.6 / 0712 1.7 / TH 1216 7.8 / 1927 1.5		**22** 0030 8.4 / 0748 0.7 / F 1246 8.4 / 2011 0.5	
8 0039 7.6 / 0742 1.7 / F 1248 7.7 / 1957 1.6		**23** 0114 8.3 / 0828 0.8 / SA 1330 8.2 / 2048 0.8	
9 0110 7.5 / 0810 1.8 / SA 1320 7.6 / 2026 1.7		**24** 0156 8.0 / 0905 1.2 / SU 1412 7.9 / 2125 1.3	
10 0142 7.4 / 0841 2.0 / SU 1354 7.4 / 2057 1.9		**25** 0237 7.6 / 0940 1.7 / M 1454 7.4 / 2159 2.0	
11 0216 7.2 / 0913 2.3 / M 1430 7.1 / 2130 2.3		**26** 0319 7.2 / 1017 2.3 / TU 1542 6.9 / ◐ 2239 2.6	
12 0255 7.0 / 0950 2.6 / TU 1513 6.8 / ◐ 2211 2.6		**27** 0412 6.8 / 1106 2.8 / W 1649 6.5 / 2335 3.1	
13 0342 6.8 / 1038 2.8 / W 1611 6.6 / 2304 2.9		**28** 0528 6.5 / 1215 3.2 / TH 1821 6.3	
14 0447 6.6 / 1139 3.0 / TH 1732 6.5			
15 0014 3.0 / 0612 6.6 / F 1304 3.0 / 1904 6.7			

MARCH

Time	m	Time	m
1 0054 3.3 / 0658 6.4 / F 1338 3.2 / 1951 6.4		**16** 0543 6.5 / 1235 3.0 / SA 1848 6.6	
2 0217 3.2 / 0815 6.6 / SA 1458 2.8 / 2054 6.7		**17** 0129 2.9 / 0718 6.8 / SU 1420 2.5 / 2007 7.1	
3 0331 2.7 / 0907 6.9 / SU 1601 2.4 / 2137 7.1		**18** 0252 2.4 / 0829 7.2 / M 1529 1.8 / 2106 7.6	
4 0425 2.3 / 0947 7.2 / M 1649 2.0 / 2212 7.4		**19** 0359 1.7 / 0924 7.7 / TU 1636 1.3 / 2155 7.9	
5 0508 1.9 / 1020 7.5 / TU 1728 1.7 / 2243 7.5		**20** 0504 1.2 / 1012 8.0 / W 1736 0.8 / 2241 8.2	
6 0545 1.5 / 1051 7.7 / W 1804 1.5 / ● 2313 7.6		**21** 0559 0.8 / 1058 8.3 / TH 1826 0.5 / ○ 2325 8.3	
7 0619 1.5 / 1122 7.7 / TH 1836 1.4 / 2343 7.7		**22** 0646 0.6 / 1142 8.4 / F 1909 0.4	
8 0650 1.5 / 1154 7.8 / F 1906 1.3		**23** 0007 8.3 / 0727 0.6 / SA 1225 8.3 / 1948 0.5	
9 0015 7.7 / 0721 1.5 / SA 1226 7.8 / 1935 1.3		**24** 0048 8.2 / 0804 0.8 / SU 1306 8.1 / 2021 0.9	
10 0046 7.7 / 0749 1.5 / SU 1258 7.7 / 2004 1.5		**25** 0127 8.0 / 0838 1.2 / M 1345 7.8 / 2054 1.5	
11 0118 7.6 / 0820 1.7 / M 1332 7.5 / 2035 1.7		**26** 0205 7.6 / 0909 1.7 / TU 1424 7.3 / 2125 2.1	
12 0151 7.4 / 0851 1.9 / TU 1408 7.2 / 2107 2.1		**27** 0242 7.2 / 0942 2.3 / W 1507 6.9 / 2200 2.7	
13 0226 7.1 / 0926 2.3 / W 1448 7.0 / 2144 2.5		**28** 0328 6.7 / 1024 2.9 / TH 1608 6.4 / ◐ 2252 3.3	
14 0309 6.9 / 1008 2.6 / TH 1543 6.6 / ◐ 2233 2.8		**29** 0440 6.3 / 1139 3.3 / F 1742 6.2	
15 0412 6.6 / 1107 2.9 / F 1706 6.5 / 2343 3.1		**30** 0012 3.5 / 0615 6.2 / SA 1253 3.3 / 1917 6.3	
		31 0140 3.4 / 0741 6.4 / SU 1419 3.0 / 2026 6.6	

APRIL

Time	m	Time	m
1 0258 2.9 / 0839 6.7 / M 1527 2.5 / 2110 7.0		**16** 0234 2.3 / 0809 7.2 / TU 1510 1.8 / 2045 7.6	
2 0355 2.4 / 0918 7.1 / TU 1617 2.1 / 2143 7.3		**17** 0341 1.7 / 0904 7.6 / W 1616 1.3 / 2134 7.9	
3 0440 2.0 / 0951 7.4 / W 1700 1.7 / 2213 7.5		**18** 0445 1.3 / 0952 7.9 / TH 1714 0.9 / 2218 8.1	
4 0519 1.7 / 1023 7.6 / TH 1736 1.5 / 2244 7.6		**19** 0538 1.0 / 1036 8.1 / F 1802 0.7 / ○ 2301 8.2	
5 0553 1.5 / 1056 7.7 / F 1809 1.4 / ● 2316 7.7		**20** 0623 0.8 / 1120 8.1 / SA 1843 0.7 / 2342 8.1	
6 0624 1.4 / 1129 7.8 / SA 1839 1.3 / 2348 7.8		**21** 0702 0.8 / 1201 8.1 / SU 1920 0.9	
7 0656 1.3 / 1203 7.8 / SU 1909 1.3		**22** 0021 8.0 / 0737 1.0 / M 1241 7.9 / 1952 1.2	
8 0020 7.8 / 0726 1.4 / M 1237 7.7 / 1941 1.4		**23** 0059 7.8 / 0810 1.3 / TU 1320 7.6 / 2025 1.7	
9 0054 7.7 / 0759 1.5 / TU 1313 7.6 / 2014 1.6		**24** 0135 7.5 / 0841 1.8 / W 1358 7.3 / 2055 2.2	
10 0129 7.5 / 0832 1.7 / W 1352 7.3 / 2047 2.0		**25** 0211 7.2 / 0911 2.3 / TH 1439 6.8 / 2129 2.8	
11 0207 7.3 / 0907 2.1 / TH 1437 7.0 / 2125 2.4		**26** 0253 6.7 / 0950 2.8 / F 1532 6.4 / ◐ 2218 3.3	
12 0253 6.9 / 0950 2.5 / F 1534 6.6 / ◐ 2214 2.8		**27** 0354 6.3 / 1049 3.2 / SA 1657 6.2 / 2329 3.6	
13 0356 6.7 / 1048 2.8 / SA 1656 6.5 / 2327 3.1		**28** 0525 6.1 / 1205 3.3 / SU 1820 6.2	
14 0527 6.5 / 1222 2.9 / SU 1834 6.7		**29** 0052 3.5 / 0643 6.2 / M 1324 3.2 / 1932 6.5	
15 0116 2.9 / 0700 6.8 / M 1400 2.4 / 1949 7.1		**30** 0208 3.1 / 0748 6.5 / TU 1435 2.7 / 2024 6.8	

Chart Datum is 4·38 metres below IGN Datum. HAT is 8·4 metres above Chart Datum.

FREE monthly updates. Register at
www.reedsnauticalalmanac.co.uk

STANDARD TIME UT –01
Subtract 1 hour for UT
For French Summer Time add
ONE hour in **non-shaded areas**

LE HAVRE   LAT 49°29'N   LONG 0°07'E
TIMES AND HEIGHTS OF HIGH AND LOW WATERS

Dates in red are SPRINGS
Dates in blue are NEAPS

YEAR **2019**

MAY

	Time	m		Time	m
1 W	0309 0836 1531 2102	2.6 6.9 2.3 7.2	**16** TH	0315 0840 1548 2109	1.8 7.5 1.5 7.7
2 TH	0358 0915 1618 2137	2.2 7.2 2.0 7.4	**17** F	0417 0929 1645 2154	1.5 7.7 1.3 7.9
3 F	0441 0951 1659 2211	1.9 7.4 1.7 7.6	**18** SA	0511 1014 1733 ○ 2237	1.3 7.8 1.2 7.9
4 SA	0518 1027 1735 ● 2246	1.7 7.7 1.6 7.7	**19** SU	0555 1058 1814 2317	1.2 7.8 1.2 7.9
5 SU	0554 1103 1809 2320	1.5 7.7 1.4 7.8	**20** M	0634 1139 1850 2356	1.2 7.8 1.4 7.8
6 M	0630 1139 1844 2355	1.4 7.7 1.4 7.8	**21** TU	0709 1219 1924	1.3 7.7 1.6
7 TU	0704 1218 1919	1.3 7.7 1.4	**22** W	0032 0743 1257 1958	7.7 1.5 7.5 1.9
8 W	0033 0740 1258 1955	7.7 1.4 7.6 1.6	**23** TH	0109 0815 1335 2030	7.5 1.8 7.2 2.3
9 TH	0113 0817 1342 2032	7.6 1.6 7.4 2.0	**24** F	0145 0847 1415 2105	7.2 2.2 6.9 2.7
10 F	0156 0855 1431 2113	7.4 1.9 7.2 2.3	**25** SA	0224 0923 1459 2148	6.9 2.6 6.6 3.1
11 SA	0246 0940 1529 2205	7.1 2.3 6.9 2.7	**26** SU	0312 1012 1602 ◗ 2247	6.5 3.0 6.4 3.4
12 SU	0349 1041 1647 ◗ 2322	6.9 2.6 6.3 2.9	**27** M	0424 1114 1719 2353	6.3 3.2 6.3 3.4
13 M	0512 1218 1813	6.7 2.6 6.9	**28** TU	0541 1221 1825	6.2 3.2 6.4
14 TU	0058 0636 1337 1922	2.7 6.9 2.3 7.2	**29** W	0104 0645 1328 1923	3.2 6.4 2.9 6.7
15 W	0210 0743 1444 2020	2.3 7.2 1.9 7.5	**30** TH	0208 0741 1431 2012	2.9 6.7 2.6 7.0
			31 F	0305 0831 1526 2056	2.5 7.0 2.3 7.2

JUNE

	Time	m		Time	m
1 SA	0354 0915 1614 2136	2.2 7.2 2.0 7.5	**16** SU	0440 0955 1702 2215	1.7 7.5 1.7 7.7
2 SU	0440 0957 1658 2215	1.9 7.4 2.0 7.6	**17** M	0526 1040 1744 ○ 2255	1.6 7.5 1.7 7.7
3 M	0523 1038 1740 ● 2254	1.6 7.6 1.6 7.7	**18** TU	0607 1121 1823 2333	1.6 7.5 1.7 7.7
4 TU	0605 1119 1821 2335	1.4 7.7 1.5 7.8	**19** W	0644 1200 1859	1.6 7.5 1.8
5 W	0646 1202 1902	1.3 7.8 1.5	**20** TH	0010 0720 1237 1936	7.6 1.6 7.4 2.0
6 TH	0017 0726 1247 1943	7.8 1.3 7.7 1.6	**21** F	0046 0754 1314 2010	7.5 1.8 7.3 2.2
7 F	0102 0808 1335 2025	7.7 1.4 7.6 1.8	**22** SA	0122 0827 1350 2044	7.3 2.1 7.1 2.5
8 SA	0149 0851 1425 2110	7.6 1.7 7.4 2.1	**23** SU	0158 0900 1427 2120	7.1 2.4 6.9 2.8
9 SU	0240 0940 1522 2205	7.4 2.0 7.2 2.4	**24** M	0236 0937 1509 2202	6.9 2.6 6.7 3.0
10 M	0340 1042 1629 ◗ 2317	7.1 2.2 7.0 2.5	**25** TU	0322 1022 1603 ◗ 2254	6.6 2.9 6.5 3.2
11 TU	0450 1159 1742	7.0 2.3 7.0	**26** W	0423 1117 1711 2352	6.5 3.0 6.5 3.2
12 W	0032 0605 1307 1850	2.5 7.0 2.2 7.2	**27** TH	0535 1217 1816	6.4 3.0 6.6
13 TH	0140 0713 1412 1951	2.3 7.1 2.1 7.3	**28** F	0055 0640 1322 1916	3.1 6.5 2.9 6.8
14 F	0244 0814 1514 2044	2.1 7.1 1.9 7.5	**29** SA	0203 0741 1431 2010	2.8 6.8 2.6 7.1
15 SA	0345 0907 1612 2131	1.9 7.4 1.8 7.6	**30** SU	0310 0837 1532 2101	2.4 7.0 2.3 7.3

JULY

	Time	m		Time	m
1 M	0404 0929 1625 2148	2.0 7.3 2.0 7.6	**16** TU	0502 1027 1721 ○ 2239	1.9 7.4 2.1 7.6
2 TU	0455 1017 1714 ● 2233	1.7 7.5 1.7 7.7	**17** W	0545 1107 1802 2315	1.8 7.4 1.9 7.7
3 W	0544 1103 1803 2319	1.4 7.7 1.5 7.9	**18** TH	0623 1143 1840 2350	1.7 7.5 1.9 7.7
4 TH	0633 1150 1851	1.2 7.9 1.4	**19** F	0700 1217 1917	1.7 7.4 1.9
5 F	0005 0721 1237 1939	8.0 1.1 7.9 1.4	**20** SA	0024 0735 1250 1951	7.6 1.7 7.4 2.0
6 SA	0052 0809 1326 2026	8.0 1.2 7.8 1.5	**21** SU	0058 0806 1324 2021	7.5 1.9 7.3 2.2
7 SU	0140 0856 1415 2114	7.9 1.3 7.7 1.7	**22** M	0131 0836 1356 2052	7.4 2.1 7.2 2.4
8 M	0230 0943 1506 2203	7.7 1.6 7.5 2.0	**23** TU	0204 0905 1430 2124	7.2 2.3 7.0 2.6
9 TU	0322 1033 1601 ◗ 2258	7.4 1.9 7.3 2.2	**24** W	0241 0939 1509 2203	7.0 2.5 6.9 2.8
10 W	0420 1129 1704	7.2 2.2 7.1	**25** TH	0327 1021 1559 ◗ 2252	6.8 2.8 6.7 3.0
11 TH	0000 0529 1233 1813	2.4 6.9 2.4 7.0	**26** F	0424 1116 1704 2353	6.6 3.0 6.6 3.1
12 F	0106 0642 1337 1921	2.5 6.9 2.5 7.1	**27** SA	0538 1221 1817	6.5 3.0 6.7
13 SA	0212 0751 1442 2021	2.4 6.9 2.4 7.2	**28** SU	0106 0654 1341 1928	3.0 6.6 2.9 6.9
14 SU	0315 0857 1542 2113	2.3 7.1 2.3 7.3	**29** M	0231 0806 1459 2032	2.7 6.9 2.5 7.2
15 M	0412 0942 1635 2159	2.1 7.2 2.2 7.5	**30** TU	0336 0908 1600 2127	2.2 7.3 2.1 7.5
			31 W	0433 1001 1656 2217	1.7 7.6 1.7 7.8

AUGUST

	Time	m		Time	m
1 TH	0528 1050 1751 ● 2305	1.3 7.9 1.4 8.1	**16** F	0607 1122 1824 2329	1.7 7.5 1.8 7.8
2 F	0623 1137 1845 2353	1.0 8.0 1.2 8.2	**17** SA	0641 1153 1857	1.6 7.6 1.8
3 SA	0715 1224 1935	0.8 8.1 1.1	**18** SU	0001 0714 1224 1930	7.8 1.6 7.5 1.8
4 SU	0039 0802 1310 2021	8.2 0.8 8.1 1.1	**19** M	0033 0743 1255 1956	7.7 1.7 7.5 1.9
5 M	0125 0847 1356 2104	8.1 0.9 8.0 1.4	**20** TU	0104 0810 1325 2024	7.6 1.8 7.4 2.1
6 TU	0211 0928 1441 2145	7.9 1.3 7.7 1.7	**21** W	0135 0838 1357 2054	7.4 2.0 7.3 2.3
7 W	0257 1008 1529 2229	7.6 1.7 7.4 2.1	**22** TH	0209 0907 1432 2127	7.2 2.3 7.1 2.6
8 TH	0348 1054 1624 2322	7.2 2.2 7.1 2.5	**23** F	0249 0943 1514 ◗ 2209	6.9 2.6 6.8 2.9
9 F	0451 1151 1733	6.8 2.7 6.8	**24** SA	0340 1030 1611 2305	6.6 2.9 6.6 3.1
10 SA	0029 0612 1303 1852	2.8 6.6 2.9 6.8	**25** SU	0452 1134 1731	6.5 3.2 6.6
11 SU	0142 0734 1414 2004	2.8 6.7 2.9 6.9	**26** M	0020 0621 1303 1857	3.1 6.5 3.1 6.7
12 M	0251 0842 1520 2101	2.6 6.9 2.7 7.1	**27** TU	0201 0746 1437 2012	2.8 6.8 2.7 7.1
13 TU	0353 0933 1619 2146	2.3 7.1 2.4 7.4	**28** W	0315 0852 1542 2112	2.2 7.3 2.1 7.6
14 W	0446 1015 1707 2224	2.1 7.3 2.2 7.6	**29** TH	0414 0945 1642 2202	1.6 7.7 1.5 7.9
15 TH	0529 1051 1747 ○ 2257	1.9 7.5 2.0 7.7	**30** F	0515 1033 1741 ● 2250	1.2 8.0 1.2 8.2
			31 SA	0612 1119 1835 2336	0.8 8.2 1.0 8.4

Chart Datum is 4·38 metres below IGN Datum. HAT is 8·4 metres above Chart Datum.

>> FREE monthly updates. Register at <<
www.reedsnauticalalmanac.co.uk

N France

STANDARD TIME UT –01
Subtract 1 hour for UT
For French Summer Time add
ONE hour in **non-shaded areas**

LE HAVRE LAT 49°29'N LONG 0°07'E
TIMES AND HEIGHTS OF HIGH AND LOW WATERS

Dates in red are **SPRINGS**
Dates in blue are **NEAPS**

YEAR 2019

SEPTEMBER

Day	Time m	Day	Time m
1 SU	0702 0.6 / 1205 8.3 / 1922 0.8	16 M	0648 1.5 / 1156 7.7 / 1904 1.7
2 M	0021 8.4 / 0745 0.6 / 1249 8.3 / 2004 0.9	17 TU	0007 7.8 / 0716 1.5 / 1226 7.7 / 1930 1.7
3 TU	0105 8.3 / 0825 0.8 / 1332 8.1 / 2043 1.2	18 W	0037 7.7 / 0743 1.7 / 1256 7.6 / 1959 1.9
4 W	0148 8.0 / 0903 1.2 / 1413 7.8 / 2120 1.6	19 TH	0109 7.6 / 0812 1.9 / 1327 7.4 / 2029 2.1
5 TH	0230 7.6 / 0938 1.8 / 1455 7.4 / 2157 2.2	20 F	0144 7.3 / 0842 2.2 / 1401 7.2 / 2102 2.4
6 F	0316 7.2 / 1016 2.5 / 1544 7.0 / 2243 2.7	21 SA	0223 7.0 / 0916 2.6 / 1441 6.9 / 2141 2.8
7 SA	0416 6.7 / 1109 3.0 / 1654 6.6 / 2350 3.1	22 SU	0313 6.7 / 1001 3.0 / 1538 6.6 / 2233 3.1
8 SU	0545 6.4 / 1229 3.3 / 1825 6.5	23 M	0427 6.5 / 1103 3.3 / 1702 6.5 / 2349 3.2
9 M	0112 3.2 / 0719 6.5 / 1351 3.2 / 1947 6.7	24 TU	0604 6.5 / 1240 3.3 / 1838 6.7
10 TU	0230 2.9 / 0831 6.7 / 1506 2.9 / 2047 7.0	25 W	0148 2.8 / 0733 6.9 / 1421 2.7 / 1957 7.1
11 W	0337 2.5 / 0919 7.1 / 1605 2.5 / 2129 7.3	26 TH	0256 2.2 / 0836 7.5 / 1526 2.1 / 2055 7.6
12 TH	0428 2.1 / 0956 7.4 / 1650 2.1 / 2204 7.6	27 F	0358 1.5 / 0927 7.9 / 1627 1.5 / 2145 8.1
13 F	0509 1.8 / 1027 7.6 / 1728 1.9 / 2235 7.7	28 SA	0459 1.0 / 1014 8.2 / 1725 1.1 / 2231 8.3
14 SA	0545 1.6 / 1057 7.7 / 1802 1.7 / 2305 7.8	29 SU	0553 0.7 / 1057 8.4 / 1816 0.8 / 2315 8.5
15 SU	0618 1.5 / 1126 7.7 / 1833 1.7 / 2336 7.9	30 M	0640 0.6 / 1141 8.4 / 1901 0.7 / 2359 8.5

OCTOBER

Day	Time m	Day	Time m
1 TU	0722 0.6 / 1224 8.4 / 1941 0.9	16 W	0648 1.5 / 1158 7.8 / 1905 1.6
2 W	0042 8.3 / 0759 0.9 / 1305 8.2 / 2017 1.2	17 TH	0014 7.8 / 0718 1.6 / 1229 7.7 / 1936 1.7
3 TH	0123 8.0 / 0834 1.4 / 1344 7.8 / 2051 1.7	18 F	0048 7.6 / 0750 1.9 / 1303 7.6 / 2009 1.9
4 F	0204 7.6 / 0907 2.0 / 1424 7.4 / 2125 2.3	19 SA	0126 7.4 / 0823 2.2 / 1340 7.4 / 2043 2.3
5 SA	0248 7.1 / 0941 2.7 / 1509 7.0 / 2207 2.8	20 SU	0209 7.1 / 0858 2.6 / 1423 7.0 / 2123 2.6
6 SU	0345 6.6 / 1032 3.3 / 1617 6.5 / 2311 3.3	21 M	0303 6.8 / 0943 3.0 / 1523 6.7 / 2215 3.0
7 M	0517 6.3 / 1153 3.6 / 1752 6.4	22 TU	0416 6.6 / 1046 3.3 / 1646 6.6 / 2331 3.1
8 TU	0036 3.4 / 0650 6.4 / 1322 3.5 / 1917 6.5	23 W	0552 6.7 / 1229 3.2 / 1822 6.7
9 W	0200 3.1 / 0806 6.7 / 1439 3.0 / 2021 6.8	24 TH	0128 2.7 / 0714 7.1 / 1402 2.6 / 1937 7.2
10 TH	0306 2.6 / 0853 7.1 / 1536 2.5 / 2103 7.2	25 F	0236 2.1 / 0815 7.6 / 1507 2.0 / 2034 7.7
11 F	0356 2.1 / 0928 7.4 / 1620 2.1 / 2136 7.5	26 SA	0338 1.5 / 0905 8.0 / 1607 1.5 / 2124 8.0
12 SA	0438 1.9 / 0957 7.6 / 1659 1.8 / 2207 7.7	27 SU	0436 1.1 / 0951 8.2 / 1703 1.1 / 2210 8.3
13 SU	0516 1.7 / 1026 7.7 / 1734 1.7 / 2238 7.8	28 M	0529 0.9 / 1035 8.3 / 1753 0.9 / 2254 8.4
14 M	0549 1.6 / 1057 7.8 / 1805 1.6 / 2310 7.9	29 TU	0614 0.8 / 1117 8.4 / 1836 0.9 / 2337 8.3
15 TU	0619 1.5 / 1127 7.8 / 1837 1.6 / 2342 7.9	30 W	0655 0.9 / 1159 8.3 / 1915 1.0
		31 TH	0019 8.2 / 0731 1.2 / 1239 8.1 / 1951 1.3

NOVEMBER

Day	Time m	Day	Time m
1 F	0100 7.9 / 0806 1.7 / 1317 7.8 / 2025 1.8	16 SA	0034 7.7 / 0731 1.8 / 1246 7.7 / 1954 1.8
2 SA	0141 7.5 / 0839 2.2 / 1356 7.4 / 2058 2.3	17 SU	0116 7.5 / 0808 2.1 / 1329 7.5 / 2031 2.1
3 SU	0224 7.1 / 0914 2.8 / 1439 7.0 / 2138 2.8	18 M	0203 7.3 / 0847 2.5 / 1417 7.2 / 2113 2.4
4 M	0318 6.7 / 1002 3.3 / 1541 6.6 / 2235 3.2	19 TU	0258 7.0 / 0934 2.8 / 1516 7.0 / 2206 2.7
5 TU	0439 6.4 / 1113 3.6 / 1708 6.3 / 2350 3.4	20 W	0407 6.9 / 1037 3.0 / 1631 6.8 / 2323 2.8
6 W	0600 6.4 / 1237 3.6 / 1825 6.4	21 TH	0532 6.9 / 1213 3.0 / 1757 6.9
7 TH	0109 3.2 / 0713 6.6 / 1352 3.2 / 1933 6.7	22 F	0102 2.5 / 0647 7.2 / 1336 2.6 / 1910 7.2
8 F	0219 2.8 / 0809 6.9 / 1453 2.7 / 2023 7.0	23 SA	0210 2.1 / 0748 7.5 / 1442 2.1 / 2010 7.6
9 SA	0313 2.4 / 0848 7.3 / 1541 2.3 / 2101 7.3	24 SU	0312 1.7 / 0841 7.8 / 1542 1.6 / 2102 7.8
10 SU	0400 2.1 / 0922 7.5 / 1624 2.0 / 2136 7.6	25 M	0410 1.4 / 0928 8.0 / 1638 1.4 / 2149 8.0
11 M	0440 1.8 / 0954 7.7 / 1702 1.8 / 2210 7.7	26 TU	0502 1.2 / 1012 8.1 / 1728 1.2 / 2235 8.1
12 TU	0516 1.7 / 1027 7.8 / 1735 1.7 / 2245 7.8	27 W	0548 1.2 / 1055 8.2 / 1811 1.2 / 2318 8.1
13 W	0549 1.6 / 1100 7.8 / 1810 1.6 / 2319 7.8	28 TH	0628 1.3 / 1136 8.1 / 1850 1.2
14 TH	0622 1.6 / 1133 7.9 / 1842 1.5 / 2355 7.8	29 F	0000 8.0 / 0705 1.5 / 1216 8.0 / 1927 1.4
15 F	0656 1.7 / 1208 7.8 / 1918 1.6	30 SA	0041 7.8 / 0742 1.8 / 1254 7.7 / 2002 1.8

DECEMBER

Day	Time m	Day	Time m
1 SU	0122 7.5 / 0818 2.2 / 1333 7.5 / 2037 2.2	16 M	0110 7.7 / 0801 1.9 / 1323 7.7 / 2028 1.7
2 M	0203 7.2 / 0854 2.7 / 1414 7.1 / 2115 2.6	17 TU	0158 7.5 / 0844 2.1 / 1412 7.5 / 2112 2.0
3 TU	0249 6.8 / 0937 3.1 / 1503 6.7 / 2201 3.0	18 W	0251 7.3 / 0932 2.4 / 1507 7.3 / 2204 2.2
4 W	0350 6.6 / 1031 3.4 / 1612 6.5 / 2258 3.2	19 TH	0351 7.2 / 1032 2.6 / 1611 7.1 / 2310 2.4
5 TH	0501 6.5 / 1134 3.5 / 1725 6.4	20 F	0501 7.1 / 1146 2.7 / 1725 7.0
6 F	0002 3.3 / 0607 6.5 / 1243 3.4 / 1830 6.5	21 SA	0028 2.4 / 0613 7.2 / 1303 2.5 / 1839 7.1
7 SA	0110 3.1 / 0706 6.7 / 1354 3.1 / 1928 6.7	22 SU	0137 2.3 / 0719 7.3 / 1412 2.3 / 1944 7.3
8 SU	0217 2.8 / 0757 7.0 / 1454 2.7 / 2018 7.0	23 M	0242 2.0 / 0816 7.5 / 1515 2.0 / 2042 7.5
9 M	0313 2.5 / 0841 7.3 / 1543 2.3 / 2102 7.3	24 TU	0343 1.8 / 0908 7.7 / 1614 1.7 / 2133 7.7
10 TU	0400 2.2 / 0921 7.5 / 1625 2.0 / 2143 7.5	25 W	0437 1.7 / 0954 7.8 / 1705 1.6 / 2221 7.8
11 W	0441 1.8 / 0959 7.7 / 1706 1.8 / 2222 7.7	26 TH	0525 1.6 / 1038 7.9 / 1750 1.5 / 2305 7.8
12 TH	0521 1.8 / 1036 7.8 / 1746 1.8 / 2302 7.8	27 F	0607 1.6 / 1119 7.9 / 1831 1.4 / 2346 7.8
13 F	0600 1.7 / 1114 7.9 / 1826 1.5 / 2342 7.8	28 SA	0645 1.7 / 1158 7.9 / 1908 1.5
14 SA	0639 1.6 / 1154 7.9 / 1905 1.5	29 SU	0025 7.7 / 0724 1.8 / 1235 7.8 / 1944 1.7
15 SU	0025 7.8 / 0720 1.7 / 1237 7.8 / 1946 1.5	30 M	0103 7.5 / 0801 2.1 / 1312 7.6 / 2019 1.9
		31 TU	0140 7.3 / 0835 2.3 / 1349 7.3 / 2054 2.2

Chart Datum is 4·38 metres below IGN Datum. HAT is 8·4 metres above Chart Datum.

》》 **FREE** monthly updates. Register at 《《
www.reedsnauticalalmanac.co.uk

4.16 HONFLEUR

Calvados 49° 25'·69N 00° 13'·90E ✿✿✿✿♦♦♦ ✿✿✿

CHARTS AC 2146, 2990, 2879, 1349; SHOM 7418, 6683, 7420; Navi 1012; Imray C31

TIDES –0135 Dover; ML 5·0; Duration 0540

Standard Port LE HAVRE (←)

Times				Height (metres)			
High Water		Low Water		MHWS	MHWN	MLWN	MLWS
0000	0500	0000	0700	7·9	6·6	2·8	1·2
1200	1700	1200	1900				
LE HAVRE-ANTIFER (49°39'·5N 00°06'·8E)							
+0025	+0015	+0005	–0005	+0·2	+0·1	+0·1	0·0
CHENAL DE ROUEN (49°26'N 00°07'E)							
0000	0000	0000	+0015	+0·1	0·0	+0·1	–0·1
HONFLEUR							
–0150	–0135	+0025	+0040	+0·1	+0·1	+0·1	+0·3

In the Seine a HW stand lasts about 2hrs 50 mins. The HW time differences refer to the beginning of the stand.

SHELTER Excellent in the Vieux Bassin and Bassin de l'Est.

NAVIGATION WPT No 2 PHM buoy, QR, 49°27'·70N 00°00'·60E, 9M

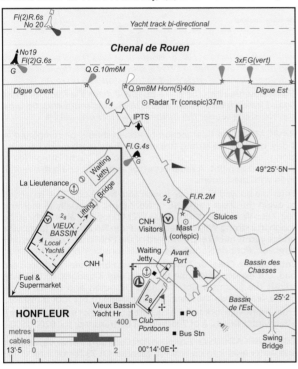

HONFLEUR

to Honfleur. Begin appr between LW+1½ to HW–1½ to avoid foul tide. From WPT keep to the N of the Chenal de Rouen between PHM buoys and posts marking Digue basse du Nord; there is no room for E-bound yachts to navigate S of the chan. Beware risk of collision in low vis with W-bound yachts. This rule applies from No 2 buoy to Tancarville bridge.

After No 20 PHM buoy cross to the ent. Caution: strong stream across ent, shoal patches reported at 'shoulders' of lock chamber.

Rouen is 60M and Paris 192M upriver.

LIGHTS AND MARKS See chartlet and Lights, buoys and waypoints. The radar tower by the lock and the Pont de Normandie 1·6M upriver are highly conspic; slightly less so are the wooded hills SW of Honfleur.

COMMUNICATIONS CROSS 02·33·52·16·16; Ⓛⓑ 02·31·89·20·17; Honfleur Radar Ch **73** (H24); Auto (local) 08·92·68·08·14; Auto (regional) 08·92·68·08·76; Auto (offshore) 08·92·68·08·08; Police 02·31·14·44·45; ⊜ 02·31·14·44·30; Brit Consul 01·44·51·31·00; Dr 02·31·89·34·05; Ⓗ 02·31·89·89·89; Port HM 02·31·49·19·19, VHF Ch **17**; Aff Mar 02·31·53·66·50; Lock VHF **17** (H24); Vieux Bassin VHF Ch 09.

THE LOCK operates H24 and opens at H, every hour for arrivals; and every H +30 for departures. Advisable to check in with lock keeper, VHF Ch17, prior to arrival. 23m wide it has recessed floating bollards, no pontoons. **IPTS** in force. Freeflow operates from 2hrs before HW to HW Le Havre.

THE ROAD BRIDGE at the Vieux Bassin lifts at times (LT) below. Inbound yachts can wait on E side of the jetty immediately outside the Vieux Bassin. Departing yachts take priority over arrivals; be ready 5 mins in advance.

1 May to 31 Aug
Every day: 0830, 0930, 1030, 1130, 1530, 1630, 1730, 1830.

1 Jan to 30 Apr and 1 Sep to 31 Dec
Mon-Fri: 0830, 1130, 1430, 1730.
Sat, Sun and public hols, 0830, 0930, 1030, 1130, 1430, 1730.

FACILITIES **Cercle Nautique d'Honfleur (CNH)** ☎02·31·98·87·13. www.cnh-honfleur.net cnh14@wanadoo.fr controls all Ⓥ pontoon berthing. **Avant Port,** NW wall, Public gardens walk ashore pontoon max 15 Ⓥ<LOA 20m, €3·00 inc ⚓ (get access code when you pay). **Vieux Bassin** 120 +15Ⓥ €3·00 inc ⚓ max LOA 16m; multihulls and large motor yachts +50%. Ⓥ raft on pontoons on NW side; or in a CNH berth as assigned by CNH dory which meets and directs all arrivals. The waiting jetty in the Avant port at ent to Vieux Bassin, €2·50, max stay 24hrs, is handy for locking out before the Vieux Bassin's 0830 bridge lift or to await next lock opening. ⚓ pontoon on W side of transit pier at Bassin de l'Est ent lock.

Bassin de l'Est can be used by yachts >20m LOA rafting alongside quay; see HM. ⊼ ✕ Ⓔ ⚓(10t and for masting) ◣.

Bassin Carnot, accessed through Bassin de l'Est E swing bridge has 70 berths on new pontoons (2018), some available for long stay visitors. See HM for details and bridge times.

Town ✕ ⌂ ⬚ ⓞ ⓟ ⓣ ⚓ @ ⊠ Ⓑ ⇌ ✈ Deauville.

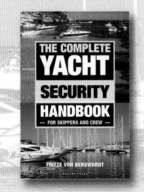

4.17 RIVER SEINE

CHARTS AC 2146, 2990, 2879; SHOM 6683, 6796, 6117; Imray C31; *Fluviacarte No 1.*

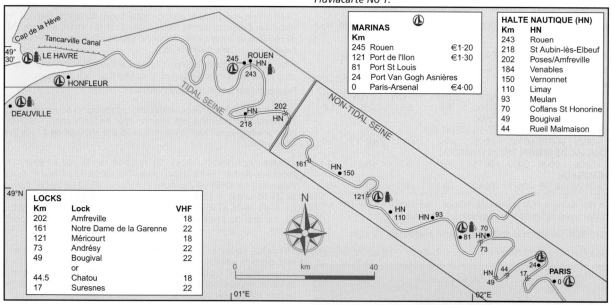

MARINAS		
Km		
245	Rouen	€1·20
121	Port de l'Ilon	€1·30
81	Port St Louis	
24	Port Van Gogh Asnières	
0	Paris-Arsenal	€4·00

HALTE NAUTIQUE (HN)	
Km	**HN**
243	Rouen
218	St Aubin-lès-Elbeuf
202	Poses/Amfreville
184	Venables
150	Vernonnet
110	Limay
93	Meulan
70	Coflans St Honorine
49	Bougival
44	Rueil Malmaison

LOCKS		
Km	**Lock**	**VHF**
202	Amfreville	18
161	Notre Dame de la Garenne	22
121	Méricourt	18
73	Andrésy	22
49	Bougival	22
	or	
44.5	Chatou	18
17	Suresnes	22

TIDES Standard Port LE HAVRE (⟵)

Times				Height (metres)			
High Water		Low Water		MHWS	MHWN	MLWN	MLWS
0000	0500	0000	0700	7·9	6·6	2·8	1·2
1200	1700	1200	1900				

Differences TANCARVILLE* *(Km 338. 49°28'·35N 00°27'·88E)*
| −0135 | −0120 | +0030 | +0145 | +0·2 | +0·2 | +0·4 | +1·0 |

QUILLEBEUF* *(Km 332. 49°28'·25N 00°32'·00E)*
| −0055 | −0110 | +0105 | +0210 | +0·2 | +0·2 | +0·6 | +1·4 |

VATTEVILLE* *(Km 317. 49°28'·89N 00°40'·01E)*
| −0015 | −0040 | +0205 | +0240 | 0·0 | 0·0 | +1·0 | +2·2 |

CAUDEBEC* *(Km 309. 49°31'·35N 00°43'·66E)*
| −0005 | −0030 | +0220 | +0300 | 0·0 | 0·0 | +1·1 | +2·3 |

HEURTEAUVILLE* *(Km 295. 49°25'·70N 00°48'·17E)*
| +0055 | +0005 | +0250 | +0330 | −0·1 | 0·0 | +1·3 | +2·6 |

DUCLAIR* *(Km 278. 49°28'·73N 00°52'·38E)*
| +0210 | +0145 | +0350 | +0410 | −0·1 | +0·1 | +1·6 | +3·2 |

ROUEN *(Km 245. 49°26'·50N 01°03'·13E)*
| +0305 | +0240 | +0505 | +0515 | 0·0 | +0·3 | +1·7 | +3·3 |

*HW time differences refer to the start of a stand lasting about 2¾hrs up to Vatteville, decreasing to about 1¾hrs at Duclair.

SHELTER Marinas/Haltes Nautiques (HN)/fuel, see above. Masts can be lowered at Dives, Deauville, Le Havre, Honfleur and Rouen.

TIDAL STREAMS require careful study. ▶At Ratier NW SHM buoy the flood starts at LW Le Havre +1 and at La Roque (Km 342) at LW+2, and progressively later further upriver. So even a 5kn boat which left Le Havre at LW (or Honfleur at LW+1) and passed the W end of Digue du Ratier at LW+15, should carry the flood to Rouen. If delayed, try another day rather than stopping en route.

Going downriver on the ebb a boat will meet the flood. In a fast boat it is worth continuing, rather than ⚓ for about 4hrs, because the ebb starts sooner the further downriver one gets.◀

NAVIGATION The tidal section to Rouen should be completed in 1 day, because the strong stream and ships' wash make berthing alongside dangerous and there are no ⚓s. Yacht navigation is prohibited at night. A radar reflector and VHF aerial are needed, even with mast down.

Enter the buoyed Chenal de Rouen abm No 2 buoy. Care is needed to avoid the shoal banks NE of No 2 buoy, especially in morning fog; keep W of Rade de la Carosse WCM buoy. The approach can be rough in a strong W'ly over ebb tide. From No 2 buoy to Tancarville bridge yachts must keep outboard of the PHM chan buoys, in either direction. Pont de Normandie (52m) is conspic 1·7M E of Honfleur.

Entry via the Canal de Tancarville is possible if sea conditions are bad, but expect delays due to 2 locks and 8 bridges. At Le Havre transit the Bassins Bellot, Vétillart, Despujols and the Canal du Havre, leading via the Canal de Tancarville to the lock (Km 338).

From Amfreville lock to Paris current is about 1kn in summer, more after winter floods. 6 of 7 locks must be transited; foc 0700-1900.

COMMUNICATIONS Ch 73 *(Rouen Port Control)* gives nav, vis, weather and water level info on request. Monitor Ch 73 for tfc info. Ch 82 broadcasts water levels from Le Havre to Caudebec every 5 mins.

ROUEN, Seine Maritime, **49°26'·07N 01°06'·24E**. AC 2879; SHOM 6796, 6117. Yacht navigation prohib SS+½ to SR−½. ⌒ in the Halte Nautique, NE side of Île Lacroix, Km 241 (lat/long above) or **Marina Bassin St Gervais**, N bank, Km 245, see below. A one-way system in the chan NE of Île Lacroix only allows navigation *against* the stream. For example, if leaving on the young flood to go upriver, you must first go downriver to the NW tip of the island, then turn upriver via the SW chan. *Rouen Port* VHF Ch **73** 68. HM 02·35·52·54·56; Aff Mar 02·35·98·53·98; ⊜ 02·35·98·27·60; ⚓ 02·35·80·11·44.

Facilities: **Marina Bassin St Gervais**, ☎02·35·08·30·59, www.la-crea.fr/port-de-plaisance-de-la-crea.html; €1·20. Mast down/up at pontoon SE side; ⛟(25t,3t).

Halte Nautique ☎02·35·07·33·94, www.rouen.fr/adresse/halte-nautique 50 ⌒ ⚓ ⛽ ⚓ ▷ ⚓ ✕ Ⓔ ⛟(3t). ⚓ ⚓ from barge at Km 239.

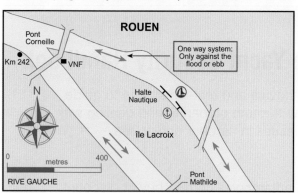

ROUEN

Pont Corneille
Km 242
VNF
One way system: Only against the flood or ebb
Halte Nautique
Île Lacroix
RIVE GAUCHE
metres 0 400
Pont Mathilde

4.18 PARIS

Île-de-France, **48°50'·84N 02°22'·02E** (Paris-Arsenal marina).

CHARTS AC & SHOM charts do not cover the Seine E of Rouen, but Fluviacarte No 1 covers Le Havre to Paris in detail.

SHELTER Complete in Paris-Arsenal Marina.

NAVIGATION Between Eiffel Tower and Notre Dame the wash of many *vedettes* can raise quite a chop.

A traffic system (chartlet) around Île de la Cité and Île St Louis must be understood and obeyed. Going upriver, the simpler route is: Hug the S bank in the narrowish one-way Bras de la Monnaie, leaving Notre Dame to port. Or take the wider, two-way N bank chan, noting that this route is only available H to H+20; is regulated by R/G tfc lts on Pont au Change; and requires two cross-overs where shown below.

Going downriver, make good a window H+35 to H+50, obey R/G tfc lts on Pont de Sully, then leave Île St Louis to stbd and Notre Dame to port. Cross-over twice where indicated.

By 2020 Paris should be joined to Cambrai in N France by the 66-mile long Seine-Nord Europe canal, which will be open to leisure craft.

LIGHTS AND MARKS Eiffel Tower and Notre Dame. Check off other landmarks and all bridges carefully; obey cross-overs.

COMMUNICATIONS Brit Consul 01·44·51·31·00; Monitor VHF Ch 10 on the river (mandatory); at the lock to Arsenal talk direct to HM via intercom, CCTV monitors; Marina VHF Ch 09; Canal St Martin VHF Ch 20; Taxi 01·45·85·85·85.

FACILITIES

Paris-Arsenal Marina 11 Blvd de la Bastille, 75012 Paris, ☎01·43·41·39·32. www.fayollemarine.fr No reservations 1/6-30/9, but contact marina several days before arrival. Enter 0800-1900 (2000: w/e, and Jul/Aug) via small lock, with waiting pontoon

Step by step advice on handling a traditional craft, from setting up a rig to steering using sails alone.

Visit www.adlardcoles.com to buy at discount

outside; obey R/G tfc lts signed 'Plaisance'. Air clearance in lock is 5·2m, least depth 1·9m. Inside, HM and office are to stbd; berth as directed. 176 €4·00; max LOA 25m. (plus on marina trolley, 750m SE) (7t). Metro stns: Quai de la Rapée (Line 5) at S end of marina and Bastille (Lines 1, 5 & 8) at the N end. Numerous buses from Bastille. Tourist Office at Gare de Lyon (550m).

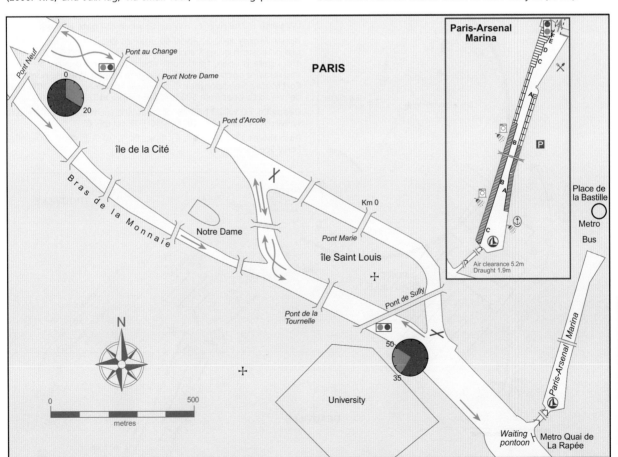

4.19 DEAUVILLE/TROUVILLE

Calvados 49° 22'·38N 00° 04'·15E ❀❀❀♦♦♦❀❀❀❀

CHARTS AC 2613, 2146, 1349; SHOM 7418, 7420; Navi 526; Imray C32

TIDES –0130 Dover; ML 5·1; Duration 0510

Standard Port LE HAVRE (←)

Times				Height (metres)			
High Water		Low Water		MHWS	MHWN	MLWN	MLWS
0000	0500	0000	0700	8·0	6·7	2·9	1·2
1200	1700	1200	1900				
Differences TROUVILLE							
–0100	–0010	0000	+0005	+0·4	+0·3	+0·3	+0·1

NOTE: There is a double HW at Trouville. The HW time differences, when referred to the time of HW Le Havre, give the times of the first HW.

SHELTER Access difficult in NW/N winds >F5. Best shelter in Bassin Morny (3m) Port Deauville outer basin (2·8m) can be uncomfortable during freeflow, again in strong winds from the N quadrant. Only FVs berth at Trouville.

NAVIGATION WPT 49°22'·94N 00°03'·62E,145°/0·65M to abeam W training wall lt, Fl WG 4s, beware drying shoal extending E from the training wall, keep well to stbd. Do not appr from E of N due to Les Ratelets drying shoal and Banc de Trouville. Trouville SW WCM buoy is 1M WNW of ent. Chan and river dry 2·4m; no access LW±2½, for 2m draft, less water than charted reported in Avant Port.

LIGHTS AND MARKS Trouville's Casino and Deauville's Royal Hotel/ Casino (off chartlet to the SW) are conspic. Front 145° ldg lt not visible when brg >150°. Bassin Morny ent gate: IPTS, sigs 2-4.

COMMUNICATIONS LB 02·31·88·31·70; CROSS 02·33·52·16·16; Auto (local) 08·92·68·08·14; Auto (regional) 08·92·68·08·76; Auto (offshore) 08·92·68·08·08; Police 02·31·88·13·07; ⊖ 02·31·88·35·29; Brit Consul 01·44·51·31·00; Dr & ⊞ 02·31·14·33·33; Port Deauville and Bassin Morny VHF Ch 09; Aff Mar 02·31·88·36·21; Deauville Tourist office 02·31·14·40·00.

FACILITIES inbound: **Port Deauville Marina** Lock ☎02·31·98·30·01 free flow approx HW–2 to +2. Marina ☎02·31·98·30·01. port-deauville-sa@wanadoo.fr www.deauville.org 700 + 70 Ⓥ €3·05(Jul, Aug). ⓟ €2·00, ⓙ (H24 with cr/card) ✗ ⊡ ⚒ ⓔⓓ ⚓ ⏃ ⚐ ⌻(45t).

Port Morny Entry gate (☎02·31·98·50·40) portdeplaisance@ deauville.fr G IPTS signals when open, about HW ±3. Visit www. dyc14.com portdeplaisance@deauville.fr for actual local times.

Port Morny Marina ☎02·31·98·50·40 (0830-1200, 1330-1800) www. deauville.org 320 ⌇ inc 70 Ⓥ, €3·70 inc ⓦⓒ ⓟ at hbr office, wi-fi. ⚓ ⓣⓣ (1km). Water taxi to Trouville runs frequently from the LW berth in the river or HW berth when the gate is open.

Bassin Morny ☎02·31·98·50·40, max 20 Ⓥ ⌇ €3·90 inc ⓟ wi-fi. raft on eastern quay; S end, residents only. From 0700-2300 when entry gate is open, footbridge opens for 30 mins every 50 mins, and remains open 2300-0700 when gate is open. Waiting pontoon, Port Morny, S end pontoon C, by bridge. For cheap petrol (30% less) delivered to boat (min 500 ltrs) call *Taupin* ☎02·31·79·21·02.

Deauville YC ☎02·31·88·38·19. deauville.yacht.club@wanadoo.fr www.dyc14.com; ⊡ ⚒ ⓔⓓ ⚒ ⚓(6t). Hotel Ibis @.

Both towns ✗ ⊡ ⛴ ⓘ ⚑ ⚓ ⊠ Ⓑ ⚓ ⚒ ⚒ ⓔⓓ ⓧ Ⓔ ≷ ✈ Deauville.
Ferry See Ouistreham.

4.20 DIVES-SUR-MER

Calvados **49°17'·68N 00°05'·63W** (entry gate)

CHARTS AC 2146, 1349; SHOM 7418, 7420 (the preferred chart); Navi 526; Imray C32

TIDES –0135 Dover; ML 5·1m

Standard Port LE HAVRE (←)

Times				Height (metres)			
High Water		Low Water		MHWS	MHWN	MLWN	MLWS
0000	0500	0000	0700	8·0	6·7	2·9	1·2
1200	1700	1200	1900				
Differences DIVES-SUR-MER							
–0100	–0010	0000	0000	+0·3	+0·2	+0·2	+0·1
OUISTREHAM							
–0045	–0010	–0005	0000	–0·3	–0·3	–0·2	–0·3

NOTE: There is a double HW at Dives. HW time differences, referred to the time of HW Le Havre, give times of the first HW.

SHELTER Excellent inside. Appr can be difficult in NW/NE >F5 when a bad sea can build quickly.

NAVIGATION WPT WPT 49°19'·18N 00°05'·85W, SWM buoy 'DI' (off chartlet) Iso 4s, 168°/7ca to Nos 1/2 buoys; thence follow the buoyed chan between sandbanks (drying 4·2m). Appr dries to 1M offshore. See facilities for access times.

4 PHM and SHM buoy/bns are moved to suit shifting chan; Nos 1 (SHM) and 2 (PHM) are the only lit lateral buoys; Nos 3 and 5 SHMs are fixed lt bns which should be given a wide berth. Caution: Buoys/bns may lack numbers.

After last chan buoys (7 & 8) keep to port for the marina ent (400m W of lt ho). Or fork stbd to follow drying Dives river to Cabourg pontoon (2 ❶ berths) E of footbridge.

LIGHTS AND MARKS See Lights, buoys and waypoints and the chartlet. Wooded hills E of Dives and Houlgate church spire help locate the marina.

COMMUNICATIONS CROSS/LB 02·31·88·11·13; via HM; Auto 08·92·68·08·14; Police 02·31·28·23·00; Brit Consul 01·44·51·31·00; Dr 02·31·91·60·17; HM VHF Ch 09 (Office hours).

FACILITIES Marina ☎02·31·24·48·00. portguillaume@caen.cci.fr www.port-dives-cabourg.com Access via entry gate HW±3; HW±2½ for draft >1·5m; HW±1½ in bad conditions. Note: unless recently dredged, craft with draft >1·8m should only attempt entry HW±2. Sill below gate is 2·5m above CD; gate opens when tide 4·5m above CD and at springs may create a 4–5kn current for the first 45 mins. 600+25 ❶ €2·85, max LOA 15m, draft 2·3m, on 2 pontoons, short finger berths. Office hrs: daily, Jul/Aug, 0800-2000,

Apr,May,Sep,Oct, 0830-1830, Nov-Mar 0900-1200,1400-1700; H24, WC (also Tourist office, near Cabourg pontoon) free wi-fi, Defib. bike hire. (30t) demasting (1·5t).

Societé des Régates Dives Houlgate (SRDH), www.srdh.fr

Town ✕ ⬚ ⛟ ⬚ ✉ Ⓑ ✈ (Deauville). **Ferry** Ouistreham.

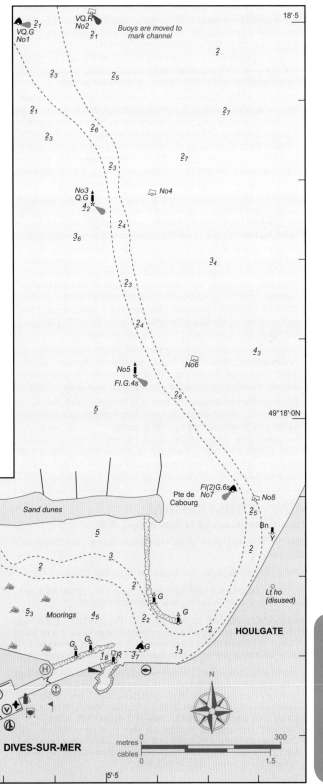

DEAUVILLE TO GRANDCAMP

(AC 2146, 2136). Since access to all the hbrs along this coast is to some extent tidally limited, it is best to cruise this area from W to E, so as to take advantage of the main Channel flood. ▶*Thus you can leave one harbour at the start of the tidal window and reach the next hbr before the window has closed; in the opposite direction this is difficult and may be impossible unless you have a fast boat. Between Le Havre and Ouistreham there is a useful stand at HW of between 2-3hrs.*◀

Le Havre, possibly Ouistreham, and St Vaast (in all but S/SE winds) are the only ports of refuge. Strong onshore winds render the approaches to other hbrs difficult or even dangerous.

Deauville, 8M S of Cap de la Hève, is an important yachting hbr. The sands dry more than 5ca offshore; dangerous in strong W or N winds, the sea breaks between the jetties. In such conditions it is best attempted within 15 mins of HW, at slack water. To the NE beware Banc de Trouville (1·7m) and Les Ratelets (0·9m) to the N at the mouth of the Seine. 7M SW is **River Dives** with marina. The banks dry for 1M to seaward, and entry is only possible from HW ± 2½; do not attempt in fresh onshore winds.

6M W lies **Ouistreham** ferry hbr and marina, the latter accessed by lock. The waiting pontoon N of the locks is for boats intending to transit the locks, not for yachts on passage pausing to await a fair tide. Here a canal leads 7M inland to a marina in the centre of Caen. 2-6M W the drying rocky ledges, Roches de Lion and Les Essarts de Langrune, extend up to 2·25M offshore in places. Roches de Ver (drying rks) lie near the centre of Plateau du Calvados, extending 1M offshore about 2M W of **Courseulles-sur-Mer** which is dangerous in strong onshore winds. During 2018/19, anticipate increasing commercial activity around the construction of the Calvados offshore wind farm (75 turbines) 9M north of Courseulles, the project shorebase.

5M W of Courseulles are the remains of the wartime hbr of **Arromanches**, where there is a fair weather anchorage, usually affected by swell. Rochers du Calvados (1·6m) lies close NE of the ruined Mulberry hbr. Beware many wrecks and obstructions.

Between **Port-en-Bessin** and Pte de la Percée a bank lies up to 3ca offshore, with drying ledges and the wrecks and obstructions of Omaha Beach marked by 3 NCM buoys.

▶*A small race forms off Pte de la Percée with wind against tide. Off Port-en-Bessin the E-going stream begins about HW Le Havre –0500, and the W-going at about HW Le Havre +0050, sp rates 1·25kn.*◀

4.21 OUISTREHAM

Calvados 49°16'·82N 00°14'·89W (E lock) ⚓⚓◊◊◊⚜⚜

CHARTS AC 2613, 2146, 2136, 1349; SHOM 7418, 7421, 7420; Navi 526; Imray C32

TIDES –0118 Dover; ML 4·6; Duration 0525. See Dives-sur-Mer.

SHELTER Approach is difficult/dangerous in strong NW-NE winds, and the lock waiting pontoon is exposed to the N.

NAVIGATION WPT 49° 19'·18N 00° 14'·54W (abm Nos 1 & 2 buoys) 184·5°/2·1M to front ldg lt. In outer channel, the harbour and canal to Caen, ships and ferries have priority. Outer chan is marked by lt buoys 1-6 and 3 pairs of lt bns; see half-scale inset on chartlet.

The only waiting pontoon on E side of Avant Port dries at LWS on a steeply shelving bank. Yachts usually use smaller E lock. Beware current when gates open. Timings (0700 -2100 LT, Jun 15 -Sept 15): Seaward lock opens: HW-2h15,-1h15,HW+0h30,+2h15, +3h15. Canal lock opens: HW-2h45,-1h45, HW, HW+1h45, +2h45. Detailed timings are posted on the waiting pontoon, at Ouistreham Marina (Ch 09) and www.ouistreham-plaisance.com. Warning: *All crew must wear lifejackets within lock*, failure to comply can result in a fine.

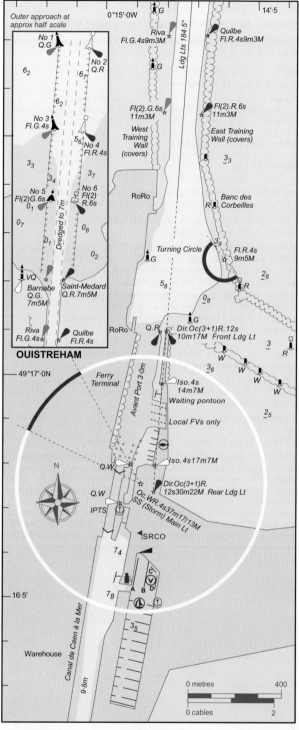

OUISTREHAM

LIGHTS AND MARKS See chartlet and Lights, buoys and waypoints. The 184·5° synch ldg lts are easily seen even in poor vis or bright sun. The huge square block of Caen hospital is conspic from afar; ditto the main lt ho, W + R top.

IPTS (full code) shown from lock control tower, 3 FG (vert) lts on the E or W side of the twr indicate which lock to enter. Two Iso 4s, two QW and Fl R 4s define the ferry turning area.

COMMUNICATIONS CROSS/🆔 02·33·52·16·16; Police 02·31·36·22·70; ⊖ 09·70·27·44·50; Dr 02·31·97·18·45; Ⓗ(Caen) 02·31·27·27·27;

British Consul 01·44·51·31·00; Port HM and Lock 02·31·36·22·00; Call *Ouistreham Port* or *Ouistreham Lock* VHF Ch 74 (HW -2 to HW+3); Marina VHF Ch 09; Sté des Régates VHF Ch 09 office hours; ☏ 08·99·71·02·14; Auto 08·92·68·08·77; Aff Mar 02·31·53·66·50; Taxi 02·31·97·39·41; Ferry terminal 02·31·36·36·36.

FACILITIES
Marina ⚓ ☎02·31·96·91·37. www.ouistreham-plaisance.com 650 inc 100 Ⓥ €3·10. Ⓥ raft on pontoon D, in 3·5m. 🛢🛠 (H24) 🔧 ⚓ Defib. 🕭 ⛽ 🚲 bike hire.🅿 ⚖ Ⓔ SHOM wi-fi 🔩 ✕ ⚓ 📧 (14t).

Société des Régates de Caen-Ouistreham (SRCO) www.srco.fr ☎02·31·96·13·30. 🖂. **Town** ✕ 🖵 🍴 🖂 Ⓑ 🛒 🚇 (bus to Caen) ✈ (Caen). Ferry to Portsmouth.

HARBOUR AT SOUTH END OF OUISTREHAM CANAL
CAEN 49°10'·95N 00°21'·17W. Charts: AC 1349, SHOM 7420.

NAVIGATION The 8M canal passage is simple and takes approx 1½hrs. Depths 2·5m to 9·8m; max speed 7kn; no overtaking. The 3 opening bridges are: Bénouville (Pegasus) (2½M S of Ouistreham) Colombelles (5M) and La Fonderie, Caen (8M). Tfc lts are not used; N-bound vessels have right of way at bridges.

Free passage through bridges is available for vessels at the following times: S-bound, depart Bénouville bridge 1010, 1330,

1630 and during summer 2030 (Allow ½hr from Ouistreham lock/marina). N-bound, depart La Fonderie bridge 0845, 1200, 1500 and summer 1915. Times can vary; check with marina or the website below.

R/T Monitor *Caen Port* Ch **74** in the canal.

Marina ⚓ ☎02·31·95·24·47. www.caen-plaisance.com After La Fonderie bridge turn 90° stbd for marina (4m) at Bassin St Pierre in city centre. Ⓥ berths on pontoon E or as directed. 76+16Ⓥ €2·27, max LOA 25m, draft 3.8m. ⚓ 🏴 🕭.

City ✕ 🛏 🍴 🖵 🏚 🏚 Ⓗ 🖂 Ⓑ @ 🚇 ✈ Caen-Carpiquet.

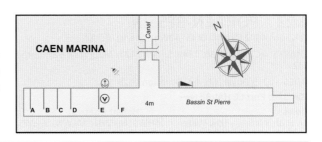

4.22 COURSEULLES-SUR-MER
Calvados **49°20'·40N 00°27'·32W** ⚓⚓🌊🌊🌊🌊🌸🌸

CHARTS AC 2613, 2136, 1349; SHOM 7421, 7420; Navi 526; Imray C32

TIDES –0145 Dover; ML 4·6; Duration No data. See Port-en-Bessin.

SHELTER Good in Bassin Joinville (3m) but appr becomes difficult in strong N to NE winds. Exposed offshore ⚓ (3-5m) at L'Anneau de la Marguerite, 134°/0·4M from SWM buoy.

NAVIGATION WPT 49° 21'·28N 00° 27'·69W (SWM buoy Iso 4s). From waypoint steer 134° 5ca, (use outer ldg marks, front, Bernières ✠ tower ≠ 134° with rear, twin spires of Douvres-La-Déliverande church (partly obsc'd by trees and below skyline close in) then steer 198° 5ca direct to E Jetty hd. At night the two bkwtr lights ≠ 186° form a useful leading line. Beware: Plateau du Calvados extending 2M seaward of the hbr, with drying banks for >0·6M; also oyster beds to the west marked by a line of unlit Y bys running parallel to the shore between Ver-sur-Mer and Asnelles.

The Avant Port dries 2·5m; best ent at HW –1, keep to the E side. Entry gate into Bassin Joinville stays open approx HW ±2. (A detailed timetable of gate opening/closing times for every day of the year can be downloaded at www.courseulles-sur-mer. com/leport.) The Control Twr opens the swing bridge above it on request Ch 09 or when an arriving yacht is spotted, usually quite quickly. **IPTS** in use. There is space at the SW end of the Bassin to turn a boat round.

LIGHTS AND MARKS See chartlet and Lights, buoys and waypoints. Both drying bkwtrs and their perches are easily seen, as is a conspic crucifix at root of the W bkwtr. A Cross of Lorraine is 4ca, and Pte de Ver lt ho is 2·5M W of the hbr ent.

COMMUNICATIONS CROSS 02·33·52·16·16,(Emergency 196); 🛟 02·31·37·45·47; ☏ 08·92·68·08·14; Auto 08·92·68·32·50; Police 02·31·29·55·20 or 17; ⊖02·31·21·71·09 @ Port-en-Bessin; Brit Consul 01·44·51·31·00 (Paris); Ⓗ 02·31·27·27·27 (Caen); Dr 02·31·37·45·24; Ambulance 02·31·44·88·88 or 15; Fire 18; Control twr 02·31·37·46·03, VHF Ch 09 HW±2; Aff Mar (@Caen) 02·31·53·66·50; Tourist Info 02·31·37·46·80.

FACILITIES Bassin Plaisance local shoal draft boats in 1m; no Ⓥ. **Bassin Joinville** ☎02·31·37·51·69. Marina hours: All year, Mon-Sat 0900-1200, + Sun and hols, April-Aug, + July-Aug, 1400-2000 at HW times. 730+20Ⓥ berth as directed by HM, or if absent, on pontoon X on E Quay. €3·10 and €0·60pp/day tax. 14·49m max LOA. 🔧 🏴 🏚 (600m) 🍴 🅿 🔩 🚇 Ⓔ ✕ ⚖ 🛒(25t).

Société des Régates de Courseulles (SRC) ☎02.31.37.47.42, 🖂.

Town ✕ 🖵 🍴 🏚 🛒 @ 🖂 Ⓑ 🚇 (via bus to Caen) ✈ (Caen-Carpiquet). Taxi ☎02·31·37·90·37. Ferry Ouistreham.

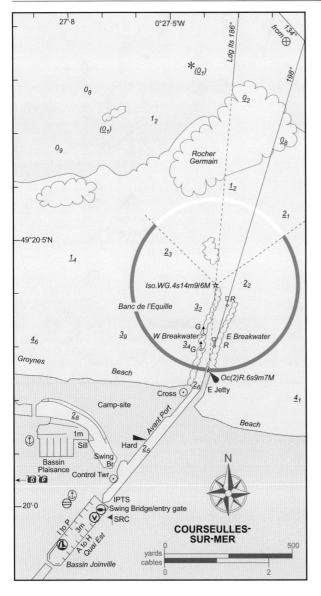

MULBERRY ANCHORAGE 6·5M WEST OF COURSEULLES

ARROMANCHES, Calvados, **49°21'·36N 00°37'·25W**. AC 2613, 2136; SHOM 7421; Navi 527. Tides, see Port-en-Bessin. Strictly a fair weather anchorage, with limited shelter from the ruined WW II Mulberry caissons which are conspicuous especially at LW. Swell intrudes and can make anchoring uncomfortable. Rochers du Calvados, 7ca ENE of the entrance, dries 1·6m but can be crossed near HW.

There are many obstructions within the harbour and offshore: 3 to the W and N are marked by a WCM and two ECM buoys. From the more N'ly ECM buoy (Roseberry, 49°23'·10N 00°36'·49W) the entrance (lat/long as line 1) bears approx 196°/1·8M; it is marked by a small unlit SHM and PHM buoy.

Enter on about 245° for charted anchorage in 3-5m to S of the caissons, or sound closer inshore with caution. Entry from E or W is not advised due to uncharted obstructions. Caution: rocky plateau (3·4m) and further obstructions W of the dinghy landing area.

Facilities: YC Port Winston ☎02·31·22·31·01, 2 dinghy slips. Worth going ashore for D-Day museums and the panoramic view from the clifftops, but very crowded in season.

4.23 PORT-EN-BESSIN

Calvados **49°21'·16N 00°45'·40W** (Hbr ent) ❄◈✿

CHARTS AC 2613, 2136, 1349; SHOM 7421, 7420; Navi 527; Imray C32

TIDES –0215 Dover; ML 4·4; Duration 0520

Standard Port LE HAVRE (⟵)

Times				**Height (metres)**			
High Water		Low Water		MHWS	MHWN	MLWN	MLWS
0000	0500	0000	0700	8·0	6·7	2·9	1·2
1200	1700	1200	1900				
Differences COURSEULLES-SUR-MER							
–0100	–0055	–0015	–0025	–0·5	–0·5	–0·1	–0·1
PORT-EN-BESSIN							
–0055	–0030	–0030	–0035	–0·7	–0·7	–0·1	+0·1
ARROMANCHES							
–0055	–0025	–0025	–0035	–0·5	–0·5	–0·1	–0·2

SHELTER Good. Ent difficult in strong N/NE'lies and dangerous in F8+. No ⚓ in outer hbr nor 1ca either side of 204° ldg line.

NAVIGATION WPT 49°21'·62N 00°45'·09W, 204°/0·5M to hbr ent. Avoid G sector of W mole hd Fl WG 4s which covers the inshore dangers off Omaha Beach, 3 to 6M to the WNW. In the outer hbr beware submerged jetty (marked by R bcn) to E of ent chan.

LIGHTS AND MARKS Conspic marks: a water twr (104m) 1·7M ESE of the hbr; a Sig Stn 7ca W of hbr. Ldg lts 204°: Front, W pylon, G top, at bottom of hill; rear, W house/grey roof near top of the hillside. See chartlet and Lights, buoys and waypoints.

IPTS control entry to the basins. Entry gate opens HW ±2. While gate stays open, swing bridge opens whenever possible to suit yachts and FVs; there is little road traffic. When shut the bridge shows FR in the middle.

COMMUNICATIONS CROSS 02·33·52·16·16; Sig Stn/LB 02·31·21·81·51; Auto 08·92·68·08·14; Police 02·31·21·70·10; ⊖02·31·21·71·09; Brit Consul 01·44·51·31·00; Ⓗ 02·31·51·51·51; Dr 02·31·21·74·26; HM 02·31·21·70·49, VHF Ch 18 (HW ±2) for gate opening H24; Aff Mar 02·31·21·71·52; Mairie 02·31·21·72·12.

FACILITIES A major FV port. **Outer Hbr** dries up to 3·7m; ▬ L. Drying ⛵ on E side Quai de L'Epi, no charge.

Bassin 1: Inside the entry gate, tight turn 90° stbd, Ⓥ berth/raft alongside 25m pontoon in NNW corner. €0·77 + €0·20pp/day tax, 🚿 WC €3·00 pp/day, code access (free, children <10yrs) at Tourist Office opp on E quay. Yachts may stay only 24-48hrs, due to limited space. No berthing on sloping E wall. Call Gatemaster ☎02·31·21·71·77 (HW ±2) to pre-book gate time if leaving early.

Bassin 2, no entry. **Services** ▬ ⛵ Ⓔ ✎ ⚒ (4t).

Town ✕ 🛒 ✉ Ⓑ ☎ ⛽ ⇌ (bus to Bayeux) ✈ (Caen, ☎ 02·31·26·58·00). **Ferry** see Ouistreham.

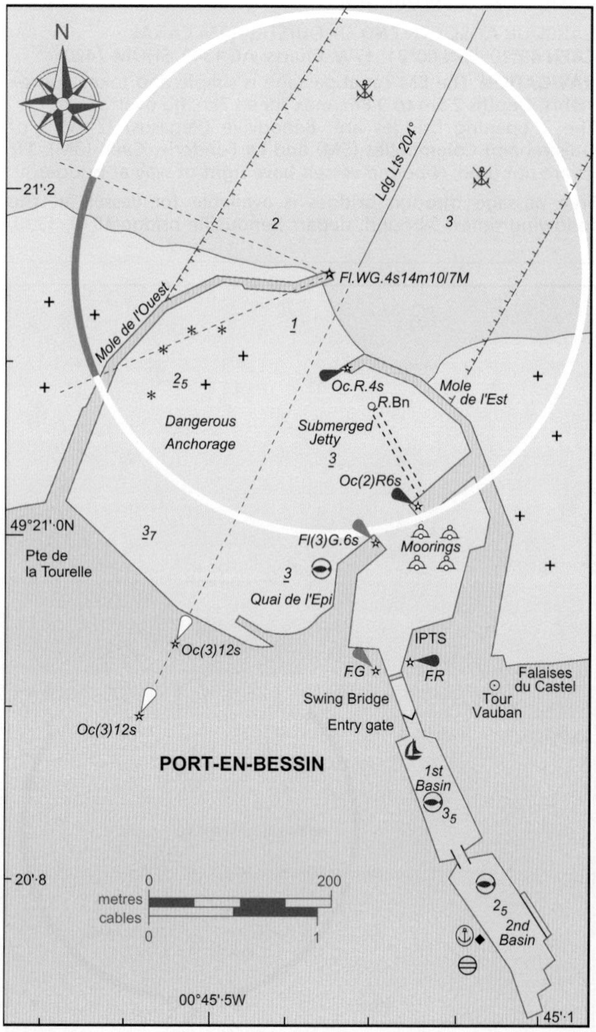

4.24 GRANDCAMP

Calvados **49°23'·51N 01°02'·99W** (Pierheads) ※❀♨♨✿✿

CHARTS AC 2613, 2135, 1349; SHOM 7422, 7420; Navi 527; Imray C32

TIDES –0220 Dover; ML Rade de la Capelle 4·4; Duration 0510

Standard Port CHERBOURG (→)

Times				Height (metres)			
High Water		Low Water		MHWS	MHWN	MLWN	MLWS
0300	1000	0400	1000	6·4	5·1	2·6	1·1
1500	2200	1600	2200				
Differences RADE DE LA CAPELLE (3M to NW)							
+0125	+0050	+0125	+0110	+0·8	+0·9	+0·2	+0·2
ILES SAINT MARCOUF							
+0130	+0055	+0120	+0110	+0·5	+0·6	–0·1	0·0

SHELTER Good, but approach is difficult in NW-NE winds >F6.

NAVIGATION WPT 49°24'·82N 01°04'·33W, 146°/1·6M to E pier head. Nos 1, 3 & 5 NCM buoys mark the seaward limit of Les Roches de Grandcamp, a large plateau of flat rock extending about 1½M offshore and drying approx 1·5m; heavy kelp can cause echo sounders to under-read. It can be crossed from most directions, given adequate rise of tide.

Watch out for FVs in the narrow access chan. 14·3m wide gate into wet basin stays open approx HW ±2½; entry is controlled by IPTS. HW –2½ is the same as LW Dunkerque and HW +2½ is the same as HW Dunkerque.

LIGHTS AND MARKS Nos 1, 3 and 5 unlit NCM buoys are 1·4-1·7M NE to NW of the hbr. The 221° ldg lts are primarily for FVs. Maisy's modern church twr is conspic 6ca SSW of hbr ent. See chartlet and Lights, buoys and waypoints and Carentan for other data.

COMMUNICATIONS CROSS 02·33·52·16·16; ⒧ 02·31·22·67·12; ⚓ 02·21·33·25·26; Auto 08·92·68·08·14; Police 02·31·22·00·18; Brit Consul 01·44·51·31·00; Dr 02·31·22·60·44; Hbr/marina VHF Ch 09; Aff Mar 02·31·22·60·65.

FACILITIES Marina ☎02·31·22·63·16. 240 + 10❶ €2·00 (no cr/cards). ❶ berths in 2·5m at E end of pontoon C, either side; or as directed. ⚓ ⌖ ✖ ⛽ ⚓ ⌕(5t) ⚓ YC ☎02.31.22.14.35.
Town ✕⌂☲🍴☎👕🏧✉Ⓑ⚞(Carentan)✈(Caen). **Ferry** Cherbourg, Ouistreham.

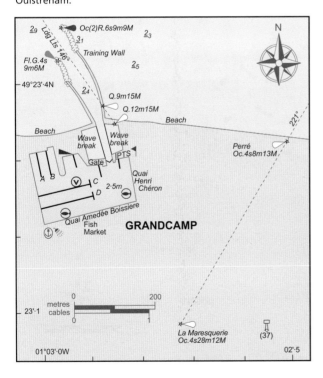

GRANDCAMP TO POINTE DE BARFLEUR

(AC 2135) On E side of B du Grand Vey, Roches de Grandcamp (dry) extend more than 1M offshore, N and W of **Grandcamp**, but they are flat and can be crossed from the N in normal conditions HW±1½. Three NCM buoys mark the N edge of this ledge. Heavy kelp can give false echo-sounder readings.

At the head of B du Grand Vey, about 10M S of Îles St Marcouf, are the (very) tidal hbrs of Carentan and Isigny; entry is only possible near HW and should not be attempted in strong onshore winds. The **Carentan** chan is well buoyed and adequately lit. It trends SSW across sandbanks for about 4M, beyond which it runs between two training walls leading to a lock, and thence into the canal to Carentan's popular marina. The **Isigny** chan is slightly deeper, but the hbr dries to steepish mudbanks.

Îles St Marcouf lie 7M SE of St Vaast-la-Hougue, about 4M offshore, and consist of Île du Large (lt) and Île de Terre about ¼M apart. The Banc du Cardonnet, with depths of 5·2m and many wks, extends for about 5M ESE from the islands. Banc de St Marcouf, with depths of 2·4m and many wks, extends 4·5M NW from the isles, and the sea breaks on it in strong N/NE winds.

Approach **St Vaast-la-Hougue** S of Île de Tatihou, but beware drying rks: La Tourelle (unmarked) Le Gavendest and La Dent, which lie near the approaches and are buoyed. 2M N is Pte de Saire with rks and shoals up to 1M offshore. From Pte de Saire the coast runs 4M NNW to drying **Barfleur**.

Raz de Barfleur, in which the sea breaks heavily, must be avoided in bad weather, particularly with NW or SE winds against the tide, by passing 5-7M seaward of Pte de Barfleur.

▶*In calmer weather it can be taken at slack water (HW Cherbourg –4½ and +2) via an inshore passage, 3ca NE of La Jamette ECM beacon. Expect some turbulence.*◀

Pte de Barfleur marks the W end of Baie de Seine, 56M west of Cap d'Antifer. There are no obstructions on a direct course across the Bay, but a transhipment area for large tankers is centred about 10M ESE of Pte de Barfleur. Be advised: The W end of the Baie de Seine has at times been affected by floating Sargassum weed; in harbours from Grandcamp to St Vaast propellers may be fouled.

POINTE DE BARFLEUR TO CAP DE LA HAGUE

(AC 1114) The N coast of the Cotentin peninsula runs E/W for 26M, bordered by rks which extend 2·5M offshore from Pte de Barfleur to C Lévi, and 1M offshore between Pte de Jardeheu and C de la Hague. ▶*Tidal streams reach 5kn at sp, and raise a steep, stopping sea with wind against tide.*◀ Between Pte de Barfleur and Cherbourg, yachts route via Les Equets and Basse du Rénier NCM buoys and La Pierre Noire WCM buoy; keep a lookout for oncoming traffic using same Wpts.

The inshore passage between Pte de Barfleur and C Lévi, well S of the above 3 cardinal buoys, is not advised except in fair weather and good visibility. Numerous depths were reported (2017) to be less than those shown on AC 1114, 2135 and 2613, the most significant, 1·8m at 49°44'·08N 01°21'·83W. SHOM 7444 and due caution are essential. Port Lévi and Port de Becquet are two small drying hbrs E of Cherbourg. ▶*Tidal streams run strongly with considerable local variations. Off C Lévi a race develops with wind against tide, and extends nearly 2M N. Off Cherbourg the stream is E-going from about HW –0430 and W-going from HW +0230.*◀

Cherbourg is a port of refuge, hugely popular with British yachtsmen which will become increasingly busy during 2018/19 as the assembly hub for offshore wind farms and tidal energy turbines along the N French coast.

If not catered for, the tidal streams are strong enough to set you downtide of Cherbourg. Do your sums and bias the approach to be uptide of the harbour. At night positively identify the leading lights through either entrance, *before* entering.

▶*Close inshore between Cherbourg and C de la Hague a back eddy runs W.*◀ West of **Omonville** and about 2M E of C de la Hague there is an anchorage off Port Racine in Anse de St Martin, guarded by at least 3 groups of rocks. It is open to the N, but useful to await a fair tide through the Alderney Race.

4.25 CARENTAN

Manche **49°19'·09N 01°13'·53W** 🌊🌊🌊🌊🌊🌺🌺🌺

CHARTS AC 2613, 2135; SHOM 7056, 7422; Navi 527; Imray C32

TIDES –0225 Dover; ML Rade de la Capelle 4·4; Duration 0510

Standard Port CHERBOURG (→) Use differences **RADE DE LA CAPELLE** (Grandcamp). HW Carentan = HW Cherbourg +0110.

SHELTER Complete shelter in the land-locked marina (3kn max) but do not attempt to approach in onshore winds >F5.

NAVIGATION WPT 49°25'·45N 01°07'·07W (C-I SWM buoy) 210°/1·8M to Nos 1/2 buoys, not to be confused with the unlit Isigny chan bys 1/2 to the ESE. Start from WPT at HW –2 to –1½. Graph inset gives approx availability. Shallowest water is typically in the area between bys 7-10, but channels in the bay are broad and fairly flat. Once inside trng walls chan deepens.

Appr protected from prevailing S to W winds. Drying out on the hard sands of the estuary is not advised; nor is an early approach, as a bore (*mascaret*) may occur in the river chans between HW –3 and HW–2½ especially at springs. Chan is liable to vary in both depth and direction; buoys are moved accordingly. All buoys have R or G reflective panels; 6 are lit as on chartlet. After about 2·2M the chan lies between 2 training walls, marked by lt bns.

3·8M SW a small pool is formed where R Taute and Douve meet. Ahead, the little lock (room for about 6 boats) opens HW –2 to HW+3. Lock sigs: ● = open, ● = shut (use phone at lock office if lock-keeper absent). Small waiting pontoons are on E side, down- and upstream of the lock. A magnetic anomaly has been reported in the vicinity of the N13 underpass. Electronic autopilots should not be relied upon for steering in this area.

LIGHTS AND MARKS See chartlet and Lights, buoys and waypoints. Maisy Ch twr is a good landmark initially.

COMMUNICATIONS CROSS 02·33·52·16·16; Auto 08·92·68·08·50; Police 02·33·42·00·17; ⊖ 02·31·21·71·09; Brit Consul 01·44·51·31·00; Ⓗ 02·33·42·14·12; Dr 02·33·42·33·21; Lockmaster 02·33·71·10·85; *Écluse de Carentan* VHF Ch 09 (0800-1200, 1400-1600LT & lock opening hrs); Aff Mar 02·33·44·00·13.

FACILITIES Marina ☎02·33·42·24·44, mob 06·89·13·15·56. port-carentan@orange.fr www.cc-carentan-cotentin.fr 310 + 50🅥 fees based on m², so approx fee €2·07/m. Max/min depths 3·5/3m. 🅥 berths normally on pontoon K. 🛠🛥 (daily: 0900-1000, 1500-1600) ▬ ⛟(35t) 🏗(50t) ⚓ 🅿 🔧 🅰 ⓧ ⛽. Additional space for wintering ashore at Top Nautique ☎02·33·42·23·87.

YC Croiseurs Côtiers de Carentan ☎02·33·42·06·61, 🖂.

Town ⤬ 🍴 🛒 🖂 ⊠ Ⓑ @ in ⊠ ⇌ ✈ and ferry Cherbourg.

ISIGNY-SUR-MER 49°19'·31N 01°06'·24W 🌊🌊🌊🌺

SHELTER Good in town, but the drying river bed shelves steeply there, soft mud. Cautions in outer approaches as for Carentan, though the Isigny chan is somewhat deeper.

NAVIGATION Leave C-I WPT at HW –2½ tracking 158°/1·25M to IS, a very small unlit NCM buoy at 49°24'·29N 01°06'·37W, near ent to unlit buoyed chan.

From IS buoy track about 220°/0·9M to first unlit chan buoys. Pairs of buoys are spaced at approx 3ca intervals, but between Nos 9/10 and 11/12 the interval is 6ca, with an IDM bn midway.

From Pte du Grouin, ldg lts lead 172·5°/1·9M (as chartlet) between training walls to a Y-junction. Here turn port into R l'Aure for 4ca to town quays on SW bank.

COMMUNICATIONS Ch 09 (0900-1200 & 1400-1800LT).

FACILITIES HM ☎02·31·51·24·01; **Quay** (dries) ⚓ 40+5 🅥. Drying pontoons on SW bank; berth bows-in/stern to buoys. Pay fees at town-hall; 🚿 🅿 🔧 ▬ 🅰 ⛽ 🔧 🏗(8t). **Town** ⤬ 🍴 🖂 ⊠ Ⓑ.

ADJACENT ANCHORAGE

ILES ST MARCOUF, Manche, **49°29'·78N 01°08'·92W**, the charted ⚓. AC 2613, 2135; SHOM 7422. Tides, Grandcamp. A fair weather ⚓ only, no shelter in S'lies. The two islands, Île du Large and SW of it, Île de Terre, look from afar like ships at ⚓. The former is a fort with a small dinghy hbr on its SW side; the latter is a bird sanctuary, closed to the public. ⚓ SW or SE of Île du Large or SE or NE of Île de Terre. Holding is poor on pebbles and kelp. Île du Large lt, VQ (3) 5s 18m 8M, on top of the fort. Both islands are surrounded by drying rks, uninhabited and have no facilities.

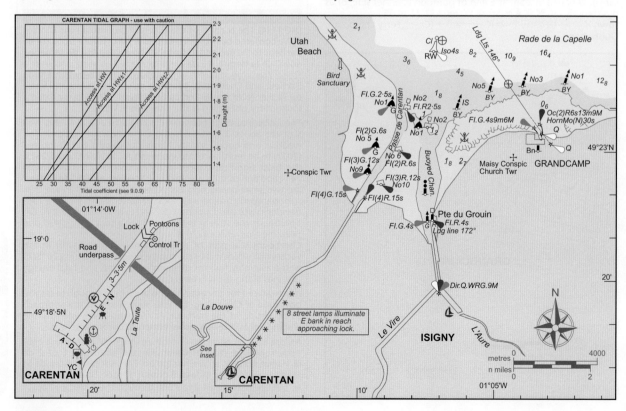

4.26 ST VAAST-LA-HOUGUE

Manche **49°35'·17N 01°15'·41W** (Jetty head) ✹❀♨♨♨ ✿✿✿

CHARTS AC 2613, 2135; SHOM 7422, 7120, 7090; Navi 527, 528; Imray C32

TIDES –0240 Dover; ML 4·1; Duration 0530

Standard Port CHERBOURG (→)

Times				Height (metres)			
High Water		Low Water		MHWS	MHWN	MLWN	MLWS
0300	1000	0400	1000	6·4	5·1	2·6	1·1
1500	2200	1600	2200				
Differences ST VAAST-LA-HOUGUE							
+0115	+0045	+0120	+0110	+0·3	+0·4	–0·3	–0·1

SHELTER Excellent in marina especially in W'lies. If full, or awaiting entry, ⚓ off between brgs of 330° and 350° on jetty lt, but this ⚓ becomes untenable in strong E-S winds.

NAVIGATION WPT 49°34'·36N 01°13'·88W (Le Gavendest buoy) 310°/1·3M to main jetty lt, Oc (2) WRG 6s; appr in its wide, well marked W sector, dries 1·8m. Beware boats at ⚓, cross-currents and numerous unlit pot markers bween La Dent and La Manquet bys. 'Le Run' appr (over extensive oyster beds) is only feasible if draft <1·2m and when the entry gate is open, HW –2¼ to HW +3½. If barometer is high and Coeff 40-50, gate may close 30 mins early; and 1-1¼hrs early if Coeff <40. Gate times are on the website.

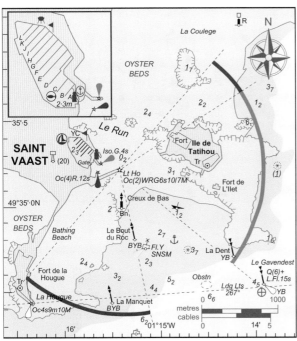

LIGHTS AND MARKS By day do not confuse similar twrs on Ile de Tatihou and Fort de la Hougue. From N, Pte de Saire is a squat white lt ho/G top. From E, ldg lts 267°: front La Hougue; rear, Morsalines in W sector. Main jetty hd lt is a conspic white twr/R top; W sector covers the bay. See chartlet, Lights, buoys and waypoints.

COMMUNICATIONS CROSS 02·33·52·72·13; ⚓ 08·99·71·08·08; Police 17 or 02·33·23·31·90; ⊖ 02·33·23·34·01; British Consul 01·44·51·31·00; Ambulance (SAMU) 15; Aff Mar 02·33·54·51·74; Semaphore stn (Fort de la Hougue) VHF Ch 16; Marina VHF Ch 09; Tourist Office 02·33·23·19·32.

FACILITIES Marina ☎02·33·23·61·00. www.saint-vaast-reville.com port-st-vaast@saint-vaast-reville.com Season 0900-1200, 1400-1700. 604 + 100 ♥ in 2·3m, €2·83 + €0·22pp/day tax, multihulls +50%. ♥ berths by LOA: B >14m; C 12-14m; D no ♥; E (outer part) <12m. 🚿♿ (pontoon A) 🛢🍴🛢⚒🔧⚓🗲(45t).

YC de St Vaast ☎02·33·44·07·87, ✕ 🛒 ⊠(3t). Town ✕ (closed Mondays) 🛒🛒🛢🔲⊠ Ⓑ 🚈 (bus to Valognes) Cherbourg ✈ ferry.

4.27 BARFLEUR

Manche **49°40'·34N 01°15'·48W** ✹❀♨ ✿✿✿✿

CHARTS AC 2613, 2135, 1114; SHOM 7422, 7120; Navi 528; Imray C32

TIDES –0208 Dover; ML 3·9; Duration 0550

Standard Port CHERBOURG (→)

Times				Height (metres)			
High Water		Low Water		MHWS	MHWN	MLWN	MLWS
0300	1000	0400	1000	6·4	5·1	2·6	1·1
1500	2200	1600	2200				
Differences BARFLEUR							
+0115	+0050	+0045	+0050	+0·1	+0·3	0·0	0·0

SHELTER Excellent, except in fresh NE/E winds when ent is difficult and berths uncomfortable. Hbr dries. Rks/shoals in SE of hbr. Safe to ⚓ outside hbr in offshore winds.

NAVIGATION WPT 49°41'·13N 01°14'·41W, 219·5°/1·0M to hbr ent. Positively identify Roche à l'Anglais SHM and Le Hintar PHM buoys before closing the hbr any further. Beware cross-currents.

In rough weather, esp wind against tide, keep 5M off Pte de Barfleur to clear the Race; see Passage information. If using the inshore passage, identify La Jamette ECM bn and La Grotte SHM buoy. From the S, keep seaward of Pte Dranguet and Le Moulard, both ECM bns.

LIGHTS AND MARKS Pte de Barfleur lt ho is conspic 1½M NNW of hbr, a 72m high, grey twr/B top. The squat church twr is a conspic daymark. Ldg lts 219·5°: both W ☐ twrs, not easy to see by day. Light details as chartlet and Lights, buoys and waypoints.

COMMUNICATIONS CROSS 02·33·52·72·13; 🆔 02·33·23·10·10; ⚓ 02·33·10·16·10; Auto 08·92·68·02·50; Police 17; ⊖ 02·33·23·34·01; Brit Consul 01·44·51·31·00; SAMU 15; Dr 02·33·43·73·55; Dentist 02·33·54·49·31; HM 02·33·54·08·29; Aff Mar 02·33·23·36·00.

FACILITIES
NW Quay Yachts dry out on firm, level sand/mud or at SW end of quay, but space may be restricted by FVs. 6 x ♥⊶ €15·00 night, max 14.99m; 6 x ♥ rafting €13·00 night, max 9.99m. 🛟 🔧(needs long cable), 🛢🚻, nearby at La Blanche Nef camping/caravan site on presentation of berthing fee receipt. ⚓🗲 SC ☎02·33·54·79·08.

Town www.ville-barfleur.fr 🛢🔧 (Montfarville: 1km S) ✕🛢🛒⊠ Ⓑ no wi-fi, bus to Cherbourg for 🚈 ✈ ferry.

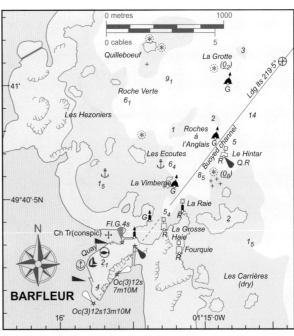

4.28 CHERBOURG

Manche **49°38′·94N 01°37′·11W** (Marina ent) ❀❀❀♨♨♨❀❀

CHARTS AC 2656, 2669, 1114, 1112; SHOM 7120, 7092, 7086; Navi 528, 1014; Imray C32, C33A

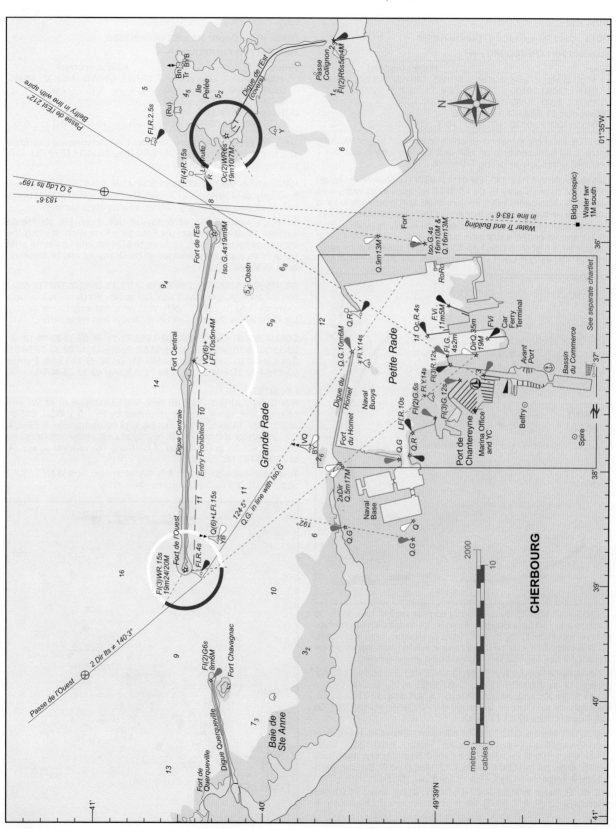

TIDES Cherbourg is a Standard Port (→). –0308 Dover; ML 3·8; Duration 0535

SHELTER Excellent; a port of refuge available H24 in all tides and weather. Small craft ⚓ N of marina bkwtr in 2·3m, keeping clear of the charted limits of the military port, close N and W.

NAVIGATION For coastal features from Pte de Barfleur to Cap de la Hague see Passage Information. Enter the Grande Rade via:

- **Passe de l'Ouest** (W ent). WPT 49°41'·05N 01°39'·86W, 140°/0·85M to abeam Fort de l'Ouest. Rks, marked by a PHM buoy, Fl R 4s, off Fort de l'Ouest, extend about 80m off each bkwtr. From W, the white sector of Fort de l'Ouest lt (or by day bearing >122°) keeps clear of offlying dangers E of Cap de la Hague. From CH1 SWM buoy, L Fl 10s, Fort de l'Ouest bears 145°/3·5M.

- **Passe de l'Est** (E ent, least depth 8m). WPT 49°40'·91N 01°35'·55W, 189°/0·65M to abeam Fort de l'Est. Keep to W side of chan (but at least 80m off Fort de l'Est) to avoid dangers W and NW of Ile Pelée marked by two PHM buoys. N/NE of Ile Pelée an extensive drying area is marked by 2 unlit bn trs.

- **Passe Collignon** is a shallow (2m) chan, 93m wide, through Digue de l'Est (covers) near the shore. Tidal stream reaches 3·4kn. Only advised in good conditions, near HW.

No anchoring in the Passe de L'Ouest and Passe de L'Est. Speed limits: Grande Rade 14kn; Petite Rade 8kn; Marina 3kn. No entry to: the area S of the Digue Centrale due to fish farms, nor to the area east of the Port Militaire. Keep clear of ferries and shipping using the large N and E extension of the commercial port, the base for offshore wind farm turbine and tidal turbine assembly.

LIGHTS AND MARKS Daymarks include the chimney of Jobourg nuclear plant, 10M W; the cranes of the naval base; and higher ground S of the city. Five powerful lights assist a good cross-Channel landfall:

- 13M E, Pte de Barfleur, Fl (2) 10s 29M
- 5M ENE, Cap Lévi, Fl R 5s 22M
- 12M WNW, Cap de la Hague, Fl 5s 23M
- 28M W, Casquets, Fl (5) 30s 24M

See also Lights, buoys and waypoints (LBW) and Area 19.

Passe de l'Ouest ldg lts 140·3° are a Dir Q rear ldg lt aligned between two Q front ldg lts (horizontally disposed) at base of Digue du Homet. Other lts as on chartlet and LBW.

On entering the Petite Rade make good 200° for marina ent, Fl (3) R 12s and Fl (3) G 12s. Shore lights may confuse and mask navigational lights.

COMMUNICATIONS CROSS 02·33·52·16·16 or 112; *Jobourg Traffic* Ch **13** 80 (H24) provides radar surveillance of the Casquets TSS and ITZ and Mont St Michel to Cap d'Antifer. Radar assistance available on request, Ch 80, to vessels in the 270° sector from S clockwise to E, radius 40M from CROSS Jobourg at 49°41'·1N 01°54'·5W. Jobourg broadcasts nav, weather and traffic info in English and French Ch 80 at H+20 & H+50; ⚓ 08·92·68·32·50; Police 02·33·92·70·00; ⊖ 02·33·23·34·02; Brit Consul 01·44·51·31·00; Ⓗ 02·33·20·70·00; Dr 02·33·01·58·58; HM (Port & Lock) 02·33·20·41·25; Aff Mar 02·50·79·15·00; Marina: *Chantereyne* VHF Ch 09; lock into Bassin du Commerce (long stay/>25m) VHF Ch 06.

FACILITIES Port Chantereyne Marina ☎ 02·33·87·65·70. www.portchantereyne. fr portchantereyne@cherbourg.fr Access H24. 1380 + 220 ❹ (large detached waiting

pontoon E side marina) €2·77 on pontoons J,N,P, Q, N side of pontoon H and W side of K; max LOA 25m; max draught 3m: ▮ ▮ (HO) or (H24) cr/debit card. max fill €120·00. ⚏ ⚏ @ ⚓ ⚒ 🛒 Ⓔ ✕ ⚓ SHOM, ⚓ ⚓(40t).

YC de Cherbourg ☎02·33·53·02·83, ✕ 🍽.

Bassin du Commerce is for permanent residents only/>25m LOA.

City ✕ 🍽 🛒 🏥 ✉ 🖂 Ⓑ ⚓. Cherbourg (Maupertus) ✈ ☎ 02·33·88·57·60. Ferry: Portsmouth, Poole.

MINOR HARBOURS EAST OF CHERBOURG

PORT DE LÉVI, Manche, **49°41'·23 N 01°28'·46W.** AC 1114; SHOM 7120, 5609, 7092; HW –0310 on Dover (UT); +0024 on Cherbourg. HW ht +0·2m on Cherbourg. Shelter good except in SW to N winds. By day appr on 094° keeping the white wall and lt between the white marks on each pier hd. Lt is Oc (2) WRG 6s; see Lights, buoys and waypoints. Keep in W sector (083°-105°) but night entry not advised due to many pot floats. Berth bows on to NE side below white wall, amongst small FVs. Dries to clean sand. Facilities: Peace. Fermanville (1·5M) has ✕ 🍽 🛒.

PORT DU BECQUET, Manche, **49°39'·304N 01°32'·81W.** AC 1114; SHOM 7120, 7092. Tides as Cherbourg. Ldg lts/marks 186·3°: Both Dir Oc (3) 12s (front lt White; rear lt Red) W 8-sided twrs; see LBW. Large PHM bcn twr, La Tounette, marks rks close E. Shelter good except from N to E when a strong scend enters. Dries, clean sand. Secure to S of the E/W jetty (little space). Facilities: very few; all facilities at Cherbourg 2·5M.

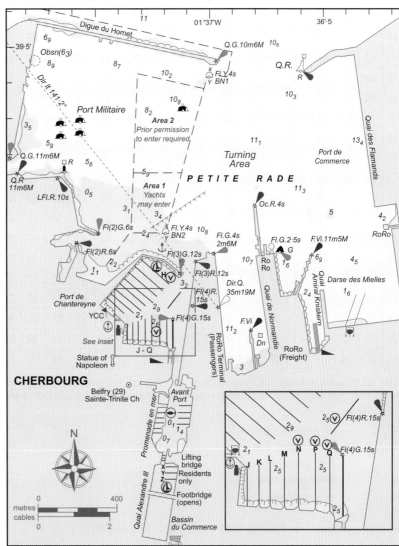

CHERBOURG TIDAL COEFFICIENTS 2019

Date	Jan am	Jan pm	Feb am	Feb pm	Mar am	Mar pm	Apr am	Apr pm	May am	May pm	June am	June pm	July am	July pm	Aug am	Aug pm	Sept am	Sept pm	Oct am	Oct pm	Nov am	Nov pm	Dec am	Dec pm
1	56	57	53	57	37	40	50	56	56	61	69	73	72	76	92	96	113	113	112	107	82	75	68	
2	60	63	62	66	45	50	61	67	66	71	77	81	81	85	100	103	111			102	67		62	56
3	66	69	69	73	55	60	71	76	75	79	84	86	88	91	104	104	107	102	94	86	59	51	50	45
4	71	74	76	78	65	70	80	83	83	86	88	89	93	94		103	96	88	77	69	44	39	41	38
5	76	78	80	82	74	78	86	88	88	89	89	89	94	93	100	96	80	71	59	51	34	33	36	35
6	79	80	83	83	81	83	90	90	90	90	87			92	91	85	62	54	43	37	34	36	37	39
7	80	80	83	82	86	87	90	90	89	87	85	82	89	86	79	72	47	42	34	33	41	46	43	47
8	79	78	81	79	88	88	88	86	84		78	74	82	77	65	58	38	38	36	40	51	56	52	57
9	76	74	77	74	88	86		83	81	77	70	66	72	68	53	49	40	44	45	51	61	66	61	66
10	72	69	70		85	82	78	74	71	66	62	60	64	60	46	46	49	55	57	62	71	75	70	74
11	66		66	62	79	74	68	63	61	56	58	58	58	56	48	51	60	65	67	72	78	81	78	81
12	62	59	58	54		70	57	51	53	50	59	61	56	57	54	59	69	74	76	80	84	85	83	85
13	55	52	49	46	65	59	47	44	50	52	63	67	58	61	63	66	77	80	82	85	87	87	86	87
14	48	45	43	42	54	49	44	46	56	62	70	73	63	66	70	73	82	84	86	87	87	86	87	86
15	43	42	43	47	44	42	52	59	67	73	76	78	69	71	76	78	85	85	88	87	84	82	85	82
16	43	45	52	59	41	44	67	76	79	84	80	82	73	75	79	80	85	85	86	85	79	75	79	76
17	48	53	67	76	50	58	84	91	88	92	82	83	76	77	81	81	83	82	82	79	70	66		72
18	59	65	84	92	66	76	97	102	94	95	82	81	78	77	80	79	79			75		61	69	65
19	72	79	99	105	85	93	105	107	95	95	80	78	77	76	78		76	72	71	66	56	52	62	59
20	86	92	110	113	100	107	108	107	93	91	76	73	74		75	73	68	63	61	55	49	48	58	57
21	97	102	115	115	111	114	105	101	87	84		70	72	70	70	66	58	52	50	45	49	53	58	61
22	105	107	113	110	115	115	97			79	66	62	67	64	62	58	47	42	41	40	57	63	64	67
23	108	108	104		113	109	91	85	74	68	59	55	61	57	53	49	39	37	42	47	70	76	71	75
24	106	102	98	91	104	97	78	70	63	57	51	47	54	50	45	41	39	44	54	62	83	88	79	82
25	98		83	74	90		63	55	52	47	44	42	47	44	39	39	51	60	71	80	92	96	84	86
26	92	86	65	57	82	73	48	41	42	38	40	39	42	41	41	45	69	78	88	95	98	99	87	88
27	79	71	49	43	64	55	36	32	36	35	40	41	41	42	51	59	87	96	101	106	99	98	87	87
28	64	58	38	36	47	40	31	32	35	37	44	48	45	50	67	75	103	108	109	111	96	93	85	83
29	52	48			35	32	35	39	40	45	52	57	55	61	83	91	113	115	111	109	89	85	80	77
30	46	45			31	34	45	50	49	54	62	67	67	74	98	104	116	115	106	102	80	74	73	69
31	47	50			39	44			59	64			80	86	108	112			96	90			65	

Tidal coefficients indicate the magnitude of the tide on any particular day, without having to look up and calculate the range, and thus determine whether it is springs, neaps or somewhere in between. This table is valid for all areas covered by this Almanac. Typical values are:

120	Very big spring tide
95	**Mean spring tide**
70	Average tide
45	**Mean neap tide**
20	Very small neap tide

See also coefficients for Calais (p 169), St Malo (p 215) and Brest (p 277).

STANDARD TIME UT –01
Subtract 1 hour for UT
For French Summer Time add
ONE hour in **non-shaded areas**

CHERBOURG   LAT 49°39'N   LONG 1°38'W
TIMES AND HEIGHTS OF HIGH AND LOW WATERS

Dates in red are **SPRINGS**
Dates in blue are **NEAPS**

YEAR 2019

JANUARY

Time	m		Time	m
1 TU	0458 5.4 / 1149 2.3 / 1723 5.4		**16** W	0340 5.1 / 1037 2.7 / 1613 5.1 / 2310 2.5
2 W	0020 2.2 / 0600 5.6 / 1252 2.2 / 1825 5.5		**17** TH	0457 5.2 / 1152 2.5 / 1732 5.3
3 TH	0116 2.0 / 0653 5.8 / 1345 1.9 / 1919 5.7		**18** F	0019 2.2 / 0605 5.5 / 1256 2.1 / 1838 5.6
4 F	0205 1.9 / 0740 6.0 / 1431 1.8 / 2005 5.8		**19** SA	0120 1.9 / 0702 5.9 / 1354 1.7 / 1934 6.0
5 SA	0248 1.8 / 0821 6.1 / 1512 1.6 / 2045 5.9		**20** SU	0216 1.5 / 0754 6.3 / 1449 1.3 / 2027 6.3
6 SU ●	0326 1.7 / 0858 6.2 / 1549 1.5 / 2121 6.0		**21** M ○	0309 1.3 / 0844 6.6 / 1541 0.9 / 2118 6.5
7 M	0403 1.7 / 0932 6.2 / 1624 1.5 / 2156 6.0		**22** TU	0400 1.0 / 0933 6.8 / 1631 0.7 / 2208 6.6
8 TU	0437 1.7 / 1006 6.2 / 1658 1.5 / 2229 6.0		**23** W	0449 0.9 / 1022 6.9 / 1719 0.6 / 2256 6.6
9 W	0511 1.8 / 1040 6.2 / 1731 1.6 / 2302 5.9		**24** TH	0537 1.0 / 1109 6.8 / 1806 0.7 / 2342 6.5
10 TH	0544 1.9 / 1113 6.0 / 1804 1.7 / 2335 5.7		**25** F	0623 1.1 / 1155 6.6 / 1851 1.0
11 F	0617 2.1 / 1146 5.8 / 1837 1.9		**26** SA	0027 6.2 / 0709 1.4 / 1240 6.3 / 1937 1.3
12 SA	0009 5.5 / 0652 2.3 / 1222 5.6 / 1914 2.1		**27** SU ◑	0112 5.9 / 0756 1.8 / 1327 5.9 / 2025 1.8
13 SU	0048 5.4 / 0733 2.5 / 1303 5.4 / 1957 2.3		**28** M	0200 5.6 / 0849 2.2 / 1420 5.5 / 2120 2.2
14 M ◐	0133 5.2 / 0822 2.6 / 1354 5.2 / 2049 2.5		**29** TU	0300 5.2 / 0952 2.5 / 1527 5.1 / 2228 2.5
15 TU	0231 5.1 / 0923 2.8 / 1457 5.0 / 2155 2.6		**30** W	0413 5.1 / 1109 2.6 / 1647 5.0 / 2344 2.5
			31 TH	0528 5.2 / 1222 2.4 / 1804 5.1

FEBRUARY

Time	m		Time	m
1 F	0051 2.4 / 0632 5.4 / 1324 2.2 / 1906 5.3		**16** SA	0539 5.3 / 1232 2.2 / 1821 5.4
2 SA	0147 2.2 / 0724 5.6 / 1415 1.9 / 1953 5.6		**17** SU	0101 2.1 / 0644 5.7 / 1338 1.7 / 1922 5.9
3 SU	0233 2.0 / 0806 5.9 / 1457 1.7 / 2033 5.8		**18** M	0203 1.6 / 0740 6.2 / 1437 1.2 / 2016 6.3
4 M ●	0313 1.8 / 0843 6.1 / 1534 1.5 / 2108 5.9		**19** TU ○	0259 1.2 / 0833 6.6 / 1530 0.8 / 2108 6.6
5 TU	0349 1.7 / 0918 6.2 / 1608 1.4 / 2140 6.0		**20** W	0351 0.9 / 0923 6.8 / 1620 0.5 / 2156 6.8
6 W	0422 1.6 / 0951 6.2 / 1641 1.3 / 2212 6.0		**21** TH	0438 0.7 / 1011 7.0 / 1706 0.4 / 2242 6.8
7 TH	0454 1.6 / 1022 6.3 / 1712 1.3 / 2242 6.0		**22** F	0523 0.7 / 1056 6.9 / 1749 0.5 / 2324 6.7
8 F	0525 1.6 / 1053 6.2 / 1742 1.4 / 2312 5.9		**23** SA	0605 0.8 / 1137 6.7 / 1830 0.8
9 SA	0555 1.7 / 1125 6.1 / 1813 1.6 / 2344 5.8		**24** SU	0003 6.4 / 0646 1.2 / 1217 6.4 / 1909 1.2
10 SU	0626 1.8 / 1157 5.9 / 1845 1.7		**25** M	0040 6.0 / 0726 1.6 / 1256 5.9 / 1950 1.8
11 M	0016 5.6 / 0701 2.0 / 1231 5.6 / 1921 2.0		**26** TU ◑	0120 5.6 / 0811 2.1 / 1340 5.4 / 2036 2.3
12 TU ◐	0052 5.4 / 0742 2.3 / 1311 5.4 / 2006 2.3		**27** W	0209 5.2 / 0906 2.5 / 1440 4.9 / 2138 2.7
13 W	0137 5.2 / 0835 2.5 / 1405 5.1 / 2105 2.5		**28** TH	0321 4.9 / 1023 2.7 / 1609 4.7 / 2304 2.9
14 TH	0240 5.0 / 0946 2.7 / 1523 4.9 / 2224 2.6			
15 F	0408 5.0 / 1113 2.6 / 1704 5.0 / 2349 2.4			

MARCH

Time	m		Time	m
1 F	0453 4.9 / 1150 2.7 / 1747 4.8		**16** SA	0335 4.9 / 1046 2.5 / 1651 5.0 / 2329 2.6
2 SA	0027 2.7 / 0611 5.1 / 1301 2.4 / 1853 5.1		**17** SU	0520 5.2 / 1214 2.2 / 1810 5.4
3 SU	0129 2.4 / 0705 5.4 / 1355 2.1 / 1938 5.4		**18** M	0047 2.1 / 0629 5.6 / 1324 1.6 / 1910 5.8
4 M	0217 2.1 / 0747 5.7 / 1437 1.8 / 2014 5.7		**19** TU	0151 1.6 / 0726 6.1 / 1423 1.1 / 2003 6.3
5 TU	0256 1.8 / 0824 6.0 / 1514 1.5 / 2048 5.9		**20** W	0246 1.2 / 0819 6.5 / 1515 0.7 / 2052 6.6
6 W ●	0330 1.6 / 0858 6.1 / 1547 1.3 / 2120 6.1		**21** TH ○	0336 0.8 / 0907 6.8 / 1602 0.5 / 2137 6.8
7 TH	0402 1.5 / 0931 6.3 / 1618 1.2 / 2150 6.1		**22** F	0421 0.6 / 0953 6.9 / 1645 0.4 / 2219 6.8
8 F	0432 1.4 / 1001 6.3 / 1648 1.2 / 2219 6.2		**23** SA	0503 0.6 / 1035 6.9 / 1726 0.5 / 2258 6.7
9 SA	0502 1.3 / 1031 6.3 / 1718 1.2 / 2248 6.1		**24** SU	0542 0.8 / 1114 6.6 / 1803 0.9 / 2334 6.4
10 SU	0532 1.4 / 1102 6.2 / 1748 1.3 / 2318 6.0		**25** M	0619 1.1 / 1150 6.3 / 1839 1.3
11 M	0602 1.5 / 1133 6.0 / 1818 1.5 / 2349 5.9		**26** TU	0008 6.0 / 0656 1.5 / 1226 5.8 / 1915 1.9
12 TU	0635 1.8 / 1206 5.8 / 1852 1.8		**27** W	0043 5.6 / 0736 2.0 / 1305 5.3 / 1957 2.4
13 W	0022 5.6 / 0713 2.0 / 1243 5.4 / 1934 2.2		**28** TH ◑	0127 5.2 / 0825 2.5 / 1359 4.8 / 2053 2.8
14 TH ◐	0102 5.3 / 0802 2.3 / 1333 5.1 / 2031 2.5		**29** F	0232 4.8 / 0935 2.8 / 1528 4.6 / 2222 3.0
15 F	0201 5.1 / 0912 2.6 / 1454 4.9 / 2153 2.7		**30** SA	0409 4.7 / 1109 2.8 / 1749 4.6 / 2355 2.9
			31 SU	0538 4.9 / 1226 2.5 / 1827 5.0

APRIL

Time	m		Time	m
1 M	0100 2.6 / 0635 5.2 / 1323 2.2 / 1910 5.3		**16** TU	0031 2.1 / 0611 5.6 / 1305 1.6 / 1851 5.9
2 TU	0149 2.2 / 0718 5.5 / 1406 1.9 / 1945 5.6		**17** W	0133 1.6 / 0707 6.1 / 1403 1.2 / 1942 6.3
3 W	0228 1.9 / 0756 5.8 / 1444 1.6 / 2019 5.9		**18** TH	0227 1.2 / 0758 6.4 / 1453 0.8 / 2029 6.5
4 TH	0303 1.6 / 0832 6.1 / 1517 1.4 / 2052 6.1		**19** F ○	0315 0.9 / 0846 6.7 / 1539 0.7 / 2113 6.7
5 F ●	0335 1.4 / 0905 6.2 / 1549 1.2 / 2123 6.2		**20** SA	0359 0.8 / 0930 6.7 / 1621 0.7 / 2152 6.7
6 SA	0406 1.3 / 0936 6.3 / 1620 1.2 / 2152 6.3		**21** SU	0439 0.8 / 1010 6.6 / 1659 0.8 / 2229 6.6
7 SU	0437 1.2 / 1007 6.3 / 1652 1.2 / 2222 6.3		**22** M	0517 0.9 / 1048 6.4 / 1735 1.1 / 2304 6.3
8 M	0509 1.3 / 1039 6.2 / 1723 1.3 / 2254 6.2		**23** TU	0553 1.2 / 1123 6.1 / 1809 1.5 / 2338 6.0
9 TU	0541 1.4 / 1113 6.1 / 1755 1.5 / 2327 6.0		**24** W	0628 1.5 / 1159 5.7 / 1844 2.0
10 W	0615 1.6 / 1149 5.8 / 1831 1.8		**25** TH	0013 5.6 / 0706 2.0 / 1238 5.3 / 1924 2.4
11 TH	0003 5.8 / 0655 1.9 / 1230 5.5 / 1915 2.2		**26** F ◑	0054 5.2 / 0751 2.4 / 1327 4.9 / 2016 2.8
12 F ◐	0046 5.5 / 0745 2.2 / 1324 5.2 / 2014 2.5		**27** SA	0151 4.9 / 0851 2.7 / 1443 4.6 / 2133 3.1
13 SA	0146 5.1 / 0856 2.4 / 1449 4.9 / 2139 2.7		**28** SU	0315 4.7 / 1015 2.8 / 1623 4.6 / 2306 3.0
14 SU	0321 5.0 / 1030 2.4 / 1640 5.0 / 2314 2.5		**29** M	0443 4.7 / 1136 2.6 / 1739 4.9
15 M	0502 5.2 / 1157 2.1 / 1754 5.4		**30** TU	0015 2.7 / 0549 5.0 / 1236 2.3 / 1827 5.2

Chart Datum is 3·33 metres below IGN Datum. HAT is 7·0 metres above Chart Datum.

N France

STANDARD TIME UT –01
Subtract 1 hour for UT
For French Summer Time add ONE hour in **non-shaded areas**

CHERBOURG   LAT 49°39′N   LONG 1°38′W
TIMES AND HEIGHTS OF HIGH AND LOW WATERS

Dates in red are SPRINGS
Dates in blue are NEAPS

YEAR 2019

MAY

Day	Time	m	Day	Time	m
1 W	0107 / 0637 / 1324 / 1906	2.4 / 5.3 / 2.0 / 5.5	**16** TH	0109 / 0642 / 1337 / 1916	1.7 / 5.9 / 1.4 / 6.1
2 TH	0149 / 0719 / 1404 / 1943	2.0 / 5.6 / 1.8 / 5.8	**17** F	0203 / 0734 / 1428 / 2003	1.4 / 6.2 / 1.1 / 6.3
3 F	0226 / 0757 / 1440 / 2018	1.7 / 5.9 / 1.5 / 6.0	**18** SA	0251 / 0822 / 1513 / ○2046	1.2 / 6.3 / 1.1 / 6.4
4 SA	0301 / 0834 / 1515 / ●2052	1.5 / 6.1 / 1.4 / 6.2	**19** SU	0335 / 0905 / 1554 / 2125	1.2 / 6.4 / 1.1 / 6.4
5 SU	0336 / 0909 / 1550 / 2124	1.3 / 6.2 / 1.3 / 6.3	**20** M	0415 / 0945 / 1632 / 2201	1.1 / 6.3 / 1.2 / 6.4
6 M	0411 / 0943 / 1626 / 2157	1.2 / 6.2 / 1.2 / 6.3	**21** TU	0452 / 1023 / 1708 / 2237	1.2 / 6.1 / 1.4 / 6.2
7 TU	0447 / 1019 / 1701 / 2232	1.2 / 6.2 / 1.3 / 6.3	**22** W	0528 / 1100 / 1743 / 2313	1.4 / 5.9 / 1.7 / 6.0
8 W	0524 / 1057 / 1738 / 2310	1.3 / 6.1 / 1.5 / 6.1	**23** TH	0604 / 1137 / 1819 / 2349	1.6 / 5.7 / 2.1 / 5.7
9 TH	0603 / 1139 / 1819 / 2352	1.5 / 5.9 / 1.8 / 5.9	**24** F	0641 / 1215 / 1858	1.9 / 5.3 / 2.4
10 F	0647 / 1225 / 1908	1.7 / 5.6 / 2.1	**25** SA	0028 / 0722 / 1259 / 1944	5.4 / 2.2 / 5.0 / 2.7
11 SA	0040 / 0742 / 1324 / 2010	5.6 / 2.0 / 5.3 / 2.4	**26** SU	0116 / 0811 / 1356 / ☽2044	5.1 / 2.5 / 4.8 / 2.9
12 SU	0142 / 0851 / 1446 / ☽2129	5.3 / 2.2 / 5.1 / 2.5	**27** M	0219 / 0914 / 1508 / 2200	4.8 / 2.7 / 4.7 / 3.0
13 M	0309 / 1015 / 1618 / 2255	5.2 / 2.2 / 5.0 / 2.4	**28** TU	0332 / 1029 / 1625 / 2314	4.8 / 2.7 / 4.8 / 2.8
14 TU	0437 / 1135 / 1728	5.3 / 2.0 / 5.5	**29** W	0443 / 1136 / 1728	4.9 / 2.5 / 5.1
15 W	0008 / 0545 / 1241 / 1825	2.1 / 5.6 / 1.6 / 5.8	**30** TH	0012 / 0543 / 1231 / 1817	2.6 / 5.1 / 2.3 / 5.4
			31 F	0101 / 0633 / 1317 / 1900	2.2 / 5.4 / 2.0 / 5.7

JUNE

Day	Time	m	Day	Time	m
1 SA	0144 / 0718 / 1400 / 1941	1.9 / 5.7 / 1.7 / 5.9	**16** SU	0227 / 0800 / 1448 / 2021	1.5 / 6.0 / 1.5 / 6.1
2 SU	0225 / 0801 / 1441 / 2019	1.7 / 5.9 / 1.5 / 6.1	**17** M	0312 / 0845 / 1530 / ○2101	1.4 / 6.0 / 1.5 / 6.2
3 M	0306 / 0841 / 1522 / ●2057	1.5 / 6.1 / 1.4 / 6.3	**18** TU	0353 / 0925 / 1609 / 2138	1.4 / 6.0 / 1.6 / 6.2
4 TU	0347 / 0922 / 1603 / 2136	1.3 / 6.2 / 1.3 / 6.3	**19** W	0431 / 1003 / 1646 / 2215	1.4 / 5.9 / 1.7 / 6.1
5 W	0429 / 1002 / 1645 / 2216	1.2 / 6.2 / 1.4 / 6.3	**20** TH	0507 / 1040 / 1722 / 2251	1.5 / 5.8 / 1.8 / 6.0
6 TH	0512 / 1047 / 1728 / 2300	1.2 / 6.1 / 1.5 / 6.3	**21** F	0542 / 1116 / 1757 / 2327	1.6 / 5.7 / 2.0 / 5.8
7 F	0557 / 1133 / 1814 / 2346	1.3 / 6.0 / 1.7 / 6.1	**22** SA	0618 / 1152 / 1833	1.8 / 5.5 / 2.2
8 SA	0645 / 1223 / 1906	1.5 / 5.8 / 1.9	**23** SU	0004 / 0654 / 1230 / 1913	5.6 / 2.0 / 5.3 / 2.5
9 SU	0037 / 0740 / 1320 / 2006	5.9 / 1.7 / 5.5 / 2.1	**24** M	0043 / 0735 / 1312 / 1959	5.6 / 2.3 / 5.1 / 2.7
10 M	0137 / 0842 / 1430 / ☽2114	5.6 / 1.9 / 5.4 / 2.3	**25** TU	0129 / 0822 / 1403 / ☽2054	5.1 / 2.4 / 5.0 / 2.8
11 TU	0249 / 0952 / 1545 / 2228	5.4 / 2.0 / 5.3 / 2.3	**26** W	0225 / 0919 / 1504 / 2159	5.0 / 2.6 / 4.9 / 2.8
12 W	0405 / 1105 / 1654 / 2339	5.4 / 2.0 / 5.5 / 2.1	**27** TH	0328 / 1025 / 1611 / 2308	4.9 / 2.6 / 5.0 / 2.7
13 TH	0513 / 1211 / 1754	5.5 / 1.8 / 5.7	**28** F	0436 / 1131 / 1717	5.0 / 2.5 / 5.2
14 F	0041 / 0614 / 1309 / 1848	1.9 / 5.7 / 1.7 / 5.9	**29** SA	0009 / 0541 / 1230 / 1814	2.5 / 5.2 / 2.2 / 5.5
15 SA	0137 / 0710 / 1401 / 1937	1.7 / 5.8 / 1.6 / 6.0	**30** SU	0103 / 0639 / 1322 / 1904	2.2 / 5.5 / 2.0 / 5.7

JULY

Day	Time	m	Day	Time	m
1 M	0152 / 0730 / 1411 / 1950	1.8 / 5.7 / 1.7 / 6.0	**16** TU	0253 / 0830 / 1512 / ○2043	1.7 / 5.8 / 1.8 / 6.0
2 TU	0240 / 0819 / 1459 / ●2035	1.5 / 6.0 / 1.5 / 6.2	**17** W	0335 / 0910 / 1551 / 2120	1.6 / 5.8 / 1.8 / 6.1
3 W	0328 / 0906 / 1546 / 2120	1.3 / 6.2 / 1.4 / 6.4	**18** TH	0412 / 0946 / 1628 / 2156	1.5 / 5.9 / 1.7 / 6.1
4 TH	0416 / 0953 / 1633 / 2206	1.1 / 6.3 / 1.3 / 6.5	**19** F	0448 / 1020 / 1702 / 2231	1.5 / 5.9 / 1.8 / 6.1
5 F	0504 / 1040 / 1721 / 2252	1.0 / 6.3 / 1.3 / 6.5	**20** SA	0521 / 1054 / 1736 / 2305	1.5 / 5.8 / 1.9 / 6.0
6 SA	0551 / 1128 / 1809 / 2341	1.0 / 6.2 / 1.4 / 6.4	**21** SU	0554 / 1127 / 1808 / 2338	1.6 / 5.7 / 2.0 / 5.8
7 SU	0640 / 1217 / 1859	1.1 / 6.1 / 1.6	**22** M	0626 / 1159 / 1841	1.8 / 5.6 / 2.2
8 M	0030 / 0730 / 1308 / 1952	6.2 / 1.4 / 5.8 / 1.8	**23** TU	0012 / 0700 / 1234 / 1918	5.7 / 2.0 / 5.4 / 2.4
9 TU	0122 / 0823 / 1403 / ☽2050	5.9 / 1.6 / 5.6 / 2.1	**24** W	0048 / 0738 / 1313 / 2001	5.4 / 2.2 / 5.2 / 2.5
10 W	0221 / 0923 / 1506 / 2155	5.6 / 1.9 / 5.4 / 2.2	**25** TH	0132 / 0823 / 1401 / ☽2055	5.2 / 2.4 / 5.1 / 2.7
11 TH	0328 / 1029 / 1614 / 2306	5.4 / 2.1 / 5.3 / 2.2	**26** F	0226 / 0920 / 1501 / 2203	5.0 / 2.6 / 5.0 / 2.8
12 F	0439 / 1138 / 1722	5.3 / 2.2 / 5.4	**27** SA	0334 / 1032 / 1614 / 2319	5.0 / 2.6 / 5.1 / 2.6
13 SA	0013 / 0548 / 1242 / 1823	2.2 / 5.4 / 2.1 / 5.6	**28** SU	0453 / 1146 / 1731	5.0 / 2.5 / 5.3
14 SU	0114 / 0650 / 1339 / 1916	2.0 / 5.5 / 2.0 / 5.8	**29** M	0027 / 0607 / 1251 / 1834	2.3 / 5.3 / 2.2 / 5.6
15 M	0207 / 0744 / 1428 / 2002	1.8 / 5.6 / 1.9 / 5.9	**30** TU	0126 / 0708 / 1349 / 1928	1.9 / 5.7 / 1.9 / 6.0
			31 W	0221 / 0801 / 1442 / 2018	1.6 / 6.0 / 1.6 / 6.3

AUGUST

Day	Time	m	Day	Time	m
1 TH	0314 / 0853 / 1534 / ●2107	1.2 / 6.3 / 1.3 / 6.6	**16** F	0353 / 0927 / 1609 / 2137	1.5 / 6.0 / 1.7 / 6.2
2 F	0405 / 0943 / 1624 / 2156	0.9 / 6.5 / 1.1 / 6.7	**17** SA	0427 / 0959 / 1641 / 2209	1.4 / 6.0 / 1.6 / 6.2
3 SA	0454 / 1031 / 1712 / 2244	0.7 / 6.6 / 1.0 / 6.8	**18** SU	0458 / 1029 / 1711 / 2240	1.4 / 6.0 / 1.7 / 6.2
4 SU	0541 / 1118 / 1758 / 2330	0.7 / 6.5 / 1.1 / 6.7	**19** M	0528 / 1058 / 1741 / 2310	1.5 / 6.0 / 1.8 / 6.1
5 M	0626 / 1202 / 1844	0.9 / 6.4 / 1.3	**20** TU	0557 / 1128 / 1810 / 2341	1.6 / 5.9 / 1.9 / 5.9
6 TU	0015 / 0711 / 1246 / 1930	6.4 / 1.1 / 6.1 / 1.6	**21** W	0627 / 1159 / 1842	1.8 / 5.7 / 2.1
7 W	0100 / 0757 / 1331 / ☽2020	6.1 / 1.6 / 5.8 / 2.0	**22** TH	0013 / 0659 / 1232 / 1919	5.7 / 2.0 / 5.5 / 2.3
8 TH	0149 / 0848 / 1424 / 2119	5.7 / 2.0 / 5.4 / 2.3	**23** F	0050 / 0739 / 1312 / ☽2006	5.4 / 2.3 / 5.3 / 2.6
9 F	0249 / 0950 / 1532 / 2231	5.3 / 2.4 / 5.2 / 2.5	**24** SA	0138 / 0830 / 1407 / 2110	5.3 / 2.6 / 5.1 / 2.8
10 SA	0405 / 1106 / 1650 / 2348	5.1 / 2.6 / 5.1 / 2.5	**25** SU	0246 / 0943 / 1524 / 2236	4.9 / 2.8 / 5.0 / 2.8
11 SU	0528 / 1220 / 1803	5.0 / 2.5 / 5.3	**26** M	0420 / 1113 / 1700	4.9 / 2.7 / 5.1
12 M	0056 / 0639 / 1323 / 1901	2.3 / 5.2 / 2.3 / 5.6	**27** TU	0001 / 0549 / 1230 / 1815	2.5 / 5.2 / 2.4 / 5.5
13 TU	0152 / 0734 / 1414 / 1948	2.0 / 5.5 / 2.1 / 5.8	**28** W	0108 / 0653 / 1333 / 1912	2.0 / 5.7 / 1.9 / 6.0
14 W	0238 / 0818 / 1458 / 2028	1.8 / 5.7 / 1.9 / 6.0	**29** TH	0206 / 0747 / 1429 / 2004	1.5 / 6.1 / 1.5 / 6.4
15 TH	0318 / 0854 / 1535 / ○2103	1.6 / 5.9 / 1.8 / 6.1	**30** F	0300 / 0839 / 1521 / ●2054	1.0 / 6.5 / 1.1 / 6.8
			31 SA	0350 / 0928 / 1610 / 2142	0.7 / 6.7 / 0.9 / 7.0

Chart Datum is 3·33 metres below IGN Datum. HAT is 7·0 metres above Chart Datum.

》》 FREE monthly updates. Register at 《《
www.reedsnauticalalmanac.co.uk

198

CHERBOURG   LAT 49°39′N   LONG 1°38′W

TIMES AND HEIGHTS OF HIGH AND LOW WATERS

STANDARD TIME UT –01
Subtract 1 hour for UT
For French Summer Time add ONE hour in **non-shaded areas**

Dates in red are SPRINGS
Dates in blue are NEAPS

YEAR **2019**

SEPTEMBER

Day	Time m	Day	Time m
1 SU	0438 0.5 / 1014 6.8 / 1656 0.8 / 2229 7.0	**16** M	0430 1.3 / 1001 6.2 / 1645 1.5 / 2213 6.4
2 M	0522 0.5 / 1058 6.8 / 1740 0.9 / 2312 6.9	**17** TU	0459 1.4 / 1029 6.2 / 1714 1.6 / 2242 6.3
3 TU	0604 0.8 / 1139 6.6 / 1822 1.1 / 2353 6.6	**18** W	0528 1.5 / 1058 6.1 / 1742 1.7 / 2313 6.1
4 W	0645 1.1 / 1218 6.2 / 1904 1.5	**19** TH	0556 1.7 / 1127 5.9 / 1813 1.9 / 2344 5.8
5 TH	0034 6.1 / 0727 1.7 / 1257 5.8 / 1949 2.0	**20** F	0628 2.0 / 1159 5.7 / 1847 2.2
6 F	0117 5.6 / 0812 2.2 / 1344 5.4 / 2042 2.4	**21** SA	0019 5.5 / 0705 2.3 / 1236 5.4 / 1932 2.5
7 SA	0212 5.2 / 0911 2.7 / 1450 5.1 / 2156 2.7	**22** SU	0105 5.2 / 0756 2.6 / 1329 5.2 / 2034 2.7
8 SU	0335 4.8 / 1035 2.9 / 1620 5.0 / 2324 2.7	**23** M	0217 4.9 / 0910 2.9 / 1452 5.0 / 2207 2.8
9 M	0515 4.9 / 1202 2.8 / 1745 5.1	**24** TU	0407 4.9 / 1051 2.8 / 1642 5.1 / 2341 2.5
10 TU	0038 2.5 / 0631 5.1 / 1308 2.5 / 1845 5.4	**25** W	0538 5.3 / 1215 2.4 / 1759 5.6
11 W	0134 2.2 / 0720 5.5 / 1358 2.2 / 1930 5.8	**26** TH	0052 1.9 / 0638 5.8 / 1319 1.9 / 1856 6.1
12 TH	0218 1.9 / 0758 5.8 / 1439 1.7 / 2007 6.0	**27** F	0150 1.4 / 0731 6.3 / 1414 1.4 / 1947 6.5
13 F	0256 1.6 / 0831 6.0 / 1514 1.7 / 2042 6.2	**28** SA	0242 0.9 / 0820 6.6 / 1504 1.0 / 2036 6.9
14 SA	0329 1.5 / 0903 6.1 / 1546 1.6 / 2114 6.3	**29** SU	0331 0.6 / 0907 6.9 / 1551 0.8 / 2123 7.1
15 SU	0401 1.4 / 0933 6.2 / 1616 1.5 / 2144 6.4	**30** M	0416 0.5 / 0951 6.9 / 1635 0.7 / 2208 7.1

OCTOBER

Day	Time m	Day	Time m
1 TU	0459 0.6 / 1033 6.9 / 1717 0.8 / 2249 6.9	**16** W	0431 1.4 / 1000 6.3 / 1648 1.5 / 2217 6.3
2 W	0538 0.9 / 1111 6.6 / 1757 1.1 / 2328 6.3	**17** TH	0501 1.5 / 1030 6.3 / 1719 1.6 / 2249 6.2
3 TH	0617 1.3 / 1148 6.3 / 1836 1.5	**18** F	0531 1.7 / 1102 6.1 / 1751 1.8 / 2324 5.9
4 F	0006 6.1 / 0655 1.9 / 1225 5.9 / 1918 2.0	**19** SA	0605 2.0 / 1136 5.9 / 1827 2.1
5 SA	0048 5.6 / 0738 2.4 / 1309 5.4 / 2007 2.5	**20** SU	0003 5.6 / 0645 2.3 / 1217 5.6 / 1913 2.4
6 SU	0141 5.1 / 0834 2.9 / 1412 5.0 / 2117 2.8	**21** M	0053 5.3 / 0738 2.7 / 1313 5.3 / 2018 2.6
7 M	0304 4.8 / 1001 3.1 / 1545 4.9 / 2252 2.9	**22** TU	0208 5.0 / 0855 2.9 / 1437 5.1 / 2149 2.7
8 TU	0454 4.8 / 1136 3.0 / 1717 5.0	**23** W	0359 5.0 / 1034 2.8 / 1625 5.2 / 2322 2.4
9 W	0009 2.6 / 0608 5.1 / 1242 2.7 / 1818 5.3	**24** TH	0521 5.4 / 1157 2.4 / 1739 5.6
10 TH	0104 2.3 / 0652 5.5 / 1331 2.3 / 1902 5.7	**25** F	0032 1.9 / 0619 5.9 / 1300 1.9 / 1836 6.1
11 F	0148 2.0 / 0728 5.8 / 1410 2.0 / 1939 6.0	**26** SA	0129 1.4 / 0710 6.3 / 1354 1.4 / 1927 6.5
12 SA	0225 1.7 / 0801 6.0 / 1445 1.8 / 2014 6.4	**27** SU	0220 1.0 / 0757 6.7 / 1444 1.1 / 2015 6.8
13 SU	0259 1.5 / 0833 6.2 / 1517 1.6 / 2047 6.3	**28** M	0308 0.8 / 0843 6.8 / 1530 0.9 / 2101 6.9
14 M	0330 1.4 / 0904 6.3 / 1547 1.5 / 2118 6.4	**29** TU	0352 0.7 / 0925 6.9 / 1613 0.9 / 2144 6.9
15 TU	0401 1.4 / 0932 6.3 / 1618 1.5 / 2147 6.4	**30** W	0433 0.9 / 1005 6.8 / 1654 1.0 / 2225 6.7
		31 TH	0512 1.1 / 1043 6.6 / 1732 1.2 / 2304 6.4

NOVEMBER

Day	Time m	Day	Time m
1 F	0550 1.5 / 1120 6.3 / 1811 1.6 / 2343 6.0	**16** SA	0514 1.7 / 1045 6.2 / 1738 1.7 / 2313 6.0
2 SA	0627 2.0 / 1157 5.9 / 1851 2.0	**17** SU	0552 1.9 / 1125 6.0 / 1820 1.9 / 2358 5.7
3 SU	0024 5.5 / 0709 2.5 / 1240 5.5 / 1937 2.5	**18** M	0637 2.2 / 1211 5.8 / 1909 2.1
4 M	0115 5.1 / 0801 2.9 / 1338 5.1 / 2036 2.8	**19** TU	0051 5.4 / 0733 2.5 / 1308 5.5 / 2013 2.4
5 TU	0228 4.8 / 0916 3.2 / 1459 4.9 / 2200 2.9	**20** W	0204 5.2 / 0846 2.7 / 1425 5.3 / 2133 2.4
6 W	0402 4.8 / 1050 3.1 / 1626 4.9 / 2321 2.8	**21** TH	0337 5.2 / 1012 2.7 / 1559 5.4 / 2256 2.2
7 TH	0521 5.0 / 1159 2.8 / 1734 5.2	**22** F	0454 5.5 / 1131 2.4 / 1712 5.7
8 F	0021 2.5 / 0610 5.3 / 1251 2.5 / 1822 5.5	**23** SA	0006 1.9 / 0553 5.9 / 1236 2.0 / 1811 6.0
9 SA	0108 2.2 / 0650 5.7 / 1334 2.2 / 1903 5.8	**24** SU	0105 1.5 / 0645 6.2 / 1331 1.6 / 1905 6.3
10 SU	0148 1.9 / 0726 5.9 / 1411 1.9 / 1941 6.0	**25** M	0157 1.3 / 0733 6.5 / 1422 1.3 / 1954 6.5
11 M	0224 1.7 / 0800 6.2 / 1445 1.7 / 2017 6.2	**26** TU	0245 1.1 / 0819 6.6 / 1509 1.1 / 2041 6.6
12 TU	0258 1.5 / 0833 6.3 / 1518 1.5 / 2051 6.3	**27** W	0329 1.1 / 0901 6.7 / 1552 1.1 / 2124 6.6
13 W	0331 1.5 / 0904 6.4 / 1552 1.5 / 2123 6.3	**28** TH	0410 1.2 / 0941 6.6 / 1633 1.2 / 2205 6.4
14 TH	0405 1.5 / 0935 6.4 / 1627 1.5 / 2157 6.3	**29** F	0449 1.4 / 1019 6.5 / 1712 1.3 / 2244 6.2
15 F	0439 1.5 / 1008 6.4 / 1702 1.5 / 2233 6.2	**30** SA	0527 1.7 / 1057 6.3 / 1750 1.6 / 2323 5.9

DECEMBER

Day	Time m	Day	Time m
1 SU	0605 2.0 / 1136 6.0 / 1828 1.9	**16** M	0548 1.7 / 1119 6.2 / 1818 1.6 / 2355 5.9
2 M	0004 5.6 / 0645 2.4 / 1217 5.6 / 1910 2.2	**17** TU	0635 1.9 / 1207 6.0 / 1908 1.7
3 TU	0049 5.3 / 0730 2.7 / 1305 5.3 / 1958 2.5	**18** W	0048 5.7 / 0730 2.2 / 1302 5.8 / 2005 1.9
4 W	0144 5.0 / 0827 3.0 / 1405 5.0 / 2058 2.8	**19** TH	0150 5.5 / 0832 2.4 / 1407 5.6 / 2111 2.1
5 TH	0252 4.9 / 0939 3.1 / 1516 4.9 / 2211 2.8	**20** F	0304 5.4 / 0943 2.4 / 1524 5.5 / 2223 2.1
6 F	0407 4.9 / 1056 3.0 / 1628 5.0 / 2321 2.7	**21** SA	0418 5.5 / 1058 2.3 / 1639 5.5 / 2335 2.0
7 SA	0513 5.1 / 1158 2.7 / 1730 5.2	**22** SU	0523 5.7 / 1208 2.1 / 1745 5.7
8 SU	0017 2.4 / 0603 5.4 / 1249 2.4 / 1821 5.5	**23** M	0038 1.8 / 0620 5.9 / 1309 1.8 / 1843 5.9
9 M	0105 2.2 / 0646 5.7 / 1332 2.1 / 1906 5.7	**24** TU	0135 1.7 / 0712 6.1 / 1403 1.7 / 1937 6.1
10 TU	0146 1.9 / 0726 6.0 / 1412 1.9 / 1947 6.0	**25** W	0225 1.5 / 0800 6.3 / 1452 1.4 / 2026 6.2
11 W	0226 1.7 / 0804 6.2 / 1451 1.7 / 2026 6.1	**26** TH	0311 1.5 / 0843 6.4 / 1536 1.3 / 2110 6.2
12 TH	0305 1.6 / 0840 6.3 / 1530 1.5 / 2105 6.2	**27** F	0353 1.5 / 0923 6.4 / 1617 1.3 / 2150 6.2
13 F	0344 1.5 / 0916 6.4 / 1610 1.4 / 2143 6.3	**28** SA	0432 1.6 / 1001 6.4 / 1655 1.4 / 2228 6.1
14 SA	0424 1.5 / 0954 6.4 / 1651 1.4 / 2224 6.2	**29** SU	0510 1.7 / 1039 6.3 / 1732 1.5 / 2305 5.9
15 SU	0505 1.6 / 1035 6.4 / 1733 1.4 / 2308 6.1	**30** M	0546 1.9 / 1117 6.1 / 1808 1.7 / 2343 5.7
		31 TU	0623 2.1 / 1154 5.8 / 1844 1.9

Chart Datum is 3·33 metres below IGN Datum. HAT is 7·0 metres above Chart Datum.

〉〉 FREE monthly updates. Register at 〈〈
www.reedsnauticalalmanac.co.uk

N France

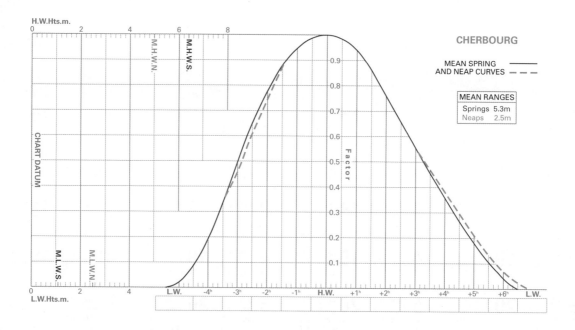

CHERBOURG

MEAN SPRING —
AND NEAP CURVES ----

| MEAN RANGES |
| Springs 5.3m |
| Neaps 2.5m |

4.29 OMONVILLE-LA-ROGUE

Manche **49°42′·29N 01°49′·84W** ✿✿⚓⚓✿✿

CHARTS AC 2669, 1114; SHOM 7120, 7158, 5636; Navi 528, 1014; Imray C33A

TIDES −0330 Dover; ML 3·8; Duration 0545

Standard Port CHERBOURG (←)

Times				Height (metres)			
High Water		Low Water		MHWS	MHWN	MLWN	MLWS
0300	1000	0400	1000	6·4	5·1	2·6	1·1
1500	2200	1600	2200				
Differences OMONVILLE							
−0015	−0010	−0015	−0015	−0·3	−0·2	−0·2	−0·1
GOURY							
−0100	−0040	−0105	−0120	+1·7	+1·6	+1·0	+0·3

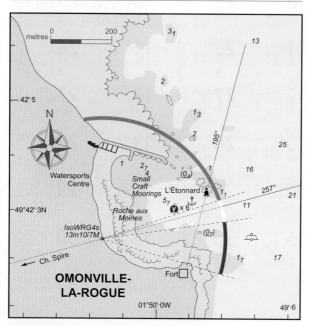

OMONVILLE-
LA-ROGUE

SHELTER Good, except in moderate/fresh N to SE winds when it may become untenable. Swell tends to work in.

NAVIGATION WPT 49°42′·47N 01°48′·66W, 257°/1·0M to Omonville dir lt. From W or N, keep clear of Basse Bréfort (depth 1m, marked by NCM buoy, VQ) 0·6M N of Pte Jardeheu. Appr on 195° transit (below) passing 100m E of L'Étonnard and into W sector of lt before turning stbd 290° for ⚓s.

From E, appr on 257° transit with church spire and in W sector of lt, until S of L'Étonnard.

Harbour entrance is 100m wide: rocks extend N from Omonville Fort, and ESE from bkwtr to L'Étonnard, G bcn twr.

LIGHTS AND MARKS Omonville lt, on W framework tr with R top, Iso WRG 4s 13m 10/7M, vis 180°-G-252°-W-262°-R-287°. Lt in transit 257° with ✠ steeple (hard to see) 650m beyond, leads S of L'Étonnard, 200m SE of which is a large unlit mooring buoy. From N, L'Étonnard in transit 195° with fort just clears Les Tataquets rocks. Street lights adequately illuminate the hbr area.

COMMUNICATIONS CROSS 02·33·52·78·23; ⚓ 02·33·10·16·10; Auto 08·92·68·02·50; Police (Beaumont Hague) 02·33·01·89·30 or 17; Brit Consul 01·44·51·31·02; Samu (emergency medical) 15; Dr 02·33·52·73·07/02·33·01·87·00; Fire (pompiers, inc. divers) 18; Aff Mar 02·50·79·15·00.

FACILITIES **Harbour** There are 6 large W conical ⚓s or ⚓ inside bkwtr; beware rocks off outer end. ↘⚓.

Village ✗◻🛒▷ at l'Association du Camping in village centre and at Watersports Centre; 📷📷🔋 (at Beaumont-Hague 5km) ✉Ⓑ ⇌ (bus to Cherbourg) ✈. Free historical tour www. visiteomonvillelarogue.fr **Ferry** see Cherbourg.

MINOR HARBOUR 2M EAST OF CAP DE LA HAGUE

ANSE DE ST MARTIN, Manche, **49°42′·72N 01°53′·78W**. AC 1114, 3653; SHOM 7120, 5636. Tides as Omonville-la-Rogue. From N, appr with conspic chy (279m) at Jobourg nuclear plant brg 175°; or from NE via Basse Bréfort NCM buoy, VQ, on with Danneville spire brg 237°. Both lines clear La Parmentière rks awash in centre of bay and Les Herbeuses and Le Grun Rk (1·9m) to W and SE respectively. Care needed, especially in poor visibility.

This bay, 2M E of Cap de la Hague, has ⚓s sheltered from all but onshore winds. Port Racine (said to be the smallest hbr in France) is in the SW corner of Anse de St Martin. ⚓ or moor off the hbr which is obstructed by lines; landing by dinghy.

PORT CHANTEREYNE
AN IRRESISTIBLE PORT OF CALL

**For information and news
have a look at the Port Chantereyne
website on www.portchantereyne.fr
Port Chantereyne - Cherbourg**

Tel: +33 (0)2 33 87 65 70
portchantereyne@cherbourg.fr
VHF Channel 9

Located in the biggest artificial harbour of Europe and accessible 24 hours, Port Chantereyne offers you a pleasant and warm welcome in its modern and functional facilities. Everything has been thought through to offer you a pleasant and efficient stay. Labelled Pavillon Bleu since 2002, Port Chantereyne was recently awarded the prestigious Gold Anchors label by winning the 5 anchors award, the highest level attainable.

In the heart of an exceptional nautical basin, close to the city center and its many shops and restaurants, Port Chantereyne offers many opportunities for sailing, as well as activities ashore. Enjoy your stopover to discover Cherbourg-en-Cotentin: visit the oceanographic park of La Cité de la Mer, the remarkable gardens including the Emmanuel-Liais Park, the Ravalet castle, the museums, the lively markets and stroll on the new Place de Gaulle. Also take advantage of the many activities scheduled throughout the year in and around the harbour and the city center.

PORT CHANTEREYNE
CHERBOURG

CHERBOURG
en Cotentin

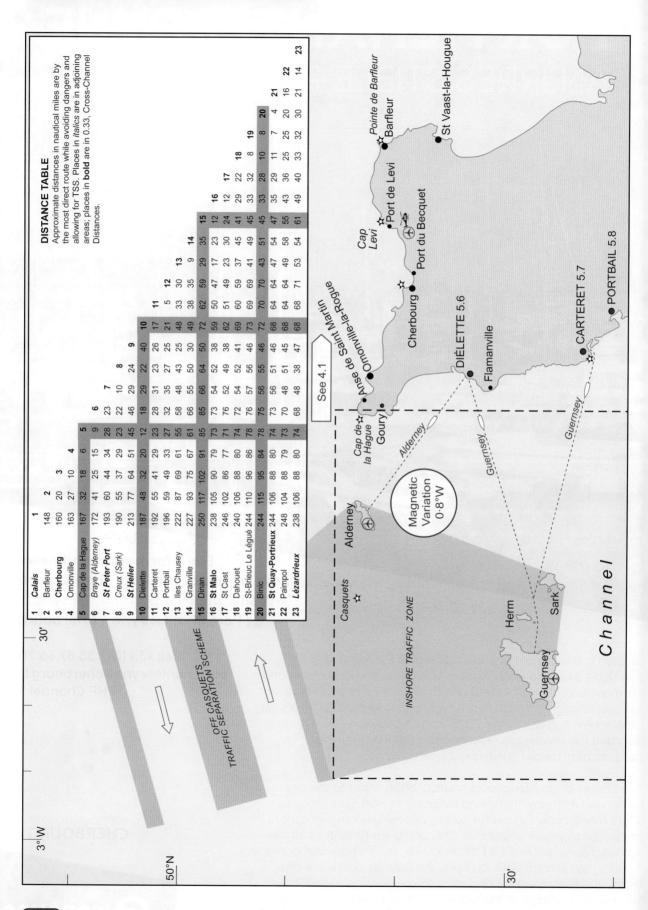

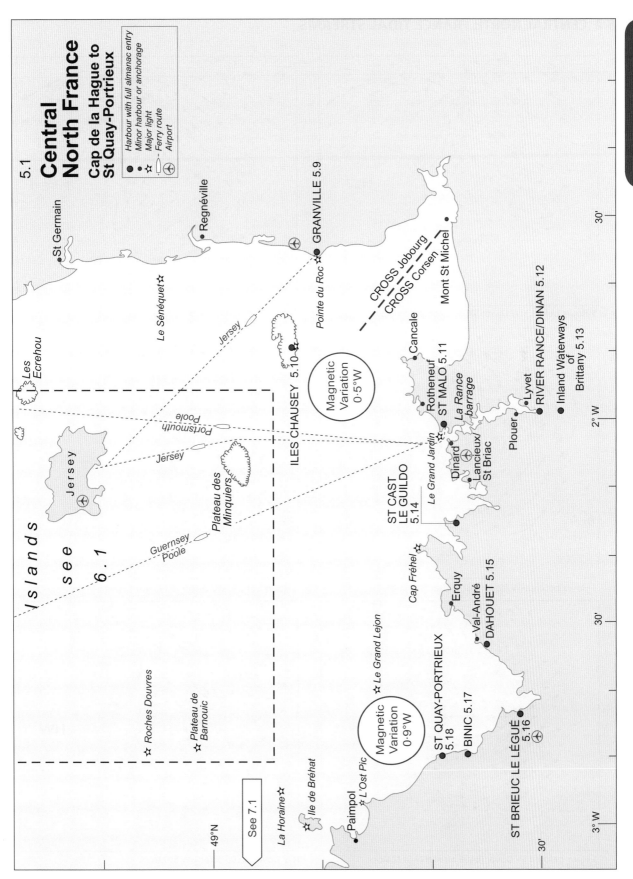

5.1

Central North France

Cap de la Hague to St Quay-Portrieux

● Harbour with full almanac entry
● · ★ Minor harbour or anchorage
★ Major light
✈ Airport

St Germain

Regnéville

GRANVILLE 5.9

Le Sénéquet ★

Les Écrehou

Jersey

Pointe du Roc

CROSS Jobourg
CROSS Corsen

Magnetic Variation 0·5°W

Cancale

Mont St Michel

ILES CHAUSEY 5.10

Rotheneuf
ST MALO 5.11
La Rance barrage

Portsmouth Poole

Jersey

Islands see 6.1

Jersey

Plateau des Minquiers

Dinard
Lancieux
St Briac

Lyvet
RIVER RANCE/DINAN 5.12

Inland Waterways of Brittany 5.13

Plouer

Le Grand Jardin

Guernsey Poole

Roches Douvres ★

Plateau de Barnouic ★

ST CAST LE GUILDO 5.14

Le Grand Léjon ★

Cap Fréhel ★

Erquy

Val-André
DAHOUET 5.15

Magnetic Variation 0·9°W

See 7.1

La Horaine ★

Ile de Bréhat ★

L'Ost Pic ★

Paimpol

ST QUAY-PORTRIEUX 5.18

BINIC 5.17

ST BRIEUC LE LÉGUÉ 5.16

49°N

30'

30'

30'

3°W

2°W

5.2 CENTRAL NORTH FRANCE TIDAL STREAMS

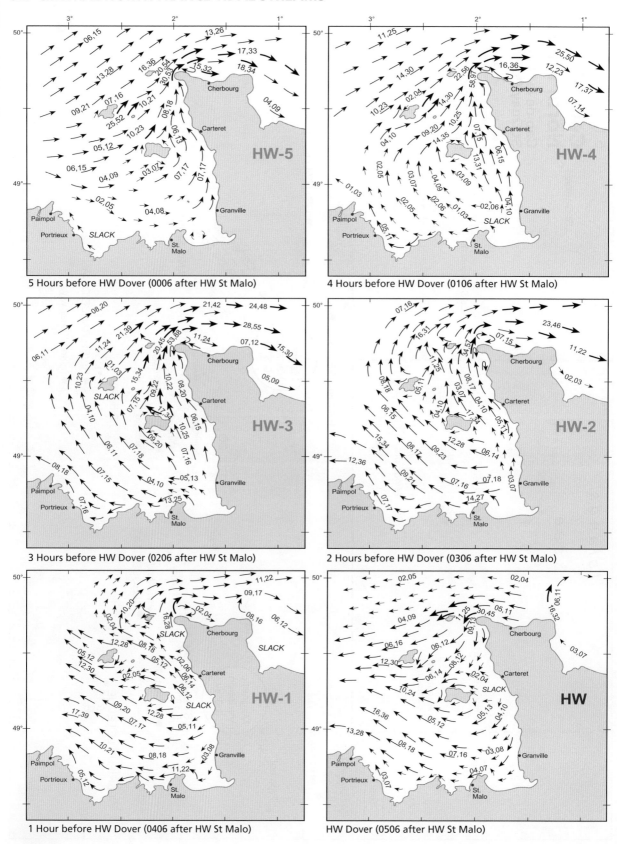

5 Hours before HW Dover (0006 after HW St Malo)

4 Hours before HW Dover (0106 after HW St Malo)

3 Hours before HW Dover (0206 after HW St Malo)

2 Hours before HW Dover (0306 after HW St Malo)

1 Hour before HW Dover (0406 after HW St Malo)

HW Dover (0506 after HW St Malo)

Channel Islands 6.2   English Channel 0.26

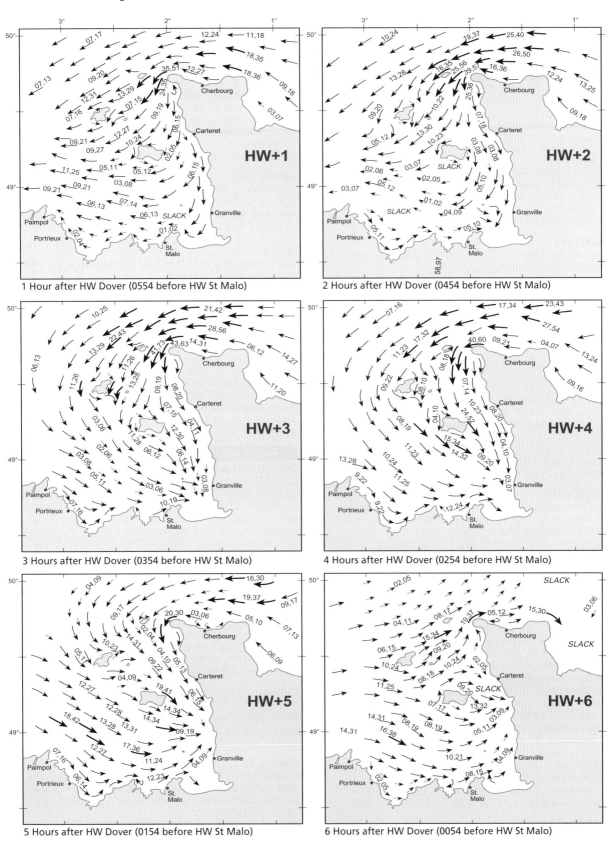

1 Hour after HW Dover (0554 before HW St Malo)

2 Hours after HW Dover (0454 before HW St Malo)

3 Hours after HW Dover (0354 before HW St Malo)

4 Hours after HW Dover (0254 before HW St Malo)

5 Hours after HW Dover (0154 before HW St Malo)

6 Hours after HW Dover (0054 before HW St Malo)

5.3 LIGHTS, BUOYS AND WAYPOINTS

Bold print = light with a nominal range of 15M or more. CAPITALS = place or feature. *CAPITAL ITALICS* = light-vessel, light float or Lanby. *Italics* = Fog signal. ***Bold italics*** = Racon. Many marks/buoys are fitted with AIS (<u>MMSI No</u>); see relevant charts.

CAP DE LA HAGUE TO ST MALO

GOURY

La Foraine ♠ VQ (9) 10s 12m 6M; 49°42'·90 N 01°58'·32W.
Ldg lts 065·2°: Front, QR 5m 7M; R □ in W □ on pier, 49°42'·89N 01°56'·70W. Rear, 116m from front, Q 11m 7M; W pylon/hut. Hervieu ⚓ 49°42'·77N 01°56'·92W.

DIELETTE

W bkwtr Dir lt 140°, Iso WRG 4s 12m W10M, R/G7M; 070°-G-135°-W-145°-R-180°; W twr, G top ;49°33'·18N 01°51'·81W.
E bkwtr ⚡ Fl R 4s 6m 2M; 49°33'·21N 01°51'·78W.
Inner N jetty, Fl (2) R 6s. Inner S jetty, Fl (2) G 6s; bth 6m 1M.
Banc des Dious ♠ Q (9) 15s; 49°32'·58N 01°54'·02W.

CARTERET

Cap de Carteret ☆ Fl (2+1) 15s 81m **26M**; Gy twr, G top; 49°22'·41N 01°48'·41W. <u>992271219</u> .
W bkwtr ⚡ Oc R 4s 7m 7M; W post, R top; 49°22'·07N 01°47'·32W.
E training wall ⚡ Fl G 2·5s 4m 2M; W post, G top; 49°22'·17N 01°47'·30W.
Channel bend ⚡ Fl (2) R 6s 5m 1M; R pylon; 49°22'·58N 01°47'·23W. Inside the bend: ⚡ Fl (2) G 6s 5m 1M; G pylon; 49°22'·55N 01°47'·20W.
Marina entry sill: ⚡ Fl (3) R 12s and Fl (3) G 12s; R & G pylons.

PORTBAIL

PB ⚓ 49°18'·37N 01°44'·75W.
Ldg lts 042°: Front, Q 14m 10M; W pylon, R top, 49°19'·75N 01°42'·50W. Rear, 870m frm front, Oc 4s 20m 10M; ch belfry.
⚓ 49°19'·32N 01°43'·16W.
▲ 49°19'·20N 01°43'·00W.
Training wall head ⚓ Fl (2) R 5s 5m 4M; W mast, R top; 49°19'·43N 01°42'·99W.

REGNÉVILLE

La Catheue ♠ Q (6) + L Fl 15s, 48°57'·67N 01°42'·23W.
Le Ronquet ⚓ Fl (2) WR 6s, W6M, R4M; 100°-R-293°-W-100°; 49°00'·11N 01°38'·07W.
Pte d'Agon ⚡ Oc (2) WR 6s 12m, W10M, R7M; 063°-R-110°-W-063°; W twr, R top, W dwelling; 49°00'·18N 01°34'·63W.
Dir lt 028°, Oc WRG 4s 9m, W12M, R/G9M; 024°-G-027°-W-029°-R-033°; house; 49°00'·63N 01°33'·36W.

PASSAGE DE LA DÉROUTE

Les Trois-Grunes ♠ Q (9) 15s, 49°21'·84N 01°55'·21W.
Écrevière ♠ Q (6) + L Fl 15s, *Bell;* 49°15'·27N 01°52'·16W.
Basse Jourdan ♠ Q (3) 10s; 49°06'·85N 01°43'·96W.
Le Boeuf ⚓ 49°06'·56N 01°47'·17W.
Les Boeuftins ⚓ 49°07'·03N 01°45'·96W.
La Basse du Sénéquet ⚓ 49°05'·96N 01°41'·11W.
Le Sénéquet ⚡ Fl (3) WR 12s 18m W13M, R10M; 083·5°-R-116·5°-W-083·5°; W twr; 49°05'·48N 01°39'·73W.
Les Nattes ⚓ 49°03'·46N 01°41'·81W.
International F ⚓ 49°02'·16N 01°42'·98W.
International E ⚓ 49°02'·08N 01°47'·21W.
Basse le Marié ♠ Q 15s; 49°01'·79N 01°48'·57W.
NE Minquiers ♠ VQ (3) 5s; *Bell;* 49°00'·85N 01°55'·30W.
Les Ardentes ♠ Q (3) 10s; 48°57'·89N 01°51'·53W.
SE Minquiers ♠ Q (3) 10s; *Bell;* 48°53'·42N 02°00'·09W.
S Minquiers ♠ Q (6) + L Fl 15s; 48°53'·09N 02°10'·10W.

ÎLES CHAUSEY

La Pointue ⚓ 48°54'·44N 01°50'·92W.

L'Enseigne, W twr, B top; 48°53'·67N 01°50'·37W.
L'Etat, BW ⚓, 48°54'·67N 01°46'·21W.
Anvers wreck ⚓ 48°53'·91N 01°41'·07W.
Le Founet ♠ Q (3) 10s; 48°53'·25N 01°42'·34W.
Le Pignon ⚡ Fl (2) WR 6s 10m, W9M, R6M; 005°-R-150°-W-005°; B twr, W band; 48°53'·49N 01°43'·36 W.
La Haute Foraine ⚓ 48°52'·89N 01°43'·66W.
Grande Île ☆ Fl 5s 39m **23M**; Gy □ twr, G top; *Horn 30s;* 48°52'·17N 01°49'·34W.
Channel ♠ Fl G 2s; 48°52'·07N 01°49'·08W.
La Crabière Est ⚓ Dir Oc(3)WRG 12s 5m, W9M, R/G6M;079°-W-291°-G-329°-W-335°-R-079°; B beacon, Y top; 48°52'·46N 01°49'·39W.
La Cancalaise ⚓ 48°51'·90N 01°51'·11W.

GRANVILLE

Le Videcoq ♠ VQ (9) 10s; 48°49'·66N 01°42'·06W.
La Fourchie ⚓ 48°50'·15N 01°37'·00W.
Pointe du Roc ☆ Fl (4) 15s 49m **23M**; 48°50'·06N 01°36'·78W.
Le Loup ♠ Fl (2) 6s 8m 11M; 48°49'·57N 01°36'·24W.
Avant Port, E jetty ⚡ Fl G 2·5s 11m 4M; 48°49'·93N 01°36'·19W.
W jetty ⚡ Fl R 2·5s 12m 4M; 48°49'·86N 01°36'·23W.
Marina S bkwtr ⚡ Fl (2) R 6s 12m 5M; W post, R top; *Horn (2) 40s;* 48°49'·89N 01°35'·90W.
N bkwtr ⚡ Fl (2) G 6s 4m 5M; 48°49'·93N 01°35'·90W.
Sill, W side, QR 2M, E side QG 2M; Gy pylons topped R (W) and G (E).

CANCALE

La Fille ⚓ 48°44'·16N 01°48'·46W.
Pierre-de-Herpin ☆ Oc (2) 6s 20m 13M; W twr, B top and base; 48°43'·77N 01°48'·92W.
Ruet ⚓ *Bell;* 48°43'·44N 01°50'·11W.
Grande Bunouze ⚓ 48°43'·17N 01°50'·95W.
Barbe Brûlée ⚓ 48°42'·11N 01°50'·57W.
Jetty hd ⚡ Oc (3) G 12s 9m 3M; W pylon B top; 48°40'·10N 01°51'·11W.

ST MALO AND RIVER RANCE

CHENAL DE LA BIGNE

Basse Rochefort (aka Basse aux Chiens) ⚓ 48°42'·69N 01°57'·31W.
La Petite Bigne ⚓ 48°41'·66N 01°58'·72W.
La Crolante ⚓; 48°41'·01N 01°59'·51W, off Pte de la Varde.
Les Létruns ♠; *Bell;* 48°40'·71N 02°00'·61W.
Roches-aux-Anglais ♠ Fl G 4s; 48°39'·65N 02°02'·27W.
Les Crapauds-du-Bey ⚓ Fl R 4s; 48°39'·37N 02°02'·57W.

CHENAL DES PETITS POINTUS

Dinard ch spire (74m) brg 203° to right of Le Petit Bé rocks.
La Saint-Servantine ♠ Fl G 2·5s; *Bell;* 48°41'·93N 02°00'·94W.
Les Petits Pontus ⚓ 48°41'·30N 02°00'·86W.

CHENAL DE LA GRANDE CONCHÉE

Villa Brisemoulin 181·5°, on with LH edge of Le Petit Bé rocks.
La Plate ♠ Q WRG 11m, W10M, R/G7M; 140°-W-203°-R-210°-W-225°-G-140°; 48°40'·78N 02°01'·91W.
Le Bouton ⚓ 48°40'·59N 02°01'·85W.

CHENAL DU BUNEL

St Énogat ldg lts 158·2°; both Iso 4s 3/85m 7/10M, sync. Front 48°38'·29N 02°04'·11W, vis 126°-236°. Rear, on white water twr, Dinard, vis 143°-210°; 48°36'·97N 02°03'·31W.
Bunel ♠ Q (9) 15s; 48°40'·84N 02°05'·38W.

CHENAL DE LA PETITE PORTE

Outer ldg lts 129·7°: **Front, Le Grand Jardin** ☆ Fl (2) R 10s 24m **15M**, 48°40'·20N 02°04'·97W. <u>992271314</u>. Rear, **La Balue** ☆ FG 20m **22M**; 3·1M from front; intens 128°-129·5°; Gy □ twr; 48°38'·16N 02°01'·30W.
Vieux-Banc E ♠ Q; 48°42'·38N 02°09'·12W.
Vieux-Banc W ♠ VQ (9) 10s; 48°41'·84N 02°10'·20W.

St Malo Atterrisage (Fairway) ⚓ Iso 4s; 48°41'·39N 02°07'·28W.
Les Courtis ⚓ Fl G 4s 14m 7M; 48°40'·46N 02°05'·80W.

Nearing Le Grand Jardin, move stbd briefly onto Ch du Bunel 152·8° ldg lts to pick up: Inner ldg lts 128·6°, both Dir FG 20/69m **22/25M**; H24. Front, **Les Bas Sablons** ☆, intens 127·2°-130·2°; W □ twr, B top; 48°38'·16N 02°01'·30W. 992271312. Rear, **La Balue** ☆, 0·9M from front; intens 128°-129·5°; Gy □ twr; 48°37'·60N 02°00'·24W. 992271313.
Basse du Nord No 5 ⚓ 48°39'·98N 02°05'·04W.
Les Pierres-Garnier No 8 ⚓ 48°39'·98N 02°04'·41W.
Les Patouillets ⚓ Fl (3) G 12s, 48°39'·68N 02°04'·30W.
Clef d'Aval No 10 ⚓ 48°39'·72N 02°03'·91W.
Basse du Buron No 12 ⚓ Fl (4) R 15s, 48°39'·42N 02°03'·51W.
Le Buron ⚓ Fl (4) G 15s 15m 7M; G twr, 48°39'·32N 02°03'·66W.
Les Grelots ⚓ VQ (6) + L Fl 10s, 48°39'·16N 02°03'·03W.

CHENAL DE LA GRANDE PORTE
Banchenou ⚓ VQ, 48°40'·44N 02°11'·48W.
Outer ldg lts 089·1°: **Front, Le Grand Jardin** ☆ Fl (2) R 10s 24m **15M**, 48°40'·20N 02°04'·97W.
Rear, **Rochebonne** ☆ Dir FR 40m **24M**; intens 088·2°-089·7°; Gy □ twr, R top, 4·2M from front; 48°40'·26N 01°58'·71W. 992271311.
Buharats W No 2 ⚓ Fl R 2·5s, 48°40'·22N 02°07'·50W.
Buharats E No 4 ⚓ *Bell*; 48°40'·24N 02°07'·20W.
Bas du Boujaron No 1 ⚓ Fl (2) G 6s, 48°40'·17N 02°05'·97W.
Le Sou ⚓ VQ (3) 5s; 48°40'·11N 02°05'·30W.
Continue on inner 128·6° ldg line: see Chenal de la Petite Porte.

RADE DE ST MALO
Plateau Rance Nord ⚓ VQ, 48°38'·64N 02°02'·35W.
Plateau Rance Sud ⚓ Q (6) + L Fl 15s, 48°38'·43N 02°02'·28W.
Crapaud de la Cité ⚓ QG, 48°38'·34N 02°02'·01W.

ST MALO
Môle des Noires hd ⚓ VQ R 11m 6M; W twr, R top; *Horn (2) 20s*; 48°38'·52N 02°01'·91W.
Écluse du Naye ldg lts 070·4°, both FR 7/23m 3/7M. Front, 48°38'·58N 02°01'·48W. Rear, 030°-120°.
Ferry jetty hd, ⚓ VQ G 6M; 48°38'·44N 02°01'·82W (128·6° ldg line). Ferry jetty, ⚓ Fl R 4s 3m 1M; 260°-080°; 48°38'·44N 02°01'·76W.
Bas-Sablons marina, mole head ⚓ Fl G 4s 7m 5M; Gy mast; 48°38'·42N 02°01'·70W.

LA RANCE BARRAGE
La Jument ⚓ Fl G 4s 6m 4M; G twr, 48°37'·44N 02°01'·76W.
ZI 12 ⚓ Fl R 4s, 48°37'·47N 02°01'·62W.
NE dolphin ⚓ Fl (2) R 6s 6m 5M; 040°-200°, 48°37'·09N 02°01'·71W.
Barrage lock, NW wall ⚓ Fl (2) G 6s 6m 5M,191°-291°; G pylon, 48°37'·06N 02°01'·73W.
Barrage lock, SW wall, ⚓ Fl (3) G 12s, 48°37'·00N 02°01'·70W.
SE dolphin ⚓ Fl (3) R 12s, 48°36'·97N 02°01'·66W.
ZI 24 ⚓ Fl (2) R 6s, 48°36'·63N 02°01'·33W.

ST MALO TO ST QUAY-PORTRIEUX

ST BRIAC
R. Frémur mouth. Dir lt 125° Iso WRG 4s 10m, W13M, R/G 11M; 121·5°-G-124·5°-W-125·5°-R-129·5°; W mast on hut, 48°37'·07N 02°08'·20W.

ST CAST
Les Bourdinots ⚓ 48°39'·01N 02°13'·48W.
St Cast môle ⚓ Iso WG 4s 12m, W9M, G6M; 180°-G-206°-W-217°-G-235°-W-245°-G-340°; G&W structure; 48°38'·41N 02°14'·61W.
Laplace ⚓, 48°39'·73N 2°16'·45W; 5ca SE of Pte de la Latte.
Cap Fréhel ☆ Fl (2) 10s 85m **29M**; Gy □ twr, G lantern; 48°41'·05N 02°19'·13W. Reserve lt range **15M**.

CHENAL and PORT D'ERQUY
Les Justières ⚓ Q (6) + L Fl 15s; 48°40'·56N 02°26'·48W.
Basses du Courant ⚓ VQ (6) + L Fl 10s; 48°39'·21N 02°29'·16W.
L'Evette ⚓ 48°38'·51N 02°31'·45W.
S môle ⚓ Fl (2) WRG 6s 11m W10M, R/G7M; 055°-R-081°-W-094°-G-111°-W-120°-R-134°; W twr; 48°38'·06N 02°28'·68W.
Inner jetty ⚓ Fl (3) R 12s 10m 3M; R/W twr; 48°38'·09N 02°28'·39W.

DAHOUET
Petit Bignon ⚓ 48°36'·82N 02°35'·06W.
Le Dahouet ⚓ 48°35'·15N 02°35'·43W.
La Petite Muette ⚓ Fl WRG 4s 10m W9M, R/G6M; 055°-G-114°-W-146°-R-196°; W twr, G band; 48°34'·82N 02°34'·30W.
Entry chan, Fl (2) G 6s 5m 1M; 156°-286°; 48°34'·71N 02°34'·19W.

BAIE DE ST BRIEUC
Grand Léjon ☆ Fl (5) WR 20s 17m **W18M**, R14M; 015°-R-058°-W-283°-R-350°-W-015°; R twr, W bands; 48°44'·91N 02°39'·87W.
Petit Léjon ⚓ ; 48°41'·80N 02°37'·55W.
Les Landas ⚓ Q, 48°41'·43N 02°31'·29W.
Le Rohein ⚓ Q (9) WRG 15s 13m, W8M, R/G5M; 072°-R-105°-W-180°-G-193°-W-237°-G-282°-W-301°-G-330°-G-072°; Y twr; B band; 48°38'·80N 02°37'·77W.

SAINT-BRIEUC LE LÉGUÉ
Tra-Hillion ⚓ 48°33'·38N 02°38'·50W.
Le Légué ⚓ Mo (A)10s; 48°34'·32N 02°41'·15W.
No 1 ⚓ Fl G 2·5s; 48°32'·42N 02°42'·51W.
No 2 ⚓ Fl R 2·5s; 48°32'·37N 02°42'·40W.
No 3 ⚓ Fl (2) G 6s; 48°32'·27N 02°42'·78W.
No 4 ⚓ Fl (2) R 6s; 48°32'·23N 02°42'·70W.
No 5 ⚓ Fl (3) G 12s; 48°32'·18N 02°42'·92W.
No 6 ⚓ Fl (3) R 12s; 48°32'·14N 02°42'·90W.
NE jetty ⚓ VQ R 4M; 48°32'·12N 02°42'·88W.
Pte à l'Aigle jetty ⚓ VQ G 13m 8M; 160°-070°; W twr, G top; 48°32'·12N 02°43'·11W.
No 7 ⚓ Fl (4) G 15s; 48°32'·11N 02°43'·08W.
No 8 ⚓ Fl R 2·5s; 48°32'·01N 02°43'·16W.
No 9 ⚓ Fl (2) G 6s; 48°31'·96N 02°43'·29W.
No 10 ⚓ Fl (2) R 6s; 48°31'·91N 02°43'·31W.
Jetty ⚓ Iso G 4s 6m 2M; W cols, G top; 48°31'·90N 02°43'·43W.
No 11 ⚓ Fl (3) G 12s; 48°31'·90N 02°43'·43W.
No 13 ⚓ Fl (4) G 15s; 48°31'·76N 02°43'·56W.
No 14 ⚓ Fl (4) R 15s; 48°31'·70N 02°43'·61W.

BINIC
N Môle hd ⚓ Fl (4) G 15s12m 6M; unintens 020°-110°; W twr, G lantern; 48°36'·07N 02°48'·92W.

ST QUAY-PORTRIEUX
Les Hors ⚓ 48°39'·60N 02°44'·04W.
Caffa ⚓ Q (3) 10s; 48°37'·82N 02°43'·08W.
La Longue ⚓ 48°37'·88N 02°44'·68W.
La Roselière ⚓ VQ (6) + L Fl 10s; 48°37'·31N 02°46'·19W.
Herflux ⚓ Dir ⚓ 130°, Fl (2) WRG 6s 10m, W 8M, R/G 5M; 115°-G-125°-W-135°-R-145°; 48°39'·07N 02°47'·95W.
Île Harbour (Roches de Saint-Quay) ⚓ Fl WRG 4s 16m, W9M, R/G 6M; 011°-R-133°-G-270°-R-306°-G-358°-W-011°; W twr & dwelling, R top; 48°39'·99N 02°48'·49W.
Madeux ⚓ 48°40'·41N 02°48'·81W.
Grandes Moulières de St Quay ⚓ 48°39'·76N 02°49'·91W.
Moulières de Portrieux ⚓ 48°39'·26N 02°49'·21W.
Les Noirs ⚓ 48°39'·09N 02°48·46W.
Marina, **NE mole elbow**, Dir lt 318·2°: Iso WRG 4s 16m **W15M**, R/G11M; W159°-179°, G179°-316°, W316-320·5°, R320·5°-159°; Reserve lt ranges 11/8M; 48°38'·99N 02°49'·09W.
NE môle hd ⚓ Fl (3) G 12s 10m 2M; 48°38'·84N 02°48'·91W.
S môle hd ⚓ Fl (3) R 12s 10m 2M; 48°38'·83N 02°49'·03W.
Old hbr ent: N side, Fl G 2.5s 11m 2M; 48°38'·71N 02°49'·36W.
S side, Fl R 2.5s 8m 2M; 48°38'·67N 02°49'·35W.

5.4 PASSAGE INFORMATION

More passage information is threaded between successive harbours in this Area. **Bibliography**: *Channel Islands, Cherbourg Peninsula and North Brittany* (Imray/RCC/Carnegie); *Shell Channel Pilot* (Imray/Cunliffe); *North Brittany and Channel Islands CC* (Fernhurst/Cumberlidge); *Channel Pilot* (NP 27); *Channel Havens* (ACN/Endean).

THE WEST COAST OF THE COTENTIN PENINSULA

This coast is exposed and often a lee shore (AC 3653, 3655, 3656, 3659). North of Carteret it is mostly rocky. Southward to Mont St Michel and W to St Malo the coast changes to extensive offshore shoals, sand dunes studded with rocks and a series of drying hbrs; there is little depth of water, so that a nasty sea can build.

▶*The seas around this coast are dominated by powerful tidal streams rotating anti-clockwise; and a very large tidal range. In the Alderney Race the main English Channel tidal streams are rectilinear NE/SW. Tidal streams need to be worked carefully; neaps are best, particularly for a first visit. The Admiralty tidal stream atlas NP 264 covers the French coast and Channel Islands. The equivalent SHOM 562-UJA gives more details.*◀

CAP DE LA HAGUE TO MONT ST MICHEL

Cap de La Hague is low-lying, but rises steeply to the south. Between the coast and Gros du Raz lt ho a difficult and narrow passage leads S to Goury. It is sheltered when the Alderney Race is fully exposed, but not for the inexperienced. Pylons and Jobourg nuclear plant are conspic 3·7M SE. CROSS Jobourg is adjacent.

5M S of Cap de la Hague beware Les Huquets de Jobourg, an extensive unmarked bank of drying (2·1m) and submerged rks, and Les Huquets de Vauville (5·4m) close SE of them.

The drying hbrs at **Goury, Carteret and Portbail** are more readily accessible if cruising from S to N on the flood. The marinas at **Diélette** and **Carteret** are non-tidal. The former has H24 access for 1·5m draught if coefficient is <80.

Déroute de Terre and Passage de la Déroute are coastal chans, poorly marked, shallow in places and not advised at night.

Déroute de Terre (not shown on AC 2669 and 3655) passes W of Plateau des Trois Grunes; Basses de Portbail; La Basse du Sénéquet WCM buoy; Les Nattes WCM buoy; Roches d'Agon; E of La Catheue SCM lt buoy; E of Iles Chausey; and W of Pte du Roc (Granville). Between Granville and Îles Chausey it carries <1m in places, but with adequate rise of tide this is not a problem for small craft; the *Channel Pilot* gives directions.

Passage de la Déroute is used by ferries from the UK to St Malo. It passes W of Les Trois Grunes; between Basses de Taillepied and Les Écrehou; Chaussée des Boeufs and Plateau de l'Arconie; E of Les Minquiers and Les Ardentes; thence NW of Îles Chausey via Entrée de la Déroute, a side channel.

Granville has an excellent but very crowded half-tide marina. **Iles Chausey**, 8·3M WNW of Granville, is a popular and attractive archipelago. The Sound can be entered from the S or N with careful pilotage and sufficient rise of tide.

The drying expanse of Baie du Mont St Michel should not be entered by sea, but is an enjoyable visit from the mainland.

5.5 SPECIAL NOTES FOR FRANCE: see Area 17

MINOR HARBOUR CLOSE SOUTH OF CAP DE LA HAGUE

GOURY, Manche, **49°42'·85N 01°56'·83W**. AC 1114, 3653, 5604.2; SHOM 7158, 5636, 7133 (essential). HW –0410 on Dover (UT); ML 5·1m. See Omonville-la-Rogue. For visitors, appr at slack water nps with no swell and good vis; a fair weather hbr only, dries to flattish shingle. At tidal diamond M, 1·3M W, the N-S stream reaches 9·7kn at sp, 5·8kn at nps.

Cap de la Hague (Gros du Raz) lt ho is 0·5M NW of hbr; La Foraine WCM lt bn is 1·0M to the W. 065° ldg lts, by day: Front W patch with R ■ at end of bkwtr; rear, W pylon, lead between Diotret to S and Les Grios to N. ⚓ W of the 2 🅛🅑 slips in 1·7m or dry out on shingle banks SE of the bkwtr.

☎02·33·52·85·92. Facilities: ✕ ☐ at Auderville (0·5M).

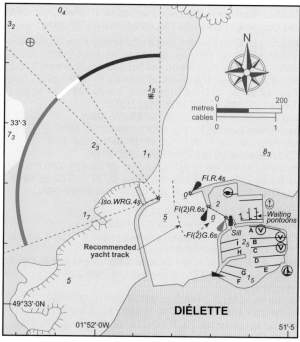

DIÉLETTE

5.6 DIÉLETTE

Manche, **49°33'·20N 01°51'·79W** ❀❀⚓⚓⚓⚓🏵🏵

CHARTS AC 2669, 3653; SHOM 7158, 7133; Navi 528, 1014; Imray C33A

TIDES HW –0430 on Dover (UT); ML 5·4m

Standard Port ST MALO (→)

Times				Height (metres)			
High Water		Low Water		MHWS	MHWN	MLWN	MLWS
0100	0800	0300	0800	12·2	9·3	4·2	1·5
1300	2000	1500	2000				
Differences DIÉLETTE							
+0045	+0035	+0020	+0035	–2·5	–1·9	–0·7	–0·3

SHELTER Good in marina, but when retaining wall covers, boats surge fore and aft; rig good springs. No entry in strong W'lies.

NAVIGATION WPT 49°33'·45N 01°52'·16W, 140°/650m to W bkwtr lt. Appr is exposed to W'ly winds/swell; 2kn cross tide at hbr ent. Caution: drying rks 3ca E of ⊕, marked by unlit WCM buoy. 1·5M WSW, keep seaward of WCM lt buoy off Flamanville power stn. Outer hbr ent dredged CD +0.5m; with Coefficient <55 accessible for 2m draft. W side of outer hbr dries approx 5m.

LIGHTS AND MARKS The power stn is conspic from afar 1·2M to SW, with pylons on the skyline. W bkwtr Lt ho, W twr/G top. Lts as chartlet.

COMMUNICATIONS CROSS 02·33·52·16·16; 🅛🅑 02·33·04·93·17; ⚓ 08·92·68·08·50; Emergency 15; ⊖ 02·33·23·34·00; Brit Consul 01·44·51·31·00; Dr 02·33·52·99·00; HM VHF Ch 09, summer 0800-1300, 1330-2000LT, winter 0800-1200, 1330-1700LT; Aff Mar 02·33·23·36·00; YC 02·33·04·14·78; Tourist office ☎02·33·52·81·60.

FACILITIES Bassin de Commerce, ⓥ waiting pontoons, E side; fuel berth 🅜 H24 cr/debit cards, W side by Marina sill, both in 2m. **Marina** ☎02·33·53·68·78. portdielette@cc-lespieux.com www.cc-lespieux.fr 420+70 ⓥ €3·50 inc 🔌, wi-fi. Enter about HW±3, tide gate over sill (CD+3·5m) drops when tide reaches CD+5m, to show 1·5m depth on gauge by sill; ⓥ berths pontoons A and E, ends of B (S side) and C (N side). If surging around HW, rig springs. 🍴 grocery/wines, 🛒 €1·70, 🗑 €4·10, 🖉 ⚓ ⚓ 🚾 🏗(40t). Ferry to CI. **Village** basic foods, mkt Sun am, ✕ ☐. **Flamanville**, where third 1750MW reactor is being built, 15 mins' walk. 🛒 ✉. **Les Pieux** ✕ ☐ 🛒 ✉ 🅑. Free daytime shuttle bus, Marina/ Flamanville/ Les Pieux, Jul/Aug, every 2 or 2½hrs.

5.7 CARTERET
Manche **49°22'·08N 01°47'·33W** ❄❀◊◊◊❀❀

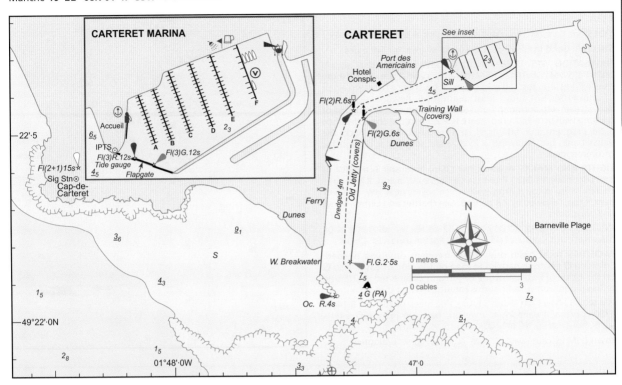

CHARTS AC 2669, 3655; SHOM 7157, 7158, 7133; Navi 1014; Imray C33A

TIDES –0440 Dover; ML 5·9; Duration 0545

Standard Port ST MALO (→)

Times				Height (metres)			
High Water		Low Water		MHWS	MHWN	MLWN	MLWS
0100	0800	0300	0800	12·2	9·3	4·2	1·5
1300	2000	1500	2000				
Differences CARTERET							
+0030	+0020	+0015	+0030	–1·6	–1·2	–0·7	–0·3
PORTBAIL							
+0030	+0025	+0025	+0030	–0·8	–0·6	–0·2	–0·1
ST GERMAIN-SUR-AY 49°13'·6N 01°39'·3W							
+0025	+0025	+0035	+0035	–0·7	–0·5	0·0	+0·1
LE SÉNÉQUET 49°05'·9N 01°41'·0W							
+0015	+0015	+0025	+0025	–0·3	–0·3	+0·1	+0·1

SHELTER Good in marina, but very crowded at weekends and in season. No safe ⚓s offshore.

NAVIGATION WPT 49°21'·86N 01°47'·36W, 006°/420m to W bkwtr lt. From N/NW, keep well off shore on appr to avoid rks 1M N of Cap de Carteret extending about 7ca from coast. From W beware Trois Grune Rks (dry 1·6m) about 4M offshore, marked by WCM lt buoy. Appr dries ½M offshore and is exposed to fresh W/SW winds which can make ent rough. Caution: strong cross-streams on the flood.

Best appr at HW–1 to avoid max tidal stream, 4½kn sp. The outer end of W bkwtr covers at big springs. Bar, at right angles to W bkwtr, dries 4m; a SHM buoy east of the W bkwtr marks a shifting shoal patch. Best water is to port of a mid-channel course. The chan dries progressively to firm sand, and is dredged to 4m and 4·5m just W of the marina.

LIGHTS AND MARKS Cap de Carteret, grey lt twr, G top, and conspic Sig stn are 8ca 295° from the ent. Breakwater and channel lts as chartlet. Marina sill is marked by PHM/SHM lt bcns, and Y poles on the retaining wall.

COMMUNICATIONS CROSS/⒧ 02·33·52·16·16; ⚓ 02·33·22·91·77; Auto 08·92·68·02·50; Police 02·33·53·80·17; ⊖ 02·33·04·90·08; Brit Consul 01·44·51·31·00; Ⓗ (Valognes) 02·33·40·14·39; Signal station VHF 09, 16; Marina VHF Ch 09; Tourist Office 02·33·04·90·58.

FACILITIES
Marina ☎02·33·04·70·84. www.barneville-carteret.net barneville-carteret @wanadoo.fr 311 + 60Ⓥ €2·95 (Apr-Sep; discounts possible Mon-Thu). Access HW ±2½ for 1·5m draft over sill 5m; lifting flapgate retains 2·3m within. IPTS N side of marina ent control ent/exit (sigs 2 & 4). Cross the sill squarely, ie heading NE, to clear the concrete bases each side. Ⓥ �/rafted on E side only of pontoon F (no need to stop at accueil pontoon).

⚓ ⚓ at accueil pontoon (limited hrs, displayed on the pumps) ⛽ ⚒⛟(35t) ⛴(24t). YC ☎02·33·52·60·73, ▷ ▬ ⚓ ⌂.

West quay Possible waiting berth clear of ferry, if too late for the marina. A 1·5m draft boat can stay afloat for 6hrs np, 9hrs sp. � free for 6hrs then at 50% of marina rates. ⚓ ▬.

Other options: Dry out in the tiny Port des Américains and basin close W of marina (up to 5m at HW); or on fine sand SW of marina.

Town ✕ ⌂ ☐ 🎦 🏧 🛢 ⚒✉ Ⓑ ⇌ (Valognes) ✈ (Cherbourg). **Ferry** Cherbourg, Jersey.

5.8 PORTBAIL

Manche **49°19'·72N 01°42'·42W** (Basin ent) ✿✿◊◊✿✿

CHARTS AC 2669, 3655; SHOM 7157, 7133; Navi 1014; Imray C33A

TIDES See Carteret. HW –0440 on Dover; ML 6·3m; Duration 0545

SHELTER Good in all winds. Occasional swell after SW gales.

NAVIGATION WPT 49°18'·25N 01°44'·58W (off chartlet, abeam unlit 'PB' SWM buoy) 042°/1·5M to training wall lt bcn. 7ca offshore the 042° ldg line crosses banks drying 5·3m, thence between a pair of unlit PHM/SHM buoys, least depth 8·1m. Beware very strong tidal stream, esp during sp flood (4-5k) at basin ent. When base of training wall bcn is covered, there is > 2·5m in the chan, dredged 5·2m. Frequent inundations of sand reported, Y buoy marks the worst bank, but following a local vessel (of similar draught) is recommended.

LIGHTS AND MARKS A water twr (38m) is conspic 6ca NNW of ent. Ldg marks/lts 042°: Front (La Caillourie) W pylon, R top; rear, church belfry in town. Trng wall, which covers near HW, head: W mast/R top, followed by 2 R perches. See chartlet and Lights, buoys and waypoints.

COMMUNICATIONS CROSS 02·33·52·16·16; ⊖ 02·33·04·90·08 (Carteret); Brit Consul 01·44·51·31·00; Port/marina VHF Ch 09.

FACILITIES Yacht basin dries 7·0m, frequent dredging needed. Enlarged ♥ pontoon parallel to road on NW side of basin; also ♥ on the E side of adjacent stone FV jetty; or as directed. At NE side of basin pontoon with fingers is for locals <7·5m LOA.

HM ☎02·33·04·83·48 (15 Jun-31 Aug, tide times; 0830-1200 & 1400-1700). portbail@wanadoo.fr www.portbail.com 160 + 30♥ €1·41. ◣╳⚓⛽▯◔(5t). YC ☎02·33·04·86·15, ╳▯◣

Town (½M by causeway) ╳▯▦◙◙⊠Ⓑ⇌ (Valognes).

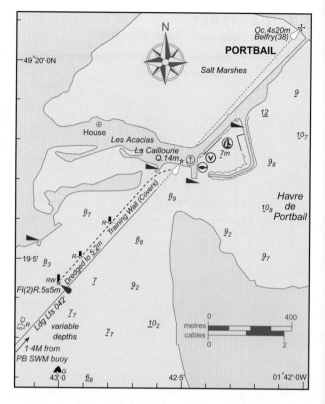

5.9 GRANVILLE

Manche **48°49'·94N 01°35'·93W** (Marina ent) ✿✿◊◊◊✿✿

See chartlet opposite

CHARTS AC 3656, 3659; SHOM 7156, 7341; Navi 534, 535; Imray C33B

TIDES –0510 Dover; ML 7·1; Duration 0525

Standard Port ST MALO (→)

Times				Height (metres)			
High Water		Low Water		MHWS	MHWN	MLWN	MLWS
0100	0800	0300	0800	12·2	9·3	4·2	1·5
1300	2000	1500	2000				
Differences REGNÉVILLE-SUR-MER							
+0010	+0010	+0030	+0020	+0·5	+0·4	+0·2	0·0
GRANVILLE							
+0010	+0005	+0025	+0015	+0·7	+0·4	+0·3	0·0
CANCALE							
0000	0000	+0010	+0010	+0·8	+0·6	+0·3	+0·1

SHELTER Good in the marina. Appr is rough in strong W winds. ⚓ in 2m about 4ca WSW of Le Loup to await the tide.

NAVIGATION WPT 48°49'·60N 01°36'·30W (abm Le Loup IDM) 042°/0·40M to S bkwtr lt. Beware rks seaward of La Fourchie WCM bcn twr, pot markers off Pte du Roc and 0·4m patches on Banc de Tombelaine, 1M SSW of Le Loup lt. At night best to keep at least 8ca W of Pte du Roc to avoid these dangers.

At bkwtr hd turn port wide into marina ent to avoid yachts leaving; cross the sill between R/G piles. Ent/exit under power; speed limit 5kn in the near approach, 3kn in hbr. The 5 R piles, Fl Y 2.5s, mark the covering wall of a windsurfing/dinghy area to stbd.

LIGHTS AND MARKS Pte du Roc is a conspic headland with large old bldgs and lt ho, grey twr, R top; the Sig stn and church spire are also obvious. The twin domes of St Paul's church in transit 034° with Le Loup lead towards the marina ent. No ldg lts, but S bkwtr hd on with TV mast leads 057° to ent; hbr lts are hard to see against town lts. 3·5M W of Pte du Roc, Le Videcoq rock

drying 0·8m, is marked by a WCM lt buoy. See chartlet and Lights, buoys and waypoints for lt details.

COMMUNICATIONS CROSS 02·33·52·16·16; Ⓛ 02·33·61·26·51; Auto 08·92·68·08·50; Police 02·33·91·27·50; ⊖02·33·50·19·90; Brit Consul 01·44·51·31·00; Ⓗ 02·33·91·50·00; Dr 02·33·50·00·07; Port HM 02·33·50·20·06, VHF Ch 09, 16 (HW±1½); Aff Mar 02·33·91·31·40; Marina VHF Ch 09, 0800-2200 Jul/Aug.

FACILITIES

Hérel Marina ④ ☎02·33·50·20·06. www.port.granville.cci.fr herel@granville.cci.fr. Depth over sill shown on digital display at marina S bkwtr elbow: eg 76 = 7·6m; 00 = no entry. 850+150 ♥ Jun-Sep, €3·34 in least depth 2.5m as indicated on SW side pontoon G & NW side pontoon H; in 1.5m pontoon V; (vessels >15m +€4·05 p/d, per additional metre); multihulls +50%. ▲♣▦◔◣▯⚓╲◭▯ ╳⚓Ⓔ SHOM, free wi-fi, ▯(12t) ◔(1·5t) ◣

YC de Granville ☎02·33·50·04·25, ╳▯ Port á Sec, St Nicolas ☎02·33·61·81·53; STL Nautisme ☎02·33·69·22·75.

Town ╳▯▦◙◙▮⊠Ⓑ╲⇌✈ (Dinard). **Ferry** UK via Jersey or Cherbourg.

MINOR HARBOUR 10M NORTH OF GRANVILLE

REGNÉVILLE, Manche, **48°59'·72N 01°34'·05W** (SHM buoy abeam Pte d'Agon). AC 2669, 3656; SHOM 7156, 7133. HW –0500 on Dover (UT); ML 7·0m; Duration 0535. See Granville. A seriously drying hbr on the estuary of R. La Sienne; few yachts visit, a magnet for hardy adventurers seeking sand and solitude.

From 48°57'·65N 01°38'·86W (2·2M E of La Catheue SCM buoy) identify Pte d'Agon lt ho, W twr/R top, and Regnéville's 028° dir lt, both 4M NE at the river mouth (see Lights, buoys and waypoints). Thence track 056°/3·8M across the drying estuary to the SHM buoy (Lat/Long in line 1) marking a drying mole extending SW from the sandspit with landing stages to stbd; a small pontoon at Regnéville is 1·2M NNE. Drying heights are from 9m-12·7m off Pte d'Agon.

Access HW –1 to +2 for 1·3m draft. Approx 80 moorings inc ☸s. YC ☎02·33·46·36·76 cnregnevillais@free.fr. Facilities: Quay, ◔(25t).

Town ╳▯▦◙◙⊠Ⓑ Tourist office ☎02·33·45·88·71.

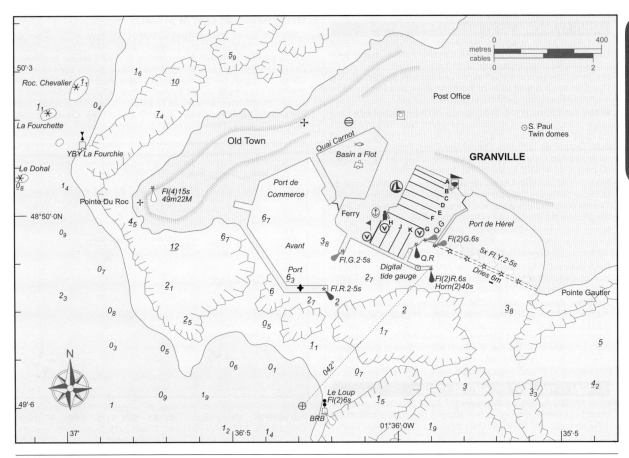

5.10 ÎLES CHAUSEY

Manche **48°52'·08N 01°49'·08W** SHM By, S ent ✸❀⚓✿✿✿

CHARTS AC 3656, 3659; SHOM 7156, 7155, 7161, 7134; Navi 534, 535; Imray C33B

TIDES –0500 Dover; ML 7·4; Duration 0530

Standard Port ST MALO (→)

Times				Height (metres)			
High Water		Low Water		MHWS	MHWN	MLWN	MLWS
0100	0800	0300	0800	12·2	9·3	4·2	1·5
1300	2000	1500	2000				
Differences ÎLES CHAUSEY (Grande Île)							
+0005	+0005	+0015	+0015	+0·8	+0·7	+0·6	+0·4
LES ARDENTES (48°58′N 01°52′W, 6M NNW of Grande Île)							
+0010	+0010	+0020	+0010	0·0	–0·1	0·0	–0·1

SHELTER Good except in strong NW or SE winds. Grande Île is privately owned, but may be visited; it is not a Port of Entry.

NAVIGATION WPT 48°51'·43N 01°48'·55W, 332°/1·1M to Crabière lt. In transit with L'Enseigne, W bn tr, B top, leads 332° to Sound. 1 SHM lt buoy, thence between unlit cardinal bns.

The N chan needs adequate ht of tide (max drying ht 5·3m) SHOM 7134, a good Pilot and/or local knowledge, plus careful pilotage. L'Enseigne ≠ Grande Île lt ho leads 156° to the N ent; thence follow charted dogleg. No access 1/4–30/6 to a bird sanctuary, ie all E of line from Grande Île lt ho to L'Enseigne bcn twr (except Aneret).

LIGHTS AND MARKS Grande Île lt ho is conspic. La Crabière lt is on blackish stilts with Y top. See chartlet and Lights, bys & w'points.

COMMUNICATIONS CROSS 02·33·52·16·16; [LB] 02·33·50·28·33; Auto 08·92·68·08·50; Police 02·33·52·72·02.

FACILITIES W ⚓ trots drying sp up to CD+1·0m, most S'ly, CD+0·5m. Max 3 boats rafted per trot, free, but donations welcome at office. Crowded w/ends; drying out at Port Homard (W side of Grande Île) discouraged. Tidal streams mod. **Village** ⚓ ✕ ☐ (limited)🛒 ▮.

MONT ST MICHEL TO ST MALO

West from Mont St Michel (AC 3659) the drying harbour of **Cancale** lies 4M S of Pte du Grouin. There are many oyster beds and a fair weather anchorage SE of Île des Rimains, N of which the many dangers are marked by La Pierre-de-Herpin (lt, fog sig). The large drying inlet of **Rothéneuf**, with a fair weather anchorage outside, is 4M NE of St Malo.

St Malo approaches (AC 2700) include 6 channels between the many islets, rocks and shoals across which tidal streams set at up to 4kn sp. Approaching from E or N, with sufficient rise of tide, Chenal de la Bigne, Chenal des Petits Pointus or Chenal de la Grande Conchée can be used in good visibility, but they all pass over or near to shoal and drying patches.

From the W and NW, Chenal de la Grande Porte and Chenal de la Petite Porte are the principal routes and are well marked/lit. Chenal du Décollé is a shallow, ill-marked inshore route from the W, and no shorter than Chenal de la Grande Porte.

▶ *Here the E-going stream begins at HW St Helier –0510, and the W-going at HW St Helier +0030, max sp rates 3·75kn. The streams set across the Chenal between Grand Jardin and Buron.*◀

Both marinas at St Malo have tidal restrictions due to either a lock or a sill. **Dinard** has a small basin 2·0m which is usually full of local boats, but it is pleasant to visit by ferry.

The **River Rance** barrage (SHOM 4233) gives access via a lock to many delightful anchorages and marinas at Plouër and Lyvet. It helps to know how the barrage affects water levels upriver. Continue to the lock at Châtelier, thence on to Dinan – subject to draught. Here enter the Canal d'Ille et Rance (Inland waterways) which leads to the S Brittany coast and provides a fascinating alternative to the offshore passage around NW Brittany.

ST MALO TO L'OST PIC AND BEYOND

From **St Malo to St Quay-Portrieux** (AC 2028, 2029) the coast has deep bays (often drying), a few rugged headlands and many offlying rocks. 6M NW of St Malo beware Le Vieux-Banc (1·2m) buoyed.

▶ *Here the ESE-going stream begins at HW St Helier –0555, and the NW-going at HW St Helier –0015, sp rates 2·5kn. There are W-going eddies very close inshore on the E-going stream.*◀

Between St Malo and Cap Fréhel there are anchorages S of Île Agot in the apprs to the drying hbr of St Briac and in Baie de l'Arguenon, SE of **St Cast's** excellent all-tide marina. More anchorages may be found in Baie de la Fresnaye and off Pte de la Latte. From Cap Fréhel, a bold steep headland, to Cap d'Erquy there are plenty of inshore rocks but no worthwhile hbrs.

Between Cap d'Erquy and the various rocky patches close to seaward, Chenal d'Erquy runs WSW/ENE into the E side of the Baie de St Brieuc. **Erquy** is a pleasant, drying hbr much used by fishing boats. There are several rocky shoals within the Baie itself, some extending nearly 2M offshore.

Baie de St Brieuc is a 20M wide bight between Cap d'Erquy and L'Ost Pic. Principal features are: to the N, Grand Léjon a rocky shoal extending 2½ca W and 8ca NNE of its lt ho. Petit Léjon (3·1m) lies 3·5M SSE. Rohein lt bcn, 3M further S, is in the centre with rocky plateaux to E and W. A 3M wide chan leads W of these features and S into the shallow, partly drying S end of the Bay and the interesting hbrs of **Dahouet, St-Brieuc Le Légué** and **Binic**.

On the W side of Bay de St-Brieuc, Roches de St Quay and offlying rocky patches extend 4M east from **St Quay-Portrieux** which can only be approached from NNW or SE. St Quay Marina is a good all-tide base from which to explore the drying hbrs around Baie de St Brieuc.

The next 2 paras are covered in greater detail in Area 20:

To the N and NE of L'Ost-Pic lt ho extensive offshore shoals guard the approaches to Paimpol, Ile de Bréhat and the Trieux river. Further N, the Plateau de Barnouic and Plateau des Roches Douvres (lt, fog sig) must be given a wide berth.

▶ *Here the E-going stream begins at about HW St Malo –0400 and the W-going at about HW St Malo +0100 with Spring rates exceeding 4kn.*◀

MINOR HARBOURS EAST OF ST MALO

CANCALE, Ille-et-Vilaine, **48°40'·10N 01°51'·11W**. AC 3659; SHOM 7155, 7131. HW –0510 on Dover (UT); ML 7·2m; Duration 0535; see Granville. La Houle is the drying hbr of Cancale just inside Bay of Mont St Michel, 2·6M S of Pte du Grouin; exposed to winds SW to SE. Drying berths may be available. Pier hd lt, obsc'd from the N, is on a mast. Avoid oyster beds at all costs. Area dries to about 1M offshore. ⚓ in 4m SE of Île des Rimains or further N off Pte Chatry. Facilities: **Quay** 🅿 ⚓ 🔧 ⚒ 🏪(1·5t). **CN de Cancale** ☎02·99·89·90·22. **Town** (famous for oysters) ✕ 🛒 🚂 🏨 🏦 ✉ Ⓑ.

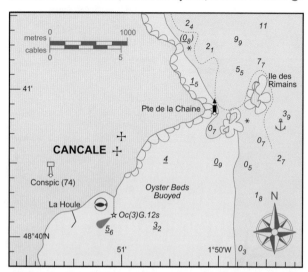

ROTHENEUF, Ille-et-Vilaine, **48°41'·38N 01°57'·64W**. AC 3659; SHOM 7155, 7131. HW –0510 on Dover (UT); Tides as for St Malo; ML 7·0m; Duration 0540. Complete shelter in quiet, delightful sandy hbr which dries up to 8·6m. ⚓ outside in 4m just N of SHM bn marking ent. Rks on both sides of ent which is barely 230m wide. Safest to enter when rks uncovered. Ldg line at 163°, W side of Pte Benard and old windmill (white twr). There are no lts. Facilities: ⚓ ⚓. **Village** ✕ 🛒 🚂 🏨 🏦.

5.11 ST MALO

Ille et Vilaine 48°38'·48N 02°01'·91W
Bassin Vauban and Bas Sablons Marinas ✿✿✿✿✿✿✿✿✿

CHARTS AC 2669, 3659, 2700; SHOM 7155, 7156, 7130; Navi 535; Imray C33B

TIDES –0506 Dover; ML 6·8; Duration 0535

SHELTER Good at Bas Sablons Marina, but in fresh W/NW'lies ♥ berths at outer ends of A & B pontoons are very uncomfortable. Excellent shelter in Bassin Vauban near the walled city.

NAVIGATION Care is needed due to strong tidal streams and many dangerous rocks close to the appr chans, which from W to E (see chartlet and Lights, buoys and waypoints) are:

> 1. Grande Porte, from the W, lit. WPT 48°40'·17N 02°08'·49W.
> 2. Petite Porte, the main lit chan from NW. WPT 48°42'·44N 02°09'·04W (abm Vieux-Banc Est NCM buoy, Q).
> 3. Bunel, from NNW, lit. WPT 48°42'·29N 02°06'·51W.

The above chans converge at or near Le Grand Jardin lt ho, 48°40'·20N 02°04'·97W. Thence continue in the main fairway 129°/2·8M to Môle des Noires.

> 4. Grande Conchée, from N, unlit. WPT 48°42'·15N 02°02'·12W.
> 5. Petits Pointus, from NNE, unlit. WPT 48°42'·34N 02°00'·38W.
> 6. La Bigne, from NE, unlit. WPT 48°42'·67N 01°57'·25W.

The above chans converge at or near Roches aux Anglais SHM buoy, 48°39'·66N 02°02'·22W. Thence continue 221°/8ca to enter the main 129° fairway.

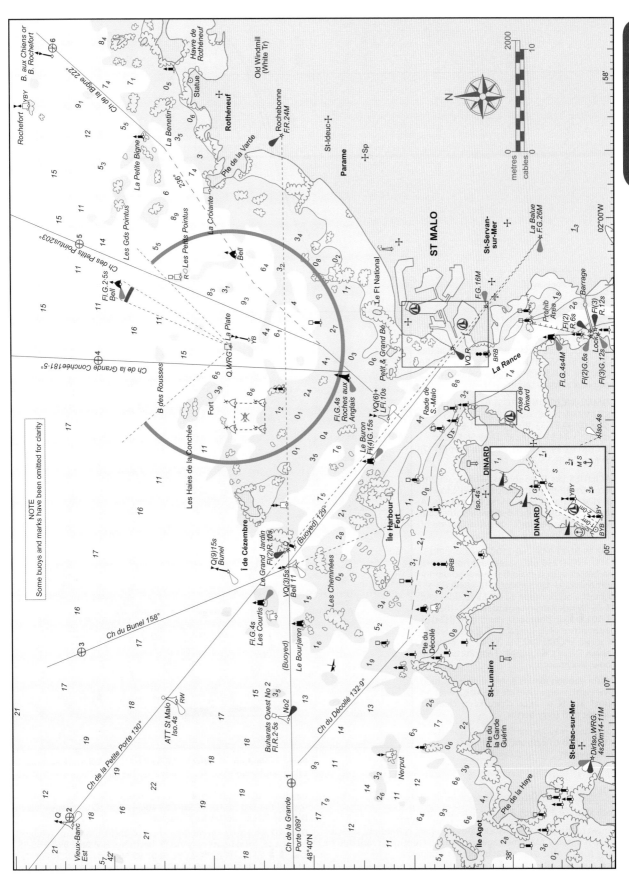

NOTE
Some buoys and marks have been omitted for clarity

ST MALO

DINARD

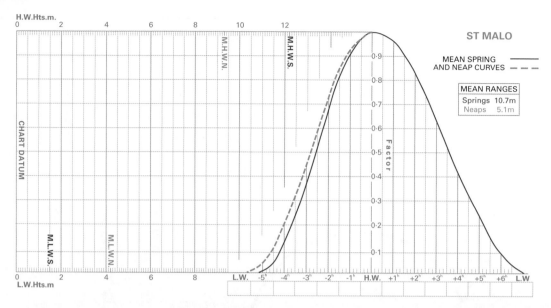

ST MALO

MEAN SPRING ——————
AND NEAP CURVES – – – –

MEAN RANGES
Springs 10.7m
Neaps 5.1m

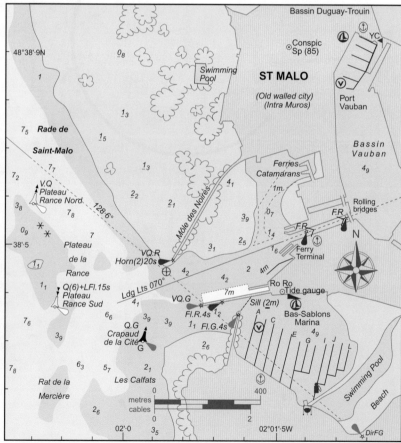

LIGHTS AND MARKS See chartlets/Lights, buoys and waypoints for ldg lts/buoys. Conspic daymarks: Île de Cézembre, Le Grand Jardin lt ho, Le Buron SHM twr, Petit and Grand Bé islets, the W lt twr/R top on Mole des Noires head and St Malo cathedral spire.

COMMUNICATIONS CROSS 02·98·89·31·31; ☎ 02·99·46·10·46; Auto 08·92·68·08·35; Police 02·99·81·52·30; ⊖ 02·99·81·65·90; Brit Consul 01·44·51·31·00; Ⓗ 02·99·56·56·19; Port HM 02·99·20·63·01, *St Malo Port* Ch **12**; Aff Mar 02·99·56·87·00; Marinas VHF Ch 09.

FACILITIES Les Bas-Sablons Marina, sill 2m. Access for 1·5m draft approx HW –3½ to +4½ sp; H24 at nps. 2 W waiting bys outside. A gauge on S bkwtr head shows depths <3m over sill. A large digital gauge on N side of marina is visible from all berths.

☎02·99·81·71·34.(0700-2100 in season, 0815-1215, 1345-1730 off season and hols). port. plaisance@ville-st-malo.fr www.guide-du-port.com/port-saint-malo/ 1216 + 64Ⓥ €3.15. Ⓥ berths on pontoon A: 32-66 (E side) and 43-75 (W side) and on pontoon B: 92-102 and 91-101.🛢 ⛽ on pontoon I; use cr/card (H24) or pay at office. ✕ Ⓟ ▷ ⚓ 🛢 ⛽ Ⓔ ⚒ ⛴ SHOM ✕ ⚓ ⬚ Ⓟ(20t) ⬚(2·5t). YC ☐. St Malo/St Servan ✕ ☐ ⬚ ✉ Ⓑ ⇌. Ferry Portsmouth, Poole, Weymouth.

Bassin Vauban (6m) is entered by lock; help given with warps. Outside the lock 3 waiting buoys are N of appr chan; keep clear of vedette and ferry berths.

Lock times may vary with traffic & tides. Lock is scheduled to operate 5 or 6 times in each direction, ie:

Inward	–2½	–1½	–½	+½	+1½	+2½
Outward		–2	–1	HW	+1	+2

Lock sigs are IPTS Nos 2, 3 and 5. In addition:

◔ next to the top lt = both lock gates are open, ie freeflow, beware current initially. Main message is the same. Freeflow is rare due to busy road traffic over retracting roller bridges.

● ● over ● = all movements prohib, big ship is outbound.

Port Vauban ☎02·99·56·51·91 portplaisancevauban@ saintmalofougeres.cci.fr www.guide-du-port.com/port-saint-malo-vauban; 175 + 50Ⓥ 🛥 on S pontoon,€46·50 (10m <12m/day Jul/Aug) inc locking, ⚓ ⋅Ⓟ ▷ 🚿 🚻, multihulls +50%. ⚓ in basins; speed limit 3kn. 🛢(1t). YC ☎02·99·40·84·42, ☐ (Ⓥ welcome). Bassin Duguay-Trouin better for long stay.

DINARD Access by 1m marked chan to yacht basin 2m; see inset prev page, local boats only, no Ⓥ berthing. HM ☎02·99·46·65·55, ⚓ ⬚ 🛢 ▷ ⚒ Ⓔ Ⓑ ✕ Ⓟ. YC ☎ 02·99·46·14·32, ✕ ☐. Town ✕ ☐ 🛒 🛢 ✉ Ⓑ Ⓗ 02·99·46·18·68. ⇌ ✈. Ferries to St Malo.

ST MALO TIDAL COEFFICIENTS 2019

Date	Jan am	Jan pm	Feb am	Feb pm	Mar am	Mar pm	Apr am	Apr pm	May am	May pm	June am	June pm	July am	July pm	Aug am	Aug pm	Sept am	Sept pm	Oct am	Oct pm	Nov am	Nov pm	Dec am	Dec pm
1	56	57	53	57	37	40	50	56	56	61	69	73	72	76	92	96	113	113	112	107	82	75	68	62
2	60	63	62	66	45	50	61	67	66	71	77	81	81	85	100	103	111	107	102	94	67	59	56	50
3	66	69	69	73	55	60	71	76	75	79	84	86	88	91	104	104	102	96	86	77	51	44	45	41
4	71	74	76	78	65	70	80	83	83	86	88	89	93	94	103	100	88	80	69	59	39			38
5	76	78	80	82	74	78	86	88	88	89	89	89	94	93	96	91		71		51	34	33	36	35
6	79	80	83	83	81	83	90	90	90	90	87	85	92	89	85		62	54	43	37	34	36	37	39
7	80	80	83	82	86	87	90	90	89	87	82	78	86	82	79	72	47	42	34	33	41	46	43	47
8	79	78	81	79	88	88	88	86	84	81	74	70	--	77	65	58	38	38	36	40	51	56	52	57
9	76	74	77	74	88	86	83	78	77	71		66	72	68	53	49	40	44	45	51	61	66	61	66
10	72	69	70	66	85	82	74	68	66	61	62	60	64	60	46	46	49	55	57	62	71	75	70	74
11	66	62	62	58	79	74	63	57		56	58	58	58	56	48	51	60	65	67	72	78	81	78	81
12	59	55	54	49	70	65		51	53	50	59	61	56	57	54	59	69	74	76	80	84	85	83	85
13	52	48		46	59	54	47	44	50	52	63	67	58	61	63	66	77	80	82	85	87	87	86	87
14	45		43	42	49		44	46	56	62	70	73	63	66	70	73	82	84	86	87	87	86	87	86
15	43	42	43	47	44	42	52	59	67	73	76	78	69	71	76	78	85	85	88	87	84	82	85	82
16	43	45	52	59	41	44	67	76	79	84	80	82	73	75	79	80	85	85	86	85	79	75	79	76
17	48	53	67	76	50	58	84	91	88	92	82	83	76	77	81	81	83	82	82	79	70	66	72	69
18	59	65	84	92	66	76	97	102	94	95	82	81	78	77	80	79	79	76	75	71	61	56	65	62
19	72	79	99	105	85	93	105	107	95	95	80	78	77	76	78	75	72	68	66	61	52			59
20	86	92	110	113	100	107	108	107	93	91	76	73	74	72	73	70	63	58	55	50	49	48	58	57
21	97	102	115	115	111	114	105	101	87	84	70	66	70	67	66	62	52			45	49	53	58	61
22	105	107	113	110	115	115	97	91	79	74	62	59	64	61	58	53	47	42	41	40	57	63	64	67
23	108	108	104	98	113	109	85	78	68	63	55	51	57	54		49	39	37	42	47	70	76	71	75
24	106	102	91	83	104	97	70	63	57	52		47		50	45	41	39	44	54	62	83	88	79	82
25	98	92	74	65	90	82	55	48	47		44	42	47	44	39	39	51	60	71	80	92	96	84	86
26	86	79	57		73	64		41	42	38	40	39	42	41	41	45	69	78	88	95	98	99	87	88
27	71		49	43	55	47	36	32	36	35	40	41	41	42	51	59	87	96	101	106	99	98	87	87
28	64	58	38	36	40		31	32	35	37	44	48	45	50	67	75	103	108	109	111	96	93	85	83
29	52	48			35	32	35	39	40	45	52	57	55	61	83	91	113	115	111	109	89	85	80	77
30	46	45			31	34	45	50	49	54	62	67	67	74	98	104	116	115	106	102	80	74	73	69
31	47	50			39	44			59	64			80	86	108	112			96	90			65	60

Tidal coefficients indicate the magnitude of the tide on any particular day, without having to look up and calculate the range, and thus determine whether it is springs, neaps or somewhere in between. This table is valid for all areas covered by this Almanac. Typical values are:

120	Very big spring tide
95	**Mean spring tide**
70	Average tide
45	**Mean neap tide**
20	Very small neap tide

See also coefficients for Calais (p 169), Cherbourg (p 196) and Brest (p 277).

STANDARD TIME UT −01
Subtract 1 hour for UT
For French Summer Time add
ONE hour in **non-shaded areas**

ST MALO LAT 48°38'N LONG 2°02'W
TIMES AND HEIGHTS OF HIGH AND LOW WATERS

Dates in red are **SPRINGS**
Dates in blue are **NEAPS**

YEAR 2019

JANUARY

#	Time	m	#	Time	m
1 TU	0301 / 0948 / 1530 / 2225	9.8 / 3.8 / 9.9 / 3.6	16 W	0142 / 0827 / 1423 / 2111	9.1 / 4.5 / 9.2 / 4.2
2 W	0408 / 1056 / 1634 / 2327	10.1 / 3.5 / 10.2 / 3.3	17 TH	0302 / 0951 / 1540 / 2228	9.4 / 4.1 / 9.6 / 3.7
3 TH	0504 / 1154 / 1728	10.5 / 3.1 / 10.6	18 F	0412 / 1103 / 1646 / 2334	10.0 / 3.4 / 10.3 / 3.0
4 F	0020 / 0552 / 1244 / 1815	2.9 / 11.0 / 2.7 / 10.9	19 SA	0512 / 1206 / 1745	10.8 / 2.6 / 11.1
5 SA	0105 / 0635 / 1327 / 1856	2.7 / 11.3 / 2.5 / 11.1	20 SU	0034 / 0608 / 1306 / 1840	2.2 / 11.6 / 1.9 / 11.8
6 SU	0144 / 0713 / 1405 / 1934	2.5 / 11.5 / 2.3 / 11.3	21 M	0132 / 0659 / 1404 / 1931	1.6 / 12.3 / 1.3 / 12.3
7 M	0220 / 0749 / 1441 / 2008	2.4 / 11.6 / 2.2 / 11.3	22 TU	0226 / 0749 / 1457 / 2020	1.2 / 12.8 / 0.9 / 12.6
8 TU	0254 / 0821 / 1515 / 2041	2.4 / 11.6 / 2.2 / 11.2	23 W	0316 / 0836 / 1546 / 2107	0.9 / 13.0 / 0.7 / 12.7
9 W	0327 / 0853 / 1548 / 2113	2.5 / 11.4 / 2.4 / 11.1	24 TH	0403 / 0922 / 1632 / 2151	0.9 / 12.9 / 0.8 / 12.5
10 TH	0358 / 0925 / 1619 / 2144	2.7 / 11.2 / 2.7 / 10.8	25 F	0447 / 1006 / 1715 / 2234	1.2 / 12.5 / 1.3 / 11.9
11 F	0428 / 0957 / 1649 / 2217	3.1 / 10.8 / 3.0 / 10.4	26 SA	0529 / 1049 / 1756 / 2316	1.8 / 11.9 / 2.0 / 11.2
12 SA	0458 / 1030 / 1722 / 2252	3.4 / 10.4 / 3.4 / 10.0	27 SU	0610 / 1133 / 1837	2.6 / 11.0 / 2.8
13 SU	0533 / 1106 / 1801 / 2332	3.8 / 9.9 / 3.8 / 9.5	28 M	0001 / 0654 / 1222 / 1924	10.4 / 3.3 / 10.2 / 3.6
14 M	0616 / 1153 / 1849	4.2 / 9.4 / 4.2	29 TU	0056 / 0749 / 1326 / 2024	9.7 / 4.0 / 9.5 / 4.1
15 TU	0027 / 0713 / 1300 / 1952	9.2 / 4.5 / 9.1 / 4.3	30 W	0210 / 0901 / 1449 / 2140	9.2 / 4.3 / 9.1 / 4.3
			31 TH	0334 / 1020 / 1609 / 2255	9.3 / 4.2 / 9.3 / 4.0

FEBRUARY

#	Time	m	#	Time	m
1 F	0442 / 1129 / 1711 / 2357	9.8 / 3.7 / 9.8 / 3.6	16 SA	0342 / 1036 / 1628 / 2311	9.5 / 3.7 / 9.8 / 3.3
2 SA	0535 / 1223 / 1800	10.3 / 3.2 / 10.4	17 SU	0455 / 1148 / 1734	10.4 / 2.8 / 10.8
3 SU	0046 / 0620 / 1309 / 1842	3.1 / 10.9 / 2.7 / 10.8	18 M	0018 / 0556 / 1254 / 1830	2.4 / 11.4 / 1.9 / 11.7
4 M	0128 / 0658 / 1349 / 1919	2.7 / 11.3 / 2.4 / 11.1	19 TU	0120 / 0649 / 1354 / 1921	1.6 / 12.3 / 1.0 / 12.5
5 TU	0206 / 0734 / 1426 / 1953	2.4 / 11.5 / 2.1 / 11.4	20 W	0216 / 0738 / 1447 / 2008	0.9 / 13.0 / 0.5 / 13.0
6 W	0240 / 0806 / 1500 / 2025	2.2 / 11.7 / 2.0 / 11.5	21 TH	0307 / 0825 / 1535 / 2052	0.5 / 13.3 / 0.2 / 13.1
7 TH	0313 / 0837 / 1532 / 2055	2.2 / 11.7 / 2.0 / 11.5	22 F	0351 / 0908 / 1618 / 2133	0.4 / 13.3 / 0.4 / 12.9
8 F	0343 / 0907 / 1602 / 2125	2.2 / 11.6 / 2.1 / 11.3	23 SA	0432 / 0948 / 1656 / 2211	0.7 / 12.9 / 0.9 / 12.3
9 SA	0411 / 0937 / 1630 / 2154	2.4 / 11.4 / 2.4 / 11.0	24 SU	0508 / 1026 / 1731 / 2248	1.4 / 12.1 / 1.7 / 11.5
10 SU	0439 / 1006 / 1659 / 2224	2.7 / 11.0 / 2.8 / 10.6	25 M	0542 / 1103 / 1804 / 2325	2.3 / 11.2 / 2.7 / 10.6
11 M	0509 / 1036 / 1731 / 2256	3.1 / 10.5 / 3.2 / 10.1	26 TU	0618 / 1143 / 1840	3.2 / 10.1 / 3.7
12 TU	0544 / 1111 / 1810 / 2337	3.6 / 9.9 / 3.8 / 9.6	27 W	0008 / 0702 / 1237 / 1931	9.6 / 4.1 / 9.2 / 4.5
13 W	0629 / 1201 / 1903	4.1 / 9.3 / 4.2	28 TH	0115 / 0810 / 1405 / 2051	8.9 / 4.7 / 8.6 / 4.9
14 TH	0037 / 0733 / 1323 / 2020	9.2 / 4.5 / 9.0 / 4.4			
15 F	0209 / 0906 / 1504 / 2152	9.0 / 4.4 / 9.1 / 4.1			

MARCH

#	Time	m	#	Time	m
1 F	0259 / 0944 / 1547 / 2226	8.7 / 4.7 / 8.7 / 4.7	16 SA	0133 / 0836 / 1446 / 2129	8.9 / 4.5 / 8.9 / 4.3
2 SA	0422 / 1104 / 1654 / 2335	9.2 / 4.1 / 9.3 / 4.0	17 SU	0323 / 1018 / 1617 / 2254	9.3 / 3.9 / 9.7 / 3.5
3 SU	0517 / 1202 / 1742	9.9 / 3.4 / 10.0	18 M	0441 / 1135 / 1723	10.3 / 2.8 / 10.8
4 M	0026 / 0600 / 1248 / 1822	3.3 / 10.6 / 2.8 / 10.7	19 TU	0005 / 0542 / 1242 / 1817	2.5 / 11.4 / 1.8 / 11.8
5 TU	0109 / 0638 / 1329 / 1858	2.7 / 11.1 / 2.4 / 11.1	20 W	0107 / 0634 / 1340 / 1905	1.5 / 12.3 / 0.9 / 12.6
6 W	0147 / 0713 / 1406 / 1932	2.3 / 11.5 / 2.0 / 11.5	21 TH	0202 / 0722 / 1431 / 1949	0.8 / 13.0 / 0.4 / 13.1
7 TH	0222 / 0746 / 1440 / 2003	2.1 / 11.8 / 1.8 / 11.7	22 F	0250 / 0806 / 1516 / 2031	0.4 / 13.3 / 0.2 / 13.2
8 F	0254 / 0817 / 1511 / 2033	1.9 / 11.9 / 1.7 / 11.8	23 SA	0332 / 0847 / 1555 / 2109	0.3 / 13.3 / 0.4 / 12.9
9 SA	0324 / 0846 / 1541 / 2102	1.8 / 11.9 / 1.7 / 11.7	24 SU	0409 / 0925 / 1630 / 2144	0.7 / 12.8 / 1.0 / 12.4
10 SU	0352 / 0915 / 1609 / 2130	1.9 / 11.7 / 2.0 / 11.5	25 M	0442 / 1000 / 1700 / 2217	1.4 / 12.0 / 1.9 / 11.5
11 M	0420 / 0944 / 1637 / 2159	2.2 / 11.4 / 2.4 / 11.1	26 TU	0513 / 1034 / 1729 / 2250	2.3 / 11.1 / 2.9 / 10.6
12 TU	0448 / 1012 / 1707 / 2229	2.7 / 10.8 / 2.9 / 10.6	27 W	0544 / 1109 / 1800 / 2328	3.2 / 10.0 / 3.8 / 9.6
13 W	0520 / 1044 / 1742 / 2306	3.2 / 10.2 / 3.5 / 9.9	28 TH	0622 / 1154 / 1844	4.1 / 9.0 / 4.7
14 TH	0601 / 1130 / 1831	3.8 / 9.5 / 4.2	29 F	0026 / 0723 / 1316 / 2002	8.8 / 4.8 / 8.3 / 5.2
15 F	0000 / 0701 / 1247 / 1946	9.3 / 4.4 / 8.9 / 4.6	30 SA	0215 / 0901 / 1515 / 2151	8.4 / 5.0 / 8.3 / 5.1
			31 SU	0351 / 1031 / 1627 / 2305	8.8 / 4.5 / 9.0 / 4.3

APRIL

#	Time	m	#	Time	m
1 M	0448 / 1131 / 1714 / 2357	9.5 / 3.7 / 9.8 / 3.6	16 TU	0423 / 1119 / 1703 / 2347	10.3 / 2.8 / 10.8 / 2.5
2 TU	0531 / 1217 / 1753	10.3 / 3.0 / 10.5	17 W	0522 / 1223 / 1755	11.3 / 1.8 / 11.7
3 W	0040 / 0609 / 1259 / 1829	2.9 / 10.9 / 2.5 / 11.1	18 TH	0047 / 0613 / 1318 / 1842	1.6 / 12.2 / 1.1 / 12.4
4 TH	0118 / 0645 / 1337 / 1903	2.4 / 11.4 / 2.1 / 11.5	19 F	0140 / 0700 / 1407 / 1925	1.1 / 12.7 / 0.7 / 12.8
5 F	0154 / 0719 / 1412 / 1936	2.1 / 11.7 / 1.8 / 11.8	20 SA	0226 / 0743 / 1450 / 2005	0.8 / 12.9 / 0.7 / 12.8
6 SA	0228 / 0751 / 1445 / 2007	1.8 / 11.9 / 1.6 / 11.9	21 SU	0307 / 0822 / 1527 / 2042	0.8 / 12.8 / 0.9 / 12.6
7 SU	0300 / 0822 / 1517 / 2037	1.7 / 12.0 / 1.6 / 12.0	22 M	0343 / 0859 / 1600 / 2116	1.1 / 12.4 / 1.5 / 12.1
8 M	0331 / 0852 / 1547 / 2107	1.7 / 11.9 / 1.8 / 11.8	23 TU	0414 / 0933 / 1629 / 2148	1.7 / 11.7 / 2.2 / 11.4
9 TU	0401 / 0923 / 1616 / 2137	2.0 / 11.5 / 2.2 / 11.4	24 W	0444 / 1006 / 1656 / 2221	2.4 / 10.9 / 3.0 / 10.6
10 W	0431 / 0954 / 1648 / 2210	2.4 / 11.0 / 2.8 / 10.8	25 TH	0515 / 1041 / 1726 / 2257	3.3 / 10.0 / 3.9 / 9.7
11 TH	0505 / 1031 / 1725 / 2250	3.0 / 10.3 / 3.4 / 10.2	26 F	0551 / 1123 / 1807 / 2348	4.1 / 9.1 / 4.7 / 8.9
12 F	0547 / 1119 / 1815 / 2348	3.7 / 9.5 / 4.1 / 9.5	27 SA	0644 / 1230 / 1913	4.7 / 8.4 / 5.3
13 SA	0648 / 1242 / 1932	4.2 / 8.9 / 4.5	28 SU	0117 / 0806 / 1415 / 2057	8.4 / 5.0 / 8.3 / 5.3
14 SU	0122 / 0824 / 1435 / 2114	9.1 / 4.3 / 9.0 / 4.3	29 M	0257 / 0938 / 1538 / 2218	8.6 / 4.7 / 8.7 / 4.7
15 M	0306 / 1003 / 1601 / 2238	9.4 / 3.7 / 9.8 / 3.5	30 TU	0402 / 1044 / 1631 / 2313	9.2 / 4.1 / 9.5 / 4.0

Chart Datum is 6·29 metres below IGN Datum. HAT is 13·6 metres above Chart Datum.

Central N France

STANDARD TIME UT –01
Subtract 1 hour for UT
For French Summer Time add
ONE hour in **non-shaded areas**

ST MALO   LAT 48°38′N   LONG 2°02′W
TIMES AND HEIGHTS OF HIGH AND LOW WATERS

Dates in red are **SPRINGS**
Dates in blue are **NEAPS**

YEAR 2019

MAY

Day	Time m	Day	Time m
1 W	0449 9.9 / 1134 3.4 / 1713 10.2 / 2358 3.3	16 TH	0455 11.1 / 1156 2.2 / 1728 11.4
2 TH	0531 10.5 / 1218 2.8 / 1752 10.8	17 F	0021 2.1 / 0547 11.7 / 1251 1.7 / 1815 11.9
3 F	0040 2.7 / 0610 11.1 / 1300 2.3 / 1829 11.3	18 SA	0113 1.7 / 0634 12.0 / 1339 1.5 / ○ 1858 12.2
4 SA	0120 2.3 / 0647 11.5 / 1339 2.0 / ● 1904 11.7	19 SU	0159 1.5 / 0718 12.1 / 1421 1.5 / 1938 12.2
5 SU	0158 2.0 / 0722 11.8 / 1416 1.8 / 1938 11.9	20 M	0239 1.5 / 0758 12.0 / 1457 1.7 / 2015 12.1
6 M	0234 1.8 / 0757 11.9 / 1452 1.7 / 2012 12.0	21 TU	0314 1.7 / 0835 11.8 / 1529 2.0 / 2050 11.8
7 TU	0310 1.8 / 0831 11.8 / 1526 1.8 / 2046 11.9	22 W	0347 2.0 / 0910 11.3 / 1600 2.5 / 2123 11.3
8 W	0345 1.9 / 0907 11.6 / 1600 2.2 / 2122 11.6	23 TH	0419 2.6 / 0944 10.8 / 1630 3.1 / 2157 10.7
9 TH	0420 2.3 / 0946 11.1 / 1637 2.7 / 2201 11.1	24 F	0451 3.2 / 1019 10.1 / 1701 3.8 / 2234 10.0
10 F	0459 2.8 / 1030 10.5 / 1719 3.1 / 2248 10.4	25 SA	0526 3.8 / 1058 9.4 / 1738 4.4 / 2318 9.3
11 SA	0547 3.4 / 1126 9.8 / 1813 3.9 / 2350 9.8	26 SU	0610 4.4 / 1150 8.9 / 1829 4.9 / ◐
12 SU	0650 3.8 / 1243 9.3 / 1928 4.2 / ◐	27 M	0021 8.8 / 0709 4.7 / 1302 8.6 / 1942 5.1
13 M	0114 9.5 / 0815 3.9 / 1414 9.4 / 2056 4.0	28 TU	0141 8.7 / 0824 4.7 / 1422 8.7 / 2106 4.9
14 TU	0241 9.7 / 0941 3.5 / 1533 10.0 / 2215 3.4	29 W	0255 8.9 / 0938 4.4 / 1529 9.1 / 2213 4.4
15 W	0355 10.4 / 1054 2.8 / 1636 10.7 / 2322 2.7	30 TH	0354 9.5 / 1038 3.8 / 1622 9.8 / 2306 3.7
		31 F	0444 10.1 / 1129 3.3 / 1708 10.4 / 2354 3.1

JUNE

Day	Time m	Day	Time m
1 SA	0529 10.6 / 1217 2.7 / 1751 11.0	16 SU	0044 2.4 / 0610 11.2 / 1309 2.3 / 1834 11.5
2 SU	0041 2.6 / 0612 11.1 / 1303 2.3 / 1831 11.5	17 M	0131 2.2 / 0656 11.4 / 1352 2.3 / ○ 1915 11.6
3 M	0126 2.2 / 0653 11.5 / 1347 2.0 / ● 1911 11.8	18 TU	0212 2.1 / 0737 11.4 / 1429 2.3 / 1954 11.6
4 TU	0209 1.9 / 0734 11.7 / 1429 1.9 / 1950 12.0	19 W	0249 2.2 / 0815 11.3 / 1504 2.4 / 2029 11.5
5 W	0252 1.8 / 0815 11.8 / 1510 1.9 / 2031 12.0	20 TH	0324 2.3 / 0850 11.1 / 1537 2.7 / 2104 11.3
6 TH	0334 1.8 / 0858 11.7 / 1551 2.0 / 2113 11.9	21 F	0358 2.6 / 0924 10.8 / 1609 3.0 / 2138 10.9
7 F	0417 2.0 / 0943 11.4 / 1634 2.4 / 2159 11.5	22 SA	0431 2.9 / 0959 10.4 / 1641 3.4 / 2212 10.4
8 SA	0502 2.4 / 1032 10.9 / 1720 2.9 / 2249 11.0	23 SU	0504 3.4 / 1034 10.0 / 1714 3.9 / 2250 9.9
9 SU	0552 2.9 / 1127 10.4 / 1814 3.3 / 2347 10.4	24 M	0539 3.8 / 1114 9.5 / 1752 4.3 / 2334 9.4
10 M	0650 3.3 / 1230 10.0 / 1918 3.7 / ◐	25 TU	0621 4.2 / 1203 9.1 / 1840 4.6 / ◑
11 TU	0054 10.1 / 0758 3.5 / 1342 9.8 / 2030 3.7	26 W	0030 9.1 / 0714 4.4 / 1305 8.9 / 1943 4.8
12 W	0208 10.0 / 0910 3.4 / 1456 10.0 / 2143 3.5	27 TH	0139 9.0 / 0820 4.5 / 1416 9.0 / 2058 4.6
13 TH	0321 10.2 / 1020 3.1 / 1602 10.4 / 2250 3.1	28 F	0249 9.2 / 0932 4.2 / 1523 9.4 / 2209 4.2
14 F	0425 10.6 / 1124 2.8 / 1659 10.8 / 2351 2.7	29 SA	0352 9.6 / 1038 3.7 / 1621 10.0 / 2309 3.6
15 SA	0521 11.0 / 1220 2.5 / 1749 11.2	30 SU	0448 10.2 / 1136 3.2 / 1714 10.6

JULY

Day	Time m	Day	Time m
1 M	0004 3.0 / 0540 10.7 / 1230 2.7 / 1802 11.2	16 TU	0108 2.7 / 0639 10.8 / 1329 2.8 / ○ 1859 11.2
2 TU	0057 2.4 / 0629 11.2 / 1321 2.2 / ● 1849 11.7	17 W	0151 2.5 / 0721 11.0 / 1409 2.6 / 1937 11.4
3 W	0149 2.0 / 0718 11.6 / 1411 1.9 / 1935 12.1	18 TH	0229 2.4 / 0758 11.1 / 1445 2.6 / 2013 11.6
4 TH	0239 1.7 / 0805 11.9 / 1500 1.7 / 2021 12.3	19 F	0305 2.3 / 0833 11.1 / 1519 2.6 / 2046 11.4
5 F	0328 1.5 / 0852 12.0 / 1546 1.7 / 2108 12.3	20 SA	0339 2.4 / 0905 11.0 / 1551 2.7 / 2117 11.2
6 SA	0416 1.5 / 0939 11.9 / 1633 1.8 / 2155 12.1	21 SU	0410 2.6 / 0937 10.9 / 1620 3.0 / 2149 11.0
7 SU	0502 1.8 / 1027 11.5 / 1718 2.2 / 2242 11.6	22 M	0440 2.9 / 1008 10.6 / 1649 3.3 / 2221 10.5
8 M	0549 2.2 / 1115 11.1 / 1806 2.7 / 2332 11.1	23 TU	0510 3.3 / 1041 10.2 / 1719 3.7 / 2253 10.1
9 TU	0637 2.7 / 1206 10.5 / 1857 3.2 / ◑	24 W	0543 3.7 / 1116 9.7 / 1756 4.1 / 2333 9.6
10 W	0027 10.5 / 0730 3.2 / 1304 10.1 / 1957 3.6 / ◑	25 TH	0623 4.0 / 1200 9.3 / 1843 4.4 / ◑
11 TH	0131 10.0 / 0832 3.6 / 1413 9.8 / 2105 3.8	26 F	0027 9.2 / 0716 4.4 / 1303 9.1 / 1947 4.7
12 F	0243 9.8 / 0942 3.7 / 1526 9.9 / 2217 3.6	27 SA	0142 9.0 / 0827 4.5 / 1422 9.1 / 2110 4.5
13 SA	0355 9.9 / 1050 3.5 / 1632 10.2 / 2322 3.3	28 SU	0303 9.2 / 0949 4.2 / 1538 9.5 / 2230 4.0
14 SU	0458 10.2 / 1152 3.3 / 1728 10.6	29 M	0414 9.7 / 1101 3.6 / 1643 10.2 / 2335 3.2
15 M	0019 3.0 / 0552 10.5 / 1244 3.0 / 1816 11.0	30 TU	0517 10.4 / 1203 2.9 / 1741 11.0
		31 W	0036 2.5 / 0613 11.1 / 1302 2.3 / 1834 11.7

AUGUST

Day	Time m	Day	Time m
1 TH	0134 1.9 / 0706 11.8 / 1359 1.7 / ● 1924 12.3	16 F	0211 2.3 / 0740 11.3 / 1428 2.5 / 1954 11.7
2 F	0230 1.3 / 0756 12.2 / 1451 1.3 / 2012 12.7	17 SA	0246 2.2 / 0812 11.4 / 1500 2.4 / 2025 11.7
3 SA	0322 1.0 / 0843 12.5 / 1540 1.1 / 2059 12.9	18 SU	0318 2.1 / 0842 11.4 / 1530 2.4 / 2055 11.7
4 SU	0409 0.9 / 0928 12.5 / 1625 1.2 / 2143 12.7	19 M	0347 2.2 / 0911 11.3 / 1558 2.5 / 2124 11.4
5 M	0453 1.1 / 1012 12.2 / 1707 1.6 / 2226 12.2	20 TU	0415 2.5 / 0940 11.1 / 1624 2.8 / 2152 11.1
6 TU	0534 1.7 / 1053 11.6 / 1748 2.2 / 2309 11.5	21 W	0441 2.8 / 1008 10.7 / 1650 3.2 / 2219 10.6
7 W	0614 2.4 / 1136 10.9 / 1830 3.0 / ◑ 2355 10.7	22 TH	0510 3.3 / 1037 10.3 / 1721 3.7 / 2249 10.0
8 TH	0657 3.2 / 1225 10.1 / 1920 3.7	23 F	0544 3.8 / 1112 9.7 / 1801 4.2 / ◑ 2332 9.4
9 F	0051 9.9 / 0750 3.9 / 1329 9.5 / 2025 4.2	24 SA	0630 4.3 / 1202 9.2 / 1856 4.6
10 SA	0205 9.3 / 0900 4.3 / 1452 9.3 / 2145 4.2	25 SU	0039 8.9 / 0736 4.6 / 1325 8.9 / 2020 4.8
11 SU	0331 9.3 / 1021 4.3 / 1612 9.6 / 2300 3.9	26 M	0224 8.9 / 0909 4.6 / 1505 9.2 / 2200 4.3
12 M	0443 9.6 / 1131 3.8 / 1713 10.2	27 TU	0354 9.4 / 1036 3.9 / 1624 10.0 / 2316 3.4
13 TU	0001 3.4 / 0539 10.2 / 1226 3.4 / 1802 10.7	28 W	0503 10.3 / 1145 3.0 / 1727 11.0
14 W	0051 2.9 / 0625 10.7 / 1312 2.9 / 1843 11.2	29 TH	0021 2.4 / 0601 11.3 / 1248 2.2 / 1821 11.9
15 TH	0133 2.5 / 0704 11.0 / 1352 2.6 / ○ 1920 11.5	30 F	0122 1.6 / 0653 12.1 / 1347 1.4 / ● 1911 12.7
		31 SA	0218 0.9 / 0741 12.7 / 1440 0.9 / 1958 13.2

Chart Datum is 6·29 metres below IGN Datum. HAT is 13·6 metres above Chart Datum.

》 FREE monthly updates. Register at 《
www.reedsnauticalalmanac.co.uk

217

STANDARD TIME UT –01
Subtract 1 hour for UT
For French Summer Time add ONE hour in **non-shaded areas**

ST MALO LAT 48°38'N LONG 2°02'W
TIMES AND HEIGHTS OF HIGH AND LOW WATERS

YEAR **2019**

Dates in red are **SPRINGS**
Dates in blue are **NEAPS**

SEPTEMBER

Day	Time m	Time m	Time m	Time m	Day	Time m	Time m	Time m	Time m
1 SU	0308 0.5	0827 13.0	1527 0.6	2043 13.4	16 M	0253 1.9	0816 11.8	1506 2.1	2029 11.9
2 M	0353 0.5	0909 13.0	1609 0.8	2125 13.2	17 TU	0322 2.0	0844 11.7	1534 2.2	2057 11.8
3 TU	0434 0.8	0949 12.6	1648 1.3	2204 12.5	18 W	0348 2.2	0911 11.5	1559 2.5	2124 11.4
4 W	0510 1.5	1026 11.9	1724 2.1	2242 11.6	19 TH	0415 2.5	0938 11.2	1626 2.9	2150 10.9
5 TH	0544 2.5	1104 11.0	1759 3.0	2322 10.6	20 F	0442 3.1	1006 10.7	1655 3.4	2219 10.3
6 F	0620 3.5	1146 10.1	1842 3.9		21 SA	0515 3.7	1038 10.1	1732 4.0	2300 9.6
7 SA	0012 9.6	0706 4.4	1246 9.2	1944 4.6	22 SU	0558 4.3	1125 9.4	1825 4.6	
8 SU	0130 8.8	0820 4.9	1423 8.8	2116 4.8	23 M	0002 8.9	0703 4.8	1247 8.9	1949 4.8
9 M	0313 8.8	0958 4.8	1556 9.2	2241 4.3	24 TU	0205 8.7	0844 4.8	1446 9.1	2141 4.4
10 TU	0431 9.3	1114 4.2	1657 9.9	2342 3.6	25 W	0343 9.4	1019 4.0	1609 10.0	2301 3.3
11 W	0523 10.0	1208 3.5	1743 10.6		26 TH	0451 10.5	1131 3.0	1711 11.1	
12 TH	0030 2.9	0605 10.7	1252 2.9	1822 11.2	27 F	0007 2.2	0546 11.6	1233 2.0	1805 12.2
13 F	0111 2.5	0642 11.2	1331 2.5	1857 11.6	28 SA	0106 1.3	0635 12.4	1330 1.2	1853 13.0
14 SA	0148 2.2	0715 11.5	1405 2.3	1930 11.9	29 SU	0200 0.6	0721 13.0	1421 0.7	1939 13.4
15 SU	0222 2.0	0747 11.7	1437 2.1	2001 12.0	30 M	0248 0.3	0804 13.3	1507 0.5	2022 13.5

OCTOBER

Day	Time m	Time m	Time m	Time m	Day	Time m	Time m	Time m	Time m
1 TU	0331 0.4	0845 13.2	1547 0.7	2102 13.1	16 W	0255 1.9	0816 11.9	1510 2.1	2032 11.9
2 W	0408 0.9	0922 12.7	1623 1.3	2139 12.5	17 TH	0324 2.1	0845 11.8	1538 2.3	2101 11.6
3 TH	0442 1.7	0958 11.9	1656 2.2	2215 11.5	18 F	0352 2.4	0913 11.5	1607 2.7	2129 11.1
4 F	0512 2.7	1033 11.0	1729 3.1	2252 10.4	19 SA	0422 3.0	0944 11.0	1638 3.2	2203 10.4
5 SA	0544 3.7	1111 10.0	1808 4.1	2337 9.4	20 SU	0456 3.6	1021 10.3	1717 3.8	2245 9.7
6 SU	0626 4.7	1207 9.1	1906 4.8		21 M	0541 4.2	1111 9.6	1812 4.4	2356 9.0
7 M	0054 8.6	0739 5.3	1350 8.6	2040 5.1	22 TU	0649 4.7	1237 9.1	1938 4.7	
8 TU	0249 8.5	0929 5.2	1531 8.9	2213 4.6	23 W	0155 8.9	0830 4.7	1429 9.3	2125 4.2
9 W	0407 9.1	1047 4.5	1631 9.7	2313 3.8	24 TH	0326 9.6	1002 3.9	1549 10.2	2244 3.2
10 TH	0457 9.9	1139 3.7	1715 10.5		25 F	0431 10.6	1112 2.9	1651 11.2	2347 2.2
11 F	0000 3.1	0536 10.6	1222 3.0	1753 11.1	26 SA	0524 11.6	1213 1.9	1743 12.2	
12 SA	0040 2.6	0612 11.2	1300 2.6	1828 11.6	27 SU	0045 1.3	0612 12.4	1308 1.2	1831 12.8
13 SU	0118 2.2	0645 11.6	1336 2.3	1902 11.9	28 M	0136 0.8	0657 12.9	1358 0.9	1916 13.1
14 M	0152 2.0	0717 11.8	1409 2.1	1933 12.0	29 TU	0222 0.7	0739 13.1	1442 0.8	1958 13.1
15 TU	0225 1.9	0747 12.0	1440 1.9	2003 12.0	30 W	0304 0.9	0818 12.9	1522 1.0	2038 12.8
					31 TH	0340 1.4	0856 12.5	1557 1.6	2115 12.1

NOVEMBER

Day	Time m	Time m	Time m	Time m	Day	Time m	Time m	Time m	Time m
1 F	0412 2.1	0931 11.8	1630 2.3	2150 11.3	16 SA	0336 2.4	0857 11.7	1556 2.5	2121 11.2
2 SA	0442 2.9	1006 10.9	1703 3.2	2227 10.3	17 SU	0411 2.8	0935 11.2	1633 3.0	2201 10.6
3 SU	0514 3.8	1044 10.0	1740 4.0	2310 9.4	18 M	0450 3.4	1018 10.6	1717 3.5	2252 9.9
4 M	0554 4.7	1135 9.2	1832 4.7		19 TU	0539 3.9	1114 10.0	1814 4.0	
5 TU	0015 8.7	0657 5.3	1300 8.6	1950 5.1	20 W	0002 9.4	0646 4.3	1232 9.5	1933 4.2
6 W	0155 8.4	0836 5.3	1441 8.7	2121 4.8	21 TH	0133 9.3	0813 4.3	1402 9.7	2101 3.9
7 TH	0321 8.9	1001 4.8	1548 9.3	2228 4.2	22 F	0256 9.8	0936 3.8	1520 10.3	2217 3.2
8 F	0416 9.6	1057 4.1	1636 10.0	2318 3.5	23 SA	0402 10.6	1046 3.0	1624 11.0	2321 2.4
9 SA	0458 10.3	1142 3.4	1716 10.7		24 SU	0458 11.4	1147 2.2	1718 11.7	
10 SU	0001 2.9	0536 10.9	1223 2.9	1754 11.2	25 M	0019 1.8	0547 12.0	1243 1.7	1808 12.2
11 M	0041 2.5	0612 11.4	1301 2.5	1830 11.6	26 TU	0110 1.4	0633 12.4	1333 1.4	1854 12.4
12 TU	0119 2.2	0646 11.7	1338 2.2	1905 11.8	27 W	0156 1.4	0716 12.5	1417 1.3	1937 12.4
13 W	0155 2.0	0719 11.9	1414 2.1	1938 11.9	28 TH	0237 1.5	0756 12.4	1457 1.5	2017 12.2
14 TH	0229 2.0	0751 12.0	1448 2.1	2011 11.8	29 F	0313 1.8	0833 12.1	1533 1.8	2055 11.7
15 F	0303 2.1	0823 11.9	1522 2.2	2045 11.6	30 SA	0346 2.3	0910 11.7	1608 2.4	2131 11.1

DECEMBER

Day	Time m	Time m	Time m	Time m	Day	Time m	Time m	Time m	Time m
1 SU	0419 2.9	0946 11.0	1642 3.0	2208 10.4	16 M	0409 2.4	0934 11.6	1637 2.4	2205 11.1
2 M	0452 3.6	1023 10.3	1718 3.7	2248 9.7	17 TU	0453 2.8	1021 11.2	1724 2.8	2255 10.6
3 TU	0529 4.3	1107 9.6	1800 4.3	2336 9.1	18 W	0542 3.2	1114 10.7	1817 3.2	2353 10.1
4 W	0617 4.8	1204 9.0	1855 4.7		19 TH	0640 3.6	1216 10.3	1919 3.5	
5 TH	0042 8.7	0722 5.1	1321 8.8	2005 4.8	20 F	0101 9.8	0748 3.8	1328 10.2	2029 3.6
6 F	0203 8.7	0844 5.1	1439 8.9	2120 4.6	21 SA	0216 9.9	0902 3.7	1444 10.1	2142 3.4
7 SA	0314 9.1	0956 4.6	1541 9.4	2223 4.1	22 SU	0327 10.2	1014 3.3	1553 10.5	2250 3.0
8 SU	0408 9.7	1052 4.0	1632 10.0	2315 3.5	23 M	0430 10.7	1120 2.8	1654 10.9	2352 2.6
9 M	0454 10.3	1140 3.4	1716 10.6		24 TU	0524 11.2	1218 2.4	1748 11.3	
10 TU	0001 3.0	0536 10.9	1225 2.9	1758 11.0	25 W	0045 2.2	0613 11.6	1310 2.1	1837 11.6
11 W	0045 2.5	0616 11.4	1308 2.5	1838 11.4	26 TH	0133 2.1	0658 11.9	1356 1.9	1922 11.7
12 TH	0127 2.2	0654 11.7	1350 2.2	1917 11.6	27 F	0215 2.1	0739 11.9	1437 1.9	2002 11.6
13 F	0207 2.0	0732 11.9	1431 2.0	1956 11.7	28 SA	0252 2.1	0818 11.9	1514 2.0	2040 11.5
14 SA	0248 2.0	0810 12.0	1512 2.0	2037 11.7	29 SU	0327 2.3	0854 11.6	1550 2.2	2116 11.2
15 SU	0328 2.1	0850 11.9	1553 2.1	2119 11.5	30 M	0401 2.7	0929 11.3	1624 2.6	2150 10.8
					31 TU	0434 3.1	1004 10.8	1657 3.1	2224 10.3

Chart Datum is 6·29 metres below IGN Datum. HAT is 13·6 metres above Chart Datum.

5.12 RIVER RANCE

Ille-et-Vilaine 48°37'·04N 02°01'·71W (Barrage) ❀❀🌢🌢❁❁❁

CHARTS AC 3659, 2700; SHOM 7130, 4233; Imray C33B

TIDES Standard Port ST MALO (←)
Water levels upriver of the Rance hydro-electric tidal barrage are strongly affected by the operation of the sluice gates and occasional use of the turbines as pumps. On most days from 0700–2100LT, 4m above CD is maintained. There is generally 8·5m above CD for a period of 4hrs between 0700–2000LT. A French language pamphlet, issued by Électricité de France (EDF) should be obtained from HMs at St Malo or Bas Sablons, or from the office at the barrage lock. It gives forecasts for the summer months of when heights of 4m and 8·5m above CD will occur in the period 0700–2000LT. The local daily paper *Ouest France* gives a forecast for the next day of HW and LW upstream of the barrage, under the heading *Usine Marémotrice de la Rance*.

SHELTER Good shelter upriver dependent on wind direction. The principal ⚓s/moorings on the E bank are at St Suliac and Mordreuc, and at La Richardais, La Jouvente, Le Minihic and La Pommeraie on the W bank. Marinas at Plouër, Lyvet (E bank, beyond Chatelier lock) and Dinan: see next column.

NAVIGATION From St Malo/Dinard, appr the lock (at the W end of the barrage) between Pte de la Jument to stbd and a prohib sluicing zone to port, marked by PHM buoys linked by wire cables. 4 W waiting buoys lie on the E side of the appr chan, close to the prohibited zone, with 4 W waiting buoys upstream of the lock.

> **Barrage lock** opens day/night (on request 2030–0430) on the hour, every hour provided the level is at least 4m above CD on both sides of the barrage see: *www.plouer-sur-rance.fr/fr/horairesdesmarees*. Yachts should arrive at H–20 mins. An illuminated display board gives access times in French/English. Lock entry sigs are **IPTS** sigs 2, 3 and 5. Masted boats entering from sea should berth at the S end of the lock so that the bridge can close astern of them. Vertical wires assist berthing/warp-handling. The lifting road-bridge across the lock opens only between H and H +15. Upstream of the lock a further prohib area to port is marked as above.

The chan upriver is marked by perches; binoculars needed. The 3M chan to St Suliac has min depth of 2m. The next 6M to the Chatelier lock to access Lyvet marina partially dries and is buoyed; keep to the outside of bends, (note: chan reptd increasingly silted 2018, notably in approach to lock). The suspension bridge and road bridge at Port St Hubert have 20m clearance. A viaduct 1M beyond Mordreuc has 19m clearance. Allow 2–3hrs from the barrage to Chatelier.

Chatelier lock and swing bridge operate 0600-2100LT, provided there is at least 8·5m rise of tide. HW Chatelier is 2–3hrs after HW St Malo depending on the barrage. Entry is controlled by CEVNI sigs (variations on a R and G theme).

The final 3M to Dinan has a published min depth in the marked chan of 1·4m; check with lock-keeper. Dinan gives access to Ille et Rance Canal and River Vilaine to Biscay (see Inland waterways).

LIGHTS AND MARKS Approaching the barrage from seaward: Pte de la Jument bn tr, Fl G 4s 6m 4M; PHM buoy opposite, Fl R 4s (prohib zone). The lock control twr is conspic. NW side of lock, Fl (2) G 6s, with G ▲ on W □. First dolphin, Fl (2) R 6s, R ■ on W □.

Approaching from Dinan: PHM buoy, Fl (2) R 6s, at S end of prohib zone; leave to stbd. First dolphin, Fl (3) R 12s, with R ■ on W □. SW side of lock, Fl (3) G 12s, G ▲ on W □.

COMMUNICATIONS ⚕ 02·99·46·10·46; Auto 08·92·68·08·35; Police 02·99·81·52·30; Brit Consul 01·44·51·31·00; Aff Mar 02·96·39·56·44; Water levels/navigation 02·99·46·14·46; Barrage/lock info 02·99·46·21·87, *Barrage de la Rance* (lock) VHF Ch 13; Chatelier lock 02·96·39·55·66, VHF Ch 14.

FACILITIES St Suliac ✕ ☐ 🛒 🔧 ⚓ ⚓ Divers (Convoimer). **Mordreuc** ✕ ☐ ⚓ 🔧 🛥.

La Richardais HM ☎02·99·46·24·20, ✕ ☐ 🛒 🛢 🛢 ✉ Ⓑ ⊞ ⚒ ✕.
La Jouvente ⚓ ✕ ☐. **Le Minihic** ⚓ 🔧 ⚓ ⚓ ⚒ ⊞ ✕.

La Cale de Plouër ⚓ 🔧 SC ✕.

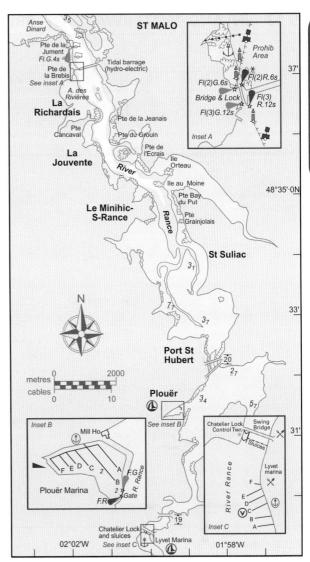

MARINAS ON THE RIVER RANCE
(NB: Berthing space limited, visitors are advised to call ahead.)
PLOUËR, Côtes d'Armor, 48°31'·54N 01°58'·95W. ❀❀🌢🌢❁❁. On W bank of R Rance, 6M S of the barrage and 2M N of Chatelier lock. Tide 8m above CD gives 1·5m water over rising gate. Appr with Plouër ⊕ spire brg 285°. Off S end bkwtr unlit PHM and SHM perches are 30m from narrow ent between 2 metal lattice pylons marking sill; access when FG/FR lts lit, day/night. Floodlit depth gauge (hard to read).
Marina plouer.portplaisance@wanadoo.fr www.plouer-sur-rance.fr/fr/portdeplaisance ☎02·96·86·83·15; 🖷02·96·89·11·00. Ch 09. HM, Mlle Liliane Faustin is very helpful. 230 ⚓ +10 Ⓥ pontoon B, €2·53. ✕ ☐ ⚒ ⛽ ⚓ 🛒 ⚓ elec bike hire 🏗(14t) 🛠. **Village** 🚆 1M.

LYVET, 48°29'·40N 02°00'·00W, on E bank just S of Chatelier lock. ❀❀🌢🌢. **La Vicomté Marina** ☎02·96·83·35·57. 210+25Ⓥ €2·00. Berth in 2m on pontoon D, or as directed. ✕ ☐ limited 🛒. Draught restricted trot moorings at La Hisse (opp Lyvet), contact La Mairie at Sampson-sur-Rance.

DINAN ❀❀🌢🌢❁❁❁. **Marina** ☎02·96·39·56·44. ⚓ line the W bank of the river, with fingers (€1·50) near to the port in about 1·5m. 🛢 ⚓ ✕ ☐ 🛠 for de-masting. Low bridge beyond port has 2·5m headroom, giving access to the Ille et Rance canal. **Town** (75m above river) ✕ 🚆 🛒 ✉ Ⓑ ⚲ ✈ (Dinard). Also ⚓ on Quai du Tallard, Lanvally (opp Dinan).

Ille et Rance Canal (opposite): Tolls (see Special notes) are not levied on Breton canals. A CEVNI is required, unless <5m LOA *and not capable* of more than 20kph/11kn.

5.13 INLAND WATERWAYS OF BRITTANY

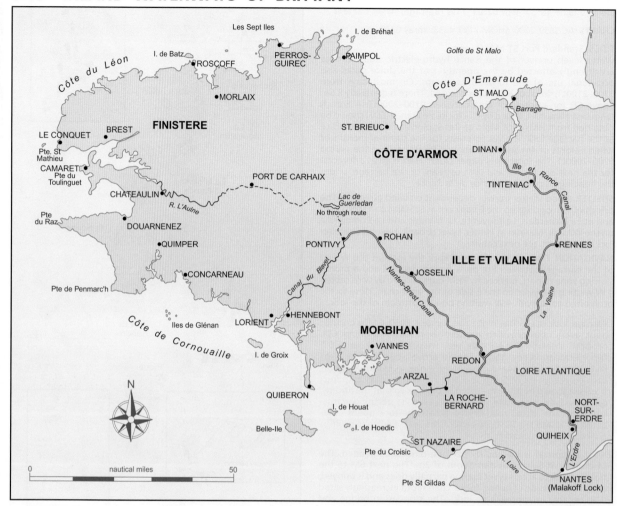

NAVIGATION Canals and rivers across and within Brittany enable boats of limited water and air draughts to go from the English Channel to the Bay of Biscay avoiding the passage around Finistère. Distances, number of locks, boat size and speed limits are summarised below. Dinan-Arzal takes about 5 days. Despite the many locks it can be a thoroughly enjoyable trip through unspoiled countryside and some interesting towns. Most overnight stops are free or at nominal cost.

LOCKS From Apr to Sept locks are worked 7 days a week 0800–1930LT, closing for lunch 1230-1330 approx. All locks are attended, but a fair measure of self-help is the order of the day. In Jul/Aug, in order to conserve water, locks may open on the hour only (and at H+30 if traffic demands).

ACCESS For prior estimate of max possible draught, contact: Equipement, Ille et Vilaine, 1 Avenue de Mail, 35000 Rennes, ☎02·99·59·20·60; or obtain recorded information update on ☎02·99·59·11·12. For latest info on the Ille et Rance Canal/R Vilaine, contact: Rennes ☎02·99·59·20·60 or Redon ☎02·99·71·03·78. For the Lorient-Nantes Canal, contact: Lorient ☎02·97·21·21·54; Hennebont ☎02·97·85·15·15; Pontivy ☎02·97·25·55·21; Nantes ☎02·40·71·02·00. Closures *(Chômages)* for maintenance are scheduled every Wed from approx first week in Nov to last week in March.

INFORMATION *Cruising French Waterways* (McKnight/ACN) *Through the French Canals* (Jefferson/ACN) and *Navicarte Guide No 12 (Bretagne).*

TOLLS may be due on the R Loire only; see Area 4, Special Notes for France, for rates.

SUMMARY	Length km	No of locks	Max draft m	Max air draft m	Max LOA m	Max beam m	Speed limit kn
St MALO–ARZAL (Ille et Rance Canal and La Vilaine river)							
R Rance-Dinan	29·0	1	1·3	19	25	–	5·4
Ille et Rance Canal							
Dinan-Rennes	79·0	48	1·2	2·5	25	4·5	4·3
Rennes-Redon	89·0	13	1·2	3·2/2·6*	25	4·5	4·3
Redon-Arzal	42·0	1	1·3	–			
*Depending on water level							
LORIENT–NANTES							
Canal du Blavet (See Lorient)							
Lorient-Pontivy	70	28	1·4	2·6	25	4·6	4·3
Nantes-Brest Canal							
	184·3	106	–	3	25	4·6	4·3
Pontivy-Rohan			0·8 (possible closure)				
Rohan-Josselin			1·0				
Josselin-Redon			1·4				
Redon-Quiheix			1·1				
L'Erdre River	27·6	1	1·4	3·8	400	6·2	13·5
R Loire, above Nantes (St Nazaire) may be navigable to Angers.							
R L'AULNE (See Brest)							
Brest-Chateaulin	42	1	3·0	N/A	25	–	–
Chateaulin-Carhaix							
	72	33	1·1	2·5	25	4·6	4·3

MINOR HARBOUR BETWEEN ST MALO AND ST CAST LE GUILDO
ST BRIAC 48°37´·31N 02°08´·71W. AC 3659, 2700; SHOM 7155, 7130, 7129. HW −0515 on Dover (UT); ML 6·8m; Duration 0550. Tides as Dahouet; use Ile des Hébihens. Drying hbr (6m) open to SW-NW. A Dir Iso WRG 4s lt leads 125° between offlying drying and beaconed rocks/islets. The last 6ca are marked by 4 PHM and 3 SHM perches. 10 ❻s in Le Bechet cove or ‡ on sand W of Ile du Perron in 3-5m. HM ☎02·99·88·01·75, ⚓ ⚓. YC de St Briac ☎02·99·88·31·45. Town ✕ ☐ ☷.

5.14 ST CAST LE GUILDO

Côtes d'Armor **48°38´·39N 02°14´·76W** ✿✿✿❁❁❁❁✿✿✿

CHARTS AC 2669, 3659; SHOM 71, 7129, 7155; Navi 535/6; Imray C33B

TIDES −0520 Dover; ML 6·3; Duration 0550

Standard Port ST MALO (◄—)

Times				Height (metres)			
High Water		Low Water		MHWS	MHWN	MLWN	MLWS
0100	0800	0300	0800	12·2	9·3	4·2	1·5
1300	2000	1500	2000				
Differences SAINT CAST							
0000	0000	−0005	−0005	−0·2	−0·2	−0·1	−0·1
ERQUY							
0000	−0005	−0020	−0010	−0·8	−0·6	−0·2	−0·2

SHELTER Good shelter in marina, access all states of tide, but exposed shoaling approach in strong NE/E/SE conditions. No anchoring.

NAVIGATION. WPT 48°39´·00N 02°13´·25W E of ECM Les Bourdinots (¾M NE of Pte de St Cast, marks 2m patch) then 225° for 1M to marina ent chan, R & G bys. Beware west mole, submerged HW.

LIGHTS AND MARKS Iso WG 4s 11/8M (appr on W sectors to clear Les Bourdinots). Bright shore lights confusing until within 2M.

COMMUNICATIONS CROSS Corsen 02·98·89·31·31; Ⓛ 02·96·41·76·61; Police 02·99·81·52·30; British Consul 01·44·51·31·00; Ⓗ St Malo 02·99·56·56·19; HM 02·96·70·88·30, VHF Ch 09, 16; Semaphore 02·96·41·85·30; Affaires Maritime 02·96·68·30·70.

FACILITIES **Port d'Armor Marina** ☎02·96·81·04·43 www.cotesdarmor.cci.fr stcast.plaisance@cotesdarmor.cci.fr; 825 inc 500❻ €3·20 inc ▮ and visitor tax; pontoon A arrivals, ❻▱ D or Z, according to LOA indicated or as instructed. ⛽ ⛽ (H24 cr/card only) ✕ ⌁ ▣ ▮(20t) ⛴(2·5t, masts). YC ☎02·96·41·71·71, ▮.

Town (10 mins) ✕ ☐ ☷ ⬟ Ⓑ ⚒ ✕ ✉ Ⓑ ⇌ (Lamballe) ✈ Pleurtuit/Dinard, Rennes). **Ferry** St Malo.

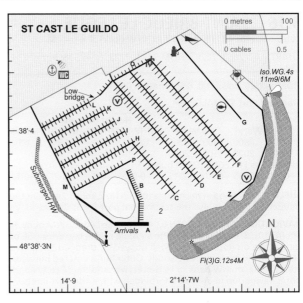

MINOR HARBOUR BETWEEN ST CAST LE GUILDO AND DAHOUET
ERQUY (pronounced Erky) **48°38´·04N 02°28´·69W**. AC 2669, 2029; SHOM 7154, 7310. HW −0515 on Dover (UT); ML 6·5m; Duration 0550; see St Cast Le Guildo. Sheltered from E, but open to SW/W'lies. An active FV hbr, but yachts can dry out E of the inner jetty hd Fl (3) R 12s; or ‡ on sand 3ca SW of the outer mole. Beware Plateau des Portes d'Erquy (dry) about 2M to W. Beware rks off Pte de Lahoussaye. Mole hd lt Fl (2) WRG 6s 11m W10M, R/G7M; approach in either W sector (see Lights, buoys and waypoints). HM & ⊖ ☎02·96·72·19·32. YC ☎02·96·72·32·40. Facilities: ❻ (drying) €0·56, **Quay** ⛴(3·5t) ⚓ ⛽ ⛽. **Town** ✕ ☐ ☷ ⬟ Ⓑ ⚒ ✕.

5.15 DAHOUET

Côtes d'Armor **48°34´·79N 02°34´·39W** ✿✿❁❁✿

CHARTS AC 2669, 2029; SHOM 7154, 7310; Navi 536; Imray C33B, C34

TIDES −0520 Dover; ML 6·3; Duration 0550

Standard Port ST MALO (◄—)

Times				Height (metres)			
High Water		Low Water		MHWS	MHWN	MLWN	MLWS
0100	0800	0300	0800	12·2	9·3	4·2	1·5
1300	2000	1500	2000				
Differences ÎLE DES HÉBIHENS (7M W of St Malo)							
0000	0000	−0005	−0005	−0·2	−0·2	−0·1	−0·1
DAHOUET							
−0005	0000	−0025	−0010	−0·8	−0·7	−0·2	−0·1

SHELTER Good, but ent (dries 4m) has strong currents and is unsafe in fresh/strong NW'lies, when a bar may form.

NAVIGATION WPT 48°35´·15N 02°35´·44W, unlit NCM buoy, 117°/0·8M to La Petite Muette (LPM) lt tr. Appr in W sector, crossing into the R until LPM bears 160°. Enter the narrow break in the cliffs on that track leaving LPM to stbd, 1st R pole to port, then 3 remaining R poles to port, to clear slipway/wall below the shrine on headland. It is dangerous to enter S of LPM, due to rocks. At SHM bcn the chan turns E, then SE to the marina sill marked by PHM/SHM perches. Night entry not advised without prior day experience.

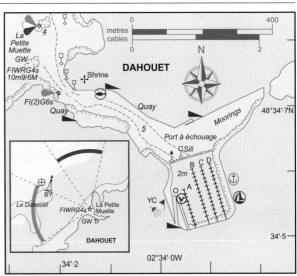

LIGHTS AND MARKS La Petite Muette (see Lights, buoys and waypoints) is the key feature. Conspic daymarks: the wide beach NE at Val André and, in the hbr ent, a pagoda-like shrine. See also St Brieuc.

DAHOUET *continued*

COMMUNICATIONS CROSS 02·96·70·42·18; ☎ 02·36·65·02·22; Auto 08·92·68·02·22; Police 02·96·72·22·18; ⊖ 02·96·74·75·32; British Consul 01·44·51·31·00; Ⓗ 02·96·45·23·28; HM VHF Ch 09, 16; Affaires Maritimes 02·96·72·31·42.

5.16 SAINT-BRIEUC LE LÉGUÉ

Côtes d'Armor **48°31′·68N 02°43′·97W** (lock) ✹✹◊◊◊❀

CHARTS AC 2669, 2029; SHOM 7154, 7128; Navi 536; Imray C34, C33B

TIDES –0520 Dover; ML 6·5; Duration 0550

Standard Port ST MALO (←)

Times				Height (metres)			
High Water		Low Water		MHWS	MHWN	MLWN	MLWS
0100	0800	0300	0800	12·2	9·3	4·2	1·5
1300	2000	1500	2000				
Differences LE LÉGUÉ (SWM buoy)							
–0010	–0005	–0020	–0015	–0·8	–0·5	–0·2	–0·1

SHELTER Very good in the marina (Bassin No 2, min 3·2m) near the viaduct. Le Légué is also the commercial port for St Brieuc.

NAVIGATION WPT 48°34′·33N 02°41′·15W, Le Légué SWM buoy, Fl Mo (A) 10s, 210°/2·6M to abm Pte à l'Aigle lt. The bay dries E/SE of conspic Pte du Roselier. Chan well buoyed/lit, dredged 5m above CD; but avoid in strong N/NE winds. Outer hbr is entered between Pte à l'Aigle, on NW bank of Le Gouet river and the SE jetty head. Note: navigation under sail is forbidden within the port limits.

Lock opens, relative to height of tide at St Malo, *at local HW*:

FACILITIES Outer hbr (FVs) dries 5·5m. **Marina**, min depth 2m, is entered over 5·5m sill, with depth gauge. Very crowded in season. ☎02·96·72·82·85, 318+25 ♥ berths on 'O' pontoon in about 1·6-2·0m, €2·25. ⬚⬚⬚⬚⬚⬚⬚⬚(8t mobile) ⚓ **YC du Val-André** ☎02·96·72·21·68, ⬚.
Town ✕◻⬚⬚ ⇌ (Lamballe) ✈ (St Brieuc). **Ferry** St Malo.

(MHWN 9·3m)	<9·4m	±1
	9·4-10·3m	±1¼
	10·3-10·8m	±1½
(MHWS 12·2m)	>10·8m	–2 to +1½

The lock is 85m x 14m; sill dries 5·0m. Lock staff help with warps. Commercial traffic takes priority and the outer Bassin, No 1, is reserved for shipping only. A low swing bridge at the ent to Bassin No 2 (marina) opens when the lock is in use. A sliding footbridge gives access further upstream (no ♥).

LIGHTS AND MARKS Conspic marks: Le Verdelet Is (42m) from E; Rohein twr, from N; Pte du Roselier in near appr; St Brieuc.

COMMUNICATIONS CROSS 02·98·89·31·31; ⬚ 02·96·88·35·47; ☎ 02·99·46·10·46; Auto 08·92·68·08·22; Police 02·96·61·22·61; ⊖ 02·96·74·75·32; Brit Consul 01·44·51·31·00; Dr 02·96·61·49·07; Port HM 02·96·33·35·41, *Le Légué Port* VHF Ch 12 (approx HW–2 to +1½); Aff Mar 02·96·68·30·70.

FACILITIES Marina ☎02·96·77·49·85, mob 06·75·91·67·63, legue. plaisance@cotesdarmor.cci.fr/ www.cotesdarmor.cci.fr 250 + 20♥ €2·10. Berth by HM's office on S quay for directions. ✕◻⬚⬚ ⬚⬚⬚⬚⬚◊ⒺⒸ YC⬚(20t) ⬚(2t for de-masting).

Ship Repair Yard ☎06·76·33·56·66, ⬚(350t).

Town ⬚ ⬚⬚⬚ @ ⇌. **Ferry** St Malo, Roscoff.

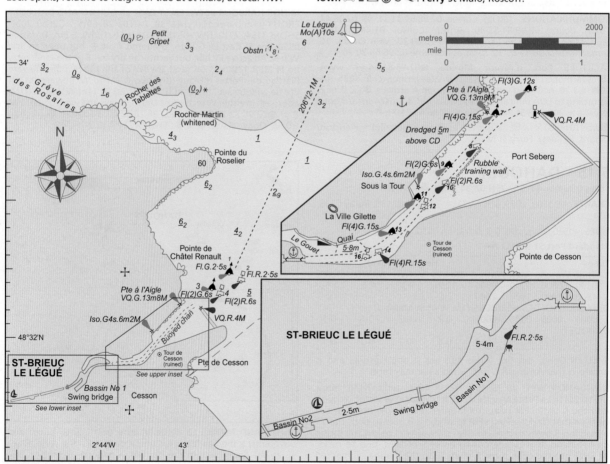

5.17 BINIC

Côtes d'Armor **48°36´·06N 02°48´·99W** ❀❀⚓⚓☆☆☆

CHARTS AC 2648, 2669, 2029; SHOM 7154, 7128; Navi 536; Imray C33B, C34

TIDES –0525 Dover; ML 6·3; Duration 0550

Standard Port ST MALO (←—)

Times				Height (metres)			
High Water		Low Water		MHWS	MHWN	MLWN	MLWS
0100	0800	0300	0800	12·2	9·3	4·2	1·5
1300	2000	1500	2000				
Differences BINIC							
–0005	–0005	–0025	–0015	–0·8	–0·6	–0·2	–0·1

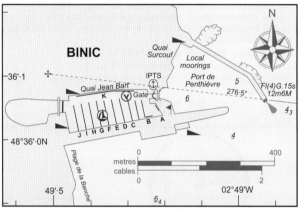

SHELTER Good in marina. Appr not advised in strong E/NE'lies.

NAVIGATION WPT 48°35´·86N 02°46´·68W, 276·5°/1·5M to ent. Appr dries 7ca offshore. From E, appr via Caffa ECM buoy, whence ent bears 246°/4·2M; or from N through Rade de St Quay-Portrieux.

Avant Port dries 5·5m, is easy day/night, except in E'lies. IPTS, sigs 2 & 3, control entry H24 to the marina via hydraulic flapgate and retracting bridge (office manned only when tide is sufficient). Gate opens when height of tide >8·5m. Times vary (see www.ville-binic.fr) with coefficient/barometer:

Coeff 40	<8·5m	=	No entry; no exit, ie neaped
40-50	8·5m	=	HW –1 to HW +1
50-60	9·15m	=	HW –1½ to HW + 1½
60-80	9·5m	=	HW –2 to HW + 2
>80	10·6m	=	HW –2¼ to HW + 2¼

Outbound vessels have priority. Beware strong current (up to 5kn) in the ent for 20 mins after the gate first opens.

LIGHTS AND MARKS Ldg line 276·5°: N mole lt, W twr, G gallery, ≠ church spire. A water twr is conspic 8ca N of hbr.

COMMUNICATIONS CROSS 02·98·89·31·31; ⊜ 02·96·74·75·32; 🚑 08·92·68·08·08; Ⓗ 02·96·94·31·71; SAMU 15; Fire 18; Call Port/marina VHF Ch 09 to check tide height and to advise of boat size and length of stay; Aff Mar 02·96·68·30·70.

FACILITIES Marina ☎02·96·73·61·86. English spoken. port@ville-binic.fr www.ville-binic.fr 350 + 50Ⓥ €2·05. Ⓥ berth/raft on pontoon K, to stbd on entry (1·5-2·5m).⫶⧽ ⚓ 🛢 🛢 SHOM ━ Ⓔ SC ⛴(20t mobile), hire car available. At St Quay-Portrieux 🅟 🅟 and nearby ⚠ ⚒ 🛢 ✖.

Town ✖ 🛏 🛒 🛢 🛢 @ ✉ Ⓑ. Bus to St-Brieuc ✈. **Ferry** St Malo.

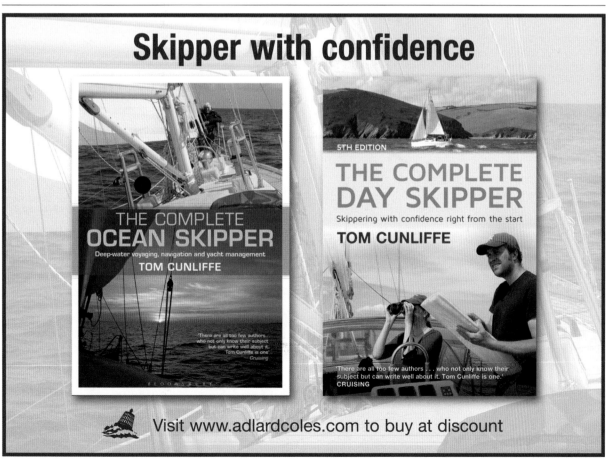

5.18 ST QUAY-PORTRIEUX

Côtes d'Armor 48°38′·84N 02°48′·97W ❄❄❄♨♨♨❀❀

CHARTS AC 2648, 2669, 3674; SHOM 7154, 7128; Navi 536, 537; Imray C33B, C34

TIDES –0520 Dover; ML 6·3; Duration 0550

Standard Port ST MALO (←)

Times				Height (metres)			
High Water		Low Water		MHWS	MHWN	MLWN	MLWS
0100	0800	0300	0800	12·2	9·3	4·2	1·5
1300	2000	1500	2000				
Differences ST QUAY-PORTRIEUX							
–0010	–0010	–0030	–0010	–0·9	–0·7	–0·2	–0·1

SHELTER Excellent in marina. ⚓ in the Rade de St Quay-Portrieux, sheltered by Roches de St Quay, but open to NW & SE winds.

NAVIGATION From E and SE appr via Caffa ECM lt buoy to WPT 48°37′·08N 02°46′·18W (2ca S of La Roselière SCM buoy VQ (6) + L Fl 10s); thence 314°/2·5M to NE mole head.

From the N, WPT 48°40′·94N 02°49′·69W, 169°/2·0M to NE mole elbow. At night, to avoid rocks/shoals, follow the white sectors of 4 Dir lts (see chartlet and tinted box) tracking 169°, 130°, 185° (astern) and 318° in sequence to the marina ent:

- NE mole elbow lt, **W159°-179°**. White concrete twr.
- Herflux lt, **W125°-135°**. A white bcn twr on drying rock.
- Ile Harbour lt, **W358°-011°**. Short W twr/R top on an islet.
- NE mole elbow lt (as first • above) **W316°-320·5°**.

LIGHTS AND MARKS See chartlet and Lights, buoys and waypoints. Signal stn at Pointe de St Quay, NNW of the marina, is conspic. N of the marina Moulières de Portrieux, ECM bcn twr is unlit. E of the marina an unlit WCM buoy marks Les Noirs rks (2·4m).

COMMUNICATIONS CROSS 02·98·89·31·31; ⚓ 02·96·76·76·80; Auto 08·92·68·08·22; Police 02·96·70·61·24; ⊖ 02·96·33·33·03; Brit Consul 01·44·51·31·00; Dr 02·96·70·41·31; HM (Old Hbr) 02·96·70·95·31; Aff Mar 02·96·68·30·73; Marina VHF Ch 09 (H24).

FACILITIES Marina ④ ☎02·96·70·81·30. welcome@port-armor.com www.port-armor.com 900+100❶ Jul/Aug, €3·00, less other months. ❶ berths on No 7 pontoon (no fingers NW side). Jul/Aug pre-call for a berth if LOA >12m. 🛉 🏋 ✕ 🗑 🛢⚓ 🔧 Ⓔ ✖ ⚓(40t) 🗑(5t).

SN de St Quay-Portrieux ☎02·96·70·93·34.

Cercle de la Voile de Portrieux ☎02·96·70·41·76.

Old Hbr ☎02·96·70·95·31, dries approx 5·3m. 500+8 ⚓ €1·50. ⚓ ⛽ 🗑(1·5t).

Town ✕ 🗑 🏪 ✉ Ⓑ 🛢. Bus to St-Brieuc 🚃 ✈. **Ferry** St Malo-Portsmouth.

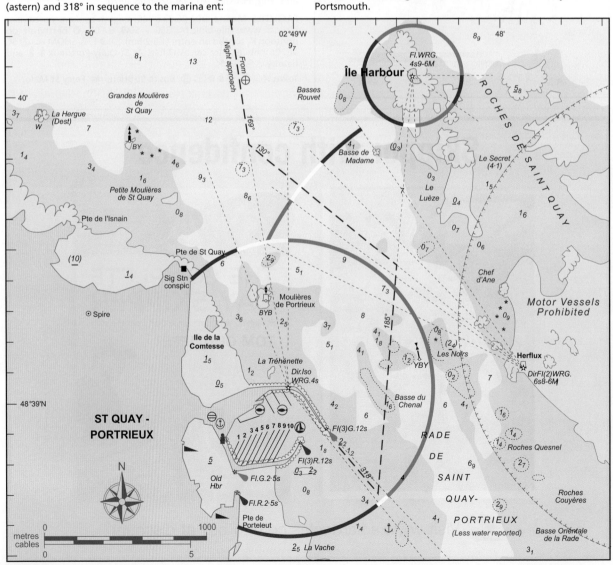

Gripping true stories from Adlard Coles

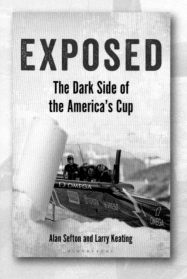

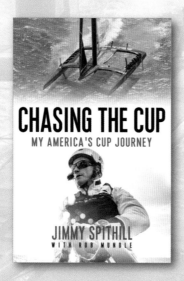

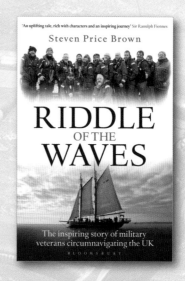

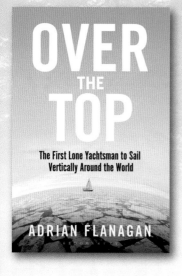

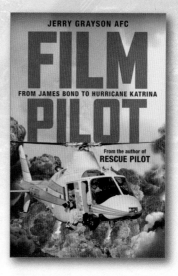

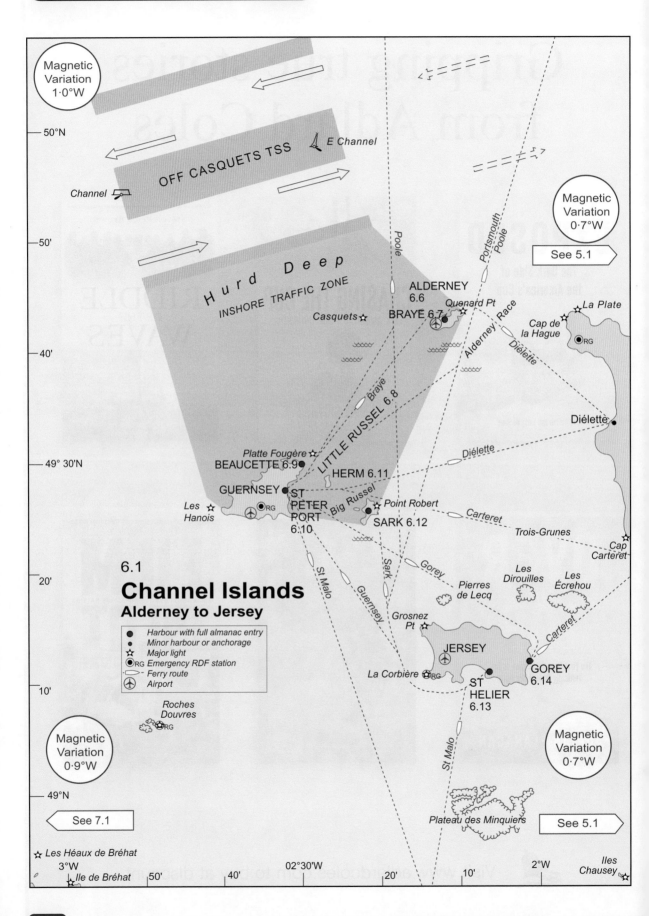

Magnetic Variation 1·0°W

50°N

OFF CASQUETS TSS

E Channel

Channel

50'

Hurd Deep

INSHORE TRAFFIC ZONE

Poole

Portsmouth Poole

ALDERNEY 6.6

BRAYE 6.7

Quenard Pt

Magnetic Variation 0·7°W

See 5.1

La Plate

Cap de la Hague

RG

40'

Casquets ☆

Alderney Race

Diélette

Braye

Diélette

LITTLE RUSSEL 6.8

Diélette

49° 30'N

Platte Fougère ☆

BEAUCETTE 6.9

GUERNSEY

ST PETER PORT 6.10

HERM 6.11

Big Russel

Point Robert

SARK 6.12

Carteret

Trois-Grunes

Cap Carteret

Les Hanois ☆

RG

6.1

Channel Islands
Alderney to Jersey

● *Harbour with full almanac entry*
• *Minor harbour or anchorage*
☆ *Major light*
◉RG *Emergency RDF station*
⌐ *Ferry route*
⊕ *Airport*

20'

St Malo

Sark

Guernsey

Gorey

Pierres de Lecq

Les Dirouilles

Les Écrehou

Carteret

Grosnez Pt

JERSEY

GOREY 6.14

10'

La Corbière

RG

ST HELIER 6.13

Roches Douvres

RG

Magnetic Variation 0·9°W

Magnetic Variation 0·7°W

49°N

See 7.1

See 5.1

Plateau des Minquiers

☆ Les Héaux de Bréhat

3°W

Ile de Bréhat

50'

40'

02°30'W

20'

10'

2°W

Iles Chausey

#	Place																				
1	*L'Aberwrac'h*	145	131	126	128	116	115	122	107	103	93	109	110	123	103	91	88	84	72	32	**1**
2	*Roscoff*	117	103	95	96	84	87	94	77	73	63	79	80	93	71	59	58	54	41	**2**	
3	*Tréguier*	94	80	72	72	60	66	72	56	52	42	58	53	63	58	30	29	22	**3**		
4	*Lézardrieux*	88	74	68	54	49	65	68	52	48	42	38	47	55	33	21	14	**4**			
5	*Paimpol*	91	77	65	56	42	67	70	54	50	45	50	45	53	24	24	**5**				
6	*St Quay-Portrieux*	88	74	64	54	35	71	73	55	56	48	51	46	52	12	**6**					
7	*Dahouet*	88	74	62	44	28	70	72	62	58	57	53	41	47	**7**						
8	Gorey (Jersey)	47	33	16	29	38	36	35	32	29	35	20	13	**8**							
9	**St Helier**	59	45	28	30	38	43	46	33	29	32	24	**9**								
10	Creux (Sark)	37	23	23	50	52	18	22	11	10	16	**10**									
11	Les Hanois	49	35	37	58	56	23	29	14	10	**11**										
12	**St Peter Port**	42	28	31	55	54	18	23	4	**12**											
13	Beaucette	39	25	34	59	56	15	19	**13**												
14	**Braye (Alderney)**	23	9	26	66	73	8	**14**													
15	Casquets	31	17	32	63	70	**15**														
16	*St Malo*	87	73	49	23	**16**															
17	*Granville*	75	61	38	**17**																
18	*Carteret*	41	23	**18**																	
19	*Cap de la Hague*	14	**19**																		
20	*Cherbourg*	**20**																			

DISTANCE TABLE
Approximate distances in nautical miles are by the most direct route while avoiding dangers and allowing for TSS. Places in *italics* are in adjoining areas; places in **bold** are in 0.33, Cross-Channel Distances.

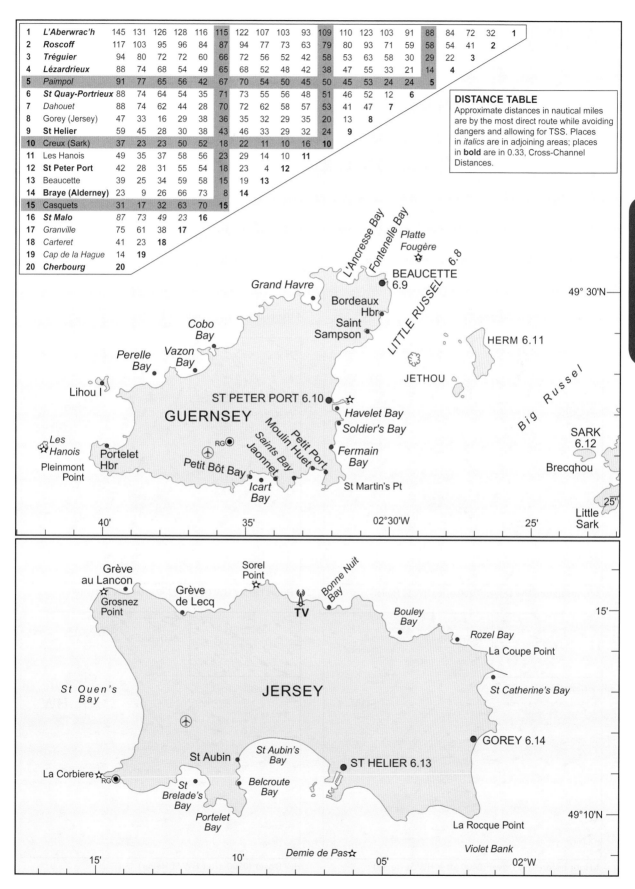

Channel Islands

L'Ancresse Bay
Fontenelle Bay
Platte Fougère
6.8
BEAUCETTE 6.9
49° 30'N
Grand Havre
Bordeaux Hbr
Saint Sampson
LITTLE RUSSEL
HERM 6.11
Cobo Bay
JETHOU
Perelle Bay
Vazon Bay
Big Russel
Lihou I
ST PETER PORT 6.10
SARK 6.12
GUERNSEY
Havelet Bay
Soldier's Bay
Les Hanois
Moulin Huet
Petit Port
Saints Bay
Fermain Bay
Brecqhou
Portelet Hbr
Jaonnet
Pleinmont Point
Petit Bôt Bay
Icart Bay
St Martin's Pt
Little Sark
25"
40'
35'
02°30'W
25'

Grève au Lancon
Sorel Point
Bonne Nuit Bay
Grève de Lecq
Bouley Bay
Grosnez Point
TV
15'
Rozel Bay
La Coupe Point
St Catherine's Bay
St Ouen's Bay
JERSEY
GOREY 6.14
St Aubin
St Aubin's Bay
ST HELIER 6.13
La Corbiere
St Brelade's Bay
Belcroute Bay
Portelet Bay
La Rocque Point
49°10'N
Demie de Pas
Violet Bank
15'
10'
05'
02°W

6.2 CHANNEL ISLANDS TIDAL STREAMS

More UK/Irish tides at www.reedstides.co.uk

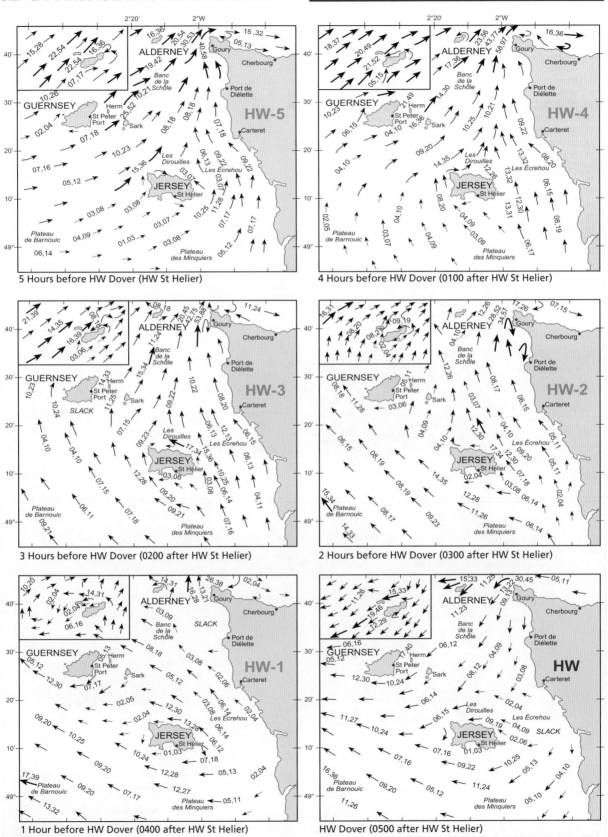

5 Hours before HW Dover (HW St Helier)

4 Hours before HW Dover (0100 after HW St Helier)

3 Hours before HW Dover (0200 after HW St Helier)

2 Hours before HW Dover (0300 after HW St Helier)

1 Hour before HW Dover (0400 after HW St Helier)

HW Dover (0500 after HW St Helier)

English Channel 0.26   Central North France 5.2

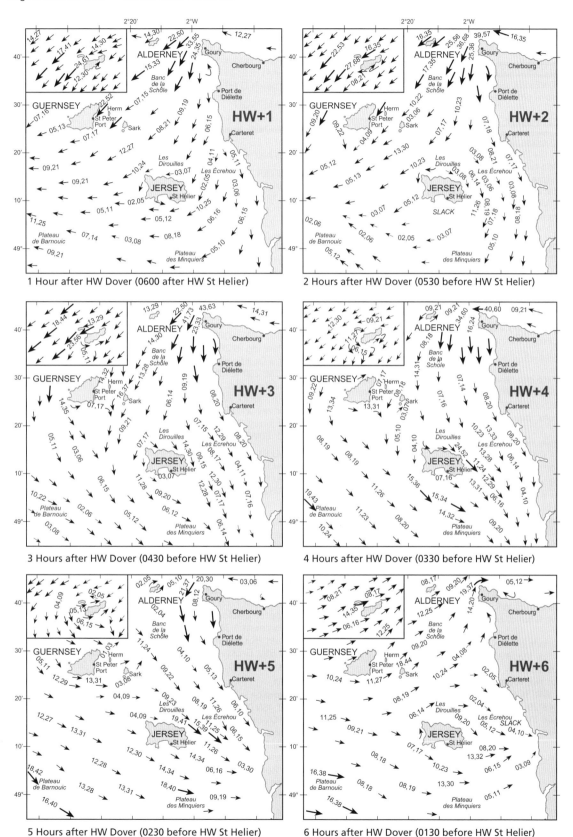

1 Hour after HW Dover (0600 after HW St Helier)

2 Hours after HW Dover (0530 before HW St Helier)

3 Hours after HW Dover (0430 before HW St Helier)

4 Hours after HW Dover (0330 before HW St Helier)

5 Hours after HW Dover (0230 before HW St Helier)

6 Hours after HW Dover (0130 before HW St Helier)

Channel Islands

6.3 LIGHTS, BUOYS AND WAYPOINTS

Bold print = light with a nominal range of 15M or more. CAPITALS = place or feature. *CAPITAL ITALICS* = light-vessel, light float or Lanby. *Italics* = Fog signal. ***Bold italics*** = Racon. Many marks/buoys are fitted with AIS (<u>MMSI No</u>); see relevant charts.

MID-CHANNEL MARKS

CHANNEL LT VESSEL ⌐ Fl 15s 12m **15M**; R hull with lt twr amidships; *Horn (20s)*; ***Racon O, 15M***; 49°54'·46N 02°53'·74W; <u>992351028</u>.
E Channel ⌐ Fl Y 5s 6M; ***Racon T, 10M***; 49°58'·67N 02°29'·01W.

COTENTIN PENINSULA (NW COAST)

La Plate ⌑ Fl (2+1) WR 10s 11m; W9M, R6M; 115°-W-272°-R-115°; Y 8-sided twr, with B top; 49°43'·98N 01°55'·76W.
Cap de la Hague (Gros du Raz) ☆ Fl 5s 48m **23M**; Gy twr, W top; *Horn 30s*; 49°43'·31N 01° 57'·27W.
La Foraine ⌑ VQ (9) 10s 12m 6M; 49°42'·90 N 01°58'·32W.
Cap de Carteret ☆ Fl (2+1) 15s 81m **26M**; Gy twr, G top; 49°22'·41N 01°48'·41W.

THE CASQUETS AND ALDERNEY

Casquets ☆ Fl (5) 30s 37m **18M**, H24; W twr, 2 R bands; NW'most of three; ***Racon T, 25M***; 49°43'·32N 02°22'·63W. <u>992351121</u>.
Ortac rock, unlit; 49°43'·40N 02°17'·44W.
Pierre au Vraic, unmarked rk 1·2m ⊛; 49°41'·61N 02°16'·94W.
Quenard Pt (Alderney) ⌇ Fl (4) 15s 37m 12M; 085°-027°; W ○ twr, B band; 49°43'·75N 02°09'·86W.
Château à L'Étoc Pt ⌇ Iso WR 4s 20m W10M, R7M; 071·1°-R-111·1°-W-151·1°; 49°43'·94N 02°10'·63W.

BRAYE

Ldg bns 142° (to clear the submerged Adm'ty bkwtr). Front, W ⌇, 49°43'·90N 02°10'·97W. Rear, BW ⌇; 720m from front.
Ldg lts 215°: both Q 8/17m 9/12M, synch; 210°-220°; orange △s. Front, old pier elbow, 49°43'·40N 02°11'·91W. Rear, 215°/335m.
Admiralty bkwtr head ⌇ L Fl 10s 7m 5M; 49°43'·82N 02°11'·67W.
Fairway No 1 ▲ QG; 49°43'·72N 02°11'·72W.
No 2 ◪ QR; 49°43'·60N 02°11'·75W.
Inner fairway ▲ Q (2) G 5s; 49°43'·58N 02°11'·98W.
Braye quay ⌇ 2 FR (vert) 8m 5M; 49°43'·53N 02°12'·00W.
Little Crabby hbr ent ⌇ FG & FR 5m 2M; 49°43'·45N 02°12'·12W.

GUERNSEY, NORTHERN APPROACHES

LITTLE RUSSEL CHANNEL

Grande Amfroque, two unlit bcn twrs: larger, BW-banded; smaller, white; 49°30'·57N 02°24'·62W.
Tautenay ⌇ Q (3) WR 6s 7m W7M, R6M; 050°-W-215°- R-050°; B & W striped bcn 49°30'·11N 02°26'·84W.
Platte Fougère ☆ Fl WR 10s 15m **16M**; 155°-W-085°-R-155°; W 8-sided twr, B band; *Horn 45s TD 2017*; ***Racon P***; 49°30'·83N 02°29'·14W.
Corbette d'Amont ⌑ Y bcn twr, topmark; 49°29'·64N 02°29'·38W.
Roustel ⌇ Q 8m 7M; BW chequered base, W framework col; 49°29'·23N 02°28'·79W.
Rousse, Y bcn twr, topmark ⌗; 49°28'·98N 02°28'·36W.
Platte ⌇, Fl WR 3s 6m, W7M, R5M; 024°-R-219°-W-024°; G conical twr; 49°29'·08N 02°29'·57W.
Vivian bcn twr, BW bands, 49°28'·45N 02°30'·66W.
Brehon ⌇ Iso 4s 19m 9M; bcn on ○ twr, 49°28'·28N 02°29'·28W.
Demie Flieroque, Y bcn twr, topmark F; 49°28'·13N 02°31'·38W.

BIG RUSSEL

Noire Pute ⌇ Fl (2) WR 15s 8m 6M; 220°-W-040°-R-220°; on 2m high rock; 49°28'·21N 02°25'·02W; unreliable.
Fourquies ⌑ Q; 49°27'·34N 02°26'·47W.
Lower Heads ⌑ Q (6) + L Fl 15s; *Bell*; 49°25'·85N 02°28'·55W.

GUERNSEY, HERM AND SARK

BEAUCETTE MARINA

Petite Canupe ⌑ Q (6) + L Fl 15s; 49°30'·20N 02°29'·14W.
Ldg lts 277°: Both FR. Front, W □, R stripe; 49°30'·19N 02°30'·23W. Rear, R □, W stripe; 185m from front.
Appr chan buoys: SWM L Fl 10s 49°30'·15N 02°29'·66W. SHM Fl G 5s, PHM Fl R 5s & Fl (3) R 5s. NCM perch Q 49°30'·165N 02°30'·06W. SHM perch Q (3) G 5s at ent; 49°30'·18N 02°30'·24W.

ST SAMPSON

Ldg lts 286°: Front, FR 3m 5M; 230°-340°; tfc sigs; 49°28'·90N 02°30'·74W. Rear, FG 13m; clock twr, 390m from front.
N Pier ⌇ FG 3m 5M; 230°-340°; 49°28'·92N 02°30'·71W.
Crocq pier ⌇ FR 11m 5M; 250°-340°; 49°28'·98N 02°31'·00W.

ST PETER PORT

Outer ldg lts 220°: **Front**, Castle bkwtr, Al WR 10s 14m **16M**; 187°-007°; dark ○ twr, W on NE side; *Horn 15s*; 49°27'·31N 02°31'·45W.
Rear 220° ldg lt, Belvedere, Oc 10s 61m 14M; 179°-269°; intens 217°-223°; W □ on W twr; 980m from front. The lts are synchronised so that the rear lt is on when the front is W, and the off when the front is R.
Queen Elizabeth II marina, 270° Dir Oc ⌇ WRG 3s 5m 6M; 258°-G-268°-W-272°-R-282°; 49°27'·73N 02°31'·87W.
Reffée ⌑ Q (6) + L Fl 15s; 49°27'·74N 02°31'·27W.
Appr buoys: outer pair ▲ QG; 49°27'·83N 02°31'·54W. ◪ QR; 49°27'·71N 02°31'·52W. Inner pair: ▲ QG; 49°27'·76N 02°31'·74W. ◪ QR; 49°27'·72N 02°31'·74W.
The Pool ldg lts 265° (*not* into moorings): Front, Victoria marina, S Pier, ⌇ Oc R 5s 10m 14M; 49°27'·32N 02°32'·03W. Rear, ⌇ Iso R 2s 22m 3M; 260°-270°; on Creasey's bldg, 160m from front.
White Rock pier ⌇ Oc G 5s 11m 14M; intens 174°-354°; ○ twr; tfc sigs; 49°27'·38N 02°31'·59W.
S Fairway ▲ QG; 49°27'·30N 02°31'·76W; N of fuel pontoon W of buoy 3 x ◪ Fl R; 49°27'·28N 02°31'·74W. 49°27'·27N 02°31'·80W. 49°27'·27N 02°31'·86W.

HAVELET BAY

Oyster Rock ⌑ Y bcn, topmark 'O'; 49°27'·09N 02°31'·46W.
Oyster Rock ▲ QG; 49°27'·04N 02°31'·47W.
Moulinet ◪ QR; 49°26'·97N 02°31'·54W.
Moulinet ⌑ Y bcn, topmark 'M'; 49°26'·95N 02°31'·58W.

GUERNSEY, SOUTH-EAST and SOUTH COASTS

Anfré, Y bcn, topmark 'A'; 49°26'·45N 02°31'·48W.
Longue Pierre, Y bcn, topmark 'LP'; 49°25'·36N 02°31'·48W.
St Martin's Pt ⌇ Fl (3) WR 10s 15m 14M; 185°-R-191°-W-011°-R-061·5°; flat-topped, W bldg. *Horn (3) 30s;* 49°25'·30N 02°31'·70W.

GUERNSEY, NORTH-WEST COAST

Les Hanois ☆ Fl (2) 13s 33m **20M**; 294°-237°; Gy ○ twr, B lantern, helicopter platform; *Horn (2) 60s;* 49°26'·10N 02°42'·15W.
4 FR on mast 1·3M ESE of Les Hanois lt ho.
Portelet Hbr, bkwtr bcn, 49°26'·16N 02°39'·84W.
Cobo Bay, Grosse Rock, B bcn 11m; 49°29'·02N 02°36'·19W.
Grand Havre, Rousse Point bkwtr, B bcn; 49°29'·92N 02°33'·05W.

HERM

Corbette de la Mare, Wh disc (W side) and R disc (E side) on Y pole, 49°28'·48N 02°28'·72W.
Petit Creux ⌇ QR; red 'C' on red pole; 49°28'·09N 02°28'·72W.
Alligande ⌇ Fl (3) G 5s; B pole, Or 'A'; 49°27'·86N 02°28'·78W.
Épec ⌇ Fl G 3s; black 'E' on G mast; 49°27'·98N 02°27'·89W.
Vermerette ⌇ Fl (2) Y 5s; Or 'V' on bcn; 49°28'·12N 02°27'·75W.
Gate Rock (Percée Pass) ⌑ Q (9) 15s; 49°27'·88N 02°27'·54W.
Hbr ldg lts 078°: White drums. ⌇ 2F occas; 49°28'·25N 02°27'·11W.
Hbr pier, N end ⌇ 2 FG; G □ on G bcn; 49°28'·21N 02°27'·26W.

SARK

Courbée du Nez ⌇ Fl (4) WR 15s 14m 8M; 057°-W-230°-R-057°; W structure on rock; 49°27'·09N 02°22'·17W.

Point Robert ☆ Fl 15s 65m **18M**; vis 138°-353°; W 8-sided twr; *Horn (1) 30s;* 49°26'·19N 02°20'·75W.
Founiais ⊥, topmark 'F'; 49°26'·02N 02°20'·36W.
Blanchard ⌇ Q (3) 10s; *Bell;* 49°25'·36N 02°17'·42W.
Pilcher monument (070° appr brg); 49°25'·71N 02°22'·44W.

JERSEY, WEST AND SOUTH COASTS

Desormes ⌇ Q (9) 15s; 49°18'·94N 02°17'·98W.
Grosnez Point ☆ Fl (2) WR 15s 50m **W19M, R17M**; 081°-W-188°-R-241°; W hut; 49°15'·50N 02°14'·80W.
L a Rocco twr (conspic) 15m; 49°11'·90N 02°14'·05W.
La Frouquie ⌁ (seasonal); 49°11'·30N 02°15'·38W.
La Corbière ☆ Iso WR 10s 36m **W18M, R16M**; shore-W-294°-R-328°-W-148°-R-shore; W ○ twr; *Horn Mo (C) 60s;* 49°10'·79N 02°15'·01W.
Pt Corbière ⌁ FR; R ☐, W stripe; 49°10'·87N 02°14'·38W.

WESTERN PASSAGE

Ldg lts 082°. Front, La Gréve d'Azette Oc 5s 23m 14M; 034°-129°; 49°10'·16N 02°05'·09W. Rear, Mont Ubé, Oc R 5s 46m 12M; 250°-095°; 1M from front.
Passage Rock ⌇ VQ; 49°09'·54N 02°12'·26W.
Les Fours ⌇ Q; 49°09'·59N 02°10'·16W.
Noirmont Pt ⌁ Fl (4) 12s 18m 10M; B twr, W band; 49°09'·91N 02°10'·08W.
Pignonet ⊥ 49°09'·88N 02°09'·70W.
Ruaudière Rock ⬥ Fl G 3s; *Bell;* 49°09'·74N 02°08'·62W.

ST AUBIN'S BAY and HARBOUR

Les Grunes du Port ⌂ 49°10'·02N 02°09'·14W.
Diamond Rock ⌂ Fl (2) R 6s; 49°10'·12N 02°08'·64W.
Castle pier ⌁ Fl R 4s 8m 1M; 49°11'·13N 02°09'·634W.
North pier, Dir lt 254° ⌁ F WRG 5m, 248°-G-253°-W-255°-R-260°; 49°11'·22N 02°10'·03W. Same col, Iso R 4s 12m 10M.
Beach Rock ⌂ 49°11'·29N 02°08'·42W, (Apr-Oct).
Rocquemin ⊥ 49°10'·64N 02°07'·95W.
Baleine ⬥ 49°10'·41N 02°08'·23W.

ST HELIER

Elizabeth marina, west appr: Dir ⌁ 106°: F WRG 4m 1M; 096°-G-104°-W-108°-R-119°; R dayglo ☐, B stripe; 49°10'·76N 02°07'·12W.
La Vrachiére ⊥ 49°10'·90N 02°07'·59W, Fl (2) 5s 1M.
Fort Charles North ⊥ 49°10'·81N 02°07'·50W.
Marina ent ⌁ Oc G & Oc R, both 4s 2M; 49°10'·83N 02°07'·13W.
Red & Green Passage, ldg lts 022·7° on dayglo R dolphins: Front, Elizabeth E berth Dn ⌁ Oc G 5s 10m 11M; 49°10'·63N 02°06'·94W. Rear, Albert Pier root ⌁ Oc R 5s 18m 12M; synch; 230m SSW.
East Rock ⬥ QG; 49°09'·96N 02°07'·28W.
Dog's Nest Rk ⌇ Fl Y 3s 3M; Y cross on bcn; 49°09'·99N 02°06'·95W.
Oyster Rocks, R/W bcn, topmark 'O'; 49°10'·10N 02°07'·49W.
Platte Rock ⌐ Fl R 1·5s 6m 5M; R col; 49°10'·16N 02°07'·34W.
Small Road No 2 ⌂ QR; 49°10'·39N 02°07'·24W.
No 4 ⌂ QR; 49°10'·53N 02°07'·12W.
Elizabeth marina, S appr: E1 ⬥ Fl G 3s; 49°10'·59N 02°07'·09W.
E2 ⌂ Fl R 2s; 49°10'·58N 02°07'·13W.
E5 ⬥ Fl G 5s; 49°10'·70N 02°07'·17W.
E6 ⌂ Fl R 2s; 49°10'·69N 02°07'·21W.
Fort Charles East ⊥ Q (3) 5s 2m 1M; 49°10'·74N 02°07'·26W.
La Collette basin ⌂ QR; 49°10'·54N 02°06'·91W. ⌂ 49°10'·52N 02°06'·90W.
St Helier Hbr, ldg lts 078°, both FG on W cols. Front, 49°10'·62N 02°06'·66W. Rear, 80m from front.
Victoria pier hd, Port control twr; IPTS; 49°10'·57N 02°06'·88W.

JERSEY, SOUTH-EAST AND EAST COASTS

Hinguette ⌂ QR; 49°09'·33N 02°07'·32W.
Hettich ⌇ Fl Y 5s; 49°08'·10N 02°09'·00W.
South Pier Marine ⌇ Fl Y 5s; 49°09'·10N 02°06'·30W.

Demie de Pas ⬥ Mo (D) WR 12s 11m, WR 8M; 130°-R-303°-W-130°; *Horn (3) 60s;* **Racon T, 10M**; B bn twr, Y top; 49°09'·01N 02°06'·15W. Icho Tower (conspic, 14m) 49°08'·89N 02°02'·90W.
Canger Rock ⌇ Q (9) 15s; 49°07'·35N 02°00'·38W.
La Conchière ⌇ Q (6) + L Fl 15s 2M; 49°08'·22N 02°00'·17W.
Frouquier Aubert ⌇ Q (6) + L Fl 15s; 49°06'·08N 01°58'·84W.
Violet ⌇ L Fl 10s; 49°07'·81N 01°57'·14W.
Petite Anquette; SCM ⊥; 49°08'·47N 01°56'·29W.
Grande Anquette ⌇ VQ(9) 10s; 49°08'·32N 01°55'·20W.
Le Cochon ⌂ 49°09'·77N 01°58'·80W.
La Noire ⊥ 49°10'·13N 01°59'·23W.
Le Giffard ⌂ Fl R 3s 49°10'·59N 01°58'·99W.

GOREY

Pier Hd ⌁ Occ WRG 5s 8m 8M; 228°-G-296·5°-W-299·5°-R-299·5°-008°; W twr on pierhead; 49°11'·81N 02°01'·33W.
Horn Rock ⊥, topmark 'H'; 49°10'·96N 01°59'·85W.
Les Burons, R bcn, topmark 'B'; 49°11'·33N 02°00'·81W.
Fairway ⬥ QG; 49°11'·50N 02°00'·34W.
Écureuil Rock ⊥ 49°11'·67N 02°00'·78W.
Equerrière Rk, bcn 'fishtail' topmark; 49°11'·80N 02°00'·67W.
Les Arch ⊥, BW bcn, 'A' topmark; 49°12'·02N 02°00'·60W.

ST CATHERINE BAY

St Catherine Bay, Le Fara ⌇ Q (3) 10s 3M; 49°12'·85N 02°00'·48W.
Archirondel Tower (conspic, 16m) 49°12'·72N 02°01'·42W.
Verclut bkwtr ⌁ Fl 1·5s 18m 13M; 49°13'·34N 02°00'·64W. In line 315° with unlit turret, 49°13'·96N 02°01'·57W, on La Coupe Pt.

JERSEY, NORTH COAST

Rozel Bay Dir lt 245°, F WRG 11m 5M; 240°-G-244°-W-246°-R-250°; W col; 49°14'·21N 02°02'·76W.
Bonne Nuit Bay ldg lts 223°: both FG 7/34m 6M. Front, Pier 49°15'·10N 02°07'·17W. Rear, 170m from front.
Demie Rock ⬥ 49°15'·56N 02°07'·36W.
Sorel Point ☆ L Fl WR 7·5s 50m **15M**; 095°-W-112°-R-173°-W-230°-R-269°-W-273°; W ○ twr, only 3m high; 49°15'·60N 02°09'·54W.

OFFLYING ISLANDS

LES ÉCREHOU

Écrevière ⌇ Q (6) + L Fl 15s; 49°15'·26N 01°52'·15W.
Mâitre Ile ⊥ 49°17'·08N 01°55'·60W.

PLATEAU DES MINQUIERS

N Minquiers ⌇ Q; 49°01'·64N 02°00'·58W.
NE Minquiers ⌇ VQ (3) 5s; *Bell;* 49°00'·85N 01°55'·30W.
SE Minquiers ⌇ Q (3) 10s; *Bell;* 48°53'·42N 02°00'·09W.
S Minquiers ⌇ Q (6) + L Fl 15s; 48°53'·09N 02°10'·10W.
SW Minquiers ⌇ Q (9) 15s 5M; *Bell;* 48°54'·34N 02°19'·38W.
NW Minquiers ⌇ Q 5M; *Whis;* 48°59'·63N 02°20'·59W.
Refuge ⊥; B/W bcn; 49°00'·13N 02°10'·16W.
Demie de Vascelin ⬥ Fl G 3s, 49°00'·81N 02°05'·17W.
Grand Vascelin, BW bcn ⌂ 48°59'·97N 02°07'·26W.
Maitresse Ile, Puffin B&W bcn twr ⊥ 48°58'·33N 02°03'·65W.
Le Coq Reef, ⌇ Q (3) 10s 6m 3M; 48°57'·88N 02°01'·29W.

FRENCH MARKS NORTH OF ILE DE BRÉHAT

Roches Douvres ☆ Fl 5s 60m **24M**; pink twr on dwelling with G roof; 49°06'·30N 02°48'·87W. (16M NNE of Ile de Bréhat).
Barnouic ⌇ VQ (3) 5s 15m 7M; 49°01'·64N 02°48'·41W.
Roche Gautier ⌇ VQ (9) 10s; 49°02'·013N 02°54'·36W.
Les Héaux de Bréhat ☆ Fl (4) WRG 15s 48m, **W15M**, R/G11M; 227°-R-247°-W-270°-G-302°-W-227°; Gy twr; 48°54'·50N 03°05'·18W (4·7M WNW of Ile de Bréhat).

6.4 PASSAGE INFORMATION

More passage information is threaded between successive harbours in this Area. **Bibliography:** *Channel Islands, Cherbourg Peninsula and North Brittany* (Imray/RCC/Carnegie); *N Brittany & Channel Islands CC* (Fernhurst/Cumberlidge); *Channel Pilot* (NP27); *Channel Havens* (ACN/Endean); *Shell Channel Pilot* (Imray/Cunliffe).

THE CHANNEL ISLANDS

In an otherwise delightful cruising area (AC 2669) the main problems include fog and poor visibility, the very big tidal range, strong tidal streams, overfalls and steep seas which get up very quickly. The shoreline is generally rugged with sandy bays and many offlying rks. It is important to use large scale AC, and recognised leading marks (of which there are plenty) when entering or leaving many of the harbours and anchs. Several passages are marked by beacons and perches identified by an alphabetical letter(s) in lieu of topmark. Beware high speed ferries.

From the N, note the Casquets TSS and ITZ. Soundings of Hurd Deep can help navigation. The powerful lights at the Casquets, Alderney, Cap de la Hague, Cap Levi, Barfleur and Les Hanois greatly assist a night or dawn landfall. By day Alderney is relatively high and conspicuous. Sark is often seen before Guernsey which slopes down from S to N. Jersey is low-lying in the SE. The islands are fringed by many rky dangers. In poor visibility stay in harbour.

▶*Over a 12hr period tidal streams broadly rotate anti-clockwise around the Islands, particularly in open water and in wider chans. The E-going (flood) stream is of less duration than the W-going, but is stronger. The islands lie across the main direction of the streams, so eddies are common along the shores. The range of tide is greatest in Jersey (9·6m sp, 4·1m np), and least in Alderney (5·3m sp, 2·2m np). Streams run hard through the chans and around headlands and need to be worked carefully; neaps are easier, particularly for a first visit. Strong W'lies cause a heavy sea, usually worst HW ±3.◀*

As well as the main harbours, described in some detail, there are also many attractive minor harbours and anchorages. In the very nature of islands a lee can usually be found somewhere. Boats which can take the ground can better explore the quieter hbrs. Avoid lobster pots and oyster beds.

THE CASQUETS AND ORTAC CHANNEL

Casquets lt ho is conspic on the largest island of this group of rks 5·5M W of Alderney (AC 60). Offlying dangers extend 4ca W and WSW (The Ledge and Noire Roque) and 4ca E (Pt Colote). The tide runs very hard around and between these obstructions. A shallow bank, on which are situated Fourquie and l'Equêt rks (dry) lies from 5ca to 1M E of Casquets, and should not be approached. Ortac Rk (24m) is 3·5M E of Casquets. Ortac Chan runs NNW/SSE 5ca W of Ortac.

▶*Here the stream starts to run NE at HW St Helier −2½, and SW at HW St Helier +0355; sp rates up to 5½kn (7kn reported).◀*

Ortac Channel should not be used in bad weather due to tremendous overfalls; these and violent eddies also occur over Eight Fathom Ledge (8½ca W of Casquets) and over the SW, SSW and SSE Casquet Banks where least depths of 7·3m occur.

ALDERNEY AND THE SWINGE

Braye Harbour (AC 2845) together with pleasant bays and anchs around the island, offer shelter from different wind/sea directions. The sunken NE extremity of Admiralty Breakwater should only be crossed in calm conditions, outside LW±2 and about 50m off the head of the Breakwater where there is 2·3m.

The **Swinge** lies between Burhou with its bordering rks, and the NW coast of Alderney. It can be a dangerous chan, and should only be used in reasonable vis and fair weather. On the N side of the

Swinge the main dangers are Boues des Kaines, almost awash at LW about 7½ca ESE of Ortac, and North Rk 2½ca SE of Burhou. On S side of the Swinge beware Corbet Rk (0·5m high) with drying outliers, 5ca N of Fort Clonque, and Barsier Rk (0·9m) 3½ca NNW of Fort Clonque.

▶*The SW-going stream begins at HW St Helier +0340, and the NE stream at HW St Helier −0245, sp rates 7-8kn. On the NE-going stream, beware the very strong northerly set in vicinity of Ortac.◀*

The tide runs very hard, and in strong or gale force winds from S or W there are very heavy overfalls on the SW-going stream between Ortac and Les Etacs (off W end of Alderney).

In strong E winds, on the NE-going stream, overfalls occur between **Burhou** and Braye breakwater. These overfalls can mostly be avoided by choosing the best time and route (see below) but due to the uneven bottom and strong tides broken water may be met even in calm conditions.

▶*It is best to pass SW through the Swinge is about HW St Helier +4, when the SW-going stream starts; hold to the SE side of the chan since the stronger stream runs on the Burhou side. But after HW St Helier +5, to clear the worst of the overfalls keep close to Burhou and Ortac, avoiding North Rk and Boues des Kaines.◀*

Pierre au Vraic (dries 1·2m) is an unmarked pinnacle rock at 49°41'·61N 02°16'·94W, 1·8M S of Ortac and 1·8M WSW of Les Étacs, almost in the fairway to/from the Swinge. Arriving on a fair tide from Guernsey it will be well covered, but it is a serious hazard if leaving the Swinge on a SW-going spring tide close to local LW. Carefully study the clearing bearings shown on AC 60.

Heading NE at about HW St Helier −2, Great Nannel in transit with E end of Burhou clears Pierre au Vraic to the E, but passes close W of Les Etacs. On this transit, when Roque Tourgis fort is abeam, alter slightly to stbd to pass 1ca NW of Corbet Rk; keep near SE side of the channel.

THE ALDERNEY RACE

This race, characterised by very strong tidal streams, runs SW/NE between Cap de la Hague and Alderney, but its influence extends at least as far SW as 02° 20'W (AC 3653). The fairway, approx 4M wide, is bounded by Race Rk and Alderney S Banks to the NW, and to the SE by rky banks 3M E of Race Rk, and by Milieu and Banc de la Schôle (least depth 2·7m). These dangers which cause breaking seas and heavy overfalls should be carefully avoided. In bad weather and strong wind-against-tide the seas break in all parts of the Race and passage is not advised. Conditions are worst at springs.

▶*In mid-chan the SW-going stream starts at HW St Helier +4½ (HW Dover) and the NE-going stream at HW St Helier −2¼ (HW Dover +5½), sp rates both 5·5kn. The times at which the stream turns do not vary much in various places, but the rates do; for example, 1M W of Cap de la Hague sp rates reach 7-8kn.◀*

For optimum conditions, timing is of the essence. As a rule of thumb the Race should be entered on the first of the fair tide so as to avoid peak tidal streams with attendant overfalls/seas.

▶*Thus SW-bound, arrrive off Cap de la Hague at around HW St Helier +4½ (HW Dover −½ to HW Dover) when the stream will be slack, whilst just starting to run SW off Alderney. A yacht leaving Cherbourg at HW Dover −3 will achieve the above timing by utilising the inshore W-going tidal eddy.*

Conversely, NE-bound, leave St Peter Port, say, at approx local HW St Helier −4½ (HWD+3) with a foul tide so as to pass Banc de la Schôle as the fair tide starts to make.◀ A later departure should achieve a faster passage, but with potentially less favourable conditions in the Race. On the NE-going stream the worst overfalls are on the French side.

6.5 SPECIAL NOTES: CHANNEL ISLANDS

The Channel Islands (Alderney, Guernsey, Sark, Jersey and other smaller islands) lie not in the Channel, but in the Bay of St Malo. Alderney, Herm, Jethou, Sark and Brecqhou are all part of the Bailiwick of Guernsey and the States (Parliament) of Alderney have seats in the States of Guernsey.

The Islands are self-governing with their own laws, judiciary and Customs, but conduct their foreign affairs mostly through the English crown. They are not part of either the UK or the EU. The French call the CI Les Îles Anglo-Normandes (Aurigny, Guernsey, Sercq et Jersey). Standard Time is UT (Zone 0).

Charts Admiralty Leisure folio SC5604 (£44·30, 7th ed. published Feb 2013) contains 17 sheets of the CIs. These and other charts are listed under individual ports.

Ports of entry Braye, Beaucette, St Sampson, St Peter Port, St Helier and Gorey.

Customs Yachts arriving in **Guernsey** *must fly a Q flag until clearance has been obtained.* Customs declaration form is provided on arrival, post it in yellow Customs box, or hand to a Customs Official. This does not apply to **Jersey** unless you arrive from outside the European Economic Area, (EEA), have non-EEA nationals on board (with valid passports), or have goods to declare. A Q flag must then be flown. On return to UK Customs Formalities apply. Yachts going to France need usual documents, passports etc.

Speed limits, Jersey, 5kn in hbrs, and within 200m of water's edge in any bay; inside 10m depth contour, Les Écrehou and Les Dirouilles; inside 5m depth contour, Les Minquiers. **Guernsey**, 6kn within bays, 0·5M of Herm and the Humps, and the area defined by the coastlines of Jethou, Fauconnière and Crevichon. St Peter Port outer hbr 6kn, inner hbr 4kn (see chartlet).

Medical Jersey has reciprocal medical arrangements with the UK, but the other CI do not and visitors will have to pay for costly inpatient / outpatient care. Medical insurance recommended.

Weather forecasts prepared by Jersey Met Office for the CI area and adjacent coasts of Normandy and Brittany are available ☎0900 669 0022 (premium rate) throughout the CI and UK. From France replace the first 0 by 00 44. See also 0.16.2.

SAR operations are co-ordinated by Guernsey Coastguard in the N area and Jersey Coastguard in the S (see 0.15). Major incidents are co-ordinated with CROSS Jobourg and Falmouth CGOC. LBs at Braye, St Peter Port, St Helier and St Catherine's Bay (Jersey).

Telephones CI phones are UK integrated but charges, especially on mobiles, may vary. Toll-free and premium numbers may be blocked when roaming with non-CI mobiles. For an online telephone directory visit www.theguernseydirectory.com

Courtesy flags A courtesy flag in CI ports is a mark of politeness but not essential. Local flags are: *Jersey* R ensign with Jersey Royal Arms (three lions passant with gold crown above); see Fig 5(9).

Guernsey R ensign with Duke William's cross in the fly. Vessels owned by Guernsey residents may wear this ensign; see Fig 5(9).

Sark The English (St George's) flag, Normandy arms in the canton.

Alderney The English (St George's) flag and in the centre a green disc charged with a gold lion.

Herm The English (St George's) flag, and in the canton the Arms of Herm (three cowled monks on a gold diagonal stripe between blue triangles containing a silver dolphin).

Cars can be hired in Jersey, Guernsey and Alderney. In Sark cars are prohib, but bikes and horse-drawn carriages can be hired.

Animals may be landed off boats from UK, Ireland, Isle of Man or other CI, but not if the boat has visited France. No vessel with an animal on board may lie alongside a pontoon or quay without a Revenue Officer's permission.

UK currency can be used but not vice versa. Postage stamps, issued by Jersey, Guernsey and Alderney, must be used in the appropriate Bailiwick; Guernsey and Alderney stamps are interchangeable.

Gas Some UK Gas retailers may not accept Channel Island bottles.

9.19.7 ALDERNEY

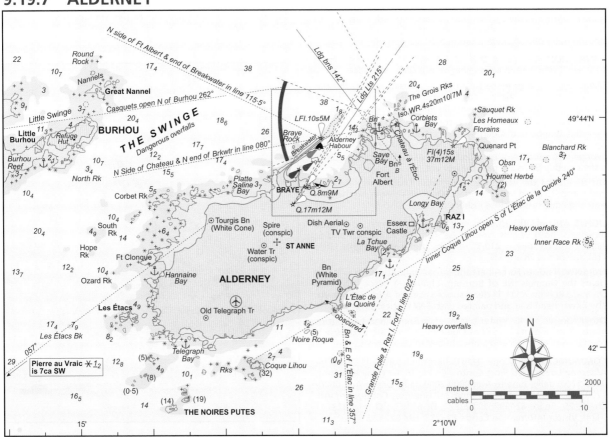

ANCHORAGES AROUND ALDERNEY There are several ⚓s, all picturesque but only safe in settled weather and offshore winds; most are very small and many have offlying rocks. Entry to these anchorages is not advised without local knowledge, advice from the HM, a good Pilot book plus AC 60 and/or preferably 2845. None provide any facilities. Clockwise from Braye they are:

Saye Bay 49°43´·85N 02°10´·87W. Small sandy bay 4ca E of ent to Braye Hbr. Appr on transit 142° of ldg bns; move E to open up the ent. Ent is 100m wide between Homet des Pies and Homet des Agneaux. Exposed to N'lies. Château à l'Étoc ☆ is 300m ENE. Speed limit 4kn.

Longy Bay 49°43´·06N 02°10´·29W. Wide drying bay with good holding on sand. Speed limit 4kn. Appr on N between Queslingue (a 14m high stack) and a rk 0̲·6m to stbd. ⚓ in 3·5m closer to Essex Castle than to Raz Island to await fair tide in the Race. Whilst the stream is setting SW offshore, a NE-going eddy develops along the island's southern shore and this would assist a straightforward arrival.

La Tchue 49°42´·78N 02°10´·75W. Good holding in small bay surrounded by cliffs. La Rocque Pendante to the E and smoking rubbish tip to the NW are both conspic.

Telegraph Bay 49°42´·01N 02°13´·51W. Pleasant sandy bay on SW tip of the island but ringed by rks. Appr 035° between Noires Putes and Coupé. Old Telegraph Twr (93m) is conspic until obsc'd close inshore, when a pillar on the cliff edge offers brgs, as does the astern transit of Coupé (7m) and and Orbouée (0.5m).

Hannaine Bay 49°42´·73N 02°13´·98W. A good place to await the flood tide. Appr Tourgis Bn △ ≠ 057° SE side of Fort Clonque. Beware rks either side. ⚓ on sand in 3m, 100m S of Fort.

Platte Saline Bay 49°43´·37N 02°12´·65W. Good shelter from E'lies. Les Jumelles and Outer Fourchie guard the ent. Beach is unsafe for swimming.

Burhou 49°43´·64N 02°15´·49W. Temp'y ⚓ in bay SW of Burhou, only on SW stream; exposed at HW on the NE-going stream. Appr on 010° for the gap between Burhou and Little Burhou.

6.7 BRAYE HARBOUR

Alderney 49°43´·77N 02°11´·51W ✸✸✸✸♨♨♨☆☆☆

CHARTS AC 2669, 3653, 60, 2845, 5604.7/.8; SHOM 7158, 6934; Navi 1014; Imray C33A, 2500

TIDES –0400 Dover; ML 3·5; Duration 0545

Standard Port ST HELIER (→)

Times				Height (metres)			
High Water		Low Water		MHWS	MHWN	MLWN	MLWS
0300	0900	0200	0900	11·0	8·1	4·0	1·4
1500	2100	1400	2100				
Differences BRAYE							
+0050	+0040	+0025	+0105	–4·8	–3·4	–1·5	–0·5

SHELTER Good, but exposed to strong N/NE'lies. ⚓s as charted; good holding on sand, but only fair on rock or weed. Keep clear of the fairway and jetty due to steamer traffic. No berthing/landing on Admiralty bkwtr; beware submerged 3ca extension.

NAVIGATION WPT 49°44´·08N 02°11´·18W, 215°/0.82M to front 215° ldg lt (see Lts and Marks below). Hbr speed limit 4kn.

The safest appr is from the NE. In fog or limited visibility the HM can give small craft radar assistance and RDF bearings HO. Caution: Strong tidal streams, eddies, overfalls and many rocks.

To avoid dangerous overfalls in certain wind and tide conditions take the Swinge and Alderney Race at/near slack water; this is approx HW Braye +2½ & LW +2½ (or HW St Helier +3½ and –3). The Swinge is often calmest near Corbet Rk. At mid-flood (NE-going) a strong eddy flows SW past the hbr ent.

Off NE end of the island give Brinchetais Ledge and Race Rk a wide berth to avoid heavy overfalls. On the S side during the ebb, a strong eddy sets NE close inshore of Coque Lihou.

LIGHTS AND MARKS Hbr ldg lts 215°, both Q synch; co-located daymarks, both W cols, orange △s. Old Hbr white bcn and St Anne's church spire ≠ 210·5° remains useful in daylight. Little Crabby Hbr has FR and FG.

From NW, N side of Fort Albert and hd of Admiralty Bkwtr ≠ 115·5° clears the Nannels (NE of Burhou). Château à l'Etoc lt ho and Quenard Pt lt ho ≠ 111·1° clears sunken ruins of Admiralty Bkwtr. The Bkwtr head is lit and painted with B/W vertical stripes.

Other conspic marks: Water twr W of St Anne's and two lattice masts (R lts) E of it. 2ca SW of Quenard Pt lt ho a blockhouse is conspic. See chartlets and Lights, buoys and waypoints.

COMMUNICATIONS (Code 01481) Recorded forecasts see 0.14; Alderney Coastguard /Port Control Ch **74**, 16, 67 (May and Sept, 0800-1700, daily; June to Aug 0800-2000 daily; Oct to Apr, 0800-1700, Mon-Fri: all LT) at other times call Guernsey Coastguard VHF Ch 16, 20; Police 824999; Ⓗ 822822; Dr 822077; ♥ Info 823737 (H24); Visitor Information Centre 823737; HM 822620.

FACILITIES 70 Y ⚓s (£20·00) with pick-up lines lie parallel to the Admiralty bkwtr and in blocks E of Braye Jetty and SW of Toulouse Rock. Orange buoys are for locals only. If >12m LOA, check suitability of ⚓ with HM. ⚓s (£5·00 inc ⛟) as on chartlet.

HM (aka CG & ☻): ☎822620. mark.gaudion@gov.gg www.sailalderney.com/harbour.html www.alderney.gov.gg

Commercial Quay Visitors dinghy landing pontoon on W side, N of slipway. ⚓ ⛟ ⊙ ⊸ ⊞ Ⓔ ⚒(25t).

Little Crabby Hbr Access at the top of the tide only for **Mainbrayce**, VHF Ch 37 Mainbrayce, ☎822772, mob 07781 415420: ⚓ ⛟ ⚖ ⚒ ACA ⚓ ⛟. Water taxi call Mainbrayce Ch M (0800-2359) ☎ as above, £1·50 each way till 2130, £2·50 after 2130.

Alderney SC ☎822758, ⌂ visitors welcome.

Town (St Anne) ✕ ⌂ ⛏ 🛒 ✉ Ⓑ. ✈ direct to Guernsey, Southampton; other connections via Guernsey. **Ferries** via Guernsey to Jersey, Poole and St Malo.

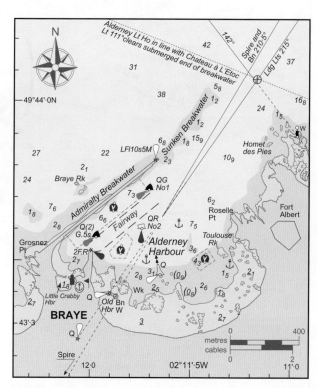

Channel Islands *(side tab)*

APPROACHES TO GUERNSEY

From the N/NE, The Little Russel Channel (AC 3654, 808) gives the most convenient access to Beaucette Marina and St Peter Port (AC 3140). But it needs care, due to rks which fringe the chan and appr, and the strong tide which sets across the ent.

▶*In mid-chan, S of Platte and NW of Bréhon, the NE-going stream begins at HW St Peter Port –0245, and the SW stream at HW St Peter Port +0330, sp rates both 5·25kn which can raise a very steep sea with wind against tide.*◀

With lts on Platte Fougère, Tautenay, Roustel, Platte and Bréhon, plus the 220° ldg lts for St Peter Port, the Little Russel can be navigated day or night in reasonable vis, even at LW.

The Big Russel is wider and easier. In bad weather or poor vis it may be a better approach to St Peter Port, via Lower Heads SCM lt buoy, though may be lumpy on a rising sp tide. From the NW, Doyle Passage, which is aligned 146°/326° off Beaucette, can be used but only by day with local knowledge. From the S or W, the natural route is around St Martin's Pt, keeping 1·5ca ENE of Longue Pierre bcn (LP) and a similar distance off Anfré bcn (A).

In onshore winds keep well clear of the W coast, where in bad weather the sea breaks on dangers up to 4M offshore. But in settled weather it is worth exploring the rock-strewn bays.

6.8   LITTLE RUSSEL CHANNEL

See notes above and AC 5604.9/10, 807, 808

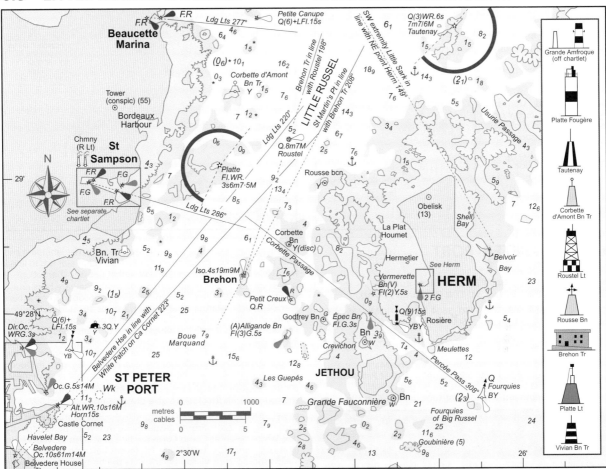

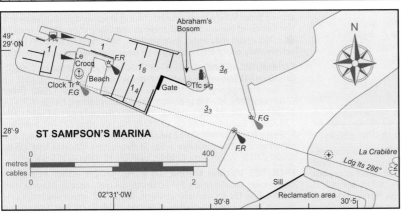

St Sampson's Hbr 49°28'·91N 02°30'·73W. Dries 3·6m. Port of entry. Good shelter but strictly a working hbr. Visit only by prior arrangement for repair work. Entry via lifting flapgate HW -1 to +2. Marina tfc sigs SW cnr Abraham's Bosom: steady ● = ent/exit; Fl ◐ or steady ● = No ent/exit. WPT 49°28'·70N 02°29'·65W. 2 Chy's NW of hbr conspic. Ldg lts 286°: Front on S Pier, FR; rear on clocktower, FG. Other lts as chartlet and LBW. On S Pier a 360° Fl R above FR (shown landward) = large vessel under way in the hbr; no ent/exit for other craft without Dockmaster's approval on Ch 12 (H24) or ☎01481 720229. Speed limit 6kn.⚓ in hbr. Services: ☍ (Marine & General 01481 243048, VHF Ch 17) ◧(75t) ⚒ ▣ ⚒ ▱ ⛟. ▮ by bowser at Abraham's Bosom (E side) must pre-book ☎01481 200800/245858.

6.9 BEAUCETTE
Guernsey **49°30′·19N 02°30′·20W** ❄❄☀☗☗☗❀❀❀

CHARTS AC 3654, 807, 808, 5604.9; SHOM 7159, 6904, 6903; Navi 1014; Imray C33A, 2500

TIDES –0450 Dover; ML 5·0; Duration 0550. See Facilities and St Peter Port for access times.

SHELTER Excellent in marina. Entry not advised in strong N–SE onshore winds.

NAVIGATION WPT 49°30′·15N 02°28′·85W, 277°/8½ ca to ent. Appr from Little Russel to mid-way between Platte Fougère lt tr (W with B band, 25m) and Roustel lt tr. Pick up the ldg marks/lts, SWM L Fl 10s, a SHM Fl G 5s, 2 PHM buoys Fl R 5s & Fl (3) R 5s, a NCM perch Q and a SHM perch Q (3) G 5s at the ent. Beware stream setting across the appr chan and drying rocks either side.

LIGHTS AND MARKS See chartlet and Lights, buoys and waypoints. Petite Canupe is a very spindly SCM lt bcn. 277° ldg lts: both FR. Front, R arrow on W background, stbd side of ent; rear, W arrow on R background, on roof of W bldg, with windsock.

COMMUNICATIONS (Code 01481); Police 725111;⊖ 245000; St John Ambulance 725211; HM & water taxi VHF Ch 80 (0700-2200).

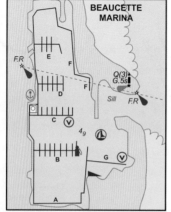

FACILITIES Marina ④ sill dries 2·37m. 6 Y waiting buoys lie outside and a tide gauge is inside the 8m wide ent channel. For entry sequence and a berth it is essential to pre-call HM with boat details: ☎245000, mob 07781102302. www.beaucettemarina.com info@beaucettemarina.com

115 inc approx 45❶ £2·25, min charge £15·75/boat; £0·75 for <4hrs. ⚓ on pontoon B, ⚓✕🛒🍴⊘🛢 ⚒ 🗑(16t) 🛢(12t) ☎247066, bike hire, ♿ access ramp is steep @ LW.

Village Basic food shop. Good coastal walks. For further information see St Peter Port; 20 mins by bus 6/6A (but bus stop is nearly a mile away).

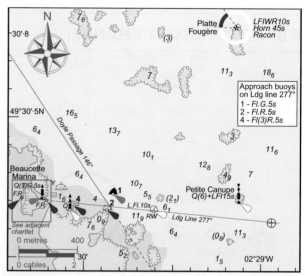

6.10 ST PETER PORT
Guernsey ❄❄❄☗☗☗☗❀❀❀

N appr WPT: 49°30′·23N 02°27′·68W (Little Russel)
S appr WPT: 49°25′·00N 02°31′·17W (St Martin's Point)

CHARTS AC 3654, 808, 807, 3140, 5604.10/11; SHOM 7159, 6903, 6904; Navi 1014; Imray C33A, 2500

TIDAL ACCESS –0439 Dover; ML 5·2; Duration 0550
NOTE: St Peter Port is a Standard Port.

To find depth of water over the sill into Victoria Marina:
1 Look up predicted time and height of HW St Peter Port.
2 Enter table below on the line for height of HW.
3 Extract depth (m) of water for time before/after HW.

Ht (m) of HW St Peter Port	Depth of water in metres over the sill (dries 4·2 m)						
	HW	±1hr	±2hrs	±2½hrs	±3hrs	±3½hrs	±4hrs
6·20	1·85	1·67	1·30	1·03	0·75	0·47	0·20
·60	2·25	2·00	1·50	1·13	0·75	0·37	–
7·00	2·65	2·34	1·70	1·23	0·75	0·27	–
·40	3·05	2·67	1·90	1·33	0·75	0·27	–
·80	3·45	3·00	2·10	1·43	0·75	0·07	–
8·20	3·85	3·34	2·30	1·53	0·75	–	–
·60	4·25	3·67	2·50	1·63	0·75	–	–
9·00	4·65	4·00	2·70	1·73	0·75	–	–
·40	5·05	4·34	2·90	1·83	0·75	–	–
·80	5·45	4·67	3·10	1·93	0·75	–	–

SHELTER Good, but Victoria Marina is exposed to strong E'lies.

NAVIGATION WPT 49°27′·82N 02°30′·78W, 227°/0·68M to hbr ent. Offlying dangers, big tidal range and strong tidal streams demand careful navigation. From N the Little Russel is most direct appr, but needs care especially in poor visibility; see Passage Information and Little Russel Channel chartlet. Big Russel between Herm and Sark, passing S of Lower Hds SCM lt buoy, is an easier appr. From W and S of

Guernsey, give Les Hanois a wide berth. Beware ferries , shipping and unlit racing marks. Hbr speed limits: 6kn from outer pier hds to line S from New Jetty to Castle pier; 4kn W of that line (see chartlet).

Access via buoyed/lit chan along S side of hbr. Beware busy cruise ship tenders using same chan and landing pontoon at E end Albert Pier (Apr-Oct). Marina boat directs arrivals to marina, waiting pontoon (colour coded) or, N of it, Swan ❶ pontoons nos 1–5, also the best options if arriving when marina unmanned (2200-0600). All have ⚓, nos 1–4 are walk-ashore (seasonal only). Pontoons for tenders each side of marina ent. Local moorings, centre of hbr, with a secondary fairway N of them. ⚓ prohib. ❶ berths in Queen Elizabeth II (in 3·0m) and Albert Marinas only by prior arrangement.

GY RDF beacon, 304·50 kHz, on Castle Bkwtr is synchronised with the co-located horn (15s) to give distance finding; see LBW.

LIGHTS AND MARKS See chartlet and Lights, buoys and waypoints. Outer ldg lts 220°: Front, Castle bkwtr hd; rear, Belvedere. By day, white patch at Castle Cornet in line 223° with Belvedere Ho (conspic). Inner ldg lts 265° are for ferries berthing at New Jetty. Yachts should appr Victoria marina via the buoyed/lit S channel.

Traffic Signals When a large vessel is under way a single FR ● is shown from White Rock pier hd, facing landward and/or seaward; the S ends of New Jetty and Inter-Island Quay show Q Fl Y over FR ● = 'No vessel may enter or leave hbr', but boats <15m LOA under power may proceed, keeping well clear commercial vessel berths.

COMMUNICATIONS (Code 01481) *Guernsey Coastguard* 720672, Ch 20 (H24) for safety traffic, VHF Ch 16/67 for DF brgs, VHF Ch 62 for emergency link calls; Guernsey Met office 12080 (from Guernsey only, lowest premium rate) for weather; Police 725111; ⊖ 741410; Dr 711237 (H24); St John Ambulance 725211; HM 720229; *St Sampson* VHF Ch 12 (H24); Port Control 720481, monitor *St Peter Port Control* VHF Ch 12 (H24) call only if absolutely necessary, when within the pilotage area; *St Peter Port Marinas* VHF Ch 80 (0600-2200); water taxi 424042, VHF Ch 10 (0800-2359LT).

FACILITIES Victoria Marina, sill 4·2m; gauges either side show depth over sill (on entry, PH accurate, SH over-reads). Marina staff and/or R/G tfc lts control ent/exit. ☎725987. guernsey.harbour@gov.gg www.guernseyharbours.gov.gg 400❶ £3·00, per 24H or part

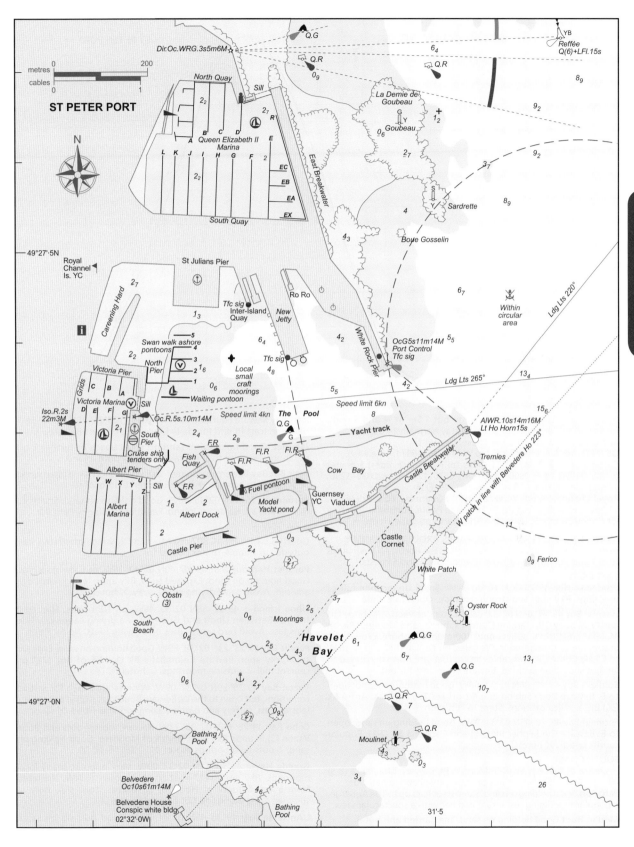

ST PETER PORT

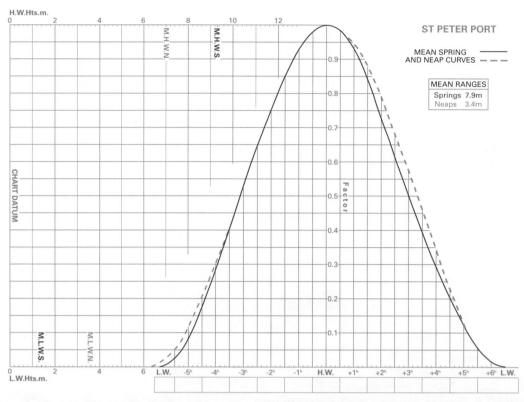

ST PETER PORT

MEAN SPRING ———
AND NEAP CURVES – – –

| MEAN RANGES |
| Springs 7.9m |
| Neaps 3.4m |

thereof. Max LOA/draft, 12·8m/2.1m but 1·5m at N and S ends, (deeper draft, enquire ahead); max stay normally14 days; ⬛ ⬛ ⬛ ⚓ but you must use your own hose; free wi-fi.

QE II Marina 🚶 🚶 times vary, esp Winter, ☎726071, free wi-fi.

Castle Pier ⚓ 🚶 🚶 ☎726071 (dries springs, check tides) Summer (01/05–01/09) M–F, 0800–1730. Sat, 0830–1730 Sun, 0830–1230. Winter, closed Sun. ⬛ 🔋 ACA, ⬛ ⬛ ⬛ ⬛ Ⓔ Ⓦ.

Royal Channel Islands YC ☎723154, ⬛. **Guernsey YC** ☎725342, ✕ ⬛. **Town** ✕ ⬛ ⬛ ⬛ ⬛ @ ✉ Ⓑ. Ferries to Poole, Portsmouth, Jersey (St Helier/Gorey), St Malo, Diélette, Carteret, Sark, Herm; ✈.

MINOR HARBOURS AND ANCHORAGES AROUND GUERNSEY

Hbrs and ⚓ages are listed clockwise from the north. All have buses to St Peter Port (from cliff-top level on the E and S coasts).

EAST COAST

Bordeaux Hbr 49°29'·31N 02°30'·37W. Small drying hbr full of moorings. Rky appr and strong crosstide. Open to E. Café.

Havelet Bay 49°27'·06N 02°31'·68W. Enter between SHM buoy QG and PHM buoy QR, marking Oyster Rk, bn 'O', and Moulinet Rk, bn M. SHM and PHM lt buoys about 100m closer inshore. Crowded ⚓ in summer but sheltered in SW gales; no 🛥s. Landing slip close W of Castle Cornet. Power cables on chartlet are normally buried to 1m but should be avoided. Beware swimmers far out.

Soldier's Bay 49°27'·06N 02°31'·68W. Good holding on sand; open to E. Beware Boue Sablon and rks off Les Terres Point. Anfré Rk (3₃) bn A, is 4ca offshore. Steps to cliff path.

Fermain Bay 49°26'·05N 02°31'·95W. Good holding on sand; open to E. Beware Gold Fisher Rk (3·6) and Gabrielle Rk (2·1). Popular tourist beach. ✕ hotel ⬛.

SOUTH COAST

In centre of first bay W of St Martin's Pt, beware Mouillière (8·5) 49°25'·28N 02°32'·86W. S'lies can bring swell above half tide.

Petit Port Good holding on sand. Steep steps lead up to The Auberge, a popular fine dining restaurant and bar with a spectacular view.

Moulin Huet Good holding on sand. Tea garden and hotel.

Saints Bay Good holding on sand. Below half tide beware uncharted rock in middle and unburied telephone cable on E side. ⚓ outside moorings with trip line. Café, hotel, bar.

Icart Bay Beware Fourquie de la Moye (3·1) 49°25'·28N 02°32'·86W between Icart Pt and Pte de la Moye. Within Icart Bay are:

Jaonnet. Good holding on sand off small beach or further E off rky shore. Exposed to S. No facilities. Cliff path inland.

Petit Bôt Bay Beware drying reef on E side. A short swell often works in. ✕ up hill to hotel, ⬛ ✈ ✉ ⬛. Better ⚓ close W at:

Portelet 49°25'·11N 02°35'·28W. Good holding on sand. Sheltered from N and W. No access inland. Facilities via dinghy/Petit Bôt.

WEST COAST

Good visibility and AC 807 essential; an E'ly wind is desirable. Pass outside Les Hanois, unless bound for:

Portelet Harbour ⚓ at 49°26'·40N 02°40'·26W, outside moorings. Good holding on sand. Exposed to W. Rky appr; local knowledge advised. Avoid small drying stone quay. ✕ hotel ⬛.

Lihou Island ⚓ 49°27'·88N 02°39'·46W, off NE corner. Sheltered from E. Between Lihou and Guernsey is a drying causeway. Tide runs fast. Avoid bird sanctuaries on offlying islets. No facilities.

Perelle Bay 49°28'·23N 02°38'·89W. Good holding on sand. Exposed to W; rky appr. Beware Colombelle Rk (1₅) NE of brg 128° on cement mill. ⚓ outside moorings. ⬛ hotel ⬛ ⬛.

Vazon Bay 49°28'·29N 02°37'·09W. Wide sandy beach for settled conditions, but open to the W. Beware Boue Vazon (3) in appr, many lobster pots, surfers and bathers. Long surf line. Hotel, ✕ ⬛.

Cobo Bay 49°28'·66N 02°36'·33W. Beware lobster pots and Boue Vazon (3) in appr. Rky appr from S of Moulière. Good holding on sand; ⚓ outside moorings. Hotel, ✕ ⬛ ✉ ⬛ ⬛ ⬛.

Grande Havre 49°30'·11N 02°33'·09W. Very popular, many local moorings. Rky appr 171° as charted; exposed to NW, sheltered from S'lies. ⚓ to W of Hommet de Grève. Stone slip, busy in summer; lying alongside not advised. Hotel ⬛.

L'Ancresse Bay 49°30'·45N 02°31'·38W. Good holding on sand. Exposed to the N, but good shelter from S/SW. Hotel ⬛ café.

Fontenelle Bay 49°30'·48N 02°30'·69W. Good holding on sand. Exposed to the N. Beware drying rks on E of ent. No facilities.

ST PETER PORT   LAT 49°27'N   LONG 2°32'W

TIMES AND HEIGHTS OF HIGH AND LOW WATERS

STANDARD TIME (UT)
For Summer Time add ONE hour in **non-shaded areas**

Dates in **red** are SPRINGS
Dates in blue are NEAPS

YEAR **2019**

JANUARY

Time	m		Time	m
1 TU	0234 7.5 / 0900 3.2 / 1500 7.6 / 2133 3.0	**16** W	0108 7.0 / 0740 3.7 / 1349 7.1 / 2019 3.4
2 W	0338 7.7 / 1003 2.9 / 1601 7.8 / 2230 2.8	**17** TH	0229 7.3 / 0900 3.4 / 1505 7.4 / 2132 3.0
3 TH	0431 8.1 / 1056 2.6 / 1653 8.0 / 2319 2.5	**18** F	0339 7.7 / 1008 2.8 / 1612 7.9 / 2236 2.5
4 F	0517 8.4 / 1143 2.3 / 1739 8.3	**19** SA	0439 8.3 / 1109 2.2 / 1711 8.4 / 2333 2.0
5 SA	0002 2.3 / 0558 8.7 / 1225 2.1 / 1820 8.5	**20** SU	0534 8.9 / 1204 1.6 / 1805 8.9
6 SU ●	0042 2.1 / 0637 8.9 / 1304 1.9 / 1859 8.6	**21** M ○	0026 1.5 / 0625 9.4 / 1257 1.1 / 1855 9.3
7 M	0118 2.0 / 0714 8.9 / 1340 1.9 / 1934 8.6	**22** TU	0117 1.1 / 0713 9.8 / 1346 0.8 / 1943 9.6
8 TU	0151 2.1 / 0748 8.9 / 1413 2.0 / 2008 8.5	**23** W	0205 0.8 / 0800 9.9 / 1433 0.6 / 2029 9.6
9 W	0222 2.2 / 0821 8.7 / 1445 2.1 / 2040 8.4	**24** TH	0250 0.8 / 0846 9.9 / 1518 0.7 / 2113 9.5
10 TH	0252 2.4 / 0853 8.5 / 1516 2.3 / 2112 8.1	**25** F	0333 1.1 / 0930 9.6 / 1601 1.1 / 2157 9.1
11 F	0323 2.7 / 0925 8.2 / 1548 2.6 / 2144 7.8	**26** SA	0417 1.5 / 1014 9.1 / 1645 1.6 / 2241 8.5
12 SA	0355 3.0 / 0959 7.9 / 1621 2.9 / 2219 7.5	**27** SU ◑	0502 2.1 / 1100 8.4 / 1731 2.3 / 2329 7.9
13 SU	0432 3.3 / 1038 7.5 / 1702 3.2 / 2302 7.3	**28** M	0552 2.7 / 1152 7.7 / 1824 3.0
14 M ◑	0518 3.6 / 1128 7.3 / 1753 3.4 / 2357 7.1	**29** TU	0027 7.3 / 0653 3.3 / 1259 7.2 / 1930 3.4
15 TU	0620 3.7 / 1231 7.1 / 1901 3.5	**30** W	0145 7.0 / 0812 3.6 / 1422 7.0 / 2052 3.6
		31 TH	0306 7.1 / 0933 3.5 / 1538 7.1 / 2205 3.4

FEBRUARY

Time	m		Time	m
1 F	0409 7.4 / 1036 3.1 / 1636 7.4 / 2300 3.0	**16** SA	0309 7.4 / 0945 3.0 / 1553 7.6 / 2216 2.8
2 SA	0459 7.9 / 1126 2.7 / 1723 7.9 / 2346 2.6	**17** SU	0422 8.0 / 1054 2.3 / 1658 8.2 / 2320 2.1
3 SU	0542 8.3 / 1209 2.3 / 1805 8.2	**18** M	0521 8.7 / 1153 1.6 / 1754 8.9
4 M ●	0026 2.3 / 0621 8.6 / 1248 2.0 / 1843 8.5	**19** TU ○	0015 1.4 / 0613 9.4 / 1246 0.9 / 1844 9.4
5 TU	0103 2.0 / 0658 8.9 / 1324 1.8 / 1918 8.6	**20** W	0106 0.8 / 0702 9.9 / 1334 0.4 / 1931 9.8
6 W	0136 1.9 / 0733 9.0 / 1357 1.7 / 1952 8.7	**21** TH	0152 0.5 / 0747 10.2 / 1419 0.2 / 2014 9.9
7 TH	0207 1.8 / 0805 8.9 / 1428 1.7 / 2023 8.6	**22** F	0235 0.4 / 0830 10.2 / 1500 0.3 / 2055 9.8
8 F	0236 1.9 / 0836 8.8 / 1457 1.9 / 2052 8.5	**23** SA	0316 0.6 / 0911 9.9 / 1539 0.7 / 2134 9.4
9 SA	0305 2.1 / 0905 8.5 / 1525 2.1 / 2120 8.3	**24** SU	0354 1.1 / 0950 9.3 / 1617 1.4 / 2211 8.7
10 SU	0333 2.4 / 0934 8.3 / 1554 2.4 / 2150 8.0	**25** M	0432 1.8 / 1028 8.5 / 1655 2.2 / 2250 8.0
11 M	0404 2.7 / 1005 8.0 / 1627 2.7 / 2226 7.7	**26** TU ◑	0513 2.6 / 1110 7.7 / 1737 3.0 / 2334 7.3
12 TU ◑	0440 3.0 / 1049 7.6 / 1709 3.1 / 2311 7.4	**27** W	0603 3.3 / 1205 6.9 / 1833 3.7
13 W	0530 3.4 / 1143 7.2 / 1805 3.4	**28** TH	0041 6.7 / 0715 3.8 / 1333 6.5 / 1956 4.0
14 TH	0013 7.1 / 0641 3.6 / 1258 7.0 / 1925 3.5		
15 F	0138 7.0 / 0818 3.5 / 1430 7.1 / 2058 3.3		

MARCH

Time	m		Time	m
1 F	0226 6.6 / 0857 3.9 / 1514 6.6 / 2138 3.9	**16** SA	0104 7.0 / 0750 3.5 / 1409 6.9 / 2036 3.5
2 SA	0346 7.0 / 1015 3.5 / 1617 7.0 / 2241 3.4	**17** SU	0247 7.2 / 0930 3.0 / 1540 7.5 / 2202 3.0
3 SU	0439 7.5 / 1106 2.9 / 1704 7.5 / 2326 2.9	**18** M	0407 7.9 / 1041 2.2 / 1646 8.2 / 2306 2.1
4 M	0522 8.0 / 1148 2.4 / 1744 8.0	**19** TU	0506 8.7 / 1138 1.4 / 1739 8.9
5 TU	0006 2.4 / 0600 8.5 / 1227 2.0 / 1821 8.4	**20** W	0000 1.3 / 0557 9.4 / 1229 0.8 / 1827 9.5
6 W ●	0042 2.0 / 0637 8.8 / 1302 1.7 / 1857 8.7	**21** TH ○	0049 0.7 / 0644 9.9 / 1315 0.3 / 1911 9.9
7 TH	0116 1.7 / 0712 9.0 / 1335 1.5 / 1930 8.9	**22** F	0133 0.3 / 0728 10.2 / 1357 0.1 / 1953 10.0
8 F	0147 1.6 / 0744 9.1 / 1405 1.4 / 2001 8.9	**23** SA	0215 0.2 / 0809 10.2 / 1437 0.3 / 2031 9.9
9 SA	0216 1.6 / 0814 9.0 / 1433 1.5 / 2029 8.8	**24** SU	0253 0.5 / 0847 9.8 / 1513 0.7 / 2107 9.5
10 SU	0243 1.7 / 0842 8.8 / 1501 1.7 / 2056 8.6	**25** M	0329 1.0 / 0923 9.2 / 1548 1.4 / 2141 8.8
11 M	0311 1.9 / 0911 8.6 / 1528 2.0 / 2124 8.3	**26** TU	0403 1.7 / 0958 8.4 / 1620 2.3 / 2214 8.1
12 TU	0340 2.3 / 0943 8.2 / 1559 2.4 / 2158 8.0	**27** W	0439 2.5 / 1034 7.6 / 1656 3.1 / 2251 7.4
13 W	0414 2.7 / 1021 7.8 / 1638 2.9 / 2240 7.6	**28** TH ◑	0522 3.3 / 1120 6.9 / 1744 3.8 / 2345 6.7
14 TH ◑	0459 3.1 / 1105 7.4 / 1731 3.3 / 2339 7.2	**29** F	0629 3.9 / 1237 6.3 / 1902 4.2
15 F	0607 3.5 / 1227 6.9 / 1850 3.6	**30** SA	0126 6.4 / 0804 4.0 / 1436 6.3 / 2053 4.2
		31 SU	0310 6.7 / 0940 3.7 / 1548 6.8 / 2209 3.6

APRIL

Time	m		Time	m
1 M	0408 7.2 / 1035 3.1 / 1635 7.4 / 2256 3.0	**16** TU	0346 8.0 / 1021 2.1 / 1626 8.3 / 2246 2.0
2 TU	0452 7.8 / 1117 2.5 / 1715 7.9 / 2336 2.5	**17** W	0445 8.7 / 1116 1.4 / 1717 8.9 / 2338 1.4
3 W	0532 8.3 / 1155 2.0 / 1753 8.4	**18** TH	0536 9.3 / 1205 0.9 / 1804 9.5
4 TH	0012 2.0 / 0608 8.7 / 1232 1.7 / 1828 8.8	**19** F ○	0025 0.8 / 0622 9.7 / 1250 0.5 / 1847 9.8
5 F ●	0047 1.7 / 0644 9.0 / 1305 1.4 / 1902 9.0	**20** SA	0110 0.5 / 0705 9.9 / 1332 0.4 / 1928 9.9
6 SA	0120 1.5 / 0718 9.1 / 1337 1.3 / 1934 9.0	**21** SU	0151 0.5 / 0746 9.8 / 1411 0.6 / 2005 9.7
7 SU	0151 1.4 / 0749 9.1 / 1407 1.4 / 2003 9.0	**22** M	0229 0.7 / 0823 9.5 / 1447 1.1 / 2040 9.3
8 M	0221 1.5 / 0819 9.0 / 1437 1.5 / 2032 8.9	**23** TU	0304 1.2 / 0858 9.0 / 1519 1.7 / 2113 8.8
9 TU	0250 1.7 / 0850 8.7 / 1506 1.9 / 2103 8.6	**24** W	0337 1.8 / 0932 8.3 / 1551 2.4 / 2145 8.1
10 W	0321 2.0 / 0924 8.4 / 1539 2.3 / 2138 8.2	**25** TH	0411 2.6 / 1007 7.6 / 1623 3.1 / 2221 7.5
11 TH	0357 2.4 / 1005 7.9 / 1619 2.8 / 2222 7.8	**26** F ◑	0452 3.2 / 1050 7.0 / 1707 3.7 / 2309 6.9
12 F ◑	0445 2.9 / 1058 7.4 / 1714 3.3 / 2322 7.3	**27** SA	0552 3.8 / 1154 6.5 / 1817 4.2
13 SA	0555 3.3 / 1214 7.0 / 1836 3.6	**28** SU	0027 6.5 / 0713 4.0 / 1333 6.4 / 1948 4.2
14 SU	0048 7.1 / 0738 3.4 / 1356 7.0 / 2021 3.4	**29** M	0210 6.6 / 0840 3.7 / 1457 6.7 / 2114 3.8
15 M	0229 7.3 / 0914 2.9 / 1524 7.6 / 2144 2.8	**30** TU	0321 7.0 / 0946 3.3 / 1552 7.2 / 2211 3.3

Chart Datum is 5·06 metres below Ordnance Datum (Local). HAT is 10·3 metres above Chart Datum.

FREE monthly updates. Register at www.reedsnauticalalmanac.co.uk

239

STANDARD TIME (UT)
For Summer Time add ONE hour in **non-shaded areas**

ST PETER PORT LAT 49°27'N LONG 2°32'W
TIMES AND HEIGHTS OF HIGH AND LOW WATERS

Dates in red are **SPRINGS**
Dates in blue are **NEAPS**

YEAR 2019

MAY

Time	m	Time	m
1 0411	7.5	**16** 0419	8.5
1034	2.7	1049	1.7
W 1636	7.8	TH 1651	8.7
2255	2.7	2312	1.6
2 0454	8.0	**17** 0510	9.0
1115	2.2	1139	1.3
TH 1716	8.2	F 1738	9.1
2334	2.2		
3 0534	8.5	**18** 0000	1.2
1154	1.8	0557	9.2
F 1754	8.6	SA 1225	1.1
		○ 1821	9.4
4 0012	1.8	**19** 0045	1.0
0611	8.8	0641	9.4
SA 1231	1.6	SU 1307	1.0
● 1830	8.9	1902	9.5
5 0049	1.5	**20** 0127	1.0
0648	9.0	0723	9.3
SU 1307	1.4	M 1346	1.2
1904	9.1	1941	9.3
6 0125	1.4	**21** 0205	1.2
0723	9.1	0801	9.0
M 1342	1.4	TU 1422	1.5
1938	9.1	2016	9.1
7 0159	1.4	**22** 0241	1.5
0758	9.0	0837	8.7
TU 1416	1.5	W 1455	2.0
2012	9.0	2050	8.6
8 0233	1.5	**23** 0315	2.0
0834	8.8	0911	8.2
W 1450	1.8	TH 1527	2.5
2048	8.8	2123	8.2
9 0310	1.8	**24** 0350	2.5
0913	8.5	0946	7.7
TH 1528	2.2	F 1600	3.0
2128	8.4	2159	7.7
10 0351	2.2	**25** 0428	3.1
0959	8.0	1028	7.2
F 1613	2.6	SA 1640	3.5
2216	8.0	2242	7.2
11 0444	2.7	**26** 0519	3.5
1056	7.5	1120	6.8
SA 1711	3.1	SU 1736	3.9
2318	7.6	◑ 2341	6.8
12 0555	3.0	**27** 0625	3.7
1209	7.2	1230	6.6
SU 1830	3.4	M 1850	4.0
◑			
13 0037	7.4	**28** 0058	6.7
0726	3.0	0736	3.7
M 1338	7.3	TU 1347	6.7
2002	3.2	2005	3.9
14 0205	7.6	**29** 0215	6.9
0849	2.7	0842	3.4
TU 1457	7.7	W 1453	7.1
2119	2.7	2110	3.5
15 0319	8.0	**30** 0316	7.3
0955	2.2	0939	3.0
W 1559	8.2	TH 1546	7.5
2220	2.1	2204	3.0
		31 0407	7.7
		1028	2.6
		F 1633	8.0
		2251	2.5

JUNE

Time	m	Time	m
1 0453	8.1	**16** 0535	8.6
1114	2.2	1201	1.8
SA 1716	8.4	SU 1758	8.9
2336	2.1		
2 0537	8.5	**17** 0023	1.6
1157	1.8	0621	8.8
SU 1757	8.7	M 1245	1.7
		○ 1840	9.0
3 0018	1.7	**18** 0106	1.6
0619	8.8	0703	8.8
M 1239	1.6	TU 1325	1.7
● 1837	9.0	1920	9.0
4 0100	1.5	**19** 0145	1.6
0700	9.0	0742	8.7
TU 1320	1.4	W 1402	1.9
1916	9.2	1957	8.8
5 0142	1.4	**20** 0222	1.8
0742	9.0	0819	8.5
W 1400	1.5	TH 1436	2.1
1957	9.0	2031	8.6
6 0223	1.4	**21** 0257	2.1
0824	8.9	0854	8.2
TH 1441	1.6	F 1508	2.5
2039	9.0	2105	8.3
7 0306	1.6	**22** 0331	2.4
0909	8.7	0928	7.9
F 1525	1.9	SA 1540	2.9
2124	8.8	2139	7.9
8 0353	1.9	**23** 0406	2.8
0958	8.3	1005	7.5
SA 1613	2.3	SU 1615	3.2
2215	8.4	2217	7.6
9 0447	2.3	**24** 0445	3.1
1053	7.9	1047	7.2
SU 1710	2.7	M 1657	3.5
2313	8.0	2302	7.2
10 0550	2.6	**25** 0534	3.4
1157	7.7	1136	7.0
M 1818	3.0	TU 1751	3.7
◐		◐ 2357	7.0
11 0020	7.8	**26** 0634	3.5
0703	2.7	1237	6.9
TU 1311	7.6	W 1858	3.8
1934	3.0		
12 0135	7.7	**27** 0103	7.0
0817	2.7	0739	3.5
W 1424	7.7	TH 1345	7.0
2048	2.8	2008	3.6
13 0248	7.9	**28** 0213	7.1
0924	2.4	0843	3.3
TH 1528	8.0	F 1449	7.3
2152	2.5	2112	3.3
14 0351	8.1	**29** 0317	7.4
1022	2.2	0942	2.9
F 1623	8.3	SA 1547	7.7
2247	2.1	2209	2.8
15 0446	8.4	**30** 0413	7.8
1113	1.9	1036	2.5
SA 1713	8.7	SU 1639	8.1
2337	1.8	2302	2.3

JULY

Time	m	Time	m
1 0505	8.2	**16** 0006	2.1
1127	2.1	0605	8.3
M 1728	8.6	TU 1228	2.2
2353	1.9	○ 1823	8.7
2 0555	8.6	**17** 0049	1.9
1216	1.8	0647	8.5
TU 1815	8.9	W 1308	2.0
●		1902	8.8
3 0042	1.5	**18** 0129	1.8
0643	8.9	0726	8.5
W 1304	1.5	TH 1345	2.0
1901	9.2	1939	8.8
4 0130	1.2	**19** 0205	1.8
0731	9.1	0802	8.5
TH 1351	1.3	F 1419	2.1
1947	9.4	2014	8.7
5 0217	1.1	**20** 0239	1.9
0818	9.2	0836	8.4
F 1436	1.3	SA 1450	2.2
2033	9.4	2046	8.5
6 0304	1.1	**21** 0310	2.2
0904	9.1	0908	8.2
SA 1522	1.5	SU 1519	2.5
2119	9.2	2118	8.3
7 0351	1.4	**22** 0341	2.4
0951	8.8	0939	7.9
SU 1609	1.8	M 1550	2.8
2207	8.9	2150	8.0
8 0440	1.7	**23** 0412	2.7
1041	8.4	1012	7.6
M 1659	2.2	TU 1622	3.1
2258	8.5	2225	7.6
9 0533	2.1	**24** 0447	3.1
1135	8.0	1049	7.4
TU 1755	2.6	W 1701	3.4
◑ 2355	8.0	2307	7.3
10 0633	2.6	**25** 0532	3.3
1237	7.7	1136	7.1
W 1900	2.9	TH 1752	3.6
		◐	
11 0101	7.7	**26** 0000	7.1
0740	2.8	0631	3.5
TH 1347	7.5	F 1237	7.0
2012	3.0	1902	3.7
12 0215	7.5	**27** 0110	7.0
0851	2.9	0745	3.5
F 1457	7.6	SA 1351	7.1
2123	2.9	2022	3.5
13 0326	7.6	**28** 0229	7.2
0956	2.8	0859	3.3
SA 1559	7.8	SU 1505	7.4
2225	2.7	2134	3.1
14 0426	7.8	**29** 0340	7.5
1053	2.6	1005	2.9
SU 1653	8.1	M 1610	7.9
2319	2.4	2238	2.6
15 0519	8.1	**30** 0443	8.0
1143	2.4	1105	2.4
M 1740	8.4	TU 1707	8.4
		2336	2.0
		31 0539	8.5
		1200	1.8
		W 1800	9.0

AUGUST

Time	m	Time	m
1 0030	1.4	**16** 0111	1.8
0631	9.0	0706	8.6
TH 1253	1.4	F 1327	1.9
● 1849	9.4	1920	8.9
2 0121	1.0	**17** 0145	1.7
0721	9.4	0741	8.7
F 1342	1.0	SA 1359	1.9
1937	9.8	1953	8.9
3 0209	0.7	**18** 0217	1.7
0808	9.6	0813	8.7
SA 1428	0.9	SU 1428	1.9
2023	9.9	2024	8.8
4 0254	0.6	**19** 0245	1.9
0852	9.5	0842	8.5
SU 1512	0.9	M 1455	2.1
2107	9.7	2053	8.6
5 0338	0.9	**20** 0312	2.1
0936	9.3	0910	8.3
M 1555	1.2	TU 1522	2.4
2151	9.4	2121	8.3
6 0421	1.3	**21** 0339	2.4
1020	8.8	0938	8.0
TU 1639	1.8	W 1550	2.7
2235	8.8	2151	8.0
7 0506	1.9	**22** 0409	2.8
1106	8.3	1009	7.7
W 1722	2.4	TH 1622	3.1
◑ 2324	8.1	2226	7.6
8 0557	2.6	**23** 0445	3.2
1159	7.7	1049	7.4
TH 1821	3.0	F 1704	3.4
		◑ 2313	7.3
9 0023	7.5	**24** 0536	3.5
0658	3.2	1144	7.1
F 1307	7.2	SA 1805	3.7
1932	3.4		
10 0142	7.0	**25** 0020	7.0
0817	3.5	0650	3.7
SA 1429	7.1	SU 1301	7.0
2058	3.5	1937	3.7
11 0308	7.0	**26** 0151	7.0
0937	3.4	0825	3.6
SU 1541	7.3	M 1433	7.2
2211	3.2	2110	3.4
12 0414	7.3	**27** 0320	7.3
1039	3.1	0945	3.1
M 1638	7.7	TU 1550	7.8
2306	2.8	2223	2.7
13 0506	7.7	**28** 0430	8.0
1129	2.8	1051	2.5
TU 1725	8.2	W 1652	8.5
2352	2.4	2323	1.9
14 0550	8.1	**29** 0527	8.6
1212	2.4	1148	1.8
W 1806	8.5	TH 1746	9.1
15 0033	2.1	**30** 0017	1.2
0630	8.4	0618	9.2
TH 1251	2.1	F 1240	1.2
○ 1844	8.8	● 1836	9.7
		31 0107	0.7
		0706	9.7
		SA 1327	0.7
		1922	10.1

Chart Datum is 5·06 metres below Ordnance Datum (Local). HAT is 10·3 metres above Chart Datum.

》》 **FREE** monthly updates. Register at 《
www.reedsnauticalalmanac.co.uk 《

ST PETER PORT   LAT 49°27'N   LONG 2°32'W
TIMES AND HEIGHTS OF HIGH AND LOW WATERS

Dates in red are **SPRINGS**
Dates in blue are **NEAPS**

YEAR 2019

Channel Islands

SEPTEMBER

Time	m		Time	m
1 0153	0.4	**16**	0149	1.6
0750	9.9		0745	9.0
SU 1412	0.5	M 1402	1.7
2006	10.2		1958	9.1
2 0236	0.3	**17**	0217	1.7
0833	9.9		0814	8.8
M 1454	0.4	TU 1429	1.9
2048	10.1		2026	8.9
3 0317	0.6	**18**	0243	1.9
0913	9.6		0840	8.6
TU 1534	1.0	W 1455	2.2
2128	9.6		2053	8.6
4 0356	1.2	**19**	0309	2.2
0952	9.1		0907	8.3
W 1613	1.6	TH 1522	2.5
2208	8.9		2122	8.2
5 0435	2.0	**20**	0338	2.6
1032	8.3		0937	8.0
TH 1654	2.4	F 1552	2.9
2250	8.0		2155	7.8
6 0518	2.8	**21**	0412	3.1
1118	7.6		1014	7.6
F 1743	3.2	SA 1632	3.3
◐ 2342	7.2		2241	7.4
7 0615	3.6	**22**	0500	3.5
1222	7.0		1108	7.2
SA 1852	3.8	SU 1731	3.7
		◓ 2348	6.9
8 0105	6.6	**23**	0614	3.9
0739	4.1		1228	6.9
SU 1401	6.7	M 1908	3.8
2034	3.9		
9 0254	6.6	**24**	0128	6.8
0921	3.9		0803	3.8
M 1525	7.0	TU 1412	7.1
2158	3.6		2055	3.4
10 0402	7.1	**25**	0309	7.3
1025	3.5		0932	3.2
TU 1621	7.5	W 1535	7.8
2251	3.1		2210	2.7
11 0450	7.6	**26**	0417	8.1
1112	3.0		1037	2.5
W 1706	8.1	TH 1637	8.6
2333	2.5		2309	1.8
12 0530	8.1	**27**	0512	8.8
1152	2.5		1131	1.7
TH 1745	8.5	F 1729	9.3
13 0011	2.1	**28**	0000	1.1
0606	8.5		0600	9.5
F 1229	2.1	SA 1221	1.0
1821	8.9	● 1817	9.9
14 0047	1.8	**29**	0047	0.6
0641	8.9		0645	9.9
SA 1302	1.8	SU 1308	0.6
○ 1856	9.1		1902	10.2
15 0119	1.6	**30**	0131	0.3
0714	8.9		0728	10.1
SU 1333	1.7	M 1351	0.4
1928	9.1		1944	10.3

OCTOBER

Time	m		Time	m
1 0213	0.4	**16**	0148	1.6
0808	10.1		0745	9.0
TU 1431	0.5	W 1403	1.8
2025	10.1		2000	9.0
2 0251	0.7	**17**	0216	1.8
0847	9.7		0813	8.9
W 1509	1.0	TH 1431	2.0
2103	9.5		2029	8.7
3 0328	1.4	**18**	0244	2.1
0923	9.1		0841	8.6
TH 1546	1.7	F 1500	2.3
2139	8.8		2100	8.4
4 0403	2.2	**19**	0315	2.6
0959	8.4		0914	8.2
F 1624	2.5	SA 1533	2.7
2217	7.9		2136	8.0
5 0442	3.1	**20**	0351	3.0
1039	7.6		0954	7.8
SA 1708	3.3	SU 1615	3.2
◓ 2303	7.1		2224	7.5
6 0532	3.9	**21**	0441	3.5
1136	6.9		1050	7.4
SU 1814	4.0	M 1717	3.6
		◓ 2334	7.0
7 0022	6.5	**22**	0557	3.9
0655	4.4		1211	7.1
M 1343	6.6	TU 1854	3.7
1956	4.2		
8 0229	6.5	**23**	0114	6.9
0854	4.3		0746	3.8
TU 1458	6.8	W 1353	7.3
2131	3.8		2038	3.3
9 0338	6.9	**24**	0252	7.4
1000	3.7		0914	3.2
W 1554	7.4	TH 1515	7.9
2223	3.3		2151	2.6
10 0423	7.5	**25**	0358	8.2
1045	3.2		1017	2.4
TH 1638	7.9	F 1615	8.6
2304	2.7		2247	1.8
11 0501	8.1	**26**	0450	8.9
1123	2.6		1110	1.7
F 1716	8.4	SA 1707	9.3
2341	2.2		2337	1.2
12 0537	8.5	**27**	0537	9.5
1158	2.2		1158	1.1
SA 1752	8.8	SU 1755	9.8
13 0015	1.9	**28**	0023	0.8
0611	8.9		0621	9.9
SU 1232	1.9	M 1245	0.8
○ 1827	9.1	● 1839	10.0
14 0048	1.6	**29**	0107	0.6
0643	9.1		0703	10.0
M 1304	1.7	TU 1328	0.6
1900	9.2		1922	10.0
15 0119	1.6	**30**	0148	0.7
0716	9.1		0743	9.9
TU 1334	1.7	W 1408	0.8
1931	9.1		2002	9.8
		31	0226	1.1
			0821	9.6
		TH 1446	1.2
			2039	9.2

NOVEMBER

Time	m		Time	m
1 0302	1.7	**16**	0227	2.1
0857	9.0		0824	8.8
F 1522	1.9	SA 1448	2.1
2115	8.6		2048	8.5
2 0336	2.4	**17**	0303	2.4
0932	8.4		0902	8.5
SA 1559	2.6	SU 1526	2.5
2152	7.8		2129	8.1
3 0411	3.2	**18**	0344	2.9
1010	7.7		0947	8.1
SU 1641	3.3	M 1613	2.9
2235	7.1		2221	7.7
4 0456	3.9	**19**	0436	3.3
1100	7.0		1045	7.7
M 1739	3.9	TU 1716	3.3
◐ 2339	6.6	◑ 2328	7.3
5 0608	4.4	**20**	0549	3.6
1223	6.7		1200	7.4
TU 1902	4.2	W 1841	3.4
6 0126	6.4	**21**	0055	7.2
0749	4.4		0723	3.6
W 1405	6.7	TH 1327	7.5
2034	4.0		2011	3.1
7 0252	6.8	**22**	0223	7.6
0913	4.0		0846	3.1
TH 1511	7.2	F 1446	7.9
2138	3.5		2123	2.6
8 0343	7.3	**23**	0330	8.1
1004	3.5		0951	2.5
F 1559	7.7	SA 1549	8.5
2223	3.0		2221	2.0
9 0424	7.9	**24**	0424	8.7
1045	2.9		1046	2.0
SA 1640	8.1	SU 1643	9.0
2302	2.5		2312	1.6
10 0501	8.3	**25**	0512	9.2
1122	2.5		1135	1.5
SU 1718	8.5	M 1732	9.3
2339	2.1		2359	1.3
11 0537	8.7	**26**	0558	9.5
1158	2.1		1222	1.2
M 1755	8.8	TU 1818	9.5
		●	
12 0014	1.9	**27**	0044	1.2
0612	9.0		0640	9.7
TU 1233	1.9	W 1306	1.1
○ 1830	9.0		1901	9.5
13 0048	1.7	**28**	0126	1.3
0646	9.1		0721	9.6
W 1307	1.7	TH 1348	1.2
1904	9.1		1942	9.3
14 0122	1.7	**29**	0204	1.5
0718	9.1		0800	9.3
TH 1340	1.8	F 1427	1.5
1938	9.0		2021	8.9
15 0154	1.8	**30**	0241	2.0
0750	9.0		0836	8.9
F 1413	1.9	SA 1504	2.0
2011	8.8		2057	8.4

DECEMBER

Time	m		Time	m
1 0315	2.5	**16**	0259	2.1
0911	8.4		0859	8.9
SU 1541	2.5	M 1528	2.0
2133	7.9		2128	8.4
2 0349	3.1	**17**	0344	2.4
0949	7.9		0946	8.6
M 1620	3.1	TU 1616	2.3
2213	7.4		2219	8.1
3 0429	3.6	**18**	0435	2.8
1032	7.4		1040	8.2
TU 1706	3.5	W 1712	2.7
2303	6.9		2317	7.8
4 0521	4.0	**19**	0537	3.1
1130	7.0		1142	7.9
W 1807	3.9	TH 1818	2.9
◐		◑	
5 0010	6.7	**20**	0026	7.6
0633	4.2		0651	3.2
TH 1246	6.8	F 1255	7.7
1917	3.9		1933	2.9
6 0131	6.7	**21**	0144	7.6
0752	4.1		0810	3.1
F 1402	6.9	SA 1411	7.8
2027	3.7		2048	2.8
7 0241	7.0	**22**	0256	7.9
0901	3.8		0921	2.8
SA 1504	7.3	SU 1520	8.0
2126	3.4		2153	2.5
8 0335	7.5	**23**	0357	8.2
0954	3.3		1022	2.4
SU 1555	7.7	M 1620	8.3
2215	2.9		2249	2.2
9 0420	7.9	**24**	0450	8.6
1040	2.9		1115	2.0
M 1639	8.1	TU 1713	8.6
2259	2.5		2339	1.9
10 0501	8.4	**25**	0538	8.9
1122	2.4		1204	1.8
TU 1722	8.4	W 1801	8.8
2340	2.2		
11 0541	8.7	**26**	0026	1.8
1203	2.1		0622	9.1
W 1802	8.7	TH 1250	1.6
		● 1846	8.9
12 0020	1.9	**27**	0109	1.7
0619	8.9		0704	9.2
TH 1244	1.8	F 1333	1.6
○ 1842	8.9		1928	8.9
13 0100	1.8	**28**	0149	1.8
0657	9.1		0744	9.1
F 1324	1.7	SA 1412	1.7
1922	9.0		2007	8.7
14 0139	1.7	**29**	0225	2.0
0736	9.2		0821	8.9
SA 1403	1.7	SU 1449	1.9
2002	8.9		2042	8.5
15 0218	1.8	**30**	0259	2.3
0816	9.1		0855	8.6
SU 1444	1.8	M 1524	2.2
2043	8.7		2117	8.2
		31	0331	2.7
			0930	8.2
		TU 1558	2.6
			2151	7.8

Chart Datum is 5·06 metres below Ordnance Datum (Local). HAT is 10·3 metres above Chart Datum.

》》 FREE monthly updates. Register at 《
www.reedsnauticalalmanac.co.uk

241

6.11 HERM

Herm **49°28'·22N 02°27'·26W** ✳❀⚓☆☆☆

CHARTS and TIDES As for St Peter Port.

SHELTER Good shelter E or W of island depending on winds.

NAVIGATION Access is not difficult, if properly planned. Herm is reached from the Little Russel Channel via any of 7 passages all of which require reasonable visibility and care with tidal streams. AC 807 & 808 list the marks and bearings of all the passages.

The Alligande Passage is the simplest and most direct from St Peter Port. Approach with the Vermerette bcn (topmark 'V') in transit 074° with W patch on Hbr quay. Leave Alligande 'A', Godfrey 'GB' and Epec 'E' bcns to starboard; skirt close N of Vermerette. When its base is awash, there is 1m at hbr ent. The tide will be setting N. Sand build-up W of Vermerette affects craft NW-bound in the Percée Passage.

The appr from the Big Russel is more open and leads easily to pleasant ⚓s at Belvoir Bay and Shell Bay.

LIGHTS AND MARKS Ldg lts, both FW (occas) and W drums at 078°. 2FG (vert) on quay hd. Night appr not advised for visitors.

COMMUNICATIONS (Code 01481) Island Admin ☎750000 for overnight stay in hbr.

FACILITIES

Herm Harbour, options:

• Lie to mooring lines secured to N and S walls inside hbr.
• Dry out on the beach to the E, moored fore/aft to chains.
• 5 ⚓s to N and 6 to W of hbr dry out; 6 more to NW: rky bottom to the N; isolated boulders/sand to W.

www.herm.com for navigation details and webcams of hbr & buoys. No fees; donations welcome. ⚓ ⛟ ✕ ⌷ limited 🛒. Can be busy at summer weekends, better mid-week.

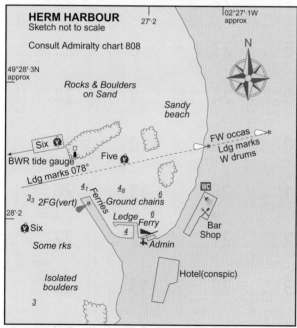

Rosière Steps Easiest appr is from the Big Russel via Percée passage; avoid Fourquies (2₇, NCM lt buoy) Meulettes (1₇) and Tinker (2₇). ⚓ 49°28'·22N 02°27'·22W NW of Rosière steps; good holding on sand, but exposed to S and SW. Access for landing only; do not linger alongside. Buoys are for ferries and Herm-owned boats only; hourly ferries by day. The bottom step is about 1.5m above the seabed which may be inconvenient at LWS. Caution: From just before HW to HW+2 tide sets hard onto the steps.

Belvoir Bay 49°28'·40N 02°26'·34W and **Shell Bay** are good ⚓s on sand, sheltered from W. Easy access from E; from S keep 400m offshore. Beach café or walk 800m to Harbour village. Note: Jethou, Crevichon and Grande Fauconnière are private. No landing.

6.12 SARK

Sark **49°25'·81N 02°20'·45W** Creux ❀✳⚓☆☆☆

CHARTS AC 808, 5604.12; SHOM 7159, 6904; Navi 1014; Imray C33A, 2500

TIDES −0450 Dover; ML 5·3; Duration 0550
Standard Port ST HELIER (➝)

Times				Height (metres)			
High Water		Low Water		MHWS	MHWN	MLWN	MLWS
0300	0900	0200	0800	11·0	8·1	4·0	1·4
1500	2100	1400	2100				
Differences SARK (MASELINE PIER)							
+0005	+0015	+0005	+0010	−2·1	−1·5	−0·6	−0·3

Tidal streams Beware large tidal range, strong streams and many lobster pots. In Gouliot (W coast) and Goulet Passages the streams reach 6-7kn at springs. Note that at about half-tide the streams are slack around Sark. At HW the stream sets hard to the N, ie onto Little Sark. At LW the stream sets hard to the S, ie onto Bec du Nez (N tip). If bound from Guernsey to Sark's E coast, go N-about at HW and S-about at LW; conversely on the return.

SHELTER Sark is fringed by rocks, but the centres of the bays are mainly clear of dangers. A safe ⚓age or ⚓s can usually be found sheltered from offshore winds. But, depending on wind and tide, they may be uncomfortable, except in settled weather; see also Facilities.

NAVIGATION From the West, WPT 49°25'·27N 02°24'·30W, 070°/1·29M towards the Pilcher monument (Sark mill is obsc'd by trees) for Havre Gosselin or La Grande Grève.
From the N or after rounding Bec du Nez, the WPT is 49°27'·30N 02°21'·42W, on the 153° charted transit (aka the outside passage) towards Grève de la Ville and Pécheresse. Noire Pierre rk is unlit but marked with a Y post & radar reflector. The inside passage, W of drying Pécheresse, is used by locals but ill advised for visitors.

LIGHTS AND MARKS Point Robert lt ho and Courbée du Nez are the only navigational lights on Sark; see chartlet and Lights, buoys and waypoints.

COMMUNICATIONS (Code 01481) Police (Guernsey) 725111; ⊖ (Guernsey) 726911; Dr 832045; HM 832323, VHF Ch 10; Maseline Hbr VHF Ch 13, season only; Tourist Office 832345.

ANCHORAGES, ⚓s, HARBOURS AND FACILITIES
Anti-clockwise from Bec du Nez, the ⚓s below (all unlit) are safe in offshore winds; in other conditions they can be exposed and sometimes dangerous. Some are only suitable around LW.

⚓s at Havre Gosselin and Grève de la Ville are free (courtesy Sark Moorings ☎832260, 07781 106065). Donations welcome in local boxes or c/o Le Grand Fort, Sark, Channel Islands, GY9 0SF. Water taxi Ch 10. All other moorings are private; use only in emergency.

WEST COAST (all ⚓s are exposed to W'lies)
Saignie Bay 49°26'·53N 02°22'·10W. Sand and shingle with fair holding. Picturesque rock formations.

Port à la Jument 49°26'·17N 02°22'·42W. Sand and shingle with fair holding. Difficult shore access.

Brecqhou Island is strictly private; landing prohibited.

Havre Gosselin 49°25'·77N 02°22'·68W. Popular deep (4-9m) ⚓. 20 Y ⚓s (see above). Beware of drying rk at extreme NW of bay. Crowded in summer. 299 steps to cliff top and panoramic views.

Port és Saies 49°25'·41N 02°22'·28W. Sandy inlet, steep cliff path.

La Grande Grève 49°25'·38N 02°22'·59W. Wide sandy bay, subject to swell, but popular day ⚓age. Beware two rks (drying 0·3m and ⊛) in the appr. Temporary steps to cliff-top panoramic views.

LITTLE SARK
Port Gorey 49°24'·60N 02°22'·72W. ⚓ or pick up Y ⚓ (free) in centre of deep, weedy bay over LW only; heavy swell begins near half-flood. Rocky appr from just NW of Grande Bretagne (18m high) then 045° into bay. Rocks must be positively identified. Remains of quay with ladder. Cliff walk past silver mine ruins to hotel.

Rouge Terrier 49°24'·78N 02°21'·87W. Sandy with some local moorings under high cliffs. Exposed to E. Landing with cliff path to hotel. Also ⚓ 49°25'·09N 02°21'·70W, 4ca NNE in Baleine Bay.

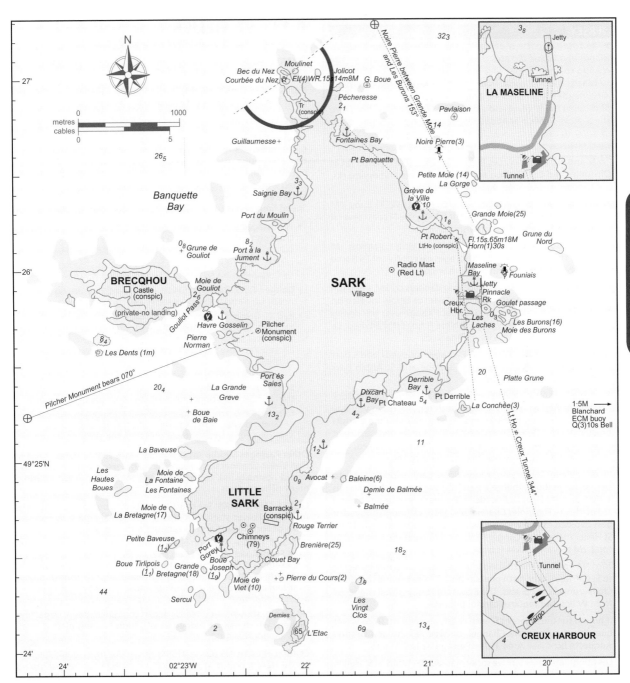

EAST COAST

Dixcart Bay 49°25´·33N 02°21´·48W. Popular sandy bay with good holding, but open to S'lies. Drying rocks either side of the bay but not in the centre. Cliff path and walk to hotels. Beware rockslides.

Derrible Bay 49°25´·41N 02°21´·07W. Sandy bay with good holding. Open to the S. No dangers in the bay, but keep off SW tip of Derrible Pt. Picturesque caves and steep climb ashore.

Creux Harbour 49°25´·80N 02°20´·61W dries completely, fair weather only, prone to surge/swell. Local transits: Pinnacle Rk on with E edge of Grand Moie 001°; or Creux tunnel (white arch) on with Pt Robert lt ho 344°. The lt ho dips behind the cliffs when close inshore. In the Goulet Passage (E of Creux) S-bound navigation is prohibited to commercial vessels (except with HM's permission) but not to yachts in either direction.

Dry out bow to NE wall, stern to ⚓. ⌇ possible only at inner part of the S pier, clear of steps and cargo berth at outer end. ☎832025

(kiosk) ⚓(free) ⚓ ⚓ ⚓ 🛆 🛆 via HM; walk or tractor up steep hill to village ✕ 🗆 🛒 🛆 🛆 🛆 ✉ ⑧. Bikes and horse-drawn carriages for hire.

Maseline Harbour 49°25´·96N 02°20´·60W is prone to surge/swell. Busy ferry hbr with no yacht berths (except to land people). Call HM Ch 13 for approval to ⚓. Moorings are private. ☎832070 (kiosk) ⌇(3t). Condor catamaran to Jersey/St Malo.

Grève de la Ville 49°26´·35N 02°21´·06W. ⚓ close in out of tide. Fair holding sand/shingle. Open to E. Seven Y ⚓s (free, see p 840). Easy walk ashore.

Les Fontaines 49°26´·78N 02°21´·62W. Sand and shingle with fair holding. Reef drying 4·5m extends 1ca N from shore. ⚓ between reef and Eperquerie headland. Exposed to the E.

With grateful acknowledgements to John Frankland, author of *Sark, Round the Island*.

▶*The rotatory pattern of tidal streams affecting the Channel Islands as a whole dictates that when streams are slack on the N and S coasts of Jersey, they are running strongly on the E and W coasts; and vice versa. If approaching Jersey from Guernsey/ Alderney at HW St Helier +4, a fair tide can be carried for at least 6hrs down the W coast and along the S coast to St Helier. From the S, leave St Malo at about HW, keeping E of the Minquiers, in order to carry a fair tide for 6hrs to St Helier. Follow similar tidal tactics when coasting around the island.*◀

To N and NE of Jersey, Les Pierres de Lecq (Paternosters) Les Dirouilles and Les Écrehou are groups of islets and drying rks, 2–4M offshore (AC *3655*, 1136, 1137, 1138).

From the N, a convenient landfall is Desormes WCM buoy, 4M NNW of Grosnez Pt (conspic lookout tr). On the N coast several bays offer anchs sheltered in offshore winds.

In St Ouen Bay on the W coast, which has drying rks almost 1M offshore, there are no good anchorages except NW of La Rocco tower which is sheltered in offshore winds.

Rounding the SW tip, to clear offlying dangers by 1M, keep the top of La Corbière lt ho (conspic) level with or below the FR lt on the clifftops behind. The inshore Boat Passage over the drying causeway between Jersey and Corbière lt ho is not advised. La Frouquie rock (9.8m) is no longer marked.

Along the S coast the **NW and W Passages** (buoyed) lead E past Noirmont Pt towards St Helier. St Brelade and St Aubin Bays provide some shelter from W'lies.

From the SW and S St Helier can be approached via **Danger Rock Passage** or **Red & Green Passage**. Both require good visibility to identify the transit marks and care to maintain the transits exactly. Only the Red & Green Passage is lit, but it needs sufficient water to pass over Fairway Rk (1.2m). Although charted, the marks for the South Passage 341° are hard to see.

SE of St Helier the drying, rocky Violet Bank extends 1M S to Demie de Pas lt bcn, thence E past Icho Twr (conspic). It extends 1.7M S and 2M SE of La Rocque Pt. Further rky plateaux extend 1M offshore. The **Violet Channel** (AC 1138) although buoyed, is best avoided in bad weather, wind-over-tide or poor vis. From St Helier make good Canger Rk WCM lt buoy, thence track 078° for 2.2M to Violet SWM lt buoy.

Turn N to pick up the charted ldg lines towards Gorey (dries) or to St Catherine Bay, both popular hbrs. The safe width of Violet Chan is only 5ca in places. The E coast of Jersey is well sheltered from W'lies, but requires careful pilotage.

If bound for the adjacent French coast, proceed NE from Violet buoy via the **Anquette Channel**, between Petite and Grande Anquette bcns. See Area 18 for Passage de la Déroute.

Les Minquiers, an extensive rocky plateau 10–18M S of St Helier, can be left to port, ie via NW and SW Minquiers buoys, if making for St Malo; or to stbd via NE and SE Minquiers buoys. A more direct route via N and SE Minquiers buoys requires sufficient height of tide to clear rocks drying 2m in the northern part. For a first visit the plateau should only be entered in settled weather, with exquisite care and a good Pilot book – not to mention *The Wreck of the Mary Deare* by Hammond Innes. The anchorage off Maîtresse Île is the principal attraction. From Jersey approach via the Demie de Vascelin SHM buoy with bcns in transit 167°. See St Helier.

6.13 ST HELIER

Jersey 49°10'·57N 02°06'·98W ✲✲✲✲♦♦♦♦✿✿

CHARTS AC 3655, 1137, 3278, 5604.14/15; SHOM 7160, 7161, 6938; Navi 534, 1014; Imray C33B, 2500

TIDES −0455 Dover; ML 6·1; Duration 0545. St Helier is a Standard Port. The tidal range is very large, 9·6m at springs.

SHELTER Excellent in 2 marinas and La Collette basin (no ⓥ).

NAVIGATION WPT (where all apprs converge) 49°09'·95N 02°07'·38W, 023°/0·74M to front 023° ldg lt. WPT is near the centre of a Precautionary Area extending up-hbr to the ferry terminal. Caution: Rule 9 (narrow channels) applies; vessels under sail not fitted with a working engine must notify St Helier VTS VHF Ch 14 before entering Precautionary area. ⚓ in Small Road discouraged. Speed limit 5kn N of La Collette Yacht Basin. Approach channels:

- **W Passage** 082° ldg lts and Dog's Nest Rock bcn, Fl Y 3s 3M (Y cross on top) lead N of Les Fours and Ruaudière Rock SHM lt buoys, to WPT. Beware race off Noirmont Pt.
- **NW Passage** (095°/110°, much used by yachts) passes 6ca S of La Corbière lt ho to join W Passage abeam Noirmont Pt.
- **Danger Rock Passage** (044°) unlit; past rky, drying shoals; needs care and good vis. Beware overfalls S of Danger Rk.
- **Red & Green Passage** Ldg line 022·7° passes over Fairway Rk 1·2m, 8ca SSW of the WPT, with drying rocks either side; thence via Small Road to the ferry terminal, where daymarks are red dayglow patches on front dolphin and rear lt twr. The fairway's W side is beaconed and buoyed. Outer pier hds and dolphin are painted white and floodlit.
- **S Passage** (341°) unlit marks are hard to see. Demie de Pas in transit 350° with floodlit power stn chy is easier to see D/N.
- **E Passage** 290° from Canger Rock WCM lt buoy passes close SW of Demie de Pas, B tr/Y top (at night stay in W sector); thence 314° past Hinguette buoy and 341° to the WPT.

Other Passages Middle Passage, unlit, leads 339° towards St Aubin Hbr. Violet Channel skirts round SE tip of Jersey to/from the N and would normally link to the E Passage to/from St Helier. Caution: many offlying reefs.

LIGHTS AND MARKS See chartlet and Lights, buoys and waypoints for lights. Power station chy (95m, floodlit) Sig mast, W concave roofs of Fort Regent and Elizabeth Castle are conspic.

IPTS Sigs 1-4 are shown from the VTS tower and are easily seen from Small Road and the Main Hbr. An Oc Y 4s, shown above Sigs 1-4, exempts power-driven craft <25m LOA from the main signals. All leisure craft should: *Keep to starboard, well clear of shipping; maintain a sharp all-round lookout and monitor VHF Ch 14 during arrival/departure.*

Sig 2 is also shown when vessels depart the Tanker Basin. Departing tankers, which can be hidden at LW, sound a long blast if small craft are approaching.

COMMUNICATIONS (Code 01534) *Jersey Coastguard* 447705; Ch 82, 16 H24; monitor *St Helier VTS* Ch 14 H24 for ferry/shipping movements; *St Helier Pierheads* on Ch 18 broadcasts recorded wind direction and speed; Police 612612; ⊖ 448000; Ⓗ 622000. No marina VHF, but call *VTS* if essential; Marine Leisure Centre (enquiries) 447708.

FACILITIES Port Control ☎447708 or 447788, ⚓ ⌫(32t) ⚓ Grids. www.jerseymarinas.je marinas@gov.je

Fuel ♦ ♦ at the ent to La Collette (HW +4; 07700 347313); a fuel barge by ent to Old Hbr (H24, self service, cr/debit card; ☎525247 07829 876543); Elizabeth marina (HW ±3; ☎07797 723090) self service, cr/debit card.

La Collette Basin, good shelter in 1·8m. Access H24. Caution: Ent narrow at LWS, least depth 1·1m; keep close to W side; PHM buoys mark shoal on E side. 130 ⌷ locals only, no ⓥ berths, access H24. ♦ ♦ (at top of tide) ⚑(65t and 25t, ☎447773) ⚓.

St Helier YC ☎721307, ✕ ⌸.

Continued on page 246

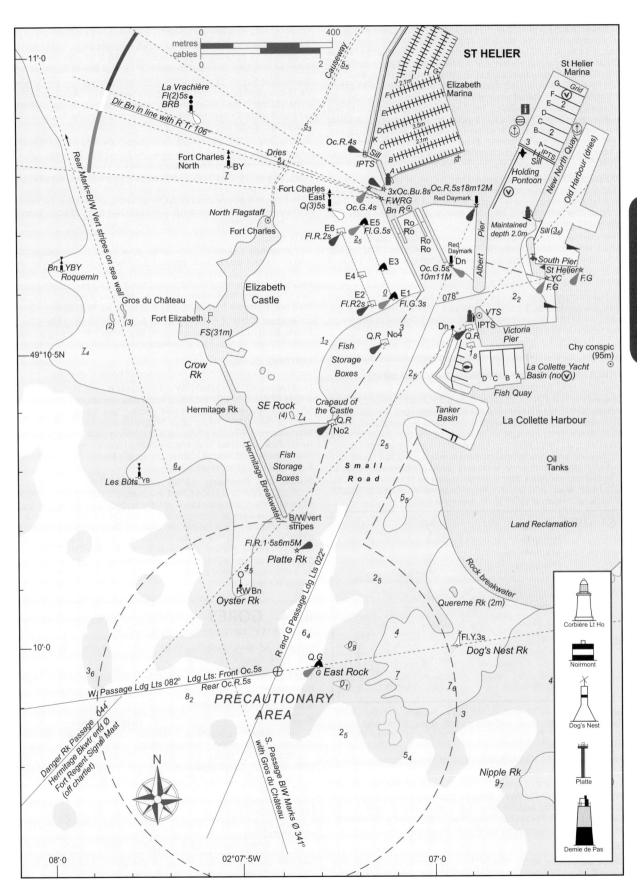

St Helier Marina ⚓☎447708. IPTS and digital depth gauges indicate when entry and exit is possible, approx (HW± 3) slight diffs sp or nps. A 3·6m sill with 1·4m flap gate maintains a depth of 5m above CD at S end of marina basin; LW depths in marina are mostly 2m-3m. When the rising tide reaches 5m, the flap gate drops, and1·4m will begin flashing on the depth gauges until there is more than 2m showing over the sill when they stop flashing. On a falling tide the gauges begin to flash at 2·5m; once the flap gate rises they will display zero and IPTS will show 3 vert R lts = do not enter. A 150m long waiting pontoon is SW of marina ent, with ⚓☇; rafting and overnight stays are permitted. ❶ berths: E, F and G in 2m; yachts >14m or >2·1m draught call 447730/447708, £2·85. ☐▮▯☐☐☐☐ Ⓔ☓△ Grid, www.portsofjersey.je marinas@ports.je

Elizabeth Marina Preferred appr, when height of tide >7m, is from the W on 106°. Ldg marks, just S of ent, are orange ☐s with B vert line; at night stay in the W sector of Dir lt at front daymark. Pass S of La Vrachière IDM lt bn, cross the causeway 5·3m, turn ENE then NNE into the ent.

The marina is mainly for locals, or large visiting yachts by prior arrangement. Access over sill/flapgate; max LOA 20m, drafts 2·1-3·5m. At ent a digital gauge reads depth over sill. IPTS (sigs 2 & 3) control one-way ent/exit, usually 10 mins in each direction. ☎447708, 564☐▮▯ (H24) ☇.

The secondary appr chan 338° from Small Road (ie N of No 4 PHM buoy) is marked by 3 pairs of PHM and SHM buoys, 1st and 3rd pairs are lit. 5 Y can waiting buoys in about 1·0m are outboard of the lateral buoys (3 to the W, 2 to the E).

Town ☓△⊠☐☐☐Ⓑ✈. Fast ferries (Mar-Nov): Poole, St Malo, Granville, Sark, Guernsey, Carteret, RoRo Ferry: Portsmouth.

OTHER HARBOURS AND ANCHORAGES AROUND JERSEY
SOUTH COAST (clockwise from St Helier)

St Aubin 49°11′·21N 02°10′·02W. Dries 6·8m. From seaward Middle Passage (339°, Mon Plaisir Ho ≠ twr on St Aubin Fort) leads towards St Aubin Fort. Final appr N pier head bearing 254° lies N of St Aubin Fort; at night in W sector of the pier head lt – see chartlet and Lights, buoys and waypoints. Castle pier head is also lit. Beware very strong tidal streams in ent during full flood. Dry out alongside N pier or on wide grid. With offshore winds, quiet ⚓ in bay (partly dries). Hbr is administered by St Helier. Facilities: ⚓£14·50, ☇☐(1t 5t) Grid. **Royal Channel Islands YC** ☎741023, ☓△▮☐. **Town** ☓△☐☐☐, bus to St Helier.

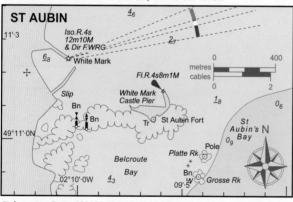

ST AUBIN

Belcroute Bay 49°10′·43N 02°09′·60W. Beware shellfish beds. Excellent shelter from W/SW'lies, but dries 3·5m and many moorings; ⚓ off Pt de Bût in 2·7m. Platte and Grosse Rks are marked by poles.

Portelet Bay 49°10′·13N 02°10′·61W. W of Noirmont Pt. Good ⚓ in N'lies, either side of Janvrin Twr.

St Brelade's Bay 49°10′·64N 02°11′·84W. Good shelter from N and W, but open to SW'ly swell. Beware many drying rks, especially Fournier Rk (0·9m) and Fourché (3·4m) in centre. A quiet ⚓ is in Beau Port. Bouilly Port in NW corner has local moorings and a small stone jetty. Very popular sandy tourist beach with hotels.

NORTH COAST

Grève au Lancon 49°15′·50N 02°14′·07W. Between Grosnez Pt's squat lt ho and Plémont Pt a wide, part-drying sandy bay, suitable

for short stay on calm days. Open to swell.

Grève de Lecq 49°14′·98N 02°12′·12W. Ldg line 202°: W Martello tr on with W hotel with grey roof. ⚓ in 5m N of pier and W of submarine cables. Exposed to swell. ☓△, and bus to St Helier.

Bonne Nuit Bay 49°15′·19N 02°07′·09W. To the NNW beware Demie Rk (5·2m; SHM buoy); and to the E Les Sambues 5·5m. Conspic TV mast (232m) is ½M W of the bay. Ldg lts 223°. ⚓ in 5m NNE of pier and W of Chevel Rk. Hbr dries to sand/shingle. Many local moorings. ☓△.

Bouley Bay 49°14′·50N 02°04′·70W. Good ⚓ on sand in 2·5m SE of pier (rocky footings). Exposed to NE, see Gorey for tides. Local moorings. ☓△.

Rozel Bay 49°14′·25N 02°02′·43W. Appr with pierhead brg 245°, in W sector of Dir lt on shore. Conspic bldg close N on Nez du Guet. Pass between pierhead and WCM bn. Hbr, dries 1·5m to sand/shingle, and is full of moorings; ⚓ outside in 4-5m S of appr. Shops ☓△, bus to St Helier.

EAST COAST

St Catherine's Bay 49°13′·30N 02°00′·90W. Local moorings off W end of Verclut bkwtr. Pass bkwtr hd by over 50m due to strong tidal streams and rock armour extending 15m seaward. ⚓ S of bkwtr in 3–7m, but not in S/SE winds. Land on slip at root of bkwtr. See Gorey for tides, beware rocky St Catherine Bank, 3·1m (ECM bcn) and submarine cables to the S. Dinghy SC, RNLI Ⓛ☓.

La Rocque 49°09′·80N 02°01′·88W. Small hbr dries 6·5m–9·6m; bkwtr, local moorings, sandy beach, slip. Appr 330° across Violet Bank requires detailed local knowledge. No facilities.

OFFLYING ISLANDS
Here, local knowledge or a detailed pilot book is essential.

Les Écrehou 49°17′·39N 01°55′·58W (States ⚓). AC 3655, 5604.15; SHOM 6937 (1:25,000). 5M NE of Rozel Bay, has about 12 cottages. ML 6·2m. Arrive at about ½ tide ebbing; see Gorey. Appr with Bigorne Rk on 022°; when SE of Maître Ile alter to 330° for FS on Marmotière Is. Beware of strong and eddying tidal streams 4–8kn. No lts. Pick up ⚓ close SE of Marmotière or ⚓ in a pool 3ca WSW of Marmotière (with houses); other islands are Maître Ile (one house) Blanche and 5 other small islets.

Plateau des Minquiers 48°58′·04N 02°03′·74W. AC 5604.17 (1:25,000); 3656 and SHOM 7161 are both 1:50,000. About 12M S of Jersey, encircled by six cardinal light buoys. See Gorey for tides. ML 6·4m. Beware of strong and eddying tidal streams. Appr by day in good vis from Demie de Vascelin SHM buoy, 49°00′·81N 02°05′·15W, on 161°: Jetée des Fontaines RW bn in transit with FS on Maîtresse Ile. Further transits skirt the W side of the islet to ⚓ due S of it; safe only in settled weather and light winds. Maîtresse Ile has about a dozen cottages. Land at the slipway NW of the States of Jersey ⚓.

6.14 GOREY
Jersey 49°11′·78N 02°01′·37W ❀❀⚓⚓⚓❀❀❀

CHARTS AC 3655, 5604./16, 1138; SHOM 7157, 7160, 6939; Navi 534, 1014; Imray C33B, 2500

TIDES –0454 Dover; ML 6·0; Duration 0545 Note the very large tidal range.

Standard Port ST HELIER (➝)

Times				Height (metres)			
High Water		Low Water		MHWS	MHWN	MLWN	MLWS
0300	0900	0200	0900	11·0	8·1	4·0	1·4
1500	2100	1400	2100				
Differences ST CATHERINE BAY							
0000	+0010	+0010	+0010	0·0	–0·1	0·0	+0·1
BOULEY BAY							
+0002	+0002	+0004	+0004	–0·3	–0·3	–0·1	–0·1
LES ECREHOU							
+0005	+0009	+0011	+0009	–0·2	+0·1	–0·2	0·0
LES MINQUIERS							
–0014	–0018	–0001	–0008	+0·5	+0·6	+0·1	+0·1

SHELTER Good in the hbr, except in S/SE winds. ⚓ about 2ca E of pier hd or in deeper water in the Roads.

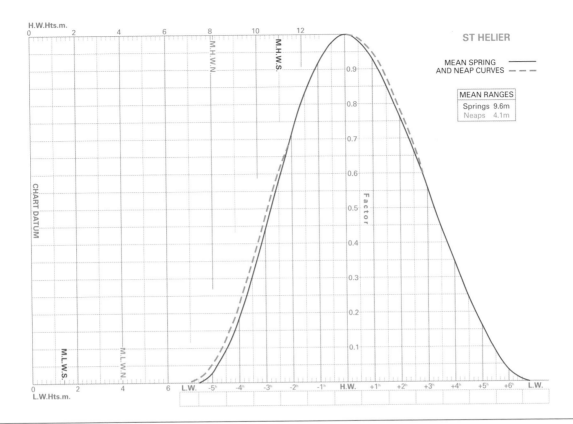

ST HELIER

MEAN SPRING
AND NEAP CURVES

MEAN RANGES	
Springs	9.6m
Neaps	4.1m

NAVIGATION WPT 49°11´·18N 01°59´·59W, 298°/1·3M to pier hd lt. On appr, keep well outside all local bcns until the ldg marks are identified, but beware Banc du Chateau (0·4m least depth) 1M offshore NE of 298° ldg line (see below); and Azicot Rk (dries 2·2m) just S of 298° ldg line, 2ca from ent. There are at least 3 approaches:

- Ldg line 298°: Gorey pierhead, W frame-work twr; rear ldg mark House (west gable end). Best for visitors.
- Pierhead ≠ ✠ spire 304° leads close to Road Rk (3·3m) and over Azicot Rk (2·2m).
- Pierhead ≠ white house/R roof 250° leads close to Les Arch bn (B/W with A topmark) and Pacquet Rk (0·3m).

See Passage information for the Violet Chan to St Helier. The Gutters and Boat Passage across Violet Bank are not advised.

LIGHTS AND MARKS See chartlet and Lights, buoys and waypoints. Mont Orgueil Castle (67m) is highly conspic.

COMMUNICATIONS (Code 01534) Marine-call 09068 969656; ⊖833833; for Dr contact Port Control St Helier 447788; *Gorey Hbr* Ch 74 (not permanently manned); Info *Jersey Coastguard* Ch 82.

FACILITIES Hbr dries to 6·9m. 12 drying ⚓s, 150m W of pier hd, 4 drying ⚓ against pier hd, all £14·50. Port Control ☎447708. 🛉 🛈 HW ±3 (HO) by hose at pierhead mob 07797 742384, ⚓ ⚓ ⚒ ⚓ ⚓(7t).

Town ✕ ⌂ ⚒ ⌂ ✉ Ⓑ bus to St Helier. Ferries to Guernsey, Carteret.

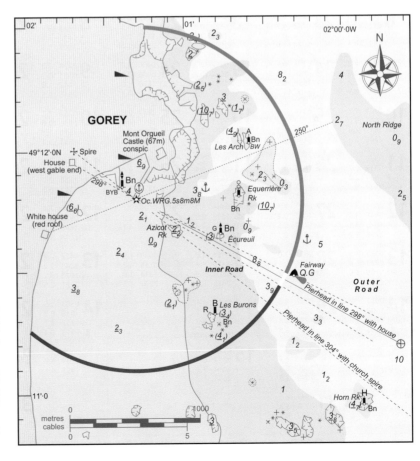

ST HELIER   LAT 49°11'N   LONG 2°07'W
TIMES AND HEIGHTS OF HIGH AND LOW WATERS

STANDARD TIME (UT)
For Summer Time add ONE hour in **non-shaded areas**

Dates in red are SPRINGS
Dates in blue are NEAPS

YEAR **2019**

JANUARY

Time m	Time m
1 TU 0217 8.7 / 0855 3.4 / 1445 8.8 / 2131 3.2	**16** W 0055 8.1 / 0738 4.2 / 1334 8.2 / 2019 3.8
2 W 0325 9.0 / 1001 3.1 / 1550 9.1 / 2232 3.0	**17** TH 0214 8.4 / 0859 3.8 / 1452 8.6 / 2134 3.4
3 TH 0421 9.4 / 1059 2.8 / 1644 9.4 / 2325 2.7	**18** F 0324 9.0 / 1010 3.1 / 1559 9.3 / 2240 2.7
4 F 0509 9.8 / 1149 2.5 / 1731 9.7	**19** SA 0426 9.8 / 1112 2.4 / 1659 10.0 / 2338 2.0
5 SA 0010 2.5 / 0551 10.1 / 1232 2.3 / 1812 10.0	**20** SU 0522 10.5 / 1209 1.7 / 1755 10.6
6 SU 0049 2.3 / 0629 10.3 / 1309 2.1 / 1849 10.1	**21** M 0034 1.5 / 0615 11.1 / 1305 1.2 / 1846 11.1
7 M 0124 2.2 / 0704 10.4 / 1344 2.1 / 1924 10.1	**22** TU 0126 1.1 / 0705 11.5 / 1356 0.8 / 1935 11.4
8 TU 0156 2.2 / 0738 10.4 / 1416 2.1 / 1957 10.1	**23** W 0216 0.9 / 0752 11.7 / 1444 0.6 / 2022 11.4
9 W 0227 2.3 / 0810 10.3 / 1448 2.2 / 2028 9.9	**24** TH 0302 0.9 / 0838 11.6 / 1530 0.7 / 2107 11.2
10 TH 0258 2.5 / 0841 10.0 / 1519 2.4 / 2100 9.6	**25** F 0346 1.1 / 0922 11.3 / 1613 1.1 / 2149 10.7
11 F 0329 2.8 / 0912 9.7 / 1551 2.7 / 2132 9.3	**26** SA 0428 1.6 / 1004 10.7 / 1654 1.8 / 2232 10.0
12 SA 0401 3.1 / 0945 9.3 / 1626 3.1 / 2206 8.9	**27** SU 0511 2.3 / 1047 9.9 / 1737 2.5 / 2317 9.3
13 SU 0437 3.5 / 1022 8.8 / 1705 3.5 / 2248 8.5	**28** M 0557 3.0 / 1136 9.1 / 1827 3.2
14 M 0521 3.9 / 1109 8.4 / 1755 3.9 / 2343 8.2	**29** TU 0012 8.6 / 0654 3.6 / 1239 8.4 / 1931 3.8
15 TU 0620 4.2 / 1214 8.1 / 1901 4.0	**30** W 0125 8.2 / 0809 3.9 / 1400 8.1 / 2049 3.9
	31 TH 0247 8.3 / 0928 3.8 / 1522 8.3 / 2204 3.7

FEBRUARY

Time m	Time m
1 F 0356 8.7 / 1037 3.4 / 1626 8.7 / 2305 3.0	**16** SA 0253 8.6 / 0941 3.3 / 1538 8.9 / 2217 3.0
2 SA 0450 9.2 / 1131 2.9 / 1716 9.2 / 2353 2.8	**17** SU 0408 9.4 / 1054 2.5 / 1647 9.7 / 2323 2.2
3 SU 0535 9.7 / 1215 2.5 / 1758 9.7	**18** M 0510 10.3 / 1157 1.6 / 1744 10.5
4 M 0033 2.4 / 0614 10.1 / 1255 2.2 / 1834 10.0	**19** TU 0022 1.4 / 0604 11.1 / 1254 0.9 / 1835 11.2
5 TU 0110 2.2 / 0649 10.3 / 1330 2.0 / 1908 10.2	**20** W 0116 0.8 / 0653 11.7 / 1345 0.4 / 1923 11.6
6 W 0143 2.1 / 0722 10.5 / 1402 1.8 / 1940 10.3	**21** TH 0204 0.5 / 0739 12.0 / 1432 0.2 / 2007 11.8
7 TH 0213 2.0 / 0753 10.5 / 1433 1.8 / 2010 10.3	**22** F 0248 0.4 / 0822 12.0 / 1514 0.3 / 2048 11.5
8 F 0243 2.0 / 0823 10.4 / 1502 1.9 / 2039 10.1	**23** SA 0329 0.6 / 0903 11.6 / 1552 0.8 / 2126 11.0
9 SA 0312 2.2 / 0852 10.2 / 1531 2.1 / 2108 9.9	**24** SU 0406 1.2 / 0940 10.9 / 1628 1.5 / 2202 10.3
10 SU 0341 2.5 / 0921 9.8 / 1601 2.5 / 2138 9.5	**25** M 0442 2.0 / 1017 10.0 / 1702 2.4 / 2239 9.4
11 M 0411 2.9 / 0952 9.4 / 1633 3.0 / 2210 9.1	**26** TU 0519 2.9 / 1057 9.0 / 1739 3.4 / 2323 8.6
12 TU 0445 3.3 / 1028 8.9 / 1711 3.4 / 2252 8.6	**27** W 0605 3.7 / 1149 8.2 / 1833 4.1
13 W 0531 3.8 / 1119 8.4 / 1804 3.9 / 2355 8.2	**28** TH 0029 7.9 / 0716 4.2 / 1314 7.6 / 2001 4.5
14 TH 0639 4.1 / 1237 8.1 / 1925 4.0	
15 F 0122 8.2 / 0813 3.9 / 1414 8.2 / 2058 3.7	

MARCH

Time m	Time m
1 F 0208 7.7 / 0852 4.3 / 1456 7.7 / 2137 4.2	**16** SA 0049 8.1 / 0743 3.9 / 1353 8.0 / 2034 3.9
2 SA 0332 8.2 / 1014 3.8 / 1608 8.3 / 2244 3.6	**17** SU 0233 8.4 / 0923 3.4 / 1527 8.7 / 2201 3.1
3 SU 0431 8.8 / 1110 3.1 / 1658 9.0 / 2333 3.0	**18** M 0354 9.3 / 1039 2.4 / 1635 9.3 / 2309 2.1
4 M 0516 9.5 / 1155 2.5 / 1739 9.5	**19** TU 0456 10.3 / 1142 1.5 / 1731 10.6
5 TU 0014 2.5 / 0554 10.0 / 1234 2.1 / 1814 10.0	**20** W 0007 1.3 / 0549 11.1 / 1238 0.8 / 1819 11.3
6 W 0051 2.1 / 0629 10.3 / 1310 1.8 / 1847 10.3	**21** TH 0059 0.7 / 0636 11.7 / 1328 0.3 / 1904 11.7
7 TH 0124 1.9 / 0701 10.6 / 1342 1.6 / 1918 10.5	**22** F 0146 0.4 / 0720 12.0 / 1412 0.2 / 1945 11.8
8 F 0155 1.7 / 0732 10.7 / 1413 1.5 / 1947 10.6	**23** SA 0228 0.3 / 0801 11.9 / 1451 0.4 / 2023 11.6
9 SA 0224 1.7 / 0801 10.7 / 1442 1.6 / 2016 10.5	**24** SU 0306 0.6 / 0839 11.5 / 1526 0.9 / 2059 11.1
10 SU 0253 1.7 / 0830 10.6 / 1510 1.7 / 2044 10.4	**25** M 0340 1.2 / 0914 10.8 / 1558 1.6 / 2132 10.3
11 M 0321 2.0 / 0859 10.2 / 1538 2.1 / 2112 10.0	**26** TU 0412 2.0 / 0947 9.9 / 1627 2.6 / 2204 9.5
12 TU 0350 2.4 / 0928 9.8 / 1608 2.6 / 2143 9.5	**27** W 0444 2.9 / 1022 8.9 / 1658 3.5 / 2241 8.6
13 W 0422 2.9 / 1001 9.2 / 1643 3.2 / 2221 8.9	**28** TH 0523 3.7 / 1107 8.0 / 1743 4.3 / 2339 7.8
14 TH 0503 3.5 / 1048 8.5 / 1732 3.7 / 2318 8.4	**29** F 0625 4.4 / 1229 7.4 / 1906 4.8
15 F 0606 3.9 / 1205 8.0 / 1851 4.1	**30** SA 0124 7.5 / 0807 4.5 / 1423 7.5 / 2059 4.6
	31 SU 0301 7.9 / 0939 4.0 / 1539 8.1 / 2213 3.9

APRIL

Time m	Time m
1 M 0402 8.6 / 1038 3.3 / 1630 8.8 / 2303 3.2	**16** TU 0336 9.4 / 1020 2.3 / 1617 9.8 / 2249 2.1
2 TU 0447 9.3 / 1123 2.7 / 1710 9.4 / 2344 2.6	**17** W 0437 10.3 / 1121 1.5 / 1710 10.6 / 2345 1.4
3 W 0526 9.8 / 1203 2.2 / 1746 9.9	**18** TH 0528 11.0 / 1215 1.0 / 1757 11.2
4 TH 0022 2.1 / 0601 10.2 / 1241 1.8 / 1819 10.3	**19** F 0036 0.9 / 0614 11.5 / 1303 0.6 / 1840 11.5
5 F 0058 1.8 / 0634 10.5 / 1315 1.6 / 1850 10.6	**20** SA 0122 0.6 / 0657 11.6 / 1346 0.6 / 1920 11.6
6 SA 0131 1.6 / 0706 10.7 / 1348 1.4 / 1921 10.7	**21** SU 0203 0.6 / 0737 11.5 / 1424 0.8 / 1957 11.4
7 SU 0202 1.5 / 0737 10.8 / 1418 1.4 / 1951 10.8	**22** M 0240 0.9 / 0814 11.1 / 1457 1.3 / 2031 10.9
8 M 0232 1.5 / 0808 10.7 / 1448 1.6 / 2021 10.6	**23** TU 0313 1.4 / 0848 10.5 / 1528 1.9 / 2103 10.3
9 TU 0302 1.7 / 0839 10.4 / 1518 1.9 / 2052 10.3	**24** W 0345 2.1 / 0921 9.7 / 1557 2.7 / 2135 9.5
10 W 0333 2.1 / 0911 9.9 / 1550 2.4 / 2125 9.8	**25** TH 0416 2.9 / 0955 8.9 / 1627 3.5 / 2210 8.7
11 TH 0408 2.6 / 0948 9.3 / 1627 3.0 / 2206 9.1	**26** F 0453 3.6 / 1038 8.1 / 1708 4.2 / 2301 8.0
12 F 0451 3.2 / 1039 8.6 / 1719 3.6 / 2305 8.5	**27** SA 0547 4.2 / 1147 7.5 / 1816 4.7
13 SA 0556 3.7 / 1158 8.1 / 1838 4.0	**28** SU 0032 7.5 / 0710 4.5 / 1333 7.5 / 1958 4.7
14 SU 0036 8.2 / 0731 3.7 / 1344 8.2 / 2018 3.8	**29** M 0212 7.8 / 0843 4.2 / 1453 7.9 / 2122 4.2
15 M 0219 8.6 / 0906 3.2 / 1512 8.9 / 2142 3.0	**30** TU 0318 8.3 / 0949 3.6 / 1547 8.6 / 2218 3.5

Chart Datum is 5·88 metres below Ordnance Datum (Local). HAT is 12·2 metres above Chart Datum.

STANDARD TIME (UT)
For Summer Time add ONE hour in **non-shaded areas**

ST HELIER   LAT 49°11'N   LONG 2°07'W

TIMES AND HEIGHTS OF HIGH AND LOW WATERS

Dates in red are **SPRINGS**
Dates in blue are **NEAPS**

YEAR **2019**

Channel Islands

MAY

Time	m		Time	m
1 0407	9.0		**16** 0411	10.1
1039	3.0		1054	1.8
W 1630	9.2		TH 1644	10.3
2303	2.9		2319	1.7
2 0448	9.5		**17** 0503	10.6
1123	2.4		1147	1.4
TH 1709	9.7		F 1731	10.8
2344	2.4			
3 0526	10.0		**18** 0010	1.4
1204	2.0		0550	10.9
F 1744	10.2		SA 1236	1.3
			○ 1814	11.0
4 0024	2.0		**19** 0056	1.2
0601	10.4		0633	11.0
SA 1243	1.7		SU 1318	1.3
● 1818	10.5		1854	11.1
5 0102	1.7		**20** 0137	1.2
0637	10.6		0713	10.9
SU 1319	1.5		M 1356	1.4
1852	10.6		1931	11.0
6 0137	1.5		**21** 0214	1.4
0712	10.8		0750	10.6
M 1354	1.5		TU 1430	1.8
1927	10.9		2006	10.6
7 0212	1.5		**22** 0248	1.8
0747	10.7		0825	10.2
TU 1428	1.5		W 1502	2.2
2002	10.8		2040	10.1
8 0246	1.6		**23** 0321	2.3
0824	10.5		0859	9.6
W 1503	1.8		TH 1532	2.8
2038	10.5		2113	9.6
9 0322	1.9		**24** 0354	2.8
0904	10.0		0934	9.0
TH 1540	2.3		F 1605	3.4
2118	10.0		2149	8.9
10 0403	2.4		**25** 0430	3.4
0948	9.4		1015	8.4
F 1623	2.9		SA 1643	3.9
2205	9.4		2234	8.3
11 0452	2.9		**26** 0516	3.9
1044	8.8		1110	7.9
SA 1718	3.3		SU 1737	4.3
2307	8.9		◑ 2338	7.9
12 0557	3.3		**27** 0618	4.2
1159	8.4		1226	7.7
SU 1833	3.6		M 1849	4.5
◑				
13 0029	8.6		**28** 0104	7.8
0720	3.3		0733	4.2
M 1328	8.5		TU 1347	7.9
2000	3.5		2010	4.3
14 0157	8.8		**29** 0218	8.1
0843	3.0		0845	3.8
TU 1447	9.0		W 1450	8.3
2117	2.9		2118	3.8
15 0310	9.4		**30** 0315	8.6
0953	2.4		0945	3.3
W 1550	9.7		TH 1540	8.9
2222	2.3		2213	3.3
			31 0402	9.1
			1036	2.8
			F 1624	9.4
			2301	2.7

JUNE

Time	m		Time	m
1 0445	9.6		**16** 0527	10.1
1123	2.4		1209	2.0
SA 1705	9.9		SU 1750	10.4
2347	2.2			
2 0527	10.1		**17** 0032	1.8
1208	2.0		0611	10.3
SU 1746	10.4		M 1253	1.9
			○ 1831	10.5
3 0030	1.9		**18** 0115	1.8
0608	10.5		0652	10.3
M 1251	1.7		TU 1332	2.0
● 1825	10.7		1910	10.5
4 0113	1.6		**19** 0153	1.8
0650	10.7		0731	10.2
TU 1332	1.5		W 1407	2.1
1906	10.9		1946	10.4
5 0154	1.5		**20** 0228	2.0
0732	10.7		0807	10.0
W 1412	1.5		TH 1441	2.4
1948	10.9		2021	10.1
6 0236	1.5		**21** 0302	2.3
0816	10.6		0841	9.7
TH 1453	1.7		F 1513	2.7
2032	10.7		2055	9.8
7 0318	1.7		**22** 0335	2.6
0902	10.3		0916	9.3
F 1537	2.0		SA 1545	3.1
2117	10.3		2129	9.3
8 0404	2.0		**23** 0409	3.0
0951	9.8		0952	8.9
SA 1624	2.4		SU 1620	3.4
2207	9.9		2207	8.9
9 0455	2.4		**24** 0447	3.4
1045	9.3		1033	8.5
SU 1718	2.8		M 1701	3.8
2305	9.4		2252	8.4
10 0554	2.7		**25** 0533	3.7
1148	9.0		1124	8.1
M 1822	3.1		TU 1753	4.1
◑			◑ 2350	8.1
11 0011	9.1		**26** 0629	3.9
0701	2.9		1228	8.0
TU 1300	8.9		W 1858	4.2
1934	3.2			
12 0126	9.0		**27** 0101	8.0
0813	2.9		0735	4.0
W 1413	8.9		TH 1340	8.1
2045	3.0		2010	4.1
13 0237	9.2		**28** 0211	8.2
0921	2.7		0844	3.7
TH 1518	9.4		F 1443	8.5
2151	2.6		2118	3.7
14 0341	9.6		**29** 0311	8.7
1023	2.4		0947	3.3
F 1615	9.8		SA 1537	9.0
2250	2.3		2217	3.1
15 0437	9.9		**30** 0405	9.2
1119	2.1		1043	2.7
SA 1705	10.2		SU 1628	9.6
2344	2.0		2311	2.5

JULY

Time	m		Time	m
1 0455	9.8		**16** 0014	2.3
1135	2.2		0555	9.8
M 1717	10.2		TU 1235	2.4
			○ 1814	10.2
2 0003	2.0		**17** 0058	2.1
0545	10.3		0637	10.0
TU 1226	1.8		W 1314	2.3
● 1804	10.7		1853	10.3
3 0053	1.6		**18** 0136	2.0
0634	10.6		0714	10.0
W 1315	1.6		TH 1350	2.2
1852	11.0		1929	10.4
4 0142	1.3		**19** 0211	2.0
0722	10.8		0749	10.0
TH 1402	1.4		F 1423	2.3
1939	11.1		2003	10.3
5 0229	1.2		**20** 0243	2.1
0810	10.9		0822	9.9
F 1448	1.4		SA 1454	2.4
2026	11.1		2035	10.1
6 0316	1.2		**21** 0314	2.3
0857	10.7		0854	9.7
SA 1534	1.5		SU 1524	2.6
2113	10.9		2107	9.8
7 0402	1.4		**22** 0346	2.5
0945	10.4		0925	9.4
SU 1621	1.8		M 1555	2.9
2200	10.5		2138	9.4
8 0450	1.8		**23** 0418	2.9
1033	9.9		0958	9.1
M 1709	2.2		TU 1629	3.3
2250	10.0		2212	9.0
9 0539	2.3		**24** 0454	3.3
1125	9.4		1035	8.7
TU 1802	2.7		W 1708	3.7
◑ 2344	9.4		2253	8.5
10 0634	2.7		**25** 0536	3.7
1223	9.0		1121	8.3
W 1902	3.1		TH 1758	4.0
			◑ 2347	8.2
11 0048	9.0		**26** 0632	4.0
0737	3.1		1224	8.1
TH 1332	8.8		F 1905	4.2
2010	3.3			
12 0200	8.8		**27** 0100	8.0
0846	3.2		0743	4.0
F 1443	8.9		SA 1341	8.2
2120	3.2		2024	4.0
13 0312	8.9		**28** 0220	8.3
0954	3.1		0901	3.7
SA 1548	9.2		SU 1454	8.7
2226	2.9		2138	3.5
14 0415	9.2		**29** 0331	8.8
1056	2.8		1009	3.1
SU 1644	9.5		M 1558	9.3
2324	2.6		2243	2.8
15 0509	9.5		**30** 0433	9.5
1149	2.6		1111	2.5
M 1732	9.9		TU 1656	10.0
			2342	2.1
			31 0529	10.2
			1207	1.9
			W 1750	10.7

AUGUST

Time	m		Time	m
1 0038	1.5		**16** 0119	2.0
0622	10.7		0656	10.1
TH 1302	1.4		F 1332	2.2
● 1841	11.2		1910	10.5
2 0132	1.0		**17** 0152	1.9
0712	11.1		0729	10.3
F 1352	1.1		SA 1403	2.1
1929	11.5		1942	10.5
3 0222	0.7		**18** 0222	1.9
0800	11.3		0759	10.3
SA 1440	0.9		SU 1432	2.1
2016	11.7		2012	10.5
4 0308	0.7		**19** 0251	1.9
0845	11.3		0828	10.2
SU 1525	1.0		M 1501	2.2
2100	11.5		2041	10.3
5 0352	0.9		**20** 0320	2.1
0928	11.0		0856	10.0
M 1608	1.3		TU 1530	2.5
2144	11.0		2109	9.9
6 0433	1.4		**21** 0349	2.5
1011	10.4		0924	9.6
TU 1650	1.9		W 1559	2.9
2226	10.3		2137	9.4
7 0515	2.1		**22** 0419	3.0
1054	9.7		0954	9.1
W 1734	2.6		TH 1631	3.4
◑ 2312	9.5		2209	8.9
8 0601	2.8		**23** 0454	3.5
1144	9.0		1030	8.6
TH 1826	3.2		F 1711	3.8
			◑ 2252	8.4
9 0008	8.7		**24** 0540	4.0
0657	3.5		1123	8.2
F 1248	8.4		SA 1811	4.2
1933	3.7			
10 0122	8.2		**25** 0001	8.0
0811	3.9		0651	4.2
SA 1410	8.4		SU 1246	8.0
2053	3.8		1938	4.2
11 0249	8.2		**26** 0140	8.0
0931	3.8		0823	4.0
SU 1528	8.6		M 1421	8.3
2208	3.5		2109	3.7
12 0401	8.6		**27** 0310	8.6
1040	3.4		0946	3.4
M 1628	9.1		TU 1539	9.1
2310	3.0		2224	2.9
13 0457	9.1		**28** 0419	9.4
1134	2.9		1053	2.6
TU 1717	9.6		W 1642	10.0
2359	2.5		2327	2.0
14 0542	9.6		**29** 0518	10.3
1219	2.6		1153	1.8
W 1759	10.0		TH 1737	10.9
15 0041	2.2		**30** 0025	1.3
0621	9.9		0609	11.0
TH 1258	2.3		F 1248	1.2
○ 1836	10.3		● 1827	11.5
			31 0118	0.7
			0658	11.5
			SA 1338	0.7
			1914	11.9

Chart Datum is 5·88 metres below Ordnance Datum (Local). HAT is 12·2 metres above Chart Datum.

》》 FREE monthly updates. Register at 《
www.reedsnauticalalmanac.co.uk 《

249

ST HELIER — LAT 49°11'N LONG 2°07'W
TIMES AND HEIGHTS OF HIGH AND LOW WATERS

STANDARD TIME (UT)
For Summer Time add ONE hour in **non-shaded areas**

Dates in **red** are SPRINGS
Dates in blue are NEAPS

YEAR 2019

SEPTEMBER

Time m	Time m
1 0207 0.4 / 0743 11.8 / SU 1425 0.5 / 1959 12.1	**16** 0157 1.7 / 0732 10.6 / M 1408 1.9 / 1946 10.7
2 0251 0.4 / 0825 11.7 / M 1507 0.6 / 2041 11.8	**17** 0225 1.7 / 0800 10.5 / TU 1436 1.9 / 2014 10.6
3 0331 0.7 / 0905 11.3 / TU 1547 1.1 / 2120 11.1	**18** 0253 1.9 / 0827 10.4 / W 1504 2.2 / 2041 10.2
4 0409 1.3 / 0943 10.6 / W 1625 1.8 / 2158 10.3	**19** 0321 2.3 / 0854 10.0 / TH 1533 2.6 / 2108 9.8
5 0445 2.2 / 1020 9.8 / TH 1703 2.7 / 2238 9.4	**20** 0350 2.8 / 0921 9.5 / F 1603 3.1 / 2137 9.2
6 0524 3.2 / 1103 8.9 / F 1749 3.5 / ◗ 2327 8.4	**21** 0422 3.4 / 0954 9.0 / SA 1640 3.7 / 2217 8.5
7 0614 4.0 / 1203 8.1 / SA 1855 4.2	**22** 0505 4.0 / 1043 8.3 / SU 1736 4.2 / ◗ 2325 8.0
8 0045 7.7 / 0736 4.5 / SU 1340 7.8 / 2029 4.3	**23** 0617 4.4 / 1208 8.0 / M 1909 4.3
9 0232 7.7 / 0913 4.3 / M 1512 8.2 / 2154 3.9	**24** 0116 7.8 / 0759 4.3 / TU 1401 8.2 / 2051 3.8
10 0350 8.3 / 1026 3.7 / TU 1613 8.9 / 2254 3.2	**25** 0258 8.5 / 0929 3.5 / W 1525 9.1 / 2208 2.9
11 0442 9.0 / 1117 3.1 / W 1700 9.5 / 2340 2.6	**26** 0407 9.5 / 1038 2.6 / TH 1628 10.1 / 2311 1.9
12 0524 9.6 / 1159 2.6 / TH 1739 10.1	**27** 0503 10.4 / 1136 1.7 / F 1721 11.0
13 0019 2.2 / 0600 10.0 / F 1236 2.2 / 1814 10.4	**28** 0007 1.1 / 0552 11.2 / SA 1229 1.0 / ● 1809 11.7
14 0055 1.8 / 0633 10.3 / SA 1309 2.0 / ○ 1846 10.6	**29** 0058 0.6 / 0638 11.7 / SU 1318 0.6 / 1855 12.1
15 0127 1.8 / 0703 10.5 / SU 1339 1.9 / 1917 10.7	**30** 0145 0.4 / 0720 11.9 / M 1403 0.5 / 1937 12.1

OCTOBER

Time m	Time m
1 0227 0.4 / 0801 11.8 / TU 1443 0.6 / 2017 11.8	**16** 0159 1.7 / 0732 10.8 / W 1412 1.8 / 1948 10.7
2 0305 0.8 / 0838 11.4 / W 1521 1.1 / 2054 11.1	**17** 0228 1.9 / 0800 10.6 / TH 1441 2.0 / 2017 10.4
3 0340 1.6 / 0913 10.6 / TH 1557 1.9 / 2130 10.2	**18** 0257 2.2 / 0829 10.3 / F 1511 2.4 / 2047 9.9
4 0413 2.5 / 0947 9.7 / F 1632 2.8 / 2206 9.2	**19** 0327 2.7 / 0859 9.8 / SA 1544 2.9 / 2120 9.3
5 0447 3.5 / 1025 8.8 / SA 1713 3.6 / ◗ 2251 8.2	**20** 0402 3.3 / 0936 9.2 / SU 1624 3.5 / 2204 8.7
6 0532 4.3 / 1120 8.0 / SU 1816 4.4	**21** 0448 3.9 / 1028 8.6 / M 1723 4.0 / ◗ 2315 8.1
7 0008 7.5 / 0655 4.9 / M 1305 7.6 / 1958 4.6	**22** 0601 4.3 / 1154 8.1 / TU 1854 4.1
8 0209 7.5 / 0846 4.7 / TU 1447 8.0 / 2128 4.1	**23** 0105 8.0 / 0742 4.2 / W 1343 8.4 / 2033 3.7
9 0327 8.1 / 1000 4.0 / W 1548 8.7 / 2226 3.4	**24** 0242 8.6 / 0910 3.5 / TH 1506 9.2 / 2148 2.8
10 0416 8.9 / 1049 3.3 / TH 1633 9.4 / 2310 2.8	**25** 0348 9.6 / 1017 2.5 / F 1608 10.1 / 2249 1.9
11 0456 9.6 / 1129 2.7 / F 1712 10.0 / 2348 2.3	**26** 0442 10.5 / 1114 1.7 / SA 1700 10.9 / 2343 1.2
12 0531 10.0 / 1205 2.3 / SA 1746 10.4	**27** 0530 11.2 / 1205 1.1 / SU 1748 11.5
13 0023 2.0 / 0603 10.4 / SU 1239 2.0 / ○ 1818 10.6	**28** 0033 0.8 / 0614 11.6 / M 1254 0.8 / ● 1832 11.8
14 0057 1.8 / 0634 10.6 / M 1312 1.9 / 1849 10.8	**29** 0119 0.7 / 0656 11.7 / TU 1338 0.8 / 1913 11.8
15 0128 1.7 / 0703 10.7 / TU 1342 1.8 / 1919 10.8	**30** 0200 0.9 / 0735 11.6 / W 1418 1.0 / 1953 11.4
	31 0237 1.3 / 0812 11.2 / TH 1455 1.4 / 2030 10.8

NOVEMBER

Time m	Time m
1 0312 1.9 / 0846 10.5 / F 1530 2.1 / 2105 10.0	**16** 0239 2.1 / 0814 10.5 / SA 1458 2.2 / 2037 10.1
2 0344 2.7 / 0920 9.7 / SA 1605 2.9 / 2141 9.1	**17** 0314 2.5 / 0851 10.1 / SU 1536 2.6 / 2118 9.5
3 0417 3.6 / 0957 8.9 / SU 1644 3.7 / 2224 8.3	**18** 0354 3.0 / 0934 9.5 / M 1621 3.1 / 2209 8.9
4 0458 4.3 / 1046 8.1 / M 1739 4.4 / ◖ 2330 7.6	**19** 0445 3.5 / 1030 9.0 / TU 1721 3.5 / ◗ 2318 8.5
5 0607 4.9 / 1213 7.6 / TU 1905 4.6	**20** 0554 3.9 / 1146 8.6 / W 1841 3.7
6 0116 7.5 / 0751 4.9 / W 1358 7.8 / 2036 4.4	**21** 0046 8.4 / 0720 3.9 / TH 1317 8.7 / 2007 3.4
7 0241 8.0 / 0913 4.4 / TH 1506 8.4 / 2140 3.8	**22** 0212 8.8 / 0842 3.4 / F 1436 9.2 / 2120 2.8
8 0336 8.6 / 1007 3.7 / F 1555 9.0 / 2227 3.2	**23** 0320 9.5 / 0949 2.7 / SA 1540 9.9 / 2222 2.2
9 0418 9.3 / 1050 3.1 / SA 1636 9.6 / 2308 2.6	**24** 0416 10.2 / 1047 2.0 / SU 1635 10.5 / 2317 1.7
10 0456 9.8 / 1129 2.6 / SU 1712 10.0 / 2347 2.2	**25** 0505 10.7 / 1140 1.6 / M 1724 10.9
11 0530 10.2 / 1206 2.2 / M 1747 10.4	**26** 0007 1.4 / 0550 11.1 / TU 1230 1.3 / ● 1809 11.1
12 0023 2.0 / 0602 10.5 / TU 1242 2.0 / ○ 1820 10.6	**27** 0053 1.3 / 0632 11.3 / W 1314 1.3 / 1852 11.1
13 0059 1.8 / 0634 10.7 / W 1317 1.8 / 1853 10.7	**28** 0135 1.4 / 0712 11.2 / TH 1355 1.4 / 1932 10.9
14 0133 1.8 / 0706 10.8 / TH 1350 1.8 / 1926 10.7	**29** 0213 1.7 / 0750 10.9 / F 1433 1.7 / 2010 10.5
15 0206 1.9 / 0739 10.8 / F 1423 1.9 / 2001 10.5	**30** 0248 2.2 / 0825 10.5 / SA 1509 2.2 / 2047 9.9

DECEMBER

Time m	Time m
1 0321 2.8 / 0900 9.8 / SU 1544 2.8 / 2123 9.2	**16** 0310 2.1 / 0850 10.5 / M 1536 2.1 / 2122 10.0
2 0354 3.4 / 0937 9.2 / M 1621 3.4 / 2202 8.6	**17** 0354 2.5 / 0937 10.1 / TU 1623 2.4 / 2212 9.5
3 0432 4.0 / 1020 8.6 / TU 1705 4.0 / 2252 8.1	**18** 0444 2.9 / 1029 9.6 / W 1717 2.8 / 2309 9.1
4 0522 4.4 / 1118 8.0 / W 1803 4.3 / ◖	**19** 0543 3.2 / 1130 9.2 / TH 1821 3.1 / ◗
5 0003 7.7 / 0631 4.7 / TH 1241 7.8 / 1918 4.4	**20** 0016 8.8 / 0651 3.4 / F 1242 9.0 / 1932 3.2
6 0128 7.8 / 0755 4.6 / F 1402 8.0 / 2032 4.1	**21** 0131 8.8 / 0806 3.3 / SA 1357 9.1 / 2045 3.0
7 0237 8.2 / 0906 4.2 / SA 1503 8.5 / 2132 3.7	**22** 0243 9.1 / 0917 3.0 / SU 1508 9.3 / 2151 2.7
8 0329 8.7 / 1000 3.6 / SU 1551 9.0 / 2223 3.2	**23** 0346 9.6 / 1020 2.6 / M 1610 9.7 / 2251 2.4
9 0413 9.3 / 1047 3.1 / M 1634 9.5 / 2308 2.7	**24** 0441 10.0 / 1118 2.2 / TU 1704 10.1 / 2345 2.1
10 0453 9.8 / 1131 2.6 / TU 1713 9.9 / 2350 2.3	**25** 0530 10.4 / 1210 1.9 / W 1753 10.3
11 0530 10.2 / 1213 2.2 / W 1752 10.3	**26** 0033 1.9 / 0614 10.7 / TH 1257 1.8 / ● 1836 10.5
12 0031 2.0 / 0608 10.6 / TH 1254 1.9 / ○ 1831 10.5	**27** 0117 1.9 / 0655 10.7 / F 1339 1.7 / 1917 10.4
13 0111 1.9 / 0646 10.8 / F 1333 1.8 / 1911 10.6	**28** 0156 2.0 / 0734 10.7 / SA 1418 1.9 / 1955 10.3
14 0150 1.8 / 0726 10.9 / SA 1412 1.7 / 1953 10.6	**29** 0231 2.2 / 0810 10.5 / SU 1453 2.1 / 2031 10.0
15 0229 1.9 / 0807 10.8 / SU 1453 1.9 / 2036 10.3	**30** 0304 2.5 / 0845 10.1 / M 1526 2.5 / 2105 9.6
	31 0335 2.9 / 0919 9.6 / TU 1559 2.9 / 2140 9.1

Chart Datum is 5·88 metres below Ordnance Datum (Local). HAT is 12·2 metres above Chart Datum.

〉〉 **FREE** monthly updates. Register at 〈
www.reedsnauticalalmanac.co.uk 〈

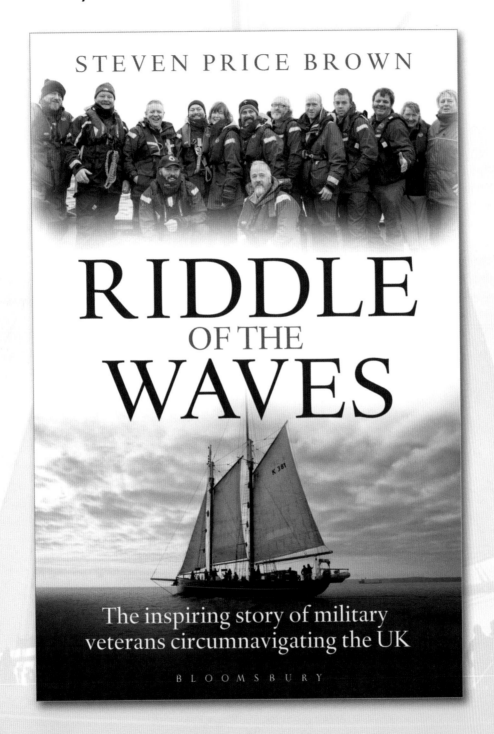

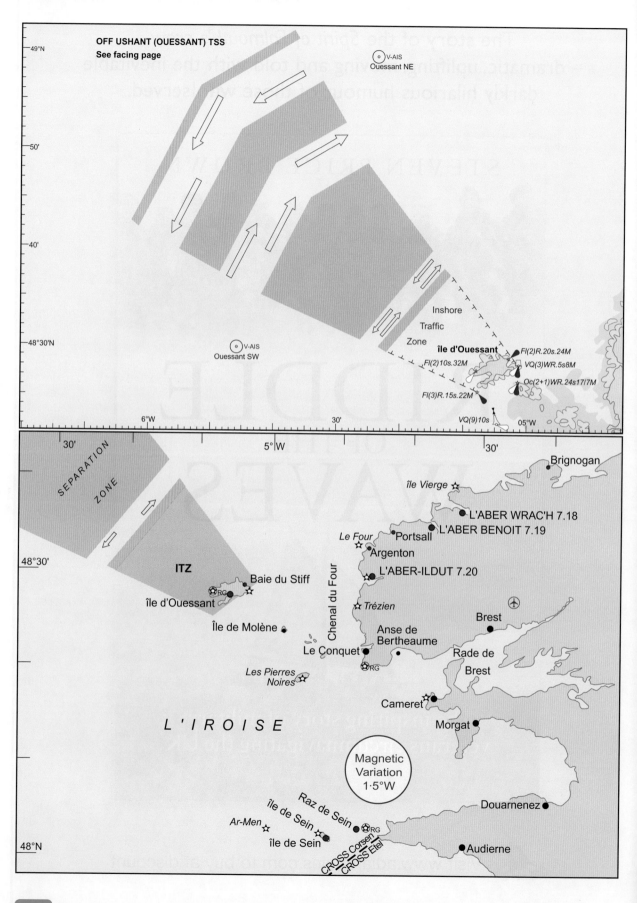

OFF USHANT (OUESSANT) TSS
See facing page

⊙ V-AIS
Ouessant NE

49°N

50'

40'

Inshore

Traffic

Zone

Île d'Ouessant

Fl(2)R.20s.24M

VQ(3)WR.5s8M

Fl(2)10s.32M

Oc(2+1)WR.24s17/7M

48°30'N

⊙ V-AIS
Ouessant SW

Fl(3)R.15s.22M

VQ(9)10s

6°W

30'

05°W

30'

5°W

30'

Brignogan

île Vierge ☆

SEPARATION

ZONE

L'ABER WRAC'H 7.18

L'ABER BENOIT 7.19

Le Four ☆

Portsall

Argenton

ITZ

48°30'

L'ABER-ILDUT 7.20

Baie du Stiff

☆

☆ Trézien

île d'Ouessant

⊛RG

Chenal du Four

Brest

Île de Molène

Anse de
Bertheaume

Rade de

Le Conquet

Brest

⊛RG

Les Pierres
Noires ☆

Cameret ☆

L ' I R O I S E

Morgat

Magnetic
Variation
1·5°W

Raz de Sein

Douarnenez

île de Sein

Ar-Men ☆

☆

⊛RG

48°N

île de Sein

CROSS Corsen
CROSS Etel

Audierne

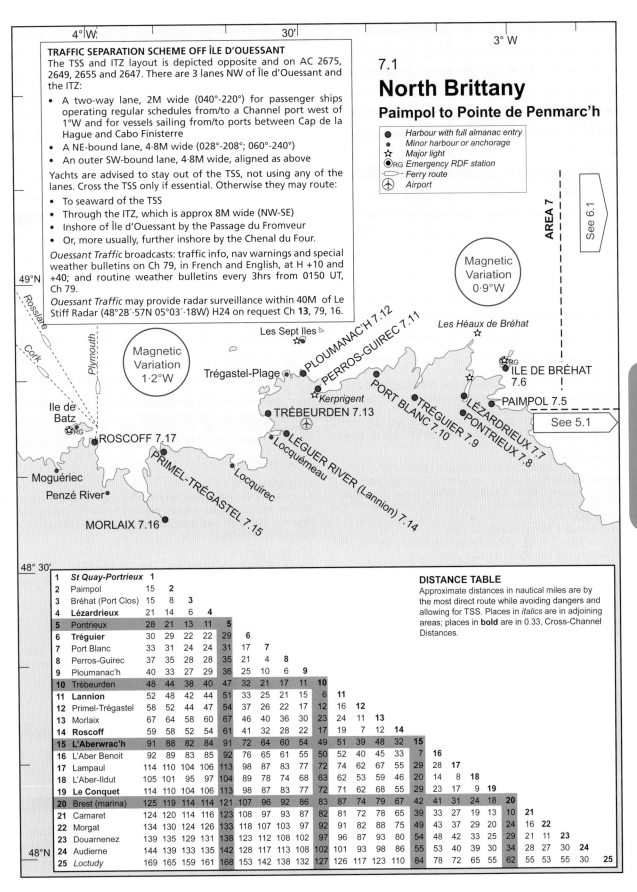

TRAFFIC SEPARATION SCHEME OFF ÎLE D'OUESSANT

The TSS and ITZ layout is depicted opposite and on AC 2675, 2649, 2655 and 2647. There are 3 lanes NW of Île d'Ouessant and the ITZ:

- A two-way lane, 2M wide (040°-220°) for passenger ships operating regular schedules from/to a Channel port west of 1°W and for vessels sailing from/to ports between Cap de la Hague and Cabo Finisterre
- A NE-bound lane, 4·8M wide (028°-208°; 060°-240°)
- An outer SW-bound lane, 4·8M wide, aligned as above

Yachts are advised to stay out of the TSS, not using any of the lanes. Cross the TSS only if essential. Otherwise they may route:

- To seaward of the TSS
- Through the ITZ, which is approx 8M wide (NW-SE)
- Inshore of Île d'Ouessant by the Passage du Fromveur
- Or, more usually, further inshore by the Chenal du Four.

Ouessant Traffic broadcasts: traffic info, nav warnings and special weather bulletins on Ch 79, in French and English, at H +10 and +40; and routine weather bulletins every 3hrs from 0150 UT, Ch 79.

Ouessant Traffic may provide radar surveillance within 40M of Le Stiff Radar (48°28'·57N 05°03'·18W) H24 on request Ch 13, 79, 16.

7.1

North Brittany

Paimpol to Pointe de Penmarc'h

- ● Harbour with full almanac entry
- ● Minor harbour or anchorage
- ☆ Major light
- ⊙RG Emergency RDF station
- ⌐Ⓐ⌐ Ferry route
- Ⓐ Airport

Magnetic Variation 0·9°W

DISTANCE TABLE

Approximate distances in nautical miles are by the most direct route while avoiding dangers and allowing for TSS. Places in *italics* are in adjoining areas; places in **bold** are in 0.33, Cross-Channel Distances.

		1	2	3	4	5	6	7	8	9	10	11	12	13	14	15	16	17	18	19	20	21	22	23	24	25
1	*St Quay-Portrieux*	**1**																								
2	Paimpol	15	**2**																							
3	Bréhat (Port Clos)	15	8	**3**																						
4	**Lézardrieux**	21	14	6	**4**																					
5	Pontrieux	28	21	13	11	**5**																				
6	**Tréguier**	30	29	22	22	29	**6**																			
7	Port Blanc	33	31	24	24	31	17	**7**																		
8	Perros-Guirec	37	35	28	28	35	21	4	**8**																	
9	Ploumanac'h	40	33	27	29	36	25	10	6	**9**																
10	Trébeurden	48	44	38	40	47	32	21	17	11	**10**															
11	**Lannion**	52	48	42	44	51	33	25	21	15	6	**11**														
12	Primel-Trégastel	58	52	44	47	54	37	26	22	17	12	16	**12**													
13	Morlaix	67	64	58	60	67	46	40	36	30	23	24	11	**13**												
14	**Roscoff**	59	58	52	54	61	41	32	28	22	17	19	7	12	**14**											
15	**L'Aberwrac'h**	91	88	82	84	91	72	64	60	54	49	51	39	48	32	**15**										
16	L'Aber Benoit	92	89	83	85	92	76	65	61	55	50	52	40	45	33	7	**16**									
17	Lampaul	114	110	104	106	113	98	87	83	77	72	74	62	67	55	29	28	**17**								
18	L'Aber-Ildut	105	101	95	97	104	89	78	74	68	63	62	53	59	46	20	14	8	**18**							
19	Le Conquet	114	110	104	106	113	98	87	83	77	72	71	62	68	55	29	23	17	9	**19**						
20	**Brest (marina)**	125	119	114	114	121	107	96	92	86	83	87	74	79	67	42	41	31	24	18	**20**					
21	Camaret	124	120	114	116	123	108	97	93	87	82	81	72	78	65	39	33	27	19	13	10	**21**				
22	Morgat	134	130	124	126	133	118	107	103	97	92	91	82	88	75	49	43	37	29	20	24	16	**22**			
23	Douarnenez	139	135	129	131	138	123	112	108	102	97	96	87	93	80	54	48	42	33	25	29	21	11	**23**		
24	Audierne	144	139	133	135	142	128	117	113	108	102	101	93	98	86	55	53	40	39	30	34	28	27	30	**24**	
25	*Loctudy*	169	165	159	161	168	153	142	138	132	127	126	117	123	110	84	78	72	65	55	62	55	53	55	30	**25**

N Brittany

7.2 LIGHTS, BUOYS AND WAYPOINTS

Bold print = light with a nominal range of 15M or more. CAPITALS = place or feature. *CAPITAL ITALICS* = light-vessel, light float or Lanby. *Italics* = Fog signal. ***Bold italics*** = Racon. Many marks/buoys are fitted with AIS (<u>MMSI No</u>); see relevant charts.

OFFSHORE MARKS

Roches Douvres ☆ Fl 5s 60m **24M**; pink twr on dwelling with G roof; 49°06'·30N 02°48'·87W.
Barnouic ⸕ VQ (3) 5s 15m 7M; 49°01'·63N 02°48'·41W.
Roche Gautier ⸕ VQ (9) 10s; 49°02'·00N 02°54'·73W.

PAIMPOL TO ÎLE DE BRÉHAT

PAIMPOL

Les Calemarguiers ⸕ 48°46'·98N 02°54'·84W.
L'Ost Pic ⸔ Fl(4)WR 15s 20m, W9M, R6M; 105°-W-116°-R-221°-W-253°- R-291°-W-329°; obsc by islets near Bréhat when brg < 162°; W twr/turret, R top; 48°46'·77N 02°56'·42W.
Les Charpentiers ⸕ 48°47'·89N 02°56'·01W.
Pte de Porz-Don ☆ Oc (2) WR 6s 13m **W15M**, R11M; 269°-W-272°-R-279°; W house; 48°47'·48N 03°01'·55W.
El Bras ⸔ Fl G 2·5s; 48°47'·21N 03°01'·50W.
⸕ Fl R 2·5s; 48°47'·17N 03°01'·49W.
Ldg lts 262·2°, both QR 5/12m 7/10M. Front, Kernoa jetty; W & R hut; 48°47'·09N 03°02'·44W. Rear, Dir QR, intens 260·2°-264·2°; W pylon, R top; 360m from front. ⸔ QG 48°47'·12N 03°02'·47W.

CHENAL DU FERLAS (277°-257°-271°)

Lel Ar Serive ⸕ 48°49'·98N 02°58'·76W.
Loguivy Dir lt 257°: Q WRG 12m 10/8M; 254°-G-257°-W-257·7°-R-260·7°; Gy twr; 48°49'·37N 03°03'·67W.
Les Piliers ⸕ 48°49'·77N 02°59'·99W.
Kermouster Dir ⸔271°: Fl WRG 2s 16m, W 10M, R/G 8M; 267°-G-270°-W-272°- R-274°; W col; 48°49'·55N 03°05'·19W (R. Trieux).

ÎLE DE BRÉHAT

Le Paon ⸔ Oc WRG 4s 22m W11M, R/G8M; 033°-W-078°-G-181°-W-196°- R-307°-W-316°-R-348°; Y twr; 48°51'·92N 02°59'·15W.
Roche Guarine ⸕ 48°51'·63N 02°57'·63W, Chenal de Bréhat.
Rosédo ☆ Fl 5s 29m **20M**; W twr; vis 006°-340° (334°); 48°51'·45N 03°00'·29W.
La Chambre ⸕ 48°50'·16N 02°59'·58W.
Men-Joliguet ⸕ Fl (2) WRG 4s 6m W11M, R/G8M; 255°-R-279°-W-283°-G-175°; 48°50'·12N 03°00'·20W.

LÉZARDRIEUX TO TRÉGUIER

LE TRIEUX RIVER to LÉZARDRIEUX

Nord Horaine ⸕ VQ, 48°54'·53N 02°55'·42W.
La Horaine ⸔ Fl (3) 12s 13m 7M; Gy 8-sided twr on B hut; 48°53'·50N 02°55'·22W.
Men-Marc'h ⸕ 48°53'·17N 02°51'·82W.
Ldg lts 224·8°: Front, **La Croix** ☆ Q 15m **18M**; intens 215°-235°; two Gy ○ twrs joined, W on NE side, R tops; 48°50'·23N 03°03'·25W. Rear **Bodic** ☆ Dir Q 55m **22M**; intens 221°-229°; W ho with G gable; 2·1M from front.
Les Sirlots ⸔ *Whis;* 48°52'·95N 02°59'·58W.
Coatmer Ldg lts 218·7°. Front, Q RG 16m R/G7M; 200°-R-250°-G-053°; W gable; 48°48'·26N 03°05'·75W. Rear, QR 50m 7M; vis 197°-242°; W gable; 660m from front.
Les Perdrix ⸔ Fl (2) WG 6s 5m, W6M, G3M; 165°-G-197°-W-202·5°-G-040°; G twr; 48°47'·74N 03°05'·79W.

CHENAL DE LA MOISIE (339·4°); PASSE DE LA GAINE (241·5°)

La Vieille du Tréou ⸔ 48°52'·00N 03°01'·09W.
An Ogejou Bihan ⸕ 48°53'·37N 03°01'·91W.
La Moisie ⸕ 48°53'·83N 03°02'·22W.
Les Héaux de Bréhat ☆ Fl (4) WRG 15s 48m, **W15M**, R/G11M; 227°-R-247°-W-270°-G-302°-W-227°; Gy ○ twr; 48°54'·50N 03°05'·17W.

Basse des Héaux ⸕ 48°54'·07N 03°05'·28W.
Pont de la Gaine ⸕ 48°53'·12N 03°07'·40W.

JAUDY RIVER TO TRÉGUIER

La Jument des Héaux ⸕ VQ; 48°55'·42N 03°08'·04W.
Grande Passe ldg lts 137°. Front, Port de la Chaine, Oc 4s 12m 11M; 042°-232°; W house; 48°51'·55N 03°07'·89W. Rear, **St Antoine** ☆ Dir Oc R 4s 34m **15M**; intens 134°-140°; R & W house; 0·75M from front. (Both marks are hard to see by day.)
Basse Crublent ⸕ QR; *Whis;* 48°54'·29N 03°11'·16W.
Le Corbeau ⸔ Fl R 4s; 48°53'·35N 03°10'·26W.
Pierre à l'Anglais ⸔ Fl G 4s; 48°53'·21N 03°10'·46W.
Petit Pen ar Guézec ⸔ Fl (2) G 6s; 48°52'·52N 03°09'·44W.
La Corne ⸔ Fl (3) WRG 12s 14m W8M, R/G6M; 052°-W-059°-R-173°-G-213°-W-220°-R-052°; W twr, R base; 48°51'·35N 03°10'·62W.

TRÉGUIER TO TRÉBEURDEN

PORT BLANC

Le Voleur Dir ⸔ 150°: Fl WRG 4s 17m, W14M, R/G 11M; 140°-G-148°-W-152°-R-160°; W twr; 48°50'·20N 03°18'·52W.
Basse Guazer ⸔; 48°51'·58N 03°20'·96W.

PERROS-GUIREC

Passe de l'Est, ldg lts 224·8°. Front, Le Colombier ⸔ Dir Q 28m 14M; intens 214·5°-234·5°; W house; 48°47'·87N 03°26'·66W. Rear, **Kerprigent** ☆, Dir Q 79m **21M**; intens 221°-228°; W twr, 1·5M from front.
Pierre du Chenal ⸕ 48°49'·29N 03°24'·67W.
Passe de l'Ouest. Kerjean ⸔ Dir lt 143·6°, Oc (2+1) WRG 12s 78m, W10M, R/G 8M; 133·7°-G-143·2°-W-144·8°- R-154·3°; W twr, B top; 48°47'·79N 03°23'·40W.
Les Couillons de Tomé ⸕ 48°50'·90N 03°25'·75W.
La Horaine ⸕ 48°49'·89N 03°27'·26W.
Roche Bernard ⸕ 48°49'·43N 03°25'·46W.
Gommonénou ⸕ VQ R 1M; 48°48'·27N 03°25'·83W.
Jetée du Linkin ⸔ Fl (2) G 6s 4m 6M; W pile, G top; 48°48'·20N 03°26'·31W.

PLOUMANAC'H

Mean-Ruz ⸔ Oc WR 4s 26m W12M, R9M; 226°-W-242°-R-226°; obsc by Pte de Trégastel when brg <080°; partly obsc by Les Sept-Îles 156°-207° and partly by Île Tomé 264°-278°; pink □ twr; 48°50'·25N 03°29'·00W.

LES SEPT ÎLES

Île-aux-Moines ☆ Fl (3) 20s 59m **24M**; obsc by Îliot Rouzic and E end of Île Bono 237°-241°, and in Baie de Lannion when brg <039°; Gy twr/dwelling; 48°52'·72N 03°29'·40W.
Les Dervinis ⸕ 48°52'·35N 03°27'·32W.

TRÉGASTEL-PLAGE

Île Dhu ⸕ 48°50'·37N 03°31'·24W.
Les Triagoz ⸔ Fl (2) WR 6s 31m W14M, R11M; 010°-W-339°-R-010°; obsc in places 258°-268° by Les Sept-Îles; Gy □ twr, R lantern; 48°52'·28N 03°38'·80W.
Bar-ar-Gall ⸕ VQ (9) 10s; 48°49'·79N 03°36'·22W.
Le Crapaud ⸕ Q (9) 15s; 48°46'·67N 03°40'·59W.

TRÉBEURDEN

Pte de Lan Kerellec ⸔ Iso WRG 4s; W8M, R/G5M; 058°-G-064°-W-069°-R-130°; Gy twr; 48°46'·74N 03°35'·06W.
Ar Gouredec ⸕ VQ (6) + L Fl 10s; 48°46'·41N 03°36'·59W.
An Ervennou ⸔ Fl (2) R 6s; 48°46'·48N 03°35'·96W.
NW bkwtr ⸔ Fl G 2·5s 8m 2M; IPTS; 48°46'·33N 03°35'·21W.

TRÉBEURDEN TO MORLAIX

LÉGUER RIVER

Beg-Léguer ⸔ Oc (4) WRG 12s 60m W12M, R/G9M; 007°-G-084°-W-098°-R-129°; west face of W house, R lantern; 48°44'·31N 03°32'·91W.
Kinierbel ⸔; *Bell;* 48°44'·15N 03°35'·18W.

LOCQUÉMEAU
Ldg lts 121°: Front, ⚡ QR 21m 7M; 068°-228°; W pylon, R top; 48°43'·41N 03°34'·44W. Rear, ⚡ QR 39m 7M; 016°-232°; W gabled house; 484m from front.
Locquémeau ⚓ Whis; 48°43'·86N 03°35'·93W.
Locquémeau approach ⚓Fl G 2.5s; 48°43'·62N 03°34'·96W.

LOCQUIREC
Gouliat Ⅰ 48°42'·57N 03°38'·97W.

PRIMEL-TRÉGASTEL
Méloine Ⅰ 48°45'·56N 03°50'·68W.
Ldg lts 152°, both ⚡ QR 35/56m 7M. Front, 134°-168°; W □, R stripe, on pylon; 48°42'·45N 03°49'·19W. Rear, R vert stripe on W wall; 172m from front.
W bkwtr ⚡ Fl G 4s 6m 7M; 48°42'·77N 03°49'·52W.

BAIE DE MORLAIX
Chenal de Tréguier ldg lts 190·5°: Front, ⚡ Île Noire Oc (2) WRG 6s 15m, W11M, R/G8M; 051°-G-135°-R-211°-W-051°; obsc in places; W □ twr, R top; 48°40'·35N 03°52'·54W. Common Rear, **La Lande** ☆ Fl 5s 85m **23M**; obsc by Pte Annelouesten when brg >204°; W □ twr, B top; 48°38'·20N 03°53'·14W.
La Pierre Noire Ⅰ 48°42'·56N 03°52'·20W.
La Chambre ⚓ 48°40'·74N 03°52'·51W.

Grande Chenal ldg lts 176·4°: Front, **Île Louet** ☆ Oc (3) WG 12s 17m **W15M**,G10M; 305°-W (except where obsc by islands)-244°-G-305°; W □ twr, B top; 48°40'·41N 03°53'·34W. Common Rear, **La Lande** as above.
Pot de Fer Ⅰ Bell; 48°44'·23N 03°54'·02W.
Stolvezen ⚓ 48°42'·64N 03°53'·41W.
Ricard ⚓ 48°41'·54N 03°53'·51W.
Vieille ⚓ 48°42'·60N 03°54'·11W. (Chenal Ouest de Ricard 188·8°)
La Noire Ⅰ 48°41'·65N 03°54'·06W.
Corbeau ⚓ 48°40'·63N 03°53'·33W.

MORLAIX RIVER
Barre de-Flot No 1 ⚓ 48°40'·18N 03°52'·95W.
No 2 ⚓ Fl R 2s; 48°39'·88N 03°52'·53W.
No 3 ⚓ Fl G 2s; 48°39'·25N 03°52'·32W.
No 4 ⚓ Fl R 2s; 48°38'·62N 03°51'·62W.
No 5 ⚓ Fl G 2s; 48°38'·04N 03°51'·34W.
No 7 ⚓ 48°37'·68N 03°51'·03W.

PENZÉ RIVER
Cordonnier ⚓ 48°42'·93N 03°56'·59W.
Guerhéon ⚓ 48°42'·73N 03°57'·18W.
Trousken ⚓ 48°42'·26N 03°56'·54W.
Pte Fourche ⚓ 48°42'·14N 03°56'·76W.
Ar Tourtu Ⅰ 48°41'·99N 03°56'·57W.
An Nehou (Caspari) Ⅰ 48°41'·56N 03°56'·45W.
Le Figuier Ⅰ 48°40'·46N 03°56'·16W.

ROSCOFF TO ÎLE VIERGE
BLOSCON/ROSCOFF
Astan ⚓ VQ (3) 5s 9m 6M; 48°44'·91N 03°57'·66W.
Le Menk ⚓ Q (9) WR 15s 6m W5M, R3M; 160°-W-188°-R-160°; 48°43'·28N 03°56'·71W.
Basse de Bloscon ⚓ VQ; 48°43'·71N 03°57'·55W.
Bloscon pier ⚡ Fl WG 4s 9m W10M, G7M (in fog Fl 2s); 200°-W-210°-G-200°; W twr, G top; 48°43'·21N 03°57'·69W.
Roscoff Marina ent, E side, ⚡ Fl (2) R 6s 4M; 48°43'·10N 03°57'·90W.
Ar Pourven ⚓ Q; 48°43'·04N 03°57'·70W.
Ar-Chaden ⚓ Q (6) + L Fl WR 15s 14m, W8M, R6M; 262°-R-289·5°-W-293°-R-326°- W-110°; YB twr; 48°43'·94N 03°58'·26W.
Men-Guen-Bras ⚓ Q WRG 14m, W9M, R/ G6M; 068°-W-073°-R-197°-W-257°-G-068°; BY twr; 48°43'·76N 03°58'·07W.
Roscoff ldg lts 209°: Front, N môle ⚡ Oc (3) G 12s 7m 7M; synch;

078°-318°; W col, G top; 48°43'·56N 03°58'·67W. **Rear** ☆ Oc (3) 12s 24m **15M**; 062°-242°; Gy □ twr, W on NE side; 430m from front.

CANAL DE L'ÎLE DE BATZ
Roc'h Zu Ⅰ 48°43'·88N 03°58'·59W.
Jetty hd (LW landing) ⚡ Q 5m 1M; BY col; 48°43'·92N 03°58'·97W.
Run Oan Ⅰ 48°44'·10N 03°59'·30W.
Perroch ⚓ 48°44'·10N 03°59'·71W.
Tec'hit Bihan Ⅰ 48°44'·02N 04°00'·88W.
La Croix Ⅰ 48°44'·19N 04°01'·30W.
L'Oignon Ⅰ 48°44'·04N 04°01'·35W.
Basse Plate ⚓ 48°44'·25N 04°02'·53W.

ÎLE DE BATZ
Île aux Moutons Ⅰ VQ (6)+ L Fl 10s 3m 7M; 48°44'·25N 04°00'·52W; landing stage, S end and ent to hbr.
Malvoch⚓ 48°44'·26N 04°00'·67W, W side of ent.
Lt ho ☆ Fl (4) 25s 69m **23M**; Gy twr; 48°44'·71N 04°01'·62W. Same twr, auxiliary lt, FR 65m 7M; 024°-059°.

MOGUÉRIEC
Ldg lts 162°: Front ⚡ Iso WG 4s 9m W11M, G6M; 158°-W-166°-G-158°; W twr, G top; jetty 48°41'·34N 04°04'·47W. Rear ⚡ Iso G 4s 22m 7M, synch; 142°-182°; W col, G top; 440m from front.

PONTUSVAL
Pointe de Pontusval Ⅰ 48°41'·42N 04°19'·32W.
Ar Peich ⚓ 48°40'·90N 04°19'·16W.
Pte de Beg-Pol ⚡ Oc (3) WR 12s 16m W10M, R7M; Shore-W-056°-R-096°-W-shore; W twr, B top, W dwelling; 48°40'·67N 04°20'·76W. QY and FR lts on towers 2·4M S.
Aman-ar-Ross ⚓ Q 7M; 48°41'·88N 04°27'·03W.
Lizen Ven Ouest ⚓ VQ (9) 10s 5M; 48°40'·53N 04°33'·63W.
Île-Vierge ☆ Fl 5s 77m **27M**; 337°-325°; Gy twr; 48°38'·33N 04°34'·05W.

ÎLE VIERGE TO L'ABER-ILDUT
L'ABER WRAC'H
Libenter ⚓ Q (9) 15s 4M; 48°37'·45N 04°38'·46W.
Outer ldg lts 100·1°: Front, Île Wrac'h ⚡ QR 20m 7M; W □ twr, Or top; 48°36'·88N 04°34'·56W. Rear, Lanvaon ⚡ Q 55m 12M; intens 090°-110°; W □ twr, Or △ on top; 1·63M from front.
Grand Pot de Beurre Ⅰ 48°37'·21N 04°36'·48W.
Petit Pot de Beurre Ⅰ 48°37'·12N 04°36'·23W.
Basse de la Croix ⚓ Fl (3) G 12s 3M; 48°36'·92N 04°35'·99W.
Breac'h Ver ⚓ Fl (2) G 2·5s 6m 3M; △ on twr; 48°36'·63N 04°35'·38W.
Dir ⚡ 128°, Oc (2) WRG 6s 5m W13M, R/G11M; 125·7°-G-127·2°-W-128·7°-R-130·2°; 48°35'·89N 04°33'·82W, root of W bkwtr .
Marina ent: QG 3m 2M; vis 186°-167° (341°); 48°35'·97N 04°33'·64W. QR 3m 2M; vis 244°-226° (342°).

L'ABER BENOÎT
Petite Fourche Ⅰ 48°36'·99N 04°38'·75W.
Rusven Est ⚓ 48°36'·30N 04°38'·63W.
Rusven Ouest Ⅰ Bell; 48°36'·07N 04°39'·43W.
Basse du Chenal Ⅰ 48°35'·81N 04°38'·53W.
Poul Orvil Ⅰ 48°35'·52N 04°38'·29W.
La Jument Ⅰ 48°35'·10N 04°37'·41W.
Ar Gazel ⚓ 48°34'·90N 04°37'·27W.
Le Chien Ⅰ 48°34'·67N 04°36'·88W.

ROCHES DE PORTSALL
Le Relec Ⅰ 48°35'·99N 04°40'·84W.
Corn-Carhai ⚡ Fl (3) 12s 19m 9M; W 8-sided twr, B top; 48°35'·19N 04°43'·94W.
Grande Basse de Portsall ⚓ VQ (9) 10s 5m 4M; 48°36'·70N 04°46'·13W.
Basse Paupian Ⅰ 48°35'·31N 04°46'·27W.

Bosven Aval ⌓ 48°33'·82N 04°44'·27W.
Men ar Pic ⚓ 48°33'·65N 04°44'·03W.
Portsall ⚟ Oc (4) WRG 12s 9m W13M, R/G10M; 058°-G-084°-W-088°-R-058°; W col, R top; 48°33'·84N 04°42'·26W.

ARGENTON to L'ABER-ILDUT
Le Taureau ⚓ 48°31'·46N 04°47'·34W.
Argenton, Île Dolvez, front ldg bcn 086° ⌓ 48°31'·25N 04°46'·23W.
Le Four ☆ Fl (5) 15s 28m **22M**; Gy ○ twr; 48°31'·38N 04°48'·32W.
L'Aber-Ildut ⚟ Dir Oc (2) WRG 6s 12m W10M, R10M, G10M; 078°-G-081°-W- 085°-R- 088° ; W bldg; 48°28'·26N 04°45'·56W.
Le Lieu ⚏Fl R 2.5s 4M; 48°28'·25N 04°46'·66W.

CHENAL DU FOUR AND CHENAL DE LA HELLE
Ldg lts 158·5°. Front, **Kermorvan** ☆ Fl 5s 20m **22M**; obsc'd by Pte de St Mathieu when brg <341°; W □ twr; *Horn 60s;* 48°21'·72N 04°47'·40W. 992271316.
Rear, **Pte de St Mathieu** ☆ Fl 15s 56m **24M**; W twr, R top; 48°19'·79N 04°46'·26W. Same twr: Dir F 54m **28M**; intens 157·5°-159·5°.
⚟ Q WRG 26m, W12M, R/G 8M; 085°-G-107°-W-116°-R-134°; W twr 54 m WNW of previous entry; 48°19'·80N 04°46'·30W.
Plâtresses N ⚐ Fl G 2·5s; 48°26'·46N 04°50'·71W.
Valbelle ⚏ Fl (2) R 2·5s; 48°26'·43N 04°50'·03W.
Plâtresses SE ⚐ Fl (2) G 6s; 48°25'·96N 04°50'·52W.
Saint Paul ⚑ Oc (2) R 6s; 48°24'·82N 04°49'·16W.
Pte de Corsen ⚟ Dir Q WRG 33m W12M, R/G8M; 008°-R-012°-W-015° (ldg sector)- G-021°; W hut; 48°24'·89N 04°47'·61W.

CHENAL DE LA HELLE
Ldg lts 137·9°: Front, **Kermorvan** ☆ see above. Rear, **Lochrist** ☆ Dir Oc (3) 12s 49m **22M**; intens 135°-140°; W 8-sided twr, R top; 48°20'·55N 04°45'·82W.
Luronne *i*; 48°26'·61N 04°53'·78W.
Ldg lts 293·5° astern (to join Ch du Four S of St Paul ⚑): Front, Le Faix ⚏ VQ 16m 8M; 48°25'·73N 04°53'·91W. Rear, **Le Stiff** ☆ (below at Ouessant); 6·9M from front.
Ldg line 142·5°, optional day only (to join Ch du Four SE of St Pierre ⚐): Front, **Kermorvan** ☆ see above. Rear, two W gables (Les Pignons de Kéravel, 48m) 48°20'·10N 04°45'·55W
Pourceaux ⚑ Q; 48°24'·00N 04°51'·31W.
Saint-Pierre ⚐ Fl (3) G 12s; 48°23'·09N 04°49'·09W.
Rouget ⚑ Fl G 4s; 48°22'·04N 04°48'·88W (Ch du Four).
Grande Vinotière ⌓ Fl R 4s 12m 5M; R 8-sided twr; 48°21'·94N 04°48'·42W.

Le Conquet, Môle Ste Barbe ⚟ Fl G 2·5s 5m 6M; 48°21'·57N 04°46'·98W.
Les Renards *i* 48°21'·00N 04°47'·48W.
Tournant et Lochrist ⚑ Fl (2) R 6s; 48°20'·64N 04°48'·12W,
Ar Christian Braz ⚓ 48°20'·68N 04°50'·14W, E of Ile de Béniguet.
Ldg line 325°: Front, Grand Courleau ⚓; rear, La Faix (above).
Ldg lts 007°: Front, **Kermorvan** ☆ above. Rear, **Trézien** ☆ Dir Oc (2) 6s 84m **20M**; intens 003°-011°; Gy twr, W on S side; 48°25'·41N 04°46'·74W. 992271309.
Les Vieux-Moines ⚑ Fl R 4s 16m 5M; 280°-133°; R 8-sided twr; 48°19'·33N 04°46'·63W, 5ca SSW of Pte de St Mathieu.
La Fourmi ⚐ Fl G 4s; 48°19'·25N 04°47'·96W.

OUESSANT AND ÎLE MOLÈNE

OFF USHANT TSS
Ouessant NE ⊙; V-AIS; 48°59'·51N 05°24'·00W.
Ouessant SW ⊙; V-AIS 48°30'·00N 05°45'·00W.

ÎLE D'OUESSANT
Le Stiff ☆ Fl (2) R 20s 85m **24M**; two adjoining W twrs; 48°28'·47N 05°03'·41W. Radar twr, conspic, 340m NE, ⚟ Q (day); ⚟ FR (night).
Port du Stiff, E môle ⚟ Dir Q WRG 11m W10M, R/G7M; 251°-G-254°-W-264°-R-267°; W twr, G top; 48°28'·12N 05°03'·25W.
Baie du Stiff: Gorle Vihan ⚓ 48°28'·32N 05°02'·60W.
Men-Korn ⚑ VQ (3) WR 5s 21m 8M; 145°-W-040°-R-145°; BYB twr; 48°27'·96N 05°01'·32W.
Men ar Froud ⚓ 48°26'·62N 05°03'·67W.
La Jument ⚟ Fl (3) R 12s 36m 10M; 241°-199°; Gy 8-sided twr, R top; 48°25'·33N 05°08'·04W.
Nividic ⚑ VQ (9) 10s 28m 10M; 290°-225°; Gy 8-sided twr, helicopter platform; 48°26'·74N 05°09'·06W.
Créac'h ☆ Fl (2) 10s 70m **32M**; obsc 247°-255°; W twr, B bands; 48°27'·55N 05°07'·76W.

ÎLE MOLÈNE and ARCHIPELAGO
Kéréon ⚟ Oc (3) WR 12s 38m 10M; 019°-W-248°- R-019°; Gy twr; 48°26'·24N 05°01'·59W; unreliable.
Les Trois-Pierres ⚟ Iso WRG 4s 15m W9M, R/G6M; 070°-G-147°-W-185°-R-191°-G-197°-W-213°-R-070°; W col; 48°24'·70N 04°56'·84W.
Molène, Old môle Dir ⚟ 191°, Fl (3) WRG 12s 6m W9M, R/G7M; 183°-G-190°-W-192°-R-203°; 48°23'·85N 04°57'·29W.
Same structure: Chenal des Laz , Dir ⚟ 261°: Fl (2) WRG 6s 9m W9M, R/G7M ; 252·5°-G-259·5°-W-262·5°-R-269·5°.
Pierres-Vertes ⚑ VQ (9) 10s 9m 5M; 48°22'·19N 05°04'·76W.
Pierres Noires *i*; 48°18'·47N 04°58'·15W.
Les Pierres Noires ☆ Fl R 5s 27m **19M**; W twr, R top; 48°18'·67N 04°54'·88W.

7.3 PASSAGE INFORMATION
More passage information is threaded between successive harbours in this Area. **Bibliography:** *Channel Islands, Cherbourg Peninsula and North Brittany* (Imray/RCC/Carnegie); *North Brittany and Channel Islands CC* (Fernhurst/Cumberlidge); *Shell Channel Pilot* (Imray/Cunliffe); *Channel Pilot* (NP 27); *Channel Havens* (ACN/ Endean); *Secret Anchorages of Brittany* (Imray/Cumberlidge).

NORTH BRITTANY
This ever-popular cruising ground (AC 2648, 2647, 2643, 2028) is not one to trifle with, but conversely it is rewarding to navigate into a remote anchorage or small fishing harbour. Here the unique character and appeal of N Brittany is all around you.

Good landfall marks must be carefully identified, as a back-up to GPS, before closing the rock-strewn coast. In rough weather, poor visibility (fog and summer haze are frequent) or if uncertain of position, it may be prudent to lie off and wait for conditions

to improve; there are few safe havens. Closer inshore the tidal streams are variable with associated overfalls. Thorough careful planning remains the key to safe pilotage.

In the E of the area the outer approaches to **Paimpol, Île de Bréhat, Lézardrieux** and **Tréguier** may be complicated by strong tidal streams. At times there may be a proliferation of landmarks which must be selectively identified. Concentrate on those which affect your pilotage and discard those which are non-essential and may even distract you from the task.

Once W of **Roscoff** there are few safe hbrs and anchorages until **L'Aberwrac'h** where many British yachts tend to pause and await good conditions for negotiating the Chenal du Four. This is rarely as difficult as it is imagined to be. The Atlantic swell can be a new experience for some, but in moderate winds it is rarely a hazard. It can however much reduce the range at which objects, especially floating marks and other small craft, are seen.

PAIMPOL TO PLOUMANAC'H

In the offing, 11-18M NNE of Île de Bréhat and on a direct track from/to Guernsey, are Plateau de Barnouic (lit) and **Plateau des Roches Douvres** (light house); both with drying and submerged rks, to be given a wide berth in poor vis (AC 2027, 2028, 3673).

Approaching L'Ost-Pic from the SE, keep to seaward of the three ECM marks E of it or enter Baie de **Paimpol** from a point about 1M E of the most N'ly ECM (Les Charpentiers); but it is safe to pass about 300m east of L'Ost Pic in moderate weather.

The Ferlas channel (AC 3673) lies E-W south of **Ile de Bréhat**, and is useful if entering/leaving R Trieux from/to the E. It is easiest to beacon-hop past the 5 SCM bcns and 1 IDM bcn marking the N side of the chan; the S side is less well marked.

For the many yachts approaching from Guernsey, Les Héaux-de-Bréhat light house is a conspic landfall day/night for either Tréguier or Lézardrieux. Closer in or from the E, La Horaine (lt beacon) is a better landfall for the latter. It marks the Plateaux de la Horaine and des Échaudés and other rocks to the SE. In poor visibility it should be closed with caution and left at least 7ca to the SE, as the flood stream sets strongly onto it.

The Grand Chenal 224·7° is the main, lit channel into the R de Trieux for **Lézardrieux** and further upriver to **Pontrieux**. From NW the unlit Chenal de La Moisie leads 159° to join the Grand Chenal off Ile de Bréhat.

The **Passe de la Gaine** 241·5° is a useful inshore route between Lézardrieux and Tréguier, avoiding a detour round Les Héaux. If taken at above half tide, it presents no problem in fair weather. But it is unlit and from the NE the distant ldg marks are hard to see in poor visibility or against a low sun. However two SHM beacons to the S and SW of Les Héaux-de-Bréhat lighthouse can be used as a handrail towards Pont de la Gaine.

The Grande Passe 137·3° into **Rivière de Tréguier** is well lit, but ldg marks are hard to see by day. Use the NE Passage with caution.

Between Basse Crublent lt buoy and Port Blanc unmarked rocks extend 2M offshore. **Port Blanc** can be difficult to identify by day until it opens on 150°. **Perros-Guirec** is approached either side of Ile Tomé from NE or NW via well lit/marked channels; take care not to be neaped. **Ploumanac'h** can only be entered by day. It boasts some of the most spectacular pink granite along the North Brittany coast.

LES SEPT ÎLES TO BAIE DE MORLAIX

(AC 2026). **Les Sept Îles** (AC 2027) consist of five main islands and several islets, through which the tide runs strongly. Île aux Moines is lit, and all the islands are bird sanctuaries. Further W, Plateau des Triagoz has offlying dangers WSW and NE of the lt, where the sea breaks heavily.

▶*Here the stream turns ENE at HW Brest −0230, and WSW at HW Brest +0330, sp rates both 3·8kn.*◀

Trégastel-Plage is a small anchorage W of Ploumanac'h. To the SW the coast as far as **Trébeurden** is not easily approached due to many offlying rks. The radome NE of Trébeurden is conspic. Further S in the Baie de Lannion is Locquémeau and anchorages near the mouth of the drying R Léguer up to **Lannion**. Locquirec is a pleasant drying harbour 3·5M SW of Locquemeau. **Primel-Trégastel**, at the E ent to Baie de Morlaix, is a useful inlet to await the tide up to Morlaix. To the N Plateau de la Méloine dries.

The Baie de Morlaix (AC 2745) is bestrewn with drying rks and shoals, all marked. Careful pilotage and adequate visibility are needed to negotiate any of the chans which are narrow in parts. Chenal de Tréguier 190·5° and the Grand Chenal 176·4° are both lit. The former should only be used HW ±3 due to shoals at the S end. Grand Chenal passes close E of Île Ricard with Île Louet and La Lande lights in transit; abeam Calhic bn tr alter to port to pass between Château du Taureau (conspic) and Île Louet. Continue SSE and upriver to lock into **Morlaix** Marina in complete shelter. The ⚓age NE of Carantec is reached from Chenal Ouest de Ricard.

ÎLE DE BATZ TO LE FOUR

▶*2M N of Île de Batz the E-going stream begins at HW Brest −0400, and the W-going stream at HW Brest +0200, sp rates 3·8kn. The flood and ebb streams deflected N of Ile de Batz may be >5kn, in opposition and violent. Against any existing sea state, the resultant breaking seas can endanger small craft which are therefore strongly advised to pass at least 2-2½M north of Batz. See the French tidal stream atlas 563-UJA.*◀

Approaching **Roscoff** from the NE, leave Astan ECM lt buoy to stbd tracking with Men Guen Bras lt bcn in transit 213° with white Chapelle St Barbe, to round Ar Chaden lt bcn for Roscoff's drying harbour. Bound for Roscoff Marina pass E of No1 pillar by and Basse de Bloscon NCM, then S towards Bloscons prominent ferry terminal pier. (AC 2026, 2025, 2745)

S of the island **Canal de L'Île de Batz**, day only and above half tide, requires careful pilotage at its narrower eastern end. From near Ar Chaden steer 275° for the Q bn at end of the conspic LW ferry pier. Pass 30m N of this bn, then alter to 300° for Run Oan SCM. Thence steer 283°, leaving Perroch NCM bcn twrs 100m to port. When clear of this rky, drying shoal alter to 270°, leaving Porz Kernok hbr bkwtrs well to stbd then midway between L'Oignon NCM and La Croix SCM bcns. With these abeam steer 281° for Basse Plate NCM bn; thence West into open waters.

Proceeding W from Île de Batz towards Le Four there are many offlying dangers, in places 3M offshore. Swell may break on shoals even further to seaward. With strong tides, in poor vis or bad weather it is a coast to avoid. But in fair weather enjoy its delightful harbours such as Moguériec, Brignogan, L'Aberwrac'h, L'Aberbenoit, Portsall and Argenton. N of L'Aberwrac'h is the conspic Île Vierge lt ho, reputedly the tallest in the world.

▶*N of Le Libenter buoy, off L'Aberwrac'h, the NE-going stream starts at HW Brest −0500, sp rate 3·8kn, and the WSW-going stream at HW Brest +0100.*◀

L'Aberwrac'h is accessible at all tides and strategically placed to catch the tide S through Chenal du Four. The marina is also well protected from NW'lies.

W of L'Aberwrac'h a daytime only/good vis chan leads inshore of Roches de Portsall to Le Four lt ho.The distant marks are not easy to identify and it is only 1M shorter than the offshore route via Basse Paupian WCM buoy. AC 1432 or SHOM 7094 and a detailed pilot book essential.

OUESSANT (USHANT)

(AC 2356) Île d'Ouessant lies 10M off NW Brittany, surrounded by five lt ho's. It is a rky island, with dangers extending 5ca to NE, 7½ca to SE, 1·5M to SW and 1M to NW; here Chaussée de Keller is a dangerous chain of unmarked drying and submerged rks running 1M W of Île de Keller and into the ITZ. There are anchorages and moorings at **Lampaul** and **Baie du Stiff**, the former exposed to the SW and the latter to the NE.

Ouessant is an important landfall, but avoid in thick weather. Ideally, stay in harbour until the vis improves. In fair weather and reasonable visibility the pilotage in the well-marked chans between it and the mainland is not too demanding. But the tide runs hard in places, causing overfalls when against wind >F5.

▶*Tidal streams are strong around the island, and in the chans between it and the mainland. Off Baie du Stiff (lt, radar twr) the stream turns NW at HW Brest +0400, and SE at HW Brest −0200, max rates 2¾kn.*◀

The routes outside Ouessant TSS or via the ITZ add much to the distance and are exposed to sea, swell and shipping. Yachts usually take the inshore chans, ie: Chenal du Four, most direct and popular; Chenal de la Helle, an alternative to N part of Chenal du Four (also gives access to Île Moléne) is less direct but better in bad weather. Passage du Fromveur, SE of Ouessant, is easiest but longer and can be extremely rough. ▶*The NE-going stream starts HW Brest −0515, max rates 8-9kn; the SW-going stream starts around HW Brest, max rates 7-8kn.*◀

7.4 Special Notes for France: See Area 4.

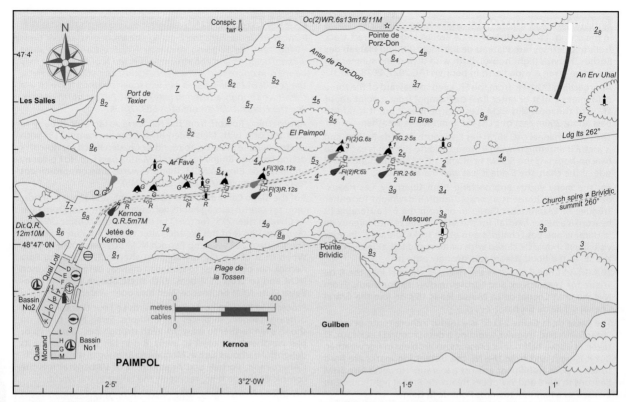

7.5 PAIMPOL

Côtes d'Armor 48°47′·00N 03°02′·56W ❀❀☀♦♦♦❁❁

CHARTS AC 2648, 2027, 2028, 3673; SHOM 7152, 7154, 7127; Navi 537; Imray C34

TIDES Dover –0525; ML 6·1; Duration 0600

Standard Port ST MALO (←)

Times				Height (metres)			
High Water		Low Water		MHWS	MHWN	MLWN	MLWS
0100	0800	0300	0800	12·2	9·3	4·2	1·5
1300	2000	1500	2000				
Differences PAIMPOL							
–0010	–0005	–0035	–0025	–1·4	–0·9	–0·4	–0·1

SHELTER Good shelter from all winds in hbr, but few ⚓s as most of the Anse de Paimpol dries, including the appr chan to hbr.

NAVIGATION Bearing in mind the very large tidal range, there is enough water in the bay for most craft from half-flood.

WPT 48°47′·82N 02°54′·58W, 262°/5·2M to Kernoa jetty hd. Chenal de la Jument 260° is the outer appr. After La Jument PHM bn tr, alter onto 262·2° inner ldg line; or ⚓ in 5m to await the tide. Small buoys/bns in final 1M of chan (dries 6-7m) may be hard to see against a low evening sun. The drying rks (El Paimpol, El Bras and Ar Fav) are close N of the ldg line.

An alternative appr from Île de Bréhat lies E of Les Piliers NCM bn tr, thence S past Pte de la Trinité. Or appr via Cadenenou NCM and Chenal du Denou 193°. SHOM 7127 is essential for these and other inshore passages which should only be attempted towards HW.

LIGHTS AND MARKS From the S L'Ost-Pic lt is a conspic ☐ W tr 4M E of Paimpol. Pte de Porz-Don, lt ho on a white house, is 7ca ENE of hbr ent; its W sector leads 270° to intercept the inner 262·2° ldg lts at La Jument bcn twr. A conspic twr (52m) is 3ca W of Porz-Don.

Chenal de la Jument, outer ldg marks 260°: Paimpol ✠ spire (the N'ly of two) on with the ill-defined summit of Pte Brividic. Inner ldg lts, both QR, 262·2°: front, Jetée de Kernoa.

COMMUNICATIONS CROSS 02·98·89·31·31; Auto 08·92·68·08·22; Police 02·96·20·80·17; ⊜ 02·96·20·81·87; Brit Consul 01·44·51·31·00; Ⓗ 02·96·55·60·00; Dr 02·96·55·15·15; Port Mgr 02·96·20·80·77; HM and Lock VHF Ch 09 (0800-1200LT and lock opening hrs); Aff Mar 02·96·55·35·00.

FACILITIES Lock opens HW ±2½, 60m x 12m, ☎02·96·20·90·02. **Marina** ☎02·96·20·47·65. port-plaisance.paimpol@wanadoo.fr www.ville-paimpol.fr **Basin No 2** 306+24 Ⓥ, May-Sep €2·62; Ⓥ berths, pontoon A, min depth 3·8m. ⛽ (hose) ▬ @ ♿. **Basin No 1** Yachts up to 40m LOA. ⛴(25t, 6t, 4t). **E Quays** ⚒ ⚒ ⚒Ⓜ SHOM Ⓔ. **Town** ⚒ ◻ 🍴 ⛽ ♨ 🏪 ⊠ Ⓑ Ⓗ ⇌ ✈ Dinard, Brest, Rennes. **Ferry** Roscoff, St Malo.

7.6 ÎLE DE BRÉHAT

Côtes d'Armor 48°51′·00N 03°00′·00W ❀❀❀♦♦❁❁❁

CHARTS AC 2648, 2027, 2028, 3673; SHOM 7152, 7154; Navi 537; Imray C34

TIDES –0525 Dover; ML 5·8; Duration 0605

Standard Port ST MALO (←)

Times				Height (metres)			
High Water		Low Water		MHWS	MHWN	MLWN	MLWS
0100	0800	0300	0800	12·2	9·3	4·2	1·5
1300	2000	1500	2000				
Differences LES HEAUX DE BRÉHAT							
–0015	–0010	–0100	–0045	–2·3	–1·6	–0·6	–0·1
ÎLE DE BRÉHAT							
–0015	–0005	–0040	–0030	–1·6	–1·2	–0·5	0·0

SHELTER Good in Port Clos and at Port de la Corderie, except in W-NW'lies. See Facilities for other ⚓s depending on wind direction.

NAVIGATION From the NE use R. Trieux WPT (Lézardrieux) and 224·8° ldg line. From the N use Ch de Bréhat 167·4° (distant ldg marks) then enter Ferlas Chan from the E via WPT 48°49′·39N 02°55′·08W, 277°/3·4M to abm Port Clos. For the R Trieux, follow Ferlas Chan transits Ⓒ – Ⓕ, as shown on AC 3673.

LIGHTS AND MARKS Chapelle St Michel, Amer du Rosedo and nearby Sig stn are all conspic. For the 3 main lts, see chartlet and Lights, buoys and waypoints.

COMMUNICATIONS CROSS 02·98·89·31·31; LB 02·96·20·00·14; Auto 08·92·68·08·22; Police 02·96·20·80·17; ⊖ 02·96·20·81·87; H 02·96·55·60·00; Dr 02·96·20·09·51; Signal station VHF Ch 16, 10, day only.

FACILITIES Port Clos the main hbr (dries) is busy with vedettes. No ⬭; ⚓ clear of fairway. SW of Port Clos ⚓ is prohib due to cables/pipe across the Ferlas Chan. So moor/⚓ in **Le Kerpont** W of Bréhat. **Port de la Corderie** dries; some ⚓s but get well out of strong tidal streams.

E of Le Bourg there are free drying private buoys near ⚓, but some reported unsafe due to lack of maintenance.

La Chambre ⚓ in upper reaches just S of the ⚓ area. ◣ can be floodlit by pressing button on lamp post at top of slip.

Guerzido off the Chenal de Ferlas has good holding, out of the strong tides.

Hbrs (no HM) full access at HW, ⛟ 🗼 ◣ from fuel barge at Port Clos, ◣; CN de Bréhat, 🗼 ⬭ ⚒.

Village ✕ 🍴 🏪 📮 ⛽ 🏠 ✉ Ⓑ. Ferry to Pte de l'Arcouest, bus to Paimpol thence ⇌ to Paris, Brest, Roscoff and St Malo. ✈ Dinard, Brest, Rennes to London. No cars on island.

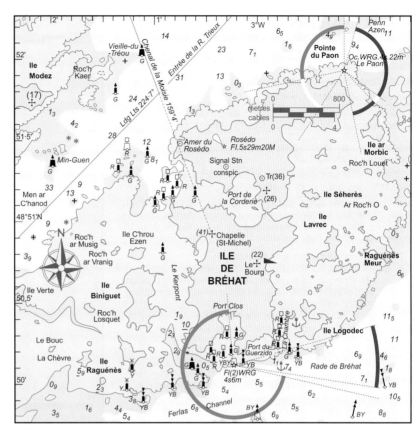

OUTER APPROACHES TO LÉZARDRIEUX AND TRÉGUIER WITH PASSE DE LA GAINE BETWEEN

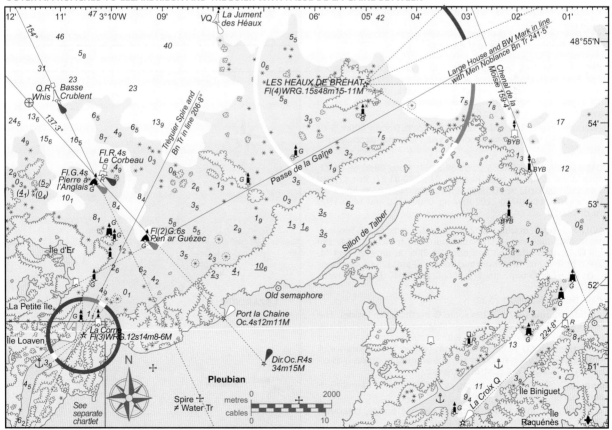

7.7 LÉZARDRIEUX

Côtes d'Armor **48°47´·34N 03°05´·89W** ✳✳✳⚓⚓⚓✿✿✿

CHARTS AC 2648, 2027, 2028, 3673; SHOM 7152/3, 7126/7; Navi 537; Imray C34

TIDES –0510 Dover; ML 5·9; Duration 0610

Standard Port ST MALO (⟵)

Times				Height (metres)			
High Water		Low Water		MHWS	MHWN	MLWN	MLWS
0100	0800	0300	0800	12·2	9·3	4·2	1·5
1300	2000	1500	2000				
Differences LÉZARDRIEUX							
–0020	–0015	–0055	–0045	–1·7	–1·3	–0·5	–0·2

SHELTER Very good in all weathers. Strong stream at half tide.

NAVIGATION WPT 48°54´·98N 02°56´·07W, 224·7°/6·7M to La Croix, front 224·8° ldg lt. Offshore dangers include Roches Douvres, Barnouic, Plateau de la Horaine and rky shoals to the W in the outer apps. Off the river ent allow for strong cross-streams.

The 3 well-marked approach channels are:
- **Grand Chenal**, main lit chan from NE, best for strangers. Ldg lts 224·7°: Front, La Croix, two double-barrelled trs, W on NE side with R tops. Rear, Bodic (2·1M from front) high up amongst the trees.
- **Ch de la Moisie**, unlit from the NW: Amer du Rosédo, W obelisk on 159·4° with St Michael's chapel (both conspic on Île de Bréhat). It also connects with Passe de la Gaine, short cut from/to Tréguier; see chartlet on previous page.
- **Ferlas Chan** (lit) from E or W, passing S of Île de Bréhat. By day follow the 5 SCM bcns on the N side of the chan. At night: W sector (281°) of Men Joliguet, WCM bcn twr. W sector (257°) of Roche Quinonec at Loguivy. W sector (271°) of Kermouster joins Coatmer 218·7° ldg line.

Within the Trieux river:
- Coatmer ldg lts 218·7°: front, low down amongst trees; rear, 660m from front, high up almost obsc'd by trees.
- W sector (200°) of Les Perdrix, a stout green bcn twr. Speed limit 5kn from Perdrix to the bridge.

LIGHTS AND MARKS See Lights, buoys and waypoints for offshore lts at Roches Douvres lt ho, Barnouic ECM bn tr, Les Héaux de Bréhat lt ho and, on Île de Bréhat, Pte du Paon and Rosédo. Outer and inner approach marks are described above. There are no navigational lights S of Perdrix.

Beware, at night, the unlit, 16m high Roc'h Donan 400m S of Perdrix. At the tidal marina floodlights on shore may dazzle yachts arriving.

COMMUNICATIONS CROSS 02·98·89·31·31; Auto 08·92·68·08·22; Police 02·96·20·18·17; ⊖ at Paimpol 02·96·20·81·87; Brit Consul 01·44·51·31·00; Dr 02·96·20·18·30; Aff Mar (Paimpol) 02·96·55·35·00; Marina VHF Ch 09 (0730-2200 Jul/Aug; 0800-1200 and 1400-1800 rest of year).

FACILITIES **Tidal marina** ☎02·96·20·14·22. http://port.3douest.com/lezardrieux/ port-lezardrieux@wanadoo.fr Access H24. 380 + 50 **Ⓥ**. Berth on the first pontoon (No 3) or as directed by HM. ⌱€2·39. ⚑ €2·00. Multihulls and boats >12·5m LOA should moor on **⚓**s €1·67 or **Ⓥ** pontoon. ⛽🛢✕⌂🖥🛎🛒🔧 ⌂ Ⓔ ⚠ ✖ Divers, ⛴🏗(50t).

YC de Trieux ☎02·96·20·10·39.

Non-tidal marina (247 ⌱) has some **Ⓥ** berths (2·4m inside). The retaining wall is marked by 5 unlit Y SPM perches, the sill by PHM/SHM perches. Access over sill **4**·9m with automatic flap. As sill covers on the flood, at CD +6·15m, flap drops to give 1·1m clearance. A depth gauge shows water over sill. The waiting pontoon is as on chartlet. IPTS sigs 2 & 4 control entry/exit.

Town ✕ ⌂ 🍴 🛢 ✉ Ⓑ 🚲 (occas bus to Paimpol & Lannion) ✈ Lannion. **Ferry** Roscoff.

Yachts can continue about 12km upriver (via bridge, clearance 17m) to lock in at Pontrieux in complete shelter.

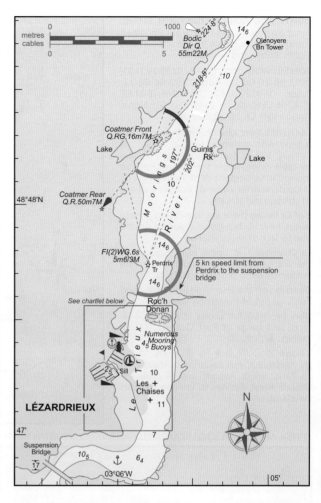

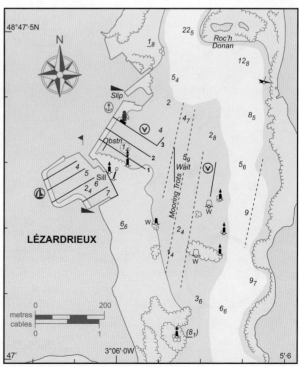

7.8 PONTRIEUX

Côtes d'Armor **48°42'·74N 03°08'·98W** ✹❀⚓⚓✿✿✿

CHARTS Navi 537; AC 3673 & SHOM 7126 end S of Lézardrieux. A useful river guide/map is available from HM Pontrieux.

TIDES HW at Pontrieux lock is also HW St Malo (Area 18).

SHELTER Complete.

NAVIGATION See Lézardrieux for the river up to it. Not before HW –3, proceed via suspension bridge (17m clearance at MHWS) for 6M upriver, keeping to high, rky bank on bends. Between river bends the best water is indicated by the alignment of reflective posts (1-2m high) near the bends. Allow 1½hrs for the passage, and aim to reach the lock at HW –1, to avoid sand coasters (sabliers) which use the lock occas at HW.

Lock (48°42'·78N 03°08'·31W) ☎02·96·95·60·70; opens HW –2¼ to +2¼, all year round, (full details/ times at www. port-de-pontrieux. fr/la-capitainerie/les-ecluses/).

LIGHTS AND MARKS River is unlit; few marks.

COMMUNICATIONS Auto 08·36·68·08·22; Lock and port VHF Ch 12, (lock hrs); ☎ link to HM Pontrieux; Public ☎ at Ch Roche Jagu.

FACILITIES Berth in 1·5-3m at quay (SE bank) about 1km WSW of lock. HM ☎ 02·96·95·36·28; www.port-de-pontrieux.fr; pontrieux. marinov@lyonnaise-des-eaux.fr 80+40 ❶ ⌣ €2·90 inc ⚓ ⛾ ⛽ wi-fi; ▢ ✕ ▢ ⬡.
Town ✕ ▢ ⛴ ⬛ ▢ ▢ ▢ @ ✉ Ⓑ ⚓ ⇌ Paimpol/Guingamp, ✈ Brest, Rennes, Dinard & Paris. Taxi 02·96·95·60·43.

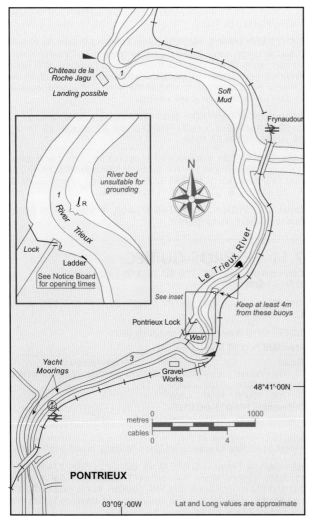

Château de la Roche Jagu

Landing possible

Soft Mud

Frynaudour

River bed unsuitable for grounding

River Trieux

Lock

Ladder

See Notice Board for opening times

See inset

Le Trieux River

Keep at least 4m from these buoys

Pontrieux Lock

Weir

Yacht Moorings

Gravel Works

48°41'·00N

metres
cables

0   1000

0   4

PONTRIEUX

03°09'·00W   Lat and Long values are approximate

7.9 TRÉGUIER

Côtes d'Armor **48°47'·21N 03°13'·27W** ✹❀❀⚓⚓⚓✿✿✿

CHARTS AC 2648, 2027, 2028; SHOM 7152, 7126; Navi 537; Imray C34

TIDES –0540 Dover; ML 5·7; Duration 0600

Standard Port ST MALO (⟵)

Times				Height (metres)			
High Water		Low Water		MHWS	MHWN	MLWN	MLWS
0100	0800	0300	0800	12·2	9·3	4·2	1·5
1300	2000	1500	2000				
Differences TRÉGUIER							
–0020	–0020	–0100	–0045	–2·3	–1·6	–0·6	–0·2
PORT-BÉNI (5ca SSE of La Corne lt twr)							
–0020	–0020	–0105	–0050	–2·4	–1·7	–0·6	–0·2

SHELTER Good. Possible ⚓s, keeping clear of fairway and fish farms. 7ca SW of La Corne lt tr, but exposed to N'lies; N and S of La Roche Jaune village; and in well sheltered pool (6m) 1ca NE of No 10 buoy (8ca N of marina).

NAVIGATION WPT 48°54'·13N 03°11'·50W (abeam Basse Crublent PHM buoy) 137°/2·1M to Pen ar Guézec SHM light buoy, a key turning point. Caution: strong tidal streams across the well-lit chan. The three approach channels, with marks, are:

- **Grande Passe** 137·3°. Ldg marks (hard to see by day): Front, Port de la Chaine, white ho; rear, St Antoine, RW ho. At Pen ar Guézec SHM buoy alter 216° in La Corne's W sector. A clearer transit from Basse Crublent is: Pleubian spire (charted) and adj water twr (on skyline) ≠ 154° to pass between Pierre à l'Anglais and Le Corbeau bys (both lit).

- **Passe de la Gaine** Unlit short cut to/from Lézardrieux. See chartlet below Île De Bréhat. Adequately marked, navigable by day with care/sufficient tide, good vis; but marks hard to see from afar in low sun/ poor vis. 241·5° ldg marks: Front, Men Noblance bcn twr, W with B band, ≠ rear mark (W wall with B vert stripe) below the skyline, right of conspic Plougrescant ✠ spire. Stay on transit to hold narrow chan. See Passage info.

- **Passe du Nord-Est** 205°/207°: unlit, dangerous with W-NW winds as sea breaks across the shallowest part of chan. On 205° keep W of La Jument NCM buoy and adjacent rky shoals; jink port onto 207° transit of Tréguier spire and Skeiviec bcn twr, for direct appr to La Corne.

In river, heed buoys and bns, eg keep E of Taureau SHM light buoy, 300m SW of La Corne lt ho, to clear adjacent drying mudbank. Speed limit 5kn south of No 3 SHM lt buoy. Less water reported from No 11 lt by to marina, approx 5M from La Corne lt hse.

LIGHTS AND MARKS See Lights, buoys and waypoints/chartlet. Important marks: La Corne, a stumpy WR lt tr; 6ca to the N is Men Noblance black/white bcn tr (for Passe de la Gaine); and 4ca SW is Skeiviec W bn tr. The spire of Tréguier cathedral is 4·6M SSW of La Corne, but obsc. by high wooded banks entering the estuary.

COMMUNICATIONS CROSS 02·96·89·31·31; Auto 08·92·68·08·22; Police 02·96·20·84·30; ⊖02·96·20·81·37; Brit Consul 01·44·51·31·00; Ⓗ 02·96·05·71·11; Dr 02·96·92·32·14; Aff Mar 02·96·92·30·38 (Paimpol); Marina VHF Ch 09, office hrs in season, 7/7: 0700-2145: Winter, M-F 0800-1200 & 1330-1730 (Sat 1630): Sun shut.

FACILITIES A 30m waiting pontoon, least depth 1·2m LWS, 500m N of marina, boats >15m LOA & multihulls can use overnight. **Marina** Arr/ dep at slack water, tide sets hard N/S diagonally through pontoons. Berth bows to tide on outer part of pontoons C, D, E. F pontoon is exclusively for vessels >14m, max LOA/Wt, 30m/100t, depth 3m LAT on outside, 2.8m inside. HM ☎02·96·92·42·37, mob 06·72·70·70·20 portplaisance.treguier@wanadoo.fr www.ville-treguier.fr/culture-sports-loisirs/port-plaisance.php 260 + 70❶ €2·80 > 20m, €57.00 + €3.00 per/m extra. ▢ ⬛ wi-fi @, ⬡ ⛾ ▢ ⚒ ✕ ⛴(40t) ⚓ ⬡ ☎02·96·92·15·15/93·55.

Le Petit Baigneur ☎02·96·11·40·69. **CN de Tréguier** ☎02·96·92·37·49, opposite marina, ▢ Sat evenings only. **Town** ✕ ▢ ⛴ (small s/ market near cathedral delivers) ▢ ▢ ✉ Ⓑ. Market Wed. ⇌ via bus to Paimpol, Guingamp, Lannion. ✈ Brest, Rennes, Dinard. **Ferry** Roscoff, St Malo.

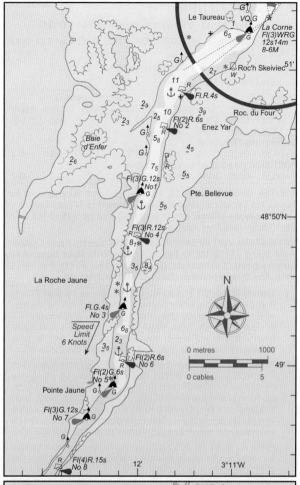

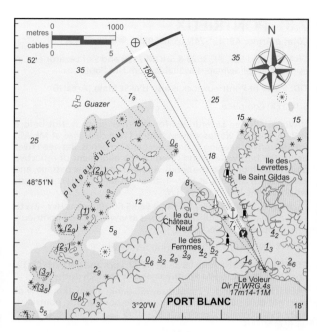

150° towards Le Voleur Dir lt; the rear ldg mark is hard to see amongst trees. From the SW approach via Basse Guazer PHM buoy, 7ca SW of the WPT. Beware ebb tide initially setting hard towards drying Plateau du Four.

LIGHTS AND MARKS 150° ldg marks: Front, Le Voleur Dir lt, low down amongst trees (between white cottage and Grand Hotel, conspic white block); rear, 5ca from front, La Comtesse Mill, unlit and obscured by trees on skyline. A white obelisk (16m) on Île du Château Neuf is conspic to stbd of the fairway. A smaller obelisk to port on Île St Gildas is less obvious.

COMMUNICATIONS CROSS 02·33·52·72·13; Auto 08·92·68·02·22; ⊖ 02·96·20·81·87; Brit Consul 01·44·51·31·00; *Port Blanc* VHF Ch 09, 16; Aff Mar 02·96·92·30·38;.

FACILITIES 5 W 🅐s, marked VISIT in approx 5m. Safe ⚓ and good holding, where charted in 7m. Or dry out alongside quays, 1·3m; best to pre-recce. HM ☎02·96·92·89·11 in sailing school. ⌂ or 🅐 €7·00 all LOA, ⚓ 🏠 🏠 🄰 🄶 🔧 ⚒ 🅗(16t) 🛢 .

Town ✕ 🏠 🍴(basic). Taxi 02·96·92·64·00, 🚂 Lannion. ✈ Brest, Rennes, Dinard. **Ferry** Roscoff.

7.11 PERROS-GUIREC
Côtes d'Armor **48° 48'·17N 03° 26'·21W** (Jetty hd)
❀❀🅐💧💧💧🅐🅐❀❀

CHARTS AC 2648, 2027; SHOM 7152, 7125; Navi 537, 538; Imray C34

TIDES –0550 Dover; ML 5·4; Duration 0605

Standard Port ST MALO (⬅)

Times				Height (metres)			
High Water		Low Water		MHWS	MHWN	MLWN	MLWS
0100	0800	0300	0800	12·2	9·3	4·2	1·5
1300	2000	1500	2000				
Differences PERROS-GUIREC							
–0035	–0035	–0115	–0100	–2·8	–2·0	–0·7	–0·1

SHELTER Very good in marina. Off Pointe du Chateau safe ⚓, except in NE'lies, in approx 3m good holding; plus 5 small W 🅐s.

NAVIGATION From the E, WPT 48°49'·86N 03°23'·66W, 224·8°/2·4M to Jetée Linkin. From the W, WPT 48°50'·53N 03°26'·45W, 143·5°/1·8M to join 224·8° ldg line. Caution: Rocks extend 7ca W and 6ca NE of Île Tomé. Local moorings E of Jetée du Linkin are marked by 2 SHM buoys.

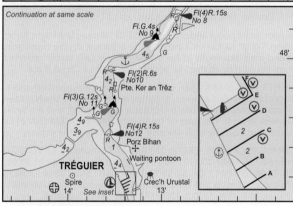

7.10 PORT BLANC
Côtes d'Armor **48°50'·54N 03°18'·89W** ❀❀🅐💧💧💧❀❀

CHARTS AC 2648, 2027; SHOM 7152, 7125/6; Navi 537, 538; Imray C34

TIDES –0545 Dover; ML 5·3; Duration 0600. Interpolate between Les Héaux de Bréhat and Perros-Guirec. HW–0040 and ht –2·0m on St Malo. MHWS 9·0m; MHWN 6·8m.

SHELTER Good in natural hbr (aka Port Bago) but very uncomfortable in winds with any N in them.

NAVIGATION WPT 48°52'·24N 03°20'·16W, 150°/1·9M to first 🅐. The fairway does not open up until at the WPT. Appr on ldg line

LIGHTS AND MARKS 225° ldg marks are hard to see by day, but bcns/buoys in the appr are adequate. Passe de l'Ouest: Kerjean Dir lt 143·6° also hard to see, but old lt ho (gable end) on the foreshore is clearer.

COMMUNICATIONS CROSS 02·98·89·31·31; LB 02·96·91·40·10; Auto 08·92·68·02·22; Police 02·96·23·20·17; 02·96·20·81·87; Brit Consul 01·44·51·31·00; Dr 02·96·23·20·01; Entry gate 02·96·23·19·03; Aff Mar 02·96·91·21·28; Marina VHF Ch 09, 16.

FACILITIES Access: 6m wide gate (no lock) opens when rise of tide reaches 7m. Sill under gate dries 3·5m, giving 3·5m water inside gateway on first opening, less water reported near fuel pontoon. Caution: strong current initially. Gate sigs: IPTS. No ⚓ in basin. R & W poles mark retaining wall; do not try to cross it near HW.

Gate opens depending on tidal coefficient (see St Malo coefficients) may open > 30 mins earlier and close slightly later:

Coeff <40	=	Shut, 4 days max
40-50	=	HW –½ to HW
50-60	=	HW –1 to +½
60-70	=	HW ±1
>70	=	approx HW ±1½

Marina ⚓ ☎02·96·49·80·50. www.port-perros-guirec.com portdeplaisance@perros-guirec.com 720+80📶 €3·30. 📶 berth on N'most 2 pontoons in 2·5m. 🛒🚿🚽📮 🔌 wi-fi, ⛽ Ⓔ SHOM △ ⚒🔧 🏗(7t). **YC** ☎02·96·91·12·65.

Town ✕🏠🛒✉Ⓑ. 🚂 Lannion. ✈ Brest. **Ferry** Roscoff.

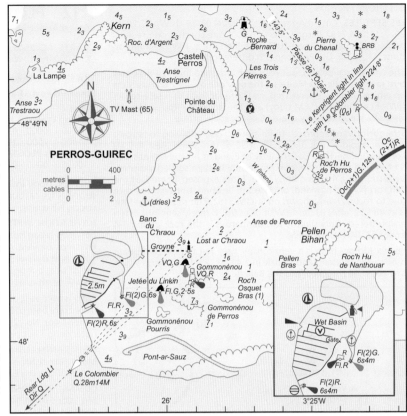

LES SEPT ÎLES *See notes on following page*

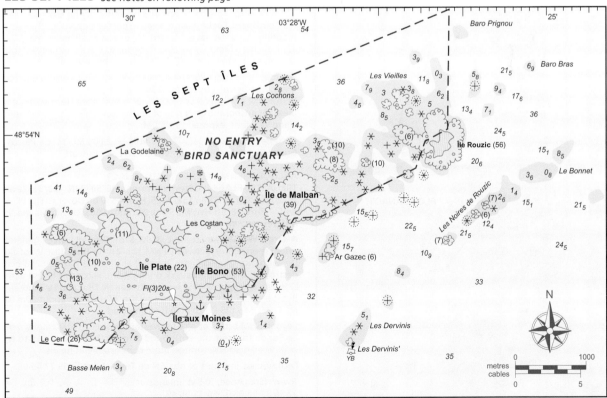

OFFSHORE ISLANDS

LES SEPT ÎLES, Côte d'Armor, **48° 52´·72N 03°29´·16W**. AC 2026, 2027; SHOM 7152 (large scale 1:20,000). See chartlet on previous page. HW–0550 on Dover (UT); +0005 on Brest. HW ht –1·8m on Brest. ML 5·2m. Use Ploumanac'h differences.

The whole archipelago, outlined with a dashed line, is a no-entry bird sanctuary. Île aux Moines is the only one of the 7 islands/islets on which landing is permitted; no shore facilities. It is safe to sail NW of Les Dervinis (SCM buoy) and Les Noires de Rouzic (always visible) a steep-to rocky chain SE of the main group.

Straightforward approach to Île aux Moines, allowing for strong stream. Avoid a rock (0·6m) 300m SE of the island. Best ⚓ (Lat/Long as on line 1) is due E of jetty at E end of Île aux Moines and S of W end of Île Bono; overnight only in calm settled weather. Vedettes use a mooring buoy here. More exposed ⚓s are S of the Old Fort (W end of island) or close S of Île Bono; both require care due to rocks, clearly shown on SHOM 7152. The most dangerous is at 48° 52´·45N 03°28´·68W; it dries 0·1, is rarely seen but often encountered. Only slightly less dangerous is the rock drying 1·4 at 48° 52´·64N 03°28´·28W.

Île aux Moines lt ho is a very conspic grey twr whose 24M beam is obscured by Bono and Rouzic in a 4° arc (237°-241°) and when brg <039°, ie if tucked into the Baie de Lannion.

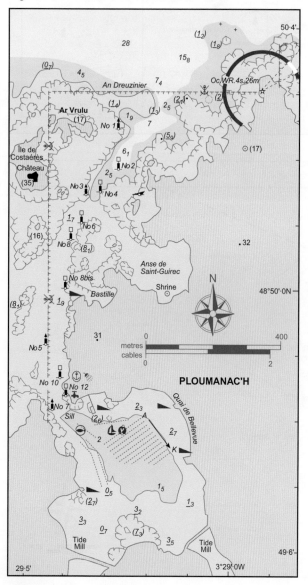

7.12 PLOUMANAC'H

Côtes d'Armor 48° 49'·78N 03° 29'·51W (Sill) 🌸🌸⚓⚓🌸🌸🌸

CHARTS AC 2648, 2027 (limited scale 1:48,700); SHOM 7125 (1:10,000 inset); Navi 537, 538; Imray C34

TIDES –0550 Dover; ML 5·5; Duration 0605

Standard Port ST MALO (←)

Times				Height (metres)			
High Water		Low Water		MHWS	MHWN	MLWN	MLWS
0100	0800	0300	0800	12·2	9·3	4·2	1·5
1300	2000	1500	2000				
Differences PLOUMANAC'H							
–0035	–0040	–0120	–0100	–2·9	–2·0	–0·7	–0·2

SHELTER Good inside. Appr is uncomfortable against the ebb in strong N'lys, favour the PHMs; dangerous during spring ebb with strong NE winds.

NAVIGATION WPT 48°51´·44N 03°29´·09W, 188°/1·2M to ent between Mean Ruz lt ho and Château Costaérès (conspic). At the entrance beware An Dreuzinier, unmarked rocks (drying 1·4m) 100m N and NE of No 1 SHM perch; not a problem with sufficient rise of tide, but near LW stick to a line 225° through the Nos 2 and 4 PHM perches. 11 unlit perches mark the chan.

Entry is over a sill, drying 2·55m, which retains 1·2m to 2·3m within. If concrete base of No 5 perch (the 3rd SHM) is covered, depth over sill is >1·4m. No 12 PHM perch right next to sill has depth gauge; note: gauge not visible when >3m over sill.

LIGHTS AND MARKS See chartlet and Lights, buoys and waypoints. Mean Ruz lt ho is a reddish square twr. Signal stn, 8ca SSE, is conspic from the E. Night entry not advised.

COMMUNICATIONS ⓁⒷ 02·96·20·00·45; Auto 08·92·68·08·22; Brit Consul 01·44·51·31·00; Dr 02·96·13·80·73; HM, Ch 09, 02·96·91·44·31.

FACILITIES Between Nos 8 and 8bis perches on the E side of chan are 2 waiting buoys marked VPG which dry at springs.

Port de Plaisance Inside the sill, are moorings for FVs and yachts >12m in a trot to stbd of ent in 3m, with 2 drying trots of small craft outboard. For best water keep to port and approach the yacht moorings from NW. ⚓s are 1st trot (A) of dumbbell buoys; max LOA 12m. Yachts >12m may be able to tie up to the fishing vessel trots perpendicular to the sill. SE & SW sides of hbr are very shallow.

HM (part time) ☎02·96·91·44·31 or 02·96·49·80·50 (Perros-Guirec). www.port-perros-guirec.com 230 + 20 ⚓s, €3·30. The full time Perros HM also oversees Ploumanac'h and Anse de Trestraou, a pleasant anchorage on N side of Perros town, well sheltered from S and W winds; exposed to NE'lies.

Quai Bellevue ⚓ ⚓ 🚿 ⛽. ⚓ ☎02·96·23·05·89. **YC de Perros-Guirec** Bus to Lannion and Perros. **Ferry** see Roscoff.

ANCHORAGE 1M WEST OF PLOUMANAC'H

TRÉGASTEL-PLAGE, Côtes d'Armor, **48°50´·04N 03°31´·29W**, AC 2026, 2027; SHOM 7152, 7125. HW –0550 on Dover (UT); +0005 and –1·8m on Brest HW; ML 5·1m; Duration 0605. Use Ploumanac'h differences.

A large W radome, conspic 3·2M S of the ent, gives general orientation. Enter between PHM bcn on Île Dhu (just W of Île Dé, with dice-shaped Pierre Pendue) and SHM bcn (or buoy if bcn destroyed) off Le Taureau, rk drying 4·5m. A conspic house with ☐ turret, brg approx 165° leads between Île Dhu and Le Taureau. But it is easier to track 183° leaving the 3 PHM bcns about 60m to port. SHOM 7125 (1: 20,000) is strongly advised.

Beyond SHM bcn, ⚓ or pick up westernmost ⚓s €10·00/night, S of Île Ronde (nearest the ent). Good ⚓ in 2m, but exposed W to N, with swell near HW springs; more comfortable at neaps.

Facilities: ⚓ in E part of hbr. **CN de Trégastel** ☎02·96·23·45·05.

Town (Ste Anne, 0·5M inland) ✕ ☐ 🛒 🏪 ⛽ 🔧 ✉ Ⓑ; Dr 02·96·23·88·08/02·96·15·96·49.

7.13 TRÉBEURDEN

Côtes d'Armor **48°46´·29N 03°35´·15W** 🌊❄️♨️♨️♨️🌸🌸

TIDES –0605 Dover; ML 5·5; Duration 0605

Standard Port BREST (→)

Times				Height (metres)			
High Water		Low Water		MHWS	MHWN	MLWN	MLWS
0000	0600	0000	0600	6·9	5·4	2·6	1·0
1200	1800	1200	1800				
Differences TRÉBEURDEN							
+0100	+0110	+0120	+0100	+2·3	+1·9	+0·9	+0·4

CHARTS AC 2026; SHOM 7151/2, 7125, 7124; Navi 537, 538; Imray C34

SHELTER Good in marina (2–2·6m). ⚓ off NE side of Île Milliau is exposed to W'lies.

NAVIGATION WPT 48°45´·34N 03°40´·00W, 067°/3M to first PHM buoy in white sector of Lan Kerellec Dir lt. From WSW, go direct to ⊕; thence enter the buoyed chan (105°/088°) to marina.

From E and N, round Bar-ar-Gall and Le Crapaud WCM buoys; continue S for 1·4M, before altering 067° towards the ⊕. If properly planned, the short-cut 183° between Le Crapaud reef and reefs N and S of Île Losket is safe in offshore winds.

LIGHTS AND MARKS See chartlet and Lights, buoys and waypoints. Lan Kerellec Dir lt leads 067° to the start of buoyed chan, aligned approx 100° with ✠ spire (conspic, 115m).

COMMUNICATIONS CROSS 02·98·89·31·31; Police 02·96·23·51·96 (Jul/Aug); ⊖ 02·96·48·45·32; Brit Consul 01·44·51·31·00; *Port Trébeurden* VHF Ch 09, 16; Aff Mar 02·96·91·21·28.

FACILITIES 10 Y waiting buoys lie S of the fairway and some 600m W of marina ent. Access times based on coefficients are:

95 (sp)	HW –3½ to +4
70	HW –3¾ to +4¼
45 (np)	HW –4¾ to +5¼

An automatic flap (2·5m) above the sill starts to open when ht of tide is 4·0m and is fully open at 4·3m (conversely on the ebb). Best to calculate tidal heights.

IPTS (sigs 2 and 4) on bkwtr and at gate show when access is possible. Do not ent/exit within 20 mins of the flap moving, due to fast, turbulent water in the entrance.

Enter between gate posts in retaining wall, which dries 3·5m and is marked by 4 Y SPM bns. A floodlit tide gauge is port side of the 15m wide gate.

Marina ☎02·96·23·64·00. portrebeurden@ wanadoo.fr www.port-trebeurden.com 625 + 70 Ⓥ on pontoon G (NE side <11m, SW side >11m) July/Aug €3·70, inc ⚡ @, ⛽€1·00. ⚓🛢️🚿🔥⊕✗🏗️(16t mobile) **YC** ☎02·96·15·45·97 (July/Aug).

Town ✗ ⬚ 🍴 ✉ Ⓑ. ✈ Brest, Rennes, Dinard. **Ferry** Roscoff or St Malo.

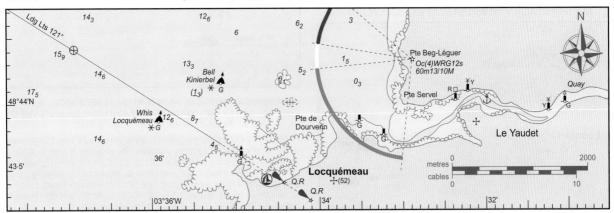

7.14 LÉGUER RIVER (Lannion)

Côtes d'Armor **48°43´·76N 03°33´·28W** (2nd G bcn) 🌊❄️♨️🌸🌸🌸

CHARTS AC 2648, 2026; SHOM 7124; Navi 537, 538; Imray C34

TIDES –0605 Dover; ML 5·4; Duration: no data

N Brittany (side tab)

Standard Port BREST (→)

Times				Height (metres)			
High Water		Low Water		MHWS	MHWN	MLWN	MLWS
0000	0600	0000	0600	6·9	5·4	2·6	1·0
1200	1800	1200	1800				
Differences LOCQUIREC							
+0100	+0110	+0125	+0100	+2·3	+1·9	+0·9	+0·4

SHELTER Good, except in strong W/NW winds. No access at LW sp, esp in strong NW'lies when seas break on the drying bar. ⚓ in non-drying pools off Le Yaudet and up to the 2nd Y perch, but very little space due to silting and many local moorings; a vacant mooring is possible or ⚓ outside and recce by dinghy.

NAVIGATION WPT 48°44'·38N 03°36'·87W, 091°/2·6M to Pte Beg-Léguer lt (also on Locquémeau ldg lts 121°). Leave Kinierbel SHM buoy close to stbd; stand on until Trébeurden spire bears 004° (to clear a drying patch) then alter 135° for the ent, passing close to two G bcn twrs. Le Yaudet pool is 8ca further E.

Chan upriver to Lannion is narrow, steep-to, dries and is unmarked by bcns; best seen at LW before attempting. Just below Lannion town, opp Quai de Loguivy, drying out on the N bank may be feasible; a recce by dinghy or on foot is advised.

LIGHTS AND MARKS See chartlet and Lights, buoys and waypoints. Pte de Beg-Léguer lt ho is a squat twr, R top, built onto a house and on high cliffs. Le Yaudet spire helps to pinpoint the river ent.

COMMUNICATIONS CROSS 02·98·89·31·31; Auto 08·92·68·08·22; Police 02·96·46·69·50; ⊖02·96·48·45·32; Brit Consul 01·44·51·31·00; Dr 02·96·47·23·15; Aff Mar 02·96·55·35·00.

FACILITIES HM ☎02·96·47·29·64, ⎯ (dries) ⚓ ⌂(1t) ⬥ ⌾ only in emergency. ✕ ⌾ in Le Yaudet, but few supplies.

Lannion, ✕ ⛽ 🛒 🔋 🔌 ⚒ 🚰 ✗ SHOM Ⓔ ✉ Ⓑ.

ADJACENT HARBOURS IN BAIE DE LANNION

LOCQUÉMEAU, Côtes d'Armor, 48°43'·60N 03°34'·96W (1st SHM perch). AC 2648, 3669; SHOM 7124. HW –0600 on Dover (UT); +0110 on Brest; HW ht +1·5m on Brest; ML 5·3m. SHOM 7124 is desirable as it shows the various perches. A small drying hbr by ent to Lannion River (above); use the same WPT. Approach leaving Locquémeau SHM buoy close to stbd. There are two quays: the outer is accessible at LW, but open to W winds. Yachts can dry out at inner quay, on S side. Ldg lts 121° to outer quay: front, W pylon + R top; rear, hard-to-see W gable and R gallery. Facilities: ⚒ 🚰 ✗. **Town** 🛒.

LOCQUIREC, Finistère, 48°41'·41N 03°38'·79W. AC 3669, 2648; SHOM 7124. Tidal data see Lannion. Small drying hbr at mouth of Le Douron R, with good shelter from W'lies. The whole estuary dries to extensive, shifting sandbanks (2·9m to 6·1m). No lights, but Gouliat NCM buoy is 7ca N of Pte du Château and a SHM bcn twr (6ca S of the hbr, marking Roche Rouge) is conspic. E of Pte du Château, off Île Vert, there are 25 W moorings +2 ⚓, €6·50p/n, in up to 4m. Beyond, Locquirec harbour and bay has 304 W drying moorings, + 5 ⚓, also €6·50, pleasant tempy ⚓, overnight if able to take the ground, or dry ⌾ on stone quay, ⚓ 🔧. Land at slip. A small, laid-back resort, inc some fine trad. hotel-restaurants and Ⓒ. Mairie (Port Authority) ☎02·98·67·42·20. Good LW cockling.

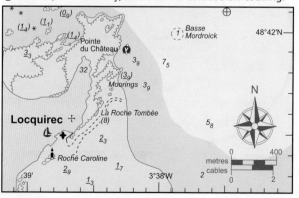

7.15 PRIMEL-TRÉGASTEL

Finistère 48°42'·79N 03°49'·46W ❀❀🐚🐚🌸🌸

CHARTS AC 2026, 2745; SHOM 7151, 7124, 7095; Navi 538; Imray C34, C35

TIDES See Morlaix for differences ANSE DE PRIMEL. –0610 Dover; ML 5·3; Duration 0600.

SHELTER Access at all tides, but exposed to N/NW'lies; if strong, seas break across ent. Drying upper reaches are well sheltered.

NAVIGATION WPT 48°43'·59N 03°50'·10W, 152°/0·9M to the hbr bkwtr. Beware drying rks NE of Pte de Primel. Enter exactly on 152° ldg marks, past a SHM bcn marking Ar Zammeguez (3m high). The ent, between a PHM and 2nd SHM bcns, is narrow. A useful hbr to await the tide up to Morlaix.

LIGHTS AND MARKS Pte de Primel is a conspic 42m high rky islet NNE of the hbr. 152° ldg marks/lights are clear but precise: three white walls with R vert stripe in serried ranks up the hillside surrounded by small pine trees; ldg lts, both QR, are on the front and rear marks.

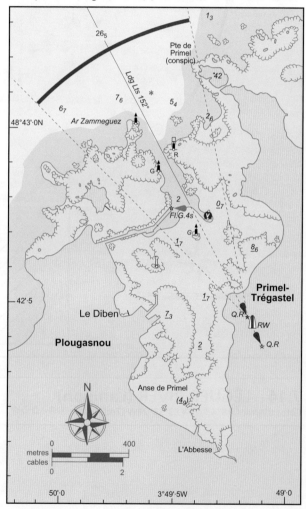

COMMUNICATIONS 🛟 06·14·38·26·81; Dr 02·98·67·30·20; Brit Consul 01·44·51·31·00; HM 06·81·79·67·49 (English speaker); VHF Ch 09, (in season); Rolland Marine B'yard 02·98·72·30·06.

FACILITIES Le Diben (Plougasnou) on W side is a FV hbr. FVs occupy most of the bkwtr; rafting possible. In summer a line is strung along the quay for yachts to tie on to, the E end is non-drying; not safe in N'lies. A small ⚓age and 10 W ⚓, in about 1m (drying), and 6 W ⚓, with R tops, non-drying, are SE of the bkwtr head. Or ⚓ to SE or SW of bkwtr hd in 2·9m. ⎯ 🔋 🔌 ⚒ ✗ Ⓔ ⌂(45t). **Primel-Trégastel** 🏨 🏨 ✕ 🛒 🛒 🛒 🔋 ✉ Ⓑ. **Ferry** Roscoff.

7.16 MORLAIX

Finistère **48°35′·31N 03°50′·30W** (Lock) 🌼🌼♢♢♢🌸🌸🌸

CHARTS AC 2026, 2745; SHOM 7151, 7095; Navi 538; Imray C34, C35

TIDES −0610 Dover; ML 5·3; Duration 0610

Standard Port BREST (→)

Times				Height (metres)			
High Water		Low Water		MHWS	MHWN	MLWN	MLWS
0000	0600	0000	0600	6·9	5·4	2·6	1·0
1200	1800	1200	1800				
Differences MORLAIX (CHÂTEAU DU TAUREAU)							
+0055	+0105	+0115	+0055	+2·0	+1·7	+0·8	+0·3
ANSE DE PRIMEL							
+0100	+0110	+0120	+0100	+2·1	+1·7	+0·8	+0·3

SHELTER Good in the bay, except in onshore winds. Strong N'lies can raise a steep sea even in the estuary. ⚓ NE of Carantec is exposed, especially to NW; landing stage dries about 4·6m. Complete shelter in the marina.

NAVIGATION The three appr channels have rocky dangers and require careful pilotage; they converge near No 1 buoy, Barre de Flot. La Lande is the rear ldg lt/mark for the first two channels.

- **Chenal de Tréguier** is best at night, but almost dries: WPT 48°42′·55N 03°51′·93W (abm La Pierre Noire SHM bcn) 190·5°/1·84M to La Chambre bcn twr. Ldg lts 190·5°: Front mark, Île Noire lt bcn.

- **Grand Chenal** is shallower but lit: WPT 48°42′·65N 03°53′·53W, (abm Stolvezen PHM buoy) 176°/2·23M to Île Louet lt, passing E of Ile Ricard. Ldg lts 176·4°: Front mark, Île Louet lt ho.

- **Chenal Ouest de Ricard** is the deepest, but unlit. It deviates 188·8° from Grand Chenal, passing W of Ile Ricard. 188·8° ldg marks: Pierres de Carantec and Kergrist, both painted white.

The Morlaix River is buoyed/bcn'd but unlit and prone to silting. ⚓ off Pen al Lann and Dourduff (dries) clear of extensive oyster beds marked by small orange buoys/stakes.

Morlaix lock, marina and town are 5·5M upriver from No 1 SHM buoy.

LIGHTS AND MARKS See Lights, buoys and waypoints and chartlet. Château du Taureau fort is conspic 4ca N of No 1 buoy. Some marks omitted for clarity.

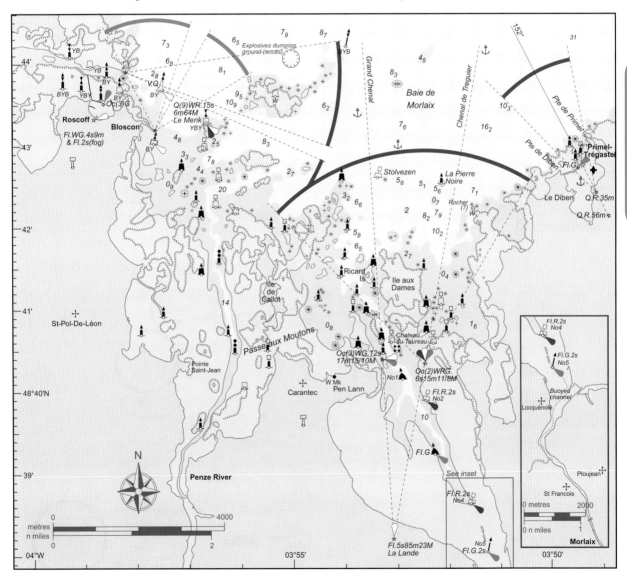

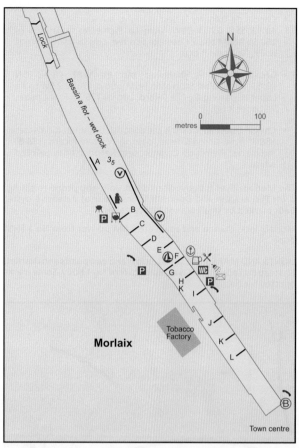

Morlaix

Town centre

COMMUNICATIONS CROSS 02·98·89·31·31; ℹ️02·98·72·35·10; Auto 08·92·68·02·29; Police 02·98·88·17·17; 🖥️ 02·98·69·70·84; British Consul 01·44·51·31·00; Ⓗ 02·98·62·62·45; Port and Marina VHF Ch 09; Aff Mar 02·98·62·10·47.

FACILITIES Morlaix lock ☎02·98·88·15·10/06·77·50·15·90 opens by day only (SR-SS) at HW –1½, HW and HW+1; wait alongside W quay.

Marina ☎02·98·62·13·14 www.plaisancebaiedemorlaix.com plaisance.morlaix@morlaix.cci.fr 150+32🅥 Jun-Sep €2·50; multihulls x150%. 🅥 pontoon (fingers) on E bank, parallel to road. In lock hours a movable footbridge across the marina is usually open. 🛢️🛠️🔧🚽✗🏗️(8t)⚓.

Town ✗🏠🍴🔲🏤🏧🛢️🏔️Ⓔ SHOM ✉️ Ⓑ. ✈ Brest, ferry Roscoff.

ANCHORAGES W AND NW OF MORLAIX

PENZÉ RIVER, Finistère, Ent **48°42´·24N 03°56´·66W**, AC 2745; SHOM 7095. HW –0610 on Dover (UT) +0105 on Brest; HW ht +1·2m Brest; ML 5·0m; Duration 0605.

The Penzé river lies W of Île Callot. From NNW, appr between Cordonnier and Guerhéon bcn twrs at mid-flood. Or, for deeper water, from the ENE pass between Les Bizeyer reef and Le Paradis bcn twr; thence S via beaconed channel.

Passe aux Moutons, between Carantec and Île Callot, is a drying (6·2m) short cut from the E into the Penzé, with adequate rise.

S of Le Figuier IDM bcn the chan narrows and is scantily marked by small oyster withies, then by mooring trots. Best shelter is SW of Pte de Lingos: moor in 1·2m off St Yves where the old ferry slips provide landing places.

S of Pont de la Corde (10m clearance) the river is buoyed and navigable on the tide for 3M to Penzé. No access 1 Mar–31 Aug to a Nature Reserve at the head of the river.

Facilities at Carantec 🛢️🛢️🛠️🏗️🔧✗🏔️Ⓔ. **Town** ✗🏠🍴✉️Ⓑ.
Penzé 🛥️ (drying) limited ✗🏠🍴.

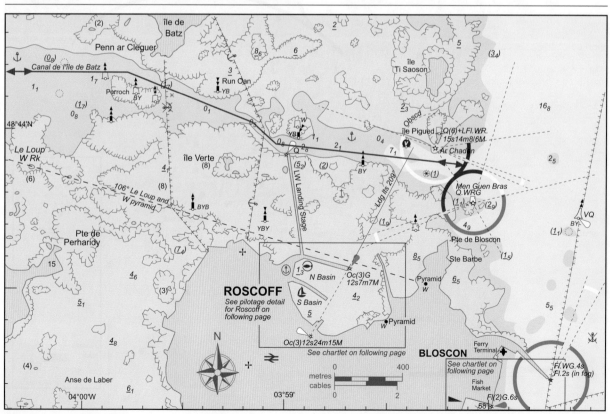

7.17 ROSCOFF

Finistère **48°43'·54N 03°58'·59W** ❄❄🌊🌊🌸🌸🌸

CHARTS AC 2026, 2745; SHOM 7151, 7095; Navi 538; Imray C35

TIDES –0605 Dover; ML 5·2; Duration 0600

Standard Port BREST (→)

Times				Height (metres)			
High Water		Low Water		MHWS	MHWN	MLWN	MLWS
0000	0600	0000	0600	6·9	5·4	2·6	1·0
1200	1800	1200	1800				
Differences ROSCOFF							
+0055	+0105	+0115	+0055	+2·0	+1·7	+0·8	+0·3
ÎLE DE BATZ							
+0045	+0100	+0105	+0055	+2·0	+1·6	+0·9	+0·4
BRIGNOGAN (48°40'·6N 04°19'·2W)							
+0040	+0045	+0100	+0040	+1·5	+1·2	+0·6	+0·2

SHELTER Good in drying Vieux Port except in strong N/E'lies. 12 W 🛟s, €8·50 all LOA, in 4-5m close W of Ar Chaden are exposed. Shelter very good in large modern marina.

NAVIGATION Roscoff Marina (S of ferry port): Appr H24 on 205° in W sector of mole hd lt. Avoid ferries; do not attempt to enter marina from south. Beware strong tide on all outer berths and close to E bkwtr, also long seaweed fronds if berthing stern to tide. **Vieux Port**: From the N, WPT 48°45'·88N 03°56'·00W, 213°/2·5M to Men Guen Bras lt. From the E approach between Plateau de la Méloine and Plateau des Duons; or pass S of the latter and N of Le Menk bcn lt twr in Ar Chaden's W sector (289·5°-293°). From the W, WPT 48°44'·27N 04°04'·06W, 090°/1M to Basse Plate bcn twr; thence via well-marked Canal de l'Île de Batz; see Passage information. Approach near HW over many large drying rks on 209° ldg line: see Lights, buoys and waypoints. N mole hd has B/W vert stripes.

LIGHTS AND MARKS See chartlet and Lights, buoys and waypoints. Île de Batz lt ho is conspic through 360°. E ent to Canal de l'Île de Batz flanked by bcn twrs: Ar Chaden YB and Men-Guen-Bras BY.

COMMUNICATIONS CROSS 02·98·89·31·31; 🆔 06·82·18·01·34; Auto 08·92·68·08·29; Police 02·98·69·00·48; ⊖ (Bloscon) 02·98·69·70·84; ⊖ (Roscoff) 02·98·69·70·15; Brit Consul 01·44·51·31·00; Ⓗ 02·98·88·40·22; Dr 02·98·69·71·18; Roscoff HM 02·98·69·76·37, mob 06·70·50·96·68, VHF Ch 09; *Bloscon* VHF Ch 09, (0700-2200) all vessels must call for authorisation to enter marina, only authorised vessels to proceed on IPTS GWG vert, no vessels to leave, IPTS 3R vert. Aff Mar 02·98·69·70·15.

FACILITIES

Vieux Port, S Basin 5m. Dry out with fender board against rough inner jetty or quay. 280+20🅥 €0·73. 🔩🍽🔧 🗄🍴. N Basin for FVs, 🏗(5t). CN de Roscoff ☎02·98·69·72·79, 🚻.

Bloscon ferry terminal. HM ☎02·98·61·27·84, www.roscoff.fr. Adjacent ⚓age may be used only with HM's approval.

Roscoff Marina ⚓ ☎02·98·79·79·49, www.plaisancebaiedemorlaix. com plaisance.roscoff@morlaix.cci.fr 625 + 40🅥Jun-Sep, €2·80, on S side pontoon B, N side pontoon D. Convenient lift from pontoons to quayside level. 🛥🛢 (H24) ✕🍽🛒🗄⛽🔩🍴(50t)🏗(2t). Shuttle bus to town, June–Aug, 0700–1700, or 10 min walk.

Roscoff Town ✕🍽🛒🔩🛒📮Ⓑ🔧🗄🍴. ⇌✈ Brest, Morlaix. Ferries to Plymouth, Rosslare, Cork.

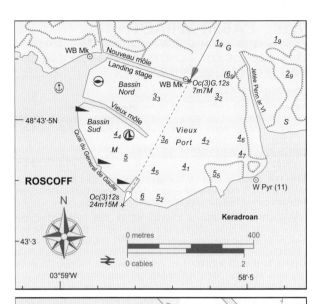

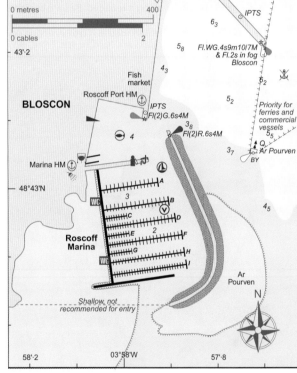

OTHER HARBOURS AND ANCHORAGES WEST OF ROSCOFF

ÎLE DE BATZ, Finistère, **48°44'·26N 04°00'·60W** (Porz-Kernok ent). AC 2026, 2745; SHOM 7151, 7095; HW +0610 Dover (UT); ML 5·2m. See Roscoff. Good shelter but hbr dries up to 5m; ⚓ where space permits. E landing stage is for ferries only. Charted ⚓ in the chan E of hbr ent, but poor holding and may be exposed. ⚓ prohib between SE tip of Batz and Roscoff LW landing jetty, due to cables. Île de Batz lt ho (gy twr, B top) the CG Stn and church spire are all conspic. Facilities: ✕🍽, basic shops, bike hire.

MOGUÉRIEC, Finistère, **48°41'·35N 04°04'·41W**, AC 2648, 2026; SHOM 7151. Tides approx as Île de Batz. A small drying fishing harbour, 3M SSW of W entrance to Canal de l'Île de Batz, open to NW swell. Ldg lts 162°, both W towers/G tops: Front on jetty, Iso WG 4s 9m, W158°-166°; rear Iso G 4s 22m. Beware Méan Névez rk,

dries 3·3m, to W of approach. ⚓ close SW of Île de Siec, or 3ca WSW of Ar Skeul WCM bn tr in 4m, or 2ca NNE of Moguériec's drying jetty. 🛟 on jetty, but clear of FVs. 🛟 also possible against jetty at Île de Siec's drying hbr, but rky bottom. Facilities: ✕🍽🛒.

BRIGNOGAN (Pontusval), Finistère, **48°40'·59N 04°19'·17W**, about 10M ENE of Île Vierge lt ho. AC 2025; SHOM 7150. HW +0605 on Dover (UT); ML 4·7m; Duration 0600; see Roscoff.
Pontusval is the port, Brignogan a small resort. App from ECM buoy (48°41'·42N 04°19'·33W) via 178° ldg marks, W bn on with Plounéour ch spire 1M S. Ent between Ar Neudenn R bn tr to E and 3 white-topped rks to W. ⚓ here in approx 4m or dry out closer in. Hbr is open to N winds and often full of FVs. Entry at night prohib as it is unlit. Pte de Beg Pol, Oc (3) WR 12s, 1M to the west. Facilities: ⚓✕🍽🛒 5 🛟's (free).

7.18 L'ABER WRAC'H

Finistère **48°35′·97N 04°33′·62W** (marina ent) ❄❄⚓🕭🕭❀❀❀

CHARTS AC 2025, 1432; SHOM 7150, 7094; Navi 539; Imray C35

TIDES +0547 Dover; ML 4·5; Duration 0600

Standard Port BREST (→)

Times				Height (metres)			
High Water		Low Water		MHWS	MHWN	MLWN	MLWS
0000	0600	0000	0600	6·9	5·4	2·6	1·0
1200	1800	1200	1800				
Differences L'ABER WRAC'H, ÎLE CÉZON							
+0030	+0030	+0040	+0035	+0·8	+0·7	+0·2	0·0

SHELTER Good in marina, but ⚓s uncomfortable in strong NW wind. No ⚓ off the marina. Paluden, 1·5M upriver (unlit/unbuoyed; beware oyster beds) is well sheltered in all winds.

NAVIGATION WPT (Grand Chenal) 48°37′·35N 04°38′·51W,abm Libenter WCM buoy, 100°/1·5M to Petit Pot de Beurre ECM bcn twr. Here pick up the inner appr chan 128°/1·7M, in W sector of Dir Lt Oc (2) WRG 6s, passing a SHM lt buoy, SHM bcn lt twr and a PHM lt buoy, to the last SHM bcn twr; thence visual for 700m to the pool/pontoon at L'Aber Wrac'h. Beware: lge unlit mooring buoys.

Chenal de la Malouine is a narrow short cut from/to N & E, only by day and in good weather: 176·2° transit of Petit Pot de Beurre with Petite Île W obelisk; precise tracking is required. Beware breakers and cross-tides.

Chenal de la Pendante 135·7° is rarely used as it crosses drying rks, saves few miles and the unlit marks are hard to see. Front, B/W

'smiley' on Île Cézon; rear, W obelisk/orange top, 650m S of hbr. **Caution**: E side of marina ent, rocks drying 0·8m marked by 4 Y SPM buoys outside the E wavebreak and 5 inside where a W cardinal bn and 2 PHM perches also define this No Entry area. These bys/marks are unlit. Speed limits 5kn in chan, 3kn in marina.

LIGHTS AND MARKS See chartlets and Lights, buoys and waypoints. Grand Chenal 100·1° ldg marks/lts: front on Île Wrac'h, W ☐ twr orange top, QR; rear, 1·63M E at Lanvaon, W ☐ twr orange △ top, Q. A Dir lt Oc (2) WRG 6s 128° and two elusive ldg twrs, both W with R tops, mark the inner chan. QG and QR lts mark the marina ent.

COMMUNICATIONS CROSS 02·98·89·31·31; Auto 08·92·68·02·29; Police 02·98·04·00·18 or just 17; ⊜ 02·98·85·07·40; Brit Consul 01·44·51·31·00; Ⓗ 02·98·22·33·33; Dr 02·98·04·91·87; HM 02·98·04·91·62, VHF Ch 09, 16 (0800-2000LT in season); Aff Mar 02·98·04·90·13.

FACILITIES

Marina 70 Ⓥ ⌇ inside W wavebreak (or outside in good weather and inside E wavebreak (tight space, prepare warps/fenders prior to entry) outer E side for trip boats to Île Vierge and FVs only (see inset). Jun-Sep, €2·88; 20 numbered W ⚓s €2·28 (+ €0.22pp/day tax) multihulls +150%. www.port-aberwrach.com aberwrach@ port.cci-brest.fr. 🛢 🛢 (H24 with cr/card) ⚓ 🔧 ⚒ 🗑 ✕ Ⓔ ▯(15t) ⌇(1t) 🔌. **YC des Abers** ☎02·98·04·92·60, ▱🚿🛁, 🗖.

L'Aber Wrac'h ✕ ▱ 🛢.

Landéda (Jul/Aug bus €1.00 rtn: M-Sat 11 times/day; Sun 3 times/day; or 20 mins up-hill walk): 🛒🏧✉Ⓑ a.m.Tues-Fri. Bus to Brest 🚆 & ✈. **Ferry** Roscoff.

Paluden. HM ☎02·98·04·63·12. 20 dumbbell ⚓s in 5m ✕ ▱ 🗖.

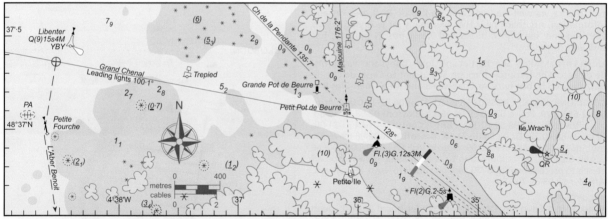

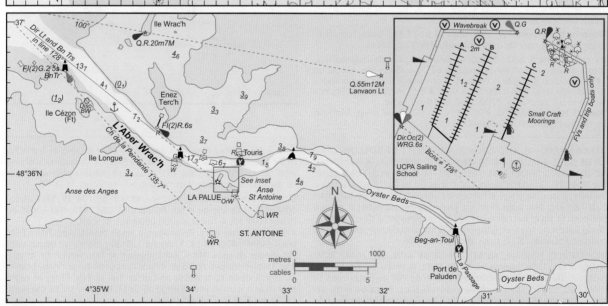

7.19 L'ABER BENOIT

Finistère **48°34'·64N 04°36'·89W** ✿✿✿✿✿✿✿

CHARTS AC 2025, 1432; SHOM 7150, 7094; Navi 539, 540; Imray C35

TIDES +0535 Dover; ML 4·7; Duration 0555

Standard Port BREST (→)

Times				Height (metres)			
High Water		Low Water		MHWS	MHWN	MLWN	MLWS
0000	0600	0000	0600	6·9	5·4	2·6	1·0
1200	1800	1200	1800				
Differences L'ABER BENOIT							
+0022	+0025	+0035	+0020	+1·0	+0·9	+0·4	+0·2
PORTSALL							
+0015	+0020	+0025	+0015	+0·6	+0·5	+0·1	0·0

SHELTER Excellent, but do not enter at night, in poor vis nor in strong WNW winds; best near LW when dangers can be seen.

NAVIGATION WPT 48°37'·00N 04°38'·84W, close W of Petite Fourche WCM buoy; thence follow the tracks shown on chartlet. Beware oyster beds. Tidal streams up to 3kn at half-tide. River is navigable on the tide to Tréglonou bridge 3M upstream.

LIGHTS AND MARKS Unlit. Vital to identify correctly buoys and beacons in the approach chan.

COMMUNICATIONS CROSS/⒧ 02·98·89·31·31; Auto 08·92·68·08·29; Police 02·98·48·10·10 or 17; ⊖ 02·98·85·07·40; British Consul 01·44·51·31·00; HM VHF Ch 09, 06·19·87·75·39 (0830–1230, 1500–1830 Jul–Aug).

FACILITIES Many moorings in lower reach, call HM who will allocate a vacant buoy, Jul-Aug, €1·50, otherwise half price. ⚓ poss upriver nps. ⬛ ⚓ ⚓ ⚒ ✕ (☎02·98·89·86·55) ⚲ ⬜. **Village** ⓟ ⬡(☎02·98·89·87·06) ⬛ ⊠ ⑧. Bus to Brest ⇌ ✈. **Ferry** Roscoff.

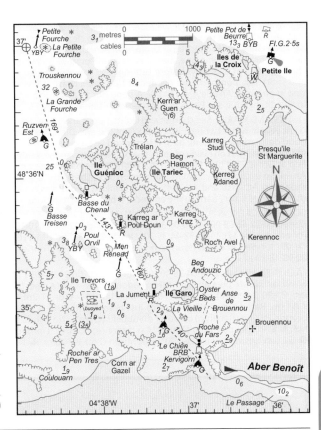

THE PASSAGE INSHORE OF ROCHES DE PORTSALL AND ROCHES D'ARGENTON

This 8·5M passage is only 1M shorter than the offshore passage from L'Aber Wrac'h towards Le Four lt ho. It requires about 6M visibility in daylight (to see the more distant marks) and is navigationally challenging because of strongish tidal streams and the risk of mis-identifying the marks. A well-illustrated Pilot and detailed study of AC 1432, or SHOM 7094, are essential.

The passage from NE to SW is summarised in 3 main sections:

- 7 ca SW of Libenter WCM buoy (see L'Aber Wrac'h) enter Ch du Relec 218·5°: Pte de Landunvez W bcn twr, R top (rear mark, 6M distant) just open left of Petit Men Louet, W bcn twr 4M distant. Leave Le Relec ECM buoy (1·5M distant) close to stbd.

- With Corn Carhai lt ho almost abeam to stbd, alter stbd into Ch du Raous: Bosven Kreiz W bcn twr in transit 249° with rks S of Le Gremm, a 12m high rk. Hold this for 7ca then alter port 228°: Le Four lt ho open left of Bosven Aval W bcn twr. 7ca later jink 7° to port to clear Bosven Aval; then pick up:

- The astern transit 036° of Bosven Aval with Bosven Kreiz. After 1·5M (Pte de Landunvez to port) intercept Chenal Méridional de Portsall, another astern transit 049°: the saddle of Le Yurc'h (7m high rk) on with Grand Men Louet W bcn twr. After 2·2M pass SE of Le Four lt ho into the Ch du Four.

HARBOURS ADJACENT TO THE INSHORE PASSAGE

PORTSALL, Finistère, **48°33'·79N 04°43'·04W**. AC 3688, 1432; SHOM 7150, 7094. HW +0535 on Dover (UT); ML 4·4m; Duration 0600. See L'Aber Benoit.

Small drying hbr at head of bay. Good shelter except in strong NW'lies. Outer 109° ldg marks: Le Yurc'h rk (7m) on with Ploudalmézeau spire. Inner 085° ldg marks (W cols; rear with R top). Front col has Oc (4) WRG 12s, W sector 084°–088°. Appr marked by 3 W bn trs to port, 1 SHM and 1 NCM bcn twr. ⚓ in >10m to W of ent, or enter inner hbr to berth on quay. Tel/Facilities: ⒧ ☎02·98·48·69·17; Police 02·98·48·10·10; Dr ☎02·98·48·75·52; Aff Mar ☎02·98·48·66·54; Taxi ☎02·98·48·68·48. **Quay** ⚲(0·5t) ⬛ ⚓ ⚒ ⬛; **CN Portsall Kersaint** ☎02·98·48·77·49; **Ancre an Eor** ☎02·98·48·73·19, ⛽. **Town** ✕ ⬜ 🛒 ⊠.

ARGENTON, Finistère, **48°31'·26N 04°46'·34W**. AC 1432, 3345; SHOM 7150, 7122. HW +0535 on Dover (UT); ML 4·6m; Duration 0600; use PORTSALL diffs L'Aber Benoit.

Small hbr drying approx 5·5m; good shelter except in W winds when swell enters. WPT 48°31'·14N 04°48'·29W (2½ca S of Le Four lt ho) 084·6°/1·4M to first of 2 W bcn twrs and a white wall with R vert stripe, all on Île Dolvez and in transit. A conspic water twr is slightly S of this ldg line and 1M inland. The N side of chan is marked by 3 PHM bcn twrs. Beware strong NE-SW tidal streams in the approach.

⚓ W of Île Dolvez in about 2m; or skirt round its N side to enter the inner hbr and dry out against the stone quay/slip on N side. 10 🛟s reported but not seen. Facilities: ⚓ ⬛ on quay. **Nautisme en Pays d'Iroise** ☎02·98·89·54·04, ⚲. **Village** ⬜ ✕ 🛒.

7.22 L'ABER-ILDUT

Finistère **48°28'·24N 04°45'·75W** (1st PHM bcn) ✿✿✿✿✿✿

CHARTS AC 2647, 2356, 3345; SHOM 7149, 7122; Navi 540; Imray C36

TIDES +0520 on Dover (UT); ML 4·2m

Standard Port BREST (→)

Times				Height (metres)			
High Water		Low Water		MHWS	MHWN	MLWN	MLWS
0000	0600	0000	0600	6·9	5·4	2·6	1·0
1200	1800	1200	1800				
Differences L'ABER-ILDUT							
+0010	+0010	+0020	+0010	+0·4	+0·3	0·0	0·0

SHELTER Good, but in the appr a high swell runs in strong W'lies; strong river current inside. In fair weather ⚓ outside, as chartlet.

NAVIGATION WPT 48°28'·07N 04°47'·95W, 083°/1·6M to Dir lt which is on conspic white gable end of house. Beware: Les Liniou rks 1·5M NNW of the WPT, Plateau des Fourches 1·2M S and strong cross tides in the approach. Drying rks are marked by Pierre de l'Aber SHM bcn and Le Lieu PHM bcn twr 5m; the latter is easier to see. At the narrow ent (2m) leave Men Tassin PHM bn and Rocher du Crapaud, a large rounded rock, 30m to port to clear drying spit on S side chan, marked by Beg ar Groas SHM bcn. Drop sails before entering under power alone, max 3kts. Beyond Porscave quay, the channel lies S of of 3 Y bys before turning NNE towards the quay and short-stay pontoon Care needed at LWS (max draft 2m) visit first in daylight. All quays dry at LW. A good hbr to await the tide in the Ch du Four; see Passage information, Chenal du Four.

LIGHTS AND MARKS See chartlet and Lights, buoys and waypoints. Glizit yellow water twr is conspic 6·5ca N of the ent. Dir lt 083° is a good daymark (see above); its 3° R sector and Le Lieu bcn, Fl R 2·5s 4M, mark dangers on N side of chan, the only lit mark on the S side, Beg ar Groas SHM bcn, Fl G 2.5s 1M, is right at the narrowing ent. An inland uncharted water twr is almost on 083°; the 078·5° transit shown on AC 2356 and AC 3345, using Lanildut and Brélès small church spires is very hard to see and obsc'd by trees when close in.

COMMUNICATIONS CROSS 02·98·89·31·31; ⓛⒷ 02·98·89·30·31; Meteo 08·92·68·08·08; Auto 08·36·68·08·08; Police 02·98·48·10·10 (or emergency 17); ⊜ 02·98·44·35·20; Dr 02·98·04·33·08; HM VHF Ch 09, Jun-Sep in office hours; Aff Mar 02·98·48·66·54; Tourism 02·98·48·12·88.

FACILITIES HM ☎02·98·04·36·40 (season) mob 06·31·93·58·71. www.lanildut.fr port.aber-ildut@wanadoo.fr **Hbr** 410 M, inc 16 ⚓s; €1·90 (average) multihull +50%; max LOA 11m, draft 2m.1 x ⚓13m. Raft on dumbbells (V trot) just before 2nd PHM bcn or beyond FV quay (landing) on outer dumbbells (G trot). No ⚓ inside hbr. ⛽ ⛽ at fuel pontoon, (2m), H24 self service, cr/cards. Limited Ⓥ ⌂ just upstream on new (2017) pontoon (2m) €2·30/m, ⚓ ⚑

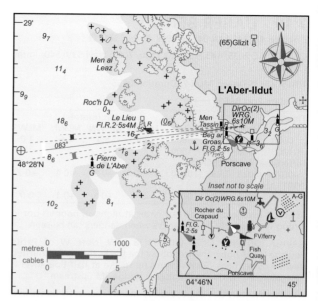

⚓ ⊕ ⊟ ⌂ ⦂ ⚒ ⊞ ⚓. Short stay (1hr) and ⚓ also available at Porscave (opp bank) on E side of pontoon.
Village ✕ ⌂ 🛒 ⊠. Bus to Brest 25km. Ferries: Ouessant, Molène.

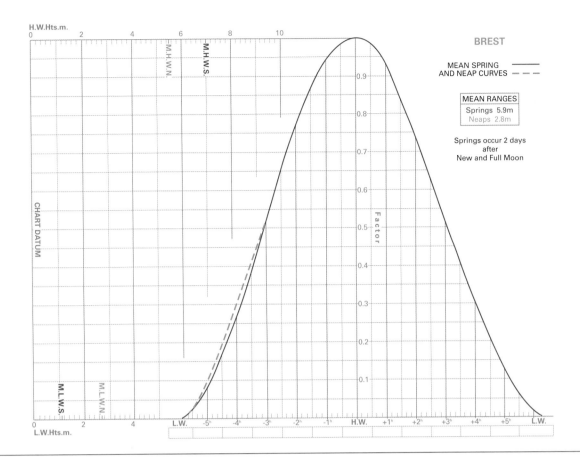

H.W.Hts.m.

BREST

MEAN SPRING —————
AND NEAP CURVES – – – –

MEAN RANGES
Springs 5.9m
Neaps 2.8m

Springs occur 2 days
after
New and Full Moon

CHART DATUM

Factor

L.W.Hts.m.

CHENAL DU FOUR

The Chenal du Four (and de la Helle; AC 3345, 2356) are unhappily split between the small scale Admiralty tidal stream atlases for the English Channel (NP 250) and W France (NP 265). Both channels and the Raz de Sein are better covered at large scale by the French atlas (560-UJA: NW Brittany, Brignogan to Penmarc'h).

Using tides to advantage it is possible for a 5½kn boat to pass through the Chenal du Four and Raz de Sein in one leg from L'Aberwrac'h to Audierne or vice versa. Anse de Bertheaume (anchorage 4M east of Pte de Ste Mathieu) Camaret or Brest are alternative interim stops.

▶ *1M W of Le Four the S-going stream begins around HW Brest (arrive a little early); the N-going stream at HW Brest – 5¾, sp rates 3·6kn. Further S, off Pte de Corsen, the stream is weaker, max 2·3kn at springs.*

The tide runs strongest at S end of Chenal du Four, off Le Conquet. Here the S-going stream starts at HW Brest + ¼, max 5kn; the N-going stream begins at HW Brest – 5¾, 5·2kn max at sp. Wind-over-tide may raise many short, steep seas. ◀

Pilotage. On the mainland coast are 6 major lights, either directional or leading. The charted transits are: St Mathieu directional lt and Kermorvan ≠ 158·5°; Pte de Corsen directional lt bearing 012° (in its white sector and on your port quarter). Thence pass between Rouget SHM lt buoy and Grande Vinotière lt bn. Maintain the 325·5° astern transit of Grand Courleau unlit NCM bcn twr ≠ Le Faix NCM lt bcn twr. Track 145·5° to pass midway between La Fourmi SHM lt buoy and Vieux-Moines lt bcn twr.

Alternatively a simple buoy-hopping sequence (below) can be used by day or night. The 3 legs are short and the speed over the ground is likely to be quite high, but other buoys between the waypoints provide a useful check on progress and any deviation from the planned track.

Leg 1: From ⊕ 48°31'·5N 04°49'·0W (5ca W of Le Four) track 188°/5M (leaving Les Liniou well to port) to Valbelle PHM buoy.

Leg 2: Thence track 168°/4·6M to abeam Grande Vinotière lt bn.

Leg 3: Finally track 172°/2·7M to La Fourmi SHM lt buoy (for Raz de Sein); or

Leg 3A: 156°/2·9M to Vieux-Moines lt bcn (for Brest/Camaret).

Homeward-bound, or along the N Brittany coast, enter the S end of Chenal du Four at LW Brest; a fair tide can then be carried through the channel and NE past Île Vierge. The reverse sequence of pilotage is followed. Two hours is par for the 12M passage.

L'Aber-Ildut, 3·5M SSE of Le Four lt ho, and **Le Conquet**, 3ca SE of Pte de Kermorvan, are the only ports on the mainland coast: the former is safe if there is no swell or W'lies; the latter has limited facilities for yachts but in offshore winds there are anchorages in Anse de Porsmoguer and Anse des Blancs-Sablons, both between Corsen and Kermorvan.

STANDARD TIME UT –01
Subtract 1 hour for UT
For French Summer Time add
ONE hour in **non-shaded areas**

BREST   LAT 48°23'N   LONG 4°30'W
TIMES AND HEIGHTS OF HIGH AND LOW WATERS

Dates in red are **SPRINGS**
Dates in blue are **NEAPS**

YEAR 2019

JANUARY

Time	m		Time	m
1 0102	5.9	**16** 0616	2.8
0722	2.4	1227	5.6
TU 1333	5.9	W 1852	2.6
1958	2.2		
2 0205	6.0	**17** 0103	5.6
0826	2.2	0728	2.6
W 1433	6.1	TH 1337	5.8
2055	2.1	1958	2.3
3 0259	6.2	**18** 0208	6.0
0921	2.0	0833	2.2
TH 1524	6.2	F 1440	6.2
2145	1.9	2059	1.9
4 0346	6.5	**19** 0306	6.4
1008	1.8	0931	1.7
F 1608	6.4	SA 1537	6.6
2228	1.8	2153	1.5
5 0428	6.6	**20** 0400	6.9
1049	1.7	1024	1.3
SA 1647	6.5	SU 1628	7.0
2307	1.7	2244	1.2
6 0506	6.7	**21** 0450	7.2
1127	1.6	1114	0.9
SU 1723	6.5	M 1718	7.2
● 2343	1.7	○ 2334	0.9
7 0541	6.8	**22** 0539	7.5
1202	1.6	1203	0.7
M 1757	6.5	TU 1805	7.4
8 0018	1.7	**23** 0022	0.8
0615	6.8	0626	7.6
TU 1237	1.6	W 1251	0.6
1830	6.5	1852	7.4
9 0052	1.8	**24** 0109	0.8
0648	6.7	0713	7.5
W 1310	1.7	TH 1338	0.7
1903	6.3	1939	7.2
10 0126	1.9	**25** 0157	1.0
0722	6.5	0759	7.3
TH 1344	1.9	F 1426	1.0
1937	6.2	2026	6.8
11 0200	2.1	**26** 0245	1.3
0756	6.3	0847	6.9
F 1419	2.1	SA 1515	1.4
2012	6.0	2115	6.4
12 0236	2.3	**27** 0335	1.7
0833	6.1	0937	6.4
SA 1458	2.3	SU 1608	1.9
2052	5.7	◗ 2208	6.0
13 0317	2.5	**28** 0430	2.2
0916	5.8	1033	6.0
SU 1542	2.5	M 1706	2.3
2141	5.5	2309	5.7
14 0406	2.7	**29** 0533	2.5
1009	5.6	1141	5.6
M 1636	2.7	TU 1813	2.6
◗ 2242	5.4		
15 0505	2.8	**30** 0022	5.6
1115	5.5	0645	2.7
TU 1741	2.7	W 1258	5.5
2352	5.4	1926	2.7
		31 0136	5.6
		0759	2.6
		TH 1411	5.6
		2033	2.5

FEBRUARY

Time	m		Time	m
1 0239	5.9	**16** 0138	5.8
0903	2.4	0806	2.3
F 1509	5.8	SA 1420	5.9
2128	2.3	2037	2.1
2 0330	6.1	**17** 0248	6.2
0953	2.1	0914	1.8
SA 1555	6.1	SU 1524	6.5
2213	2.0	2137	1.6
3 0413	6.4	**18** 0346	6.8
1035	1.8	1010	1.2
SU 1634	6.3	M 1617	7.0
2252	1.8	2231	1.1
4 0450	6.6	**19** 0438	7.3
1111	1.7	1101	0.8
M 1708	6.5	TU 1705	7.3
● 2327	1.7	○ 2320	0.7
5 0525	6.7	**20** 0526	7.6
1145	1.5	1149	0.4
TU 1740	6.6	W 1751	7.5
6 0000	1.6	**21** 0008	0.5
0557	6.8	0612	7.8
W 1217	1.5	TH 1235	0.3
1811	6.6	1835	7.6
7 0032	1.6	**22** 0053	0.5
0628	6.8	0656	7.7
TH 1248	1.5	F 1320	0.5
1842	6.6	1918	7.4
8 0103	1.6	**23** 0137	0.7
0700	6.8	0738	7.5
F 1320	1.6	SA 1403	0.8
1912	6.5	2000	7.0
9 0135	1.7	**24** 0221	1.1
0731	6.6	0820	7.0
SA 1352	1.7	SU 1447	1.3
1944	6.3	2042	6.6
10 0208	1.9	**25** 0306	1.6
0803	6.4	0903	6.5
SU 1426	1.9	M 1533	1.9
2018	6.1	2128	6.1
11 0244	2.1	**26** 0355	2.1
0840	6.1	0951	5.9
M 1505	2.2	TU 1625	2.5
2059	5.9	◗ 2224	5.6
12 0327	2.4	**27** 0452	2.6
0925	5.8	1054	5.4
TU 1552	2.5	W 1729	2.9
◑ 2151	5.6	2337	5.3
13 0419	2.6	**28** 0604	2.9
1023	5.6	1219	5.2
W 1651	2.7	TH 1850	3.0
2259	5.4		
14 0526	2.8		
1139	5.4		
TH 1805	2.7		
15 0018	5.5		
0646	2.7		
F 1303	5.6		
1925	2.5		

MARCH

Time	m		Time	m
1 0104	5.3	**16** 0619	2.7
0729	2.9	1241	5.4
F 1349	5.3	SA 1901	2.7
2010	2.8		
2 0217	5.6	**17** 0117	5.7
0842	2.6	0748	2.4
SA 1453	5.6	SU 1409	5.9
2109	2.5	2020	2.2
3 0311	5.9	**18** 0232	6.2
0933	2.3	0858	1.8
SU 1539	5.9	M 1510	6.4
2154	2.2	2122	1.6
4 0353	6.3	**19** 0331	6.8
1014	1.9	0955	1.2
M 1616	6.2	TU 1601	7.0
2231	1.9	2215	1.1
5 0430	6.5	**20** 0422	7.3
1050	1.7	1045	0.7
TU 1648	6.4	W 1648	7.4
2305	1.7	2303	0.7
6 0503	6.7	**21** 0508	7.6
1122	1.5	1131	0.4
W 1719	6.6	TH 1732	7.6
● 2337	1.5	○ 2349	0.5
7 0534	6.9	**22** 0552	7.8
1153	1.4	1215	0.4
TH 1748	6.7	F 1813	7.6
8 0008	1.4	**23** 0032	0.5
0605	6.9	0634	7.7
F 1224	1.3	SA 1257	0.5
1818	6.8	1853	7.4
9 0039	1.4	**24** 0114	0.7
0635	6.9	0713	7.4
SA 1254	1.3	SU 1337	0.9
1847	6.7	1931	7.1
10 0110	1.5	**25** 0155	1.1
0705	6.8	0751	6.9
SU 1325	1.5	M 1417	1.4
1917	6.6	2010	6.6
11 0142	1.6	**26** 0237	1.6
0737	6.6	0830	6.4
M 1358	1.7	TU 1458	2.0
1950	6.4	2051	6.1
12 0218	1.9	**27** 0322	2.1
0812	6.3	0913	5.8
TU 1436	2.0	W 1546	2.5
2028	6.1	2142	5.6
13 0259	2.1	**28** 0415	2.6
0855	6.0	1010	5.3
W 1521	2.4	TH 1646	3.0
2117	5.8	◗ 2253	5.3
14 0350	2.5	**29** 0523	3.0
0951	5.6	1135	5.0
TH 1618	2.6	F 1808	3.2
◑ 2223	5.5		
15 0455	2.7	**30** 0023	5.2
1108	5.4	0649	3.0
F 1733	2.8	SA 1315	5.1
2348	5.4	1937	3.0
		31 0144	5.4
		0808	2.8
		SU 1426	5.4
		2040	2.7

APRIL

Time	m		Time	m
1 0241	5.7	**16** 0214	6.2
0903	2.4	0840	1.7
M 1511	5.8	TU 1451	6.4
2125	2.3	2104	1.6
2 0323	6.1	**17** 0312	6.8
0944	2.0	0935	1.2
TU 1548	6.1	W 1541	6.9
2202	2.0	2156	1.2
3 0400	6.4	**18** 0401	7.2
1020	1.7	1024	0.8
W 1620	6.4	TH 1627	7.2
2236	1.7	2243	0.8
4 0434	6.6	**19** 0447	7.4
1053	1.5	1109	0.6
TH 1651	6.6	F 1709	7.4
2309	1.5	○ 2327	0.7
5 0506	6.8	**20** 0529	7.5
1125	1.3	1152	0.7
F 1721	6.8	SA 1749	7.4
● 2341	1.3		
6 0538	6.9	**21** 0010	0.7
1156	1.3	0609	7.4
SA 1751	6.9	SU 1232	0.8
		1827	7.3
7 0013	1.3	**22** 0050	0.9
0609	7.0	0647	7.1
SU 1228	1.3	M 1310	1.2
1821	6.9	1904	7.0
8 0045	1.3	**23** 0130	1.3
0641	6.9	0724	6.7
M 1300	1.4	TU 1349	1.6
1853	6.8	1941	6.6
9 0119	1.5	**24** 0210	1.7
0715	6.7	0801	6.2
TU 1335	1.6	W 1428	2.1
1928	6.6	2021	6.1
10 0157	1.7	**25** 0253	2.2
0753	6.4	0843	5.7
W 1415	1.9	TH 1512	2.6
2008	6.3	2108	5.7
11 0241	2.0	**26** 0342	2.6
0838	6.0	0935	5.3
TH 1501	2.2	F 1607	3.0
2058	5.9	◗ 2211	5.3
12 0334	2.3	**27** 0443	2.9
0936	5.7	1049	5.0
F 1600	2.6	SA 1719	3.2
◗ 2206	5.6	2331	5.2
13 0440	2.6	**28** 0558	3.0
1055	5.4	1218	5.0
SA 1716	2.8	SU 1843	3.1
2332	5.5		
14 0605	2.6	**29** 0052	5.3
1227	5.5	0715	2.9
SU 1845	2.6	M 1336	5.2
		1953	2.8
15 0102	5.8	**30** 0154	5.5
0731	2.3	0816	2.6
M 1351	5.9	TU 1428	5.6
2003	2.2	2043	2.5

Chart Datum is 3·64 metres below IGN Datum. HAT is 7·9 metres above Chart Datum.

》》 **FREE** monthly updates. Register at 《
www.reedsnauticalalmanac.co.uk 《

STANDARD TIME UT –01
Subtract 1 hour for UT
For French Summer Time add
ONE hour in **non-shaded areas**

BREST   LAT 48°23'N   LONG 4°30'W
TIMES AND HEIGHTS OF HIGH AND LOW WATERS

Dates in red are **SPRINGS**
Dates in blue are NEAPS

YEAR **2019**

MAY

Time	m	Time	m
1 0242	5.9	**16** 0248	6.6
0902	2.2	0912	1.4
W 1508	5.9	TH 1517	6.7
2124	2.1	2133	1.4
2 0322	6.2	**17** 0338	6.9
0941	1.9	1001	1.2
TH 1544	6.3	F 1603	7.0
2201	1.8	2221	1.1
3 0400	6.5	**18** 0424	7.0
1018	1.6	1046	1.1
F 1618	6.5	SA 1646	7.1
2237	1.6	○ 2305	1.0
4 0435	6.7	**19** 0506	7.0
1053	1.4	1128	1.1
SA 1651	6.8	SU 1725	7.1
● 2312	1.4	2347	1.1
5 0509	6.8	**20** 0546	6.9
1128	1.3	1208	1.2
SU 1724	6.9	M 1803	7.0
2347	1.3		
6 0544	6.9	**21** 0027	1.2
1203	1.3	0623	6.7
M 1758	6.9	TU 1246	1.5
		1840	6.8
7 0023	1.3	**22** 0106	1.5
0620	6.9	0700	6.5
TU 1239	1.4	W 1324	1.8
1834	6.9	1917	6.5
8 0102	1.4	**23** 0146	1.8
0659	6.7	0738	6.1
W 1318	1.5	TH 1402	2.1
1913	6.7	1956	6.2
9 0144	1.6	**24** 0227	2.1
0742	6.4	0818	5.8
TH 1402	1.8	F 1444	2.5
1958	6.4	2040	5.8
10 0232	1.8	**25** 0311	2.4
0832	6.1	0904	5.4
F 1452	2.1	SA 1531	2.8
2053	6.1	2133	5.5
11 0327	2.1	**26** 0402	2.7
0933	5.8	1002	5.2
SA 1552	2.4	SU 1630	3.0
2201	5.8	◑ 2237	5.3
12 0434	2.3	**27** 0504	2.9
1047	5.6	1114	5.1
SU 1705	2.6	M 1740	3.1
◑ 2320	5.7	2347	5.3
13 0552	2.3	**28** 0611	2.9
1210	5.6	1226	5.2
M 1827	2.5	TU 1849	2.9
14 0041	5.9	**29** 0053	5.4
0710	2.1	0715	2.7
TU 1326	5.9	W 1328	5.4
1940	2.1	1948	2.7
15 0150	6.2	**30** 0150	5.7
0816	1.8	0810	2.4
W 1426	6.3	TH 1419	5.7
2041	1.7	2038	2.3
		31 0238	6.0
		0857	2.1
		F 1502	6.1
		2122	2.0

JUNE

Time	m	Time	m
1 0321	6.2	**16** 0403	6.5
0940	1.8	1025	1.5
SA 1543	6.4	SU 1625	6.7
2203	1.7	2246	1.4
2 0402	6.5	**17** 0446	6.6
1021	1.6	1107	1.5
SU 1621	6.6	M 1706	6.8
2244	1.5	○ 2328	1.4
3 0443	6.7	**18** 0526	6.6
1101	1.4	1147	1.6
M 1700	6.8	TU 1744	6.7
● 2324	1.3		
4 0523	6.8	**19** 0008	1.5
1141	1.3	0604	6.5
TU 1740	6.9	W 1225	1.7
		1821	6.7
5 0006	1.2	**20** 0046	1.6
0605	6.8	0640	6.3
W 1223	1.3	TH 1302	1.8
1821	6.9	1858	6.5
6 0049	1.3	**21** 0123	1.8
0649	6.7	0716	6.1
TH 1307	1.4	F 1339	2.0
1906	6.8	1935	6.3
7 0136	1.4	**22** 0201	2.0
0737	6.5	0753	5.9
F 1354	1.6	SA 1417	2.3
1955	6.6	2013	6.0
8 0226	1.6	**23** 0240	2.2
0829	6.3	0833	5.7
SA 1446	1.9	SU 1457	2.5
2050	6.4	2055	5.8
9 0322	1.8	**24** 0322	2.5
0928	6.0	0918	5.4
SU 1544	2.1	M 1543	2.7
2153	6.1	2145	5.5
10 0424	2.0	**25** 0411	2.6
1033	5.8	1012	5.3
M 1651	2.3	TU 1638	2.9
◑ 2302	6.0	◑ 2243	5.4
11 0533	2.1	**26** 0508	2.7
1144	5.8	1117	5.2
TU 1802	2.3	W 1741	2.9
		2348	5.4
12 0013	6.0	**27** 0611	2.7
0643	2.1	1222	5.3
W 1254	5.9	TH 1846	2.8
1912	2.2		
13 0121	6.1	**28** 0051	5.5
0748	1.9	0714	2.6
TH 1357	6.1	F 1323	5.5
2015	1.9	1947	2.6
14 0222	6.3	**29** 0150	5.7
0846	1.8	0811	2.3
F 1451	6.4	SA 1418	5.8
2111	1.7	2042	2.2
15 0315	6.4	**30** 0243	6.0
0938	1.6	0903	2.0
SA 1540	6.6	SU 1508	6.2
2201	1.5	2132	1.9

JULY

Time	m	Time	m
1 0333	6.3	**16** 0432	6.3
0951	1.7	1051	1.8
M 1555	6.5	TU 1650	6.6
2219	1.6	○ 2313	1.6
2 0420	6.6	**17** 0511	6.3
1038	1.5	1130	1.7
TU 1640	6.8	W 1729	6.6
● 2305	1.3	2350	1.6
3 0507	6.8	**18** 0547	6.4
1124	1.3	1207	1.7
W 1726	7.0	TH 1804	6.6
2352	1.1		
4 0553	6.9	**19** 0026	1.6
1210	1.2	0621	6.3
TH 1812	7.1	F 1241	1.8
		1838	6.6
5 0039	1.0	**20** 0100	1.7
0641	6.9	0654	6.3
F 1257	1.2	SA 1315	1.9
1900	7.1	1911	6.5
6 0127	1.0	**21** 0134	1.8
0729	6.8	0727	6.1
SA 1345	1.3	SU 1349	2.0
1949	7.0	1945	6.3
7 0217	1.2	**22** 0208	2.0
0819	6.6	0800	6.0
SU 1436	1.5	M 1424	2.2
2040	6.7	2020	6.1
8 0309	1.4	**23** 0244	2.2
0912	6.3	0837	5.8
M 1530	1.8	TU 1501	2.4
2135	6.4	2059	5.8
9 0405	1.7	**24** 0324	2.4
1009	6.1	0919	5.6
TU 1628	2.0	W 1545	2.6
◑ 2235	6.1	2145	5.6
10 0506	2.0	**25** 0411	2.6
1111	5.9	1013	5.4
W 1733	2.2	TH 1639	2.8
2341	5.9	◑ 2244	5.4
11 0611	2.2	**26** 0509	2.7
1219	5.8	1118	5.3
TH 1841	2.3	F 1744	2.8
		2354	5.4
12 0050	5.8	**27** 0617	2.7
0718	2.2	1229	5.4
F 1327	5.9	SA 1856	2.7
1949	2.2		
13 0157	5.9	**28** 0105	5.5
0822	2.2	0727	2.6
SA 1429	6.0	SU 1337	5.7
2051	2.1	2004	2.5
14 0257	6.0	**29** 0211	5.8
0918	2.0	0830	2.2
SU 1522	6.2	M 1439	6.0
2145	1.9	2105	2.0
15 0347	6.1	**30** 0310	6.2
1008	1.9	0927	1.9
M 1609	6.4	TU 1534	6.5
2231	1.7	2159	1.6
		31 0404	6.5
		1019	1.5
		W 1625	6.9
		2250	1.2

AUGUST

Time	m	Time	m
1 0453	6.9	**16** 0528	6.4
1109	1.2	1146	1.7
TH 1714	7.2	F 1743	6.7
● 2339	0.9		
2 0541	7.1	**17** 0004	1.5
1157	0.9	0559	6.5
F 1801	7.4	SA 1218	1.6
		1814	6.7
3 0026	0.7	**18** 0035	1.5
0628	7.2	0629	6.5
SA 1244	0.8	SU 1250	1.7
1848	7.4	1845	6.7
4 0113	0.7	**19** 0106	1.6
0714	7.2	0658	6.4
SU 1331	0.9	M 1320	1.8
1934	7.3	1915	6.6
5 0200	0.9	**20** 0137	1.7
0800	6.9	0728	6.3
M 1419	1.1	TU 1352	1.9
2021	7.0	1946	6.4
6 0248	1.2	**21** 0209	1.9
0848	6.6	0800	6.1
TU 1508	1.5	W 1426	2.2
2109	6.6	2020	6.1
7 0338	1.6	**22** 0244	2.2
0938	6.2	0837	5.9
W 1601	1.9	TH 1504	2.4
◑ 2202	6.2	2101	5.8
8 0434	2.1	**23** 0327	2.5
1036	5.9	0923	5.6
TH 1700	2.3	F 1553	2.7
2305	5.7	◑ 2153	5.5
9 0537	2.4	**24** 0420	2.7
1144	5.6	1025	5.4
F 1809	2.5	SA 1655	2.9
		2305	5.3
10 0019	5.5	**25** 0530	2.8
0648	2.6	1144	5.4
SA 1301	5.6	SU 1813	2.9
1925	2.6		
11 0138	5.5	**26** 0029	5.4
0801	2.6	0650	2.7
SU 1411	5.8	M 1305	5.6
2035	2.4	1935	2.6
12 0244	5.7	**27** 0148	5.7
0904	2.3	0806	2.4
M 1508	6.0	TU 1418	6.0
2131	2.1	2045	2.1
13 0336	5.9	**28** 0255	6.1
0954	2.1	0909	1.9
TU 1555	6.3	W 1518	6.5
2217	1.8	2143	1.5
14 0419	6.2	**29** 0350	6.7
1036	1.9	1003	1.4
W 1635	6.5	TH 1610	7.0
2257	1.7	2235	1.0
15 0455	6.3	**30** 0439	7.1
1113	1.8	1053	1.0
TH 1710	6.7	F 1659	7.4
○ 2331	1.6	● 2323	0.6
		31 0525	7.4
		1141	0.7
		SA 1745	7.7

Chart Datum is 3·64 metres below IGN Datum. HAT is 7·9 metres above Chart Datum.

N Brittany

》 **FREE** monthly updates. Register at 《
www.reedsnauticalalmanac.co.uk

275

STANDARD TIME UT –01
Subtract 1 hour for UT
For French Summer Time add ONE hour in **non-shaded areas**

BREST LAT 48°23′N LONG 4°30′W

TIMES AND HEIGHTS OF HIGH AND LOW WATERS

Dates in red are SPRINGS
Dates in blue are NEAPS

YEAR **2019**

SEPTEMBER

Day	Time	m	Day	Time	m
1 SU	0009 0610 1227 1830	0.4 7.5 0.6 7.7	16 M	0008 0602 1222 1817	1.4 6.7 1.5 6.9
2 M	0054 0653 1312 1913	0.5 7.4 0.7 7.5	17 TU	0037 0630 1252 1847	1.5 6.7 1.6 6.8
3 TU	0138 0736 1356 1956	0.7 7.2 1.0 7.2	18 W	0107 0658 1323 1917	1.6 6.5 1.8 6.6
4 W	0222 0819 1442 2040	1.2 6.8 1.4 6.6	19 TH	0138 0729 1356 1949	1.8 6.4 2.0 6.3
5 TH	0308 0905 1531 2128	1.7 6.3 2.0 6.1	20 F	0213 0804 1434 2029	2.1 6.1 2.3 6.0
6 F	0359 0959 1628 2229	2.3 5.8 2.5 5.6	21 SA	0255 0848 1522 2120	2.4 5.8 2.6 5.6
7 SA	0501 1109 1739 2350	2.8 5.5 2.8 5.2	22 SU	0347 0949 1624 2234	2.7 5.5 2.8 5.3
8 SU	0620 1236 1902	3.0 5.4 2.8	23 M	0457 1113 1745	2.9 5.4 2.9
9 M	0121 0743 1353 2019	5.3 2.8 5.6 2.6	24 TU	0006 0624 1244 1915	5.3 2.9 5.6 2.6
10 TU	0231 0847 1451 2114	5.6 2.5 6.0 2.3	25 W	0133 0747 1401 2028	5.7 2.5 6.1 2.0
11 W	0321 0936 1536 2158	5.9 2.2 6.3 2.0	26 TH	0241 0852 1501 2126	6.2 1.9 6.7 1.4
12 TH	0400 1015 1613 2235	6.2 1.9 6.5 1.7	27 F	0333 0946 1553 2216	6.8 1.3 7.2 0.9
13 F	0434 1050 1647 2307	6.4 1.7 6.7 1.5	28 SA	0421 1035 1640 2303	7.3 0.9 7.6 0.5
14 SA	0504 1122 1718 2338	6.6 1.6 6.8 1.4	29 SU	0505 1121 1725 2348	7.6 0.6 7.8 0.4
15 SU	0533 1152 1748	6.7 1.5 6.9	30 M	0548 1206 1807	7.7 0.5 7.8

OCTOBER

Day	Time	m	Day	Time	m
1 TU	0031 0629 1249 1849	0.5 7.6 0.7 7.5	16 W	0010 0603 1226 1821	1.4 6.9 1.5 6.8
2 W	0113 0709 1332 1929	0.8 7.3 1.0 7.1	17 TH	0041 0633 1259 1853	1.6 6.8 1.7 6.7
3 TH	0155 0749 1416 2011	1.3 6.8 1.5 6.6	18 F	0114 0705 1334 1928	1.8 6.6 1.9 6.4
4 F	0238 0833 1503 2056	1.9 6.3 2.1 6.0	19 SA	0150 0742 1415 2009	2.0 6.3 2.2 6.0
5 SA	0327 0924 1557 2155	2.5 5.8 2.6 5.4	20 SU	0234 0827 1504 2104	2.4 6.0 2.5 5.7
6 SU	0427 1035 1707 2318	2.9 5.4 2.9 5.1	21 M	0328 0931 1608 2220	2.7 5.6 2.7 5.4
7 M	0547 1205 1832	3.2 5.3 3.0	22 TU	0439 1056 1729 2351	2.9 5.5 2.8 5.4
8 TU	0055 0716 1326 1951	5.2 3.1 5.5 2.8	23 W	0606 1227 1858	2.8 5.7 2.5
9 W	0207 0821 1424 2046	5.5 2.7 5.9 2.4	24 TH	0117 0729 1342 2008	5.8 2.4 6.2 2.0
10 TH	0255 0908 1507 2129	5.8 2.3 6.2 2.1	25 F	0220 0833 1441 2105	6.3 1.9 6.7 1.4
11 F	0332 0947 1544 2205	6.2 2.0 6.5 1.8	26 SA	0312 0926 1532 2155	6.9 1.3 7.2 1.0
12 SA	0405 1021 1618 2237	6.5 1.8 6.7 1.6	27 SU	0359 1015 1619 2242	7.3 0.9 7.5 0.7
13 SU	0436 1053 1649 2309	6.7 1.6 6.9 1.4	28 M	0443 1101 1703 2326	7.5 0.7 7.7 0.6
14 M	0505 1124 1720 2339	6.8 1.5 6.9 1.4	29 TU	0525 1145 1745	7.6 0.7 7.6
15 TU	0534 1155 1750	6.9 1.5 6.9	30 W	0008 0605 1227 1825	0.8 7.5 0.9 7.3
			31 TH	0049 0645 1309 1905	1.1 7.2 1.2 6.9

NOVEMBER

Day	Time	m	Day	Time	m
1 F	0130 0724 1352 1945	1.5 6.8 1.6 6.4	16 SA	0056 0651 1320 1917	1.7 6.7 1.8 6.5
2 SA	0212 0806 1437 2029	2.0 6.4 2.1 5.9	17 SU	0137 0732 1405 2003	1.9 6.5 2.0 6.2
3 SU	0257 0854 1528 2123	2.5 5.9 2.6 5.4	18 M	0224 0822 1457 2100	2.2 6.2 2.3 5.8
4 M	0352 0957 1630 2238	3.0 5.5 2.9 5.1	19 TU	0319 0924 1559 2211	2.5 5.9 2.5 5.6
5 TU	0504 1118 1747	3.2 5.3 3.0	20 W	0427 1041 1715 2331	2.7 5.8 2.5 5.6
6 W	0006 0627 1239 1904	5.1 3.2 5.4 2.9	21 TH	0546 1203 1834	2.7 5.9 2.3
7 TH	0121 0737 1341 2003	5.3 2.9 5.7 2.6	22 F	0050 0703 1316 1943	5.9 2.4 6.2 2.0
8 F	0214 0829 1428 2049	5.7 2.6 6.0 2.3	23 SA	0154 0808 1417 2041	6.3 1.9 6.6 1.4
9 SA	0255 0910 1509 2128	6.1 2.2 6.3 2.0	24 SU	0248 0904 1510 2132	6.7 1.5 7.0 1.2
10 SU	0330 0947 1545 2204	6.3 1.9 6.6 1.7	25 M	0337 0954 1558 2220	7.1 1.2 7.2 1.1
11 M	0404 1022 1619 2238	6.6 1.7 6.8 1.6	26 TU	0422 1041 1643 2305	7.3 1.0 7.3 1.0
12 TU	0436 1057 1653 2311	6.8 1.6 6.9 1.5	27 W	0504 1125 1725 2347	7.3 1.0 7.2 1.1
13 W	0508 1131 1726 2345	6.9 1.5 6.9 1.5	28 TH	0545 1208 1806	7.3 1.1 7.0
14 TH	0541 1205 1801	6.9 1.5 6.9	29 F	0028 0625 1250 1845	1.4 7.1 1.4 6.7
15 F	0019 0614 1241 1837	1.5 6.9 1.6 6.7	30 SA	0108 0704 1331 1925	1.7 6.8 1.7 6.4

DECEMBER

Day	Time	m	Day	Time	m
1 SU	0149 0745 1413 2006	2.0 6.5 2.0 6.0	16 M	0130 0730 1359 2000	1.7 6.8 1.6 6.4
2 M	0232 0828 1458 2053	2.4 6.1 2.4 5.6	17 TU	0218 0820 1450 2054	1.9 6.5 1.8 6.2
3 TU	0319 0919 1549 2149	2.7 5.7 2.7 5.3	18 W	0311 0917 1548 2156	2.1 6.3 2.1 6.0
4 W	0415 1021 1650 2259	3.0 5.5 2.9 5.2	19 TH	0412 1021 1654 2304	2.3 6.1 2.2 5.9
5 TH	0522 1131 1758	3.1 5.4 3.0	20 F	0520 1132 1804	2.4 6.0 2.2
6 F	0011 0632 1240 1903	5.2 3.1 5.5 2.8	21 SA	0015 0632 1245 1913	5.9 2.3 6.1 2.1
7 SA	0115 0734 1338 1958	5.4 2.8 5.7 2.6	22 SU	0123 0741 1351 2015	6.1 2.1 6.3 1.9
8 SU	0207 0825 1427 2045	5.7 2.5 6.0 2.3	23 M	0223 0842 1449 2111	6.4 1.9 6.5 1.7
9 M	0250 0910 1510 2127	6.1 2.2 6.3 2.0	24 TU	0316 0936 1541 2202	6.6 1.6 6.7 1.5
10 TU	0330 0951 1550 2207	6.4 1.9 6.5 1.8	25 W	0405 1026 1628 2248	6.9 1.4 6.8 1.5
11 W	0408 1030 1628 2246	6.6 1.7 6.7 1.6	26 TH	0449 1111 1711 2331	7.0 1.3 6.8 1.5
12 TH	0446 1109 1707 2324	6.8 1.5 6.8 1.5	27 F	0531 1153 1751	7.0 1.3 6.7
13 F	0523 1148 1746	6.9 1.4 6.8	28 SA	0011 0610 1233 1829	1.5 7.0 1.4 6.6
14 SA	0004 0603 1229 1828	1.5 7.0 1.4 6.8	29 SU	0050 0648 1311 1906	1.7 6.8 1.6 6.4
15 SU	0045 0644 1312 1912	1.5 6.9 1.5 6.6	30 M	0128 0725 1349 1943	1.9 6.6 1.8 6.2
			31 TU	0206 0802 1428 2021	2.1 6.3 2.1 5.9

Chart Datum is 3·64 metres below IGN Datum. HAT is 7·9 metres above Chart Datum.

276

BREST TIDAL COEFFICIENTS 2019

Date	Jan am	Jan pm	Feb am	Feb pm	Mar am	Mar pm	Apr am	Apr pm	May am	May pm	June am	June pm	July am	July pm	Aug am	Aug pm	Sept am	Sept pm	Oct am	Oct pm	Nov am	Nov pm	Dec am	Dec pm
1	56	57	53	57	37	40	50	56	56	61	69	73	72	76	92	96	113	113	112	107	82	75	68	62
2	60	63	62	66	45	50	61	67	66	71	77	81	81	85	100	103	111	107	102	94	67	59	56	50
3	66	69	69	73	55	60	71	76	75	79	84	86	88	91	104	104	102	96	86	77	51	44	45	41
4	71	74	76	78	65	70	80	83	83	86	88	89	93	94	103	100	88	80	69	59	39	34	38	36
5	76	78	80	82	74	78	86	88	88	89	89	89	94	93	96	91	71	62	51	43	33		35	--
6	79	80	83	83	81	83	90	90	90	90	87	85	92	89	85	79	54	47	37		34	36	37	39
7	80	80	83	82	86	87	90	90	89	87	82	78	86	82	72	65		42	34	33	41	46	43	47
8	79	78	81	79	88	88	88	86	84	81	74	70	77	72	58		38	38	36	40	51	56	52	57
9	76	74	77	74	88	86	83	78	77	71	66	62	68	64	53	49	40	44	45	51	61	66	61	66
10	72	69	70	66	85	82	74	68	66	61	60			60	46	46	49	55	57	62	71	75	70	74
11	66	62	62	58	79	74	63	57	56	53	58	58	58	56	48	51	60	65	67	72	78	81	78	81
12	59	55	54	49	70	65	51	47	50		59	61	56	57	54	59	69	74	76	80	84	85	83	85
13	52	48	46	43	59	54	44		50	52	63	67	58	61	63	66	77	80	82	85	87	87	86	87
14	45	43	42		49	44	44	46	56	62	70	73	63	66	70	73	82	84	86	87	87	86	87	86
15	42	43	43	47	42	41	52	59	67	73	76	78	69	71	76	78	85	85	88	87	84	82	85	82
16		45	52	59		44	67	76	79	84	80	82	73	75	79	80	85	85	86	85	79	75	79	76
17	48	53	67	76	50	58	84	91	88	92	82	83	76	77	81	81	83	82	82	79	70	66	72	69
18	59	65	84	92	66	76	97	102	94	95	82	81	78	77	80	79	79	76	75	71	61	56	65	62
19	72	79	99	105	85	93	105	107	95	95	80	78	77	76	78	75	72	68	66	61	52	49	59	58
20	86	92	110	113	100	107	108	107	93	91	76	73	74	72	73	70	63	58	55	50	48	49	57	--
21	97	102	115	115	111	114	105	101	87	84	70	66	70	67	66	62	52	47	45	41		53	58	61
22	105	107	113	110	115	115	97	91	79	74	62	59	64	61	58	53	42	39	40		57	63	64	67
23	108	108	104	98	113	109	85	78	68	63	55	51	57	54	49	45		37	42	47	70	76	71	75
24	106	102	91	83	104	97	70	63	57	52	47	44	50	47	41		39	44	54	62	83	88	79	82
25	98	92	74	65	90	82	55	48	47	42	42	40	44	42	39	39	51	60	71	80	92	96	84	86
26	86	79	57	49	73	64	41	36	38	36		39		41	41	45	69	78	88	95	98	99	87	88
27	71	64	43	38	55	47	32			35	40	41	41	42	51	59	87	96	101	106	99	98	87	87
28	58	52		36	40	35	31	32	35	37	44	48	45	50	67	75	103	108	109	111	96	93	85	83
29	48				32		35	39	40	45	52	57	55	61	83	91	113	115	111	109	89	85	80	77
30	46	45			31	34	45	50	49	54	62	67	67	74	98	104	116	115	106	102	80	74	73	69
31	47	50			39	44			59	64			80	86	108	112			96	90			65	60

Tidal coefficients indicate the magnitude of the tide on any particular day, without having to look up and calculate the range, and thus determine whether it is springs, neaps or somewhere in between. This table is valid for all areas covered by this Almanac. Typical values are:

120	Very big spring tide
95	**Mean spring tide**
70	Average tide
45	**Mean neap tide**
20	Very small neap tide

See also coefficients for Calais (p 169), Cherbourg (p 196), St Malo (p 215).

N Brittany

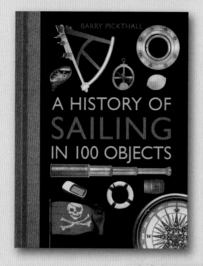

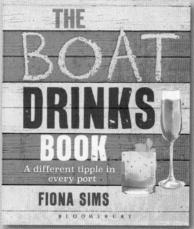

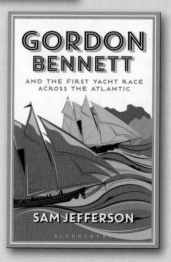

Brest
The Perfect Getaway

City center or beach-front, just pick where you'll stay !

Château

10 min walk to the City Center
and the Railway Station

- Lifting Capacity: 25t
- Bicycle Lending
- Numerous restaurants

Jérémy Kergourlay

Moulin Blanc

- 15 min drive from the Airport
- Lifting Capacity: 40t
- Ideal for Technical Stops
- Wintering
- Boat Suppliers and Services

Both marinas offer the following facilities:

- Free Wi-Fi
- Pump Station
- Toilets, Showers, Laundry

- Recycling Containers
- Careening Area
- Fuel Station 24/7

www.marinasbrest.fr/en   f @marinasbrest

TransEurope

Passeport Escales

Moulin Blanc
+33 (0) 298 02 20 02
moulinblanc@marinasbrest.fr

Château
+33 (0) 298 33 12 50
chateau@marinasbrest.fr

⇒ISO 14001

⇒Blue Flag

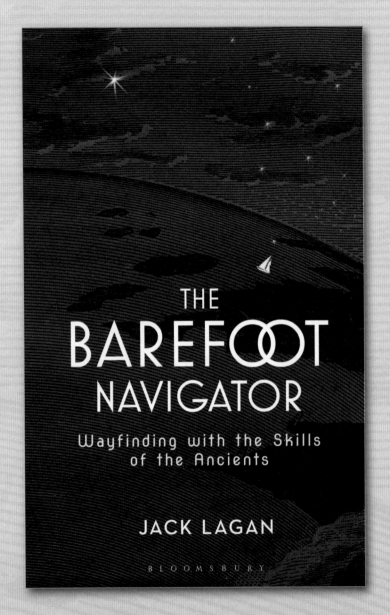

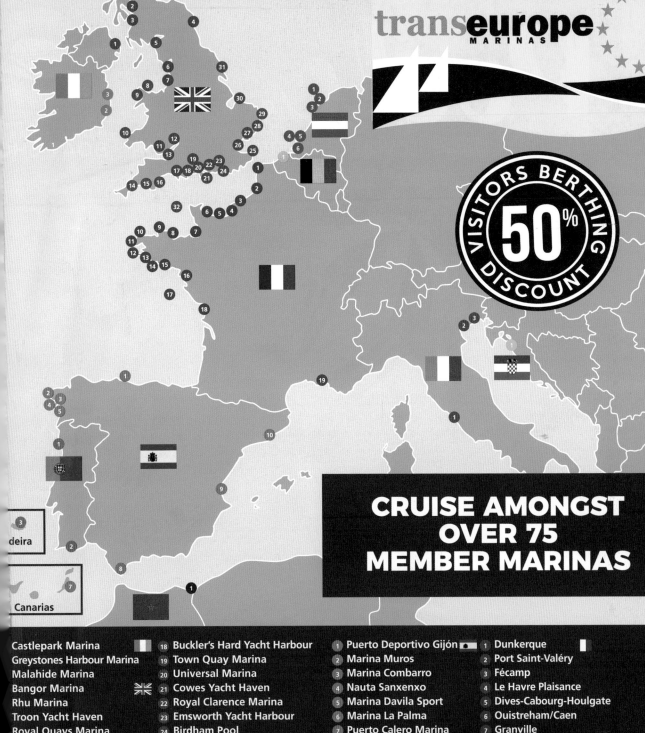

transeurope
MARINAS

VISITORS BERTHING 50% DISCOUNT

CRUISE AMONGST OVER 75 MEMBER MARINAS

Castlepark Marina	18 Buckler's Hard Yacht Harbour	1 Puerto Deportivo Gijón
Greystones Harbour Marina	19 Town Quay Marina	2 Marina Muros
Malahide Marina	20 Universal Marina	3 Marina Combarro
Bangor Marina	21 Cowes Yacht Haven	4 Nauta Sanxenxo
Rhu Marina	22 Royal Clarence Marina	5 Marina Davila Sport
Troon Yacht Haven	23 Emsworth Yacht Harbour	6 Puerto Calero Marina
Royal Quays Marina	24 Birdham Pool	7 Marina Alcaidesa
Whitehaven Marina	25 Dover Marina	8 Pobla Marina
Fleetwood Haven Marina	26 Gillingham Marina	9 Port Ginesta
Liverpool Marina	27 Fambridge Yacht Haven	10 Port Ginesta
Deganwy Marina	28 Tollesbury Marina	1 Douro Marina
Conwy Quays Marina	29 Fox's Marina	2 Marina Portimão
Neyland Yacht Haven	30 Brundall Bay Marina	3 Quinta do Lorde Marina
Penarth Quays Marina	31 Hull Marina	1 Porto Romano
Upton Marina	32 Beaucette Marina	2 Venezia Certosa Marina
Portishead Quays Marina	1 Marina Den Oever	3 Marina del Cavallino
Mylor Yacht Harbour	2 Jachthaven Waterland	1 Marina Punat
Mayflower Marina	3 Jachthaven Wetterwille	1 Kos Marina
Dart Marina	4 Marina Port Zélande	1 Marina de Saïdia
Poole Quay Boathaven	5 Jachthaven Biesbosch	VVW Nieuwpoort
	6 Delta Marina	

1 Dunkerque	
2 Port Saint-Valéry	
3 Fécamp	
4 Le Havre Plaisance	
5 Dives-Cabourg-Houlgate	
6 Ouistreham/Caen	
7 Granville	
8 Saint Quay Port d'Armor	
9 Perros-Guirec	
10 Roscoff	
11 Marinas de Brest	
12 Douarnenez-Tréboul	
13 Loctudy	
14 Port La Forêt	
15 Concarneau	
16 Ports de Nantes	
17 Île d'Yeu / Port Joinville	
18 La Rochelle	
19 Port Napoléon	

Poole....

The next page in your adventure

- The Jurassic Coast
- Brownsea Island
- Dining out
- Enterainment
- Fireworks

Plus much more!

HOME OF THE
POOLE HARBOUR
BOAT SHOW

OUR FACILITIES:

Marina of the Year 2017
UK Coastal
Under 250 Berths
WINNER
PARTNERED BY
GJW Insurance

Marina of the Year 20
UK Coastal
Under 250 Berths
WINNER
PARTNERED BY
GJW Insurance

PERMANENT BERTHS

It's in a private position that makes the most of the views and gorgeous sunsets, yet it's still close to Poole's historic quay, old town and vibrant shopping centre.

- 75 permanent berths
- Superyacht berths
- Floating docks for jet skis and RIBs up to 6.1m
- 24 hour security
- Deep water: 2.5 - 6m
- Water taxi service, parking

VISITOR MARINA

Use your boat as a holiday home; entertain family, friends, colleagues or customers onboard; sail the stunning Jurassic Coast.

Enjoy all the attractions of Poole, Bournemouth and beautiful Dorset. A warm welcome always awaits!

- 125 visitor berths all year for vessels up to 70m in length and up to 4.5m draft
- Swinging moorings

SWINGING MOORINGS

Relax with a glass of wine, on a sunny afternoon, on your own swinging mooring in Poole Harbour overlooking Brownsea Island. Away from the madding crowd, these offer you ultimate privacy, peace & tranquillity.

POOLE QUAY
Boat Haven
PORT OF POO
Marina

Poole Town Quay, Poole,
Dorset BH15 1HJ t: 01202 64948

poolequayboathaven.co.uk
VHF Channel 80 call sign "Poole Quay Boat Haven